S0-BMG-671

THE SINGLE-STRANDED DNA PHAGES

(*First row*) D. Dressler, N. Zinder/D.T. Denhardt, S. Wickner, D. Capon
(*Second row*) C. Sumida-Yasumoto/J.A. Nowak, J. Das/A. Sugino
(*Third row*) D. Loring, E. Benz, S. Sugrue, L. Reha-Krantz/S. Eisenberg, R. McMacken
(*Fourth row*) M. Takanami, H. Schaller/B. Kessler-Liebscher, C. Hours, C. Miyamoto
(*Fifth row*) D. Ray, J. Schoenmakers/P. Hofschneider/L. Makowski, R. Warner, D. Caspar
Photos courtesy of C. Hours

THE SINGLE-STRANDED DNA PHAGES

Edited by

D. T. Denhardt
McGill University

D. Dressler
Harvard University

D. S. Ray
University of California at Los Angeles

Cold Spring Harbor Laboratory
1978

COLD SPRING HARBOR MONOGRAPH SERIES

THE SINGLE-STRANDED DNA PHAGES

Printed in the United States of America
Book design by Emily Harste

Library of Congress Cataloging in Publication Data

Main entry under title:

The Single-stranded DNA phages.

(Cold Spring Harbor monograph series)
Includes index.
1. Bacteriophage. 2. Deoxyribonucleic acid.
I. Denhardt, David T. II. Dressler, David
III. Ray, Daniel S. IV. Series
QR342.S56 576'.6482 78-60445
ISBN 0-87969-122-0

Contents

Replication of Viral DNA

Recombination, Repair, and Genetic Engineering

Transcription, Morphogenesis, and Virion Structure

Appendices

Preface

The single-stranded DNA phages are among the smallest organisms known. They have been tremendously valuable in elucidating several fundamental life processes, especially DNA replication and gene organization. Indeed, overlapping genes were discovered in single-stranded phages.

This book is based on an international meeting held at Cold Spring Harbor in August 1977 and contains a comprehensive and up-to-date account of what is known about the life process of the single-stranded phages.

The meeting was made possible by generous support from the National Science Foundation and from the Cold Spring Harbor Laboratory. The tireless efforts of Nancy Ford are responsible for seeing the production of this volume through to its completion. Without the encouragement of Dr. James D. Watson the book would not have been attempted.

D. T. Denhardt
D. Dressler
D. S. Ray

THE SINGLE-STRANDED DNA PHAGES

φX—Some Recollections

Robert L. Sinsheimer
University of California
Santa Cruz, California 95064

The origins of the research in my laboratory on the bacteriophage φX174 have been recounted in the chapter "φX: Multum in Parvo" in *Phage and the Origins of Molecular Biology* (Sinsheimer 1966). The more salient events in the particular research program (in collaboration with Dr. Arthur Kornberg's laboratory) which led to the first in vitro synthesis of an infective DNA are detailed in the article "Closing the Ring" (Sinsheimer 1968).

φX Rings

One of the most significant and unanticipated discoveries concerning φX was the circularity of its DNA molecule. This first demonstration of a ring form of DNA and the accompanying development of criteria for the verification of such ring structures led quite directly to their demonstration in other viral forms (polyoma, SV40, lambda DNA during the eclipse phase, etc.) and elsewhere (in plasmids, ribosomal DNA, etc.).

The concept that φX DNA might be a circular molecule arose from the negative results of persistent experiments by Dr. Walter Fiers to identify the nucleotides at the putative ends of φX DNA (Fiers and Sinsheimer 1962a). φX DNA was the first relatively homogeneous DNA preparation available with a known size (about 5500 nucleotides) and an assayable biological activity; therefore, it seemed reasonable to examine its termini for any unusual features or regularities.

Dr. Fiers attempted to identify the assumed terminal nucleotides by controlled exonuclease digestion from both the 3′ and 5′ ends. When the 3′ or 5′ exonuclease preparations used were effectively devoid of endonuclease activity, only a very small release of mononucleotides was observed from fresh DNA preparations; there was no detectable decline in biological activity accompanying this digestion, and the mononucleotides released seemed to be a random set. We concluded that they represented the digestion of some DNA fragments which invariably accompany the active fraction. The ends could, of course, have been blocked by attachment of some non-nucleotide components. Alternatively, it occurred to us that the failure to find an end might indicate that there were no ends; i.e., that the molecule was circular.

This concept of circularity led to a possible explanation for an older observation noted in the first φX paper from my laboratory: some φX DNA preparations, when sedimented under conditions leading to molecular extension (i.e., at low ionic strength or after treatment with formaldehyde), gave rise to two discrete sedimentation components which differed in sedimentation rate by 10–12%. The proportions of the two components varied from one preparation to another, but there appeared to be some correlation between the specific infectivity of a preparation and the fraction of faster-sedimenting component.

It occurred to us that the faster-sedimenting component under these conditions might be the closed circular form of the DNA; the slower-sedimenting form might then be rings that had been opened inadvertently and which were singly cleaved and therefore biologically inactive. This concept suggested a means for verifying the circularity of φX DNA. A preparation composed almost exclusively of the faster-sedimenting (ring) form could be nicked progressively with a dilute preparation of endonuclease (or by heat). The mean number of nicks could be determined from the declining infectivity of the preparation, assuming a random distribution of nicks and that one nick inactivates. If a Poisson distribution of nicks applied, the mean number of nicks should correlate with the migration of the DNA from the faster-sedimenting component to the slower-sedimenting component and then to random-size fragments (Fiers and Sinsheimer 1962b).

Measurement of the proportion of the original DNA that remained resistant to digestion by exonuclease provided another measure of the mean number of DNA nicks per molecule.

All the data fitted together precisely and supported fully the concept of circularity. An additional facet that emerged from the data was the evident presence of a single region in the φX DNA resistant to digestion by *Escherichia coli* exonuclease I, which we interpreted as a likely hairpin structure.

Satisfied that φX single-stranded DNA did indeed have a ring structure, we decided to determine whether the previously described double-stranded replicative form of φX DNA also had a ring structure. Employing an approach analogous to ours, Dr. Burton undertook the digestion of the double-stranded replicative form using spleen deoxyribonuclease as the cleaving agent. We had shown that this enzyme preferred to cleave both strands at effectively the same nucleotide level. Her results had verified the ring hypothesis (Burton and Sinsheimer 1963) when we became acquainted with the then new Kleinschmidt method of making DNA molecules visible in the electron microscope by coating them with a basic protein (cytochrome).

Dr. Kleinschmidt was then at Berkeley, and I arranged to visit him with a small sample of purified φX replicative-form DNA prepared by Dr. Burton. Dr. Kleinschmidt prepared the specimen for microscopy and took the pictures. We eagerly awaited the emergence of the first plate from the fixative.

The first molecule examined with a small magnifying glass was indeed a ring.

Although the preparation contained more fragments of *E. coli* DNA than we had realized, numerous small rings of the dimension expected for φX DNA (1.6–1.7 μm in circumference) were evident. Visualization of our previous deduction was indeed gratifying (Kleinschmidt et al. 1963).

A few years later the elaboration of the electron microscope technique by David Freifelder (Freifelder et al. 1964) made possible the direct visualization of the ring structure of φX single-stranded DNA as well.

Lysis

The appreciable spread of the latent period of φX infection (as contrasted with that of T phages), combined with the light microscopic observation that the infected cells contained much of their original structure after the release of progeny virus, raised the suspicion that mature φX particles might gradually leak out of the bacteria rather than be released in a lytic burst.

This question remained unresolved until the performance of a very ingenious experiment by Clyde Hutchison, then a graduate student in my laboratory. Clyde hit upon the idea of depositing a small number of infected bacteria upon a Millipore filter (then newly available) with a pore size that would prevent passage of bacteria but would permit passage of the virus. When 1 ml per minute of nutrient medium was flushed through the filter, any virus released from the cells in that minute would be passed through the filter and could be assayed in the filtrate. In this way Clyde demonstrated that the filter-bound cells did indeed release their phage in bursts over time periods of less than 30 seconds—that is, they lysed (Hutchison and Sinsheimer 1963).

Infectivity

After we had demonstrated that φX DNA could be extracted from the virus in an apparently intact form, with a molecular weight accounting for the full viral content, it was plausible to inquire whether this DNA might be infective by itself. To enable the DNA to enter the cell, we proposed to use spheroplasts maintained in hypertonic medium. Early experiments by George Guthrie with φX DNA preparations and *E. coli* spheroplasts did indeed provide evidence of infectivity and viral growth in the spheroplasts. The efficiency of infectivity, however, was initially very low (approximately one infected spheroplast per 10^7 or 10^8 molecules), and therefore it was necessary to prove that the naked DNA was itself infective, not some minor component of partially degraded virus.

For this purpose the DNA was banded in cesium chloride and the contents of the centrifuge tube recovered by drops. Successive drops were assayed by infectivity to ascertain whether a peak of infectivity paralleled the observed peak of DNA.

Initial results were wholly erratic—infectivity could be found in one or another drop, with none in adjacent or intervening drops. Gradually we learned to collect the drops in a solution containing protein (to prevent absorption to test-tube walls) and versene and excess RNA (to inhibit lurking deoxyribonucleases). At the same time, George Guthrie was learning how to modify the incubation conditions and medium so as to raise the infectivity, ultimately from one infected spheroplast per 10^8 DNA molecules to one per 10^2. The combination of these developments permitted us to demonstrate a clear proportionality between infectivity and DNA content throughout the band of DNA.

As a sidelight, it may be of interest that when this work was submitted for publication to the journal *Virology*, the editor decreed that studies of the activity of viral DNA were not virology and refused to countenance its publication. The paper appeared in another journal (Guthrie and Sinsheimer 1960).

REFERENCES

Burton, A. and R. L. Sinsheimer. 1963. Process of infection with φX174: Effect of exonucleases on the replicative form. *Science* **142:** 962.

Fiers, W. and R. L. Sinsheimer. 1962a. The structure of the DNA of bacteriophage φX174. I. The action of exopolynucleotidases. *J. Mol. Biol.* **5:** 408.

———. 1962b. The structure of the DNA of bacteriophage φX174. III. Ultracentrifugal evidence for a ring structure. *J. Mol. Biol.* **5:** 424.

Freifelder, D., A. K. Kleinschmidt, and R. L. Sinsheimer. 1964. Electron microscopy of single-stranded DNA: Circularity of DNA of bacteriophage φX174. *Science* **146:** 254.

Guthrie, G. D. and R. L. Sinsheimer. 1960. Infection of protoplasts of *Escherichia coli* by subviral particles of bacteriophage φX174. *J. Mol. Biol.* **2:** 297.

Hutchison, C. A., III and R. L. Sinsheimer. 1963. Kinetics of bacteriophage release by single cells of φX174-infected *E. coli*. *J. Mol. Biol.* **7:** 206.

Kleinschmidt, A. K., A. Burton, and R. L. Sinsheimer. 1963. Electron microscopy of the replicative form of the DNA of the bacteriophage φX174. *Science* **142:** 961.

Sinsheimer, R. L. 1966. φX: Multum in parvo. In *Phage and the origins of molecular biology* (ed. J. Cairns et al.), p. 258. Cold Spring Harbor Laboratory, Cold Spring Harbor, New York.

———. 1968. Closing the ring. *Technol. Rev.* **70:** 22.

Genetic and Physical Properties of Single-stranded Genomes

The Genes of the Isometric Phages and Their Functions

Ethel S. Tessman and Irwin Tessman
Department of Biological Sciences
Purdue University
West Lafayette, Indiana 47907

The small isometric DNA phages φX174 and S13 are currently believed to encode nine genes, which specify at least 16 gene products (Fig. 1). The concept that each gene on a DNA molecule is physically distinct from every other gene has been shattered for this class of phages. The production of two proteins from one DNA sequence was reported by Linney et al. (1972), who found that translation of the gene-*A* region was initiated at two different sites in the same reading frame. The spectacular finding by Barrell et al. (1976), Sanger et al. (1977), and Weisbeek et al. (1977) of overlapping sequences that are translated in different reading frames, yielding functional proteins, provided an entirely new view of gene topology.

By the criteria of complementation tests, which measure units of genetic

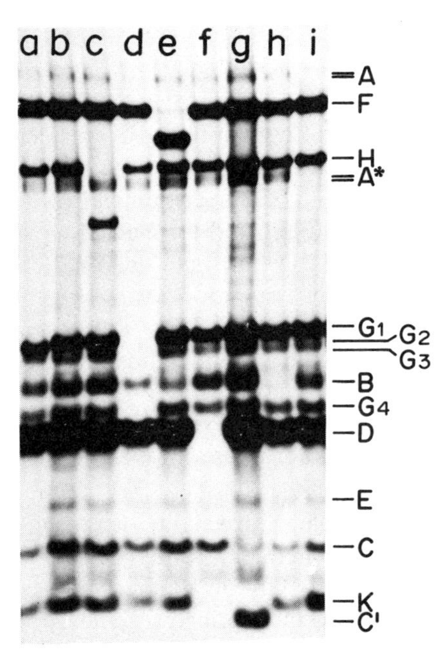

Figure 1 The proteins of φX174. Slab-gel fluorogram (T. J. Pollock, pers. comm.) of [³H]leucine-labeled extracts of UV-irradiated *E. coli* HF4704 infected with the following φX phages: (*a*) *amE*3, (*b*) *wt*, (*c*) *amH* N-1, (*d*) *amG*9, (*e*) *amF*57, (*f*) *amD*81, (*g*) *ocC*6, (*h*) *amB*210, (*i*) *amA*29. The A and A* proteins are doublets, and J does not appear in the gel.

function, there are at least eight genes in these phages (Jeng et al. 1970), designated *A* through *H*. To these must be added gene *J*, which specifies a major capsid component but is not yet represented by any known mutation.

The "one gene-one protein" hypothesis can be converted into a definition of a gene and used as a biochemical method of counting genes. Any stretch of DNA that codes for a protein is counted as one gene, whether or not some other protein is coded for somewhere in that same stretch of DNA. This method of identifying genes is implicit in current research.

We are faced with a plethora of gene products, only some of whose functions have been identified. The work of Linney et al. (1972) indicated that gene *A* codes for two proteins, A and A*, of quite different sizes, with A* initiating within gene *A*. Pollock et al. (1978) showed that A and A* each consists of two species that differ slightly in electrophoretic mobility in SDS-polyacrylamide gels. These authors also showed that gene *G* codes for four proteins of different mobilities.

For the related phage G4 (Godson 1974), extracts of irradiated cells also reveal at least 16 phage-specified protein bands but these include three high-molecular-weight proteins, labeled F*, H*, and L, not yet observed for φX (I. Tessman and B. Steffen, unpubl.).

For this discussion we will consider the function of a gene to be the function(s) that disappears when a nonpolar mutation arises in the gene. Polar mutations in genes *B*, *F*, and *G* do not cause ambiguity in this identification, because nonpolar nonsense mutations or temperature-sensitive mutations are also available.

We are also confronted by a number of phage-specified phenomena occurring after infection; for some of these, the phage gene products that may be involved have not yet been identified. Phage-related functions include adsorption of the virus particle to the bacterial cell-wall receptor, loss of infectivity of the virus particle (eclipse), penetration of phage single-stranded (SS) DNA into the cell, formation of parental complementary-strand DNA, site-specific nicking of parental replicative-form (RFI) DNA to yield the relaxed form RFII, synthesis of SS DNA, cleavage of single-stranded loops into unit genome lengths, ligation of these single-stranded pieces, conversion of progeny single strands to duplexes, synthesis of phage mRNA, cessation of host DNA synthesis, inhibition of host protein synthesis despite continued host mRNA synthesis, exclusion of superinfecting phages, formation of a phage-specified DNA methylase, aggregation of capsid proteins into small, discrete structures, formation of phage precapsid structures, formation of mature infective phage particles, and lysis of the host cells.

The molecular weights of the phage gene products presented here are derived from the φX sequence data of Sanger et al. (1977), which have been modified slightly by more recent results (see Appendix I, this volume). In the cases of the products of genes *A* (A, A*), *B*, and *C*, the molecular weights given have been determined to be the most likely ones based on inspection of the DNA sequence.

PROTEINS

The A Protein

The isolation of the first amber *A* (*amA*) mutants, using phage S13, provided the first clear example of a *cis*-acting protein, the gene-*A* protein (Tessman 1965, 1966). Sensitive measurements revealed that the *cis* action was not absolute. The yield of rescued *amA* mutants in mixed infection with an amA^+ phage always constitutes a few percent of the burst and is not due to unadsorbed particles. The number of rescued *amA* mutants never exceeds 10% of the burst, or about 20 particles per cell (Tessman et al. 1971). The few *amA* mutant phage particles that are rescued contain newly synthesized DNA and not parental DNA, as determined by density-shift experiments (A. Puga, pers. comm.). *Cis* action is also observed in single-burst experiments (E. S. Tessman, unpubl.). The *cis* action of the A protein is observed not only in S13 and φX, but also in the less closely related phage G4 (J. French, pers. comm.).

The gene-*A* product was found to be required for progeny RF DNA synthesis (Tessman 1966; Lindqvist and Sinsheimer 1967b). By infectivity measurements it was determined for a number of gene-*A* mutants altered at different loci that they all formed a small amount of RF DNA, equal to the amount formed by wild-type phage at 3 minutes in broth at 37°C (Tessman 1966). The RF DNA that was formed was assumed to be entirely parental; later this assumption was confirmed by density-shift experiments with φX (Levine and Sinsheimer 1969) and with S13 (Shleser et al. 1969). The amount of parental RF DNA formed by gene-*A* mutants was shown to be a linear function of multiplicity of infection (m.o.i.), reaching a plateau at an m.o.i. of 50 (Tessman et al. 1971).

Before the involvement of the phage A protein in progeny RF DNA synthesis was known, it had been assumed that progeny RF DNA synthesis required no phage-specified protein synthesis (Sinsheimer et al. 1962), because abundant RF DNA synthesis occurred in the presence of a dose (30 μg/ml) of chloramphenicol (Cm) known to eliminate all but a small percentage of *Escherichia coli* protein synthesis (Kurland and Maaløe 1962). Tessman (1966) found that the A protein was required for progeny RF DNA synthesis, and she was able to reconcile this requirement with the Cm results by increasing the Cm dose to 100 μg/ml. At this Cm dose progeny RF DNA synthesis is blocked but parental RF DNA synthesis is scarcely affected. (There is eventually some resumption of progeny RF DNA synthesis at 100 μg Cm/ml, however, and therefore 150 μg/ml is preferred for blocking this synthesis.) The formation of progeny RF DNA in 30 μg Cm/ml was explained by assuming either that the A protein was Cm-resistant or that only a very small amount of A protein was needed (Tessman 1966). It was later shown by Godson (1971b) and also by van der Mei et al. (1972) that synthesis of the A protein is Cm-sensitive, so it was concluded that a very small amount of A protein was sufficient to permit progeny RF DNA synthesis.

It was found that φX RF DNA sedimented on gradients as two different species: a superhelical, covalently closed duplex, RFI, and a relaxed duplex having a nick in one strand, RFII (Burton and Sinsheimer 1965; Jansz and Pouwels 1965; Roth and Hayashi 1966). In a study of the function of the gene-*A* product, Francke and Ray (1971) found that both types of RF DNA were present in the parental RF DNA synthesized after infection of a *rep* host (i.e., a host blocked in progeny RF DNA synthesis) with the control lysis mutant phage *amE*3. However, if infection was with an *amA* mutant or with *amE*3 in 150 μg Cm/ml, only the covalently closed form, RFI, was formed. They concluded that the gene-*A* product was involved in the conversion of RFI to RFII. Using alkaline gradient centrifugation to denature DNA, Francke and Ray (1971) examined the RFII DNA produced by *amE* infection of the *rep* host and found that the molecule is nicked in the viral strand only (the complementary strand remains circular). This nicking is a *cis*-dominant function, and only a small number of parental RFI molecules per cell (about four) can be converted to RFII (Francke and Ray 1972). They concluded that the gene-*A* product either is a site-specific endonuclease or else activates such an endonuclease. Their results were carried further by Henry and Knippers (1974), who purified the gene-*A* protein and showed by in vitro action that this protein is in fact a site-specific endonuclease active on φX RF and SS DNA but not on SV40 or other DNAs tested. There was some evidence for an effect on M13 DNA but none for an effect on f1 DNA. The activity is Mg^{++}-dependent and ATP-independent. The A protein shows neither exonucleolytic activity nor any activity on DNA-RNA hybrids.

Because the A protein can cleave φX SS DNA, Henry and Knippers (1974) predicted that the A protein is involved in SS DNA synthesis, presumably by cleaving the DNA into unit genome lengths. This prediction was confirmed by Fujisawa and Hayashi (1976b) for φX and by Tessman and Peterson (1977) for S13. Both groups of workers used temperature-shiftup experiments with temperature-sensitive (*ts*) mutants of gene *A* to show that the gene-*A* function is needed for SS DNA synthesis. On the basis of alkaline gradient centrifugation analysis of the pulse-labeled replicating intermediate DNA isolated during gene-*A*-mutant growth at high temperature, Fujisawa and Hayashi (1976b) concluded that this DNA consists of a double-stranded circle with a single-stranded tail of twice genome length. They suggested that *tsA* mutants are blocked in cleavage of SS DNA into pieces of unit genome length. Instead of double-length DNA tails, a continuum of oversized tails was reported from electron microscopy by Friedman et al. (1977), who analyzed the total replicative form of *tsA*-mutant-infected cells rather than the replicating intermediate alone. Further work is needed to elucidate the multiple functions of the gene-*A* protein in SS DNA synthesis.

A near-C-terminal nonsense mutant, *amA*62, studied by P. J. Weisbeek et al. (pers. comm.), provides striking evidence for the involvement of A protein in SS DNA synthesis. This mutant is blocked only in SS DNA synthesis; it forms progeny RF DNA in normal amounts. Analysis of this late-blocked gene-*A* mutant showed that the late function of the A protein is *cis*-dominant, as is true for the early function.

In both an N-terminal *tsA* mutant (in which only the A protein and not the A* protein is affected) and a C-terminal mutant (in which both the A and A* proteins are affected) SS DNA synthesis is prevented, indicating that the A protein, and not the A* protein acting alone, is involved in this stage of viral replication (Fujisawa and Hayashi 1976b). The A* protein may be involved also, but there is no evidence on this point yet.

In vivo work by Koths and Dressler (1978) has explained why the A protein is needed both early and late in phage development. Electron micrographs by these authors show that SS DNA (viral strand) is formed throughout phage development via a rolling-circle intermediate (Gilbert and Dressler 1969). Thus, A protein is needed for viral-strand synthesis at all times. At early times this SS DNA is converted to RF DNA; at late times the A protein must interact with capsid components during SS DNA synthesis (Tessman and Peterson 1976 and below) to achieve phage maturation and prevent conversion of SS DNA to RF DNA. Other aspects of gene-*A* function, including ligation of DNA, are discussed by Eisenberg et al. (this volume).

When the proper triggering signal is present, the conversion of the newly synthesized circles to duplex DNA ceases and the SS DNA is encapsidated. It seems likely that SS DNA is synthesized into an empty precapsid structure, the 108S particle (Fujisawa and Hayashi 1977a), then cleaved and ligated. Information about the signal that triggers maturation can be deduced from results of Tessman and Peterson (1976) that suggest that the signal is a recognition between an A-*rep* protein complex on the RF duplex and the F protein in its capsid configuration. These authors found that certain *E. coli* strains mutated in the *rep* gene permitted a normal amount of progeny RF DNA synthesis but were blocked in synthesis of stable SS DNA. However, mutants of phages φX and S13 designated *ogr* (*o*vercomes *g*rowth *r*estriction) were able to form both SS DNA and infective particles in these late-acting *rep* strains. Of five *ogr* mutants tested, all were mutant in gene *F*. However, other evidence suggests that *ogr* phages may be mutant in gene *A* as well as in gene *F*: (1) *ogr* mutants are *cis*-acting in their growth on *rep* strains, a result not explainable for an *F* mutant; (2) certain mutations in gene *A* are incompatible with the formation of viable *ogr* mutants. Thus, it seems likely that SS DNA synthesis at the phage maturation stage requires the formation of a complex consisting of the A, *rep*, and F proteins. It would appear that SS DNA synthesis during phage maturation must occur with the A-*rep* protein complex acting under severe stereochemical constraint imposed by the F protein in its capsid or precapsid configuration.

A convincing explanation for the *cis* action of the A protein remains to be found. The *cis* property is thought to require synthesis of the protein close to its site of attachment to DNA. The paucity of *trans* activity suggests that the protein has little success in finding its target by diffusion through the cell. Echols et al. (1976) speculated that *cis*-acting proteins in general may be trapped and lost by nonspecific binding to DNA. For the A protein, the evidence is that nonspecific binding to DNA is rare (Eisenberg et al. 1977). However, it has been postulated (van der Mei et al. 1972) that the cell membrane could be the trap for the A protein.

The A protein is involved in an unknown manner in the process of exclusion of superinfecting phages which commences 7 minutes after infection at 37°C and is completed by 15 minutes. Exclusion requires the gene-*A* function; it does not occur if the preinfecting phage is mutant in gene *A* (Tessman 1965) but does occur if it is mutant in gene *B*, *C*, *D*, *E*, *F*, *G*, or *H* (Tessman et al. 1971). In 30 μg Cm/ml, which reduces the amount of gene-*A* product but does not eliminate progeny RF DNA synthesis, exclusion does not occur (Hutchison and Sinsheimer 1971). The role of the A protein in exclusion is not known, but the finding that the membrane traps the enormous excess of A protein synthesized (van der Mei et al. 1972) suggests that the A-protein-laden membrane may be a barrier to successful infection by a superinfecting phage.

Clearly, much remains to be learned about the mechanisms by which the A protein accomplishes binding, nicking, and ligation of DNA and how it interacts with the *rep* protein and with the capsid. Isolation of mutants may yield phages having A proteins in which only one or some of these capabilities has been eliminated.

The A* Protein

The A* protein (32,000 daltons, compared with 58,650 for A) is initiated within gene *A* and in the same reading frame as the larger A protein (Linney et al. 1972). Purified extracts of A protein frequently contain A* protein (Ikeda et al. 1976; Langeveld et al. 1978), and the distinction between the functions of A and A* is ill-defined.

For one phage-specified function, shutoff of host DNA synthesis (Lindqvist and Sinsheimer 1967a), A* alone may be sufficient. Martin and Godson (1975) measured host DNA synthesis with mutants of each known phage gene except *J* and found that *B*, *C*, *D*, *E*, *F*, *G*, and *H* mutants all shut off host DNA synthesis, but two mutants (*amA*18 and *amH*90) which form neither A nor A* showed no shutoff. Three N-terminal gene-*A* amber mutants which form A* but not A could still shut off host DNA synthesis. The interpretation of the role of A* in shutoff is complicated by the fact that two gene-*A* mutants that do form A* nevertheless fail to shut off host DNA. In these two cases the A* proteins may be defective.

Martin and Godson (1975) state that *E. coli* DNA is reduced in size after wild-type infection, and they suggest that this degradation might be due to nucleolytic action of the A* protein at single-stranded regions of replication forks. No endonucleolytic activity was observed by Ikeda et al. (1976) in studies of purified A* protein in vitro. Funk and Snover (1976) reported that there is no observable DNA degradation by the time host DNA shutoff has occurred, but later in infection there is some reduction in the size of the DNA.

Funk and Snover (1976) found that two gene-*A* amber mutants that are defective in shutoff of host DNA show a great reduction in their ability to kill the host cells. At least one of these mutants is altered in the gene coding for A*. Their fate in the host cell merits further study.

At least one copy of a protein that migrates in gels close to the A* position is found in purified phage capsids (Godson 1971a; Linney and Hayashi 1974). A* is found associated with the 50S replicative structure presumed to be a precursor of mature phage particles (Fujisawa and Hayashi 1976a, 1977a). It is not present in the 108S particle, a DNA-lacking precursor of phage particles (Fujisawa and Hayashi 1977a).

The B Protein

The gene-*B* protein (13,800 daltons) is encoded within gene *A* in a second reading frame (Sanger et al. 1977; see also Godson et al.; Weisbeek and van Arkel; both this volume). Not a capsid component itself, it is involved in the formation of specific capsid subunits.

Tonegawa and Hayashi (1970) found that infected cells contain a number of discrete aggregates of phage structural proteins. By velocity sedimentation analysis they identified an aggregate of G molecules (6S), an aggregate of F molecules (9S), and an F-G complex (12S). Siden and Hayashi (1974) found that the 12S complex was missing in cells infected with gene-*B* mutants. Only mutations in genes *B*, *F*, and *G* had this effect (*J* was not tested). But T. J. Pollock (pers. comm.) found that the gene-*C* mutation *ocC*6, as well as the gene-*B* mutations *amB*210 and *amB*16, eliminates the 12S complex. Siden and Hayashi (1974) also examined the proposed precursor role of the 6S, 9S, and 12S aggregates by pulse-chase experiments and observed loss of label from all three components and its recovery in mature particles, although the loss of label from the small aggregates was only partial after a long chase. The F and G proteins in the 12S aggregate formed by the action of the B protein are not covalently bonded to each other; the aggregate is easily dissociated by urea, SDS, or chloroform (Siden and Hayashi 1974; T. J. Pollock, pers. comm.).

The mechanism of action of the B protein is unknown. If its action is only catalytic, it is surprising that it is made in relatively large amounts. Midway through an infection of nonirradiated cells the relative molar amounts of phage proteins synthesized in short pulse labelings are: B, 3.1; C, 3.1; D, 10.0; F, 2.8; G, 1.6; H, 1.0; A, 0.05 (T. J. Pollock, pers. comm.). In addition, the B protein is relatively labile compared with the other phage proteins; its half-life is 20 minutes in infected cells grown in M9 medium at 37°C (T. J. Pollock, pers. comm.).

The electrophoretic mobility of the B protein in SDS-polyacrylamide gels is quite anomalous. The B protein of φX, for example, behaves as if its molecular weight were about 19,300 daltons (Tessman et al. 1976), rather than the 13,800 daltons predicted from the nucleotide sequence, and therefore it migrates much more slowly in relation to the other phage proteins, particularly C, D, and G, than one would expect.

The C Protein

Two low-molecular-weight proteins are eliminated from gel patterns by the φX gene-*C* ochre mutation *ocC*6, and also by the S13 ochre mutation *ocC*286 (Pollock et al. 1978), leaving a smaller fragment (Borrás et al. 1971). The *ocC*6 nonsense mutation is in the correct reading frame to terminate one of these two proteins directly. This protein, designated C, is assumed to begin at base 133 on the φX sequence map of Sanger et al. (1977) and to end at base 390, leading to a calculated molecular weight of 10,069 daltons. The second protein eliminated by *ocC*6 may prove to be identical with protein K which begins at base 51 in gene *A*, in a reading frame different from those used for gene *A* and gene *B*, and ends at base 218, which is inside gene *C* (Shaw et al. 1978). Evidence that this second protein eliminated by the *ocC* mutations corresponds to the K protein includes the following: (1) its high leucine-to-lysine isotope incorporation ratio, 3.2 ± 0.5, agrees well with the ratio of 2.8 calculated from the sequence; (2) its electrophoretic mobility on a calibrated acrylamide gel agrees with the predicted molecular weight of 6400 daltons. The *ocC*6 mutation, although not in the reading frame of gene *K*, is located within gene *K* and causes a serine-to-leucine change at amino acid 49 of the K protein. The actual reason for the absence of K protein in extracts of cells infected with *ocC*6 is not yet understood. The K-protein band is also absent in extracts of φX *amA*62, *amD*81, and some *amB* mutants. These properties of the φX K protein have been reported by Pollock et al. (1978).

Fujisawa and Hayashi (1977a) have shown that the C protein functions at a late stage in phage maturation. They observed that infection with an ochre *C* mutant (φX *ocC*6) results in the abundant accumulation of a 108S protein particle lacking DNA. The 108S structure contains the F, G, and H proteins in the same proportions found in mature phage, but it lacks a major capsid protein, J. In addition, they found large amounts of the noncapsid D protein attached to these particles. They suggest that the 108S particle is a precursor to mature phage, because similar particles accumulate when SS DNA synthesis is stopped in wild-type-infected cells by thymidine deprivation late in infection. Upon restoration of thymidine, the 108S particles disappear and mature infective phage particles accumulate.

DNA synthesis by *φ*X *C* mutants differs markedly from that by other gene mutants. It was observed by Funk and Sinsheimer (1970) that in infections with an ochre *C* mutant, a DNA form accumulates which sediments as a shoulder at the leading edge of the RFI peak. Using pulse-chase experiments in a *C*-mutant infection, Fujisawa and Hayashi (1977a) observed that radioactive label incorporated into progeny RFI and RFII is chased out of these replicative forms, and although the label is chased into the region of a neutral sucrose gradient that in a wild-type infection contains the replicating intermediate, the labeled material is not a rolling circle. Instead, the label is found in a novel replicative form, RFI*, which yields no soluble counts after S1 endonuclease treatment and which therefore has no single-stranded region and is not the replicating intermediate. RFI* is dissociated into RFI by alkali or RNase treatment and consists of covalently closed RFI apparently complexed with nascent mRNA. Thus, RFI* may be a transcriptional form of RFI. RFI* is also found in cells infected with *B*, *D*, *F*, and *G* mutants, but only in the case of *C*-mutant infection did Fujisawa and Hayashi (1977a) find that label from RFII and RFI appears to be chased into RFI*. On the basis of these pulse-chase data they hypothesized that a function of the C protein is to prevent strand closure of RFII, thus allowing RFII DNA to be available as a template for SS DNA synthesis. This hypothesis remains to be proved.

T. J. Pollock and E. S. Tessman (unpubl.) found that *ocC*6 infection results in the absence of the 12S aggregate known to consist of F and G molecules (Tonegawa and Hayashi 1970). The absence of this peak (despite its presence in control [*amE*3] infections) suggests that the C protein, as well as the B protein, has a role in the formation of the F-G aggregates. It is difficult to reconcile this early function of the C protein with the C function that presumably is operative during the stage of encapsidation of SS DNA.

The D Protein

When it was found that the noncapsid D protein (16,900 daltons) is made in larger molar amount than any other phage-specified protein (Burgess and Denhardt 1969; Gelfand and Hayashi 1969), it appeared that the function of this protein was not likely to be catalytic. It was thought that the D protein of φX might act in the same manner as the major product of the filamentous phage M13, the gene-V protein, which was shown by Salstrom and Pratt (1971) to bind to progeny SS DNA, thereby blocking conversion of SS to RF DNA. By analogy, absence of D protein should cause overproduction of RF DNA. Iwaya and Denhardt (1971) found, however, that an *amD* mutant gave no increase in RF DNA beyond the initial plateau level, which indicates that the gene-V protein and the gene-*D* protein probably do not have similar functions. Later, Farber (1976) examined the D protein extensively for DNA-binding ability and found that it was not a DNA-binding protein, thus confirming the earlier results of Burgess (1969b). He suggested that it may act instead as a scaffold protein in virus assembly, perhaps in a manner analogous to the action of protein p8 of phage P22 (King and Casjens 1974).

An early indication that the D protein might be involved in capsid formation was the interaction between the D and F proteins deduced from the absence of complementation between φX *amD* and S13 *amF* mutants (Jeng et al. 1970). Poor complementation is also observed between φX *amD* mutants and S13 *amG* and *amH* mutants (E. S. Tessman, unpubl.), which indicates that there is also recognition between the D and G proteins and between the D and H proteins. The apparent involvement of D protein in capsid assembly is supported by results of Weisbeek and Sinsheimer (1974), who, using special lysis conditions, detected a peak of mature infective particles that sedimented faster (132S or 140S) than normal mature particles (114S) and that contained about 80 monomers of D protein per particle. Weisbeek et al. (1973) showed that the 132S particle can be converted to mature infective 114S particles in vitro by adding 0.1 M $MgCl_2$.

Evidence supporting the role of D protein as a scaffold protein in virus assembly was obtained by Fujisawa and Hayashi (1976a, 1977a), who found that the D protein was present in very large amounts in structures that appear to be precursors of mature phage. These precursors are (1) the DNA-lacking 108S particles, composed of the D, F, G, and H proteins, and (2) the 50S DNA-containing precursor which consists of phage DNA in its rolling-circle intermediate form associated with the phage-specified proteins A, A*, D, F, G, H, and J.

A serological approach was used by Fujisawa and Hayashi (1977a) to locate the D protein in the 132S particle. They showed that the latent infectivity of the 132S particles is not neutralized by anti-φX serum, but that after in vitro conversion to the normal 114S particle size, with release of D, there is rapid neutralization of infectivity. This indicates that the spike component (presumably the H protein) which normally reacts with anti-φX serum (Jazwinski et al. 1975a) is masked by the D protein in the 132S particle. This was confirmed by electron micrographs showing that the 132S particle is spherical rather than spiked (Fujisawa and Hayashi 1977b).

These studies, which show that the D protein is located on the outside of the 132S mature infective particle, might lead one to conclude that the capsid components (F, G, H, and J) are assembled within a shell of D protein. However, no shell of D protein has been observed in gradients, and there is no evidence of self-aggregation of D protein beyond a tetramer (Farber 1976).

An analogy between the D protein (Farber 1976) and the known scaffold protein p8 of P22 (King and Casjens 1974) should not be carried too far. In contrast to the external location of D in the 132S DNA-containing particle, the p8 protein appears to be located on the inner surface of the P22 prohead and is released when DNA enters. Although p8 protein is synthesized in small amounts, the requirement for large amounts of p8 is met by its release and reuse.

Clements and Sinsheimer (1974) reported that the D protein has a positive regulatory role in phage-specific transcription. Reexamination of this problem by Pollock (1977) showed that the D protein has no regulatory function in phage transcription or translation.

The E Protein

The gene-*E* product was first recognized as the lysis protein by Hutchison and Sinsheimer (1966), who found that one class of φX nonlysing amber mutants formed mature phage particles in the nonpermissive host but failed to lyse the cells. The average burst size (by artificial lysis) obtained 90 minutes after infection with an *amE* mutant is about 2000 particles per cell, compared with a natural burst of approximately 200 for wild-type phage. Cells infected with *amE* mutants keep increasing in length but cannot form colonies.

The E protein has recently been identified (T. J. Pollock, pers. comm.) in extracts of irradiated φX-infected cells, using φX *amE*3 to eliminate the E band. The E protein migrates on SDS gels at a position between C and D (corresponding to 10,700 daltons). This mobility is in agreement with the molecular weight of 10,600 that can be calculated for the E protein from the DNA sequence work of Barrell et al. (1976). In UV-irradiated cells, the E protein is synthesized in approximately 10% the molar amount of D. Barrell et al. (1976) located gene *E* on the φX map from its nucleotide sequence alone and found that *E* is read in a second reading frame that is entirely within gene *D*. The predicted amino acid sequence shows that the E protein is hydrophobic near the N terminus; Barrell et al. (1976) have suggested that this may allow interaction with the cell membrane. The φX lysis protein differs functionally from T4 lysozyme in that φX-infected cells containing wild-type lysis protein cannot be lysed by chloroform.

It is not known what metabolic events trigger lysis of the cell. Various treatments of the bacterial cell will prevent lysis despite the presence of some active gene-*E* product. These include UV irradiation (Gelfand and Hayashi 1969) and treatment with 300 μg mitomycin C/ml (Fukuda and

Sinsheimer 1976). It is known that phage protein synthesis decreases drastically with increasing UV dose to the cell (Burgess and Denhardt 1969; Gelfand and Hayashi 1969), so absence of lysis in UV-irradiated cells may be due to reduction of E-protein synthesis below a critical value; alternatively, results of Denhardt and Sinsheimer (1965) suggest that the lysis protein can act only if the cells are in an optimal physiological state.

For other genes mutants have been found which exhibit inhibition or delay of lysis, but in these cases no mature particles are formed. E. S. Tessman (1965, 1967) found that lysis was blocked in all *amA* mutants tested and delayed greatly in some *tsF* and *tsG* mutants; lysis was also blocked in a *tsH* mutant at 40°C (Hutchison and Sinsheimer 1966). Lysis by *amA* mutants is host-strain-dependent. The *E. coli* C strain originally used by Tessman as a nonpermissive host was not lysed by any S13 *amA* mutant, but this strain later changed to a form lysed completely by these same mutants, with lysis occurring a few minutes later than for wild-type phage. This same strain, although nonsuppressing, now actually plates S13 *amE* mutants that formerly would not plate; they show tiny, white-haloed plaques (I. Tessman, unpubl.).

Gene-*E* mutants provide the most convenient method for preparing large quantities of phage (Hutchison and Sinsheimer 1966). The *E*-mutant-infected cells are usually lysed artificially with lysozyme-EDTA.

Most amber mutants altered in the C-terminal region of S13 gene *E* tend to be very leaky, suggesting that a complete E polypeptide is not needed for lysis (E. S. Tessman, unpubl.). Despite their extreme leakiness, it was possible to map these mutants with respect to an N-terminal *F* mutant.

The F Protein

The F protein (48,400 daltons) was first identified as a capsid protein by means of the altered heat stability of various *F* mutants (Tessman and Tessman 1966). The F and J proteins together constitute the main capsid of φX, as distinct from the spikes (Edgell et al. 1969) which are composed of the G and H proteins and can be dissociated from the main capsid by 4 M urea. Despite the distinction between main capsid and spikes, the F, G, H, and J proteins are all usually termed "capsid" components. The φX capsid is in the shape of a regular polyhedron (Hall et al. 1959) that can be considered either an icosahedron or a dodecahedron; which solid is used to describe the virus is a matter of personal taste or convention. The two Platonic solids are termed reciprocal: the icosahedron has 20 faces and 12 vertices, the dodecahedron has 12 faces and 20 vertices; five three-sided faces meet at each vertex of the icosahedron, and three five-sided faces meet at each vertex of the dodecahedron. In their original work on the φX structure, Hall et al. (1959) described the 12 symmetrically arranged "knobs" of the virus, the morphological units, as if they were the 12 faces of a dodecahedron. But the knobs could just as well be considered to be at the 12 vertices of an icosahedron. This happens to be the conventional view, which we will retain.

Burgess (1969a) analyzed the molar ratios of the φX capsid components by SDS-polyacrylamide gel electrophoresis and proposed a model in which there are 60 copies of F protein per particle, grouped into 12 pentamers; each pentamer composes one vertex of an icosahedral particle and reflects the fivefold symmetry expected at each vertex. Based on the equimolar ratio of F and G proteins in the capsid and the presence of G in the spikes (Edgell et al. 1969), Burgess's model also calls for 60 copies of G protein per capsid, grouped in pentamers at each of the 12 vertices.

Mutational evidence (E. S. Tessman and Peterson 1976) suggests that the F protein in its capsid or precapsid configuration is essential to the synthesis of SS DNA at the stage of phage maturation. This evidence is discussed above in "The A Protein."

The G Protein

The amino acid sequence of the G protein (19,050 daltons) located gene *G* as a unique stretch of the nucleotide sequence (Air et al. 1976). However, in extracts of infected cells four distinct bands are identified by SDS gel electrophoresis as the products of gene *G* of φX and S13 (Pollock et al. 1978) because all four bands are eliminated by gene-*G* amber mutants of both phages. The four G proteins are also found in the mature phage capsid, in the 6S aggregate of G protein, and in the 12S F-G complex in the infected cell (Pollock et al. 1978) and therefore are functional gene products.

Infection with φX *amE*3 results in an electrophoretic mobility for each of the four G proteins different from those found with wild-type and all other φX mutants (*A*, *B*, *C*, *D*, *F*, *H*) tested, including *amE*3 *cs*70. It is assumed that this electrophoretic change in the G proteins of *amE*3 has no relation to the mutation in gene *E* but is caused by a cryptic mutation in *G* (T. J. Pollock, pers. comm.).

One might speculate that the G proteins arise by initiation and/or termination at different nucleotide sites and that the variant G proteins are useful in making a properly built capsid. Although pulse-chase experiments show no evidence of protein processing, a very rapid modification of a progenitor form remains an alternative explanation for the multiple G proteins.

The G-protein species is one of the two components of the phage coat spike (Edgell et al. 1969); the other is the H protein. The spikes can be dissociated from the main capsid in 4 M urea. The G protein is present in the capsid in an amount equimolar to that found for F (Burgess 1969a). In gradients of labeled infected-cell extracts, the G protein is found as a 6S aggregate and, in association with the F protein, as a 12S aggregate.

The G protein has been implicated as the phage adsorption protein by

Weisbeek et al. (1973), who mapped three host-range mutants of φX to gene *G* by marker-rescue of DNA restriction fragments. Controversy exists over whether G or H is the adsorption protein (see "The H Protein").

The H Protein

The H protein (34,400 daltons), one of the two components of the phage coat spike, may have multiple functions. The H protein was identified as a spike component from the finding that at 40°C the φX gene-*H* mutant *ts*4 forms nonadsorbing particles lacking one of the spike proteins (Sinsheimer 1968). In Burgess's (1969a) model of the capsid, each of the 12 spikes contains one copy of the H protein together with five copies of the G protein.

Whether G or H is the sole species of adsorption protein is difficult to answer, as mutations in genes *F*, *G*, and *H* can alter the adsorption rate (I. Tessman and S. Kumar as cited in E. S. Tessman 1965; Newbold and Sinsheimer 1970). Jazwinski et al. (1975a) used the blocking of phage adsorption by anti-H antibody to show that the H protein is required for adsorption. It was not shown, however, whether anti-F or anti-G antibody can also block phage adsorption. Similar problems arise in using host-range mutants to identify H as the adsorption protein. Three host-range mutants of φX were located in gene *H* (Sinsheimer 1968), but host-range mutants are also found in genes *F* and *G* (I. Tessman, unpubl.), and three host-range mutants have been located in gene *G* by restriction fragment mapping (Weisbeek et al. 1973).

Jazwinski et al. (1975a,c) found several types of evidence supporting the idea that the H protein acts as a pilot protein that guides infecting phage DNA through the cell membrane. It had been found earlier by Jazwinski et al. (1973) that gene-III protein of phage M13 is necessary for parental RF DNA synthesis. Jazwinski et al. (1975a) showed that when a mixture of labeled φX-specified proteins is combined in vitro with phenol-extracted φX SS DNA, only H protein binds tightly to the DNA, forming a complex that has greatly enhanced infectivity on spheroplasts. This enhancement of SS DNA infectivity is blocked by anti-H antibody and by the φX lipopolysaccharide cell receptor.

Jazwinski et al. (1975c) used labeled φX to infect host cells and found that conversion of H protein to a cell-bound form not accessible to protease was severalfold greater than for the other phage capsid proteins. They considered this to be evidence that H protein of infecting phage particles is taken up into the cell. Furthermore, after φX or S13 infection, these authors recovered H protein bound tightly to the parental RF DNA in an amount equal to about half the H protein of the infecting labeled particles. That this H protein binds in vivo to many regions of the parental RF DNA was determined by gel electrophoresis of the RF DNA after restriction endonuclease cleavage.

The possibility that the H protein might have multiple functions was first indicated in a study of the S13 mutant *tsH*1069 (Doniger and Tessman 1969), a mutant which adsorbs normally yet is noncomplementing and is blocked in formation of parental RF DNA. Jazwinski et al. (1975c) have reported that the H protein of *tsH*1069 is taken up by the cells; this phage mutant appears to be normal with respect to eclipse of the particle, but the ability of the DNA to penetrate was not measured directly. However, another gene-*H* mutant studied by Jazwinski et al. (1975c), S13*tsH*281, was found to be blocked in parental RF DNA formation at 43°C, but parental DNA and H protein did penetrate the cell. Thus, a gene-*H* mutant can be defective in SS-to-RF conversion even when there is substantial uptake of DNA and H protein.

Jazwinski et al. (1975c) summarize their results by stating that the H protein of φX and S13 is required for adsorption of the phage to the receptor region on the cell surface; for triggering the conformational change (eclipse) of the structure that abolishes infectivity of the particle; for guiding the viral DNA through the lipid bilayers of the cell membrane to its replication sites, which are believed to be on the outer cell membrane (Jazwinski et al. 1975a, b; Knippers and Sinsheimer 1968); and for conversion of the viral plus strand to parental RF DNA. Some qualifications are necessary: There is no proof that the H protein is actively involved in the transport of DNA from the adsorbed particle into the cell, nor have penetration mutants been identified. Although mutants blocked in eclipse have been isolated (Dowell 1967; Segal and Dowell 1974; Incardona, this volume), assignment of any eclipse mutant to gene *H* has not yet been established. Finally, it is not known whether the block in SS-to-RF conversion for *tsH* mutants is due to the absence of a functional H protein or to the presence of an inhibitory H protein.

The gene-*H* mutants that arise most frequently are affected at a late stage of phage development. Studies on a number of such *H* mutants have shown that SS DNA is formed (Lindqvist and Sinsheimer 1967b; Siegel and Hayashi 1969; Iwaya and Denhardt 1971). In *amH* infections, the infectivity of this SS DNA is very much less than in a wild-type infection (Tessman 1966; Siegel and Hayashi 1969). One *tsH* mutant (*ts*4) has been reported to form infectious SS DNA at 40°C (Dowell and Sinsheimer 1966). Siegel and Hayashi found that about 30% of the total SS DNA in an *amH* infection is full-size, versus 94% in a wild-type infection. This full-size SS DNA is enclosed in defective particles (105S) which do not adsorb to cells. The remaining 70% of the SS DNA, designated A-DNA, sediments as a discrete peak and consists of random, linear fragments of SS DNA enclosed in noninfectious, nonadsorbing, 70S particles.

Iwaya and Denhardt (1971) have shown that an *amH* mutant forms viral-strand DNA in large amount. They found that degradation to A-DNA occurs after synthesis of SS DNA, rather than being simultaneous with its synthesis. Thus, H protein is not essential for SS DNA synthesis during maturation. The DNA in a particle lacking H protein is vulnerable to degradation; it is surprising, then, that any full-size single-stranded molecules can be recovered from *amH* infections.

Fragmented single-stranded molecules (A-DNA) are even produced in mixed infections of *amH* and amH^+ phages (Siegel and Hayashi 1969), although the reasons for this are not yet known.

The lack of infectivity of the full-size SS DNA in *amH* infections might be explained by the absence of the H protein if it is assumed that H protein is necessary for the infectivity of SS DNA enclosed in a capsid, as well as being needed, as has been shown by Jazwinski et al. (1975a), for increasing the infectivity of purified DNA on spheroplasts.

The J Protein

The J protein (4200 daltons), for which no mutants have been isolated as yet, is a major capsid component present in an amount equimolar to that of the F protein (Burgess 1969a; T. J. Pollock, pers. comm.). The J protein is prominent in gels of purified capsid protein but is only variably seen in extracts of φX-infected cells; when it appears, it makes a prominent band (E. S. Tessman, unpubl.). Its mobility in gels is close to that of C and K (Pollock et al. 1978).

The J protein is not present in the empty 108S particle (Fujisawa and Hayashi 1977a), but it is associated with the 50S replicating intermediate structure, which is believed to represent the next stage in phage maturation (Fujisawa and Hayashi 1976a, 1977a). It is expected that directed mutagenesis of DNA fragments should lead to the isolation of gene-*J* mutants and the clarification of the function of the J protein in morphogenesis.

Other Phage Proteins (L, M, N, . . .)

A possible phage-specified function whose corresponding gene has not been identified is the DNA cytosine methylase that appears after phage infection (Razin 1973). Confirmation of the phage-specific nature of this enzymatic activity came from the induction, after φX infection, of a DNA cytosine methylase in *E. coli* B, which normally lacks such a function. Methylation of φX DNA is known to occur at a single base within gene *H* (Lee and Sinsheimer 1974). When methylation of φX DNA is inhibited by infection in nicotinamide, cleavage of single strands into unit genome lengths is inhibited, as observed in electron micrographs which show rolling circles with very long tails (Friedman et al. 1977).

As mentioned in the first section, gel analyses of infected cells reveal several protein bands not yet accounted for. These are particularly prominent in G4 infection, but corresponding bands may in fact be produced by φX and S13. It is not yet known what functions, if any, these proteins have. One of these products appears to correspond in its distinctive amino acid composition to what would be the 24,000-dalton product of a sequence in φX (nucleotides 3065–3673) that is entirely in gene *H* (I. Tessman and B. Steffen, unpubl.). This protein, labeled L, or protein K may provide the methylase activity.

It is apparent now that these phages, which are among the simplest of DNA-containing viruses, have their genetic information tightly packed into the nucleotide sequence; their coding capacity is far greater than was contemplated when these viruses were singled out, because of their small size, as an ideal subject for analyzing the genes and gene functions of one organism.

ACKNOWLEDGMENTS

The writing of this review article was supported by grant AI–11853 from the National Institute of Allergy and Infectious Diseases (to E. S. T.) and grant CA22239 from the National Cancer Institute (to I. T.). We thank Thomas J. Pollock for communicating many unpublished results and for reading the manuscript. We also thank Mrs. Sandra Pijanowski for her invaluable secretarial assistance.

REFERENCES

Air, G. M., F. Sanger, and A. R. Coulson. 1976. Nucleotide and amino sequences of gene *G* of φX174. *J. Mol. Biol.* **108:** 519.

Barrell, B. G., G. M. Air, and C. A. Hutchison III. 1976. Overlapping genes in bacteriophage φX174. *Nature* **264:** 34.

Borrás, M.-T., A. S. Vanderbilt, and E. S. Tessman. 1971. Identification of the acrylamide gel protein peak for gene *C* of phages S13 and φX. *Virology* **45:** 802.

Burgess, A. B. 1969a. Studies on the proteins of φX174. II. The protein composition of the φX coat. *Proc. Natl. Acad. Sci.* **64:** 613.

———. 1969b. "The proteins of bacteriophage φX174." Ph.D. thesis, Harvard University, Cambridge, Massachusetts.

Burgess, A. B. and D. T. Denhardt. 1969. Studies on φX174 proteins. I. Phage-specific proteins synthesized after infection of *E. coli*. *J. Mol. Biol.* **45:** 377.

Burton, A. and R. L. Sinsheimer. 1965. The process of infection with bacteriophage φX174. VII. Ultracentrifugal analysis of the replicative form. *J. Mol. Biol.* **14:** 327.

Clements, J. B. and R. L. Sinsheimer. 1974. Class of φX174 mutants relatively deficient in synthesis of viral RNA. *J. Virol.* **14:** 1630.

Denhardt, D. T. and R. L. Sinsheimer. 1965. The process of infection with bacteriophage φX174. III. Phage maturation and lysis after synchronized infection. *J. Mol. Biol.* **12:** 641.

Doniger, J. and I. Tessman. 1969. An S13 capsid mutant that makes no replicative form DNA. *Virology* **39:** 389.

Dowell, C. E. 1967. Cold-sensitive mutants of bacteriophage φX174. I. A mutant blocked in the eclipse function at low temperature. *Proc. Natl. Acad. Sci.* **58:** 958.
Dowell, C. E. and R. L. Sinsheimer. 1966. The process of infection with bacteriophage φX174. IX. Studies on the physiology of three φX174 temperature-sensitive mutants. *J. Mol. Biol.* **16:** 374.
Echols, H., D. Court, and L. Green. 1976. On the nature of *cis*-acting regulatory proteins and genetic organization in bacteriophage: The example of gene *Q* of bacteriophage λ. *Genetics* **83:** 5.
Edgell, M. H., C. A. Hutchison III, and R. L. Sinsheimer. 1969. The process of infection with bacteriophage φX174. XXVIII. Removal of the spike proteins from the phage capsid. *J. Mol. Biol.* **42:** 547.
Eisenberg, S., J. Griffith, and A. Kornberg. 1977. φX174 cistron A protein is a multifunctional enzyme in DNA replication. *Proc. Natl. Acad. Sci.* **74:** 3198.
Farber, M. B. 1976. Purification and properties of bacteriophage φX174 gene *D* product. *J. Virol.* **17:** 1027.
Francke, B. and D. S. Ray. 1971. Formation of the parental replicative form DNA of bacteriophage φX174 and initial events in its replication. *J. Mol. Biol.* **61:** 565.
———. 1972. *Cis*-limited action of the gene-*A* product of bacteriophage φX174. *Proc. Natl. Acad. Sci.* **69:** 475.
Friedman, J., A. Friedmann, and A. Razin. 1977. Studies on the role of DNA methylation. III. Role in excision of one-genome-long single-stranded φX174 DNA. *Nucleic Acids Res.* **4:** 3483.
Fujisawa, H. and M. Hayashi. 1976a. Viral DNA-synthesizing intermediate complex isolated during assembly of bacteriophage φX174. *J. Virol.* **19:** 409.
———. 1976b. Gene *A* product of φX174 is required for site-specific endonucleolytic cleavage during single-stranded DNA synthesis in vivo. *J. Virol.* **19:** 416.
———. 1977a. Functions of gene *C* and gene *D* products of bacteriophage φX174. *J. Virol.* **21:** 506.
———. 1977b. Two infectious forms of bacteriophage φX174. *J. Virol.* **23:** 439.
Fukuda, A. and R. L. Sinsheimer. 1976. Viral DNA replication of φX174 mutants blocked in progeny single stranded DNA synthesis. *J. Virol.* **18:** 218.
Funk, F. D. and R. L. Sinsheimer. 1970. Process of infection with bacteriophage φX174. XXXV. Cistron VIII. *J. Virol.* **6:** 12.
Funk, F. D. and D. Snover. 1976. Pleiotropic effects of mutants in gene *A* of bacteriophage φX174. *J. Virol.* **18:** 141.
Gelfand, D. H. and M. Hayashi. 1969. Electrophoretic characterization of φX174-specific proteins. *J. Mol. Biol.* **44:** 501.
Gilbert, W. and D. Dressler. 1969. DNA replication: The rolling circle model. *Cold Spring Harbor Symp. Quant. Biol.* **33:** 473.
Godson, G. N. 1971a. Characterization and synthesis of φX174 proteins in ultraviolet-irradiated and unirradiated cells. *J. Mol. Biol.* **57:** 541.
———. 1971b. φX174 Gene expression in u.v.-irradiated cells treated with chloramphenicol. *Virology* **45:** 788.
———. 1974. Evolution of φX174: Isolation of four new φX-like phages and comparison with φX174. *Virology* **58:** 272.
Hall, C. E., E. C. Maclean, and I. Tessman. 1959. Structure and dimensions of bacteriophage φX174 from electron microscopy. *J. Mol. Biol.* **1:** 192.

Henry, T. J. and R. Knippers. 1974. Isolation and function of the gene *A* initiator of bacteriophage φX174, a highly specific DNA endonuclease. *Proc. Natl. Acad. Sci.* **71:** 1549.

Hutchison, C. A., III and R. L. Sinsheimer. 1966. The process of infection with bacteriophage φX174. X. Mutations in a φX lysis gene. *J. Mol. Biol.* **18:** 429.

———. 1971. Requirement of protein synthesis for bacteriophage φX174 superinfection exclusion. *J. Virol.* **8:** 121.

Ikeda, J., A. Yudelevich, and J. Hurwitz. 1976. Isolation and characterization of the protein coded by gene *A* of bacteriophage φX174 DNA. *Proc. Natl. Acad. Sci.* **73:** 2669.

Iwaya, M. and D. T. Denhardt. 1971. The mechanism of replication of φX174 single-stranded DNA. II. The role of viral proteins. *J. Mol. Biol.* **57:** 159.

Jansz, H. S. and P. H. Pouwels. 1965. Structure of the replicative form of bacteriophage φX174. *Biochem. Biophys. Res. Commun.* **18:** 589.

Jazwinski, S. M., A. A. Lindberg, and A. Kornberg. 1975a. The gene *H* spike protein of bacteriophages φX174 and S13. I. Functions in phage-receptor recognition and in transfection. *Virology* **66:** 283.

———. 1975b. The lipopolysaccharide receptor for bacteriophages φX174 and S13. *Virology* **66:** 268.

Jazwinski, S. M., R. Marco, and A. Kornberg. 1973. A coat protein of the bacteriophage M13 virion participates in membrane-oriented synthesis of DNA. *Proc. Natl. Acad. Sci.* **70:** 205.

———. 1975c. The gene *H* spike protein of bacteriophages φX174 and S13. II. Relation to synthesis of the parental replicative form. *Virology* **66:** 294.

Jeng, Y., D. Gelfand, M. Hayashi, R. Shleser, and E. S. Tessman. 1970. The eight genes of bacteriophages φX174 and S13 and comparison of the phage-specified proteins. *J. Mol. Biol.* **49:** 521.

King, J. and S. Casjens. 1974. Catalytic head assembling protein in virus morphogenesis. *Nature* **251:** 112.

Knippers, R. and R. L. Sinsheimer. 1968. Process of infection with bacteriophage φX174. XX. Attachment of the parental DNA of bacteriophage φX174 to a fast-sedimenting cell component. *J. Mol. Biol.* **34:** 17.

Koths, K. and D. Dressler. 1978. An electron microscopic analysis of the φX DNA replication cycle. *Proc. Natl. Acad. Sci.* **75:** 605.

Kurland, C. G. and O. Maaløe. 1962. Regulation of ribosomal and transfer RNA synthesis. *J. Mol. Biol.* **4:** 193.

Langeveld, S. A., A. D. M. van Mansfeld, P. D. Baas, H. S. Jansz, G. A. van Arkel, and P. J. Weisbeek. 1978. Nucleotide sequence of the origin of replication in bacteriophage φX174 RF DNA. *Nature* **271:** 417.

Lee, A. S. and R. L. Sinsheimer. 1974. Location of the 5-methylcytosine group on the bacteriophage φX174 genome. *J. Virol.* **14:** 872.

Levine, A. J. and R. L. Sinsheimer. 1969. The process of infection with bacteriophage φX174. XXV. Studies with bacteriophage φX174 mutants blocked in progeny replicative form DNA synthesis. *J. Mol. Biol.* **39:** 619.

Lindqvist, B. H. and R. L. Sinsheimer. 1967a. Process of infection with bacteriophage φX174. XIV. Studies on macromolecular synthesis during infection with a lysis-defective mutant. *J. Mol. Biol.* **28:** 87.

———. 1967b. The process of infection with bacteriophage φX174. Bacteriophage DNA synthesis in abortive infections with a set of conditional lethal mutants. *J. Mol. Biol.* **30:** 69.

Linney, E. and M. Hayashi. 1974. Intragenic regulation of the synthesis of φX174 gene *A* proteins. *Nature* **249:** 345.

Linney, E., M. N. Hayashi, and M. Hayashi. 1972. Gene *A* of φX174. I. Isolation and identification of its products. *Virology* **50:** 381.

Martin, D. F. and G. N. Godson. 1975. Identification of φX174 coded protein involved in the shutoff of host DNA replication. *Biochem. Biophys. Res. Commun.* **65:** 323.

Newbold, J. E. and R. L. Sinsheimer. 1970. Process of infection with bacteriophage φX174. XXXIV. Kinetics of the attachment and eclipse steps of the infection. *J. Virol.* **5:** 427.

Pollock, T. J. 1977. Gene *D* of bacteriophage φX174: Absence of transcriptional and translational regulatory properties. *J. Virol.* **21:** 468.

Pollock, T. J., I. Tessman, and E. S. Tessman. 1978. Potential for variability through multiple gene products of bacteriophage φX174. *Nature* (in press).

Razin, A. 1973. DNA methylase induced by bacteriophage φX174. *Proc. Natl. Acad. Sci.* **70:** 3773.

Roth, T. F. and M. Hayashi. 1966. Allomorphic forms of bacteriophage φX174 replicative DNA. *Science* **154:** 658.

Salstrom, J. S. and D. Pratt. 1971. Role of coliphage M13 gene 5 in single-stranded DNA production. *J. Mol. Biol.* **61:** 489.

Sanger, F., G. M. Air, B. G. Barrell, N. L. Brown, A. R. Coulson, J. C. Fiddes, C. A. Hutchison III, P. M. Y. Slocombe, and M. Smith. 1977. Nucleotide sequence of bacteriophage φX174 DNA. *Nature* **265:** 687.

Segal, D. J. and C. E. Dowell. 1974. Cold-sensitive mutants of bacteriophage φX174. II. Comparison of two cold-sensitive mutants. *J. Virol.* **14:** 1115.

Shaw, D. C., J. E. Walker, F. D. Northrop, B. G. Barrell, G. N. Godson, and J. C. Fiddes. 1978. Gene *K*, a new overlapping gene in bacteriophage G4. *Nature* **272:** 510

Shleser, R., A. Puga, and E. S. Tessman. 1969. Synthesis of RF DNA and mRNA by gene IV mutants of bacteriophage S13. *J. Virol.* **4:** 394.

Siden, E. J. and M. Hayashi. 1974. Role of the gene *B* product in bacteriophage φX174 development. *J. Mol. Biol.* **89:** 1.

Siegel, J. E. D. and M. Hayashi. 1969. φX174 bacteriophage structural mutants which affect deoxyribonucleic acid synthesis. *J. Virol.* **4:** 400.

Sinsheimer, R. L. 1968. Bacteriophage φX174 and related viruses. *Prog. Nucleic Acid Res. Mol. Biol.* **8:** 115.

Sinsheimer, R. L., B. Starman, C. Nagler, and S. Guthrie. 1962. The process of infection with bacteriophage φX174. I. Evidence for a replicative form. *J. Mol. Biol.* **4:** 142.

Stone, A. B. 1970. Protein synthesis and the arrest of bacterial DNA synthesis by bacteriophage φX174. *Virology* **42:** 171.

Tessman, E. S. 1965. Complementation groups in phage S13. *Virology* **25:** 303.

———. 1966. Mutants of bacteriophage S13 blocked in infectious DNA synthesis. *J. Mol. Biol.* **17:** 218.

———. 1967. Gene function in phage S13. In *The molecular biology of viruses* (ed. J. S. Colter and W. Paranchych), p. 193. Academic Press, New York.

Tessman, E. S. and P. K. Peterson. 1976. Bacterial *rep*$^-$ mutations that block development of small DNA bacteriophages late in infection. *J. Virol.* **20:** 400.

———. 1977. Gene *A* protein of bacteriophage S13 is required for single-stranded DNA synthesis. *J. Virol.* **21:** 806.

Tessman, E. S., M.-T. Borrás, and I. L. Sun. 1971. Superinfection in bacteriophage S13 and determination of the number of bacteriophage particles which can function in an infected cell. *J. Virol.* **8:** 111.

Tessman, I. and E. S. Tessman. 1966. Functional units of phage S13: Identification of two genes that determine the structure of the phage coat. *Proc. Natl. Acad. Sci.* **55:** 1459.

Tessman, I., E. S. Tessman, T. J. Pollock, M.-T. Borrás, A. Puga, and R. Baker. 1976. Reinitiation mutants of gene *B* of phage S13 that mimic gene *A* mutants in blocking RF synthesis. *J. Mol. Biol.* **103:** 583.

Tonegawa, S. and M. Hayashi. 1970. Intermediates in the assembly of φX174. *J. Mol. Biol.* **48:** 219.

van der Mei, D., J. Zandberg, and H. S. Jansz. 1972. The effect of chloramphenicol on synthesis of φX174-specific proteins and the detection of the cistron A proteins. *Biochim. Biophys. Acta* **287:** 312.

Weisbeek, P. J. and R. L. Sinsheimer. 1974. A DNA-protein complex involved in bacteriophage φX174 particle formation. *Proc. Natl. Acad. Sci.* **71:** 3054.

Weisbeek, P. J., J. H. van de Pol, and G. A. van Arkel. 1973. Mapping of host range mutants of bacteriophage φX174. *Virology* **52:** 408.

Weisbeek, P. J., W. E. Borrias, S. A. Langeveld, P. Baas, and G. A. van Arkel. 1977. Bacteriophage φX174: Gene *A* overlaps gene *B*. *Proc. Natl. Acad. Sci.* **74:** 2504.

The Isometric Phage Genome: Physical Structure and Correlation with the Genetic Map

Peter J. Weisbeek and Gerard A. van Arkel
Department of Molecular Cell Biology
and Institute of Molecular Biology
State University of Utrecht
Utrecht, The Netherlands

Replication of the isometric DNA bacteriophages φX174 and S13 during the infective cycle is a very efficient and rapid event. Within 20 minutes an infection results in the synthesis of some 200 progeny phage particles and their release through lysis of the host cell. This process is rapid both because the viral genome is small, and therefore will be replicated rapidly, and because, in its life cycle, the phage makes extensive use of already existing bacterial proteins. Also, the viral gene product that is essential in the earliest stage of phage reproduction is needed only in small amounts.

These properties of the small DNA phages make them attractive and suitable organisms for experiments in molecular biology. Our knowledge of a number of physiological processes in the bacterial cell, such as DNA replication, has been enhanced by using these viruses as test systems. From detailed analyses of particle structure, the organization of the phage DNA, and the various gene functions coded for in the viral genome, a fairly complete picture has emerged as to how these viruses, in spite of their limited genome size, are able to multiply and how they interact with many bacterial processes.

The various functions of the virus will be reflected in the organization of its genetic material. Studies of the organization of the viral genome started some 15 years ago in the classical way—by the isolation and analysis of mutants. By means of complementation analysis, the mutants were grouped into genes, and subsequent genetic recombination analysis furnished the order of the genes on the genome and a rough estimate of gene sizes based on the distances between mutations.

A powerful new technique was developed when restriction enzymes became available. The method of mapping mutations by means of marker rescue with specific DNA fragments in artificially made heteroduplex DNA can now largely replace the formal recombination technique. By using mutant single-stranded DNA circles and wild-type DNA fragments with known lengths, the distances between mutations can be determined with great resolving power and high reliability. In this way the viral genome was studied in great detail. This revealed several interesting aspects of the genetic organization and clarified some older uncertainties in the genetic map. Combining the genetic map thus obtained with the recently determined base sequence has yielded a fairly complete picture of the viral genome.

Genetic Data

Complementation

The analysis of the genetic content of the genome of isometric phages started in the mid-1960s with complementation experiments utilizing conditionally lethal mutants (Tessman 1965; Hutchison and Sinsheimer 1966). The mutants were classified into complementation groups, the number of such groups being indicative of the number of biological functions directed by the viral genome. In φX, this analysis has revealed eight complementation groups (or genes) (Funk and Sinsheimer 1970); the same number was found in S13 (Jeng et al. 1970). Mutants in one of these genes (*A*) are not rescued efficiently by mutants in other genes, although gene-*A* mutants can complement other mutants quite effectively (asymmetric complementation [Tessman 1965]). For seven of these genes the corresponding gene products were identified by protein analysis of cells infected with phage mutants. From the molecular weights of the protein products as determined by SDS-polyacrylamide gel electrophoresis (summarized by Denhardt 1977; see also Table 1), it has been possible to estimate the sizes of the corresponding genes.

Complementation data give no clue as to the relative positions of genes on the genome except when one mutation blocks the expression of more than one function, as in the case of polar mutants. It was deduced from a study of polar mutants that gene *A* precedes gene *B* in φX (Sinsheimer 1968) and that gene *H* precedes gene *A* in S13 (Tessman 1965).

Table 1 φX mutants

Gene	Mutants
A	*am*86, *am*N14, *am*50, *am*8, *to*5, *am*30, *am*33, *ts*R3-1, *am*H90, *am*R8-1, *am*18, *am*35, *am*S29, *ts*56, *am*62, *op*15
B	*to*8, *am*H210, *am*16, *ts*6, *ts*173, *ts*R5-2, *ts*116
C	*oc*6
D	*am*H81, *am*42, *ts*Z7-1
E	*am*3, *am*6
J	—
F	*am*H244, *ts*27, *ts*γ, *am*H600, *ts*Z4-1, *ts*H1, *ts*41D, *am*H57
G	*am*9, *em*25, *am*H116
H	*am*H257, *ts*4, *am*N1

Recombination

Evidence for genetic recombination in the isometric single-stranded DNA phages was first obtained for S13 by Tessman and Tessman (1959) in crosses between host-range and plaque-morphology mutants. Pfeifer (1961a,b) developed a multistep host-range system for φX and found that besides a fair additivity of the distances between mutations there exists high negative interference. He was the first to suggest that the genome might have a circular structure. This assumption has been confirmed by ultracentrifugation analysis (Fiers and Sinsheimer 1962), electron microscopy (Freifelder et al. 1964), and genetic studies (Baker and Tessman 1967; Benbow et al. 1971).

Although recombination frequencies are low, ranging from 10^{-6} to 10^{-2} wild-type recombinants among progeny phage, the use of conditionally lethal mutants enables the automatic selection of recombinants and makes possible a standard genetic recombination analysis. From experiments involving two-, three-, and four-factor crosses, genetic maps of φX and S13 have been constructed (Baker and Tessman 1967; Hutchison 1969; Benbow et al. 1971; Hayashi and Hayashi 1974). A number of circumstances have influenced the accuracy of the resulting recombination maps, e.g., the relatively poor reproducibility of the low recombination frequencies, the high negative interference found in the gene-*A* region, and the effects of the choice of phage mutants and bacterial strains and other experimental conditions. However, careful experimentation has yielded recombination maps with the correct order of genes. By combining these maps with the data on the sizes of the proteins of the various genes, a reasonably accurate genetic map could be drawn (Benbow et al. 1971).

PHYSICAL STRUCTURE

The construction of a genetic map by means of purified specific DNA fragments obtained by digestion of replicative-form (RF) DNA with restriction endonucleases is based on the determination of the genetic content of each DNA fragment. The order of the genes and the distribution of the mutations on the viral DNA can be determined from the order of the DNA fragments on a cleavage map as soon as we know which fragment carries which mutations (from marker rescue) and which mutation belongs to which gene (from complementation). The accuracy of the resulting genetic map is increased considerably when DNA fragments made by many different restriction enzymes are used. Therefore, the requirements for genetic mapping with fragments are (a) cleavage maps of the DNA, obtained by using many restriction enzymes, which show the order and sizes of the fragments; (b) a good correlation of the various cleavage maps; and (c) conditionally lethal mutants that have been classified into genes by complementation.

The DNA of the isometric phages is cleaved by a large number of restriction endonucleases (Godson and Roberts 1976; Roberts 1976). The enzymes that have been used in the construction of the genetic fragment map are *Hin*dII (from *Haemophilus influenzae* R_d), *Hae*III (from *Haemophilus aegyptius*), *Alu*I (from *Arthrobacter luteus*), *Hha*I (from *Haemophilus haemolyticus*), *Hin*fI (from *Haemophilus influenzae* R_f), *Hap*II (from *Haemophilus aphrophilus*), and *Hin*HI (from *Haemophilus influenzae* H-1).

Cleavage Maps

The first attempt to construct a cleavage map of φX made use of the existing recombination map (Benbow et al. 1971). The DNA fragments were located on the recombination map by genetic fragment assay (see below); with this method, the known order of the mutants gives the order of the fragments. This is the reverse of the mutant mapping that we describe in this chapter. A limited number of *Hin*dII and *Hae*III fragments were ordered in this way (Chen et al. 1973). Owing to the large number of fragments to be ordered (13 with *Hin*dII and 11 with *Hae*III) and the limited number of mutants tested, only the large fragments could be mapped. The same problem was encountered by Hayashi and Hayashi (1974) when they constructed the *Hin*HI and *Hap*II maps in this way. Although these enzymes cleave RF DNA into only five and eight fragments, respectively, the correct localization of the smaller fragments required other methods (e.g., DNA-DNA hybridization). The limited accuracy of the recombination map is, of course, an attendant problem.

Several methods that are independent of any genetic information are available to construct cleavage maps. The ones that have been used with φX will be discussed briefly, and the resulting maps are given in Figure 1.

Partial digestion. This is the most generally applicable way of aligning fragments. It is based on analysis of DNA fragments formed during incomplete digestion of the intact DNA, and subsequent complete cleavage of the purified partials will in general yield the order of the fragments. Remaining problems can be solved by digestion of large fragments with a second enzyme. The first cleavage map of φX was determined in this way by Lee and Sinsheimer (1974) with a *Hpa*I + II enzyme mixture.

Cross-digestion. Once the fragment map of one restriction endonuclease is known, the map of a second enzyme can be determined by cleaving with the second enzyme the DNA fragments obtained with the first, and vice versa. Not only does the map of the second enzyme become known, but also the correlation between the two maps is made. The following enzymes have been analyzed in this way: *Hin*dII and *Hae*III (Lee and Sinsheimer 1974), *Alu*I (Vereijken et al. 1975), *Hin*fI and *Hha*I (Baas et al. 1976b), *Pst*I (from

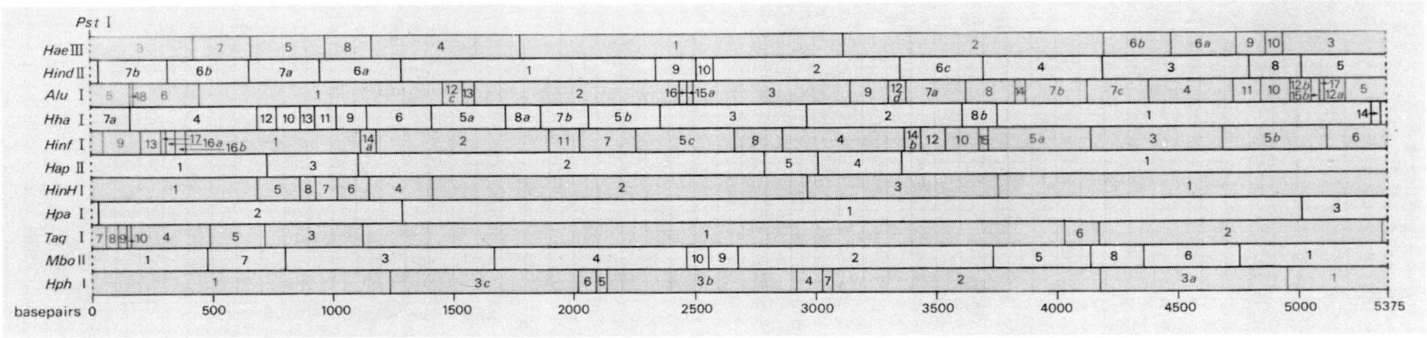

Figure 1 Linear presentation of the fragment maps of φX RF DNA. The maps start with the single *Pst*I cleavage site. The *Mbo*II and *Hph*I maps are derived from Sanger et al. (1977).

Providencia stuartii 164; Brown and Smith 1976), and *Taq*I (from *Thermus aquaticus* YTI; Sato et al. 1977).

Fragment priming. The relative order of the fragments can be determined by making use of primed DNA synthesis on single-stranded DNA. The minus strand of a given fragment is annealed to an intact plus strand and serves as a primer for DNA polymerase I action. After a short radioactive pulse followed by a chase, the newly formed double-stranded DNA will be labeled only in the DNA next to the 3′ end of the primer fragment. Cleavage of this DNA with the restriction enzyme concerned shows which DNA fragment lies adjacent to the primer fragment. A complete cleavage map can be obtained when all fragments are used successively as primer. This procedure was employed to a limited extent by Lee and Sinsheimer (1974) to support their analysis of partials. Jeppesen et al. (1976) developed the method further and used it to obtain or confirm the maps of *Hha*I, *Hap*II, *Hin*dII, *Hae*III, and *Alu*I. A limitation of this method is that it does not lead to an accurate alignment of the various maps.

Electron microscopy. The relative positions of two or more fragments are obtained by annealing the minus strands of these fragments to circular plus strands and measuring by electron microscopy the positions of the fragments and the distances between them. The *Alu*I map has been confirmed in this way (Keegstra et al. 1977). Only fragments larger than approximately 100 nucleotides can be mapped by this method.

The recently established nucleotide sequence of φX (Sanger et al. 1977; see also Appendix I, this volume) enables one to inventory quickly the possible cleavage sites of enzymes with known recognition sequences. Several other ways of obtaining a cleavage map have been described, but they have not been applied to the isometric phages and therefore will not be discussed.

Although the fragment maps described above are all derived from cleavage of double-stranded DNA, it has been found that a number of restriction enzymes are also able to cleave φX single-stranded DNA (Blakesley and Wells 1975; Godson and Roberts 1976). From the findings that *Hae*III and *Hha*I cut the single-stranded DNA into the same number of fragments produced by cleavage of RF DNA and that these fragments are the same size as the RF DNA fragments, it was concluded that the same sites are used in RF and single-stranded DNA.

Genetic Fragment Assay

φX circular DNA (either single- or double-stranded) is infectious to bacterial spheroplasts (Guthrie and Sinsheimer 1963). The information of the viral genes is expressed, and this results in DNA replication and ultimately in the formation of virus particles. When the DNA is fragmented before infection, the functioning of the viral genome is prevented and the genetic information contained in the DNA fragment is lost (Fiers and Sinsheimer 1962). No phage particles are formed and the fragments are said to be biologically inactive. However, when a fragment first is annealed to its complementary part on an intact viral DNA ring (thus forming a partial duplex DNA molecule) and then is introduced into a spheroplast, the fragment may become integrated, with a low frequency, into a newly formed RF DNA molecule. In this way a fragment can be made biologically active again and the genes or parts of genes that it contains can be determined by analyzing the progeny.

General Procedure

This method of assaying DNA fragments was first developed using non-specific fragments produced by cleavage of RF DNA with pancreatic DNase (Weisbeek and van de Pol 1970). When spheroplasts were infected with a denatured and renatured mixture of intact single-stranded DNA of mutant phage and fragmented double-stranded wild-type RF DNA, wild-type and double-mutant phages appeared in the progeny. From the kinetics of this process it was concluded that a specific nucleotide pairing of fragments with single-stranded molecules constitutes the basis of the hybridization and the formation of recombinant phages. The genetic fragment assay acquired much wider applicability in genetic analyses when it was combined with specific DNA fragments obtained by cleavage of RF DNA with restriction endonucleases (Edgell et al. 1972). It allowed a detailed correlation of the genetic and physical maps (Hayashi and Hayashi 1974; Weisbeek et al. 1976).

The approach used most often in DNA-fragment mapping studies is to anneal purified restriction enzyme fragments of wild-type double-stranded DNA to circular single-stranded DNA of a conditionally lethal mutant. The minus strand of the fragment will anneal to the viral (+) ring (Fig. 2). When the wild-type fragment covers the mutation on the circular DNA, the partial duplex (being heteroduplex at the site of the mutation) will give rise, after spheroplast infection and DNA replication, to a considerable number of wild-type particles among the otherwise mutant-phage yield. When the fragment does not cover the site of the mutation on the ring, no increase in the amount of wild-type phage will be observed (there is always a low background due to reversion).

By testing separate fragments in the assay, each mutation can be located on one of the fragments. Once the order of the fragments is known, the order of the mutations under investigation can be deduced.

When average-size fragments (200–800 base pairs) and single-stranded rings are mixed in a molar ratio of 1, the efficiency of annealing, as determined in the electron microscope, varies between 30% and 40% under standard conditions. The frequency of wild-type phage among the progeny,

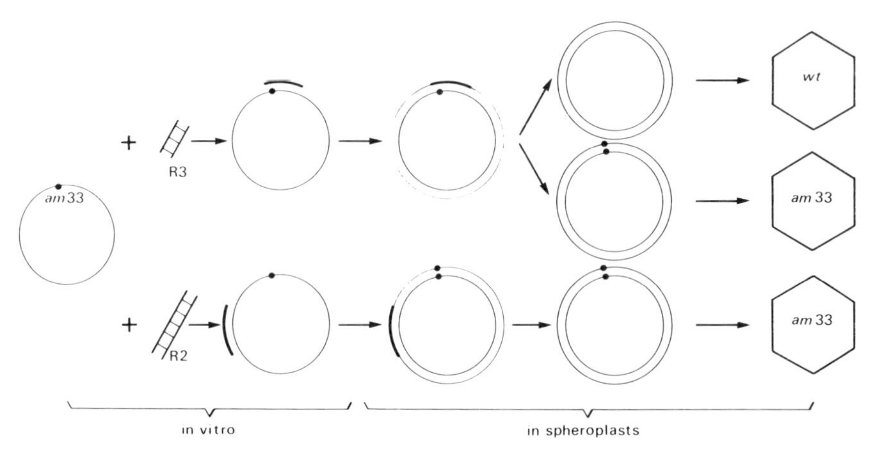

Figure 2 Procedure for genetic fragment assay of wild-type double-stranded DNA fragments and mutant single-stranded DNA. The partial heteroduplexes are made in vitro and used to infect spheroplasts. Analysis of the progeny phage shows whether or not the DNA fragment covers the mutation.

however, is much lower: 1% on the average. Our interpretation of this difference is that the probability that an annealed fragment will become integrated into the growing complementary strand (−) is much lower than 1. The frequency of rescue of the markers on the fragment will therefore depend mainly on the efficiency of fragment integration rather than on the efficiency of hybridization. This also follows from experiments in which double-mutant single-stranded rings were used (Fig. 3). When both mutations are covered by the same wild-type fragment, 90% of the recombinant progeny phage are wild-type, but when each mutation is covered by a separate fragment, 95% of the recombinants are the single mutants *am* and *ts* and only 5% are wild-type (P. Weisbeek and W. Borrias, unpubl.). If each fragment had a probability of 1 to become integrated, the two situations should have given approximately equal amounts of wild-type phage. The fact that in the situation in Figure 3b both wild-type alleles are rescued simultaneously in the majority of the partial duplexes indicates further that repair of mismatch is not an important process during the assay. Also, the

Figure 3 Progeny-phage yield when double-mutant single-stranded DNA is used in the fragment assay. (*a*) Each mutation is covered by a different DNA fragment. (*b*) Both mutations are covered by the same DNA fragment.

use of mitomycin C, which is able to reduce correction of mismatches (Baas and Jansz 1972), does not affect the marker-rescue frequencies.

Modified Procedures

In the general procedure described above, the genetic content in the minus strand of wild-type double-stranded fragments is determined. In some cases, however, it is useful or necessary to reverse the wild-type and mutant alleles in the material or to analyze the plus strand of the DNA fragment. Two examples may serve to illustrate this point. First, with certain mutants (e.g., those with an altered or extended host range) the automatic selection of recombinants in the phage yield is feasible only when mutant instead of wild-type RF DNA fragments are used and annealed to wild-type single-stranded DNA (Weisbeek et al. 1973). Second, the genetic content of fragments produced by cleavage of single-stranded DNA (Blakesley and Wells 1975) or fragments of viral DNA found in phage particles (Sinsheimer 1959) can be determined most conveniently by the analysis of the plus strand after annealing to the minus strand of double-stranded DNA. When a mixture of wild-type plus-strand fragments and mutant RFII DNA (circular duplex DNA in which at least one strand is not continuous) is denatured and renatured in this way and then assayed on spheroplasts, one observes a rescue of fragment markers to about 1% of the progeny (Weisbeek et al. 1972; Dodgson et al. 1976). Thus, the assay with mutant RFII DNA is about as efficient as the one with mutant single-stranded DNA.

Although we have not tested RFI DNA (superhelical, covalently closed, circular duplex DNA) in the fragment assay, it is likely to act as well as or even better than RFII DNA. In their studies on genetic recombination, Radding and coworkers (Holloman et al. 1975; Holloman and Radding 1976) showed that merely mixing mutant RFI DNA and wild-type plus-strand fragments suffices to bring about an efficient rescue of the wild-type allele in spheroplast infection. The negative supercoiling induces single-stranded regions in the superhelical RFI DNA that readily take up complementary pieces of single-stranded DNA. Partial triplex structures thus formed would then act as initiation points for genetic recombination.

Genetic Fragment Map

Purified fragments of wild-type double-stranded DNA produced by cleavage of RF DNA with *Hin*dII, *Hae*III, *Alu*I, *Hha*I, *Hin*fI, *Hap*II, and *Hin*HI were annealed to circular single-stranded DNA of mutants in all the viral genes. The partial duplex DNA formed was used to infect bacterial spheroplasts, and the progeny-phage yield was analyzed for parental and recombinant phage types. In this way it was determined in which fragment many of the mutations were located. The mutants tested and their gene assignments as determined by complementation analysis are listed in Table 1.

The results of the mutant mapping presented in this section are a compilation of mapping data published by Chen et al. (1973), Hayashi and Hayashi (1974), and Weisbeek et al. (1976), together with more recent unpublished information from our laboratory. More than 50% of the mutants tested belong to genes *A* and *B*. This overrepresentation of *A* and *B* mutants is caused by our special interest in this part of the genome (Weisbeek et al. 1977). All the mapping results are integrated in the seven circular fragment maps of Figure 4. The amount of overlap of the various fragments determines the accuracy of the mapping. Mutant *am*N1 is positioned between the R6c/R4 site and the F10/F15 site; these sites are separated by only 30

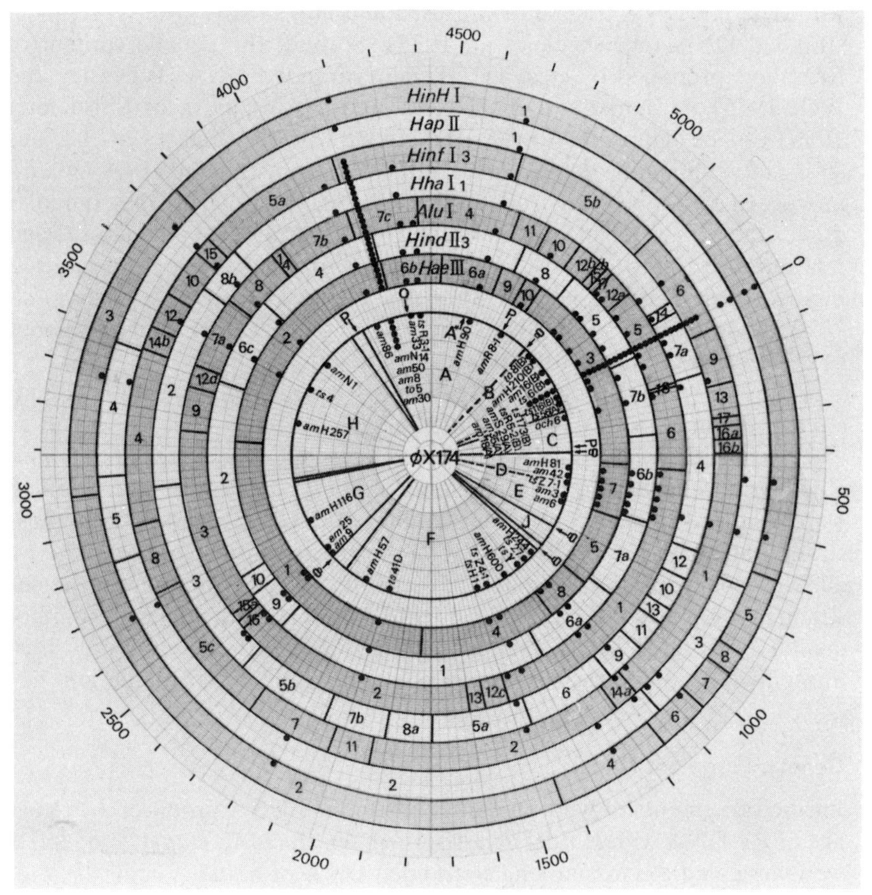

Figure 4 Correlation of the separate fragment maps with the data from the genetic fragment assays. The inner circle contains additional information, derived from sequence analysis (Sanger et al. 1977; Langeveld et al. 1978), regarding gene borders, promoters (P), ribosome-binding sites (☋), and the origin (0). The outer circle measures the map distances, in number of nucleotides, starting at the *Pst*I nick.

nucleotide pairs, whereas the possible variation in the position of *am*86 is 250 base pairs. The positions of the mutations on the DNA fragments are extrapolated toward the inner ring, in which sequence information regarding gene borders, intercistronic regions, promoters, and ribosome-binding sites (Sanger et al. 1977) is incorporated, and also toward the site at which the gene-*A* protein nicks RFI DNA (Langeveld et al. 1978). Because of the large number of mutants in the *A*-*B* region, an extended linear representation of this area is given in Figure 5.

All mutants listed in Table 1 could be mapped unambiguously, although some mutants were refractory in combination with fragments produced by certain restriction enzymes. Two different causes for a deviating behavior have been recognized. First, *ts*41D gave no reaction with each of the *Hin*fI or *Hha*I fragments, whereas with the appropriate *Hin*dII, *Hae*III, and *Alu*I fragments the frequency of recombinants was enhanced in the usual way. As *ts*41D is probably a double mutant (Hutchison 1969), we assume that both temperature-sensitive mutations map on the single fragments R1, Z1, and A2, but not on one F or H fragment. This view was later supported by results obtained with other mutants of a known double-mutant character, such as *ts*128 (Levine and Sinsheimer 1969), *ts*68, and *ts*54 (Borrias et al. 1969). In the same way, these mutants cannot be rescued by a single fragment because for each of these mutants the two mutations lie far apart in the genome (in genes *A* and *F*, *A* and *F*, and *A* and *G*, respectively).

The mutants *am*H210 (*B*) and *to*8 (*B*), which show no increase in recombinants with the *Hin*fI fragments, and *ts*116, which gives no reaction with the *Hin*dII fragments, represent another type of special behavior in the fragment assay. The best explanation for this type of behavior is that these mutations are located very close to one end of their matching fragment. The resulting partial duplex might be more prone to repair of the mismatch because of the terminal instability. It was shown by Smith et al. (1977) that *ts*116 contains two base substitutions close to the R5/R7b cleavage site (six and eight nucleotides removed from this split). Another base substitution (*am*18) at position 8 from this cleavage site does not influence the marker-rescue frequency; therefore, we may assume that the *am*H210 and *to*8 mutations map within eight base pairs of the F5b/F6 cleavage site.

The order of two mutations is determined as soon as each mutation is mapped on a different fragment. Quite often a set of mutants is mapped consistently on the same fragment, e.g., *am*9 and *em*25. In this case, no order can be determined and the mutants might even carry the same mutation. This was in fact found for *am*18 and *am*35 and for *am*3 and *am*6; each pair of mutants has an amber codon at identical positions (Barrell et al. 1976; Smith et al. 1977). Mutant *am*H57 (*F*) has not been mapped in this way. Its position within gene *F* was determined from the size of its amber protein fragment and the direction of translation.

We have found three regions in gene *A* in which relatively large numbers

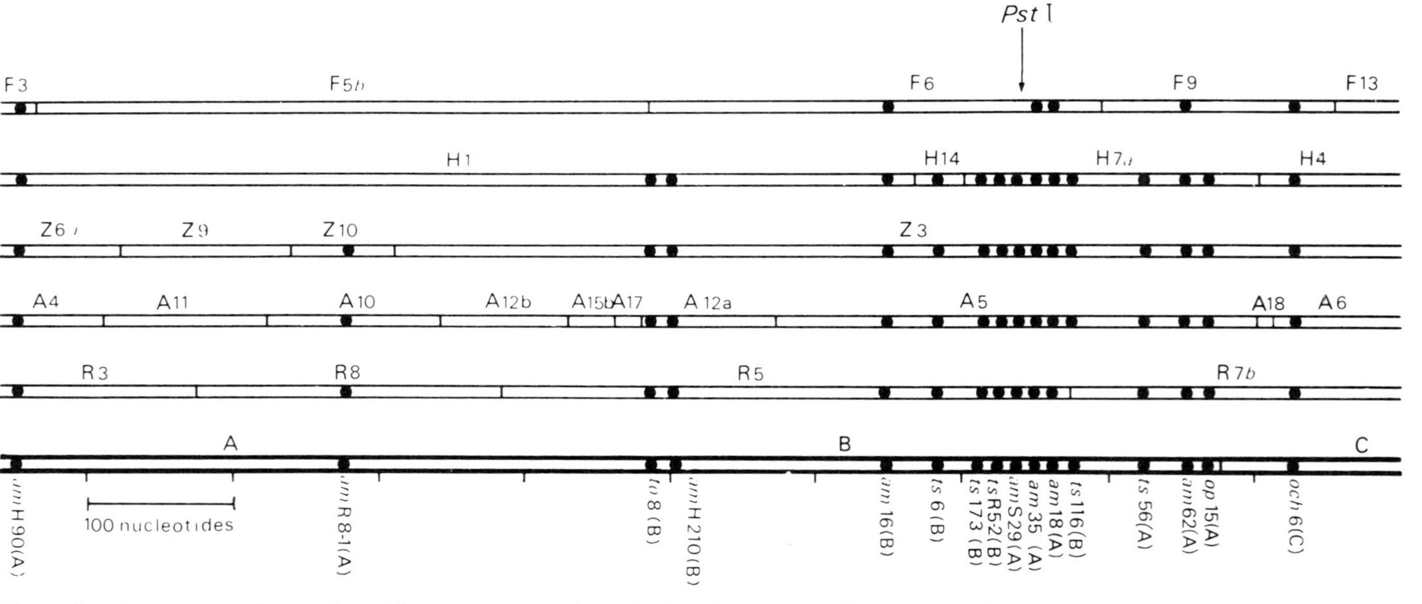

Figure 5 Genetic map (*bottom*) and fragment maps of the *A–B* region. Dots indicate the positions of mutations shown on the genetic map.

of independently isolated mutants map close together (Fig. 4; two of the regions are shown in Fig. 5). The mutants *am*N14, *am*50, *am*8, *to*5, and *am*30 map within a distance of 50 base pairs; *am*18, *am*35, *am*S29, *ts*173, and *ts*R5-2 map within 65 base pairs. Moreover, in addition to *to*8 and *am*H210, we have mapped six other amber mutants on fragment A12*a*, which has a size of 89 base pairs. Although many of these mutations might prove to be identical, the relative ease with which mutations in these regions are picked up suggests local structural differences in the DNA that influence the rate of mutation or the efficiency of isolating mutants for these regions.

Consequences of the Genetic Organization

Correlation of the physical maps of the φX DNA with the genetic map has increased our insight into the organization of the viral genome. The much more exact localization of mutants on the DNA enables one to correlate in much greater detail the genetic defects and properties of these mutants with a particular part of the DNA, and therefore also with a specific part of the protein encoded by the gene concerned. Correlation of the combined physical and genetic map with information obtained from the nucleotide sequence of the entire viral DNA (Sanger et al. 1977) adds many new aspects to the map and delineates the real gene locations and dimensions. The most interesting aspect of viral genetic organization and functioning as we know them today is the way the virus uses its DNA to store its genetic information.

Gene Overlap

Several stretches of DNA contain the information to code for two or three proteins. Gene *A* directs the synthesis of a 55,000-dalton protein (A protein), but it is also responsible for the production of a smaller protein (A* protein, 35,000 daltons) by using an internal initiator for protein synthesis (Fig. 6). The amino acid sequence of A* is identical to the sequence of the C-terminal part of the large A protein (Linney and Hayashi 1974). The reading frame in which the gene-*A* DNA is used is therefore the same for both proteins. The position of the internal initiator is within fragment A4, to the left of *am*H90 (Fig. 4).

Gene *B* is contained entirely within gene *A*; i.e., the right part of gene *A* is also used to code for the B protein. But, in contrast to the A/A* situation, the reading frame of gene *B* is different from that of gene *A*. This results in different amino acid sequences for the B and A proteins. Nonsense mutations in gene *B* are not nonsense mutations in gene *A*, and vice versa, so the genetic identities of both genes are preserved. A and A* clearly belong to one gene. The DNA region of gene *B* is therefore used for three different proteins: B, A, and A*. The position of gene *B* within gene *A* was determined in two ways: by DNA sequencing (Smith et al. 1977) and by genetic analysis of this region (Weisbeek et al. 1977).

Gene *E* is contained within gene *D* in the same way gene *B* falls in gene *A*; genes *E* and *D* use different reading frames (Barrell et al. 1976). This means that φX has found ways to obtain extra coding capacity equivalent to nearly 1500 base pairs. This is about 27% more than it would have were its DNA used only once. Moreover, the entire viral DNA does not code for protein. Intercistronic regions of different lengths are present between genes *J* and *F*, *F* and *G*, *G* and *H*, and *H* and *A* (Sanger et al. 1977). The function of these regions is unknown.

Gene J

There has been some controversy around the existence of gene *J*, which was proposed initially by Benbow et al. (1972) to explain both the aberrant behavior of *am*6 and the presence of a small protein in phage particles and infected cells. By fragment mapping we found that *am*6 is a double mutant and that one of its mutations maps very close to *am*3 (*E*). This led us to the conclusion that *am*6 does not represent a new gene (Weisbeek et al. 1976). Independent sequence analysis (Smith et al. 1977) confirmed this and showed that *am*3 and *am*6 have identical amber codons in gene *E*. Moreover, the same sequence study showed the presence of a small gene between genes *E* and *F* that codes for the small protein found in the particle. This gene has been named gene *J*; it is 114 nucleotides long. No gene-*J* mutants have been found as yet.

Gene Order

From the sequential order of the genes on the DNA, the positions of the three promoters, and the direction of transcription, it can be deduced that genes that act at different stages of the viral life cycle use different pro-

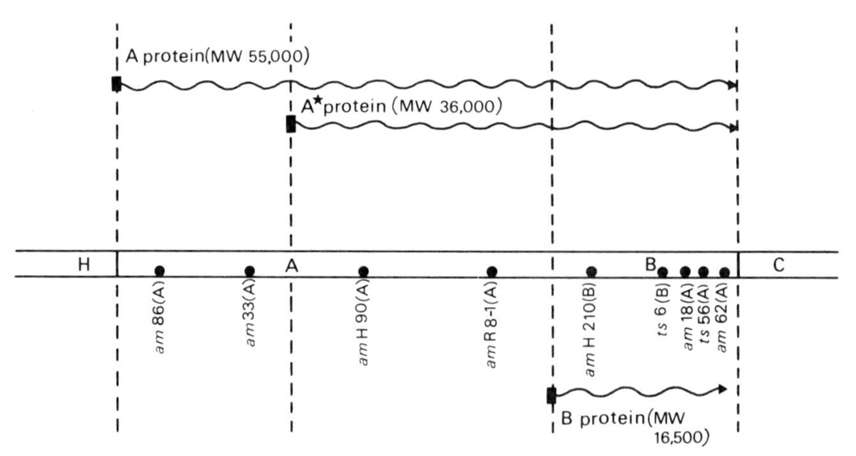

Figure 6 Organization of the *A–B* region of the φX genome and the proteins encoded.

moters. The messenger RNAs of genes involved in phage maturation and cell lysis (late functions; *D*, *E*, *J*, *F*, *G*, *H*) generally start at the same promoter, whereas gene *A*, which controls an early function, uses a different promoter (Hayashi et al. 1976). Genes *B* and *C*, although needed late in infection, have a separate promoter, perhaps because they function in late DNA synthesis (gene *C*) or are required in catalytic quantities only during phage assembly (gene *B*). However, too little is known about the regulation of viral transcription and translation to make definitive statements about the whys and wherefores of a given gene order.

Origin of Replication

Gene *A* is involved in the viral RF-to-RF DNA replication. Its gene product(s) introduces a nick in the viral strand of the parental RFI DNA molecule (Francke and Ray 1972), and this nicking initiates the semiconservative replication of the viral DNA, very likely by extending the 3′ end of the nicked viral strand. A large amount of work has been done, both in vivo and in vitro, to analyze the role of the gene-*A* product(s) in DNA replication and to determine the site of action of the A protein on the parental RF DNA. Experiments with in vivo pulse-labeled DNA showed the replication origin to be in gene *A* within the *Hin*dII fragment R3 (Godson 1974)—more precisely, within the *Hae*III fragment Z6b at approximately 100 base pairs from the Z2/Z6b split (Baas et al. 1976a). In vitro experiments with purified A protein and RFI DNA also showed the nick to be in the viral strand (Henry and Knippers 1974) in the *Hin*dII fragment R3 (Ikeda et al. 1976). Recent experiments in our laboratory placed the nick in the *Hae*III fragment Z6b. We have determined the exact position of the nick by sequence analysis. The A protein generates a free 3′-OH end, and the sequence at this 3′ end is -TGCTCCCCCAACTTG$_{OH}$ (Langeveld et al. 1978). When this sequence is compared with the φX sequence (Appendix I), the 3′ terminus at the nick site is uniquely positioned at nucleotide 4307. This position of the nick introduced in vitro in RFI DNA coincides very well with the position of the origin of replication as determined in vivo.

Previous investigations of the repair of mismatches in heteroduplex DNA in spheroplasts suggested to us that the replication origin might be located in the C-terminal part of gene *A*, between mutants *am*18 and *am*35 (Weisbeek and van Arkel 1976). However, these mutants have their amber codons at identical positions (Smith et al. 1977). Further analysis showed us that *am*35 does contain an extra mutation(s) besides its nonsense mutation. As in the experiments on heteroduplex repair, the position of the mismatch and the amount of repair of the mismatch were used to locate the position of the replication origin; the presence of more than one mismatch per heteroduplex might well have biased the results. The distances over which the repair processes extend are large (up to 2000 base pairs [Wagner and Meselson 1976]), and multiple mismatches might influence the efficiency of repair.

Evidence is increasing that many conditionally lethal mutants carry more than one mutation. These additional mutations, when not conditionally lethal, are usually not detected in standard genetic analyses. They are found by sequencing mutant DNA (e.g., in the case of *am*6 [Smith et al. 1977]) or by analyzing the restriction enzyme cleavage pattern (in the case of *to*8 [P. Weisbeek, unpubl.]).

Comparison of φX and S13

The genome of S13 has not been analyzed with the fragment assay. However, from the general properties and the detailed genetic analysis we know that the S13 genes and genome organization must be very similar to those of φX; S13 can recombine with φX and complement its mutants (Jeng et al. 1970), and S13 DNA fragments can rescue many φX markers (H. Jansz, pers. comm.). A nucleotide sequence of part of S13 (190 base pairs) shows only three differences from the homologous region in φX (Grosveld and Spencer 1977). However, there is reason to believe that this part of S13 is one of only a few regions that are highly conserved (Compton and Sinsheimer 1977) and that the differences in other parts of the genome will be more numerous. Experiments showing that S13 mutations can simultaneously affect the size of the B protein and the function of gene *A* (Tessman et al. 1976) indicate that in S13 perhaps there is also an overlap of genes *A* and *B* similar to the situation in φX (see Spencer et al., this volume).

REFERENCES

Baas, P. D. and H. S. Jansz. 1972. φX174 replicative form DNA replication, origin and direction. *J. Mol. Biol.* **63:** 569.

Baas, P. D., H. S. Jansz, and R. L. Sinsheimer. 1976a. φX174 DNA synthesis in a replication-deficient host: Determination of the origin of φX DNA replication. *J. Mol. Biol.* **102:** 633.

Baas, P. D., G. P. H. van Heusden, J. M. Vereijken, P. J. Weisbeek, and H. S. Jansz. 1976b. Cleavage map of bacteriophage φX174 RF DNA by restriction enzymes. *Nucleic Acids Res.* **3:** 1947.

Baker, R. and I. Tessman. 1967. The circular genetic map of phage S13. *Proc. Natl. Acad. Sci.* **58:** 1438.

Barrell, B. G., G. M. Air, and C. A. Hutchison III. 1976. Overlapping genes in bacteriophage φX174. *Nature* **264:** 34.

Benbow, R. M., C. A. Hutchison III, J. D. Fabricant, and R. L. Sinsheimer. 1971. Genetic map of bacteriophage φX174. *J. Virol.* **7:** 549.

Benbow, R. M., R. F. Mayol, J. C. Picchi, and R. L. Sinsheimer. 1972. Direction of translation and size of bacteriophage φX174 cistrons. *J. Virol.* **10:** 99.

Blakesley, R. W. and R. D. Wells. 1975. Single-stranded DNA from φX174 and M13 is cleaved by certain restriction endonucleases. *Nature* **257:** 421.

Borrias, W. E., J. H. van de Pol, C. van de Vate, and G. A. van Arkel. 1969. Complementation experiments between conditional lethal mutants of bacteriophage φX174. *Mol. Gen. Genet.* **105:** 152.

Brown, N. L. and M. Smith. 1976. The mapping and sequence determination of the single site in φX174 *am*3 replicative form DNA cleaved by restriction endonuclease *Pst*I. *FEBS Lett.* **65:** 284.

Chen, C.-Y., C. A. Hutchison III, and M. H. Edgell. 1973. Isolation and genetic localization of three φX174 promoter regions. *Nat. New Biol.* **243:** 233.

Compton, J. L. and R. L. Sinsheimer. 1977. Aligning the φX174 genetic map and the φX174/S13 heteroduplex denaturation map. *J. Mol. Biol.* **109:** 217.

Denhardt, D. T. 1977. The isometric single-stranded DNA phages. In *Comprehensive virology* (ed. H. Fraenkel-Conrat and R. R. Wagner), vol. 7, p. 1. Plenum Press, New York.

Dodgson, J. B., I. F. Nes, B. W. Porter, and R. D. Wells. 1976. Two new genetic assays for noninfectious fragments of φX174 DNA. *Virology* **69:** 782.

Edgell, M. H., C. A. Hutchison III, and M. Sclair. 1972. Specific endonuclease R fragments of bacteriophage φX174 DNA. *J. Virol.* **8:** 574.

Fiers, W. and R. L. Sinsheimer. 1962. The structure of the DNA of bacteriophage φX174. III. Ultracentrifugal evidence for a ring structure. *J. Mol. Biol.* **5:** 424.

Francke, B. and D. S. Ray. 1972. *Cis*-limited action of the gene *A* product of bacteriophage φX174 and the essential bacterial site. *Proc. Natl. Acad. Sci.* **69:** 475.

Freifelder, D., A. K. Kleinschmidt, and R. L. Sinsheimer. 1964. Electron microscopy of single-stranded DNA: Circularity of DNA of bacteriophage φX174. *Science* **146:** 254.

Funk, F. D. and R. L. Sinsheimer. 1970. Process of infection with bacteriophage φX174. XXXV. Cistron VIII. *J. Virol.* **6:** 12.

Godson, G. N. 1974. Origin and direction of φX174 double- and single-stranded DNA synthesis. *J. Mol. Biol.* **90:** 127.

———. 1975. Evolution of φX174. II. A cleavage map of the G4 phage genome and comparison with the cleavage map of φX174. *Virology* **63:** 320.

Godson, G. N. and R. J. Roberts. 1976. A catalogue of cleavages of φX174, S13, G4 and St-1 by 26 different restriction enzymes. *Virology* **73:** 561.

Grosveld, F. G. and J. H. Spencer. 1977. The nucleotide sequence of two restriction fragments located in the gene *A-B* region of bacteriophage S13. *Nucleic Acids Res.* **4:** 2235.

Guthrie, G. D. and R. L. Sinsheimer. 1963. Observations on the infection of bacterial protoplasts with the deoxyribonucleic acid of bacteriophage φX174. *Biochim. Biophys. Acta* **72:** 290.

Hayashi, M. N. and M. Hayashi. 1974. Fragment maps of φX174 RF DNA produced by restriction enzymes from *Haemophilus aphirophilus* and *Haemophilus influenzae* H-1. *J. Virol.* **14:** 1142.

Hayashi, M., F. K. Fujimura, and M. Hayashi. 1976. Mapping of in vivo messenger RNAs for bacteriophage φX174. *Proc. Natl. Acad. Sci.* **73:** 3519.

Henry, T. J. and R. Knippers. 1974. Isolation and function of the gene *A* initiator of bacteriophage φX174: A highly specific DNA endonuclease. *Proc. Natl. Acad. Sci.* **71:** 1549.

Holloman, W. K. and C. M. Radding. 1976. Recombination promoted by superhelical DNA and the *recA* gene of *Escherichia coli*. *Proc. Natl. Acad. Sci.* **73:** 3910.

Holloman, W. K., R. Wiegand, C. Hoessli, and C. M. Radding. 1975. Uptake of homologous single-stranded fragments by superhelical DNA: A possible mechanism for initiation of genetic recombination. *Proc. Natl. Acad. Sci.* **72:** 2394.

Hutchison, C. A., III. 1969. "Bacteriophage φX174: Viral genes and functions." Ph.D. thesis, California Institute of Technology, Pasadena, California.
Hutchison, C. A., III and R. L. Sinsheimer. 1966. The process of infection with bacteriophage φX174. X. Mutations in a φX lysis gene. *J. Mol. Biol.* **18:** 429.
Ikeda, J.-E., A. Yudelevich, and J. Hurwitz. 1976. Isolation and characterization of the protein coded by gene *A* of bacteriophage φX174 DNA. *Proc. Natl. Acad. Sci.* **73:** 2669.
Jeng, Y., D. Gelfand, M. Hayashi, R. Shleser, and E. S. Tessman. 1970. The eight genes of bacteriophages φX174 and S13 and comparison of the phage-specified proteins. *J. Mol. Biol.* **49:** 521.
Jeppesen, P. G. N., L. Sander, and P. M. Slocombe. 1976. A restriction cleavage map of φX174 DNA by pulse-chase labeling using *E. coli* DNA polymerase. *Nucleic Acids Res.* **3:** 1323.
Keegstra, W., J. M. Vereijken, and H. S. Jansz. 1977. Mapping and length measurements of restriction enzyme fragments by electron microscopy. *Biochim. Biophys. Acta* **475:** 176.
Langeveld, S. A., A. D. M. van Mansfeld, P. D. Baas, H. S. Jansz, G. A. van Arkel, and P. J. Weisbeek. 1978. Nucleotide sequence of the origin of replication in bacteriophage φX174 RF DNA. *Nature* **271:** 417.
Lee, A. S. and R. L. Sinsheimer. 1974. A cleavage map of bacteriophage φX174 genome. *Proc. Natl. Acad. Sci.* **71:** 2882.
Levine, A. J. and R. L. Sinsheimer. 1969. The process of infection with bacteriophage φX174. XXV. Studies with bacteriophage φX174 mutants blocked in progeny replicative form DNA synthesis. *J. Mol. Biol.* **39:** 619.
Linney, E. and M. Hayashi. 1974. Intragenic regulation of the synthesis of φX174 gene *A* proteins. *Nature* **249:** 345.
Pfeifer, D. 1961a. Genetic recombination in bacteriophage φX174. *Nature* **189:** 423.
———. 1961b. Genetische Untersuchungen am Bakteriophagen φX174. I. Aufbau eines selectiven Systems und Nachweis genetischer Recombination. *Z. Vererb.-Lehre* **92:** 317.
Roberts, R. J. 1976. Restriction endonucleases. *CRC Crit. Rev. Biochem.* **4:** 123.
Sanger, F., G. M. Air, B. G. Barrell, N. L. Brown, A. R. Coulson, J. C. Fiddes, C. A. Hutchison III, P. M. Slocombe, and M. Smith. 1977. Nucleotide sequence of bacteriophage φX174 DNA. *Nature* **265:** 687.
Sato, S., C. A. Hutchison III, and J. I. Harris. 1977. A thermostable sequence-specific endonuclease from *Thermus aquaticus*. *Proc. Natl. Acad. Sci.* **74:** 542.
Sinsheimer, R. L. 1959. Purification and properties of bacteriophage φX174. *J. Mol. Biol.* **1:** 37.
———. 1968. Bacteriophage φX174 and related viruses. *Prog. Nucleic Acid Res. Mol. Biol.* **8:** 115.
Smith, M., N. L. Brown, G. M. Air, B. G. Barrell, A. R. Coulson, C. A. Hutchison III, and F. Sanger. 1977. DNA sequence at the C termini of the overlapping genes *A* and *B* in bacteriophage φX174. *Nature* **265:** 702.
Tessman, E. S. 1965. Complementation groups in phage S13. *Virology* **25:** 303.
Tessman, E. S. and I. Tessman. 1959. Genetic recombination in bacteriophage S13. *Virology* **7:** 465.
Tessman, I., E. S. Tessman, T. J. Pollock, M.-T. Borrás, A. Puga, and R. Baker.

1976. Reinitiation mutants of gene *B* of phage S13 that mimic gene *A* mutants in blocking RF synthesis. *J. Mol. Biol.* **103:** 583.

Vereijken, J. M., A. D. M. van Mansfeld, P. D. Baas, and H. S. Jansz. 1975. *Arthrobacter luteus* restriction endonuclease cleavage map of φX174 RF DNA. *Virology* **68:** 221.

Wagner, R., Jr. and M. Meselson. 1976. Repair tracts in mismatched DNA heteroduplexes. *Proc. Natl. Acad. Sci.* **73:** 4135.

Weisbeek, P. J. and G. A. van Arkel. 1976. On the origin of bacteriophage φX174 replicative form DNA replication. *Virology* **72:** 72.

Weisbeek, P. J. and J. H. van de Pol. 1970. Biological activity of φX174 replicative form DNA fragments. *Biochim. Biophys. Acta* **224:** 328.

Weisbeek, P. J., J. H. van de Pol, and G. A. van Arkel. 1972. Genetic characterization of the DNA of the bacteriophage φX174 70S particle. *Virology* **48:** 456.

———. 1973. Mapping of host range mutants of bacteriophage φX174. *Virology* **52:** 408.

Weisbeek, P. J., J. M. Vereijken, P. D. Baas, H. S. Jansz, and G. A. van Arkel. 1976. The genetic map of bacteriophage φX174 constructed with restriction enzyme fragments. *Virology* **72:** 61.

Weisbeek, P. J., W. E. Borrias, S. A. Langeveld, P. D. Baas, and G. A. van Arkel. 1977. Bacteriophage φX174: Gene *A* overlaps gene *B*. *Proc. Natl. Acad. Sci.* **74:** 2504.

Comparative DNA Sequence Analysis of the G4 and φX174 Genomes

G. Nigel Godson,* John C. Fiddes,† Barclay G. Barrell, and Frederick Sanger
MRC Laboratory of Molecular Biology
Hills Road, Cambridge, CB2 2QH, England

G4 was isolated in 1974 (Godson 1974b) and was quickly shown to be different from φX174 in some aspects of its physiology and DNA replication (see below). At this time DNA sequencing methods employing φX DNA as a model system were being developed in this laboratory, which resulted in the determination of the entire 5386-nucleotide sequence of the φX genome (Sanger et al. 1977b; an updated version of this sequence is given in this volume in Appendix I). A natural next step was to apply these methods to a closely related virus such as G4 and to compare the sequences.[1] We will present the results of this comparison after first reviewing the biological differences between G4 and φX.

Biological Differences and Similarities between G4 and φX

G4 has the same host range as φX; it forms plaques on *Escherichia coli* C strains but not on K or B strains (Godson 1974b). Its growth is restricted by the *E. coli* K (*Eco*K) and B (*Eco*B) restriction systems, and it forms plaques with low efficiency on *E. coli* K and B spheroplasts and on some K/C hybrid cells (L. Dumas, pers. comm.). Growth of G4 is also restricted in cells containing the plasmid RY13 because G4 has a single *Eco*RI cleavage site (Godson and Boyer 1974). Multiplication of φX is not restricted in any of these systems.

The G4 genome is 5577 nucleotides long, compared with 5386 for φX, and it has a different G+T content. This causes the G4 viral and complementary strands to have densities in alkaline CsCl opposite to those of φX. The densities of the G4 viral and complementary strands are 1.753 g/cc and 1.765 g/cc, respectively, compared with 1.767 g/cc and 1.756 g/cc for the

*On sabbatical leave from Radiobiology Laboratories, Yale University School of Medicine, 333 Cedar Street, New Haven, Connecticut 06510.

†Present address: Department of Biochemistry and Biophysics, University of California, San Francisco, California 94143.

[1] A gene-by-gene comparison of the G4 and φX sequences is given in Appendix II to this volume.

equivalent φX DNA strands (Ray and Dueber 1975). However, there is enough similarity between the two DNAs to align the G4 and φX restriction enzyme cleavage maps, by means of the Southern filter hybridization techniques, to detect homology between G4 and φX restriction enzyme fragments (Grindley and Godson 1978).

G4 DNA codes for the same series of ten proteins as φX DNA (A–H, J, and K), but since they migrate slightly differently from their φX equivalents in SDS-polyacrylamide gels, the molecular weights may not be identical (Godson 1974b; Shaw et al. 1978). G4 proteins F, G, H, and J are present in the virion and the functions of these and the other G4 proteins are considered to be the same as those of their φX counterparts. G4 can recombine with φX at a low frequency. The G4/φX DNA heteroduplex map (Godson 1974b) has now been aligned with the φX genetic map (B. L. Smiley and R. C. Warner, pers. comm.).

The first real indication of a fundamental difference between φX and G4 came with the observation that G4 requires only *E. coli dnaG* protein, DNA-binding protein, and DNA polymerase III holoenzyme to initiate and synthesize its complementary strand in vitro (Zechel et al. 1975), whereas φX requires the *dnaB*, *dnaC/D*, i and n proteins, as well as the *dnaG* protein, DNA-binding protein, and DNA polymerase III holoenzyme (Wickner and Hurwitz 1974; Schekman et al. 1975). In this system, G4 has a single, unique origin of complementary-strand synthesis (Zechel et al. 1975) compared with multiple, possibly randomly located origins of the φX complementary-strand synthesis (Tabak et al. 1974). Initiation of G4 complementary-strand synthesis requires synthesis of an RNA primer (Bouché et al. 1975), and this molecule has now been isolated and its nucleotide sequence determined (Bouché et al. 1978). G4 also has a single, unique origin of complementary-strand synthesis in vivo (Godson 1975b; Martin and Godson 1977; Godson 1977b; J. Sims et al., pers. comm.), which probably is located at the same site as the in vitro origin.

The origin of G4 viral-strand synthesis is also unique, and its location is approximately 100° away, on the circular map, from that of the complementary strand (Ray and Dueber 1975, 1977; Martin and Godson 1977). This results in G4 replicative-form (RF) DNA synthesis taking place by a D-loop mechanism with the two origins acting independently, and a model of G4 DNA synthesis has been proposed which is different from those accepted for φX DNA (Godson 1977b).

Determination of the G4 DNA Nucleotide Sequence

The complete φX DNA base sequence (Sanger et al. 1977a) was obtained mainly by the Sanger and Coulson (1975) enzymatic plus-minus method of DNA sequencing. Where possible, the accuracy of the base sequence was checked by amino acid sequence analysis (see Barrell 1977 for a general account). The G4 sequence was determined by a combination of the plus-minus method and the new enzymatic chain-termination method (Sanger et al. 1977a). The initial alignment of the G4/φX restriction enzyme cleavage maps (Grindley and Godson 1978) was refined by computer matching of the G4 and φX DNA sequences, and this allowed the use of the amino acid phasing of the φX proteins to check the accuracy of the G4 DNA sequence. In the case of G4 proteins B, C, D, F, G, H, J, and K, the sequence of the first 20–30 N-terminal amino acids was obtained by using an automatic microsequence method (Walker et al. 1978) on ^{14}C-labeled proteins from G4-infected, UV-irradiated cells that had been separated on SDS-acrylamide gels. In all cases the amino acid sequence agreed with that predicted from the nucleotide sequence, thus confirming that the two phages have homologous genes located in the same order on the genetic map.

Because of the large amount of data being accumulated, the use of a computer to store and sort the sequence became essential. Computer programs were also used to match the G4 sequence against the φX sequence and in the analysis of sequence and structural characteristics (see Staden 1977).

The G4 nucleotide sequence is presented in this volume (see Appendix II) as a gene-by-gene computer display of the φX nucleotide and amino acid sequences matched against the comparable G4 nucleotide and amino acid sequences.

CODING SEQUENCES

Genes

G4 DNA codes for the same ten proteins (A–H, J, and K) as φX DNA, and from the nucleotide sequence it may be inferred that the gene order is the same in both genomes. The genes of the two phages vary in size, but only for genes *A* and *J* are there major changes in size, which involve lengthy deletions and insertions. The gene order of φX is shown in Figure 1 and the sizes of the genes of G4 and φX are given in Table 1.

The degree of conservation of the amino acid and nucleotide sequences between φX and G4 differs for each gene and over the whole genome, averaging 34.0% and 32.9%, respectively (not including untranslated regions and coding regions that do not have homologous counterparts). The nucleotide and amino acid differences between φX and G4 are listed gene by gene in Table 2.

Codon Use

The frequencies with which different codons are used in each φX and G4 gene are given in Table 3. The high frequency of T in the third position that was noted in φX (Sanger et al. 1977b) is found to a certain extent in G4, despite the lower overall thymine content of the G4 viral strand (27.1% in G4 vs

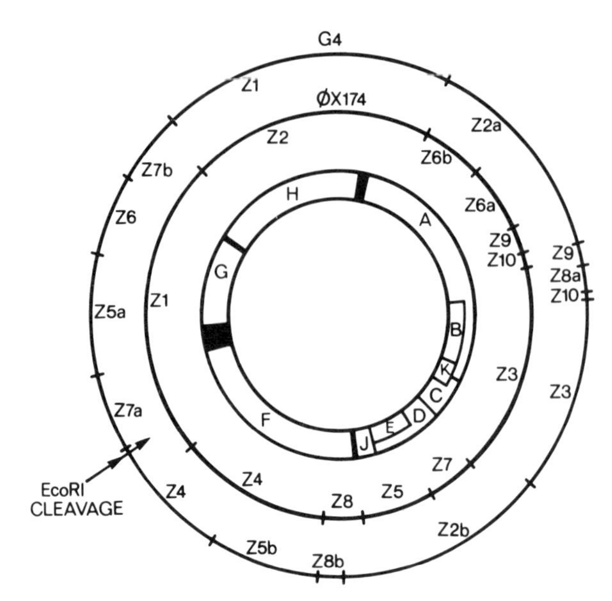

Figure 1 Aligned G4 and φX DNA *Hae*III restriction enzyme cleavage maps and φX genetic map. The φX *Hae*III cleavage and genetic maps are from Sanger et al. (1977b); the G4 *Hae*III cleavage map is from Fiddes and Godson (1978), and the alignment of the G4 and φX *Hae*III maps is from Grindley and Godson (1978).

31.2% in φX). However, this differs with the gene (Table 3). The spectrum of codon use in G4 is similar in some genes to that in φX (e.g., gene *G*), but in other genes it is rather different (e.g., gene *A*).

Some codons are used only rarely: for example, out of a total of 143

Table 1 Sizes of G4 and φX genes

Gene	φX (amino acids)	G4 (amino acids)
A	513	554
B	120	120
C	86	84
K	56	56
D	152	152
E	91	96
J	38	25
F	427	427
G	175	177
H	328	337

Sizes are calculated from the G4 and φX DNA nucleotide sequences and include the initiating methionine codon but not the termination codon.

arginine codons in φX only three are AGG and out of 149 glycine codons in φX only five are GGG. The figures are similar for G4. In general, φX and G4 codons ending in GG or GA are rare. These rare codons may have a modulating role in translation if their corresponding tRNA is also rare, as suggested by Fiers et al. (1976). The $tRNA^{Arg}$, for example, that recognizes the codon CGG is only a minor component of the bulk iso-accepting arginine tRNAs (W. Fiers, pers. comm.), and the $tRNA^{Ile}$ for AUA is only a minor component of the bulk isoleucine tRNAs (Harada and Nishimura 1974). Both of these codons are rare in φX and G4.

All of the φX and G4 genes retain the same termination codons, with the exception of genes *F* and *G*, which use TGA in φX and TAA in G4. The ATGA overlap of the termination codon of gene *A* with the initiation codon

Table 2 Conservation of φX amino acid and nucleotide sequences in the coding region of the G4 genome

Gene	Amino acid changes (%)	Nucleotide changes (%)	Nucleotide changes in third position of codon[a] (%)
A[b]	154/507 = 30.4	445/1521 = 29.3	125/445 = 28.1
B	53/120 = 44.2	86/360 = 23.9	7/86 = 8.1
K	22/56 = 39.3	40/168 = 23.4	3/40 = 7.5
C[c]	26/84 = 31.0	83/252 = 32.9	22/83 = 26.5
D	27/152 = 17.8	114/456 = 25.0	54/114 = 47.4
E[d]	40/91 = 44.0	52/273 = 19.0	5/52 = 9.6
J[b]	7/25 = 28.0	26/75 = 33.6	9/26 = 34.6
F	144/427 = 33.7	430/1281 = 33.6	124/430 = 28.8
G[b]	103/174 = 59.2	263/522 = 50.1	32/263 = 12.2
H[b]	90/323 = 27.9	309/969 = 31.9	120/309 = 38.8
A(*B*)[e]	33/120 = 27.5	86/360 = 23.9	26/86 = 30.2
A(*K*)[f]	7/29 = 24.1	17/86 = 19.8	4/17 = 23.5
C(*K*)[g]	8/29 = 27.6	23/86 = 26.7	6/23 = 26.0
D(*E*)[h]	14/90 = 15.6	52/272 = 19.1	27/52 = 51.9
Total	666/1959 = 34.0	1670/5076 = 32.9	486/1670 = 29.1

In all cases the initiating methionine codon has been included in the calculations but not the termination codon.

[a] Third-position changes that do not change the amino acid.

[b] Because amino acids are inserted or deleted in G4 as compared with φX, the calculation is based on the common amino acid codons.

[c] Gene *C* is two codons shorter in G4.

[d] Gene *E* is five codons longer in G4.

[e] Region of gene *A* which overlaps gene *B* (i.e., *A* phase).

[f] Region of gene *A* which overlaps gene *K* (i.e., *A* phase).

[g] Region of gene *C* which overlaps gene *K* (i.e., *C* phase).

[h] Region of gene *D* which overlaps gene *E* (i.e., *D* phase).

Table 3 Codon use in φX and G4

		Genes																											
		A		*A* in *B*/*K* region		A*		*B*		*K*		*C*		*D*		*E*		*J*		*F*		*G*		*H*		Total	Total
Amino acid	Codon	φX	G4	φX	G4	φX	G4	φX	G4	φX	G4	φX	G4	φX	G4	φX	G4	φX	G4	φX	G4	φX	G4	φX	G4	φX	G4
Phe	UUU	17	4	(2)	(1)	11	3	2	4	–	–	4	2	6	6	–	1	1	–	12	6	11	3	7	4	71	33
Phe	UUC	7	15	(3)	(4)	5	11	4	3	–	–	2	3	3	3	4	4	–	–	10	16	1	3	1	3	37	61
Leu	UUA	7	7	(3)	(3)	5	5	1	1	3	2	2	4	2	2	7	5	–	–	2	6	2	4	1	1	32	37
Leu	UUG	11	6	(3)	(3)	8	4	–	–	1	3	2	–	4	1	4	5	1	–	10	–	4	1	5	–	50	20
Leu	CUU	14	12	(4)	(–)	9	5	2	1	4	4	3	2	1	5	1	2	1	–	17	10	3	1	12	10	67	52
Leu	CUC	5	14	(2)	(3)	3	4	1	–	2	1	–	4	2	1	2	4	–	1	4	6	3	5	–	7	22	47
Leu	CUA	3	5	(2)	(2)	3	5	–	–	1	1	4	1	–	–	–	–	–	–	–	1	–	2	–	1	11	16
Leu	CUG	10	11	(7)	(6)	7	10	–	–	3	4	1	2	2	2	7	6	–	–	2	8	2	1	1	–	35	44
Ile	AUU	12	10	(4)	(3)	6	5	3	2	2	1	2	–	8	4	1	1	–	1	19	11	7	7	15	9	75	51
Ile	AUC	7	16	(3)	(4)	5	6	1	3	2	–	–	–	1	8	1	3	–	–	4	8	–	5	1	7	22	56
Ile	AUA	3	5	(2)	(2)	2	3	–	–	1	–	1	1	–	–	–	–	–	–	–	–	–	–	–	1	7	10
Met	AUG	22	17	(8)	(6)	16	14	1	1	1	1	2	3	3	3	3	2	1	1	12	17	3	3	12	10	76.	72
Val	GUU	15	9	(2)	(3)	9	6	3	3	1	–	3	–	9	7	–	–	–	–	16	11	15	8	12	6	83	50
Val	GUC	–	5	(–)	(1)	–	3	2	3	1	2	1	–	3	5	3	2	–	–	2	4	3	6	2	4	17	34
Val	GUA	4	6	(2)	(–)	3	4	–	–	–	–	–	–	3	2	1	–	1	1	2	7	–	2	2	4	16	26
Val	GUG	5	10	(1)	(4)	4	9	1	–	–	1	–	2	2	–	2	4	–	–	2	1	1	1	2	2	19	30
Ser	UCU	9	14	(1)	(4)	2	10	2	1	1	2	3	1	3	7	–	–	2	2	17	22	9	7	17	19	65	85
Ser	UCC	5	5	(3)	(1)	3	3	1	–	–	–	2	1	1	–	–	2	–	–	3	9	–	–	4	7	19	27
Ser	UCA	5	10	(1)	(4)	4	7	–	1	1	1	3	2	–	2	4	6	–	1	3	2	2	5	1	3	23	40
Ser	UCG	6	4	(5)	(2)	6	4	–	1	1	1	1	3	–	–	2	4	–	–	4	1	4	1	1	2	25	21
Pro	CCU	9	10	(1)	(1)	6	5	1	4	1	–	–	–	3	3	2	3	1	–	15	17	4	6	4	1	45	49
Pro	CCC	6	8	(1)	(1)	3	4	–	1	–	–	–	–	1	2	–	–	–	–	–	1	1	1	–	2	11	19
Pro	CCA	1	4	(–)	(–)	–	1	2	1	1	2	1	–	–	–	–	–	–	–	2	7	2	4	–	2	9	21
Pro	CCG	2	1	(1)	(1)	2	1	1	1	–	–	–	–	1	–	3	6	2	–	11	6	3	1	–	–	25	16
Thr	ACU	9	15	(4)	(1)	5	5	4	3	1	2	1	2	4	3	1	1	1	–	21	14	7	5	12	8	66	58
Thr	ACC	3	10	(2)	(7)	3	9	1	2	–	–	1	1	3	3	1	–	–	–	9	6	3	5	3	8	27	44
Thr	ACA	7	6	(4)	(2)	7	4	–	1	–	–	–	2	–	3	3	4	–	1	1	7	3	5	–	5	21	38
Thr	ACG	9	5	(4)	(4)	7	5	–	1	–	1	2	–	4	–	3	2	–	–	3	4	3	1	4	1	35	20
Ala	GCU	28	16	(4)	(6)	20	12	7	3	–	1	1	1	10	12	1	1	2	–	17	16	7	11	30	24	123	97
Ala	GCC	4	9	(1)	(2)	3	6	1	–	–	–	–	1	5	6	–	1	–	1	4	2	4	5	11	8	32	39
Ala	GCA	5	8	(2)	(2)	4	5	4	2	–	–	–	2	2	1	–	–	–	–	1	4	1	4	3	7	20	33
Ala	GCG	5	3	(3)	(3)	5	3	1	1	2	–	2	–	2	–	3	2	–	–	4	3	–	2	2	2	26	16

Tyr	UAU	13	14	(1)	(–)	9	7	3	2	–	–	1	–	3	3	–	–	1	1	19	13	6	3	5	7	60	50
Tyr	UAC	7	11	(1)	(2)	4	7	2	1	1	1	1	3	1	1	2	1	–	1	3	8	–	2	–	1	21	37
Ochre	UAA	–	–	(–)	(–)	–	–	–	–	–	–	–	–	1	1	–	–	1	1	–	1	–	1	1	1	3	5
Amber	UAG	–	–	(–)	(–)	–	–	–	–	–	–	–	–	–	–	–	–	–	–	–	–	–	–	–	–	–	–
His	CAU	5	9	(–)	(–)	3	4	1	3	–	–	3	4	1	1	–	–	–	–	10	7	3	4	3	2	29	34
His	CAC	4	6	(–)	(1)	2	3	2	1	–	–	–	1	–	1	–	1	–	–	3	7	–	2	–	1	11	23
Gln	CAA	6	13	(2)	(2)	3	6	8	12	–	3	2	1	4	4	–	1	2	1	8	6	–	–	15	16	48	63
Gln	CAG	12	6	(4)	(2)	9	4	2	1	2	1	–	–	2	1	2	–	1	–	14	9	6	3	12	10	62	35
Asn	AAU	16	17	(3)	(4)	12	10	6	7	2	2	1	2	1	1	3	–	–	–	11	9	7	6	14	10	73	64
Asn	AAC	10	18	(6)	(8)	7	11	2	4	2	2	–	5	5	5	–	3	–	–	9	12	4	8	4	12	43	80
Lys	AAA	21	32	(10)	(12)	17	18	6	2	3	5	4	7	3	4	4	4	5	3	8	16	4	4	16	14	91	109
Lys	AAG	18	13	(7)	(5)	9	10	1	1	2	–	3	1	1	–	4	4	1	1	7	1	3	1	7	12	56	44
Asp	GAU	19	13	(3)	(2)	9	8	5	1	1	–	3	–	4	2	1	–	–	–	14	12	7	4	16	8	79	48
Asp	GAC	15	22	(2)	(2)	10	10	5	6	1	–	4	5	4	6	–	1	–	–	14	16	3	2	6	12	62	80
Glu	GAA	7	17	(–)	(3)	1	6	6	9	3	4	–	3	9	7	1	2	–	–	6	11	–	3	3	12	36	74
Glu	GAG	19	6	(3)	(2)	7	3	4	2	2	2	5	2	3	4	1	–	–	–	9	6	2	1	13	5	65	31
Cys	UGU	5	1	(1)	(–)	2	1	2	2	–	–	–	–	1	1	1	2	–	–	1	2	3	–	–	–	15	9
Cys	UGC	4	4	(1)	(–)	2	1	–	–	1	1	–	–	2	1	2	–	–	–	2	1	2	1	–	–	15	9
Opal	UGA	1	1	(1)	(1)	1	1	1	1	1	1	1	1	–	–	1	1	–	–	1	–	1	–	–	–	8	6
Trp	UGG	7	10	(2)	(2)	6	6	1	–	–	–	2	1	1	1	3	2	1	1	6	9	1	3	3	1	31	34
Arg	CGU	12	9	(–)	(–)	8	6	2	2	–	1	1	1	7	4	1	–	4	1	19	17	1	3	2	3	57	47
Arg	CGC	12	11	(2)	(2)	7	7	4	3	–	–	2	1	4	7	1	–	1	2	6	7	4	4	4	6	45	48
Arg	CGA	3	5	(1)	(3)	3	4	2	2	3	1	–	–	–	–	–	–	1	–	–	–	–	–	–	–	12	12
Arg	CGG	2	2	(–)	(–)	2	2	1	2	–	–	–	1	–	–	4	–	–	–	–	–	–	–	1	1	10	8
Ser	AGU	3	2	(1)	(2)	3	2	1	2	2	1	–	–	2	–	1	3	–	–	1	–	1	1	2	1	16	12
Ser	AGC	3	2	(1)	(–)	3	1	1	2	–	–	2	–	–	–	1	–	–	–	–	–	–	–	–	1	10	6
Arg	AGA	6	2	(4)	(1)	5	2	1	1	–	–	4	–	–	–	–	–	–	–	–	–	–	1	–	1	16	7
Arg	AGG	1	1	(1)	(1)	1	1	–	–	–	–	–	2	–	–	–	–	–	–	–	–	–	–	1	–	3	4
Gly	GGU	14	16	(1)	(2)	12	10	–	–	–	–	2	2	1	4	–	1	5	2	14	14	8	2	19	13	75	64
Gly	GGC	8	6	(2)	(–)	4	5	2	3	–	1	1	1	2	–	–	–	3	2	12	10	1	3	12	16	45	47
Gly	GGA	4	5	(3)	(4)	4	4	2	4	1	1	1	1	3	2	–	–	–	1	2	1	1	1	5	3	23	23
Gly	GGG	2	1	(1)	(–)	1	1	2	1	–	–	–	–	–	1	–	–	–	–	–	–	–	–	–	1	5	5
Total codons		513	554	(148)	(148)	341	341	120	120	56	56	86	84	152	152	91	96	38	25	427	427	175	177	328	337	2327	2369
Percentage with 3rd U		38.9	30.9	(21.6)	(19.6)	36.9	29.0	36.7	33.3	26.8	25.0	32.5	20.2	42.1	41.4	14.3	15.6	47.4	28.0	52.2	42.4	56.6	40.1	51.8	37.1	43.0	33.9

of gene *C* is present in both genomes, but the other two overlapping φX stop/start codons (genes *C*/*D* and *D*/*J*) are not present in G4 (Table 4).

Gene A

In φX the gene-*A* protein is the largest virus-encoded protein: it can be predicted from the nucleotide sequence to be 513 amino acids long (see Appendix I). The reading frame of the φX gene-*A* protein was fixed by locating the C→T changes in φX gene-*A* mutants *am*86, at position 4116, and *am*33, at position 4380 (Slocombe 1977).

The initiation codon and reading phase of G4 gene *A* have been identified by comparison with the φX sequence, as no amino acid sequence information is available. Over the 3′-terminal two-thirds of the gene-*A* coding region there is considerable homology, but in the 5′-terminal one-third there are major structural changes. The G4 gene-*A* coding region has a deletion of 6 codons and two insertions of 1 and 46 codons when compared with φX (see Appendix II). The net result is that the G4 gene-*A* coding region is 41 codons longer than that of φX. The larger size of the G4 gene-*A* coding region is in agreement with the larger size of the G4 gene-*A* protein when it

Table 4 Overlapping termination and initiation codons in φX and G4 DNA

Gene	φX	G4
K – *B*	Met - Ser T-T-C-T-G-A-T-G-A-G-T * * * Phe	Met - Lys T-T-C-T-G-A-T-G-A-A-A * * * Phe
C – *A*	Met - Arg A-A-A-T-G-A-G-A * * * Lys	Same as φX
D – *C*	Met - Ser T-C-A-T-G-A-G-T * * * Ser	Met - Ser A-A-C-T-G-A-A-A-T-G-A-G-T * * * Asn
J – *D*	Met - Ser A-T-G-T-A-A-T-G-T-C-T * * * Met	See Fig. 8

is compared with φX gene-*A* protein on SDS-polyacrylamide gels (Godson 1974; Shaw et al. 1978).

The A* protein results from a translational start within φX gene *A* without change in reading phase (Linney and Hayashi 1974). The start of A* is most likely the ATG 516 nucleotides from the 5′ start of the φX gene-*A* coding region (i.e., nucleotide 4497 in the φX sequence in Appendix I), as this is preceded by the sequence GGAGG, which is a potential ribosome-binding site. Proposed ribosome-binding sites are given in Table 5. This is different from the A* start suggested previously (Sanger et al. 1977b) and results from a correction in the φX sequence which changes A to T at position 4498 (see Appendix I for the φX sequence). This would make the A* protein 37,160 daltons, in good agreement with the size observed on SDS-polyacrylamide gels. An A* protein is observed on SDS-polyacrylamide gels of G4-infected cells (Godson 1974; Shaw et al. 1978), and as the ATG and GGAGG sequence is conserved in G4 at the same position as in φX, it is most likely that the G4 A* protein also starts here (i.e., at G4 nucleotide 698, see Appendix II).

Gene B

It has been shown by DNA sequence studies (Smith et al. 1977) and by genetic studies (Weisbeek et al. 1977) that φX gene *B* is encoded completely within gene *A*. The reading frame of φX gene-*B* protein was confirmed by sequencing the mutants *am*18 (*A*), *am*35 (*A*), and *ts*116 (*B*) (Smith et al. 1977). The gene-*B* protein is 120 amino acids long and has a calculated molecular weight of 13,845 daltons. It migrates on SDS-polyacrylamide gels anomalously, which it has been suggested may be due to attached carbohydrate (Sanger et al. 1977b).

G4 gene *B* is the same size as that of φX, but its amino acid sequence is considerably different, with 44.2% of the amino acids different. This is seen in the gene-*B* sequence (Appendix II) and in Table 3. The start and reading frame of G4 gene *B* have been confirmed by sequencing the first 15 N-terminal amino acids of the G4 gene-*B* protein (J. E. Walker et al., in prep.).

Gene C

Two possible reading frames for φX gene *C* have been postulated (Sanger et al. 1977b). In one case the initiation codon for gene *C* would overlap with the termination codon for gene *A* and in the other case it would overlap with the termination codon of gene *B*. The reading frame consistent with the gene-*A* overlap has now been confirmed for gene *C* by sequencing the C→T transition in the φX gene-*C* mutant *oc*6 (B. G. Barrell, in prep.). The φX gene-*C* protein is thus 86 amino acids long.

G4 gene *C* can be identified in the DNA sequence because of its considerable homology with that of φX. The G4 gene-*C* protein is two amino acids

Table 5 G4 and φX proposed ribosome-binding sites

Gene		Nucleotide sequence
A	φX	A-A-T-C-T-T-[G-G-A-G-G]-C-T-T-T-T-T-T-[A-T-G]
	G4	A-T-C-A-A-A-C-[G-G-A-G-G]-C-T-T-T-T-C-[A-T-G]
*A**	φX	T-T-G-C-T-[G-G-A-G-G]-C-C-T-C-C-A-C-T-[A-T-G]
	G4	A-T-C-T-T-[G-G-A-G-G]-A-G-T-C-A-A-C-T-[A-T-G]
B	φX	G-G-T-C-T-[A-G-G-A-G]-C-T-A-A-A-G-A-A-[A-T-G]
	G4	A-G-G-C-T-[A-G-G]-G-A-A-T-A-A-A-G-A-A-[A-T-G]
K	φX	C-A-C-T-T-T-C-[G-G-A]-T-A-T-T-T-C-T-G-[A-T-G]
	G4	C-A-A-G-T-A-C-[G-G-A]-T-A-T-T-T-C-T-G-[A-T-G]
C	φX	G-T-G-G-A-C-T-G-C-T-G-G-C-[G-G-A]-A-A-[A-T-G]
	G4	G-T-G-G-A-C-T-G-C-T-G-G-T-[G-G-A]-A-A-[A-T-G]
D	φX	A-C-T-A-A-T-[A-G-G-T]-A-A-G-A-A-A-T-C-[A-T-G]
	G4	A-C-C-A-C-A-[A-A-G-G-A]-A-A-C-T-G-A-A-[A-T-G]
E	φX	G-C-G-T-T-[G-A-G-G]-C-T-T-G-C-G-T-T-T-[A-T-G]
	G4	C-T-A-T-C-[G-A-G-G]-C-T-T-G-C-G-T-A-T-[A-T-G]
J	φX	T-C-G-G-G-[A-A-G-G-A]-G-T-G-A-T-G-T-A-[A-T-G]
	G4	C-T-T-T-[T-A-A-G-G-A-G]-T-T-A-T-G-T-A-[A-T-G]
F	φX	C-T-T-A-C-T-T-G-[A-G-G-A]-T-A-A-A-T-T-[A-T-G]
	G4	C-T-A-T-T-[T-A-A-G-G-A]-T-A-C-A-A-A-A-[A-T-G]
G	φX	C-T-G-C-T-T-[A-G-G-A-G]-T-T-T-A-A-T-C-[A-T-G]
	G4	A-G-C-C-A-A-A-[A-G-G-A]-C-T-A-A-C-A-T-[A-T-G]
H	φX	A-C-T-T-A-A-G-T-[G-A-G-G-T-G-A]-T-T-T-[A-T-G]
	G4	T-G-A-A-A-[T-A-A-G-G-A]-T-T-A-T-C-C-T-[A-T-G]
16S rRNA 3′ end		HO A-U-U-C-C-U-C-C-A-C-U-A-G

shorter, since the penultimate AAA codon in φX is TGA in G4. In φX, the termination codon of gene *C* overlaps the initiation codon of gene *D*, but the shorter gene *C* in G4 results in a single untranslated A nucleotide between the two genes (see Table 4).

Radioactively labeled gene-*C* protein from G4 has been purified from SDS-polyacrylamide gels, and the sequence of the N-terminal 30 amino acids has been determined (J. E. Walker et al., in prep.). The results confirm the phasing and codons derived from the DNA sequence.

Gene D

Gene-*D* proteins are the most highly conserved proteins between φX and G4. Both gene-*D* proteins are 152 amino acids long, and despite a 24.8% base-sequence difference between the two gene-*D* coding regions, only 17.8% of the D-protein amino acids are different. This is possible because of a high incidence of third-position codon differences (47.8%) between the two phages.

φX gene *D* ends with a termination codon that overlaps the initiation codon of the following gene *J* (TAATG). This is not so in G4 because the last 14 nucleotides of G4 gene *D* appear to be duplicated and a new untranslated region is introduced between G4 genes *D* and *J* (see section on "Chromosome Rearrangements").

The start and reading frame of G4 gene *D* have been confirmed by sequencing the first 20 N-terminal amino acids of the G4 D protein (J. E. Walker et al., in prep.).

Gene E

φX gene *E* was the first overlapping gene to be recognized. The mutations *am*3, *am*27, and *am*N11 of gene *E* were found by marker rescue to lie within the coding sequences of gene *D*. When base sequences were determined, the mutations were found to be termination codons introduced into a second reading frame within gene *D* (Barrell et al. 1976). The φX gene-*E* protein has not yet been identified on SDS-polyacrylamide gels, but its calculated size is 91 amino acids.

The G4 gene-*E* protein also has not been identified on SDS-polyacrylamide gels, but from the conservation of the ATG start and the ribosome-binding site and the amino acid homology with φX, it also is presumed to be encoded as a second phase of G4 gene *D*. This is shown in Table 5 and in Appendix II. G4 gene *E* is five codons longer than φX gene *E* and it terminates beyond the end of gene *D*.

The large number of third-position codon differences in the *D* phase of the φX gene-*D* coding region relative to the *D* phase of the G4 gene-*D* coding region are second-position codon changes in the *E* phase. As a result, the G4 E protein differs considerably from the φX E protein in sequence, particularly in the C-terminal half. The N-terminal half of the protein, however, is fairly well conserved, and the run of leucine residues close to the N terminus is conserved completely on both proteins. This supports the idea that the hydrophobic N terminus of the E protein is important for its membrane-related function, as discussed further in the section on the "Evolution of Overlapping Genes."

Gene J

φX gene *J* codes for a basic protein 37 amino acids long which is packaged in the virion. Freymeyer et al. (1977) isolated this protein and determined its amino acid sequence, confirming the reading frame of gene *J* and the overlap of its initiation codon with the gene-*D* termination codon. No unambiguous φX gene-*J* mutants have been isolated, and the mutant *am*6 that was reported to be in gene *J* (Benbow et al. 1971) has now been shown to be a double mutant with one change in gene *A* and another in gene *E* (Smith et al. 1977; Weisbeek et al. 1977).

In G4, gene *J* is smaller and codes for a J protein only 24 amino acids long, but it still has a basic character. When the amino acid homologies between the G4 and φX J proteins are considered, it appears that in G4 there is a deletion of 39 nucleotides (13 amino acids) from the center of the gene. This is seen in the gene-*J* sequences (Appendix II). The deletion in G4 gene *J* has been verified by the isolation of G4 J protein and elucidation of its entire amino acid sequence (J. E. Walker et al., in prep.).

Gene F

In both phages gene *F* codes for a virion coat protein that is 427 amino acids long. The conservation of the nucleotide and amino acid sequences between the two phage genes is average (33.6% and 33.7% differences, respectively) and the amino acid differences appear to be scattered randomly. This is seen in the gene-*F* sequences (Appendix II). The reading frame of gene *F* has been fixed both in φX (Air 1976; Air et al. 1976) and in G4 (J. E. Walker et al., in prep.) by amino acid sequencing.

Gene G

Of all the viral proteins, G4 G protein (virion spike protein) is the most different from its φX counterpart, and only 40% of the amino acid sequence is common to both phage proteins. The G4 G protein is two amino acids longer than the φX G protein (177 vs 175 amino acids) because of three small insertions and deletions (see Appendix II). The reading frames of φX gene *G* (Air et al. 1975) and G4 gene *G* (J. E. Walker et al., in prep.) have been confirmed by amino acid sequencing of parts of both G proteins.

Gene H

The G4 gene-*H* coding region, which codes for a virion spike protein, is 9 codons longer than the comparable φX gene-*H* coding region (337 codons in G4 and 328 codons in φX). This difference is accounted for by an insertion of

14 codons near the 5′ end and two deletions of 1 and 4 codons near the 3′ end of the G4 gene-*H* coding region. The amino acid sequences of the G4 and φX H proteins are relatively highly conserved, and only 27.9% of the amino acids are different.

The coding regions of both φX and G4 gene *H* are unique in that they contain sequences near the 5′ end that can form fairly large, stable hairpin loops (Fig. 2). In φX the base of the loop is 66 nucleotides (2997 on the φX sequence in Appendix II) from the ATG initiation codon of gene *H* and in G4 it is 69 nucleotides (4632 on the G4 sequence in Appendix II) from the initiation codon. In G4 there is also a second loop whose base is 28 nucleotides (4591 on the G4 sequence in Appendix II) from the initiation codon. These are the only large hairpin loops that occur in coding regions of the φX and G4 genomes—all others occur in untranslated regions (see section on "Intergenic Untranslated Regions").

Protein K: A New Overlapping Gene

Examination of the G4 and φX nucleotide sequences reveals in both cases a considerable number of potential extra genes utilizing the second or third phase of preexisting coding regions. These potential reading frames are characterized by an ATG codon preceded by a ribosome-binding-site sequence and are shown in Table 6 and discussed in the section on "Control Sequences."

Some of the additional second- or third-phase proteins that are possible in φX are not possible in G4 because of differences in the initiation and termination codons. Other additional proteins are possible in G4 and not in φX for the same reason. However, as judged from the nucleotide sequences, some of the potential φX genes appear to be strongly conserved in G4, and this has led to a search to see whether any of these proteins are synthesized in infected cells.

Using a radioactive amino acid microsequencing technique on proteins extracted from SDS-polyacrylamide gels (Walker et al. 1978), a series of five small proteins that are synthesized by G4 in UV-irradiated *uvrA* cells have been characterized. One of these has the amino acid composition of a G4 extra-phase protein (Shaw et al. 1978). This protein (which has been renamed K in accordance with the φX gene-letter convention) is made in large amounts and migrates on 15% polyacrylamide-SDS gels with the gene-*J* product. It is 56 amino acids long and has a strongly hydrophobic N terminus similar to those of the G4 and φX gene-*E* proteins (see gene-*E* and gene-*K* sequences in Appendix II). Two N-terminal blocks of leucine amino acids in the K protein are in positions characteristic of the "signal sequences" of some eukaryote and prokaryote secretory proteins (Devillers-Theiry et al. 1975), which suggests that the K protein has a function involved with the *E. coli* cell membrane.

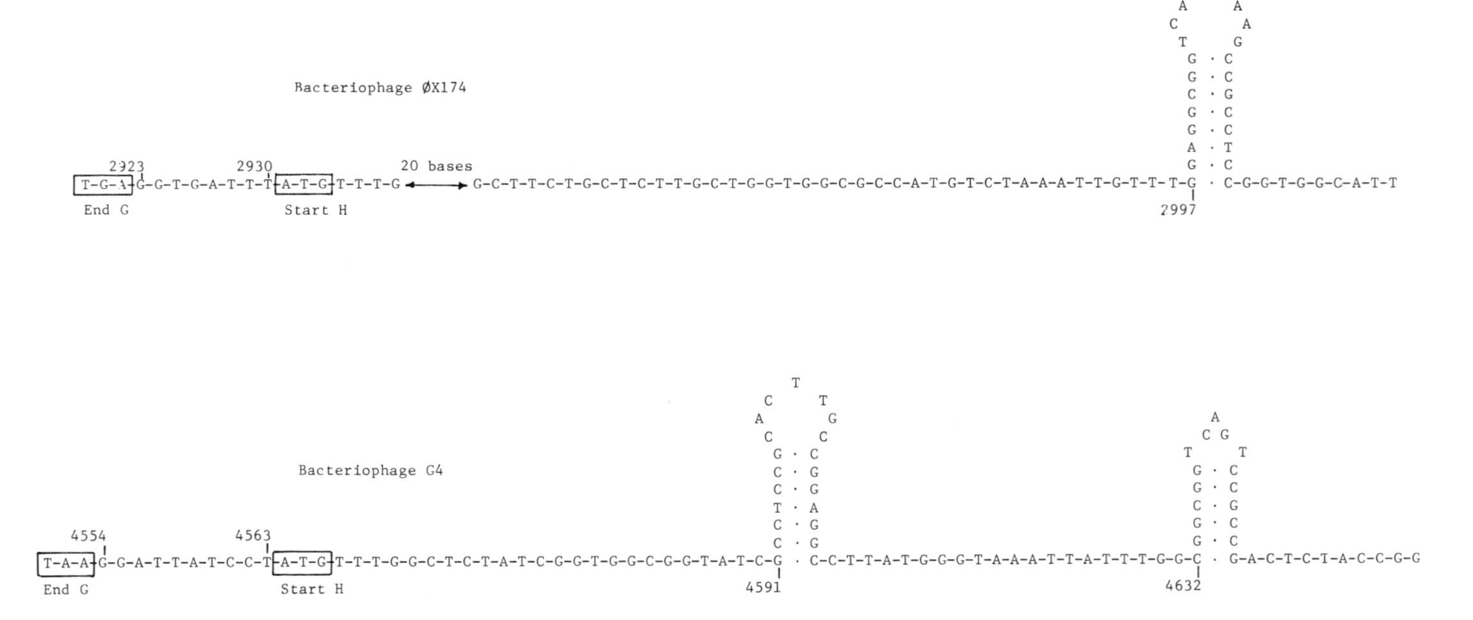

Figure 2 Untranslated regions between genes *G* and *H* and the gene-*H* hairpin loops. The φX sequence is from Sanger et al. (1977b) and the G4 DNA sequence is from data of the authors (in prep.). The boxed sequences are the gene-*G* and gene-*H* termination and initiation codons, respectively.

The K protein is encoded partly as a second phase of gene *A* and partly as a second phase of gene *C* (see Fig. 1). Its ATG initiation codon overlaps the TGA termination codon of gene *B*. Five nucleotides are therefore used in all three reading frames. The A of the TGATG overlapping the gene-*B*/*K* termination and initiation codon is used to code for Met (ATG) in the *K* phase, Asp (GAT) in the *A* phase, and as part of the TGA termination codon in the *B* phase. The T of the ATGA overlapping gene-*A* termination and gene-*C* initiation codons is part of an AAT (Asn) codon in the *K* phase, part of an ATG (Met) codon in the *C* phase, and part of the TGA termination codon in the *A* phase. The *G* of the ATGA is part of the GAG (Glu) codon in the *K* phase, part of an ATG (Met) codon in the *C* phase, and part of the TGA termination codon in the *A* phase. The last A of the ATGA is part of a GAG (Glu) codon in the *K* phase, part of an AGG (Arg) codon in the *C* phase, and part of the TGA termination codon in the *A* phase. The first A of the ATGA, however, requires tRNA in all three phases and is part of the AAT (Asn) codon in the *K* phase, part of an ATG (Met) codon in the *C* phase, and part of an AAA (Lys) codon in the *A* phase. This is shown schematically in Figure 3.

The K protein has not yet been identified in extracts of φX-infected cells, but there may be some genetic evidence for its existence (Shaw et al. 1978).

INTERGENIC UNTRANSLATED REGIONS

In φX, five consecutive genes (*J*, *F*, *G*, *H*, and *A*) are separated by untranslated intergenic spaces which vary in size. The untranslated sequences between genes *J* and *F*, *F* and *G*, *G* and *H*, and *H* and *A* are 36, 110, 8, and 63 nucleotides, respectively, and are given in Figures 4, 5, 2, and 6, respectively. The three large untranslated regions contain sequences that can form stable secondary structures, and the gene *A*/*H* (Fig. 6) and *F*/*G* (Fig. 5; and Fiddes 1976) intergenic loops are stable enough in viral single-stranded (SS) DNA to be resistant to single-strand-specific nucleases (Bartok et al. 1975; H. Schaller, pers. comm.).

The complete function of the intergenic untranslated regions is not known, but all contain a ribosome-binding site (i.e., complementarity with the 3′ end of 16S ribosomal RNA) for the proximal gene. The *A*/*H* and *J*/*F* intergenic spaces contain mRNA termination signals, and the *A*/*H* space contains the gene-*A* promoter (see below).

It is instructive to compare the G4 untranslated regions with their φX counterparts. The differences between the two viral DNAs are greatest in these intergenic spaces. Not only are they different in size (the G4 *J*/*F*, *F*/*G*, *G*/*H*, and *H*/*A* untranslated regions are 45, 136, 10, and 59 nucleotides, respectively), but they also have considerably different nucleotide sequences. It is difficult to recognize any residual sequence similarity between the φX and G4 intergenic spaces, as is shown in Figures 2, 4, 5, and 6.

Table 6 Possible extra protein initiation sites in bacteriophage φX and G4 DNA

Name[a]	Position[b]	Sequence[c]	Length of protein[d]
		φX	
A1a	4186	G-G-C-G-T-T-[G-A-G]-T-T-C-G-A-T-A-A-T-G	25
A1b	4198	G-A-T-A-A-T-[G-G-T]-G-A-T-A-T-G-T-A-T-G	21
A2	4429	C-T-T-[A-A-G-G-A]-T-A-T-T-C-G-C-G-A-T-G	23
A3a	4621	C-G-A-T-T-A-[G-A-G-G]-C-G-T-T-T-T-A-T-G	78
A3b	4699	G-A-G-G-G-T-C-G-C-[A-A-G-G]-C-T-A-A-T-G	52
F1	1038	G-C-C-[G-A-G]-C-G-T-A-T-G-C-C-G-C-A-T-G	52
F2a	1272	A-T-G-[A-A-G-G-A]-T-G-G-T-G-T-T-A-A-T-G	38
F2b	1317	A-C-T-[G-G-T]-T-A-T-A-T-T-G-A-C-C-A-T-G	23
F3a	1449	A-C-C-[G-A-G-G]-C-T-A-A-C-C-C-T-A-A-T-G	34
F3b	1464	A-A-T-[G-A-G]-C-T-T-A-A-T-C-A-A-G-A-T-G	29
F4	1686	[G-G-A-G-G-T]-A-A-A-A-C-C-T-C-T-T-A-T-G	29
G1a	2543	G-A-T-A-G-T-T-T-G-A-C-[G-G-T]-T-A-A-T-G	62
G1b	2552	A-C-[G-G-T]-T-A-A-T-G-C-T-G-G-T-A-A-T-G	59
H1a	3076	A-C-T-G-T-[A-G-G]-C-A-T-G-G-G-T-G-A-T-G	196
H1b	3109	G-C-C-A-T-T-C-[A-A-G-G]-C-T-C-T-A-A-T-G	187
H1c	3316	G-C-A-T-T-T-C-C-T-[G-A-G]-C-T-T-A-A-T-G	116
H1d	3340	G-A-G-C-G-T-G-C-T-[G-G-T]-G-C-T-G-A-T-G	108
H1e	3439	A-T-T-G-C-C-[G-A-G]-A-T-G-C-A-A-A-A-T-G	75
H1f	3508	A-C-G-A-A-A-G-A-C-C-[A-G-G-T]-A-T-A-T-G	52
H1g	3517	C-[A-G-G-T]-A-T-A-T-G-C-A-C-A-A-A-A-T-G	49
H2a	3784	G-C-A-[A-A-G-G-A]-T-A-T-T-T-C-T-A-A-T-G	48
H2b	3826	G-T-[G-G-T]-T-G-A-T-A-T-T-T-T-T-C-A-T-G	34
	16S rRNA	3′ $_{HO}$A-[U-U-C-C-U-C-C-A]-C-U-A-G 5′	

Table 6 (continued)

Name[a]	Position[b]	Sequence[c]	Length of protein[d]
		G4	
A1	564	C-A-T-T-[G-G-A]-A-C-C-T-C-C-T-C-T-A-T-G	45
A2a	822	C-G-C-T-T-A-[A-A-G-G-A]-T-T-T-C-T-A-T-G	78
A2b	900	G-A-[A-G-G-T]-C-G-C-T-C-G-G-T-G-C-A-T-G	52
F1	2637	G-C-[G-G-A]-C-C-G-T-G-T-A-C-C-T-C-A-T-G	74
F2	2871	A-T-G-[A-A-G-G-A]-T-G-G-C-G-T-T-A-A-T-G	38
F3	3285	G-G-C-[G-G-T]-C-A-T-A-C-A-T-C-C-T-A-T-G	21
G1	3970	C-T-C-[G-A-G]-T-C-T-C-C-G-A-T-A-C-A-T-G	42
G2	4192	C-A-T-T-C-[A-G-G]-C-T-T-G-T-G-C-C-A-T-G	28
G3a	4309	C-G-T-T-T-T-[G-A-G-G-T]-C-G-C-T-G-A-T-G	68
G3b	4444	G-T-C-[G-G-T]-A-A-T-C-G-C-G-T-T-T-A-T-G	24
H1	4706	A-A-T-G-T-T-G-T-[A-G-G-T-G]-C-T-A-A-T-G	101
H2	5126	A-C-T-[A-A-G-G-A]-T-A-C-T-G-T-C-T-A-T-G	78
H3	5444	G-C-T-[A-A-G-G-A]-C-G-T-G-T-C-C-A-A-T-G	36
	16S rRNA	3′ $_{HO}$A-[U-U-C-C-U-C-C-A]-C-U-A-G 5′	

[a] Sites are named according to the gene and the order within the gene in which they occur. Sites labeled a, b, etc., are in the same reading frame. There is not necessarily any correlation between the names of the φX and G4 sites.

[b] Numbers refer to the A of the ATG initiation codon and correspond to the corrected φX sequence given in Appendix I and to the complete G4 sequence to be published elsewhere (G. N. Godson et al., in prep.).

[c] Only those sequences that have at least three consecutive nucleotides complementary to the boxed sequence at the 3′ end of 16S ribosomal RNA are included. Other restrictions are that these complementary sequences must not be closer than two nucleotides and not farther than 14 nucleotides from the ATG and must have at least two potential G/C base pairs with the 16S sequence shown boxed. Possible extra initiation sites in the same reading frames as known genes are not included, nor are sites containing GTG initiation codons.

[d] The length of the polypeptide (no. of amino acids) that could be produced starting at the ATG initiation codon. Only those sites that could specify a polypeptide of greater than 20 amino acids have been considered.

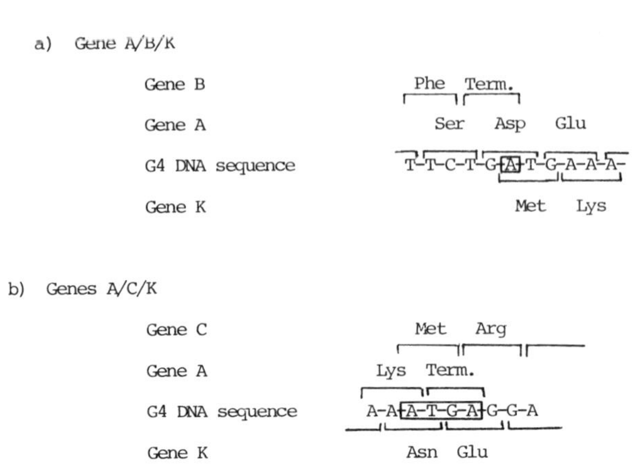

Figure 3 Triple overlaps in G4. The evidence for the triple overlaps in G4 DNA is given by Shaw et al. (1978). The nucleotides that are used in all three reading frames are boxed.

However, the sequences of the G4 *J/F*, *F/G*, and *A/H* untranslated regions all form secondary-structure hairpin loops, similar to their φX counterparts (Figs. 4, 5, and 6). This suggests that it is only the secondary structure that is important in these intergenic spaces. However, if these secondary structures are specific signals for mRNA termination, processing, or recognition of other functional proteins involved in phage synthesis or assembly, the flanking nucleotide sequences most likely also play some part in their specificity (see section on "Control Sequences"). It is noteworthy that despite the differences in base sequence, the hairpin loops in the G4 and φX *J/F* intergenic spaces have the same size (7 vs 6 base pairs in the stem), as have the large hairpin loops in the G4 and φX *F/G* intergenic spaces (20 vs 19 base pairs in the stem). The extra hairpin loop in the G4 *F/G* intergenic space (Fig. 5) has no φX equivalent and appears to be involved in a function specific to G4, i.e., the initiation of complementary-strand synthesis (see section on "Origin of DNA Replication").

G4 has two intergenic spaces that are not present in φX: a single-nucleotide space between the end of G4 gene *C* and the start of G4 gene *D* (Table 4) and a 32-nucleotide intergenic space between G4 genes *E* and *J* which may be caused by the duplication of the end of gene *D* (see section on "Chromosome Rearrangements").

CONTROL SEQUENCES

Ribosome-binding Sites

A comparison of φX and G4 ribosome-binding sites is of interest because the homologous genes presumably serve identical functions in the two phages

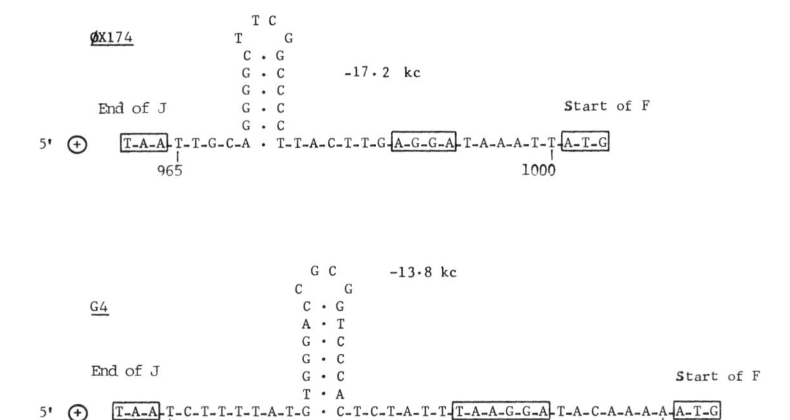

Figure 4 Untranslated region between genes *J* and *F* in G4 and φX DNAs. The φX sequence is from Sanger et al. (1977b), and the G4 sequence is from G. N. Godson and J. C. Fiddes (in prep.). The boxed sequences are the termination codons of gene *J*, the initiation codons of gene *F*, and the gene-*F* ribosome-binding sites. The stability of the hairpin loops was calculated from values for RNA loop stabilities based on Tinoco et al. (1973).

and would therefore be expected to be translated with equal efficiencies. Indeed, apart from the high yield of K protein in G4 (Shaw et al. 1978), the relative amounts of the G4 and φX proteins appear similar.

The sequences preceding the initiation sites of the 11 G4 and φX genes are shown in Table 5. In most cases the sequence preceding the ATG is considerably different, although the complementarity of the 3′ end of 16S rRNA, which is a feature common to ribosome-binding sites (Shine and Dalgarno 1974), is always present. This complementarity is usually about the same distance away from the ATG initiation codon. Homologous genes in G4 and φX do not necessarily have the same degree of 16S rRNA complementarity, although there is a general correspondence. In this and other systems no correlation has yet been observed between the degree of complementarity and the efficiency of translation. Possibly, the yield of proteins is largely dependent on other factors that are limiting, such as mRNA availability.

When the sequences preceding all of the ATG triplets in φX are examined, many are found to have partial complementarity with the 3′ end of 16S ribosomal RNA, and some of these have more complementarity than the ribosome-binding sites that are known to be used. These are shown in Table 6. Each of these potential coding regions theoretically could be used to synthesize a protein containing more than 20 amino acids, but in a second phase of an already functional coding region. The existence of these potential ribosome-binding sites has already led to the prediction of the new

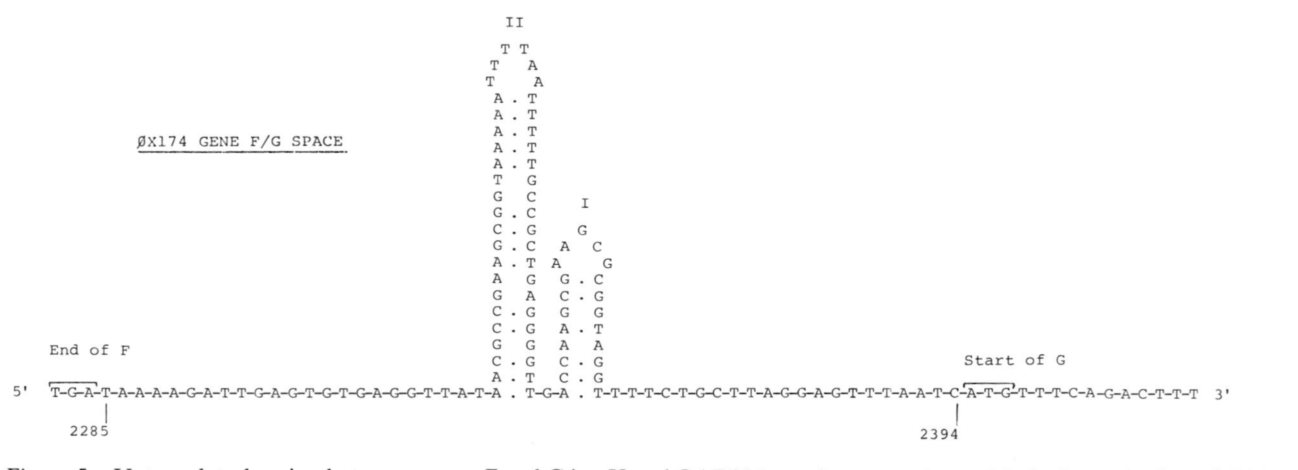

Figure 5 Untranslated region between genes *F* and *G* in φX and G4 DNAs and a comparison with the bacteriophage λ DNA *ori* sequence. The sequence of the φX DNA untranslated region is from Fiddes (1976), with two extra bases inserted as described by Fiddes et al. (1978). The G4 DNA sequence is from Fiddes et al. (1978); the G4 RNA primer sequence is from Bouché et al. (1978); and the sequence of the origin of replication in λ DNA is from Denniston-Thompson et al. (1977).

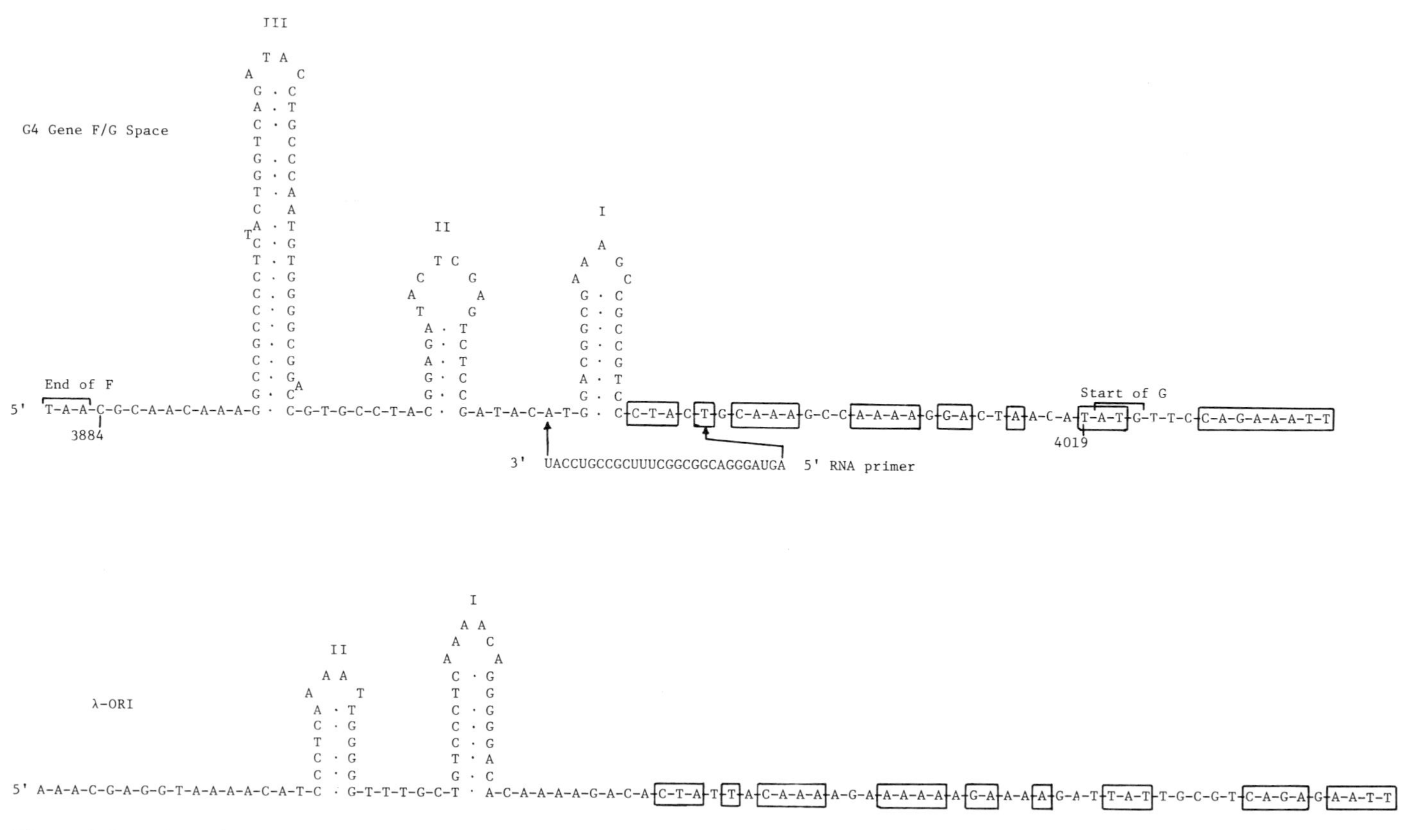

Figure 5 (continued)

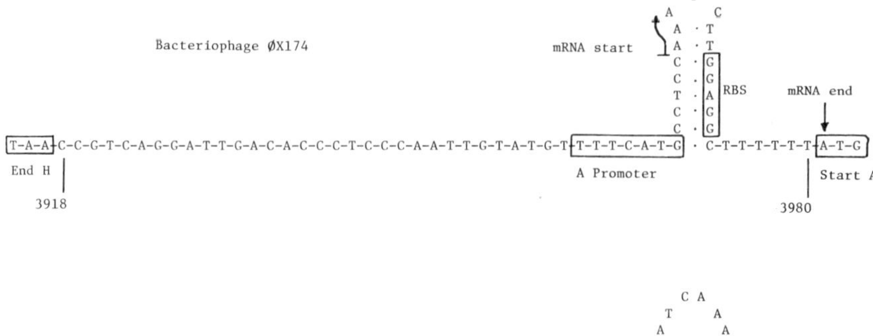

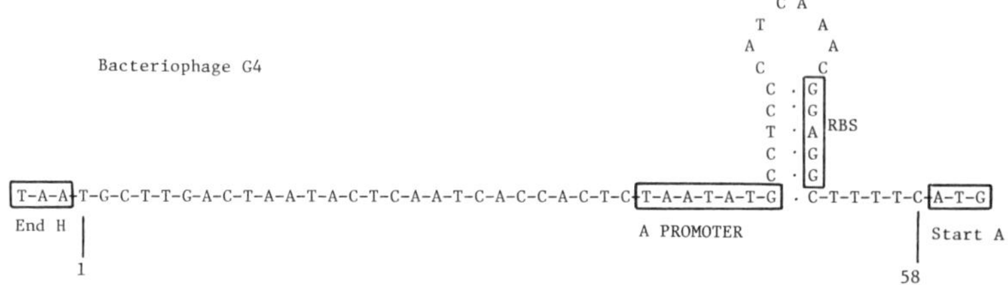

Figure 6 Untranslated region between genes *H* and *A*. The φX DNA sequence is from Sanger et al. (1977b) and the G4 DNA sequence is from G. N. Godson et al. (in prep.). The location of the *A* promoter sequence, the ribosome-binding site (RBS), and the mRNA initiation and termination sites are discussed in the text.

overlapping G4 gene *K*. It is difficult to see why some of the potential ribosome-binding sites are not used, and it is possible they are but that their products have not yet been identified.

Similar potential ribosome-binding sites are also present in G4. They are fewer than in φX and are mostly in different parts of the genome (see Table 6). Two such sites are present in both phages and could specify the same length proteins in the same part of the genome. These are the φX A3a (and A3b) and the G4 A2a (and A2b) sites and the φX F2a and the G4 F2 sites. The amino acid sequences specified by the φX A3a and G4 A2a sites are 78 amino acids long and are 63% conserved between the two phages. The amino acid sequences specified by the φX F2a and G4 F2 sites are 38 amino acids long and are 34% conserved between the two phages. In particular, the hydrophobic nature of the amino acid sequences is highly conserved.

Promoter Sites, mRNA Termination Sites, and Control of Transcription

φX has three promoters, preceding genes *A*, *B*, and *D* (Sanger et al. 1977b). Initially, two of these promoters, those for genes *A* and *D*, were identified by binding RNA polymerase to purified restriction enzyme fragments (Chen et al. 1973). This method also indicated a third binding site internal to gene *G*, but this is not believed to correspond to an mRNA initiation site (Fujimura and Hayashi, this volume). In analogous experiments with G4 (Godson 1975b) the same three RNA-polymerase-binding sites were observed, including the gene-*G* site (see also Spencer et al., this volume).

Three major and many minor species of φX mRNA are synthesized in vitro (Smith and Sinsheimer 1976a; Axelrod 1976a), and these correspond to the mRNA species synthesized in vivo (Hayashi and Hayashi 1971; Clements and Sinsheimer 1975). That the three mRNAs come from the promoters preceding genes *A*, *B*, and *D* is clear since the mRNA transcripts have been mapped with respect to φX restriction enzyme fragments (Smith and Sinsheimer 1976b,c; Axelrod 1976b). The sequences at the 5′ ends of these messages have been determined (Grohmann et al. 1975; Smith et al. 1974) and located on the φX DNA sequence (Sanger et al. 1977b; J. Sims and D. Dressler, unpubl.). They are shown in Table 7 along with the comparable regions of the G4 DNA sequence.

There is no direct evidence for the precise location of the G4 promoters (other than the RNA-polymerase-binding studies mentioned above) since neither have the 5′ termini of the G4 mRNA species been identified nor have RNA-polymerase-protected fragments been isolated. The identification of the promoters for G4 genes *A*, *B*, and *D* (shown in Table 7) is based on homology with φX and with the common features identified in all *E. coli* promoters. These common features are (a) a sequence similar to TATPuATPu centered about 8 nucleotides from the mRNA initiation point (Pribnow 1975; Schaller et al. 1975) and (b) a sequence similar to TGTTGACAATT centered about 35 nucleotides from the mRNA initiation point (Maniatis et al. 1975; Takanami et al. 1976; J. Majors, unpubl.). The presence of both of these structural features in the G4 regions analogous to the φX promoters supports the identification of the G4 promoters. The 5′ ends of the major G4 mRNA species can therefore be predicted as AUCAA (gene *A*), GUCAAC (gene *D*), and AUCGCA (gene *B*).

In φX the "leader" sequence from the start of the mRNA to the initiation codon of the first gene is very long for gene *B* (176 nucleotides) compared with the leader sequences for gene *A* (18 nucleotides) and gene *D* (32 nucleotides).

J. Drouin of this laboratory has identified the termination sites of φX mRNAs synthesized in vitro (pers. comm.) by annealing purified mRNA species to complementary-strand DNA, using the 3′ OH of the mRNA to prime DNA synthesis, and then analyzing the origin of this synthesis by DNA sequencing techniques. The main mRNA termination site is, as expected, at the 3′ end of the TTTTTTA sequence at the base of the secondary-structure loop in the φX gene-*A*/*H* untranslated region (see Fig. 6). Similar T_6A sequences have been found at the 3′ end of bacteriophage lambda 4S and 6S transcripts (Rosenberg et al. 1976), and hairpin loops before mRNA termination sites have been implicated in termination in both filamentous phage fd (Sugimoto et al. 1977) and bacteriophage λ (Rosenberg et al. 1977). The presence of a *rho*-dependent termination site in this approximate position has also been suggested by Hayashi et al. (1976) and Axelrod (1976b). If the T_6A sequence is the mRNA termination site, then

Table 7 Promoter sequences of φX and G4 DNA

Gene	Phage	Promoter sequence
A	φX	A G G A T T G A C A C C C T C C C A A T T G T A T G T T T T C A T G C C T C C A A A T C T
	G4	T G C T T G A C T A A T A C T C A A T C A C C A C T C T A A T A T G C C T C C C A T C A A
D	φX	T G T T G A C A T T T T A A A A G A G C G T G G A T T A C T A T C T G A G T C C G A T G C
	G4	G C T T G A C A T A C T G A A A G A A C G T G G C C T A T T A T C C A C A T C G T C A A C
B	φX	A G C T T G C A A A A T A C G T G G C C T T A T G G T T A C A G T A T G C C C A T C G C A
	G4	A G C T T G C A A A A C A C G T G G C C T T A T G G T T A C T C T A T G C C C A T C G C A

Boxed regions indicate (a) homology with TATPuATPu (Pribnow 1975; Schaller et al. 1975) and (b) TGTTGACAATT (J. Majors, unpubl.). The locations of the φX mRNA starts are shown with arrows.

it is located after the gene-*A* mRNA start, and one would expect a small, 20-nucleotide mRNA to be produced, but no such product has been identified. However, mRNAs initiated at gene-*B* and gene-*D* promoters contain the *A*/*H* secondary-structure loop at the 3′ end, whereas mRNAs initiated at the gene-*A* promoter start within this loop and would therefore not contain the secondary-structure loop at the 5′ end of the mRNA but would contain it at the 3′ end. This supports the idea that hairpin loops in the mRNA prior to the T_6A sequence are important in termination. The G4 *H*/*A* intercistronic space, although of a different base sequence, retains the main feature of the φX *H*/*A* space—a secondary-structure loop near the 3′ end which leads into a T_4C sequence rather than a T_6A sequence (see Fig. 6). However, at the time of this writing there is no information on the sequences of the 3′ and 5′ ends of the G4 mRNA species.

Thus, the *H*/*A* site acts as the main φX mRNA terminator. The secondary-structure loop in the gene-*J*/*F* untranslated region (Fig. 4) also acts as a terminator of a minor species of mRNA synthesized in vitro (J. Drouin, pers. comm.). Other termination sites in gene *C* (Smith and Sinsheimer 1976b) and near the beginnings of genes *F*, *G*, and *H* (Hayashi et al. 1976; Axelrod 1976a,b) have been reported. Although there are hairpin loops in these regions, the termination sites have not been located on the DNA sequence. It is interesting that in G4 a similar hairpin loop is observed in the *J*/*F* untranslated region, although there is little sequence homology with the φX *J*/*F* untranslated region (see section on "Intergenic Untranslated Regions").

ORIGINS OF DNA REPLICATION

Origin of Viral-strand Synthesis

The φX origin of viral-strand synthesis has been shown by genetic experiments to be in cistron *A* (Baas and Jansz 1972) and by pulse-label experiments to be in the *Haemophilus aegyptius* restriction enzyme cleavage fragment *Hae*III 6b (Godson 1974a; Baas et al. 1976). Its exact location has now been identified by the observation that the φX cistron-*A* protein (which is required for viral-strand synthesis) cleaves φX RFI DNA (superhelical, covalently closed, circular duplex DNA) between bases 4305 (G) and 4306 (A) on the φX DNA sequence (Langeveld et al. 1978). The G4 origin of viral-strand synthesis has also been located by pulse-label experiments (Ray and Dueber 1975, 1977; Martin and Godson 1977; Godson 1977b) and is at the same site as the origin of the viral-strand DNA tail in G4 late rolling-circle replicating intermediates (Godson 1977a). On the aligned φX/G4 restriction enzyme cleavage map (Fig. 1) the φX and G4 viral-strand origins appear to be in the same regions. The comparative φX and G4 nucleotide sequences across this region have been described (Fiddes et al. 1978) and are shown in Figure 7.

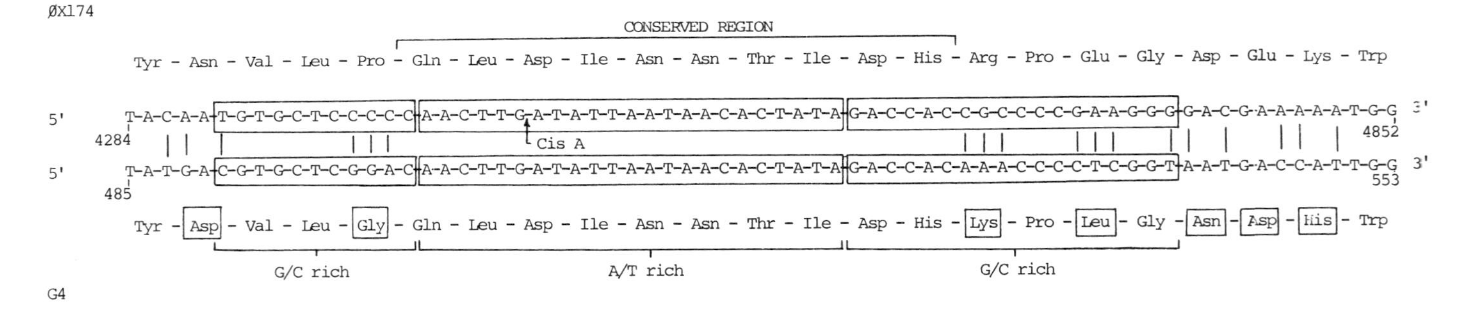

Figure 7 The origins of G4 and φX viral-strand synthesis. The φX DNA sequence is from Sanger et al. (1977b) and the G4 DNA sequence is from Fiddes et al. (1978). The gene-*A* amino acids that are different in the two bacteriophages are boxed, and the nucleotide differences are indicated with a line.

Within the φX sequence, in and around the cistron-*A* cleavage site, there is a stretch of 30 nucleotides (positions 4299–4328) that is completely conserved in the two genomes, which suggests that the conserved sequence represents a protein-recognition site common to both origins of replication. Whether this protein is a host-cell protein or the cistron-*A* protein is not clear, but by means of electron microscopy Eisenberg et al. (1977) have shown that the φX cistron-*A* protein binds to φX DNA between nucleotides 4298 and 4338, a region that spans the conserved sequence. This would suggest that the G4 and φX cistron-*A* proteins interact with the same nucleotide sequence and that experiments with φX cistron-*A* protein and G4 RFI DNA (and vice versa) would provide an interesting insight into the DNA interaction of these proteins.

The DNA sequence of this region, even outside the fully conserved portion, preserves in both phages the arrangement of an A-T-rich (low T_m) region at the origin flanked by G-C-rich (high T_m) regions on either side (see Fig. 7). It is interesting that the C_6T pyrimidine tract (4293–4299) that Eisenberg et al. (1975) observed to be present in the gap in the viral strand of late φX RFII (circular duplex DNA in which at least one strand is not continuous) molecules is absent in G4. This suggests that C_6T does not have a specific function but is simply the last part of the φX genome to be synthesized.

Origin of Complementary-strand Synthesis

The origins of complementary-strand synthesis in φX are multiple and almost randomly located, whereas G4 has a single, unique origin of complementary-strand synthesis. At present, there is no evidence for sequence specificity in the initiation of φX complementary-strand synthesis, and for this reason only the unique G4 origin will be discussed.

From in vivo experiments the origin of synthesis of the G4 complementary strand has been placed in the restriction enzyme fragment *Hae*III 5a (Godson 1975b; Martin and Godson 1977), at or close to the in vitro SS → RF origin of complementary-strand synthesis (Zechel et al. 1975). This origin has been defined by the sequence of the RNA primer that is synthesized in vitro by the *dnaG* protein during the initiation process (Bouché et al. 1975, 1978). This RNA sequence corresponds to a G4 DNA sequence contained in the untranslated region between genes *F* and *G* (see Fig. 5), with the 3′ end of the 28-nucleotide RNA primer sequence located at a T nucleotide (3997) 23 bases away from the ATG initiation codon of gene *G* and 5 nucleotides away from the first loop (loop I, Fig. 5). The primer RNAs are heterogeneous in size (Bouché et al. 1978), the largest ending in an A nucleotide (3970) before loop II (Fig. 5). Loop II may be a signal for the change to deoxyribonucleotide precursors, even though they can be added closer to the origin. (For additional details, see Dressler et al., this volume.)

The structural difference between the G4 (and φX) origin of viral-strand synthesis and the G4 origin of complementary-strand synthesis probably reflects different initiation processes. Initiation of DNA synthesis at the G4 viral-strand origin requires specific nicking of the viral strand by the cistron-*A* protein and unwinding of the DNA with *rep* protein and DNA-binding protein (Eisenberg et al. 1977), whereas initiation of DNA synthesis at the G4 complementary-strand origin requires DNA-binding protein and *dnaG* protein to synthesize an RNA primer.

Similarity between the Sequences of the G4 Complementary-strand Origin and the Bacteriophage λ Origin

It is of interest to note that while there are no obvious similarities between the G4 (and φX) origins and those of ColE1 (Tomizawa et al. 1977) and the filamentous phage f1 (Ravetch et al. 1977), both of which use RNA polymerase to synthesize a primer RNA, there is considerable sequence homology between the G4 complementary-strand origin and that of λ (Denniston-Thompson et al. 1977), both of which use *dnaG* protein to initiate DNA synthesis (Fig. 5). The region of homology is on the 5′ side of the G4 RNA primer sequence and may represent the recognition or binding site in the DNA for the *dnaG* protein.

CHROMOSOME REARRANGEMENT

Duplication of the End of Gene *D* in G4

The region around the end of φX gene *D* is complex: as well as containing the termination codon of gene *D*, it also contains the termination codon of the overlapping gene *E* in a second reading frame, the ATG start of gene *J* overlapping the gene-*D* termination codon in a third reading frame, and the gene-*J* ribosome-binding-site sequence within the *D*/*E* coding region. In G4 it appears as though the last 14 nucleotides of φX gene *D* containing all of these functions are duplicated (Fig. 8).

G4 retains the TAA termination codon of φX gene *D*, and the D-protein amino acid sequence in this region is similar to that in φX. The TGA termination codon of φX gene *E* is replaced in G4 by TCA, and another TGA, which acts as the gene-*E* terminator in G4, occurs 12 nucleotides away. This results in a larger G4 E protein (five amino acids) and adds to the great degree of difference in the C-terminal half of the G4 E protein compared with the φX E protein (see gene-*E* sequences in Appendix II). The overlapping gene-*D* termination and gene-*J* initiation codon TAATG in φX is changed in G4 to TAAAG, and another ATG initiation codon, 41 nucleotides farther on, is used for gene *J* in G4. Thus, G4 contains an intercistronic space between genes *E* and *J* that is not present in φX.

Inspection of the sequence of the new G4 gene-*E*/*J* untranslated region shows a duplication of the last 14 nucleotides of gene *D*, which are present at

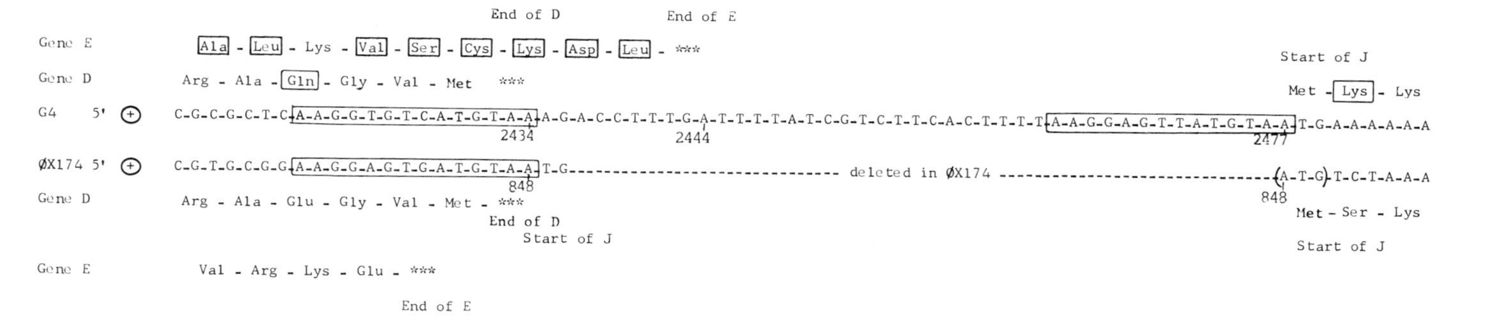

Figure 8 G4 gene-*D* duplication (from G. N. Godson and J. C. Fiddes, in prep.). The amino acids that are different in G4 and φX are boxed and the 14 bases that are duplicated in G4 DNA are also boxed. The ATG start of φX gene *J* is repeated for clarity, and the repeated ATG is bracketed.

both ends of the region. The sequence of the second repeat, which contains the start of G4 gene *J*, has only one nucleotide different from the last 14 nucleotides of φX gene *D* and still retains the overlapping TAATG end of gene *D* (not used) and start of gene *J*, plus a ribosome-binding site for gene *J*. These 14 nucleotides, however, have three differences from the true end of G4 gene *D*, which itself retains a ribosome-binding site for gene *J* and presumably is nonfunctional.

Deletions and Insertions

Apart from the duplication of the end of gene *D* and the changes in the untranslated regions of the genome, there are several insertions and deletions in the coding regions of the G4 genes in comparison with φX genes.

G4 gene *A* has one minor and two major differences from φX gene *A*, all of which are near the 5′ end of the gene. Eighteen additional nucleotides are inserted very close to the 5′ end and 141 more (47 codons) are inserted farther on. A single extra codon is deleted between these two insertions. G4 gene *J* is one-third deleted (39 nucleotides) relative to φX. G4 gene *H* has deletions of four codons and one codon in the 5′-terminal third of the coding region and an insertion of 14 codons (42 nucleotides) in the 3′-terminal third. G4 gene *G* has insertions of two codons and one codon and a deletion of one codon.

EVOLUTION OF G4 AND φX

Although the above differences have been described for convenience as though they represent changes from the φX sequence to the G4 sequence, there is, of course, no evidence that this is the way they have evolved, and it is most probable that they have evolved independently from a common ancestor. The insertion of an *E*/*J* intergenic sequence in G4 could be described equally well as a deletion in φX. Several features of the G4 and φX DNA sequences suggest that insertions and deletions of blocks of DNA by recombination have produced many of the changes observed. These features are:

1. The frequent use of alternative serine codons (AGC/T → TCX). This occurs often in highly conserved stretches of amino acids but cannot take place in single steps without changing the amino acid.
2. The insertion and deletion of single triplet codons.
3. The insertion and deletion of large blocks of nucleotides (up to 141 in gene *A*) in coding regions of the DNA. Sometimes both insertions and deletions occur in the same gene (i.e., in genes *A* and *H*). Both of these processes take place in multimers of the triplet codon, as the amino acid phasing on both sides of the deletion and insertion appears to be relatively unchanged.

4. The different sizes of the untranslated regions of the genomes imply deletion and insertion of DNA.

It is also possible that the φX and G4 genomes evolve by a mechanism of intragenic recombination, each using the pool of its own replicating viral DNA molecules as a source of new sequences. For instance, if an essential triplet codon is changed by mutation, it could be exchanged with a region of the DNA from an adjacent viral genome that contains the correct amino acid codon. In this way, different flanking codons, or alternative amino acid codons, may be incorporated.

Not only have the amino acid sequences of the viral proteins evolved so that there is an overall 33% amino acid difference between the proteins of the two phages, but the control sequences in the DNA have also evolved. This is seen in the considerably different sequences of the untranslated intercistronic regions, the differences in ribosome-binding-site and promoter sequences, and in the absence in G4 of some of the overlapping initiation and termination codons for successive φX genes. It is reasonable to suspect, therefore, that the control of expression of some of the genes (e.g., gene *J*) is either different in the two organisms or is minimal.

Now that the complete base sequences of φX and G4 are known, it will be possible to design experiments to investigate the mechanism of diversification of the two genomes.

EVOLUTION OF OVERLAPPING GENES

Three overlapping-gene systems have now been identified in φX and G4. These are gene *B* within gene *A*, gene *K* within genes *A* and *C*, and gene *E* within gene *D*.

It was postulated previously (Barrell et al. 1976) that gene *E* evolved after gene *D*. φX has a high incidence of T nucleotides in the third position of codons, and this is particularly true for gene *D* but not for gene *E*. It was speculated that the use of the phase displaced one nucleotide to the right of the *D* phase was advantageous to φX, as this generates codons that have a high incidence of T in the second position and these codons specify hydrophobic amino acids. The hydrophobic nature of the E protein, particularly around the N terminus, is considered important in the function of E in lysing the cell. Similar arguments could be applied to the φX (and G4) K protein as it also is highly hydrophobic at the N terminus.

It was also proposed, based on the third-position T phenomenon, that gene *B* was initially separate from gene *A* and that the present gene *A* in φX resulted from a readthrough of the gene-*A* termination (Smith et al. 1977). This was based on the observation that in the *A*/*B* overlap region the high third-position T is a feature of the *B* phase rather than the *A* phase.

To a certain extent, the sequence of the overlapping genes in G4 supports these ideas. Though the G4 E protein has 44.0% amino acid differences with respect to the φX E protein, the hydrophobic N terminus of the protein is conserved. The K protein is similar in this regard: there is a 39.3% amino acid difference but no alteration in the nature of the N terminus. Also, the distribution of third-position Ts in the *A/B* overlap in G4 follows the same pattern as in φX, which also suggests that initially the genes were distinct.

Table 2 gives the percentages of amino acid and nucleotide differences between the two phages in the overlapping regions (*B+A*, *K+A/C*, and *D+E*) and the percentages of nucleotide changes that are silent since they are in the third, degenerate codon position.

For the genes that are not involved in overlaps the values for the amino acid and nucleotide changes tend to be similar (see, for example, *F*, which has 33.6% and 33.7% amino acid and nucleotide changes, respectively). Within the overlapping regions this feature is observed for *A*, *A+C*, and *D* rather than for *B*, *K*, and *E*. Conversely, B, K, and E have more amino acid changes (44.2%, 39.3%, and 44.0%, respectively) than average, and these changes are associated with a lower frequency of nucleotide changes (23.9%, 23.4%, and 19.0%, respectively). This difference between the two phases is highlighted by examining the frequency of silent third-position changes.

In gene *F*, with an overall nucleotide change of 33.6%, there are 28.8% third-position changes. However, within the overlapping regions of gene *A* (*A+B*; *A+K*), gene *C* (*C+K*), and gene *D* (*D+E*), the third-position changes are 30.2%, 23.5%, 26.0%, and 51.9%, respectively. This suggests that in these regions there is considerable selection pressure on the amino acid sequences of the A, C, and D proteins. This is especially so for gene *D*, which has a very high level of third-position changes. However, in the corresponding *B*, *K*, and *E* phases, third-position changes are very low (8.1%, 7.5%, and 9.6%, respectively), which suggests that there is less pressure on these genes. This is shown in Table 2.

This observation is consistent with the concept that genes *E* and *K* arose later, and that the E and K proteins require only a hydrophobic N terminus to function. It is not consistent with the idea that the present-day gene *A* was a readthrough into a preexisting gene *B*, as gene *B* is much less conserved between the two phages than is gene *A*.

ACKNOWLEDGMENTS

The authors wish to thank Drs. D. Shaw and J. Walker for making their G4 protein sequence data available before publication, and R. Staden for assistance in computer handling of the DNA sequence data. G.N.G. was supported by U.S. Public Health Service grant CA06519 and by an American Cancer Society Scholar in Cancer Research Award (ACS-86).

REFERENCES

Air, G. M. 1976. Amino acid sequences from the gene F (capsid) protein of bacteriophage φX174. *J. Mol. Biol.* **107:** 433.

Air, G. M., E. H. Blackburn, F. Sanger, and A. R. Coulson. 1975. The nucleotide and amino acid sequences of the N(5′) terminal region of gene G of bacteriophage φX174. *J. Mol. Biol.* **96:** 703.

Air, G. M., E. H. Blackburn, A. R. Coulson, F. Galibert, F. Sanger, J. W. Sedat, and E. B. Ziff. 1976. Gene F of bacteriophage φX174. Correlation of nucleotide sequences from the DNA and amino acid sequences from the gene product. *J. Mol. Biol.* **170:** 445.

Axelrod, N. 1976a. Transcription of bacteriophage φX174 in vitro: Selective initiation with oligonucleotides. *J. Mol. Biol.* **108:** 753.

———. 1976b. Transcription of bacteriophage φX174 in vitro: Analysis with restriction enzymes. *J. Mol. Biol.* **108:** 771.

Baas, P. D. and H. S. Jansz. 1972. φX174 replicative form DNA replication, origin and direction. *J. Mol. Biol.* **63:** 569.

Baas, P. D., H. S. Jansz, and R. L. Sinsheimer. 1976. φX174 DNA synthesis in a replication-deficient host: Determination of the origin of φX DNA replication. *J. Mol. Biol.* **102:** 633.

Barrell, B. G. 1977. Sequence analysis of bacteriophage φX174 DNA. In *International review of biochemistry* (ed. B. F. C. Clarke), vol. 17. University Park Press, Baltimore. (In press.)

Barrell, B. G., G. M. Air, and C. A. Hutchison III. 1976. Overlapping genes in bacteriophage φX174. *Nature* **264:** 34.

Bartok, K., B. Harbers, and D. T. Denhardt. 1975. Isolation and characterization of self-complementary sequences from φX174 viral DNA. *J. Mol. Biol.* **99:** 93.

Benbow, R. M., C. A. Hutchison III, J. D. Fabricant, and R. L. Sinsheimer. 1971. Genetic map of bacteriophage φX174. *J. Virol.* **7:** 549.

Bouché, J. P., L. Rowen, and A. Kornberg. 1978. The RNA primer synthesized by primase to initiate phage G4 DNA replication. *J. Biol. Chem.* **253:** 765.

Bouché, J. P., K. Zechel, and A. Kornberg. 1975. *Dna*G gene product, a rifampicin-resistant RNA polymerase initiates the conversion of a single-stranded coliphage DNA to its duplex replicative form. *J. Biol. Chem.* **250:** 5995.

Chen, C. Y., C. A. Hutchison III, and M. H. Edgell. 1973. Isolation and genetic localization of three φX174 promoter regions. *Nat. New Biol.* **243:** 233.

Clements, J. B. and R. L. Sinsheimer. 1975. Process of infection with bacteriophage φX174. XXXVII. RNA metabolism in φX174-infected cells. *J. Virol.* **15:** 151.

Denniston-Thompson, K., D. Moore, K. Kruger, M. Furth, and F. Blattner. 1977. Physical structure of the replication origin of bacteriophage lambda. *Science* **198:** 1051.

Devillers-Thiery, A., T. Kindt, G. Scheele, and G. Blobel. 1975. Homology in amino-terminal sequence of precursors to pancreatic secretory proteins. *Proc. Natl. Acad. Sci.* **72:** 5016.

Eisenberg, S., J. Griffith, and A. Kornberg. 1977. φX174 cistron A protein is a multifunctional enzyme in DNA replication. *Proc. Natl. Acad. Sci.* **74:** 3198.

Eisenberg, S., B. Harbers, C. Hours, and D. T. Denhardt. 1975. The mechanism of replication of φX174 DNA. XII. Non-random location of gaps in nascent φX174 RF II DNA. *J. Mol. Biol.* **99:** 107.

Fiddes, J. C. 1976. Nucleotide sequence of the intercistronic region between genes G and F in bacteriophage φX174 DNA. *J. Mol. Biol.* **107:** 1.

Fiddes, J. C. and G. N. Godson. 1978. A restriction endonuclease cleavage map of G4. *Virology* **89:** 322.

Fiddes, J. C., B. G. Barrell, and G. N. Godson. 1978. Nucleotide sequences of the separate origins of synthesis of bacteriophage G4 viral and complementary DNA strands. *Proc. Natl. Acad. Sci.* **75:** 1081.

Fiers, W., R. Contreras, F. Duerinck, G. Haegeman, D. Iserentant, J. Merregaert, W. Min-Jou, F. Molemans, A. Raeymaekers, A. van den Berghe, G. Volckaert, and M. Ysebaert. 1976. Complete nucleotide sequence of a bacteriophage MS2 RNA: Primary and secondary structure of the replicase gene. *Nature* **260:** 500.

Freymeyer, D. K., P. R. Shank, M. H. Edgell, C. A. Hutchison III, and T. C. Vanaman. 1977. Amino acid sequence of the small core protein from bacteriophage φX174. *Biochemistry* **16:** 4551.

Godson, G. N. 1974a. Origin and direction of φX174 double and single stranded DNA synthesis. *J. Mol. Biol.* **90:** 127.

———. 1974b. Evolution of φX174: Isolation of four new φX-like phages and comparison with φX174. *Virology* **58:** 272.

———. 1975a. Evolution of φX174. II. A cleavage map of the G4 phage genome and comparison with the cleavage map of φX174. *Virology* **63:** 320.

———. 1975b. A physical map of G4 and the origin of G4 double and single stranded DNA replication. In *DNA synthesis and its regulation* (ed. M. Goulian and P. Hanawalt), p. 386. W. A. Benjamin, Menlo Park, California.

———. 1977a. G4 DNA replication. II. Synthesis of viral progeny single-stranded DNA. *J. Mol. Biol.* **117:** 337.

———. 1977b. G4 DNA replication. III. Synthesis of replicative forms. *J. Mol. Biol.* **117:** 353.

Godson, G. N. and H. Boyer. 1974. Susceptibility of the φX-like phages G4 and G14 to *Eco*RI endonuclease. *Virology* **62:** 270.

Grindley, J. and G. N. Godson. 1978. Evolution of φX174. V. Alignment of the φX174, G4, and St-1 restriction enzyme maps. *J. Virol.* **27:** 745.

Grohmann, K., L. H. Smith, and R. L. Sinsheimer. 1975. New method for isolation and sequence determination of 5′-terminal regions of bacteriophage φX174 in vitro mRNAs. *Biochemistry* **14:** 1951.

Harada, F. and S. Nishimura. 1974. Purification and characterization of AUA specific isoleucine transfer ribonucleic acid for *Escherichia coli* B. *Biochemistry* **13:** 300.

Hayashi, Y. and M. Hayashi. 1971. Fractionation of φX174 specific mRNA. *Cold Spring Harbor Symp. Quant. Biol.* **35:** 171.

Hayashi, M., F. K. Fujimara, and M. Hayashi. 1976. Mapping of in vivo messenger RNA's for bacteriophage φX174. *Proc. Natl. Acad. Sci.* **73:** 3519.

Langeveld, S. A., A. D. M. van Mansfeld, P. D. Baas, H. S. Jansz, G. A. van Arkel, and P. J. Weisbeek. 1978. Nucleotide sequence of the origin of replication in bacteriophage φX174 RF DNA. *Nature* **271:** 417.

Linney, E. and M. Hayashi. 1974. Intragenic regulation of the synthesis of φX174 gene A proteins. *Nature* **249:** 345.

Maniatis, T., M. Ptashne, K. Backman, D. Kleid, S. Flashman, A. Jeffrey, and R. Maurer. 1975. Recognition sequences of repressor and polymerase in the operators of bacteriophage lambda. *Cell* **5:** 109.

Martin, D. M. and G. N. Godson. 1977. G4 DNA replication. I. Origin of synthesis of the viral and complementary DNA strands. *J. Mol. Biol.* **117:** 321.

Pribnow, D. 1975. Nucleotide sequence of an RNA polymerase binding site at an early T7 promoter. *Proc. Natl. Acad. Sci.* **72:** 784.

Ravetch, J. V., K. Horiuchi, and N. D. Zinder. 1977. Nucleotide sequences near the origin of replication of bacteriophage f1. *Proc. Natl. Acad. Sci.* **74:** 4219.

Ray, D. S. and J. Dueber. 1975. Structure and replication of replicative forms of the φX-related bacteriophage G4. In *DNA synthesis and its regulation* (ed. M. Goulian and P. Hanawalt), p. 370. W. A. Benjamin, Menlo Park, California.

———. 1977. Location of the origin-terminus of the viral strand of the duplex replicative form of bacteriophage G4 DNA. *J. Mol. Biol.* **113:** 651.

Rosenberg, M., B. de Crombrugghe, and R. Musso. 1976. Determination of nucleotide sequences beyond the sites of transcriptional termination. *Proc. Natl. Acad. Sci.* **73:** 717.

Sanger, F. and A. R. Coulson. 1975. A rapid method for determining sequences in DNA by primed synthesis with DNA polymerase. *J. Mol. Biol.* **94:** 441.

Sanger, F., S. Nicklen, and A. R. Coulson. 1977a. DNA sequencing with chain-terminating inhibitors. *Proc. Natl. Acad. Sci.* **74:** 5463.

Sanger, F., G. M. Air, B. G. Barrell, N. L. Brown, A. R. Coulson, J. C. Fiddes, C. A. Hutchison III, P. M. Slocombe, and M. Smith. 1977b. Nucleotide sequence of bacteriophage φX174 DNA. *Nature* **265:** 687.

Schaller, H., C. Gray, and K. Herrmann. 1975. Nucleotide sequence of an RNA polymerase binding site from the DNA of bacteriophage fd. *Proc. Natl. Acad. Sci.* **72:** 737.

Schekman, R., J. H. Weiner, A. Weiner, and A. Kornberg. 1975. Ten proteins required for conversion of φX174 single-stranded DNA to duplex form in vitro: Resolution and reconstitution. *J. Biol. Chem.* **250:** 5859.

Shaw, D. C., J. E. Walker, F. D. Northrop, B. G. Barrell, G. N. Godson, and J. C. Fiddes. 1978. Gene K, a new overlapping gene in bacteriophage G4. *Nature* **272:** 510.

Shine, J. and L. Dalgarno. 1974. The 3′-terminal sequence of *Escherichia coli* 16S ribosomal RNA: Complementary to nonsense triplets and ribosome binding sites. *Proc. Natl. Acad. Sci.* **71:** 1342.

Slocombe, P. 1977. "Sequencing φX174 DNA." Ph.D. thesis, Cambridge University, England.

Smith, L. H. and R. L. Sinsheimer. 1976a. The in vitro transcription units of φX174. I. Characterization of synthetic parameters and measurement of transcript molecular weights. *J. Mol. Biol.* **103:** 681.

———. 1976b. The in vitro transcription units of φX174. II. In vitro initiation sites of φX174 transcription. *J. Mol. Biol.* **103:** 699.

———. 1976c. The in vitro transcription units of φX174. III. Initiation with specific 5′ end oligonucleotides of in vitro φX174 RNA. *J. Mol. Biol.* **103:** 711.

Smith, L. H., K. Grohmann, and R. L. Sinsheimer. 1974. Nucleotide sequences of the 5′ termini of φX174 mRNAs synthesized in vitro. *Nucleic Acids Res.* **1:** 1521.

Smith, M., N. L. Brown, G. M. Air, B. G. Barrell, A. R. Coulson, C. A. Hutchison

III, and F. Sanger. 1977. DNA sequence at the C termini of the overlapping genes A and B in bacteriophage φX174. *Nature* **265:** 702.

Staden, R., 1977. Sequence data handling by computer. *Nucleic Acids Res.* **4:** 4037.

Sugimoto, K., H. Sugisaki, T. Okamoto, and M. Takanami. 1977. Studies on bacteriophage fd DNA. IV. The sequence of messenger RNA to the major coat protein gene. *J. Mol. Biol.* **111:** 487.

Tabak, H. F., J. Griffith, K. Geider, H. Schaller, and A. Kornberg. 1974. Initiation of deoxyribonucleic acid synthesis. VII. A unique location of the gap in the M13 replicative duplex synthesized in vitro. *J. Biol. Chem.* **249:** 3049.

Takanami, M., K. Sugimoto, H. Sugisaki, and T. Okamoto. 1976. Sequence of promoter for coat protein gene of bacteriophage fd. *Nature* **260:** 297.

Tinoco, I., P. N. Borer, B. Dengler, M. D. Levine, O. C. Uhlenbeck, D. M. Crothers, and J. Gralla. 1973. Improved estimation of secondary structure in ribonucleic acids. *Nat. New Biol.* **246:** 40.

Tomizawa, J.-I., H. Ohmori, and R. E. Bird. 1977. Origin of replication of colicin E1 plasmid DNA. *Proc. Natl. Acad. Sci.* **74:** 1865.

Walker, J. E., D. C. Shaw, F. D. Northrup, and T. Horsnell. 1978. In *Proceedings of the 2nd Symposium on Solid Phase Sequence Analysis.* North-Holland, Amsterdam. (In press.)

Weisbeek, P. J., W. E. Borrias, S. A. Langeveld, P. D. Baas, and G. A. van Arkel. 1977. Bacteriophage φX174 gene A overlaps gene B. *Proc. Natl. Acad. Sci.* **74:** 2504.

Wickner, S. and J. Hurwitz. 1974. Conversion of φX174 viral DNA to double-stranded form by purified *E. coli* proteins. *Proc. Natl. Acad. Sci.* **71:** 4120.

Zechel, K., J. P. Bouché, and A. Kornberg. 1975. Replication of phage G4: A novel and simple system for the initiation of DNA synthesis. *J. Biol. Chem.* **250:** 4684.

The S13 Genome

John H. Spencer, Eric Rassart, John S. Kaptein, Klaus Harbers, Frank G. Grosveld, and Bruce Goodchild
Department of Biochemistry
McGill University
Montreal, Quebec, Canada H3G 1Y6

Bacteriophage S13, one of the isometric phages, was first described by Burnet (1927) as one of a series of phages isolated from *Salmonella* cultures; these phages were given the prefix S followed by arabic numerals. Soon after its discovery S13 was found to have anomalous properties in comparison with the other phages. For example, it was shown to migrate to the cathode in an electric field, and this electrical behavior was strikingly affected by changes in pH (Burnet and McKie 1930). It was relatively stable to reducing dyes and diffused much more rapidly than other phages. The latter result led to the speculation that S13 was smaller than most other phages (Burnet and McKie 1930). Evidence to support this hypothesis came from early centrifugal studies by Elford (1936), and an estimate was made of its size as 15 to 17 nm—a figure in agreement with results from previous ultrafiltration analyses (Elford and Andrewes 1932). Subsequent studies concentrated mainly on the action of various compounds on S13 (S13 having been chosen as a test phage because of its small size). One study revealed the unusual result that, unlike other phages, when S13 was irradiated by ultraviolet light it failed to kill its host (Mills 1955).

In 1958, Zahler published a comparison of some of the biological properties of S13 and the more recently discovered φX174; a close serological relationship between the two phages was found. In 1959, I. Tessman showed by ^{32}P-decay studies that the DNA of phage S13 is single-stranded. Later, during an investigation of various hydroxylamine-induced mutants, E. Tessman (1966, 1967) explained the sensitivity to hydroxylamine inactivation on the basis of the DNA being single-stranded. Studies on the structure and morphogenesis of S13 went into an eclipse until the late 1960s, the general assumption being that S13 was similar to φX. Then Poljak and Suruda (1969) established the presence of three different protein components among the coat proteins of S13 and published data on the amino acid compositions of the total coat proteins and coat protein β, which is probably the gene-*F* protein (Denhardt 1975).

Studies on the structure of S13 DNA were initiated in our laboratory in 1968. These included an investigation which determined that S13 DNA is single-stranded by the criteria of reactivity with formaldehyde, dependence of optical density on ionic strength, broad temperature absorbance profile, and lack of molar equivalence of the purine and pyrimidine bases. The DNA has an S°_{20} of 24.6 in standard saline-citrate solution and a molecular weight of 1.8×10^6 daltons. It was determined by electron microscopy that the phage is 26 nm in diameter (an estimate somewhat larger than that of Elford [1936]) and that its DNA is circular. Investigation of S13 replicative-form (RF) DNA showed it to be double-stranded and to have a molecular weight of 3.5×10^6 daltons. Base-composition analyses have revealed that S13 DNA is slightly more pyrimidine-rich than φX DNA, but, overall, is very similar in composition to the DNA of other isometric phages and to that of the filamentous phages (Spencer et al. 1972).

METHODS

Sequence Analysis of S13 DNA

The gathering of sequence data on S13 started with the publication of the complete pyrimidine oligonucleotide catalog by Cerny et al. (1969). This study took advantage of the single-strandedness of the phage DNA and the double-strandedness of the RF DNA to deduce the pyrimidine oligonucleotide distributions of both strands of the RF DNA. The longest pyrimidine oligonucleotides present in S13 DNA are undecanucleotides, and there is an asymmetrical distribution of pyrimidine tracts between the two strands, particularly of the longer pyrimidine oligonucleotides. This study was followed somewhat later by a sequence analysis of the longer pyrimidine oligonucleotides which confirmed the previous distribution data (Cerny et al. 1969; Spencer and Boshkov 1973) and revealed remarkable sequence homology among the longer pyrimidine oligonucleotides present in S13 DNA (Harbers et al. 1976; Delaney and Spencer 1976; Kaptein and Spencer 1978).

Recently, the sequences of portions of the S13 DNA genome have been reported (Grosveld and Spencer 1977; E. Rassart et al., in prep.). These sequences are in the gene-*A*/*B* region and are very similar to the corresponding φX sequences. By correlation of the sequences with RNA-polymerase-binding studies (E. Rassart and J. H. Spencer, in prep.), the putative promoter site for S13 gene *B* has been localized. The amino acid sequence of a portion of the gene-*A* protein has been deduced (Grosveld and Spencer 1977). It contains a cluster of five arginine residues within a seven-amino-acid block, which in turn is surrounded by a neutral region containing a number of asparagine and glutamine residues. This highly positively charged center could be a DNA-binding site of the gene-*A* protein. The φX gene-*A* protein has a nickase activity as one of its properties (Henry and Knippers 1974) and is capable of binding to DNA.

Restriction Enzyme Analysis

The small size of S13 DNA and its assumed similarity to φX DNA made it an obvious choice for structural investigation by restriction enzyme analysis. Restriction fragment maps of *Hin*dII+III and *Hae*III cleavages (Grosveld et al. 1976), *Hin*dII, *Hae*III, *Hpa*I, and *Hpa*II cleavages (Godson 1976), and the single-site cleavages of *Pst*I, *Mbo*I, *Ava*I (Godson 1976), and *Hin*dIII (Goodchild and Spencer 1978) have been published. Hayashi and Hayashi (1974) compared cleavages of S13 and φX DNA by *Hap*II and *Hin*HI restriction enzymes but did not construct maps for S13; however, *Hap*II is an isoschizomer of *Hpa*II. In a comparative study of cleavages of S13, φX, G4, and St-1 DNA by 26 different restriction enzymes, Godson and Roberts (1976) found similarities between φX and S13 and between G4 and St-1. There were fewer similarities between the two pairs of phages. The restriction fragment maps of S13 DNA have been aligned with the genetic map of the phage (Grosveld et al. 1976) by comparison with φX data (Weisbeek and van Arkel, this volume). That comparative data can be used is based on the results of complementation tests which have shown the genetic maps of S13 and φX to be composed of eight homologous genes arranged in the same order in both genomes (Jeng et al. 1970). These tests confirmed the results of genetic recombination experiments (Tessman and Shleser 1963) and serological tests (Zahler 1958) which had pointed to a close genetic relationship between S13 and φX.

Location of Promoter Sites on the S13 Genome

In a study of the distribution of possible promoter sites on S13 DNA by mapping of *Escherichia coli* RNA-polymerase-binding sites, E. Rassart and J. H. Spencer (in prep.) determined that RNA polymerase binds to restriction fragments located at three regions on the S13 map, which correspond to the beginnings of gene *A*, gene *B*, and gene *D*/*E*. In a companion study, in which the polymerase molecules bound to the S13 genome were visualized under the electron microscope, E. Rassart et al. (in prep.) located six binding sites. Three of these correspond to the sites located on the restriction fragment map and a fourth corresponds to the beginning of gene *F*; the two other sites are in the middle of gene *F* and at the junction of genes *G* and *H*. The gene-*B* RNA-polymerase-binding data have been correlated with the sequence of the restriction fragment close to the start of gene *B*, to which *E. coli* RNA polymerase binds strongly (E. Rassart et al., in prep.). These studies place the putative promoter site for S13 gene *B* approximately 150 nucleotides to the right of the corresponding site in φX DNA proposed by Sanger et al. (1977).

In this article we present additional sequence information on S13 DNA and preliminary evidence for an overlap of genes *A* and *B* in S13, as in φX.

The similarities and differences between the DNA sequences of φX and S13 are explored and discussed in terms of the results of studies on the heteroduplexes formed between S13 and φX DNA (Godson 1973; Compton and Sinsheimer 1977).

RESULTS

Restriction fragment maps of *Hpa*I, *Hpa*II, *Hin*HI, *Hin*dII, *Hae*III, and *Alu*I cleavages and the unique cuts by *Hin*dIII, *Pst*I, *Mbo*I, and *Ava*I are presented in Figure 1. The maps have been aligned so that the G of the *Pst*I cleavage site CTGCA ↓ G is nucleotide 1 of each map. The asterisks indicate positions that differ from those on the corresponding φX map. The *Hpa*I and *Hpa*II maps are reinterpretations of the data of Godson (1976). The *Hpa*II cleavages were compared with the data from the isoschizomer *Hap*II (Hayashi and Hayashi 1974). The *Hin*HI map is based on data from Hayashi and Hayashi (1974). The *Hin*dII map is based on data from Grosveld et al. (1976) and Goodchild and Spencer (1978). The *Hae*III map is redrawn from Grosveld et al. (1976). Nomenclature for the restriction fragments is that described by Grosveld et al. (1976), the fragment numbers decreasing with size. The nomenclature coincides with that used for φX by Sanger et al. (1977; see also Appendix I).

Data for the *Alu*I map were obtained by separation of the restriction fragments by polyacrylamide gel electrophoresis and comparison with fragments from an identical *Alu*I digest of φX RF DNA (see Fig. 7 in Godson

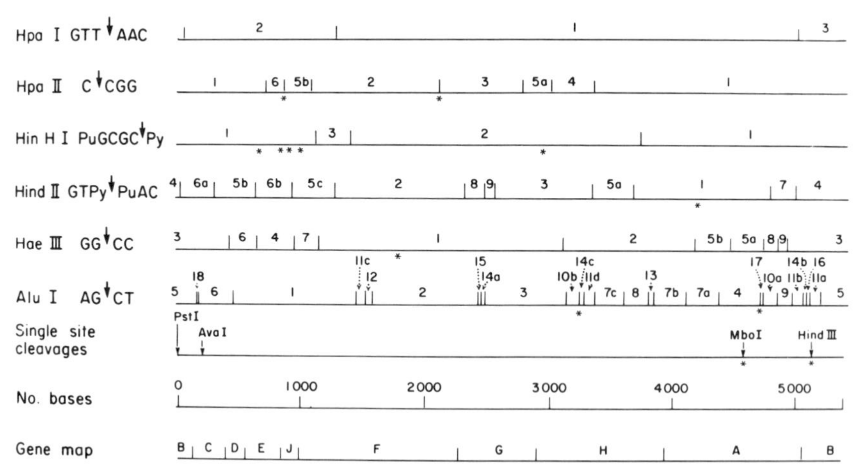

Figure 1 Restriction maps of S13 DNA. The maps are oriented to the *Pst*I cleavage site at nucleotide number 1. Asterisks indicate differences from the corresponding map of φX, i.e., either the presence or absence of certain cleavage sites. The gene map indicates only the start of each gene; gaps and overlaps have been excluded.

1976). Where the fragment sizes were identical, it was assumed that these are corresponding fragments. Confirmation of the positions of fragments 16, 14b, 11b, 10a, and 9 has been obtained from the sequence data shown in Table 1. Because of the small sizes of some of the fragments, for example, 14a, 14c, 15, and 18, and the identical sizes of others, such as 11a, 11c, and 11d, analysis by gel electrophoresis is difficult, and confirmation of these fragment positions on the S13 map has not yet been obtained. Comparison of the *Alu*I restriction fragment maps of φX and S13 reveals that cuts unique to S13 occur at S13 10b/14c and S13 4/17. However, the exact locations of 14c and 17 are not known and may be at S13 14c/10b and S13 17/4, respectively.

The position of site *Hpa*II 6/5b on the *Hpa*II map was assigned from the sequence data (sequence 6) in Table 1. The alternative position for this site is *Hpa*II 5b/6, approximately 60 bases to the right, which would be close to the *Hin*dII 6b/5c site. However, the *Hpa*II recognition sequence C↓CGG is not present in this area of S13 DNA.

The data in the *Hin*HI map do not coincide exactly with the sequence data in Table 1. A *Hin*HI sequence GGCGCT has been located at position 926–931 (sequence 6). A reinvestigation of the cleavage of S13 by *Hin*HI, or by an isoschizomer of *Hin*HI, is needed to clarify this discrepancy.

The positions of the unique cuts of *Hin*dIII and *Pst*I are from the data of Goodchild and Spencer (1978), and those of *Mbo*I and *Ava*I have been calculated from the data of Godson (1976). The *Mbo*I cut, which could be at either position ~4600 or position ~4860, was assigned to position ~4600 because the recognition sequence ↓GATC is absent at position ~4860 (Grosveld and Spencer 1977; see also Table 1, sequence 11).

Sequence Analysis of S13

The current sequence information on S13 is presented in Table 1. The sequence numbers correspond to the φX numbers (Sanger et al. 1977) to facilitate comparison of the data. However, the correspondence between the numbering of the DNA of the two phages is not exact (see sequence 2 with an insertion). The gene map location is also presented. The methods by which the various sequences have been determined are indicated and documented in the table.

Figure 2 is an autoradiograph of a plus-minus sequence analysis primed by S13 *Hae*III 6 on a viral template. The sequence is complementary to part of sequence 4, Table 1. Figure 3 is an autoradiograph of an alkylation-hydrazinolysis sequence analysis of fragment *Hin*dII/*Hae*III 75 (S13 *Hin*dII 7 cleaved by *Hae*III, resulting in a piece 75 nucleotides long) and is part of sequence 11, Table 1. Figures 2 and 3 are presented as representative of the sequence data originating in this article.

The sequences in Table 1 that differ from the corresponding φX DNA

Table 1 Nucleotide sequences of S13 DNA

Sequence	Gene	Method*	Reference†
1. [1, 10, 20, 30] g-a-g-t-t-t-t-a-t-c-g-c- - - - -a-t-g-a-c-g-c-a-g-a-a-c-a-a *Pst*I	*A/B*	a	1
2. [221, 230, 241] a-c-g-a-G-T-T-G-G-C-A-T-G-A-G-G-A-G-A-A-G	*C*	b	2
3. [323, 330, 340, 350] A-C-A-T-T-T-T-A-A-A-A-G-A-G-C-G-T-G-G-A-T-T-A-C-T-A-T-t-T-G-A-G *Hin*dII 6a/5b	*C*	c	3
4. [372, 380, 390, 400] a-c-t-a-a-t-a-g- -t-a-a- -a-a-a-t-c-a-t-g-A-G-T-C-A-A-G-t-T-A-C-T-G-A-a-C-a- *Hin*fI [410, 420, 426] A-T-C-C-G-T-A-C-G-T-T-T-C- -A-G-A	*C/D*	d	4
5. [621, 630, 640, 650] A-T-T-G-C-T-G-C-C-G-T-C-A-T-T-G-C-T-T-A-T-T-A-T-T-T-T-C-A-T-C-C- [655] C-G-T *Hin*dII 5b/6b	*D/E*	c	3
6. [924, 930, 940, 950] A-A-G-G-c-G-c-T-C-G-T-C- -T-T-c-G-T-A-T-G-T-A- -G-C-G-G-T-C-A-A- *Hha*I — *Hin*dII 6b/5c	*J*	c	3
[960, 970, 981] C-A-A-T-T-T-T-A-A-T-T-G-C-A-G-G-G-G-C-T-T-A-G-G-c-c *Hae*III 4/7		d	4
7. [1268, 1280, 1290, 1293] G-T-T-A-A-T-G-C-C-A-C-T-C-C-T-C-T-T-C-C-G-A-C-C-g-t *Hin*dII 5c/2	*F*	c	3

8. (3245) a-g-t-C-T-(3250) G-C-C-G-C-T-G-A-T-A-(3260) A-A-G-G-A-A-A-G-G-(3269) A	*H*	b	2
9. (3392) t-g- - -a- -t-g-(3400) g-a-c-a-a-T-C-A-G-A-(3410) A-A-G-A-G-(3415) A	*H*	b	2
10. (4415) g-T-G-A-T-(4420) G-A-G-T-A-T-C-A-T-T-(4430) A-C-C-C-C-A-A-A-A-A-(4440) G-A-A-A-G-(4445) G	*A*	b	2
11. (4747) G-G-C-C-G-T-C-T-T-C-A-T-T-(4760) T-C-C-A-T-G-C-G-G-T-(4770) G-C-A-T-T-T-T-A-T-G- *Hae*III 5a/8		e	5
(4780) C-G-G-A-C-A-C-T-T-C-(4790) C-T-T-C-A-G-G-T-A-G-(4800) C-G-T-T-G-A-C-C-C-T- *Hin*dII 1/7		e	5
(4810) A-A-C-T-T-T-G-G-T-C-(4820) G-T-C-G-G-G-T-A-C-G-(4830) C-A-A-T-C-G-C-C-G-C-		c,d,e	3,4,5
(4840) C-A-G-T-T-A-A-A-T-A-(4850) G-C-T-T-G-C-A-A-A-A-(4860) T-A-C-G-T-G-G-C-C-T- *Alu*I 10a/9 *Hae*III 8/9		c,d,e	3,4,5
(4870) T-A-T-G-G-T-T-A-C-A-(4880) G-T-A-T-G-C-C-C-A-T-(4890) C-G-C-A-G-T-T-C-G-C-		e	5
(4900) T-A-C-A-C-G-C-A-G-G-(4910) A-C-G-C-T-T-T-T-T-C-(4920) A-C-G-T-T-C-T-G-G-T-	*A*	e	5
(4930) T-G-G-T-T-G-T-G-G-C-(4940) C- -G-T-T-G-A-T-G-C-(4950) T-A-A-A-G-G-T-G-A-G- *Hae*III 9/3		c,e	3,5
(4960) C-C-G-C-T-T-A-A-A-G-(4970) C-T-A-C-C-A-G-T-T-A-(4980) T-A-T-G-G-C-T-G-T-T- *Alu*I 9/11b		c	3
(4990) G-G-T-T-T-C-T-A-T-G-(5000) T-G-G-C-T-A-A-A-T-A-(5010) T-G-T-T-A-A-C-A-A-A- *Hin*dII 7/4	*A/B*	c	3

(*continued*)

Table 1 (continued)

Sequence	Gene	Method*	Reference†
5020 A-A-G-T-C-A-G-A-T-A-T(5030)-G-G-A-C-c-t-t-g-C-T(5040)-G-C-T-A-A-A-g-G-T-		a,c,e	3,6
5050 C-T-A-G-G-A-G-C-T-A-A(5060)-A-G-A-A-T-G-G-A-A-C(5070)-A-A-C-T-C-A-C-T-A- *Alu*I 11b/14b		a,c,e	3,6
5080 A-A-A-A-C-C-A-A-G-C-T(5090)-G-T-C-G-C-T-A-C-T-T(5100)-C-C-C-A-A-G-A-A-G-C(5110)-T-<u>T</u> *Alu*I 14b/16 *Alu*I 16/11a *Hin*dIII		a,c,e	3,6
12. 5243 a-C-G-C-A-A-A-A-A-G-A-G-A-G-A 5257	*A*/*B*	b	2

Capital letters indicate sequences that are fully confirmed, small letters sequences to be confirmed, and spaced dashes uncertain sequences. Underlined elements indicate differences from the corresponding sequence in φX. The φX sequence numbers (Sanger et al. 1977) are used. The restriction sites conform with the nomenclature in Figure 1.

* Methods: (a) alkylation with methyl methane sulfonate; (b) plus-minus with pyrimidine oligonucleotide primer and viral DNA template; (c) alkylation and hydrazinolysis; (d) plus-minus with a restriction fragment primer and viral DNA template; (e) T4 endonuclease IV hydrolysis.

† (1) B. Goodchild and J. H. Spencer (unpubl.); (2) Kaptein and Spencer (1978); (3) K. Harbers and J. H. Spencer (unpubl.); (4) P. Darragh and J. H. Spencer (unpubl.); (5) Grosveld and Spencer (1977); (6) E. Rassart et al. (in prep.).

sequences are underlined. Sequences in small letters remain to be confirmed by a second method of analysis in which internal confirmation is not part of the method. Sequences indicated by spaced dashes have been collected but are too ambiguous for publication at this time. Restriction sites are also presented in Table 1 to aid in orienting the data.

DISCUSSION

The area of the S13 genome on which we have concentrated is the *A*/*B* region through to the gene-*J* region. In the *A*/*B* region we have sequence information on residues 4747 to 5112. E. Rassart and J. H. Spencer (in

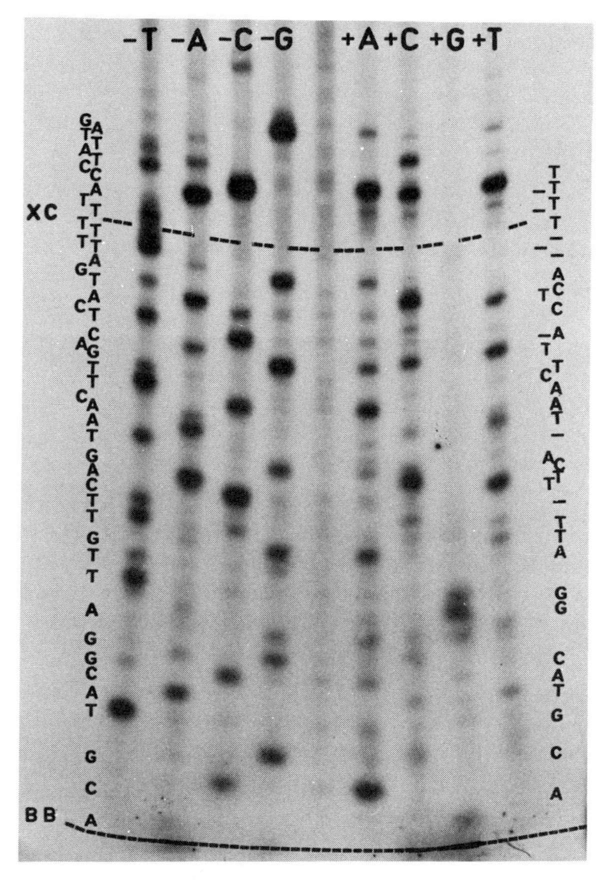

Figure 2 Autoradiograph of a plus-minus sequence analysis of S13 DNA primed by restriction fragment S13 *Hae* III 6 on a viral DNA template. *E. coli* DNA polymerase I was used for both the plus and minus reactions. The sequence is complementary to part of viral sequence 4, Table 1. XC indicates the position of the marker xylene cyanole FF, BB the position of the dye marker bromphenol blue.

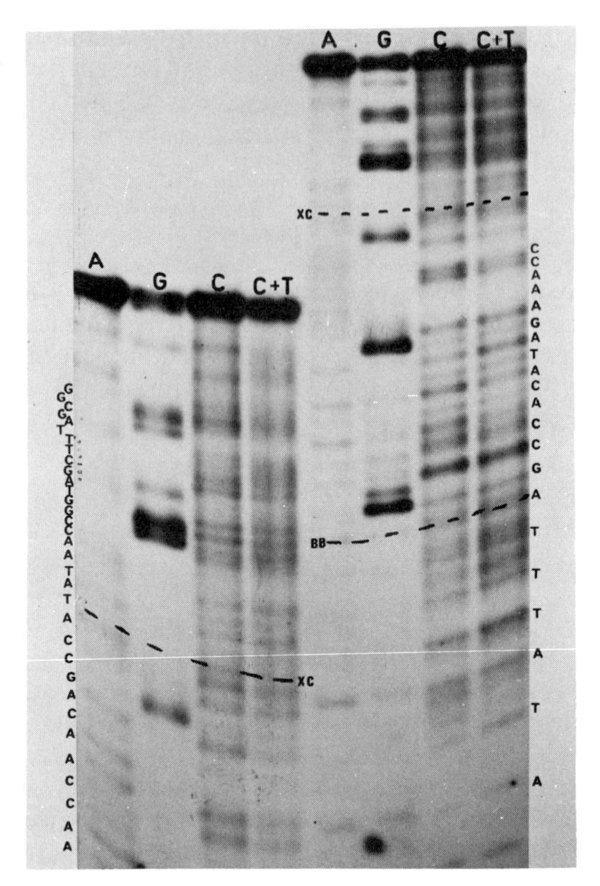

Figure 3 Autoradiograph of an alkylation-hydrazinolysis sequence analysis of S13 fragment *Hin*dII/*Hae*III 75 which was generated by *Hae*III hydrolysis of S13 *Hin*dII 7. XC indicates the position of the dye marker xylene cyanole FF, BB the position of the dye marker bromphenol blue. The alternative G and alternative strong adenine/weak cytosine cleavage methods (Maxam and Gilbert 1977) were used.

prep.) have presented evidence for RNA polymerase binding in the region from 5014 to approximately 5060 and have suggested that this is the putative promoter for gene *B*, with the RNA start position in the region 5040 to 5050. This is approximately 150 nucleotides away from the closest of the two gene-*B* start sequences suggested for φX (Sanger et al. 1977). The start of the S13 gene-*B* protein is probably at residue 5064, the position analogous to that in φX. The proposed difference in the origin of transcription of gene *B* of S13 compared to that of φX remains to be confirmed, as the same area in φX also binds RNA polymerase (E. Rassart and J. H. Spencer, in prep.). S13 transcription data are required to localize more accurately the promoter in this region.

The availability of the complete sequence of φX DNA (Sanger et al. 1977; see also Appendix I) allows a ready comparison with S13 sequence data. The differences between the two DNA sequences and the locations of the differences are listed in Table 2. At the present time we have conclusive evidence for only one insertion in S13, at position 229, in gene *C*. Although this one change alone would alter the reading frame of the gene-*C* protein, we have tentative evidence (Kaptein and Spencer 1978) for a second insertion in this region. A third insertion would maintain the φX reading frame except for the small area of difference around residue 229.

The difference at position 1285 (sequence 7, Table 1) is in the pyrimidine tract C_6T_4, which is unique to S13 (Cerny et al. 1969). This sequence difference confirms the previous suggestion that C_6T_4 in S13 and C_7T_3 in φX are related (Harbers et al. 1976).

There are five C→T, three T→C, and two G→T transitions and one G→C, one C→A, one A→C, and one A→T transition. These differences result in a slightly increased pyrimidine content for S13, in agreement with base-composition and pyrimidine-tract data indicating that S13 is more pyrimidine-rich than φX (Cerny et al. 1969; Spencer et al. 1972). The genes in which the differences occur are listed in Table 2, as are the codon positions according to the φX amino acid readout and the amino acid changes that

Table 2 Comparison of φX and S13 nucleotide sequences

Change φX → S13	Sequence number	Gene location	Position in amino acid reading frame	Amino acid change φX → S13
Insertion (C)	230	*C*	—	—
C → T	350	*C*	2nd	Ser → Phe
G → T	645	*E*	3rd	Met → Ile
		D	1st	Val → Phe
G → C	939	*J*	2nd	Trp → Ser
T → C	949	*J*	3rd	Gly → Gly
C → A	977	noncoding region between *J* and *F*		
C → T	1285	*F*	1st	Pro → Ser
T → C	1291	*F*	1st	Cys → Arg
C → T	4416	*A*	3rd	Arg → Arg
A → C	4426	*A*	1st	Asn → His
C → T	4773	*A*	3rd	His → His
A → T	4792	*A*	1st	Thr → Ser
T → C	4812	*A*	3rd	Asn → Asn
C → T	5010	*A*	3rd	Tyr → Tyr
G → T	5112	*A*	3rd	Leu → Leu
		B	1st	Val → Phe

would result from the gene differences. Seven of the variations are in the third position of the reading frame, six are in the first position, and only two are in the second position, an unusual and far from random result. These numbers include the double counts for the differences at positions 645 and 5112 which may correspond to the gene-*D*/*E* and gene-*A*/*B* overlaps. All the third-position variations except one result in no change in the amino acid residue. This conforms with expectations based on the genetic code and is also in concert with the complementation data (Jeng et al. 1970) and the close serological relationships between the products of the various genes of the two phages (Zahler 1958). All the third-position reading-frame differences except two are changes to T residues. The unusually high T distribution in the third position of codons in φX was observed by Sanger (1975; Sanger et al. 1977), and it had been suggested previously that this might be advantageous in translation (Denhardt and Marvin 1969), an advantage that may be enhanced in S13.

The sequence of residues 4747 through to 5112 covers the area of the possible overlap of genes *A* and *B* in S13. There is only one continuous reading frame throughout this region; the other frames have stop codons. The continuous frame has been assigned as the *A* reading frame and is the same as the *A* reading frame in φX. The base changes observed in this region do not result in stop codons; there are no insertions or deletions and there are only two first-position codon changes. A second reading frame starting at residue 5064 has been assigned as the *B* reading frame and is identical to the *B* frame in φX. It is the only continuous reading frame available at the 3′ end of the sequence because of stop codons elsewhere in the sequence. The G→T base change at position 5112 results in no change in the gene-*A* protein but causes a Val→Phe change in the gene-*B* protein. The evidence and conclusions described above are all consistent with an overlap of genes *A* and *B* in S13.

The sequence data and the restriction enzyme fragment data of S13 DNA allow a calculation of the percentage of base mismatch between φX and S13 DNA. The sequence data in Table 1 account for more than 13% of the total S13 genome, and the overall mismatch in relation to φX is ~2%.When the data in Figure 1 are included, and if it is assumed that there is one base change per altered restriction enzyme cleavage site, the mismatch increases to ~3%. These calculations indicate close sequence homology and are in agreement with the close biological similarities between the DNA of the two phages.

The two DNAs have been compared by heteroduplex analysis by Godson (1973) and Compton and Sinsheimer (1977). Godson (1973) reported up to a 35% base mismatch under stringent denaturating conditions of 75% formamide and found that only one small segment, 4.7 ± 1.9%, of the DNA of φX was highly homologous with that of S13. The more recent analysis by Compton and Sinsheimer (1977) made use of a technique of heteroduplex denaturation which included restriction fragments as reference points to align the homologous regions with the genetic map. Their data show that three regions of the heteroduplex are conserved as double-stranded regions under mildly denaturing conditions of 40% formamide; two regions of about 10% each in genes *A* and *H* and a third region of about 25% covering gene *E* and most of gene *F*. Under increased denaturing conditions up to 80% formamide, only two small duplex regions of 2–4% are conserved, one each in genes *E* and *F*.

The two independent studies (Godson 1973; Compton and Sinsheimer 1977) differ in the approximate amount of homology between S13 and φX. Compton and Sinsheimer (1977) observed less homology than did Godson (1973) under mild denaturing conditions and more homology under stringent conditions, conditions which were even more severe than those of Godson: 80% formamide vs 75%.

Compton and Sinsheimer (1977) recognized that the restriction enzyme data from Hayashi and Hayashi (1974) and Grosveld et al. (1976) and the pyrimidine oligonucleotide sequences reported by Harbers et al. (1976) did not fit with their heteroduplex analyses, but they pointed out that denaturation studies necessarily yield an average value for relatively large stretches of DNA. The sequence data in Table 1 accentuate this discrepancy. Also, the two regions of 2–4% conserved in genes *E* and *F* under stringent conditions exclude the region we have reported in gene *A* as a highly homologous region, yet it comprises ~6% of S13 DNA and has only five changes scattered through its length. The disagreement between the sequence divergence data and heteroduplex mapping results is not the case with other phage circular DNAs, for example φX and G4 (see Warner and Tessman, this volume).

The available sequence data are from a very limited part of the S13 genome and areas of wide divergence from φX may still exist. However, the restriction data must also be taken into account, and only one area, the *DEJ* region, reveals any indication of clustered changes (see Fig. 1), and this is part of the region of high homology delineated by Compton and Sinsheimer (1977).

ACKNOWLEDGMENTS

The authors wish to thank Pam Darragh and Pat Douglas for expert technical assistance in the sequence analyses using the Sanger-Coulson and Maxam-Gilbert techniques. The studies in our laboratory have been supported by the Medical Research Council of Canada and the National Cancer Institute of Canada, through grants to J.H.S. J.S.K. was the holder of a studentship from the Medical Research Council of Canada, and F.G.G. was the recipient of studentships from the Medical Research Council of Canada and the Conseil de la Recherche en Santé du Québec.

REFERENCES

Burnet, F. M. 1927. The relationships between heat-stable agglutinogens and sensitivity to bacteriophage in the *Salmonella* group. *Br. J. Exp. Pathol.* **8:** 121.

Burnet, F. M. and M. McKie. 1930. The electrical behaviour of bacteriophages. *Aust. J. Exp. Biol. Med. Sci.* **7:** 199.

Cerny, R., E. Cerna, and J. H. Spencer. 1969. Nucleotide clusters in deoxyribonucleic acids. IV. Pyrimidine oligonucleotides of bacteriophage S13 Su N15 DNA and replicative form DNA. *J. Mol. Biol.* **46:** 145.

Compton, J. L. and R. L. Sinsheimer. 1977. Aligning the φX174 genetic map and the φX174/S13 heteroduplex denaturation map. *J. Mol. Biol.* **109:** 217.

Delaney, A. D. and J. H. Spencer. 1976. Nucleotide clusters in deoxyribonucleic acids. XIII. Sequence analysis of the longer unique pyrimidine oligonucleotides of bacteriophage S13 DNA by a method using unlabeled starting oligonucleotides. *Biochim. Biophys. Acta* **435:** 269.

Denhardt, D. T. 1975. The single-stranded DNA phages. *CRC Crit. Rev. Microbiol.* **4:** 161.

Denhardt, D. T. and D. A. Marvin. 1969. Altered coding in single stranded DNA viruses? *Nature* **221:** 769.

Elford, W. J. 1936. Centrifugation studies. I. Critical examination of a new method as applied to the sedimentation of bacteria, bacteriophages and proteins. *Br. J. Exp. Pathol.* **17:** 399.

Elford, W. J. and C. H. Andrewes. 1932. The sizes of different bacteriophages. *Br. J. Exp. Pathol.* **13:** 446.

Godson, G. N. 1973. DNA heteroduplex analysis of the relation between bacteriophages φX174 and S13. *J. Mol. Biol.* **77:** 467.

———. 1976. Evolution of φX174. III. Restriction map of S13 and its alignment with that of φX174. *Virology* **75:** 263.

Godson, G. N. and R. J. Roberts. 1976. A catalogue of cleavages of φX174, S13, G4 and St-1 DNA by 26 different restriction endonucleases. *Virology* **73:** 561.

Goodchild, B. and J. H. Spencer. 1978. Mapping of the single cleavage site in S13 replicative form DNA of *Hemophilus influenzae* (*Hin*dIII) restriction enzyme. *Virology* **84:** 536.

Grosveld, F. G. and J. H. Spencer. 1977. The nucleotide sequence of two restriction fragments located in the gene *A-B* region of bacteriophage S13. *Nucleic Acids Res.* **4:** 2235.

Grosveld, F. G., K. M. Ojamaa, and J. H. Spencer. 1976. Fragmentation of bacteriophage S13 replicative form DNA by restriction endonucleases from *Hemophilus influenzae* and *Hemophilus aegyptius. Virology* **71:** 312.

Harbers, B., A. D. Delaney, K. Harbers, and J. H. Spencer. 1976. Nucleotide clusters in deoxyribonucleic acids. XI. Comparison of the sequences of the large pyrimidine oligonucleotides of bacteriophages S13 and φX174 deoxyribonucleic acids. *Biochemistry* **15:** 407.

Hayashi, M. N. and M. Hayashi. 1974. Fragment maps of φX174 replicative DNA produced by restriction enzymes from *Haemophilus aphirophilus* and *Haemophilus influenzae* H-1. *J. Virol.* **14:** 1142.

Henry, T. J. and R. Knippers. 1974. Isolation and function of the gene A initiator of bacteriophage φX174, a highly specific DNA endonuclease. *Proc. Natl. Acad. Sci.* **71:** 1549.

Jeng, Y., D. Gelfand, M. Hayashi, R. Shleser, and E. S. Tessman. 1970. The eight genes of bacteriophages φX174 and S13 and comparison of the phage specified proteins. *J. Mol. Biol.* **49:** 521.

Kaptein, J. S. and J. H. Spencer. 1978. Nucleotide clusters in deoxyribonucleic acids. XV. Sequence analysis of DNA using pyrimidine oligonucleotides as primers in the DNA polymerase I repair reaction. *Biochemistry* **17:** 841.

Maxam, A. M. and W. Gilbert. 1977. A new method for sequencing DNA. *Proc. Natl. Acad. Sci.* **74:** 560.

Mills, R. F. N. 1955. Properties of a small bacteriophage and the action of some compounds on it. *J. Gen. Microbiol.* **12:** 172.

Poljak, R. J. and A. J. Suruda. 1969. The coat proteins of φX174 and S13. *Virology* **39:** 145.

Sanger, F. 1975. Nucleotide sequences in DNA. *Proc. R. Soc. Lond. B* **191:** 317.

Sanger, F., G. M. Air, B. G. Barrell, N. L. Brown, A. R. Coulson, J. C. Fiddes, C. A. Hutchison III, P. M. Slocombe, and M. Smith. 1977. Nucleotide sequence of bacteriophage φX174 DNA. *Nature* **265:** 687.

Spencer, J. H. and L. Boshkov. 1973. Nucleotide clusters in deoxyribonucleic acids. VIII. The pyrimidine oligonucleotides of bacteriophage S13 wild type DNA. *Can. J. Biochem.* **51:** 1206.

Spencer, J. H., R. Cerny, E. Cerna, and A. D. Delaney. 1972. Characterization of bacteriophage S13 SuN15 single-strand and replicative-form deoxyribonucleic acid. *J. Virol.* **10:** 134.

Tessman, E. S. 1966. Mutants of bacteriophage S13 blocked in infectious DNA synthesis. *J. Mol. Biol.* **17:** 218.

———. 1967. Gene function in phage S13. In *The molecular biology of viruses* (ed. J. S. Colter and W. Paranchych), p. 193. Academic Press, New York.

Tessman, E. S. and R. Shleser. 1963. Genetic recombination between phages S13 and φX174. *Virology* **19:** 239.

Tessman, I. 1959. Some unusual properties of the nucleic acid in bacteriophages S13 and φX174. *Virology* **7:** 263.

Zahler, S. A. 1958. Some biological properties of bacteriophages S13 and φX174. *J. Bacteriol.* **75:** 310.

The Other Isometric Phages

G. Nigel Godson
Radiobiology Laboratories
Yale University School of Medicine
New Haven, Connecticut 06510

φX174, S13, and G4 are only three members of what appears to be a large and diverse family of small, tail-less, icosahedral coliphages whose genomes consist of circular single-stranded DNA of approximately 5500 bases. Twenty-one such viruses have been reported and named in the literature, but only 14 of these have been characterized well enough to place them indisputably in this family, and these are listed in Table 1. This paper will attempt to review and correlate the scattered published data on all of the small, tail-less isometric phages other than φX, S13, and G4.

These small isometric coliphages are present abundantly in sewage, have been reported from most parts of the world, and are probably present wherever *Escherichia coli* is found. They are extremely diverse, and no two isolates appear to be the same. For example, when the G phages were

Table 1 Characterized tail-less isometric coliphages

Phage	*E. coli* host strain	Temperature range (°C)	Place of isolation	First reported
S13	C	22–43		Burnet (1927)
G6	C	22–43	USA	Godson (1974)
φB	C	22–43	Japan	Taketo (1977b)
φR	C	43	England	Kay and Bradley (1962)
φX174	C	26–41	France	Sertic and Boulgakov (1935)
φA	C	25	Japan	Taketo (1974)
φC	C	22–35	Japan	Taketo (1977b)
G13	C	18–39	USA	Godson (1974)
G14	C	18–39	USA	Godson (1974)
G4	C	18–39	USA	Godson (1974)
α3	C and B	22–43	Scotland	Bradley (1962)
St-1	K12	33–43	England	Bradley (1970)
φK	K12	33–43	Japan	Taketo (1976a)
U3	K12		USA	Watson and Paigen (1971)
Others				Bradley (1962)

Only viruses that have been characterized as small icosahedra containing single-stranded DNA are listed. The upper and lower limits of the temperature range are defined by the ability to form plaques at those temperatures, and the data are from the published papers cited or from the author's unpublished data.

isolated (Godson 1974), one batch of raw sewage yielded four different φX-like phages, none of which was identical with φX or S13.

Host Range and Temperature Range

Wild-type isometric phages fall into two main categories: those that infect *E. coli* C strains and those that infect *E. coli* K12 strains (Table 1). This is a function of the host cell wall, and viral DNA from phages specific for *E. coli* C can transfect *E. coli* K12 spheroplasts and vice versa. Most of the phages have their own specific receptor sites, and it is relatively easy to pick spontaneous mutants of *E. coli* that are resistant to a chosen phage but still sensitive to other isometric phages. The chemistry of some of these cell-wall receptor sites (for S13, φR, and BR2) has been studied (Lindberg 1973).

Host-range mutants of some of the phages can be selected to grow on *E. coli* K12, C, or B strains. For instance, spontaneous mutants of φK can be isolated that form plaques on *E. coli* B and C strains with the same efficiency as on their natural host, *E. coli* K12 (Taketo 1977a), and a φX mutant has been reported that infects *E. coli* B strains (Razin 1973; D. T. Denhardt, pers. comm.). In most cases, however, it is difficult to extend the host range of these phages in this way. The only naturally occurring isometric phage that infects both *E. coli* B and *E. coli* C is *a*3 (Bradley 1970).

Wild-type isometric phages can be grouped into three classes on the basis of the temperature range at which they will grow (Table 1). S13, G6, φB, and φR will grow at any temperature between 22°C and the maximum temperature at which *E. coli* will grow (43°C). G4, G13, and G14, however, will grow and form normal-size plaques at much lower temperatures (18°C) but will not grow or form plaques above 39°C (Godson 1974). St-1 and φK will not grow at temperatures below 33°C (Bowes and Dowell 1974; Taketo 1976a), and St-1 grows much better at temperatures above 40°C than below (Grindley and Godson 1978). φC is a unique, naturally occurring temperature-sensitive phage that will not grow at temperatures above 33°C (Taketo 1977b).

Physical Properties of the Virus and DNA

φX is the best-characterized phage of this family, and most authors have used it as a standard. φX (MacLean and Hall 1962), S13 (Spencer et al. 1972), φA (Taketo 1974), φR (Kay and Bradley 1962), St-1 (Bowes and Dowell 1974), φK (Taketo 1976a), G4, G6, G13, G14 (Godson 1974), and U3 (Watson and Paigen 1971) have all been characterized under the electron microscope as small, spiked icosahedra 28–30 nm in diameter. Sedimentation studies (summarized in Table 2) have revealed that most of the isometric viruses that form plaques on *E. coli* C have sedimentation rates similar to φX (114S), whereas St-1 and φK, which grow on K12, are larger and sediment at

Table 2 Physical properties of some tail-less isometric coliphages

Phage	$S_{20,w}$ of virus	$S_{20,w}$ of viral DNA	Size of DNA (bases)
S13	—	24.6[2]	—
φX174	114[1]	23.0[1]	5386[7,8]
G4	108[3]	24–24.5[3]	5577[8]
G6	—	—	—
G13	113[3]	24–24.5[3]	—
G14	111[3]	24–24.5[3]	—
St-1	121[4]	23.8[4]	6050[9]
φK	121[5]	—	—
U3	83[6]	—	—

The data are from the following sources: [1]Sinsheimer (1968); [2]Spencer et al. (1972); [3]Godson (1974); [4]Bowes and Dowell (1974); [5]Taketo (1976a); [6]Watson and Paigen (1971); [7]Sanger et al. (1977); [8]Godson et al. (this volume); [9]Grindley and Godson (1978).

120S. U3 has an anomalously low sedimentation rate (Watson and Paigen 1971).

For all of the isometric phages shown in Table 1 their DNA has been shown unequivocally to be single-stranded, either by electron microscopy, melting curves, or reaction with hydroxylamine. Where studied, the sedimentation rate of the viral DNA of most of these phages appears to be the same, but more sensitive methods (DNA sequencing and restriction endonuclease cleavage) have shown that G4 (Godson et al., this volume) and St-1 (Grindley and Godson 1978) have genomes slightly larger than that of φX.

Virus-coded Proteins

The nine proteins encoded by φX (A to H, plus J) have been characterized extensively and their amino acid sequences, locations on the genetic map, and physiological functions have been determined. They are therefore used as standards for the proteins encoded by the other isometric phages. By means of SDS-polyacrylamide gel analysis of the virus-encoded proteins synthesized in UV-irradiated and infected cells, S13 (Jeng et al. 1970; Godson 1974), G4, G6, G13, G14 (Godson 1974), and St-1 (J. Grindley and G. N. Godson, in prep.) have been shown to code for the same series of nine proteins. These appear to vary somewhat in size from their φX counterparts (as can be seen in Fig. 1), but this is most probably due to the effect of differences in amino acid composition on their migration in SDS-polyacrylamide gels (see Godson et al., this volume). No new, major virus-encoded proteins have been identified in any of these phages. DNA sequencing studies have shown that the order of the viral genes is the same in φX,

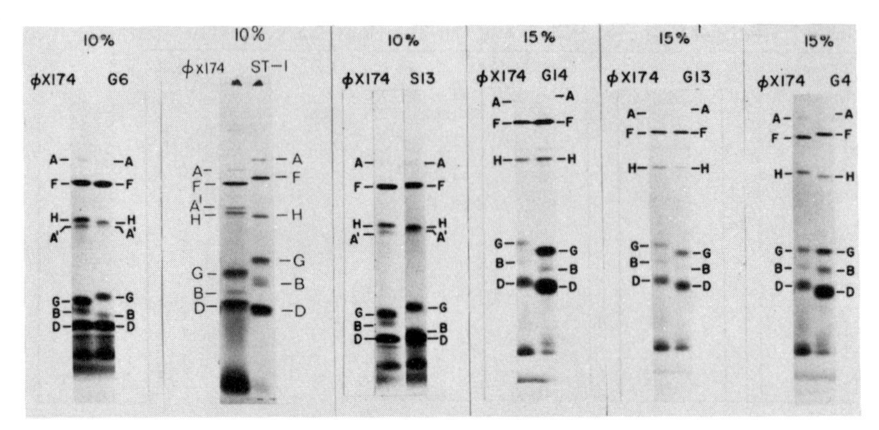

Figure 1 Proteins encoded by φX, S13, G4, G6, G13, G14, and St-1. UV-irradiated *E. coli* C *uvr* cells were infected with the phages in the presence of ^{14}C-labeled amino acids and the proteins synthesized were analyzed on SDS-polyacrylamide gels. The methods used are described by Godson (1974).

S13, and G4, and it is reasonable to expect that the genetic maps of the other isometric phages will be the same as that of φX.

Base-sequence Studies

Electron microscope studies of heteroduplex molecules made up of one strand of φX DNA and one strand of another isometric phage DNA have shown that S13 (Godson 1974; Compton and Sinsheimer 1977), G4, G6, G13, and G14 (Godson 1974) differ from φX in their DNA base sequences. However, this method of comparison suffers from the lack of a reliable method of calibration. Experiments using restriction endonucleases that cleave double-stranded DNA at specific sequences have shown that the DNA restriction fragments obtained from φX and S13 (Godson 1976; Grosveld et al. 1976) differ somewhat and that those obtained from φX, G4, and St-1 (Godson and Roberts 1976; Grindley and Godson 1978) differ greatly. The restriction enzyme fragments obtained from G6 are similar to those obtained from S13, but those obtained from G14 are very different. This is shown in Figure 2. These studies suggest that the DNA base sequences of φX, S13, and G6 are closely related, whereas the DNA base sequences of G4, G14, and St-1 are distantly related, both to φX and among themselves. Determination of the base sequences of φX, S13, and G4 (see Godson et al.; Spencer et al.; both this volume) has borne out these differences.

Host Proteins Required for DNA Replication

The isometric phages fall into three groups: those that require *E. coli dnaB* and *C/D* proteins for synthesis of the complementary strand of their parental replicative-form (RF) DNA (in vitro and in vivo) and for later stages in DNA synthesis (e.g., φX); those that do not require *dnaB* and *C/D* proteins for any stage in DNA synthesis (e.g., St-1, φK, and α3); and those that do not require *dnaB* and *C/D* proteins for synthesis of the complementary strand of their parental RF DNA (in vitro and in vivo) but do require these proteins for later stages in DNA synthesis (e.g., G4). These differences are discussed in detail by Dumas (this volume).

Antigenic Relatedness

When tested against φX antiserum, S13 (Zahler 1957), φA, φR (Taketo 1973), φB (Taketo 1977b), and G6 (Godson 1974) are inactivated at rates fairly close to the inactivation rate of φX, whereas G4, G14 (Godson 1974; Taketo 1977b), φC (Taketo 1977b), St-1 (Bradley 1970; Bowes and Dowell 1974), α3 (Bradley 1970; Taketo 1976a), and φK (Taketo 1976a) are not inactivated at all, even after prolonged incubation. G13 (Godson 1974) and φC (Taketo 1977b) are inactivated by φX antiserum to a small extent. A composite of published data is given in Figure 3A.

When tested against St-1 antiserum, φK and α3 are inactivated (Taketo

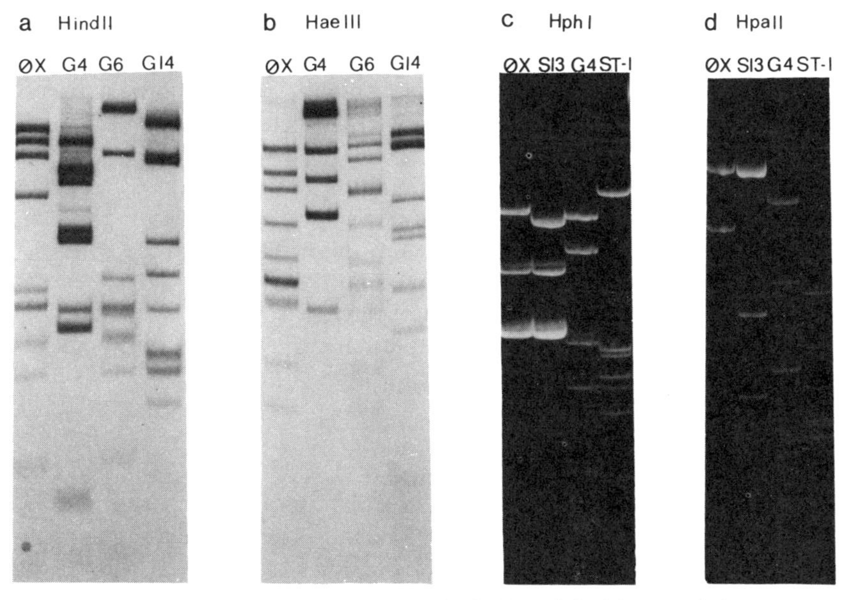

Figure 2 Comparison of φX, S13, G4, G6, G14, and St-1 by restriction enzyme cleavage analysis. RFI DNA from each of the viruses was digested with the restriction enzyme indicated and the products electrophoresed either on a 4% polyacrylamide gel (*a*, *b*) or a 2% agarose gel (*c*, *d*). The restriction endonucleases were from *Haemophilus influenzae* serotype d (*Hin*dII), *Haemophilus aegyptius* (*Hae*III), *Haemophilus parahaemolyticus* (*Hph*I), and *Haemophilus parainfluenzae* (*Hpa*II).

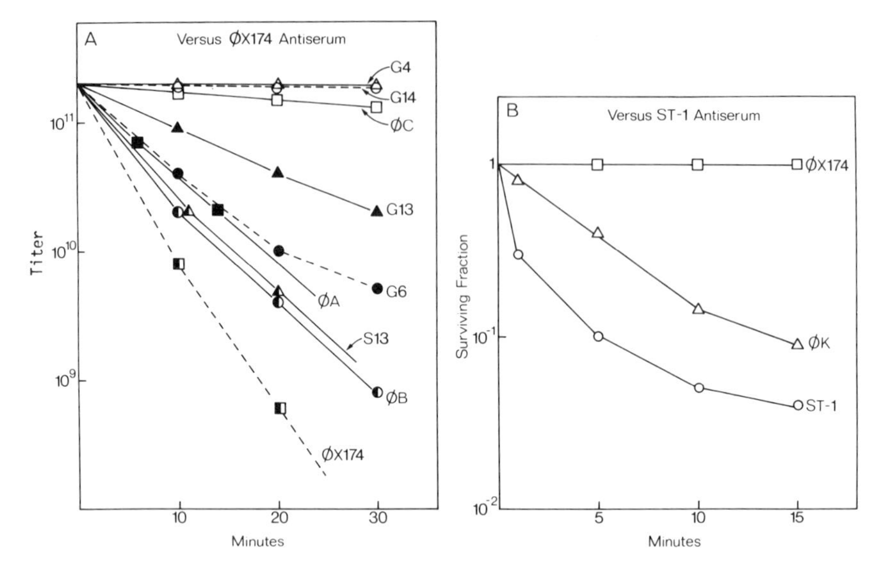

Figure 3 Antiserum specificity of the isometric phages. (Data from Godson 1974 and Taketo 1977a,b.)

1977a), but none of the other phages are. This is shown in Figure 3B and Table 3.

U3, which infects K12, does not cross-react with either St-1 or α3 antiserum (Taketo 1977b). G4 and G14 do not cross-react with φX, St-1, or α3 antiserum.

The phages thus fall into three main groups: those that cross-react well with φX antiserum (S13, G6, φA, φB, and φR) plus those that cross-react partially with this serum (G13 and φC); those that cross-react with St-1 antiserum (φK and α3); and those that do not cross-react with φX or St-1 antiserum (G4, G14, and U3). The last three phages, moreover, are completely unrelated to each other on the basis of host-cell specificity and their requirements for host-cell *dna* functions.

General Relatedness

From the criteria of antiserum cross-reaction, growth-temperature range, specificity for the same host cell-wall receptor sites, and requirements for host-cell *dna* functions, φX, S13, G6, φA, and φB are closely related, and judging from the φX, S13, and G6 restriction enzyme cleavage fragments (Fig. 2), their DNA base sequences are probably very similar. G13 and φR are also fairly closely related to this group, but not as closely related as the other members.

By the same criteria, St-1 and φK are probably very closely related to each

Table 3 Cross-reactivity of various phage antisera

Phage	Antiserum		
	φX	α3	St-1
φX174	+	–	–
S13	+	–	
G6	+	–	
φA	+	–	–
φB	+	–	
φR	+	–	
G13	partial	–	
φC	partial	–	
St-1	–	+	+
φK	–	+	+
α3	–	+	+
U3	–	–	
G4	–	–	
G14	–	–	–

The data are from Taketo (1973, 1976a, 1977a,b), Bradley (1970), and Godson (1974).

other and, together with α3, form a separate group of phages which are slightly larger than φX and require fewer host-cell proteins for DNA replication. G4 and φC may be related, as judged by their unique pattern of infecting a series of different host-cell strains (Taketo 1977b). G14 and U3 are distinct from each other and from any of the other better characterized phages. However, despite these superficial groupings, the phages are better viewed as a continuous spectrum of differences of the same basic genome structure (see below).

Tail-less Isometric Phages Specific for Other Species of Bacteria

φR was isolated originally as a *Salmonella typhimurium* phage (Kay and Bradley 1962), but it now grows normally on *E. coli* C strains and appears to be restricted on *S. typhimurium* (Burton and Yagi 1968). Likewise, S13 was isolated as a *Salmonella* phage (Burnet 1927; Spencer et al., this volume), but it now grows normally on *E. coli* C strains and behaves like a phage closely related to φX. No small isometric phages have been reported for any other bacterial species.

Evolution

The great diversity of the small, tail-less isometric coliphages is in contrast with the uniform nature of the filamentous coliphages (van den Hondel and Schoenmakers 1976), and every isolate of a small isometric phage appears to be different.

Two important features emerge from the comparative study of these phages. First, genome size is almost constant (~5500 bases), and second, although the genomes code for the same series of nine major proteins, the DNA base sequence differs considerably from phage to phage. It is remarkable that for the phages that have been analyzed using restriction enzyme fragmentation to characterize their genomes (φX, S13, G4, G6, G14, and St-1), every one is different. In some cases (φX, S13, and G6; Fig. 2) the number and sizes of different DNA fragments produced by each enzyme are similar, indicating that the DNA base sequences of these phages are closely related, but in other cases (φX, G4, G14, and St-1; Fig. 2) there are no common DNA fragments produced by any enzyme from any phage, indicating that the DNA base sequences of these phages are very dissimilar. The complete nucleotide sequences of G4 and φX (see Godson et al., this volume, and Appendix II) confirm these differences, and the coding sequences of the G4 and φX genomes have an average 33% base-sequence mismatch.

The question arises as to how this divergence of DNA base sequence is generated. There have not yet been any studies of genetic complementation and recombination between the various phages (except between φX and S13, which are very closely related). Yet whatever causes the divergence of DNA base sequence (see the comparison of G4 and φX base sequences by Godson et al., this volume, and Appendix II), the main constraint appears to be on the size of the genome, and, consequently, within narrow limits, on the sizes of the genes. Presumably this is because the amount of DNA that can be packaged into an icosahedral structure is determined by the sizes of the virion structural protein units (F, G, and H proteins). Any increase in the size of the DNA would also have to include an increase in the sizes of the structural proteins at the same time, so as to allow a larger virion to be constructed to accommodate the larger DNA. It would seem from the nucleotide sequence of G4 that the amino acid sequences of the viral proteins can change extensively without affecting their functions.

The evolution of the isometric phages, therefore, is probably not a linear series of ancestors traceable as a series of changes from phage to phage, but rather a network of converging and diverging genome base sequences that have in common only a basic genome organization (gene order) and common functions for each of their proteins. This would predict that the more divergent isometric phages might not be able to recombine efficiently but would be able to complement their protein functions efficiently, and this is testable experimentally.

ACKNOWLEDGMENT

The author has been supported by U.S. Public Health Service grant CA 06519.

REFERENCES

Bowes, J. M. and C. E. Dowell. 1974. Purification and some properties of bacteriophage St-1. *J. Virol.* **13:** 53.

Bradley, D. E. 1962. Some new small bacteriophages (φX174 type). *Nature* **195:** 622.

———. 1970. A comparative study of some properties of the φX174 type bacteriophage. *Can. J. Microbiol.* **16:** 965.

Burnet, F. M. 1927. The relationship between heat-stable agglutinogens and sensitivity to bacteriophage in the *Salmonella* group. *Br. J. Exp. Pathol.* **8:** 121.

Burton, A. J. and S. Yagi. 1968. Intracellular development of bacteriophage φR. I. Host modification of the replicative process. *J. Mol. Biol.* **34:** 481.

Compton, J. L. and R. L. Sinsheimer. 1977. Aligning the φX174 genetic map and the φX174/S13 heteroduplex denaturation map. *J. Mol. Biol.* **109:** 207.

Godson, G. N. 1974. Evolution of φX174: Isolation of four new φX-like phages and comparison with φX174. *Virology* **58:** 272.

———. 1976. Evolution of φX174. III. Restriction map of S13 and its alignment with that of φX174. *Virology* **75:** 263.

Godson, G. N. and R. J. Roberts. 1976. A catalogue of cleavages of φX174, S13, G4 and St-1 by 25 different restriction enzymes. *Virology* **73:** 561.

Grindley, J. and G. N. Godson. 1978. Evolution of φX174: Restriction map of St-1. *J. Virol.* (in press).

Grosveld, F. G., K. M. Ojamaa, and J. Spencer. 1976. Fragmentation of bacteriophage S13 replicative form DNA by restriction endonucleases from *Hemophilus influenzae* and *Hemophilus aegyptius. Virology* **71:** 312.

Jeng, Y., D. Gelfand, M. Hayashi, R. Shleser, and E. S. Tessman. 1970. The eight genes of bacteriophage φX174 and S13 and comparison of the phage-specific proteins. *J. Mol. Biol.* **49:** 521.

Kay, D. and D. E. Bradley. 1962. The structure of bacteriophage φR. *J. Gen. Microbiol.* **27:** 195.

Lindberg, A. A. 1973. Bacteriophage receptors. *Annu. Rev. Microbiol.* **27:** 205.

MacLean, E. C. and C. E. Hall. 1962. Studies on bacteriophage φX174 and its DNA by electron microscopy. *J. Mol. Biol.* **4:** 173.

Razin, A. 1973. DNA methylase induced by bacteriophage φX174. *Proc. Natl. Acad. Sci.* **66:** 646.

Sanger, F., G. M. Air, B. G. Barrell, N. L. Brown, A. R. Coulson, J. C. Fiddes, C. A. Hutchison III, P. M. Slocombe, and M. Smith. 1977. Nucleotide sequence of bacteriophage φX174 DNA. *Nature* **265:** 687.

Sertic, V. and N. Boulgakov. 1935. Classification et identification des typhi-phages. *C.R. Soc. Biol.* **119:** 1270.

Sinsheimer, R. L. 1968. Bacteriophage φX174 and related viruses. *Prog. Nucleic Acid Res. Mol. Biol.* **8:** 115.

Spencer, J. H., R. Cerny, E. Cerna, and A. Delaney. 1972. Characterization of bacteriophage S13 SuN15 single-strand and replicative form DNA. *J. Virol.* **10:** 134.

Taketo, A. 1973. Sensitivity of *E. coli* to viral nucleic acid. VI. Capacity of *dna* mutants and DNA polymerase-less mutants for multiplication of φA and φX174. *Mol. Gen. Genet.* **122:** 15.

——. 1974. Properties of bacterial virus φA and its DNA. *J. Biochem.* **75:** 951.
———. 1976a. Host factor requirements and some properties of φXtB. An evolutionary aspect of φX-type phages. *Mol. Gen. Genet.* **148:** 25.
———. 1976b. Host functions required for the replication of α3 and related phages. *Proceedings of the 1976 Molecular Biology Meeting*, Osaka, Japan. (Abstr.)
———. 1977a. Effect of *dna* mutations on replication of S13, φR and φKh-1. *J. Gen. Appl. Microbiol.* **23:** 29.
———. 1977b. Sensitivity of *Escherichia coli* to viral nucleic acid. XIII. Replicative properties of φX174-type phages. *J. Gen. Appl. Microbiol.* **23:** 85.
van den Hondel, C. A. and J. G. G. Schoenmakers. 1976. Cleavage maps of the filamentous bacteriophages M13, fd, f1, and ZJ/2. *J. Virol.* **18:** 1024.
Watson, G. and K. Paigen. 1971. Isolation and characterization of an *E. coli* bacteriophage requiring cell wall galactose. *J. Virol.* **8:** 669.
Zahler, S. A. 1957. Some biological properties of bacteriophages S13 and φX174. *J. Bacteriol.* **75:** 310.

The Filamentous Phage Genome: Genes, Physical Structure, and Protein Products

Kensuke Horiuchi, Gerald F. Vovis, and Peter Model
The Rockefeller University
New York, New York 10021

In this article we review the organization of the filamentous phage genome and the phage-specific protein products. A great deal of work has been done on the filamentous phages specific to strains of *Escherichia coli* carrying an F factor. Included in this category are f1 (Loeb 1960; Zinder et al. 1963), fd (Marvin and Hoffmann-Berling 1963), M13 (Hofschneider 1963), and ZJ/2 (Bradley 1964). These phages are so closely related that we will not usually distinguish among them. Other filamentous phages, including the I-specific phages of *E. coli* (Meynell and Lawn 1968) and the filamentous phages of *Pseudomonas* (Minamishima et al. 1968), have been described but are not included in this discussion.

We describe the properties of the various mutants of the filamentous phages, the organization of the genes on the genetic and physical maps, and the methylatable bases in the genome. We also list the gene products and describe some of their properties, as well as discuss aspects of the control of gene expression.

MUTANTS

Conditional Lethals

Initial genetic studies of the filamentous phages were carried out by Pratt and his colleagues (1966, 1969), who collected amber and temperature-sensitive mutants and classified them by standard complementation analysis into eight complementation groups. No additional genes have been identified since. In the Pratt collection, gene VIII is represented by only one conditionally lethal mutation, *am*8H1, and none of the other groups working with the filamentous phages have identified any other conditionally lethal mutations in gene VIII. Since the map is not saturated, the existence of another small gene cannot be excluded.

The physiology of cells infected with these conditionally lethal mutants under nonpermissive conditions was described initially by Pratt and subsequently by a number of other groups. Cells infected with the wild-type continue to grow, divide, and produce phage. Mutations in any gene other than gene II are lethal to the infected cell under nonpermissive conditions.

Gene II is required for all phage DNA synthesis (Pratt and Erdahl 1968). Gene III encodes a minor virion protein (Rossomando and Zinder 1968; Pratt et al. 1969), is necessary for adsorption to the pilus of the host bacterium (Pratt et al. 1969), and, furthermore, may be required for synthesis of parental replicative-form (RF) DNA (Jazwinski et al. 1973; but see Ray, this volume). The presence of the gene-III protein in infected cells appears also to be responsible for the reduced sensitivity of these cells to the action of a variety of colicins (Zinder 1973). The gene-V protein is required for synthesis of single-stranded (SS) DNA (Pratt and Erdahl 1968). The gene-VIII protein is the major structural component of the virion (Pratt et al. 1969). The functions of the products of genes I, IV, VI, and VII are not known; they are indispensable for phage synthesis but are not required for phage DNA replication. Under nonpermissive conditions, cells infected with gene-III or gene-VI amber mutants produce a small number of multiple-unit-length defective particles (Pratt et al. 1969; Beaudoin 1970). Even on suppressing hosts certain gene-III amber mutants make multiple-unit-length particles.

By genetic criteria, genes III, VI, and I form an operon, with gene-III proximal nonsense mutations polar on genes VI and I and with gene-VI nonsense mutations polar on gene I (Pratt et al. 1966). In addition, gene-V nonsense mutations are polar on gene VII (Lyons 1971).

Other Mutants

Restriction-Modification Mutants

Filamentous phages grown in *E. coli* K or C strains normally are sensitive to restriction in *E. coli* B strains. Similarly, phages grown on bacteria that do not harbor phage P1 are restricted when introduced into P1 lysogens (Arber 1966). Phage mutants that are either partially or totally insensitive to these restriction-modification systems have been isolated and characterized (Arber and Kühnlein 1967; Hartman and Zinder 1973) and have played an important role in genetic mapping and recombination analysis of the filamentous phages (Boon and Zinder 1971; Lyons and Zinder 1972; Zinder, this volume).

Mutants with Altered Coat Proteins

Braunitzer and colleagues (1970) have isolated mutants of fd that have an altered electrophoretic mobility and contain amino acid substitutions in the major coat protein, the product of gene VIII. These mutants appear to produce normal yields of viable progeny.

Deletion Mutants

After serial mass transfer, deletion mutants of the filamentous phages can be isolated. They vary in size between about 20% and 50% of the size of the wild-type phage, and all of them require helper functions from coinfecting viable phages. No deletion mutant yet isolated appears to make any functional gene product (Griffith and Kornberg 1974; Hewitt 1975; Enea and Zinder 1975). When first isolated, these mutants are heterogeneous in size, but they can be cloned by infecting cells with both a helper phage and the deletion mutant and plating at high dilution (Hewitt 1975; Enea and Zinder 1975). Several such cloned defective mutants have been studied carefully and have been mapped both genetically and physically (Enea et al. 1977; see below). These cloned defective mutants interfere with the growth of wild-type f1 to varying extents. It has been possible to isolate mutants of f1, termed IR (interference-resistant), that are resistant to growth interference by the defective particles. At least one aspect of this resistance to interference appears to be due to a mutation located within gene II (V. Enea et al., pers. comm.).

The length of the DNA contained within defective particles correlates with the length of the particle (Griffith and Kornberg 1974; Enea and Zinder 1975); thus, the length of a filamentous phage seems to be determined by the length of the DNA molecule contained within it. It is this property of the filamentous phages that has made them attractive as possible vectors for the construction of recombinant DNA molecules (Vovis et al. 1977; Vovis and Ohsumi; Messing and Gronenborn; Ray and Kook; Kaplan et al.; Nomura et al.; Herrmann et al.; all this volume).

The Genetic Map

The genetic map of bacteriophage f1 is presented in Figure 1. The maps of the other filamentous phages have not been determined independently by genetic means, but physical mapping, marker-rescue experiments, and DNA sequencing (see below and Schaller et al., this volume) suggest that their genetic maps are the same as the one shown. M13, f1, and fd complement and recombine with each other actively (Lyons and Zinder 1972).

Construction of the genetic map was seriously hampered by the fact that the filamentous phages produce many diploid or polyploid particles, which

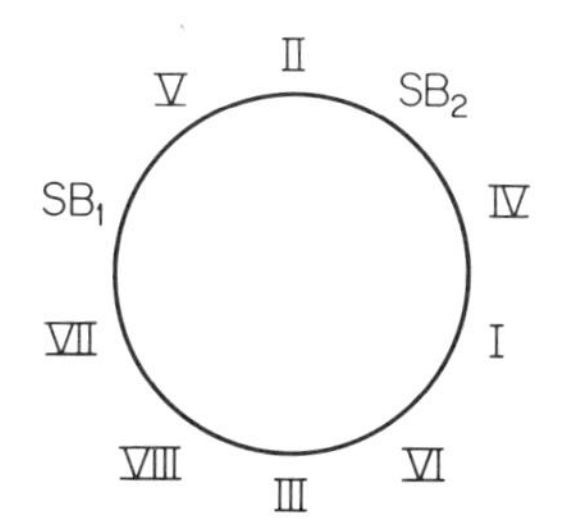

Figure 1 The genetic map of bacteriophage f1. The roman numerals refer to the genes. SB_1 and SB_2 are genetic sites which confer sensitivity to restriction by the *E. coli* B restriction-modification system. The map is modified from that of Lyons and Zinder (1972).

contain two or more unit-length DNA molecules encapsulated within a single structure (Scott and Zinder 1967; Salivar et al. 1967). When two different phage mutants are used to infect a single cell, some of the diploid particles produced contain one DNA molecule from each of the infecting parents and are therefore heterozygous. The DNA molecules in these heterozygotes complement each other, and in subsequent infections recombination may yield a wild-type phage. The frequency of heterozygote formation exceeds greatly the frequency with which the original parental molecules recombine, and as these heterozygotes plate as though they were wild-type, they make it impossible to measure the recombination frequency in intergenic crosses. Thus, to establish the gene order it was necessary to rely mainly on intragenic crosses between multiply marked parents and to score for the segregation of unselected markers (Lyons and Zinder 1972). Moreover, since the recombination frequency is very low (i.e., about the same level as the reversion frequency of single f1 mutants), only those recombinants that contain a set of unselected markers different from the set found in either of the parents could be used to construct the map. The frequency with which particular genotypes are created by this approach gives information on the order of the markers used in the cross but it does not reflect accurately the distances between the markers (see also Zinder, this volume).

Genes III, VI, and I, which form an operon (Pratt et al. 1966), are contiguous, as are genes V and VII, which form a second operon (Lyons 1971). Since the direction of transcription is counterclockwise on the map shown in Figure 1 (Model and Zinder 1974; Konings and Schoenmakers, this volume), the gradient of polarity follows the transcriptional direction, as in other operons.

PHYSICAL MAP

Construction

Initial attempts to construct a physical map of f1 used the *E. coli* B restriction enzyme, which was the first such enzyme to be characterized (Meselson and Yuan 1968; Linn and Arber 1968). Surprisingly, this enzyme was found not to cleave at specific sites (Horiuchi and Zinder 1972) even though recognition between the enzyme and the DNA occurs at discrete sites (Horiuchi et al. 1975; Brack et al. 1976). Later, the use of truly site-specific endonucleases found in other bacteria (Smith and Wilcox 1970; Kelly and Smith 1970) made physical mapping practicable (Danna et al. 1973). Detailed physical maps of f1 (Fig. 2), fd (Schaller et al., this volume [p. 148]), and M13 (Konings and Schoenmakers, this volume [p. 515]) have been constructed with the use of such enzymes. The fragments generated by these enzymes were ordered using one or a combination of three different techniques.

In one method a specific fragment generated by one enzyme is then

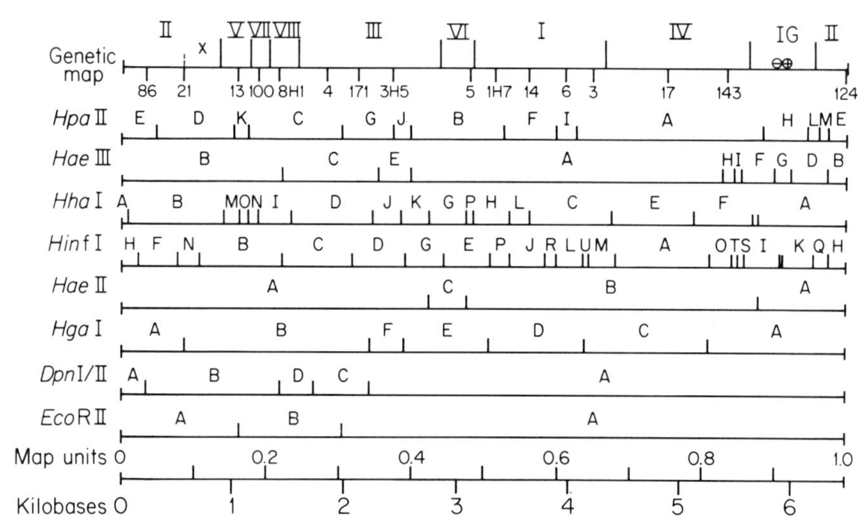

Figure 2 The physical and genetic maps of the filamentous phage f1. (For the maps of fd and M13, see, respectively, Schaller et al. [p. 148] and Konings and Schoenmakers [p. 515], both this volume.) The circular phage genome is presented in linear form as if it had been cleaved at the single *Hin*dII restriction site. The roman numerals refer to the genes. IG is the intergenic space between genes II and IV and contains the origins of plus- and minus-strand synthesis. X refers to that region of the genome which codes for the X protein (see text). The arabic numbers show the locations of various sites of amber mutations. The restriction fragments are labeled in alphabetical order according to decreasing size. The direction of transcription and translation is from left to right. This figure includes data from the following: Horiuchi et al. 1975; Vovis et al. 1975; Enea et al. 1977; Ravetch et al. 1977a,b; Vovis and Lacks 1977; K. Horiuchi et al. (unpubl.).

cleaved by another enzyme of different specificity (Takanami et al. 1975; Horiuchi et al. 1975). This allows one to determine which of the fragments produced by the second enzyme lie in a region defined by the size of the initial fragment produced by the first enzyme. Alternatively, the DNA may be digested partially with a single enzyme, and the recovered fragments may then be cleaved exhaustively with the same enzyme (van den Hondel and Schoenmakers 1975). The use of these two methods alone can generate an unambiguous map. A third method consists of using restriction fragments as primers for limited DNA synthesis and then recleaving with the same enzyme; this method also leads to an ordering of the fragments (Seeburg and Schaller 1975). More recently another method was introduced that involves terminal labeling at a unique site. The DNA is then digested for various lengths of time and subjected to gel electrophoresis. The pattern of those fragments that still contain the terminal label allows one to order the cleavage sites. In favorable cases this kind of mapping can be carried out very quickly and with a minimal number of experiments (Smith and Birnstiel 1976).

Figure 2 shows the physical map constructed for f1, and it should be compared with the physical maps of fd (Schaller et al., this volume [p. 148]) and M13 (Konings and Schoenmakers, this volume [p. 515]). In general, the maps are very similar, except for the occasional absence of a particular cleavage site from one or the other of the phage genomes (van den Hondel and Schoenmakers 1976). All three phages have a single *Hin*dII cleavage site, which serves as a reference point to relate the various maps.

Correlation of Physical and Genetic Maps

Figure 2 also shows the genetic map of f1. The genetic and physical maps were correlated by marker-rescue experiments (Hutchison and Edgell 1971; Edgell et al. 1972), in which a given restriction fragment derived from wild-type DNA is denatured and hybridized to viral SS DNA containing a particular mutation. The hybrid molecules obtained are used to transfect (Mandel and Higa 1970) *E. coli* cells under nonpermissive conditions. Phage are produced only if the mutation is located within the region included in the restriction fragment. With this procedure any mutation can be located on a given restriction fragment (Seeburg and Schaller 1975; van den Hondel et al. 1975b; Horiuchi et al. 1975), but since the number of mutants available is limited, the total genetic content of a given fragment can be defined only rarely.

Two other sorts of data have proved useful in aligning the genetic and physical maps. One is the sizes of the individual proteins and the other is the transcription-translation of individual restriction fragments (see below). One consequence of this correlation of the two maps was the discovery of a region of the genome that does not code for any of the known gene products. This region, marked IG (intergenic space) in Figure 2, is located between genes II and IV. Its existence was inferred from the following arguments: f1 gene-IV amber mutant R143 is located as shown in Figure 2. The size of the amber fragment, as estimated from acrylamide gel electrophoresis, is 40,000 daltons (P. Model, unpubl.); the size of the intact gene-IV product is estimated to be 50,000 daltons (Table 1). These data can be used to locate on the physical map the region coding for the C terminus of gene IV. Similarly, f1 gene-II amber mutant R124 maps as shown in Figure 2. The size of its amber fragment is about 10,000 daltons (Model and Zinder 1974) and thus locates on the physical map the region coding for the N terminus of gene II. Similar arguments for f1 gene-II amber mutant R86 place the N terminus of gene II at the same position. These two termini are separated by about 8% of the f1 genome. Although the existence of this region was demonstrated first in f1 (Vovis et al. 1975; Enea et al. 1977), the results of subsequent experiments are consistent with its presence in M13 also (van den Hondel et al. 1976).

Table 1 Estimated molecular weights of filamentous phage proteins

Protein	Molecular weight × 10^{-3}
Gene-I	35[a], 36[b]
Gene-II	40[a], 46[b]
Gene-III	68[a], 59[b], 70[c], 43.4[d]
Gene-IV	50[a], 48[b]
Gene-V	9.7[e]
Gene-VI	unknown
Gene-VII	3.6[f]
Gene-VIII	5.24[g]

[a] SDS electrophoresis of material synthesized in vitro (Model and Zinder 1974).
[b] SDS electrophoresis of material synthesized in vitro (Konings et al. 1975).
[c] SDS electrophoresis of material synthesized in vivo (Henry and Pratt 1969).
[d] Predicted from DNA sequence data (Schaller et al., this volume).
[e] From protein sequence data (fd data from Nakashima et al. 1974; M13 data from Cuypers et al. 1974).
[f] Predicted from RNA sequence data for fd (Takanami et al. 1976) and from RNA sequence data for M13 wild type and amber mutants (J. G. G. Schoenmakers, pers. comm.). There is no direct evidence that this protein is ever made either in vivo or in vitro.
[g] From protein sequence data (Snell and Offord 1972; Nakashima and Konigsberg 1974; Bailey et al. 1977).

Deletion Mutants

Figure 3 shows the physical maps of four cloned f1 deletion mutants (Enea et al. 1977). The techniques used for constructing these maps were similar to those used to map the parent phage. Identity between fragments derived from deletion mutants and those obtained from the parent phage was established principally by comigration in gel electrophoresis and by susceptibility to secondary digestion by restriction enzymes other than those used to generate the fragments. In some instances, moreover, the nucleotide sequences of the fragments derived from the deletion mutant and the wild-type phage were shown to be the same (J. Ravetch, unpubl.). In addition to the fragments that appear to be identical in the deletion and the parent, each deletion mutant contains at least one fragment that is different from those obtained from the parent. This is exemplified by deletion mutant MP-2 (Fig. 3) and would be expected if the mutants arose by simple deletion of a contiguous region of DNA. Electron microscopic examination of heteroduplex molecules formed from one deletion-mutant strand and one wild-type strand also leads to the hypothesis that these mutants are formed by simple deletion of contiguous DNA (Griffith and Kornberg 1974; Enea et al. 1977). Physical mapping and heteroduplex analysis have shown, however, that some of the mutants have reiterated a particular portion of the parental genome, and these mutants contain more than one fragment that has no counterpart in the wild-type parent (MP-4, MP-5, and MP-7 in Fig. 3). The patterns are still consistent with an original, simple deletion event,

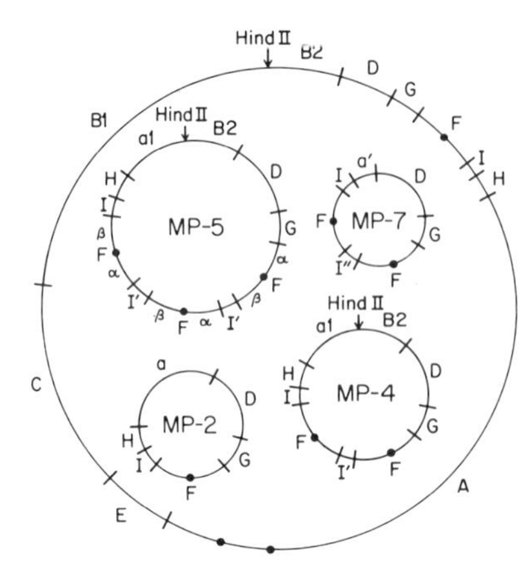

Figure 3 The physical maps of f1 and various f1 deletion mutants. The capital letters refer to *Hae*III restriction fragments, the filled circles to the *Hae*II cleavage sites, and the arrows to the *Hin*dII cleavage site. The outer circle represents the f1 map. The deletion mutants described by Enea et al. (1977) have been renamed as follows: MP-2 (formerly MB), MP-4 (formerly PII), MP-5 (formerly PIII), and MP-7 (formerly Trb).

which is then followed by reiteration of a portion of the genome. Although the mechanism by which either the deletion or the reiteration occurs remains obscure, those mutants that contain the reiterated DNA seem to grow more rapidly than those that do not (Enea et al. 1977; J. Bergelson, unpubl.).

Both genetic and physical evidence show that none of the deletion mutants contain any known complete gene (Griffith and Kornberg 1974; Enea and Zinder 1975). The deletion mutants recombine with certain f1 amber mutants, but each of the eight known genes contains at least one locus with which the deletion mutants are unable to recombine. Comparison of the physical maps of the various deletions with the map for wild-type f1 also suggests an absence of complete genes. Furthermore, if deletion-mutant double-stranded DNA is used to program protein synthesis in vitro, the polypeptide products obtained are different from any of the known authentic f1 products (P. Model and V. Enea, unpubl.).

Figure 3 shows that much of the DNA of each deletion mutant is derived from the intergenic (IG) region between genes IV and II. This region contains no known genes. A portion of the IG region is reiterated in MP-4, MP-5, and MP-7. The origin of minus-strand synthesis has been reported to be near the *Hae*III-G/*Hae*III-F border (Tabak et al. 1974; Schaller et al. 1976; Horiuchi and Zinder 1976; see also Ray, this volume), and this segment is contained within the reiterated region. The reiterated region may also contain the origin of plus-strand synthesis, which has been reported to be within *Hae*III-G (Horiuchi and Zinder 1976; Suggs and Ray 1977). When grown with wild-type f1 as helper, the deletion mutants often have a particle yield as great as or greater than that of the helper phage (Enea and Zinder 1975). Thus, they are efficient substrates for the f1 replication machinery. We note further that three independently derived deletion mutants all contain DNA derived from the IG space and that reiteration of this region is correlated with more efficient growth (Enea et al. 1977). Taken together, these data imply that the presence of some segment of the intergenic space is necessary for f1 DNA replication and that the rate of this replication may depend on the number of copies of the *cis*-acting segment present. This in turn would suggest that it is the initiation of DNA synthesis (and not propagation) that is rate-limiting.

Deletion mutants can be detected by the unique and peculiar morphology of the plaques produced by mixed infection with these mutants and wild-type helper. This morphology is probably due to interference with the growth of wild-type by the deletion mutant. To determine whether the presence of the IG region is sufficient for the successful propagation of deletion mutants, synthetic analogs of the deletion mutants were made. f1 DNA was cleaved into two pieces with endonuclease *Eco*RII. The fragments were separated and recircularized with T4 DNA ligase. One of the two fragments contained the entire IG region (Fig. 2); however, neither of these synthetic analogs gave rise to the plaques characteristic of deletion mutants. Under the same conditions transfection with DNA derived from authentic deletion mutants produces such plaques (G. F. Vovis, unpubl.). It would appear, therefore, that more is required than simply the presence of the IG region. A lack of complete genes may be a necessary rather than a fortuitous attribute of deletion mutants that can be replicated efficiently enough to give rise to interfering particles.

METHYLATED BASES

The viral DNA of filamentous phages grown in *E. coli* K12 normally contains at least six methylated bases (Smith et al. 1972; Hattman 1973). Of these six, four are 6-methyl adenine and are contributed by the major *E. coli* DNA adenine methylase, which is coded for by the *dam* gene (Marinus and Morris 1973, 1975; S. Hattman, pers. comm.). The other two methylated bases are 5-methyl cytosine (Hattman 1973) and are produced by the *E. coli* DNA cytosine methylase, which is coded for by the *mec* gene (Marinus and Morris 1973; Hattman et al. 1973). The *dam*-specified methylase recognizes the nucleotide sequence GATC (Hattman et al. 1978), and the *mec*-specified enzyme methylates within the CCAGG or CCTGG sequence (Boyer et al. 1973; Schlagman et al. 1976). These methylation sites have been mapped within specific restriction fragments (Vovis et al. 1975; Vovis and Lacks 1977). The *dam*-specified methylase shares its recognition sequence with two restriction enzymes from *Diplococcus pneumoniae*, *Dpn*I and *Dpn*II, but not with any known *E. coli* restriction enzyme (Lacks and Greenberg 1977; Vovis and Lacks 1977). The *mec*-specified methylase has the same specificity as the restriction endonuclease *Eco*RII (Schlagman et al. 1976), which is coded for by an *E. coli* plasmid (Yoshimori 1971).

E. coli mutants that are *mec*$^-$, *dam*$^-$, or *mec*$^-$ *dam*$^-$ (the double mutant) have been constructed. The viral DNA of filamentous phages grown in *mec*$^-$ mutants contains no methylated cytosine (Hattman et al. 1973); when grown in *dam*$^-$ mutants it contains no methylated adenine (Marinus and Morris 1975; S. Hattman, pers. comm.). Viral DNA synthesized in the double mutant contains no detectable methyl groups (S. Hattman, pers. comm.). Growth in these hosts, as monitored by phage production and DNA synthesis, appears to be entirely normal (G. F. Vovis, unpubl.). We conclude, therefore, that methyl groups do not normally play a role in the replication of the filamentous phages. This conclusion is in marked contrast with that drawn from the study of φX174, for in φX a methyl group appears to play an important role in DNA replication or virus maturation (Razin et al. 1970, 1973; Razin, this volume).

In addition to the six methyl groups discussed above, other methyl groups are found if certain restriction-modification systems are active in the infected cell. These systems include that found in *E. coli* B (Smith et al. 1972; Kühnlein and Arber 1972; Hattman 1973) and that specified by bacteriophage P1 (Hattman 1973; G. F. Vovis and S. Hattman, in prep.). The recognition sequences for both systems have been determined (Lautenberger et al. 1978; Ravetch et al. 1978; Hattman et al. 1978), and their locations on the physical and genetic maps of the filamentous phages have been established (Horiuchi et al. 1975; G. F. Vovis and S. Hattman, in prep.). There is no evidence that these methyl groups have any function other than that of protecting the DNA against restriction.

PROTEINS

General

The identification of eight genes in the filamentous phages led to the expectation that eight phage-specific proteins would be found. To date, seven f1-specific proteins have been detected, but only six have been assigned to known genes (Henry and Pratt 1969; Model and Zinder 1974; Konings et al. 1975). The seventh (X protein), though almost certainly f1-specific, has not been shown unequivocally to be the product of a previously identified gene (Model and Zinder 1974; Konings et al. 1975). Thus, the products of genes I, II, III, IV, V, and VIII have been detected (molecular weights are listed in Table 1), whereas the products of genes VI and VII have not yet been found. RNA from a portion of the genome known to code for gene VII has been sequenced (Takanami et al. 1976; J. G. G. Schoenmakers, pers. comm.), and it is therefore possible to predict an amino acid sequence for this protein. The sequence of RNA obtained from in vitro transcription of a gene-VII amber mutant shows a CAG-to-UAG change within the region assigned to gene VII (J. G. G. Schoenmakers, pers. comm.).

Gene Functions

The products of genes III and VIII are virion structural proteins (Rossomando and Zinder 1968; Pratt et al. 1969). The gene-VIII protein is the major coat protein of the phage (Pratt et al. 1969), and each virion contains approximately 2000 copies (Braunitzer et al. 1967). Since infected cells produce 100–200 phage particles per generation (Marvin and Hohn 1969), on the order of 2–4 $\times$ 10^5 molecules of the protein must be made per cell generation. In its mature form the gene-VIII protein consists of 50 amino acids (Snell and Offord 1972; Nakashima and Konigsberg 1974; Bailey et al. 1977). The N terminus is acidic, the C-terminal region rather basic, and the central core extremely hydrophobic (Asbeck et al. 1969; Snell and Offord 1972; Nakashima and Konigsberg 1974). In infected cells the gene-VIII protein is found only in the cell membrane (Smilowitz et al. 1972). Its properties are described more fully by Webster and Cashman (this volume). The gene-III protein is a minor coat protein (Rossomando and Zinder 1968; Pratt et al. 1969). Only a few copies are present in each virion, and these are located at one end of the phage particle (Rossomando and Zinder 1968; Marco 1975). It is extremely hydrophobic (R. E. Webster and W. Konigsberg, pers. comm.). Electrophoresis on SDS-acrylamide gels suggests a molecular weight of 60,000–70,000 daltons (Pratt et al. 1969; Model and Zinder 1974; Konings et al. 1975), whereas DNA sequence data predict a size of less than 45,000 daltons (Schaller et al., this volume). Virions containing gene-III protein with an amino acid substitution are often less stable than wild-type phage (Rossomando and Zinder 1968; Pratt et al. 1969). Gene-III protein is needed for adsorption to the host cell (Pratt et al. 1969). It has been suggested that the protein remains associated with the virion DNA and is required for proper association of the DNA with the replication site (Jazwinski et al. 1973). This finding has been challenged because of the difficulties associated with distinguishing experimentally between a failure of adsorption and a failure of parental DNA replication (D. Pratt, pers. comm.). The gene-III and gene-VIII proteins are synthesized as precursors (Pieczenik et al. 1974; Konings et al. 1975) and then cleaved to produce the mature proteins (see below).

The gene-II protein, which is required for synthesis of phage single- and double-stranded DNA (Fidanián and Ray 1972; Lin and Pratt 1972), seems to be a strand-specific, site-specific endonuclease which converts RFI (superhelical, covalently closed, circular duplex RF) to RFII (circular, duplex RF in which at least one strand is not continuous) so that SS DNA synthesis can be initiated (Fidanián and Ray 1972; Lin and Pratt 1972). It has been isolated recently (Meyer and Geider, this volume).

Gene-V protein is an SS DNA-binding protein (Alberts et al. 1972; Oey and Knippers 1972); its sequence is shown in Figure 4. Its binding properties have been examined carefully, and the nature of its interaction with DNA is the subject of intensive investigation, both by solution methods (for references, see Day and Wiseman, this volume) and by X-ray diffraction (A. Rich, pers. comm.). It seems highly likely that the gene-V protein, which is required for synthesis of SS DNA, acts by sequestering the SS DNA at the time of its synthesis, thus blocking synthesis of the complementary strand

```
Met - Ile - Lys - Val - Glu - Ile - Lys - Pro - Ser - Gln
                                                      10
Ala - Gln - Phe - Thr - Thr - Arg - Ser - Gly - Val - Ser
                                                      20
Arg - Gln - Gly - Lys - Pro - Tyr - Ser - Leu - Asn - Glu
                                                      30
Gln - Leu - Cys - Tyr - Val - Asp - Leu - Gly - Asn - Glu
                                                      40
Tyr - Pro - Val - Leu - Val - Lys - Ile - Thr - Leu - Asp
                                                      50
Glu - Gly - Gln - Pro - Ala - Tyr - Ala - Pro - Gly - Leu
                                                      60
Tyr - Thr - Val - His - Leu - Ser - Ser - Phe - Lys - Val
                                                      70
Gly - Gln - Phe - Gly - Ser - Leu - Met - Ile - Asp - Arg
                                                      80
Leu - Arg - Leu - Val - Pro - Ala - Lys
```

Figure 4 Amino acid sequence of fd (Nakashima et al. 1974) and M13 (Cuypers et al. 1974) gene-V proteins. The sequence of the f1 gene-V protein is the same as the one shown from the N-terminal Met through Ala^{57} (partial amino acid data from P. Model [unpubl.]; DNA sequence data from J. Ravetch [unpubl.]).

(Salstrom and Pratt 1971; Mazur and Model 1973). It is possible to obtain premature synthesis of SS DNA in infected cells at early times simply by manipulating the cells so as to force the accumulation of gene-V protein before DNA synthesis is allowed to proceed (Mazur and Model 1973). Complexes of SS DNA and gene-V protein can be isolated from infected cells (Webster and Cashman 1973; Pratt et al. 1974) or formed in vitro, but these complexes, though quite similar, are not identical (Pratt et al. 1974). A second, less well understood function of gene-V protein is its ability to repress the synthesis of the gene-II protein (see below).

Gene Expression In Vivo

The products of genes V and VIII are made in very large amounts and can be readily detected in infected cells by conventional electrophoretic analysis on polyacrylamide gels (P. Model, unpubl.). No other gene products are seen. Most of the gene-V protein is complexed with SS DNA (Webster and Cashman 1973; Pratt et al. 1974; Lica and Ray 1977). Virtually all of the gene-VIII protein is found as a constituent of the inner cell membrane (Smilowitz et al. 1972), and its presence there causes substantial changes in the phospholipid composition of the cell envelope (Woolford et al. 1974). If cells are heavily irradiated before infection, the gene-III protein also can be detected (Henry and Pratt 1969). The products of gene II and gene IV can be found as minor components of the membrane fraction of such irradiated cells (Webster and Cashman 1973; Lin and Pratt 1974) but not in whole-cell extracts.

A unique complex is present in nonpermissive cells infected with a gene-V amber mutant (R. E. Webster et al., in prep.). This complex contains proteins that comigrate with the gene-II, gene-VIII, and X proteins (see below) and smaller amounts of material with the mobility characteristic of gene-III and gene-IV proteins. A one-dimensional electropherogram of a tryptic digest of the putative gene-II material from the complex suggests that it is identical to the gene-II product made in vitro from f1 RFI. The other components of this complex have so far been identified only on the basis of comigration in acrylamide gels. This complex, normally not found in wild-type-infected cells, may be related to one described by Lin and Pratt (1974) in wild-type-infected cells grown at 42°C.

In summary, infected cells contain very large amounts of gene-V and gene-VIII proteins and, under normal conditions, much lesser quantities of the other gene products. Regulation of the relative amounts of these proteins is implied in this observation.

Gene Expression In Vitro

DNA Template

In a conventional coupled transcription-translation system filamentous phage DNA stimulates the synthesis of substantial amounts of the products of genes V, VIII, and IV (Model and Zinder 1974; Konings et al. 1975) (Fig. 5). Good yields of gene-II product may be obtained also, but this result is not consistent from one protein-synthesizing extract to another. These variable yields indicate that some component needed for efficient synthesis of gene-II protein may be limiting. The products of genes I and III are made also, but almost always in relatively low amounts.

Isolated restriction fragments are active as templates (their protein products are listed in Table 2). Results from such experiments help not only to define the genetic content of a given region of the genome, but also to locate the regions containing functional RNA polymerase promoters.

In addition to the above gene products, which were identified by using DNAs prepared from cells infected with phages carrying appropriate amber mutations, a protein which has been termed X is synthesized in substantial amounts (Model and Zinder 1974; Konings et al. 1975) (Fig. 5). Synthesis of X is affected only by the presence in gene II of the amber R21 mutation (P. Model and C. McGill, unpubl.). No other amber mutations in gene II affect the observed level of X synthesis. Protein synthesis carried out with R21 DNA in the presence of amber-suppressing tRNA results in the synthesis of the X protein. The simplest explanation of these data is that the X protein is the result of an initiation within gene II in phase with the rest of the gene-II protein. Because only one mutant that affects the synthesis of X protein is available, we cannot exclude more complex explanations or the

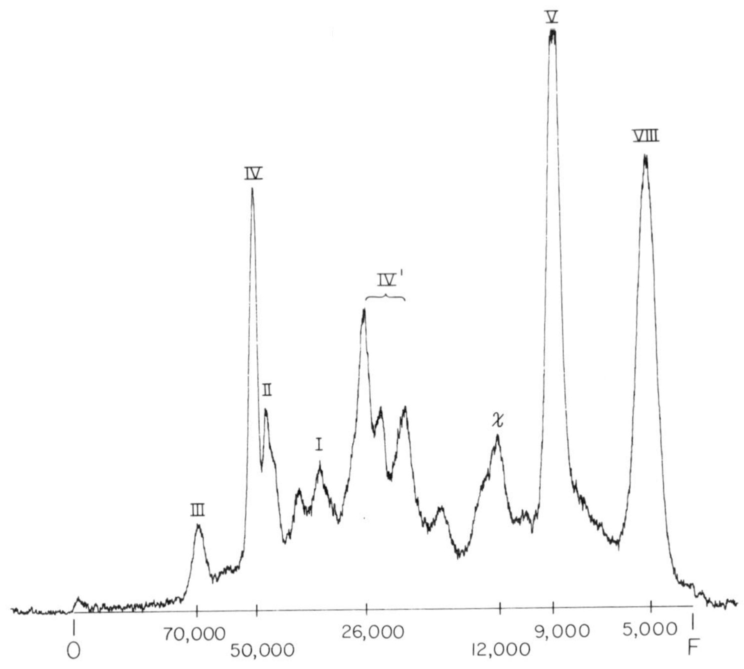

Figure 5 Protein products of an in vitro incorporation programmed by f1 covalently closed, circular DNA. Roman numerals indicate gene products, identified as in Model and Zinder (1974). Bands marked IV′ appear to be prematurely terminated gene-IV product, as argued by Model and Zinder (1974), Konings et al. (1975), and van den Hondel et al. (1975) (see text). The band marked X is not the product of an independent known gene but is encoded by specific restriction fragments (see Table 2). Synthesis of X protein is abolished when the template carries the gene-II amber mutation R21 (P. Model and C. McGill, unpubl.) (see text). The molecular-weight scale was generated by plotting the logarithms of the molecular weights of standards against their mobilities; comparison with Table 1 shows that there are discrepancies between molecular weights estimated in this fashion and those determined from other data.

possibility that X is the product of an independent gene. Synthesis of X protein in vitro from isolated restriction fragments also suggests that the X region is either contiguous to or contained within gene II (van den Hondel et al. 1975; K. Horiuchi and P. Model, unpubl.). It may be significant that two products are also encoded in the same phase by gene *A* of bacteriophage φX (Linney and Hayashi 1973, 1974) and that φX gene *A* has a function analogous to that of f1 gene II. Thus, the X protein may be the counterpart of A*.

Despite extensive efforts using both single amber mutants and multiple mutants, the products of genes VI and VII have not yet been detected in vitro (P. Model, unpubl.).

Table 2 Protein products of various restriction fragments

Fragment	Product
M13	
Hap-A[a]	IV, IV′[b]
Hap-B2[a]	V, VIII
Hap-C[a]	X protein
*Hae*III-A[a]	IV′[b]
*Hae*III-B[a]	V, X protein
f1	
*Hae*III-A[c]	(I)[e], IV′[b]
*Hae*III-B[c]	V, X protein
*Hae*III-C[c]	none
*Hae*II-A[c]	II, V, VIII, X protein
*Hae*II-B[c]	IV, IV′[b]
*Eco*RII-A[d]	II, IV, IV′[b], X protein
*Eco*RII-B[d]	VIII
RFIII/*Hin*dII[f]	I, III, IV, IV′[b], V, VIII, X protein

The restriction fragments listed were used as templates in a coupled transcription-translation system. For their locations on the genome see Fig. 2 and Konings and Schoenmakers (this volume).

[a] van den Hondel et al. (1975).

[b] IV′ appears to be a prematurely terminated product of gene IV (Model and Zinder 1974; Konings et al. 1975).

[c] K. Horiuchi and P. Model (unpubl.).

[d] Vovis et al. (1975).

[e] The band on the autoradiograph was weak.

[f] Model et al. (1975).

The in vitro products of both gene III and gene VIII are synthesized as precursors and are larger than their mature counterparts, although the differences in mobilities between precursors and mature proteins on most conventional polyacrylamide gel systems are small (Konings et al. 1975; Pieczenik et al. 1974; P. Model et al., in prep.). The existence of a precursor for the gene-VIII protein was inferred from the observation that an RNA whose only in vitro product appeared to be the gene-VIII protein contained a ribosome-binding site whose sequence did not predict the N-terminal sequence of the mature protein (Pieczenik et al. 1974). The length and sequence of the precursor can be predicted from RNA sequence data obtained in Takanami's laboratory (Sugimoto et al. 1977). The predicted precursor is shown in Figure 6. Protein sequence data suggest that the start point predicted by the ribosome-binding sequence and by Takanami's sequence is the correct one (P. Model et al., in prep.). In vitro processing of the precursor has been achieved recently with preparations derived from uninfected cells. Cleavage of the precursor requires that the cleaving activity be present during translation; the cleavage is not observed if the activity is added after the precursor has been synthesized (Chang et al. 1978). No

Met - Lys - Lys - Ser - Leu - Val - Leu - Lys - Ala - Ser

Val - Ala - Val - Ala - Thr - Leu - Val - Pro - Met - Leu

Ser - Phe - Ala - Ala - Glu - Gly - Asp - Asp - Pro - Ala

Lys - Ala...

Figure 6 Sequence of the precursor portion of the fd pre-coat protein, as predicted by the RNA sequence of Sugimoto et al. (1977). Underlined residues have been confirmed by direct radiochemical sequencing of material synthesized in vitro (P. Model et al., in prep.). The arrow marks the position of proteolytic cleavage to form the mature product (Chang et al. 1978).

evidence for a gene-VIII precursor has been obtained in vivo (R. E. Webster, pers. comm.); this is perhaps not surprising if, as in vitro, the precursor is cleaved while still a nascent chain. Less is known of the gene-III precursor, but by combining DNA sequence data (Schaller et al., this volume) with the knowledge of the N terminus of the mature gene-III protein (W. Konigsberg, pers. comm.) it is possible to predict the length and sequence of the gene-III precursor. It is not yet known whether the activity that processes the gene-III protein is the same as that which cleaves the gene-VIII protein.

In Vitro RNA Template

When RNA transcribed from DNA by purified RNA polymerase is used as the template for protein synthesis, the products are much the same as those derived from the DNA in a coupled system (Chan et al. 1975). The results are the same if the RNA is first fractionated and the largest RNA species is used. The second-largest RNA fails to encode the gene-III product. The next largest RNA encodes the gene-V and gene-VIII proteins, and the smallest RNA encodes only the product of gene VIII. These results are in accord with the model of transcription proposed by Takanami (Sugiura et al. 1969) and described by Konings and Schoenmakers (this volume; see also below) in which in vitro transcription is believed to initiate at a number of promoters and to terminate at one strong terminator located distal to the gene-VIII boundary.

In Vivo RNA Template

When total RNA extracted from infected cells is used to program in vitro translation, the products obtained are comparable to those obtained with in vivo experiments (LaFarina and Model 1978). Large amounts of the gene-V and gene-VIII products and much smaller (but detectable) amounts of the gene-III product are made. The products of genes II, IV, and I (in contrast to the results with RF-directed in vitro synthesis) are not observed, nor are the products of genes VI and VII or the X protein.

Elements That May Control Gene Expression in the Filamentous Phages

The data on protein synthesis (see above), which come from a number of different laboratories, suggest that gene expression of the filamentous phages is regulated. Regulation in vivo leads to the production of a great deal of gene-VIII protein, a substantial amount of gene-V protein, and comparatively much less of the other phage-specified proteins (Henry and Pratt 1969; Woolford et al. 1974). In coupled in vitro systems at least some of this control is lost; the amounts of gene-V and gene-VIII proteins made are more nearly equal, and the synthesis of a good deal of gene-II and gene-IV proteins can be detected (Model and Zinder 1974; Konings et al. 1975). A priori, one could expect this control to operate at the level of transcription, RNA processing, messenger decay, translation, or protein decay. Transcription of filamentous phage DNA has been studied in some detail in vitro (Konings and Schoenmakers, this volume), and the general pattern described—a series of promoters spaced around a genome to yield overlapping sets of RNAs which end at a common terminator—can afford some, but not all, of the regulatory potential observed in synthesis in vivo. Overlapping transcription can account, for example, for the preponderance of the gene-V and gene-VIII products both in vivo and in vitro, but it does not explain why there is so little gene-III synthesis on unit-length RNA or why the levels of the gene-II and gene-IV products found in vivo are low.

Some information on control mechanisms is afforded by in vitro translation of messenger RNAs and by a consideration of in vitro ribosome binding. The largest in vitro transcript, presumably a single species, encodes the products of genes IV, V, and VIII with roughly equal efficiency, and that of gene III with perhaps tenfold lower efficiency (Chan et al. 1975). Unfortunately, the yield of gene-II product is highly variable and the X protein is synthesized from RNA in relatively low amounts (Chan et al. 1975; R. Konings, pers. comm.). When the same system is used to translate total in vivo RNA, the phage-specified products consist of a massive amount of gene-VIII product, rather less gene-V protein, and a fairly minute amount of gene-III protein (LaFarina and Model 1977). Gene-IV protein is synthesized at levels so low as to be undetectable. Taken together, these data suggest that the RNA that encodes only gene-VIII protein is probably present in infected cells at higher levels than any other RNA. Furthermore, the gene-III protein is probably not made from a very long transcript, as a full-length RNA would be expected to make more gene-IV than gene-III product (see above). Thus, we may infer tentatively that in vivo there is extensive processing of the messages or that RNA transcription does not follow the simplest in vitro model. That some differential decay does occur is fairly clear; the half-life of at least a portion of the gene-VIII message is appreciably longer than that of the gene-V message (LaFarina and Model 1978). The mechanism by which differential decay or processing might occur is almost totally obscure, as it is for most *E. coli* messages. In contrast to isolated and purified T7 message, f1 RNA does not appear to be cleaved by ribonuclease III (H. D. Robertson and P. Model, unpubl.), and therefore this enzyme probably does not participate in messenger processing in the filamentous phages in vivo.

When f1 RNA synthesized in vitro is bound to ribosomes under conditions very similar to those used by Steitz and her colleagues (Steitz 1969), regions of the messenger corresponding to the sites for the initiation of synthesis of the products of genes V, VIII, and IV are protected from subsequent nuclease digestion (Pieczenik et al. 1974). Further fractionation of such ribosome-protected regions by hybridization to the minus strand of appropriate restriction fragments suggests that one more site, probably representing the ribosome-binding site for the initiation of synthesis of the gene-VI or gene-I product, is also protected (Ravetch et al. 1977a). Little or no RNA that might code for the initiation of synthesis of the gene-II or gene-III product is recovered, not even by the restriction fragment approach. Results of the same sort are obtained if the messenger RNA is bound to ribosomes under normal conditions of in vitro protein synthesis but with elongation blocked by the drug thiostrepton: the yield of ribosome-protected fragments is enhanced, and their sizes are somewhat larger, but the same catalog of fragments is obtained (Ravetch et al. 1977a). These observations suggest that the various initiation regions differ in the ability to bind to ribosomes and be protected from nuclease digestion. Of course, the conditions under which such experiments are carried out are only distantly related to the conditions that obtain in vivo. Nonetheless, these experiments are consistent with the in vitro translation experiments in that they suggest that the absence of gene-IV protein in vivo results from there being only a limited amount of RNA that codes for this protein. The relatively low level (both in vivo and in vitro) of gene-III expression may well be due, at least in part, to weak ribosome binding.

Dynamic Control

We consider dynamic control to be the interaction of one gene product with the synthesis of itself or another product, rather than a differential level of gene expression which results from essentially static elements such as promoters and ribosome-binding regions. Dynamic control of the synthesis of the gene-II product has been observed and may be central to the control of the interaction of the filamentous phages with their hosts. Regulation of the synthesis of the gene-II product will regulate the level of all phage-specific expression, as this gene product is needed both for progeny DNA synthesis and for the maintenance of the pool of RF molecules that are used as templates for phage-specific RNA synthesis.

Three phenomena related to the control of the synthesis of the gene-II product have been observed. The first, reported by Henry and Pratt (1969), is that heavily UV-treated cells infected with different gene-II amber mutants produced, in each case, large amounts of a different polypeptide that was not found in wild-type-infected or uninfected cells. Henry and Pratt inferred that these polypeptides are gene-II amber fragments. The gene-II protein was not made in amounts large enough to be detected in any of their experiments. The second phenomenon, reported by Lin and Pratt (1974), is that growth of either wild-type or gene-II temperature-sensitive mutants at 42°C leads to a markedly enhanced amount of gene-II protein in the infected cell. Pratt and his colleagues inferred from these results that a deficiency in gene-II function results directly or indirectly in elevated levels of gene-II protein synthesis. To explain enhanced synthesis of gene-II protein in wild-type cells at 42°C, Lin and Pratt (1974) suggested that even the wild-type gene-II product is somewhat thermolabile and is denatured at the higher temperature. The third phenomenon is that purified gene-V protein, when added to an in vitro protein-synthesis system, specifically and markedly reduces the amount of gene-II product made. In vivo, infection under nonpermissive conditions with gene-V mutants leads to higher levels of gene-II protein than are normally detected, and most of this protein material is probably in the unique complex that is found in cells infected with gene-V mutants (R. E. Webster et al., in prep.) (see above).

It is possible to unify these observations by one mechanism. If, on the basis of the in vitro data, we postulate that gene-V protein represses the synthesis of gene-II protein directly, then any condition that lowers either the amount or the functional efficiency of gene-V protein results in enhanced gene-II protein synthesis. Thus, infection with gene-II amber mutants would prevent progeny RF synthesis, reduce the amount of phage-specific RNA in the infected cell, and therefore also reduce the amount of gene-V protein. This has been observed experimentally (Mazur and Model 1973). The synthesis of filamentous phage SS DNA (but not double-stranded DNA) is markedly reduced at 42°C below that found at lower temperatures. Since gene-V protein is required only for SS DNA synthesis, it may be that gene-V protein is somewhat thermolabile and at 42°C not able to repress the synthesis of gene-II protein. Thus, it is quite possible that what appears to be gene-II autoregulation actually is effected via the functional levels of gene-V protein. Only further experimentation can establish whether this unifying hypothesis is correct.

SUMMARY

We know a great deal about the organization of the phage genome (including, of course, the nucleotide sequence [see Schaller et al., this volume]).

The protein products of a majority of the genes have been identified, and something is known of the functions of half of these gene products. However, we have only begun to scratch the surface in our understanding of perhaps the most interesting aspects of these phages. What are the functions of the protein products of genes I, IV, and VII and the X protein? How is the level of each protein regulated? And perhaps most interesting, how is phage replication controlled so that production of phage is maintained while the host continues to grow and multiply?

ACKNOWLEDGMENTS

We thank Norton D. Zinder for his continuing interest in and concern for our experiments and for his suggestions with regard to this manuscript. Work in this laboratory is supported by grants from the National Science Foundation and the National Institutes of Health.

REFERENCES

Alberts, B., L. Frey, and H. Delius. 1972. Isolation and characterization of gene V protein of filamentous bacterial viruses. *J. Mol. Biol.* **68:** 139.

Arber, W. 1966. Host specificity of DNA produced by *Escherichia coli.* 9. Host-controlled modification of bacteriophage fd. *J. Mol. Biol.* **20:** 483.

Arber, W. and U. Kühnlein. 1967. Mutationeller Verlust B-spezifischer Restriktion des Bakteriophagen fd. *Pathol. Microbiol.* **30:** 946.

Asbeck, F., K. Beyreuther, H. Köhler, G. von Wettstein, and G. Braunitzer. 1969. Virusproteine. IV. Die Konstitution des Hüllproteins des Phagen fd. *Hoppe-Seyler's Z. Physiol. Chem.* **350:** 1047.

Bailey, G. S., D. Gillet, D. F. Hill, and G. B. Petersen. 1977. Automated sequencing of insoluble peptides using detergent: Bacteriophage f1 coat protein. *J. Biol. Chem.* **252:** 2218.

Beaudoin, J. 1970. "Studies on coat protein mutants and on multiple-length particles of coliphage M13." Ph.D. thesis, University of Wisconsin, Madison.

Boon, T. and N. D. Zinder. 1971. Genotypes produced by individual recombination events involving bacteriophage f1. *J. Mol. Biol.* **58:** 133.

Boyer, H. W., L. T. Chow, A. Dugaiczyk, J. Hedgpeth, and H. M. Goodman. 1973. DNA substrate for the *Eco* RII restriction endonuclease and modification methylase. *Nat. New Biol.* **244:** 40.

Brack, C., H. Eberle, T. A. Bickle, and R. Yuan. 1976. Mapping of recognition sites for the restriction endonuclease from *Escherichia coli* K12 on bacteriophage PM2 DNA. *J. Mol. Biol.* **108:** 583.

Bradley, D. E. 1964. The structure of some bacteriophages associated with male strains of *Escherichia coli. J. Gen. Microbiol.* **35:** 471.

Braunitzer, G., S. Braig, F. Krug, and G. Hobom. 1970. Mutanten des Hüllproteins (B-Protein) des Bakteriophagen fd. *FEBS Lett.* **7:** 83.

Braunitzer, G., F. Asbeck, K. Beyreuther, H. Köhler, and G. von Wettstein. 1967.

Die Konstitution des Hüllproteins des Bakteriophagen fd. *Hoppe-Seyler's Z. Physiol. Chem.* **348:** 725.

Chan, T.-S., P. Model, and N. D. Zinder. 1975. In vitro protein synthesis directed by separated transcripts of bacteriophage f1 DNA. *J. Mol. Biol.* **99:** 369.

Chang, C. N., G. Blobel, and P. Model. 1978. Detection of prokaryotic signal peptidase in an *Escherichia coli* membrane fraction: Endoproteolytic cleavage of nascent f1 pre-coat protein. *Proc. Natl. Acad. Sci.* **75:** 361.

Cuypers, T., F. J. van der Ouderaa, and W. W. de Jong. 1974. The amino acid sequence of gene 5 protein of bacteriophage M13. *Biochem. Biophys. Res. Commun.* **59:** 557.

Danna, K. J., G. H. Sack, Jr., and D. Nathans. 1973. Studies of simian virus 40 DNA. VII. A cleavage map of the SV40 genome. *J. Mol. Biol.* **78:** 363.

Edgell, M. H., C. A. Hutchison III, and M. Sclair. 1972. Specific endonuclease R fragments of bacteriophage φX174 deoxyribonucleic acid. *J. Virol.* **9:** 574.

Enea, V. and N. D. Zinder. 1975. A deletion mutant of bacteriophage f1 containing no intact cistrons. *Virology* **68:** 105.

Enea, V., K. Horiuchi, B. G. Turgeon, and N. D. Zinder. 1977. Physical map of defective interfering particles of bacteriophage f1. *J. Mol. Biol.* **111:** 395.

Fidanián, H. M. and D. S. Ray. 1972. Replication of bacteriophage M13. VII. Requirement of the gene 2 protein for the accumulation of a specific RFII species. *J. Mol. Biol.* **72:** 51.

Griffith, J. and A. Kornberg. 1974. Mini M13 bacteriophage: Circular fragments of M13 DNA are replicated and packaged during normal infections. *Virology* **59:** 139.

Hartman, N. and N. D. Zinder. 1973. An f1 mutant with reduced sensitivity to P1 restriction. *Virology* **56:** 407.

Hattman, S. 1973. Plasmid-controlled variation in the content of methylated bases in single-stranded DNA phages M13 and fd. *J. Mol. Biol.* **74:** 749.

Hattman, S., J. E. Brooks, and M. Masurekar. 1978. Sequence specificity of the P1-modification methylase (M.*Eco*P1) and the DNA methylase (M. *Eco*dam) controlled by the *Escherichia coli* dam-gene. *J. Mol. Biol.* (in press).

Hattman, S., S. Schlagman, and L. Cousens. 1973. Isolation of a mutant of *Escherichia coli* defective in cytosine-specific deoxyribonucleic acid methylase activity and in partial protection of bacteriophage λ against restriction by cells containing the N-3 drug-resistance factor. *J. Bacteriol.* **115:** 1103.

Henry, T. J. and D. Pratt. 1969. The proteins of bacteriophage M13. *Proc. Natl. Acad. Sci.* **62:** 800.

Hewitt, J. A. 1975. Miniphage—A class of satellite phage to M13. *J. Gen. Virol.* **26:** 87.

Hofschneider, P. H. 1963. Untersuchungen über "kleine" *E. coli* K12 Bakteriophagen 1. und 2. Mitteilung. *Z. Naturforsch.* **18b:** 203.

Horiuchi, K. and N. D. Zinder. 1972. Cleavage of bacteriophage f1 DNA by the restriction enzyme of *Escherichia coli* B. *Proc. Natl. Acad. Sci.* **69:** 3220.

———. 1976. Origin and direction of synthesis of bacteriophage f1 DNA. *Proc. Natl. Acad. Sci.* **73:** 2341.

Horiuchi, K., G. F. Vovis, V. Enea, and N. D. Zinder. 1975. Cleavage map of bacteriophage f1: Location of the *Escherichia coli* B-specific modification sites. *J. Mol. Biol.* **95:** 147.

Hutchison, C. A., III and M. H. Edgell. 1971. Genetic assay for small fragments of bacteriophage φX174 deoxyribonucleic acid. *J. Virol.* **8:** 181.

Jazwinski, S. M., R. Marco, and A. Kornberg. 1973. A coat protein of the bacteriophage M13 virion participates in membrane-oriented synthesis of DNA. *Proc. Natl. Acad. Sci.* **70:** 205

Kelly, T. J., Jr. and H. O. Smith. 1970. A restriction enzyme from *Hemophilus influenzae* II. Base sequence of the recognition site. *J. Mol. Biol.* **51:** 393.

Konings, R. N. H., T. Hulsebos, and C. A. van den Hondel. 1975. Identification and characterization of the in vitro synthesized gene products of bacteriophage M13. *J. Virol.* **15:** 570.

Kühnlein, U. and W. Arber. 1972. Host specificity of DNA produced by *E. coli.* XV. The role of nucleotide methylation in *in vitro* B-specific modification. *J. Mol. Biol.* **63:** 9.

Lacks, S. and B. Greenberg. 1977. Complementary specificity of restriction endonucleases of *Diplococcus pneumoniae* with respect to DNA methylation. *J. Mol. Biol.* **114:** 153.

LaFarina, M. and P. Model. 1978. Transcription in bacteriophage f1-infected *E. coli.* 1. Translation of the RNA in vitro. *Virology* **86:** 368.

Lautenberger, J. A., N. C. Kan, D. Lackey, S. Linn, M. H. Edgell, and C. A. Hutchison III. 1978. Recognition site of the *Escherichia coli* B restriction enzyme on φXsB1 and SV40 DNAs: An interrupted sequence. *Proc. Natl. Acad. Sci.* **75:** 2271.

Lica, L. and D. S. Ray. 1977. Replication of bacteriophage M13. XII. In vivo cross-linking of a phage-specific DNA binding protein to the single-stranded DNA of bacteriophage M13 by ultraviolet irradiation. *J. Mol. Biol.* **115:** 45.

Lin, N. S.- C. and D. Pratt. 1972. Role of bacteriophage M13 gene 2 in viral DNA replication. *J. Mol. Biol.* **72:** 37.

———. 1974. Bacteriophage M13 gene 2 protein: Increasing its yield in infected cells, and identification and localization. *Virology* **61:** 334.

Linn, S. and W. Arber. 1968. Host specificity of DNA produced by *Escherichia coli.* X. In vitro restriction of phage fd replicative form. *Proc. Natl. Acad. Sci.* **59:** 1300.

Linney, E. and M. Hayashi. 1973. Two proteins of gene A of φX174. *Nat. New Biol.* **245:** 6.

———. 1974. Intragenic regulation of the synthesis of φX174 gene A proteins. *Nature* **249:** 345.

Loeb, T. 1960. Isolation of a bacteriophage specific for the F^+ and Hfr mating types of *Escherichia coli* K-12. *Science* **131:** 932.

Lyons, L. B. 1971. "Genetic studies of bacteriophage f1." Ph.D. thesis, The Rockefeller University, New York.

Lyons, L. B. and N. D. Zinder. 1972. The genetic map of the filamentous bacteriophage f1. *Virology* **49:** 45.

Mandel, M. and A. Higa. 1970. Calcium-dependent bacteriophage DNA infection. *J. Mol. Biol.* **53:** 159.

Marco, R. 1975. The adsorption protein in mini M13 phages. *Virology* **68:** 280.

Marinus, M. G. and N. R. Morris. 1973. Isolation of deoxyribonucleic acid methylase mutants of *Escherichia coli* K-12. *J. Bacteriol.* **114:** 1143.

———. 1975. Pleiotropic effects of a DNA adenine methylation mutation (dam^- 3). in *Escherichia coli* K12. *Mutat. Res.* **28:** 15.

Marvin, D. A. and H. Hoffmann-Berling. 1963. Physical and chemical properties of two new small bacteriophages. *Nature* **197:** 517.
Marvin, D. A. and B. Hohn. 1969. Filamentous bacterial viruses. *Bacteriol. Rev.* **33:** 172.
Mazur, B. J. and P. Model. 1973. Regulation of coliphage f1 single-stranded DNA synthesis by a DNA-binding protein. *J. Mol. Biol.* **78:** 285.
Meselson, M. and R. Yuan. 1968. DNA restriction enzyme from *E. coli*. *Nature* **217:** 1110.
Meynell, G. G. and A. M. Lawn. 1968. Filamentous phages specific for the I sex factor. *Nature* **217:** 1184.
Minamishima, Y., K. Takeya, Y. Ohnishi, and K. Amako. 1968. Physicochemical and biological properties of fibrous *Pseudomonas* bacteriophages. *J. Virol.* **2:** 208.
Model, P. and N. D. Zinder. 1974. In vitro synthesis of bacteriophage f1 proteins. *J. Mol. Biol.* **83:** 231.
Model, P., K. Horiuchi, C. McGill, and N. D. Zinder. 1975. Template activity of f1 RFI cleaved with endonucleases R·*Hin*d, R·*Eco*P1 or R·*Eco*B. *Nature* **253:** 132.
Nakashima, Y. and W. Konigsberg. 1974. Reinvestigation of a region of the fd bacteriophage coat protein sequence. *J. Mol. Biol.* **88:** 598.
Nakashima, Y., A. K. Dunker, D. A. Marvin, and W. Konigsberg. 1974. The amino acid sequence of a DNA binding protein, the gene 5 product of fd filamentous bacteriophage. *FEBS Lett.* **40:** 290 (see also erratum **43:** 125).
Oey, J. L. and R. Knippers. 1972. Properties of the isolated gene 5 protein of bacteriophage fd. *J. Mol. Biol.* **68:** 125.
Pieczenik, G., P. Model, and H. D. Robertson. 1974. Sequence and symmetry in ribosome binding sites of bacteriophage f1 RNA. *J. Mol. Biol.* **90:** 191.
Pratt, D. and W. S. Erdahl. 1968. Genetic control of bacteriophage M13 DNA synthesis. *J. Mol. Biol.* **37:** 181.
Pratt, D., P. Laws, and J. Griffith. 1974. Complex of bacteriophage M13 single-stranded DNA and gene 5 protein. *J. Mol. Biol.* **82:** 425.
Pratt, D., H. Tzagoloff, and J. Beaudoin. 1969. Conditional lethal mutants of the small filamentous coliphage M13. II. Two genes for coat proteins. *Virology* **39:** 42.
Pratt, D., H. Tzagoloff, and W. S. Erdahl. 1966. Conditional lethal mutants of the small filamentous coliphage M13. I. Isolation, complementation, cell killing, time of cistron action. *Virology* **30:** 397.
Ravetch, J., K. Horiuchi, and P. Model. 1977a. Mapping of bacteriophage f1 ribosome binding sites to their cognate genes. *Virology* **81:** 341.
Ravetch, J. V., K. Horiuchi, and N. D. Zinder. 1977b. Nucleotide sequences near the origin of replication of bacteriophage f1. *Proc. Natl. Acad. Sci.* **74:** 4219.
———. 1978. Nucleotide sequence of the recognition site for the restriction-modification enzyme of *E. coli* B. *Proc. Natl. Acad. Sci.* **75:** 2266.
Razin, A., J. W. Sedat, and R. L. Sinsheimer. 1970. Structure of the DNA of bacteriophage φX174. VII. Methylation. *J. Mol. Biol.* **53:** 251.
———. 1973. *In vivo* methylation of replicating bacteriophage φX174 DNA. *J. Mol. Biol.* **78:** 417.
Rossomando, E. F. and N. D. Zinder. 1968. Studies on the bacteriophage f1. I. Alkali-induced disassembly of the phage into DNA and protein. *J. Mol. Biol.* **36:** 387.
Salivar, W. O., T. J. Henry, and D. Pratt. 1967. Purification and properties of diploid particles of coliphage M13. *Virology* **32:** 41.

Salstrom, J. S. and D. Pratt. 1971. Role of coliphage M13 gene 5 in single-stranded DNA production. *J. Mol. Biol.* **61:** 489.

Schaller, H., A. Uhlmann, and K. Geider. 1976. A DNA fragment from the origin of single-strand to double-strand DNA replication of bacteriophage fd. *Proc. Natl. Acad. Sci.* **73:** 49.

Schlagman, S., S. Hattman, M. S. May, and L. Berger. 1976. *In vivo* methylation by *Escherichia coli* K-12 *mec*$^+$ deoxyribonucleic acid-cytosine methylase protects against *in vitro* cleavage by the RII restriction endonuclease (R·*Eco*RII). *J. Bacteriol.* **126:** 990.

Scott, J. R. and N. D. Zinder. 1967. Heterozygotes of phage f1. In *The molecular biology of viruses* (ed. J. S. Colter and W. Paranchych), p. 211. Academic Press, New York.

Seeburg, P. H. and H. Schaller. 1975. Mapping and characterization of promoters in bacteriophages fd, f1, and M13. *J. Mol. Biol.* **92:** 261.

Smilowitz, H., J. Carson, and P. W. Robbins. 1972. Association of newly synthesized major f1 coat protein with infected host cell inner membrane. *J. Supramol. Struct.* **1:** 8.

Smith, H. O. and M. L. Birnstiel. 1976. A simple method for DNA restriction site mapping. *Nucleic Acids Res.* **3:** 2387.

Smith, H. O. and K. W. Wilcox. 1970. A restriction enzyme from *Hemophilus influenzae*. I. Purification and general properties. *J. Mol. Biol.* **51:** 379.

Smith, J. D., W. Arber, and U. Kühnlein. 1972. Host specificity of DNA produced by *Escherichia coli.* XIV. The role of nucleotide methylation in *in vivo* B-specific modification. *J. Mol. Biol.* **63:** 1.

Snell, D. T. and R. E. Offord. 1972. The amino acid sequence of the B-protein of bacteriophage ZJ/2. *Biochem. J.* **127:** 167.

Steitz, J. A. 1969. Polypeptide chain initiation: Nucleotide sequences of the three ribosomal binding sites in bacteriophage R17 RNA. *Nature* **224:** 957.

Suggs, S. V. and D. S. Ray. 1977. Replication of bacteriophage M13. XI. Localization of the origin for M13 single-strand synthesis. *J. Mol. Biol.* **110:** 147.

Sugimoto, K., H. Sugisaki, T. Okamoto, and M. Takanami. 1977. Studies on bacteriophage fd DNA. IV. The sequence of messenger RNA for the major coat protein gene. *J. Mol. Biol.* **111:** 487.

Sugiura, M., T. Okamoto, and M. Takanami. 1969. Starting nucleotide sequences of RNA synthesized on the replicative form DNA of coliphage fd. *J. Mol. Biol.* **43:** 299.

Tabak, H. F., J. Griffith, K. Geider, H. Schaller, and A. Kornberg. 1974. Initiation of deoxyribonucleic acid synthesis. VII. A unique location of the gap in the M13 replicative duplex synthesized in vitro. *J. Biol. Chem.* **249:** 3049.

Takanami, M., T. Okamoto, K. Sugimoto, and H. Sugisaki. 1975. Studies on bacteriophage fd DNA. I. A cleavage map of the fd genome. *J. Mol. Biol.* **95:** 21.

Takanami, M., K. Sugimoto, H. Sugisaki, and T. Okamoto. 1976. Sequence of promoter for coat protein gene of bacteriophage fd. *Nature* **260:** 297.

van den Hondel, C. A. and J. G. G. Schoenmakers. 1975. Studies on bacteriophage M13 DNA. 1. A cleavage map of the M13 genome. *Eur. J. Biochem.* **53:** 547.

———. 1976. Cleavage maps of the filamentous bacteriophages M13, fd, f1, and ZJ/2. *J. Virol.* **18:** 1024.

van den Hondel, C. A., R. N. H. Konings, and J. G. G. Schoenmakers. 1975a.

Regulation of gene activity in bacteriophage M13 DNA: Coupled transcription and translation of purified genes and gene-fragments. *Virology* **67:** 487.

van den Hondel, C. A., L. Pennings, and J. G. G. Schoenmakers. 1976. Restriction-enzyme-cleavage maps of bacteriophage M13: Existence of an intergenic region on the M13 genome. *Eur. J. Biochem.* **68:** 55.

van den Hondel, C. A., A. Weijers, R. N. H. Konings, and J. G. G. Schoenmakers. 1975b. Studies on bacteriophage M13 DNA. 2. The gene order of the M13 genome. *Eur. J. Biochem.* **53:** 559.

Vovis, G. F. and S. Lacks. 1977. Complementary action of restriction enzymes endo R·*Dpn*I and endo R·*Dpn*II on bacteriophage f1 DNA. *J. Mol. Biol.* **115:** 525.

Vovis, G. F., K. Horiuchi, and N. D. Zinder. 1975. Endonuclease R·*Eco*RII restriction of bacteriophage f1 DNA in vitro: Ordering of genes V and VII, location of an RNA promoter for gene VIII. *J. Virol.* **16:** 674.

Vovis, G. F., M. Ohsumi, and N. D. Zinder. 1977. Bacteriophage f1 as a vector for constructing recombinant DNA molecules. In *Molecular approaches to eucaryotic genetic systems* (ed. G. Wilcox et al.), p. 55. Academic Press, New York.

Webster, R. E. and J. S. Cashman. 1973. Abortive infection of *Escherichia coli* with the bacteriophage f1: Cytoplasmic membrane proteins and the f1 DNA-gene 5 protein complex. *Virology* **55:** 20.

Woolford, J. L., Jr., J. S. Cashman, and R. E. Webster. 1974. f1 coat protein synthesis and altered phospholipid metabolism in f1 infected *Escherichia coli*. *Virology* **58:** 544.

Yoshimori, R. N. 1971. "A genetic and biochemical analysis of the restriction and modification of DNA by resistance transfer factors." Ph.D. thesis, University of California, San Francisco.

Zinder, N. D. 1973. Resistance to colicins E3 and K induced by infection with bacteriophage f1. *Proc. Natl. Acad. Sci.* **70:** 3160.

Zinder, N. D., R. C. Valentine, M. Roger, and W. Stoeckenius. 1963. f1, a rod-shaped male-specific bacteriophage that contains DNA. *Virology* **20:** 638.

Sequence and Regulatory Signals of the Filamentous Phage Genome

Heinz Schaller and Ewald Beck
Department of Microbiology
Universität Heidelberg
6900 Heidelberg, Federal Republic of Germany

Mituru Takanami
Institute for Chemical Research
Kyoto University
Kyoto, Japan

The genome of the male-specific filamentous coliphages fd, f1, and M13 is a circular single-stranded DNA of approximately 6400 bases, which codes for at least eight genes (Marvin and Hohn 1969; Ray 1977). Upon infecting a cell, a double-stranded replicative-form (RF) DNA is formed, and this RF DNA then serves as a template for mRNA synthesis as well as for progeny phage DNA production. Both reactions are catalyzed and controlled largely by host proteins. Therefore, the structure and function of the filamentous phage DNA have been studied extensively as a model system for elucidating the molecular mechanisms involved in DNA replication and gene expression.

The first nucleotide sequences of filamentous phage DNA to be determined were the pyrimidine tracts obtained by the Burton and Peterson depurination procedure; tracts of up to 20 bases in length were sequenced (Petersen and Reeves 1969; Ling 1972). Oligonucleotide-primed DNA synthesis was used by Oertel and Schaller (1972) to determine the order of pyrimidine tracts in a pyrimidine-rich segment of fd DNA (positions 5539–5630, Fig. 1) and, in combination with ribosubstitution, by Sanger et al. (1973, 1974) to deduce the sequence of 85 nucleotides in f1 DNA (positions 6303–6387, Fig. 1).

Three ribosome-binding sites on DNA-directed mRNA were sequenced by Pieczenik et al. (1974). Ribosomes bind specifically to the start point of translation on mRNA, and the bound region can be isolated and sequenced after RNase treatment. Similar protection experiments with RNA polymerase allowed the isolation and sequencing of an RNA-polymerase-binding site from fd DNA (Schaller et al. 1975; Sugimoto et al. 1975a) and the isolation and characterization of a DNA fragment from the origin of DNA replication (Schaller et al. 1976).

The introduction of restriction endonucleases greatly advanced studies of the phage genome. The construction of a physical map made it possible to locate and isolate the specific regions of the genome involved in gene

```
  1  AAC GCT ACT ACC ATT AGT AGA ATT GAT GCC ACC TTT TCA GCT CGC GCC CCA AAT GAA AAT ATA GCT AAA CAG GTT
     ASN ALA THR THR ILE SER ARG ILE ASP ALA THR PHE SER ALA ARG ALA PRO ASN GLU ASN ILE ALA LYS GLN VAL

 76  ATT GAC CAT TTG CGA AAT GTA TCT AAT GGT CAA ACT AAA TCT ACT CGT TCG CAG AAT TGG GAA TCA ACT GTT ACA
     ILE ASP HIS LEU ARG ASN VAL SER ASN GLY GLN THR LYS SER THR ARG SER GLN ASN TRY GLU SER THR VAL THR

151  TGG AAT GAA ACT TCC AGA CAC CGT ACT TTA GTT GCA TAT TTA AAA CAT GTT GAA CTA CAG CAC CAG ATT CAG CAA
     TRY ASN GLU THR SER ARG HIS ARG THR LEU VAL ALA TYR LEU LYS HIS VAL GLU LEU GLN HIS GLN ILE GLN GLN

226  TTA AGC TCT AAG CCA TCC GCA AAA ATG ACC TCT TAT CAA AAG GAG CAA TTA AAG GTA CTG TCT AAT CCT GAC CTG   II
     LEU SER SER LYS PRO SER ALA LYS MET THR SER TYR GLN LYS GLU GLN LEU LYS VAL LEU SER ASN PRO ASP LEU

301  TTG GAA TTT GCT TCC GGT CTG GTT CGC TTT GAG GCT CGA ATT AAA ACG CGA TAT TTG AAG TCT TTC GGG CTT CCT
     LEU GLU PHE ALA SER GLY LEU VAL ARG PHE GLU ALA ARG ILE LYS THR ARG TYR LEU LYS SER PHE GLY LEU PRO

376  CTT AAT CTT TTT GAT GCA ATT CGC TTT GCT TCT GAC TAT AAT AGA CAG GGT AAA GAC CTG ATT TTT GAT TTA TGG
     LEU ASN LEU PHE ASP ALA ILE ARG PHE ALA SER ASP TYR ASN ARG GLN GLY LYS ASP LEU ILE PHE ASP LEU TRY

451  TCA TTC TCG TTT TCT GAA CTG TTT AAA GCA TTT GAG GGG GAT TCA ATG AAT ATT TAT GAC GAT TCC GCA GTA TTG
     SER PHE SER PHE SER GLU LEU PHE LYS ALA PHE GLU GLY ASP SER MET ASN ILE TYR ASP ASP SER ALA VAL LEU

526  GAC GCT ATC CAG TCT AAA CAT TTT ACA ATT ACC CCC TCT GGC AAA ACT TCC TTT GCA AAA GCC TCT CGC TAT TTT
     ASP ALA ILE GLN SER LYS HIS PHE THR ILE THR PRO SER GLY LYS THR SER PHE ALA LYS ALA SER ARG TYR PHE

601  GGT TTC TAT CGT CGT CTG GTT AAT GAG GGT TAT GAT AGT GTT GCT CTT ACC ATG CCT CGT AAT TCC TTT TGG CGT   X
     GLY PHE TYR ARG ARG LEU VAL ASN GLU GLY TYR ASP SER VAL ALA LEU THR MET PRO ARG ASN SER PHE TRY ARG

676  TAT GTA TCT GCA TTA GTT GAG TGT GGT ATT CCT AAA TCT CAA TTG ATG AAT CTT TCC ACC TGT AAT AAT GTT GTT
     TYR VAL SER ALA LEU VAL GLU CYS GLY ILE PRO LYS SER GLN LEU MET ASN LEU SER THR CYS ASN ASN VAL VAL

751  CCG TTA GTT CGT TTT ATT AAC GTA GAT TTT TCC TCC CAA CGT CCT GAC TGG TAT AAT GAG CCA GTT CTT AAA ATC
     PRO LEU VAL ARG PHE ILE ASN VAL ASP PHE SER SER GLN ARG PRO ASP TRY TYR ASN GLU PRO VAL LEU LYS ILE

826  GCA TAA GGT AAT TCA AA  ATG ATT AAA GTT GAA ATT AAA CCG TCT CAA GCG CAA TTT ACT ACC CGT TCT GGT GTT
     ALA ***                 MET ILE LYS VAL GLU ILE LYS PRO SER GLN ALA GLN PHE THR THR ARG SER GLY VAL

900  TCT CGT CAG GGC AAG CCT TAT TCA CTG AAT GAG CAG CTT TGT TAC GTT GAT TTG GGT AAT GAA TAT CCG GTG CTT   V
     SER ARG GLN GLY LYS PRO TYR SER LEU ASN GLU GLN LEU CYS TYR VAL ASP LEU GLY ASN GLU TYR PRO VAL LEU
```

```
975   GTC AAG ATT ACT CTC GAC GAA GGT CAG CCA GCG TAT GCG CCT GGT CTG TAC ACC GTG CAT CTG TCC TCG TTC AAA
      VAL LYS ILE THR LEU ASP GLU GLY GLN PRO ALA TYR ALA PRO GLY LEU TYR THR VAL HIS LEU SER SER PHE LYS

1050  GTT GGT CAG TTC GGT TCT CTT ATG ATT GAC CGT CTG CGC CTC GTT CCG GCT AAG TAA C ATG GAG CAG GTC GCG GAT
      VAL GLY GLN PHE GLY SER LEU MET ILE ASP ARG LEU ARG LEU VAL PRO ALA LYS ***   MET GLU GLN VAL ALA ASP

1126  TTC GAC ACA ATT TAT CAG GCG ATG ATA CAA ATC TCC GTT GTA CTT TGT TTC GCG CTT GGT ATA ATC GCT GGG GGT   VII
      PHE ASP THR ILE TYR GLN ALA MET ILE GLN ILE SER VAL VAL LEU CYS PHE ALA LEU GLY ILE ILE ALA GLY GLY

1201  CAA AGA TGA GT GTT TTA GTG TAT TCT TTC GCC TCT TTC GTT TTA GGT TGG TGC CTT CGT AGT GGC ATT ACG TAT   IX
      GLN ARG ***

1275  TTT ACC CGT TTA ATG GAA ACT TCC TC  ATG AAA AAG TCT TTA GTC CTC AAA GCC TCC GTA GCC GTT GCT ACC CTC
                                          MET LYS LYS SER LEU VAL LEU LYS ALA SER VAL ALA VAL ALA THR LEU

1349  GTT CCG ATG CTG TCT TTC GCT GCT GAC GGT GAC GAT CCC GCA AAA GCG GCC TTT GAC TCC CTG CAA GCC TCA GCG   VIII
      VAL PRO MET LEU SER PHE ALA ALA GLU GLY ASP ASP PRO ALA LYS ALA ALA PHE ASP SER LEU GLN ALA SER ALA

1424  ACC GAA TAT ATC GGT TAT GCG TGG GCG ATG GTT GTT GTC ATT GTC GGC GCA ACT ATC GGT ATC AAG CTG TTT AAG
      THR GLU TYR ILE GLY TYR ALA TRY ALA MET VAL VAL VAL ILE VAL GLY ALA THR ILE GLY ILE LYS LEU PHE LYS

1499  AAA TTC ACC TCG AAA GCA AGC TGA TAA ACCGA TACAA TTAAA GGCTC CTTTT GGAGC CTTTT TTTTT GGAGA TTTTC AAC
      LYS PHE THR SER LYS ALA SER *** ***

1579      GTG AAA AAA TTA TTA TTC GCA ATT CCT TTA GTT GTT CCT TTC TAT TCT CAC TCC GCT GAA ACT GTT GAA AGT
          MET LYS LYS LEU LEU PHE ALA ILE PRO LEU VAL VAL PRO PHE TYR SER HIS SER ALA GLU THR VAL GLU SER

1651  TGT TTA GCA AAA CCT CAT ACA GAA AAT TCA TTT ACT AAC GTC TGG AAA GAC GAC AAA ACT TTA GAT CGT TAC GCT
      CYS LEU ALA LYS PRO HIS THR GLU ASN SER PHE THR ASN VAL TRY LYS ASP ASP LYS THR LEU ASP ARG TYR ALA

1726  AAC TAT GAG GGC TGT CTG TGG AAT GCT ACA GGC GTT GTG GTT TGT ACT GGT GAC GAA ACT CAG TGT TAC GGT ACA
      ASN TYR GLU GLY CYS LEU TRY ASN ALA THR GLY VAL VAL VAL CYS THR GLY ASP GLU THR GLN CYS TYR GLY THR

1801  TGG GTT CCT ATT GGG CTT GCT ATC CCT GAA AAT GAG GGT GGT GGC TCT GAG GGT GGC GGT TCT GAG GGT GGC GGT
      TRY VAL PRO ILE GLY LEU ALA ILE PRO GLU ASN GLU GLY GLY GLY SER GLU GLY GLY GLY SER GLU GLY GLY GLY

1876  TCT GAG GGT GGC GGT ACT AAA CCT CCT GAG TAC GGT GAT ACA CCT ATT CCG GGC TAT ACT TAT ATC AAC CCT CTC
      SER GLU GLY GLY GLY THR LYS PRO PRO GLU TYR GLY ASP THR PRO ILE PRO GLY TYR THR TYR ILE ASN PRO LEU
```

```
1951  GAC GGC ACT TAT CCG CCT GGT ACT GAG CAA AAC CCC GCT AAT CCT AAT CCT TCT CTT GAG GAG TCT CAG CCT CTT
      ASP GLY THR TYR PRO PRO GLY THR GLU GLN ASN PRO ALA ASN PRO ASN PRO SER LEU GLU GLU SER GLN PRO LEU

2026  AAT ACT TTC ATG TTT CAG AAT AAT AGG TTC CGA AAT AGG CAG GGT GCA TTA ACT GTT TAT ACG GGC ACT GTT ACT
      ASN THR PHE MET PHE GLN ASN ASN ARG PHE ARG ASN ARG GLN GLY ALA LEU THR VAL TYR THR GLY THR VAL THR

2101  CAA GGC ACT GAC CCC GTT AAA ACT TAT TAC CAG TAC ACT CCT GTA TCA TCA AAA GCC ATG TAT GAC GCT TAC TGG
      GLN GLY THR ASP PRO VAL LYS THR TYR TYR GLN TYR THR PRO VAL SER SER LYS ALA MET TYR ASP ALA TYR TRY   III

2176  AAC GGT AAA TTC AGA GAC TGC GCT TTC CAT TCT GGC TTT AAT GAG GAT CCA TTC GTT TGT GAA TAT CAA GGC CAA
      ASN GLY LYS PHE ARG ASP CYS ALA PHE HIS SER GLY PHE ASN GLU ASP PRO PHE VAL CYS GLU TYR GLN GLY GLN

2251  TCG TCT GAC CTG CCT CAA CCT CCT GTC AAT GCT GGC GGC GGC TCT GGT GGT GGT TCT GGT GGC GGC TCT GAG GGT
      SER SER ASP LEU PRO GLN PRO PRO VAL ASN ALA GLY GLY GLY SER GLY GLY GLY SER GLY GLY GLY SER GLU GLY

2326  GGC GGC TCT GAG GGT GGC GGT TCT GAG GGT GGC GGC TCT GAG GGT GGC GGT TCC GGT GAT TTT GAT TAT GAA AAA
      GLY GLY SER GLU GLY GLY GLY SER GLU GLY GLY GLY SER GLU GLY GLY GLY SER GLY ASP PHE ASP TYR GLU LYS

2401  ATG GCA AAC GCT AAT AAG GGG GCT ATG ACC GAA AAT GCC GAT GAA AAC GCG CTA CAG TCT GAC GCT AAA GGC AAA
      MET ALA ASN ALA ASN LYS GLY ALA MET THR GLU ASN ALA ASP GLU ASN ALA LEU GLN SER ASP ALA LYS GLY LYS

2476  CTT GAT TCT GTC GCT ACT GAT TAC GGT GCT GCT ATC GAT GGT TTC ATT GGT GAC GTT TCC GGC CTT GCT AAT GGT
      LEU ASP SER VAL ALA THR ASP TYR GLY ALA ALA ILE ASP GLY PHE ILE GLY ASP VAL SER GLY LEU ALA ASN GLY

2551  AAT GGT GCT ACT GGT GAT TTT GCT GGC TCT AAT TCC CAA ATG GCT CAA GTC GGT GAC GGT GAT AAT TCA CCT TTA
      ASN GLY ALA THR GLY ASP PHE ALA GLY SER ASN SER GLN MET ALA GLN VAL GLY ASP GLY ASP ASN SER PRO LEU

2626  ATG AAT AAT TTC CGT CAA TAT TTA CCT TCT TTG CCT CAG TCG GTT GAA TGT CGC CCT TAT GTC TTT AGC GCT GGT
      MET ASN ASN PHE ARG GLN TYR LEU PRO SER LEU PRO GLN SER VAL GLU CYS ARG PRO TYR VAL PHE SER ALA GLY

2701  AAA CCA TAT GAA TTT TCT ATT GAT TGT GAC AAA ATA AAC TTA TTC CGT GGT GTC TTT GCG TTT CTT TTA TAT GTT
      LYS PRO TYR GLU PHE SER ILE ASP CYS ASP LYS ILE ASN LEU PHE ARG GLY VAL PHE ALA PHE LEU LEU TYR VAL

2776  GCC ACC TTT ATG TAT GTA TTT TCG ACG TTT GCT AAC ATA CTG CGT AAT AAG GAG TCT TAA TC ATG CCA GTT CTT
      ALA THR PHE MET TYR VAL PHE SER THR PHE ALA ASN ILE LEU ARG ASN LYS GLU SER ***    MET PRO VAL LEU

2850  TTG GGT ATT CCG TTA TTA TTG CGT TTC CTC GGT TTC CTT CTG GTA ACT TTG TTC GGC TAT CTG CTT ACT TTC CTT
      LEU GLY ILE PRO LEU LEU LEU ARG PHE LEU GLY PHE LEU LEU VAL THR LEU PHE GLY TYR LEU LEU THR PHE LEU
```

```
2925  AAA AAG GGC TTC GGT AAG ATA GCT ATT GCT ATT TCA TTG TTT CTT GCT CTT ATT ATT GGG CTT AAC TCA ATT CTT
      LYS LYS GLY PHE GLY LYS ILE ALA ILE ALA ILE SER LEU PHE LEU ALA LEU ILE ILE GLY LEU ASN SER ILE LEU   VI

3000  GTG GGT TAT CTC TCT GAT ATT AGC GCA CAA TTA CCC TCT GAT TTT GTT CAG GGC GTT CAG TTA ATT CTC CCG TCT
      VAL GLY TYR LEU SER ASP ILE SER ALA GLN LEU PRO SER ASP PHE VAL GLN GLY VAL GLN LEU ILE LEU PRO SER

3075  AAT GCG CTT CCC TGT TTT TAT GTT ATT CTC TCT GTA AAG GCT GCT ATT TTC ATT TTT GAC GTT AAA CAA AAA ATC
      ASN ALA LEU PRO CYS PHE TYR VAL ILE LEU SER VAL LYS ALA ALA ILE PHE ILE PHE ASP VAL LYS GLN LYS ILE

3150  GTT TCT TAT TTG GAT TGG GAT AAA TAA  AT ATG GCT GTT TAT TTT GTA ACT GGC AAA TTA GGC TCT GGA AAG ACG
      VAL SER TYR LEU ASP TRY ASP LYS ***     MET ALA VAL TYR PHE VAL THR GLY LYS LEU GLY SER GLY LYS THR

3224  CTC GTT AGC GTT GGT AAG ATT CAG GAT AAA ATT GTA GCT GGG TGC AAA ATA GCA ACT AAT CTT GAT TTA AGG CTT
      LEU VAL SER VAL GLY LYS ILE GLN ASP LYS ILE VAL ALA GLY CYS LYS ILE ALA THR ASN LEU ASP LEU ARG LEU

3299  CAA AAC CTC CCG CAA GTC GGG AGG TTC GCT AAA ACG CCT CGC GTT CTT AGA ATA CCG GAT AAG CCT TCT ATT TCT
      GLN ASN LEU PRO GLN VAL GLY ARG PHE ALA LYS THR PRO ARG VAL LEU ARG ILE PRO ASP LYS PRO SER ILE SER

3374  GAT TTG CTT GCT ATT GGT CGT GGT AAT GAT TCC TAC GAC GAA AAT AAA AAC GGT TTG CTT GTT CTT GAT GAA TGC
      ASP LEU LEU ALA ILE GLY ARG GLY ASN ASP SER TYR ASP GLU ASN LYS ASN GLY LEU LEU VAL LEU ASP GLU CYS

3449  GGT ACT TGG TTT AAT ACC CGT TCA TGG AAT GAC AAG GAA AGA CAG CCG ATT ATT GAT TGG TTT CTT CAT GCT CGT
      GLY THR TRY PHE ASN THR ARG SER TRY ASN ASP LYS GLU ARG GLN PRO ILE ILE ASP TRY PHE LEU HIS ALA ARG

3524  AAA TTG GGA TGG GAT ATT ATT TTT CTT GTT CAG GAT TTA TCT ATT GTT GAT AAA CAG GCG CGT TCT GCA TTA GCT
      LYS LEU GLY TRY ASP ILE ILE PHE LEU VAL GLN ASP LEU SER ILE VAL ASP LYS GLN ALA ARG SER ALA LEU ALA

3599  GAA CAC GTT GTT TAT TGT CGC CGT CTG GAC AGA ATT ACT TTA CCC TTT GTC GGC ACT TTA TAT TCT CTT GTT ACT
      GLU HIS VAL VAL TYR CYS ARG ARG LEU ASP ARG ILE THR LEU PRO PHE VAL GLY THR LEU TYR SER LEU VAL THR   I

3674  GGC TCA AAA ATG CCT CTG CCT AAA TTA CAT GTT GGT GTT GTT AAA TAT GGT GAT TCA CAA TTA AGC CCT ACT GTT
      GLY SER LYS MET PRO LEU PRO LYS LEU HIS VAL GLY VAL VAL LYS TYR GLY ASP SER GLN LEU SER PRO THR VAL

3749  GAG CGT TGG CTT TAT ACT GGT AAG AAT TTA TAT AAC GCA TAT GAC ACT AAA CAG GCT TTT TCC AGT AAT TAT GAT
      GLU ARG TRY LEU TYR THR GLY LYS ASN LEU TYR ASN ALA TYR ASP THR LYS GLN ALA PHE SER SER ASN TYR ASP

3824  TCA GGT GTT TAT TCA TAT TTA ACC CCT TAT TTA TCA CAC GGT CGG TAT TTC AAA CCA TTA AAT TTA GGT CAG AAG
      SER GLY VAL TYR SER TYR LEU THR PRO TYR LEU SER HIS GLY ARG TYR PHE LYS PRO LEU ASN LEU GLY GLN LYS
```

```
3899  ATG AAA TTA ACT AAA ATA TAT TTG AAA AAG TTT TCT CGC GTT CTT TGT CTT GCG ATA GGA TTT GCA TCA GCA TTT
      MET LYS LEU THR LYS ILE TYR LEU LYS LYS PHE SER ARG VAL LEU CYS LEU ALA ILE GLY PHE ALA SER ALA PHE

3974  ACA TAT AGT TAT ATA ACC CAA CCT AAG CCG GAG GTT AAA AAG GTA GTC TCT CAG ACC TAT GAT TTT GAT AAA TTC
      THR TYR SER TYR ILE THR GLN PRO LYS PRO GLU VAL LYS LYS VAL VAL SER GLN THR TYR ASP PHE ASP LYS PHE

4049  ACT ATT GAC TCT TCT CAG CGT CTT AAT CTA AGC TAT CGC TAT GTT TTC AAG GAT TCT AAG GGA AAA TTA ATT AAT
      THR ILE ASP SER SER GLN ARG LEU ASN LEU SER TYR ARG TYR VAL PHE LYS ASP SER LYS GLY LYS LEU ILE ASN

4124  AGC GAC GAT TTA CAG AAG CAA GGT TAT TCC ATC ACA TAT ATT GAT TTA TGT ACT GTT TCA ATT AAA AAA GGT AAT
      SER ASP ASP LEU GLN LYS GLN GLY TYR SER ILE THR TYR ILE ASP LEU CYS THR VAL SER ILE LYS LYS GLY ASN

4199  TCA AAT GAA ATT GTT AAA TGT AAT TAA T TTT GTT TTC TTG ATG TTT GTT TCA TCA TCT TCT TTT GCT CAA GTA ATT
      SER ASN GLU ILE VAL LYS CYS ASN ***
          MET LYS LEU LEU ASN VAL ILE ASN  PHE VAL PHE LEU MET PHE VAL SER SER SER SER PHE ALA GLN VAL ILE

4275  GAA ATG AAT AAT TCG TCT CTG CGC GAT TTC GTG ACT TGG TAT TCA AAG CAA ACA GGT GAA TCT GTT ATT GTC TCA
      GLU MET ASN ASN SER SER LEU ARG ASP PHE VAL THR TRY TYR SER LYS GLN THR GLY GLU SER VAL ILE VAL SER

4350  CCT GAT GTT AAA GGT ACA GTG ACT GTA TAT TCC TCT GAC GTT AAG CCT GAA AAT TTA CGC AAT TTC TTT ATC TCT
      PRO ASP VAL LYS GLY THR VAL THR VAL TYR SER SER ASP VAL LYS PRO GLU ASN LEU ARG ASN PHE PHE ILE SER

4425  GTT TTA CGT GCT AAT AAT TTT GAT ATG GTT GGC TCA ATT CCT TCC ATA ATT CAG AAA TAT AAC CCA AAT AGT CAG
      VAL LEU ARG ALA ASN ASN PHE ASP MET VAL GLY SER ILE PRO SER ILE ILE GLN LYS TYR ASN PRO ASN SER GLN

4500  GAT TAT ATT GAT GAA TTG CCA TCA TCT GAT ATT CAG GAA TAT GAT GAT AAT TCC GCT CCT TCT GGT GGT TTC TTT
      ASP TYR ILE ASP GLU LEU PRO SER SER ASP ILE GLN GLU TYR ASP ASP ASN SER ALA PRO SER GLY GLY PHE PHE

4575  GTT CCG CAA AAT GAT AAT GTT ACT CAA ACA TTT AAA ATT AAT AAC GTT CGC GCA AAG GAT TTA ATA AGG GTT GTA
      VAL PRO GLN ASN ASP ASN VAL THR GLN THR PHE LYS ILE ASN ASN VAL ARG ALA LYS ASP LEU ILE ARG VAL VAL

4650  GAA TTG TTT GTT AAA TCT AAT ACA TCT AAA TCC TCA AAT GTA TTA TCT GTT GAT GGT TCT AAC TTA TTA GTA GTT
      GLU LEU PHE VAL LYS SER ASN THR SER LYS SER SER ASN VAL LEU SER VAL ASP GLY SER ASN LEU LEU VAL VAL

4725  AGC GCC CCT AAA GAT ATT TTA GAT AAC CTT CCG CAA TTT CTT TCT ACT GTT GAT TTG CCA ACT GAC CAG ATA TTG
      SER ALA PRO LYS ASP ILE LEU ASP ASN LEU PRO GLN PHE LEU SER THR VAL ASP LEU PRO THR ASP GLN ILE LEU

4800  ATT GAA GGA TTA ATT TTC GAG GTT CAG CAA GGT GAT GCT TTA GAT TTT TCC TTT GCT GCT GGC TCT CAG CGC GGC
      ILE GLU GLY LEU ILE PHE GLU VAL GLN GLN GLY ASP ALA LEU ASP PHE SER PHE ALA ALA GLY SER GLN ARG GLY   IV
```

```
4875  ACT GTT GCT GGT GGT GTT AAT ACT GAC CGT CTA ACC TCT GTT TTA TCT TCT GCG GGT GGT TCG TTC GGT ATT TTT
      THR VAL ALA GLY GLY VAL ASN THR ASP ARG LEU THR SER VAL LEU SER SER ALA GLY GLY SER PHE GLY ILE PHE

4950  AAC GGC GAT GTT TTA GGG CTA TCA GTT CGC GCA TTA AAG ACT AAT AGC CAT TCA AAA ATA TTG TCT GTG CCT CGT
      ASN GLY ASP VAL LEU GLY LEU SER VAL ARG ALA LEU LYS THR ASN SER HIS SER LYS ILE LEU SER VAL PRO ARG

5025  ATT CTT ACG CTT TCA GGT CAG AAG GGT TCT ATT TCT GTT GGC CAG AAT GTC CCT TTT ATT ACT GGT CGT GTA ACT
      ILE LEU THR LEU SER GLY GLN LYS GLY SER ILE SER VAL GLY GLN ASN VAL PRO PHE ILE THR GLY ARG VAL THR

5100  GGT GAA TCT GCC AAT GTA AAT AAT CCA TTT CAG ACG GTT GAG CGT CAA AAT GTT GGT ATT TCT ATG AGT GTT TTT
      GLY GLU SER ALA ASN VAL ASN ASN PRO PHE GLN THR VAL GLU ARG GLN ASN VAL GLY ILE SER MET SER VAL PHE

5175  CCC GTT GCA ATG GCT GGC GGT AAT ATT GTT TTA GAT ATA ACC AGT AAG GCC GAT AGT TTG AGT TCT TCT ACT CAG
      PRO VAL ALA MET ALA GLY GLY ASN ILE VAL LEU ASP ILE THR SER LYS ALA ASP SER LEU SER SER SER THR GLN

5250  GCA AGT GAT GTT ATT ACT AAT CAA AGA AGT ATT GCG ACA ACG GTT AAT TTG CGT GAT GGT CGG ACT CTT TTG CTC
      ALA SER ASP VAL ILE THR ASN GLN ARG SER ILE ALA THR THR VAL ASN LEU ARG ASP GLY ARG THR LEU LEU LEU

5325  GGA GGC CTC ACT GAT TAC AAA AAC ACT TCT CAA GAT TCT GGT GTG CCG TTC CTG TCT AAA ATC CCT TTA ATC GGC
      GLY GLY LEU THR ASP TYR LYS ASN THR SER GLN ASP SER GLY VAL PRO PHE LEU SER LYS ILE PRO LEU ILE GLY

5400  CTC CTG TTT AGC TCC CGT TCT GAT TCT AAC GAG GAA AGC ACG TTG TAC GTG CTC GTC AAA GCA ACC ATA GTA CGC
      LEU LEU PHE SER SER ARG SER ASP SER ASN GLU GLU SER THR LEU TYR VAL LEU VAL LYS ALA THR ILE VAL ARG

5475  GCC CTG TAG CGG CGC ATT AAG CGC GGC GGG TGT GGT GGT TAC GCG CGC GTG ACC GCT ACA CTT GCC AGC GCC CTA
      ALA LEU ***

5550  GCG CCC GCT CCT TTC GCT TTC TTC CCT TCC TTT CTC GCC ACG TTC TCC GGC TTT CCC CGT CAA GCT CTA AAT CGG

5625  GGG ATC CCT TTA GGG TTC CGA TTT AGT GCT TTA CGG CAC CTC GAC CTC CAA AAA CTT GAT TTG GGT GAT GGT TCA

5700  CGT AGT GGG CCA TCG CCC TGA TAG ACG GTT TTT CGC CCT TTG ACG TTG GAG TCC ACG TTC TTT AAT AGT GGA CTC

5775  TTG TTC CAA ACT GGA ACA ACA CTC ACA ACT AAC TCG GCC TAT TCT TTT GAT TTA TAA GGA TTT TTG TCA TTT TCT
```

```
5850  GCT TAC TGG TTA AAA AAT AAG CTG ATT TAA CAA ATA TTT AAC GCG AAA TTT AAC AAA ACA TTA ACG TTT ACA ATT

5925  TAA ATA TTT GCT TAT ACA ATC ATC CTG TTT TTG GGG CTT TTC TGA TTA TCA ACC GGG GTA CAT ATG ATT GAC ATG
                                                                                          MET ILE ASP MET

6000  CTA GTT TTA CGA TTA CCG TTC ATC GAA TCT CTT GTT TGC TCC AGA CTT TCA GGT AAT GAC CTG ATA GCC TTT GTA
      LEU VAL LEU ARG LEU PRO PHE ILE GLU SER LEU VAL CYS SER ARG LEU SER GLY ASN ASP LEU ILE ALA PHE VAL

6075  GAC CTC TCA AAA ATA GCT ACC CTC TCC GGC ATG AAT TTA TCA GCT AGA ACG GTT GAA TAT CAT ATT GAC GGT GAT
      ASP LEU SER LYS ILE ALA THR LEU SER GLY MET ASN LEU SER ALA ARG THR VAL GLU TYR HIS ILE ASP GLY ASP

6150  TTG ACT GTC TCC GGC CTT TCT CAC CCG TTT GAA TCT TTG CCT ACT CAT TAC TCC GGC ATT GCA TTT AAA ATA TAT   II
      LEU THR VAL SER GLY LEU SER HIS PRO PHE GLU SER LEU PRO THR HIS TYR SER GLY ILE ALA PHE LYS ILE TYR

6225  GAG GGT TCT AAA AAT TTT TAT CCC TGC GTT GAA ATT AAG GCT TCA CCA GCA AAA GTA TTA CAG GGT CAT AAT GTT
      GLU GLY SER LYS ASN PHE TYR PRO CYS VAL GLU ILE LYS ALA SER PRO ALA LYS VAL LEU GLN GLY HIS ASN VAL

6300  TTT GGT ACA ACC GAT TTA GCT TTA TGC TCT GAG GCT TTA TTG CTT AAT TTT GCT AAC TCT CTG CCT TGC TTG TAC
      PHE GLY THR THR ASP LEU ALA LEU CYS SER GLU ALA LEU LEU LEU ASN PHE ALA ASN SER LEU PRO CYS LEU TYR

6375  GAT TTA TTG GAT GTT
      ASP LEU LEU ASP VAL
```

Figure 1 A nucleotide sequence for the DNA of bacteriophage fd. The genes are shown boxed, and the amino acid sequences of gene products are included.

Note added in proof: The fd sequence in this figure requires four minor corrections: Nucleotide 2692 in gene III is A, which changes the Ser codon AGC to the Gly codon GGC. Nucleotide 4290 in gene IV is C, which changes the Ser codon TCT to the Pro codon CCT. Nucleotide 6026 in gene II is T, which changes the Glu codon GAA to the Asp codon GAT. An additional A residue is present in the intergenic region between the C in position 5520 and the G in position 5521. This changes the size of fd DNA to 6390 nucleotides. Recent work has mapped the origin of viral-strand replication as a nick produced by gene-II protein between nucleotides 5762 (T) and 5763 (A). This may be a better reference point for nucleotide positions in future work.

expression (Takanami and Okamoto 1973; Seeburg and Schaller 1975) and in DNA replication (Tabak et al. 1974; Schaller et al. 1976; Horiuchi and Zinder 1976). The genetic map was aligned precisely with the physical map by the marker-rescue technique (Seeburg and Schaller 1975; Horiuchi et al. 1975; van den Hondel et al. 1975). A strong promoter located in a region preceding gene X—P(X)—was the first to be sequenced (Schaller et al. 1975; Sugimoto et al. 1975a,b). Subsequently, the sequences of other promoters were deduced (Takanami et al. 1976; E. Beck et al. as quoted in Seeburg et al. 1977). Takanami et al. (1976) focused mainly on the determination of RNA sequences of in vivo transcripts from both strands of restriction fragments. Sugimoto et al. (1977) extended this study to obtain the sequence covering genes VII and VIII and a strong terminator of transcription immediately after gene VIII (positions 1096–1565, Fig. 1).

At this point, two new DNA sequencing methods were developed: the enzymatic "plus and minus" method of Sanger and Coulson (1975) and the chemical base-specific degradation method of Maxam and Gilbert (1977). By means of these new techniques Gray et al. (1978) have completed the sequence in the replication origin and have extended the sequenced regions to include the entire fd genome (E. Beck et al., in prep.). K. Sugimoto et al. (in prep.) have also sequenced the region covering genes X–VIII.

Nucleotide Sequence and Cleavage Map of fd DNA

The sequencing methods used required specific cleavages of the fd genome with restriction endonucleases at positions less than 250 nucleotides apart. The main enzymes employed for this purpose and their cleavage maps are given in Figure 2. They provide more than 100 reference sites throughout the fd DNA molecule for the sequencing and ordering of the various restriction fragments. For certain DNA segments, the enzyme *Mnl*I, which cleaves double-stranded fd DNA at about 60 sites, was used also. A cleavage map for this enzyme deduced from the nucleotide sequence given in Figure 1 has also been largely confirmed experimentally.

The results of the combined sequence work to date are shown in Figure 1. A sequence of 6389 nucleotides is presented in a linear fashion starting from the single *Hin*dII cleavage site as a reference point on the circular DNA molecule. Most of the sequence was obtained by partial base-specific chemical degradation of terminal-labeled restriction fragments (Maxam and Gilbert 1977). Overlaps were provided by sequencing from nearby restriction cuts produced by different enzymes or by analysis of DNA fragments that had been elongated by repair synthesis with DNA polymerase I (Gray et al. 1978). Additional sequence information was obtained by means of the "plus and minus" reaction of Sanger and Coulson (1975; positions 5500–6200), by RNA sequencing (positions 500–2300; see, e.g., Takanami et al. 1976), and by oligopyrimidine analysis (Ling 1972; Oertel and Schaller 1972; C. P.

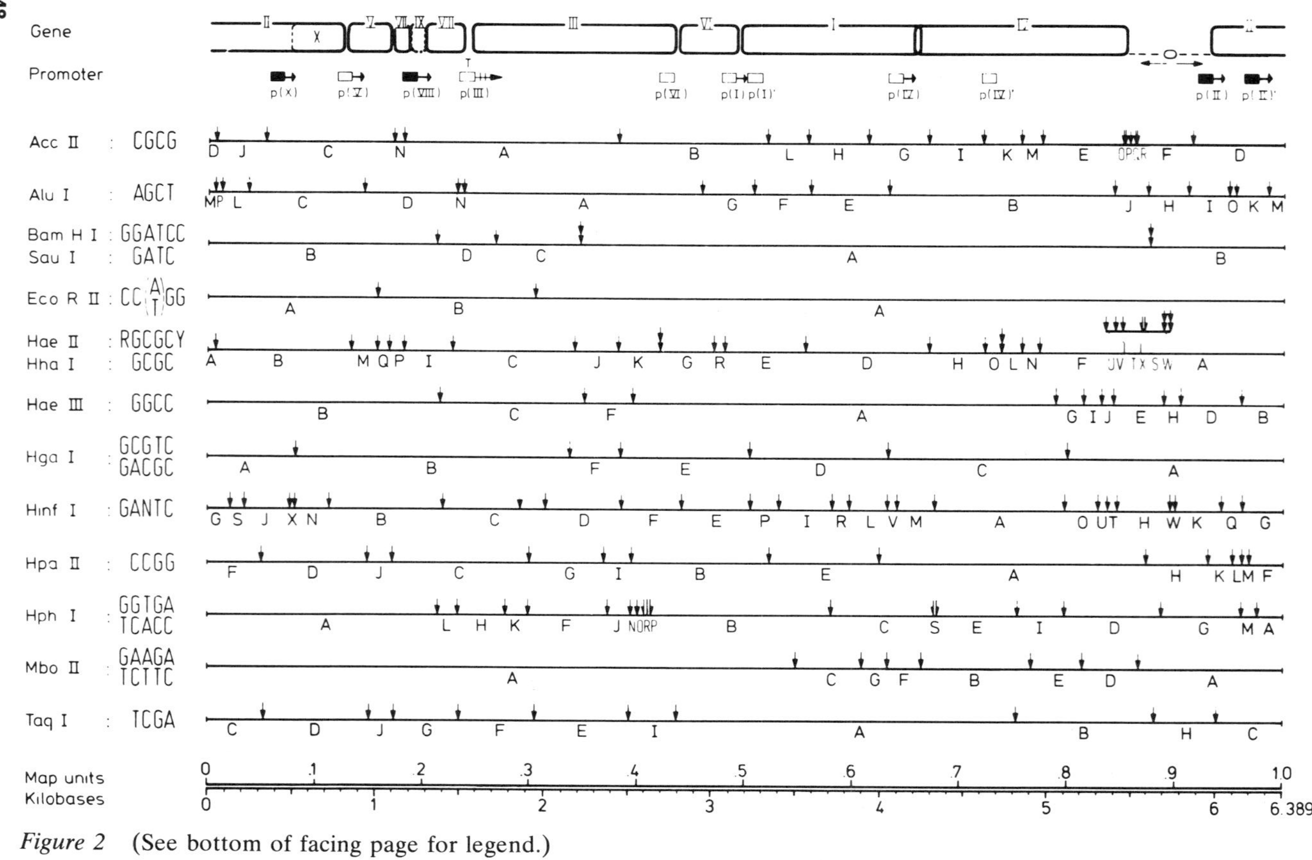

Figure 2 (See bottom of facing page for legend.)

Gray et al. 1978 and unpubl.). In addition, about 10% of the sequence is confirmed by the cleavage maps of the restriction enzymes mentioned above.

Most sections of the fd DNA sequence were determined independently from both DNA strands of RF DNA. In addition, the sequences of substantial sections were obtained both in Kyoto and in Heidelberg and more than 90% of the fd sequence has been compared with those of the related phages f1 and M13. Thus, the sequence presented in Figure 1, although probably not completely free from error, is precise enough to allow a detailed analysis of the filamentous phage genome on the nucleotide level. That the fd genome is 6389 nucleotides in size is considered to be correct within ±5 nucleotides and agrees very well with the value of 6370 ± 140 nucleotides determined with physicochemical techniques (Day and Berkowitz 1977).

Gene Structure

Genetic analysis has revealed that the filamentous phage genome consists of at least eight genes. A detailed biochemical characterization of most of the corresponding proteins has not been possible because not all of them are synthesized in large amounts in the infected cell and most cannot be obtained from the virus particle. Only the products of gene V, a DNA-binding protein, and gene VIII, the major capsid protein, have been characterized by their amino acid sequences (Asbeck et al. 1969; Nakashima and Konigsberg 1974). Approximate molecular weights are known for the products of genes I, II, III, and IV (see Table 1), whereas the existence of genes VI and VII has been deduced from genetic analysis only. A ninth small gene may have escaped detection by genetic analysis (see below).

The order of the genes and their approximate positions on the physical map of the filamentous phage genome are well established (Lyons and Zinder 1972; Seeburg and Schaller 1975; Horiuchi et al. 1975; van den Hondel et al. 1975). Analysis of the corresponding nucleotide sequences in the three possible translational reading frames for proteins of the expected sizes has allowed the identification of all known fd genes and the prediction of the start and stop sites and the amino acid sequence for each gene product.

Figure 2 Cleavage map of restriction enzymes used in the sequence analysis of fd DNA. Maps for *Hpa*II, *Hga*I, *Hae*II (*Hin*HI), *Hae*III, *Alu*I (Seeburg and Schaller 1975; Takanami and Okamoto 1975; van den Hondel et al. 1975) were confirmed and refined. Maps for *Hha*I, *Hin*fI, *Taq*I, *Bam*HI, *Sau*I (*Dpn*I, *Mbo*I), *Eco*RII, *Mbo*II, *Hph*I, and *Acc*II were newly established (E. A. Auerswald et al; M. Takanami et al.; both unpubl.). The circular phage DNA is opened at the single *Hin*dII (*Hpa*I) cleavage site. The map includes the positions of the phage genes, of promoters, and of the origin of DNA replication (o).

Table 1 Coding capacity of fd DNA

Gene	Nucleotides coding	Nucleotides noncoding	Amino acids	Protein molecular weight from DNA	Protein molecular weight from protein
II	1230		410	46,243	40,000–46,000
		14			
X	333		111	12,682	12,000
		4			
V	261		87	9690	9688
		94			
VII	99		33	3603	—
		59			
VIII	219		73	7627	5800
	(150)		(50)	(5240)	(5196)
		5			
III	1254		418	44,179	59,000
	(1200)		(400)	(42,138)	(56,000–68,000)
		5			
VI	336		112	12,352	—
		−20[a]			
I	1044		348	39,544	35,000
IV	1278		426	45,858	48,000–50,000
IG		507			

The values given in parentheses refer to the processed major and minor capsid proteins as isolated from the virus. IG is the intergenic region.
[a] The "−20 noncoding" nucleotides between genes I and IV refers to a 20-nucleotide region that is used in two different reading frames by both genes (gene overlap).

The results of this analysis are included in Figures 1 and 3 and are summarized in Table 1. In the following they will be discussed and compared briefly with the data available for the gene products obtained both in vivo and by in vitro protein synthesis (for a recent review, see Ray 1977). Although only 6% of our sequence is supported by direct amino acid sequencing of gene products (genes V and VIII), nucleotide sequences from the closely related filamentous phages f1 (E. Beck, unpubl.) and M13 (J. Schoenmakers, pers. comm.) have revealed 10 amber mutations and about 100 silent base changes within the genes. These have allowed us to determine the reading frames of the gene products with a high degree of probability.

Gene III

The only continuous translational reading frame in the gene-III region extends from nucleotide 1579 to nucleotide 2832. It starts with a GTG triplet and codes for a protein of 418 amino acids. Its amino acid composition (Table 2) is very similar to that determined for the minor coat protein (the A protein) of the fd virion (Goldsmith and Konigsberg 1977). The N-terminal amino acid sequence Ala-Glu-Thr-Val-Glu-Ser (Goldsmith and Konigsberg 1977), determined for the protein isolated from the virus, agrees with the nucleotide sequence obtained for positions 1633–1650 (amino acids 19–24). This suggests that the minor viral coat protein is formed from a precursor containing 18 additional amino acids (mainly hydrophobic) at its N terminus. The molecular weight calculated for the mature gene-III protein from the DNA sequence is 44,300 daltons of protein, which is substantially below the values of 55,000–68,000 daltons observed in SDS gels. This discrepancy, as well as the variations in molecular

```
16s RNA     3'HO AUUCCUCCACUAG--
            5'   TAAGGAGGTGATC-- ATG R-- --- ---

                                5988
GENE II          CAAACCGGGGTACAT ATG ATT GAC ATG
                                 Met Ile Asp Met

                                496
GENE X           TTTGAGGGGGATTCA ATG AAT ATT TAT
                                 Met Asn Ile Tyr

                                843
GENE V           ATAAGGTAATTCAAA ATG ATT AAA GTT
                                 Met Ile Lys Val

                                1108
GENE VII         TTCCGGCTAAGTAAC ATG GAG CAG GTC
                                 Met Glu Gln Val

                                1301
GENE VIII        AATGGAAACTTCCTC ATG AAA AAG TCT
                                 Met Lys Lys Ser

                                1579
GENE III         TTGGAGATTTTCAAC GTG AAA AAA TTA
                                 Met Lys Lys Leu

                                2838
GENE VI          TAAGGAGTCTTAATC ATG CCA GTT GCC
                                 Met Pro Val Leu

                                3179
GENE I           TTGGGATAAATAAAT ATG GCT GTT TAT
                                 Met Ala Val Tyr

                                4203
GENE IV          AAAAAGGTAATTCAA ATG AAA TTG TTA
                                 Met Lys Leu Leu

                                1206
GENE IX          CGCTGGGGGTCAAAG ATG AGT GTT TTA
                                 Met Ser Val Leu
```

Figure 3 fd DNA sequences coding for the initiation of gene products. The sequences for gene V, gene VIII, and gene IV have been isolated as ribosome-binding sites from f1-directed mRNA (Pieczenik et al. 1974). Nucleotides complementary to the 3′ terminus of 16S rRNA (Shine and Dalgarno 1974) are underlined.

Table 2 Amino acid composition of gene-III product in bacteriophage fd

	No. of residues		Mole·%	
Amino acid	precursor	mature	calculated	determined[a]
Ala	29	28	7.0	6.8
Arg	9	9	2.2	2.3
Asn	29	29	7.2	13.2
Asp	24	24	6.0	
Cys	8	8	2.0	2.1
Gln	15	15	3.7	10.4
Glu	26	26	6.5	
Gly	63	63	16.0	16.0
His	3	2	0.5	0.5
Ile	10	9	2.2	2.2
Leu	22	19	4.7	4.7
Lys	16	14	3.5	3.5
Met	7	7	1.7	1.8
Phe	22	20	5.0	4.7
Pro	27	25	6.3	6.4
Ser	33	31	7.5	7.6
Thr	26	26	6.5	6.7
Try	4	4	1.0	0.8
Tyr	22	21	5.2	5.1
Val	23	20	5.0	5.0

[a] Data from Goldsmith and Konigsberg (1977).

weights observed in different gel systems (Goldsmith and Konigsberg 1977), may be due to the unusual clustering of glycine residues encoded by a quindecanucleotide sequence which is repeated several times around position 2350. Part of this repeat sequence is deleted in an fd variant and therefore appears to be nonessential for gene-III function (E. A. Auerswald, unpubl.).

Gene VI

Gene-VI mutations have been mapped in restriction fragments around map position 3000. A protein covering this region starts at an ATG triplet at position 2838, six nucleotides beyond the C terminus of gene III, and terminates at map position 3173, six nucleotides preceding the presumed start of gene I. Its reading frame has been confirmed by the nucleotide sequences of gene-VI amber mutants from M13 and f1 which show a C → T transition at position 3048 (J. Schoenmakers, pers. comm.; E. Beck, unpubl.). The hypothetical gene-VI protein of 112 amino acids (Fig. 1) shows no apparent correlation to proteins that have been considered previously as possible products of gene VI (L. L. Bertsch et al. as cited in Kornberg 1974; K. Hodgson et al., pers. comm.).

Gene I

The product encoded by this gene has been identified as a protein of about 35,000 daltons. A nucleotide sequence coding for a protein of 348 amino acids starts at position 3179 with an ATG triplet and extends 20 nucleotides into the sequence coding for the N-terminal end of gene IV. The molecular weight predicted for the gene-I product is about 40,000 daltons

Gene IV

This gene appears to span 1278 nucleotides (426 amino acids). Its N terminus is well defined by the sequence homology of nucleotides 4187–4210 with a ribosome-binding site from f1 DNA-directed mRNA (Pieczenik et al. 1974). The reading frame used near the C-terminal end of the protein is confirmed by the nucleotide sequence of the f1 amber mutation R143, which has been located at position 5247 (E. Beck, unpubl.). The molecular weight estimated from the DNA sequence (45,800 daltons) is close to that observed experimentally (48,000 daltons).

Intergenic Region

Combined genetic and biochemical mapping of f1 mutations in genes II and IV suggested that these genes were separated by a noncoding region of several hundred nucleotides (Vovis et al. 1975; van den Hondel et al. 1975). In addition, it has been proved that the nucleotide sequences around positions 5547, 5626, and 5850 do not code for an essential gene product, as they can be interpreted by an insertion of additional DNA into the filamentous phage genome (Messing et al. 1977; Herrmann et al. 1978; Ray and Kook; Vovis and Ohsumi; both this volume).

According to our fd-sequence analysis (R. Sommer et al., in prep.) a large intergenic space is located between nucleotides 5480 and 5987. Protein synthesis is excluded in most parts of this sequence by the presence of stop codons in all three translational reading frames, except for nucleotides 5691–5739 which may code for a hypothetical peptide of 16 amino acids. However, this segment is part of a highly base-paired hairpin structure at the origin of DNA replication (see below) and as such it is not likely to allow initiation of protein synthesis in mRNA.

Gene II

This protein appears to start at either one of the closely spaced ATG triplets at positions 5988 and 5997. It is more likely that the former is the initiator codon since it is preceded by a sequence characteristic of a ribosome-binding site. This prediction has been confirmed recently by a sequence analysis of the N terminus of the gene-II protein (T. Meyer, pers. comm.). The succeeding nucleotide sequence codes for a protein of 410 amino acids with a calculated molecular weight of 46,260 daltons, which is similar to that observed experimentally (42,000–48,000 daltons).

Gene X

Gene II contains an internal translational start signal which controls in vitro the synthesis of a protein of about 12,000 daltons (Model and Zinder 1974; Konings et al. 1975), presumably a C-terminal fragment of gene II. A possible ribosome-binding site exists in the nucleotide sequence around position 496. The sequence predicts a protein of 111 amino acids (12,680 daltons) starting with fMet-Asn-Ile-Tyr.

Gene V

The nucleotide sequence from position 843 to position 1103 predicts the amino acid sequence known for gene-V protein from phage fd (Nakashima and Konigsberg 1974) and confirms the sequence for the ribosome-binding site (Pieczenik et al. 1974).

Gene VII

On the basis of the sequence information available for this region (Takanami et al. 1976) and by analysis of two amber mutations occurring in positions 1114 and 1141 (T. Hulsebos and J. Schoenmakers, unpubl.), the product of gene VII is assumed to be a peptide of 33 amino acids which is encoded in the base sequence between position 1108 and 1206.

Gene VIII

From a comparison of the nucleotide sequence in the gene-VIII region with the information on the amino acid sequence of the major capsid protein (Nakashima and Konigsberg 1974) and on a translation start site (Pieczenik et al. 1974), one may deduce that the major coat protein is formed from a precursor containing 23 extra amino acid residues at the N terminus (Sugimoto et al. 1977). The structural gene starts at position 1301, 95 nucleotides beyond the presumed end of gene VII. It terminates at position 1519, 59 nucleotides ahead of the beginning of gene III.

Gene IX

Genes VII and VIII are separated by an intergenic space of 94 nucleotides (1207–1300). This region seems to code for a noncoding "leader sequence" of the gene-VIII mRNA (Sugimoto et al. 1977). Alternatively, and perhaps more likely, it may code for an additional phage gene product (indicated as gene IX in Fig. 1). This is suggested by the fact that the region in question can be read from the ATG triplet in position 1206 in a continuous reading frame to yield a protein of 32 amino acids which overlaps the adjoining genes on either side by one nucleotide only. Furthermore, this sequence is preceded by a purine-rich sequence (1192–1202) with the features of a ribosome-binding site (Shine and Dalgarno 1974).The amino acid composition of the putative gene-IX protein (1 arginine, 6 serine, no histidine) is similar to that of a protein that has been isolated as a minor component from highly purified f1 phage (K. Hodgson et al., pers. comm.).

Coding Capacity and Genome Organization

According to our present analysis, about 89% of the filamentous phage genome codes for protein products (Table 1). Most genes are separated by only a few untranslated nucleotides. The extended intergenic spaces between genes VIII and III and between genes IV and II contain the central terminator of transcription and the region that controls DNA replication, respectively (see below). The latter region may also contain a signal for oriented packaging of the viral DNA during phage morphogenesis (J. E. Hearst, pers. comm.). As discussed above, the functional significance of an intergenic space of 94 nucleotides between genes VII and VIII is not understood.

Except for a short stretch of nucleotides at the border of genes I and IV there is no evidence for an overlap of genes in the filamentous phage genome. This is in contrast to the organization of the genome of the small icosahedral phages φX174 (Sanger et al. 1977; see also Appendix I, this volume) and G4 (Godson et al. and Appendix II, this volume) and is probably related to the different packaging mechanisms of the two phage groups. In filamentous phages the size of the DNA packaged can vary over a wide range (Herrmann et al. 1978). Thus, their genome may have grown during evolution instead of economizing and packaging a limited amount of DNA.

Promoter and Terminator Regions

On the basis of analyses of in vitro transcripts and of proteins directed by these transcripts (Sugiura et al. 1969; Okamoto et al. 1969; Chan et al. 1975; Edens et al. 1975), it has generally been accepted that transcription of the filamentous phage genome is initiated at many different promoters and that it proceeds unidirectionally to a single stop signal named the "central terminator." The resulting gradient of transcription is enhanced by the fact that the four strongest promoters are located at positions proximal to the termination site (Seeburg and Schaller 1975). Three of these G-start promoters—P(X) 0.065, P(VIII) 0.19, and P(II) 0.975 (Fig. 2)—have been analyzed in detail as to their nucleotide sequences and their interaction with RNA polymerase (see, e.g., Takanami et al. 1976; Seeburg et al. 1977). The precise locations of four additional minor promoters, including three A-starts, were determined by testing the RNA-polymerase-binding abilities as well as the template activities of restriction fragments (Seeburg and Schaller 1975; Edens et al. 1975; Okamoto et al. 1975).

Very recently, H. Schaller et al. (in prep.) analyzed all the regions in fd RF DNA at which RNA polymerase interacts specifically and identified as many as 11 binding sites for this enzyme. Several sites were sequenced directly (P[II]', P[I], P[I]', P[IV], and P[IV]'), others were located in the DNA sequence by their distance from sites of cleavage by different restriction enzymes (P[V], P[III], and P[VI]). The resulting promoter map (Fig. 2) indicates that each gene is preceded by at least one true promoter, usually within the preceding gene (see also Konings and Schoenmakers, this volume). Other promoters (P[I]' and P[II]') are located shortly after the translational start of large gene products. Their functional significance is not known.

A summary of the known fd promoter sequences is presented in Figure 4.

```
                    TGTTGACAATTT              TATAATR

P(X)     0.065  TAATCTTTTTGATGCAATTCGCTTTGCTTCTGACTATAATAGACAGGGTAAAGAC
                378
P(II)    0.93   AAACATTAACGTTTACAATTTAAATATTTGCTTATACAATCATCCTGTTTTTGGG
                5905
P(VIII)  0.19   TACAAATCTCCGTTGTACTTTGTTTCGCGCTTGGTATAATCGCTGGGGGTCAAAG
                1151
P(II)'   0.975  GTTTGAATCTTTGCCTACTCATTACTCCGGCATTGCATTTAAAATATATGAGGGT
                6176
P(IV)    0.64   TGATAAATTCACTATTGACTCTTCTCAGCGTCTTAATCTAAGCTATCGCTATGTT
                4039
P(V)     0.13   CCAACGTCCTGACTGGTATAATGAGCCAGTTCTTAAAATCGCATAAGGTAATTCA
                786
P(III)   0.24   CTGTTTAAGAAATTCACCTCGAAAGCAAGCTGATAAACCGATACAATTAAAGGCT
                1490
P(VI)    0.430  GCGCTGGTAAACCATATGAATTTTCTATTGATTGTGACAAAATAAACTTATTCCG
                2693
P(I)     0.485  TTCTCCCGTCTAATGCGCTTCCCTGTTTTTATGTTATTCTCTCTGTAAAGGCTGC
                3064
P(I)'    0.51   AAATTAGGCTCTGGAAAGACGCTCGTTAGCGTTGGTAAGATTCAGGATAAAATTG
                3203
P(IV)'   0.73   ATTAATAACGTTCGCGCAAAGGATTTAATAAGGGTTGTAGAATTGTTTGTTAAAT
                4611

                                   └ – ——RNA polymerase protected →
```

Figure 4 Nucleotide sequences of promoter sites in fd DNA. The viral DNA strand is presented in the 5 ′ to 3 ′ direction. Sequences are aligned with respect to known initiation nucleotides (→) and/or to the RNA-polymerase-binding sites. The upper four (strong G-start) promoters have been ordered according to their relative strengths (Seeburg et al. 1977). Sequences homologous to "binding" and "recognition" sequences are underlined. Map positions are given in nucleotide numbers and in fractional lengths from the single *Hin*dII cleavage site in fd RF DNA.

Although the base sequences are different, all promoters are rich in AT base pairs and contain regions partially homologous to the sequences TATAATR and TGTTGACAATTT, which have been postulated to be involved in recognition and binding of RNA polymerase (Pribnow 1975; Takanami et al. 1976; Gilbert 1976; Seeburg et al. 1977). This sequence homology appears to be less pronounced in the minor (weak) promoters presented in the lower part of Figure 4.

The central terminator at which all transcription ceases is located immediately after gene VIII. The transcripts terminate in a stretch of U residues and contain a potentially base-paired loop structure near the 3′ end (Sugimoto et al. 1977). Such a structure has also been noted near the ends of other mRNAs, including those terminated at *rho*-induced termination sites (Lebowitz et al. 1971; Pieczenik et al. 1972; Ikemura and Dahlberg 1973; Sogin et al. 1976; Lee et al. 1976). These results indicate that the presence of such a base-paired loop structure is required for the termination of transcription. An additional feature characteristic of a terminator seems to be the presence of a promoter-specific binding sequence that may help to slow down the rate of RNA polymerase movement.

The fd DNA genome has been shown to contain several *rho*-induced termination sites (Takanami et al. 1971). One such *rho* site, which acts even in high salt, was found to be located at a site immediately preceding gene V (K. Sugimoto et al., in prep.). Although the exact *rho* site has not yet been determined, a region with twofold rotational symmetry was found in the position 796–816. This region overlaps with the approximate position of the minor G-start promoter (P[V] 0.13) and seems to be involved in the positive and negative control of mRNA production for the genes downstream.

Promotion and termination of RNA synthesis at a single site has been demonstrated more clearly for the central terminator. This site overlaps with an RNA-polymerase-binding site, presumably that of the gene-III promoter (H. Schaller et al., in prep.). By analyzing transcripts of this region, K. Sugimoto et al. (in prep.) found that gene-III mRNA starts within that part of the terminator which codes for the hairpin loop at the end of gene-VIII mRNA (Fig. 5). The gene-III promoter appears to differ from other promoters in that RNA initiation at this site is promoted only in the presence of a large excess of RNA polymerase (K. Sugimoto et al., in prep.). In addition, the chain is initiated with neither A nor G, but probably with U (see Fig. 5). In φX DNA, the start site for the gene-*A* mRNA has also been assumed to overlap with a possible terminator of mRNA (Sanger et al. 1977).

Origin of DNA Replication

Conversion of the viral DNA into the double-stranded replicative form and the replication of both strands of this DNA molecule are initiated and terminated at a single site located within the intergenic region (Tabak et al.

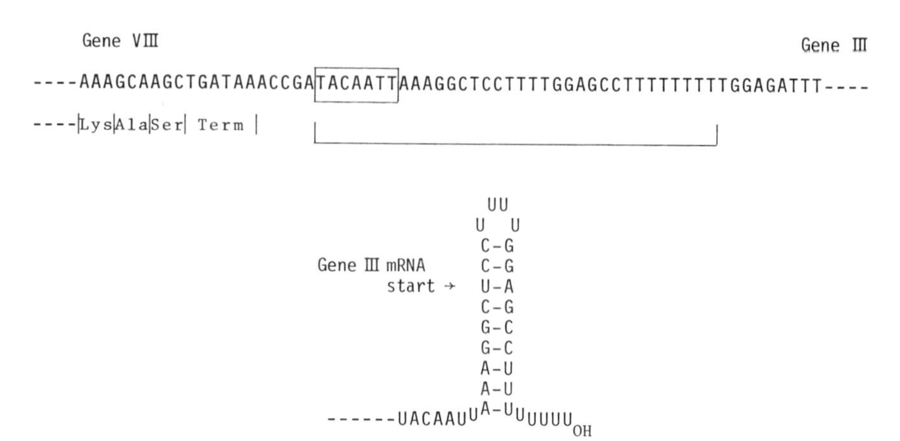

Figure 5 Sequences in the regions covering the central terminator between genes VIII and III. "Binding sequences" (Pribnow 1975) are boxed.

1974; Horiuchi and Zinder 1976; Suggs and Ray 1977). A DNA fragment containing the origin has been isolated from fd single-stranded DNA as an RNA-polymerase-protected DNA fragment of about 125 nucleotides (Schaller et al. 1976). The nucleotide sequence of this ori-DNA (nucleotides 5605–5735, Fig. 6) is folded by base-pairing of self-complementary stretches of bases into two large, stable hairpin structures (Gray et al. 1978). A third hairpin located to the left of the RNA-polymerase-protected ori-DNA (C. P. Gray and E. Beck, unpubl.) has been correlated with a self-complementary DNA sequence in the position 5480–5556, and an additional small hairpin structure can be formed from a self-complementary nucleotide sequence to the right of the ori-DNA (Fig. 6). The four self-complementary nucleotide sequences at the fd origin are separated by pyrimidine-rich nucleotide spacers. Similar features have also been noted in nucleotide sequences at the origin of DNA replication of plasmid ColE1 (Tomizawa et al. 1977) and of bacteriophage λ (Denniston-Thompson et al. 1977). They suggest a potential mechanism for the activation of the origin in double-stranded DNA by destabilization of the double helix and by conformational changes.

In fd single-stranded DNA, recognition of the origin seems to be determined mainly by the secondary structure of the origin region which prevents it from being covered by the *Escherichia coli* DNA-binding protein. However, only two of the four hairpins are protected against nuclease digestion in the preinitiation complex formed by fd DNA, *E. coli* DNA-binding protein, and RNA polymerase. A short RNA primer is transcribed selectively from the stem of one hairpin structure (Geider et al. 1978). The other two large hairpins may be less important since changes in their structures and relative positions by integration of large DNA fragments around positions 5545 and

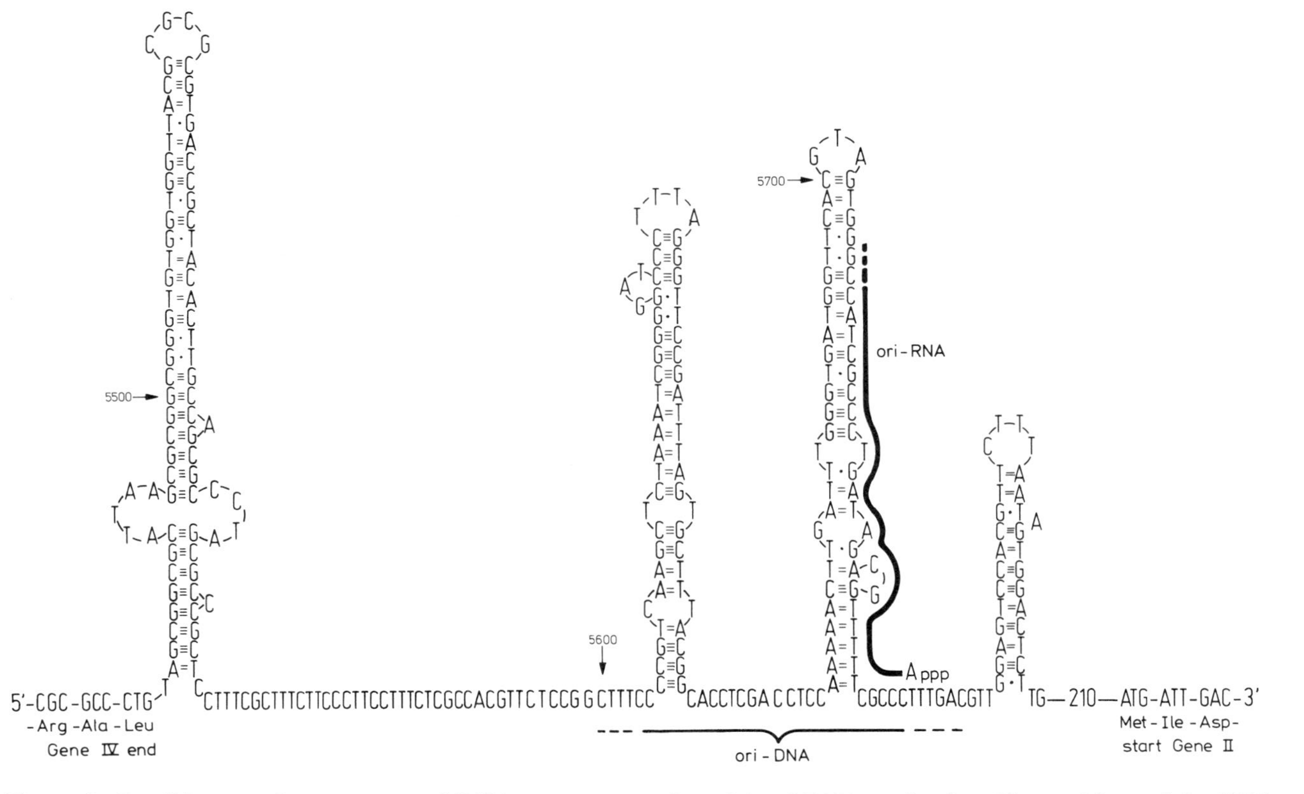

Figure 6 Possible secondary structure of DNA sequences at the origin of DNA replication. The positions of the RNA-polymerase-protected fragment (ori-DNA) and of the primer RNA (ori-RNA) are indicated.

5625 of fd DNA do not appear to interfere with phage replication (Herrmann et al., this volume).

Sequences from f1 and M13 DNA

The data used in the present analysis of the filamentous phage genome were obtained almost exclusively from work with fd DNA. Recently, sections of the DNA from the closely related phages f1 and M13 have been subjects of sequence analysis. E. Beck (unpubl.) has sequenced more than 80% of the f1 genome; sections of f1 DNA have also been sequenced by Sanger et al. (1974), by Ravetch et al. (1977, and pers. comm.), and by Korobko et al. (1977). M13 fragments covering most of the region of gene V to gene I (nucleotides 880–1575 and 1830–3440 have been sequenced by T. Hulsebos et al. (pers. comm.). In addition, the nucleotide sequence of the intergenic region in M13 has been determined by S. Suggs and D. S. Ray (pers. comm.).

All together, preliminary sequences covering about 90% of the f1 genome and close to 50% of the M13 genome have been available to be compared with the sequence given in Figure 1. The results obtained indicate that the fd sequence is conserved in f1 and M13 better than 95% in both the expressed and the silent regions of the phage genome and suggest a much closer relationship between M13 and f1. Base changes within genes only very rarely resulted in changes in the amino acid sequences of the gene products. Deletions of nucleotides were observed only in noncoding parts of the genome. The last two observations support the fd sequence given in Figure 1 and indicate that the reading frames chosen for the fd genes were correct.

ACKNOWLEDGMENTS

We thank R. Sommer and G. Osterburg for carrying out the computer storage and analysis of the sequence. This work was supported, in part, by the Deutsche Forschungsgemeinschaft.

REFERENCES

Asbeck, F., K. Beyreuther, H. Köhler, G. von Wettstein, and G. Braunitzer. 1969. Virusproteine. IV. Die Konstitution des Hüllproteins des Phagen fd. *Hoppe-Seyler's Z. Physiol. Chem.* **350:** 1047.

Chan, T., P. Model, and N. D. Zinder. 1975. In vitro protein synthesis directed by separated transcripts of bacteriophage f1 DNA. *J. Mol. Biol.* **99:** 369.

Day, L. A. and S. A. Berkowitz. 1977. The number of nucleotides and the density and refractive index increments of fd DNA. *J. Mol. Biol.* **116:** 603.

Denniston-Thompson, K., D. D. Moore, K. E. Kruger, M. E. Furth, and F. R. Blattner. 1977. Physical structure of the replication origin of bacteriophage lambda. *Science* **198:** 1051.

Edens, L., R. N. H. Konings, and J. G. G. Schoenmakers. 1975. Physical mapping of the central terminator for transcription on the bacteriophage M13 genome. *Nucleic Acids Res.* **2:** 1811.

Geider, K., E. Beck, and H. Schaller. 1978. An RNA transcribed from the origin of phage fd single-strand to replicative form conversion. *Proc. Natl. Acad. Sci.* **75:** 645.

Gilbert, W. 1976. Starting and stopping sequences for the RNA polymerase. In *RNA polymerase* (ed. R. Losick and M. Chamberlin), p. 193. Cold Spring Harbor Laboratory, Cold Spring Harbor, New York.

Goldsmith, M. E. and W. H. Konigsberg. 1977. Adsorption protein of bacteriophage fd. Isolation, molecular properties, and location in virus. *Biochemistry* **16:** 2686.

Gray, C. P., R. Sommer, C. Polke, E. Beck, and H. Schaller. 1978. Structure of the origin of DNA replication of bacteriophage fd. *Proc. Natl. Acad. Sci.* **75:** 50.

Herrmann, R., K. Neugebauer, H. Zentgraf, and H. Schaller. 1978. Transposition of a DNA sequence determining kanamycin resistance into the single-stranded genome of bacteriophage fd. *Mol. Gen. Genet.* **159:** 171.

Horiuchi, K. and N. D. Zinder. 1976. Origin and direction of synthesis of bacteriophage f1 DNA. *Proc. Natl. Acad. Sci.* **73:** 2341.

Horiuchi, K., G. F. Vovis, V. Enea, and N. D. Zinder. 1975. Cleavage map of bacteriophage f1: Location of the *Escherichia coli* B-specific modification sites. *J. Mol. Biol.* **95:** 147.

Ikemura, T. and J. E. Dahlberg. 1973. Small ribonucleic acids of *Escherichia coli*. *J. Biol. Chem.* **248:** 5024.

Konings, R. N. H., T. Hulsebos, and C. A. van den Hondel. 1975. The identification and characterization of the gene products of bacteriophage M13. *J. Virol.* **15:** 570.

Kornberg, A. 1974. *DNA synthesis.* W. H. Freeman, San Francisco.

Korobko, V. G., S. A. Grachev, and N. A. Petrov. 1977. A novel method of desoxynucleotides fingerprinting. *Bioorg. Khim.* **3:** 1423.

Lebowitz, P., S. M. Weissman, and C. M. Radding. 1971. Nucleotide sequence of a RNA transcribed in vitro from λ phage DNA. *J. Biol. Chem.* **246:** 5120.

Lee, F., C. L. Squires, C. Squires, and C. Yanofsky. 1976. Termination of transcription in vivo in the *Escherichia coli* tryptophan operon leader region. *J. Mol. Biol.* **103:** 383.

Ling, V. 1972. Fractionation and sequences of the large pyrimidine oligonucleotides from bacteriophage fd DNA. *J. Mol. Biol.* **64:** 87.

Lyons, L. B. and N. D. Zinder. 1972. The genetic map of the filamentous bacteriophage f1. *Virology* **49:** 45.

Marvin, D. A. and B. Hohn. 1969. Filamentous bacterial viruses. *Bacteriol. Rev.* **33:** 172.

Maxam, A. M. and W. Gilbert. 1977. A new method for sequencing DNA. *Proc. Natl. Acad. Sci.* **74:** 560.

Messing, J., B. Gronenborn, B. Müller-Hill, and P. H. Hofschneider. 1977. Filamentous coliphage M13 as a cloning vehicle. Insertion of a *Hin*dII fragment of the *lac* regulatory region in M13 replicative form in vitro. *Proc. Natl. Acad. Sci.* **74:** 3642.

Model, P. and N. D. Zinder. 1974. In vitro synthesis of bacteriophage f1 proteins. *J. Mol. Biol.* **83:** 231.

Nakashima, Y. and W. Konigsberg. 1974. Reinvestigation of a region of fd bacteriophage coat protein sequence. *J. Mol. Biol.* **88:** 598.

Oertel, W. and H. Schaller. 1972. A new approach to the sequence analysis of DNA. *FEBS Lett.* **27:** 316.

Okamoto, T., M. Sugiura, and M. Takanami. 1969. Length of RNA transcribed on the replicative form DNA of coliphage fd. *J. Mol. Biol.* **45:** 101.

Okamoto, T., T. Sugimoto, H. Sugisaki, and M. Takanami. 1975. Studies on bacteriophage fd DNA. II. Localization of RNA initiation sites on the cleavage map of the fd genome. *J. Mol. Biol.* **95:** 33.

Petersen, G. B. and J. M. Reeves. 1969. The occurrence of long sequences of consecutive pyrimidine deoxyribonucleotides in the DNA of bacteriophage f1. *Biochim. Biophys. Acta* **179:** 510.

Pieczenik, G., B. G. Barrell, and M. L. Gefter. 1972. Bacteriophage φ80 induced low molecular weight RNA. *Arch. Biochem. Biophys.* **152:** 152.

Pieczenik, G., P. Model, and H. D. Robertson. 1974. Sequence and symmetry in ribosome-binding sites of bacteriophage f1 RNA. *J. Mol. Biol.* **90:** 191.

Pribnow, D. 1975. Nucleotide sequence of an RNA polymerase binding site at an early T7 promoter. *Proc. Natl. Acad. Sci.* **72:** 784.

Ravetch, J. V., K. Horiuchi, and N. D. Zinder 1977. Nucleotide sequences near the origin of replication of bacteriophage f1. *Proc. Natl. Acad. Sci.* **74:** 4219.

Ray, D. S. 1977. Replication of filamentous bacteriophage. In *Comprehensive virology* (ed. H. Fraenkel-Conrat and R. Wagner), vol. 7, p. 105. Plenum Press, New York.

Sanger, F. and A. R. Coulson. 1975. A rapid method for determining sequences in DNA by primed synthesis with DNA polymerase. *J. Mol. Biol.* **94:** 441.

Sanger, F., J. E. Donelson, A. R. Coulson, H. Kössel, and H. Fischer. 1973. Use of DNA polymerase I primed by a synthetic oligonucleotide to determine a nucleotide sequence in phage f1 DNA. *Proc. Natl. Acad. Sci.* **70:** 1209.

———. 1974. Determination of a nucleotide sequence in bacteriophage f1 DNA by primed synthesis with DNA polymerase. *J. Mol. Biol.* **90:** 315.

Sanger, F., G. M. Air, B. G. Barrell, N. L. Brown, A. R. Coulson, J. C. Fiddes, C. A. Hutchison, P. M. Slocombe, and M. Smith. 1977. Nucleotide sequence of bacteriophage φX174 DNA. *Nature* **265:** 687.

Schaller, J., C. P. Gray, and K. Herrmann. 1975. Nucleotide sequence of an RNA polymerase-binding site from the DNA of bacteriophage fd. *Proc. Natl. Acad. Sci.* **72:** 737.

Schaller, H., A. Uhlmann, and K. Geider. 1976. A DNA fragment from the origin of single to double strand DNA replication of bacteriophage fd. *Proc. Natl. Acad. Sci.* **73:** 49.

Seeburg, P. H. and H. Schaller. 1975. Mapping and characterization of promoters in bacteriophage fd, f1 and M13. *J. Mol. Biol.* **92:** 261.

Seeburg, P. H., C. Nüsslein, and H. Schaller. 1977. Interaction of RNA polymerase with promoters from bacteriophage fd. *Eur. J. Biochem.* **74:** 107.

Shine, J. and L. Dalgarno. 1974. The 3′-terminal sequence of *Escherichia coli* 16S ribosomal RNA: Complementarity to nonsense triplets and ribosome binding sites. *Proc. Natl. Acad. Sci.* **71:** 1342.

Sogin, M. L., N. R. Pace, M. Rosenberg, and S. M. Weissman. 1976. Nucleotide sequence of a 5S ribosomal RNA precursor of *Bacillus subtilis*. *J. Biol. Chem.* **251:** 3480.

Suggs, S. and D. S. Ray. 1977. Replication of bacteriophage M13. XI. Localization of the origin of M13 single strand synthesis. *J. Mol. Biol.* **110:** 147.

Sugimoto, K., T. Okamoto, H. Sugisaki, and M. Takanami. 1975a. The nucleotide sequence of an RNA polymerase binding site on bacteriophage fd DNA. *Nature* **253:** 410.

Sugimoto, K., H. Sugisaki, T. Okamoto, and M. Takanami. 1975b. Studies on bacteriophage fd DNA. III. Nucleotide sequence preceding the RNA start site on a promoter-containing fragment. *Nucleic Acids Res.* **2:** 2091.

———. 1977. Studies on bacteriophage fd DNA. IV. The sequence of messenger RNA for the major coat protein gene. *J. Mol. Biol.* **111:** 487.

Sugiura, M., T. Okamoto, and M. Takanami. 1969. Starting nucleotide sequences of RNA synthesized on the replicative form DNA of coliphage fd. *J. Mol. Biol.* **43:** 299.

Tabak, H. F., J. Griffith, K. Geider, H. Schaller, and A. Kornberg. 1974. Initiation of deoxyribonucleic acid synthesis. VII. A unique location of the gap in the M13 replicative duplex synthesized in vitro. *J. Biol. Chem.* **249:** 3049.

Takanami, M. and T. Okamoto. 1975. Physical mapping of transcribing regions on coliphage fd DNA by the use of restriction endonucleases. In *Control of transcription* (ed. B. B. Biswas et al.), p. 145. Plenum Press, New York.

Takanami, M., T. Okamoto, and M. Sugiura. 1971. Termination of RNA transcription on the replicative form DNA of bacteriophage fd. *J. Mol. Biol.* **62:** 81.

Takanami, M., K. Sugimoto, H. Sugisaki, and T. Okamoto. 1976. Sequence of promoter for coat protein gene of bacteriophage fd. *Nature* **260:** 297.

Tomizawa, J.-I., H. Ohmori, and R. E. Bird. 1977. Origin of replication of colicin E1 plasmid. *Proc. Natl. Acad. Sci.* **74:** 1865.

van den Hondel, C. A., C. A. A. Weijers, R. N. H. Konings, and J. G. G. Schoenmakers. 1975. Studies on bacteriophage M13 DNA. II. The gene order of the M13 genome. *Eur. J. Biochem.* **53:** 559.

Vovis, G. F., K. Horiuchi, and N. D. Zinder. 1975. Endonuclease R·*Eco*RII restriction of bacteriophage f1 DNA in vitro. Ordering of genes V and VII, location of an RNA promoter for gene VIII. *J. Virol.* **16:** 674.

Methylated Bases in the Single-stranded DNA Phages

Aharon Razin
Department of Cellular Biochemistry
The Hebrew University-Hadassah Medical School
Jerusalem, Israel 91000

The widespread occurrence of methylated bases in the DNA of living organisms and the species-specific pattern of DNA methylation (Hall 1971) strongly suggest that these methyl groups play an important biological role. At the present time, however, few biological processes can be correlated with methylated bases in DNA.

Most of the methylated bases in the bacterial chromosome are not related to the modification activity in these cells, since bacteria that are devoid of a restriction-modification (R-M) system still methylate their own DNA. The function of these methyl groups, as well as of those found in eukaryotic cell DNA, is obscure.

The single-stranded DNA bacteriophages, which possess a few methyl groups per genome, are ideal tools for studying the role of DNA methylation. Thus, the filamentous bacteriophages f1, fd, and M13 have been investigated with respect to the function of the methylated bases in the R-M phenomena (Arber 1968). The isometric phage φX174, which is not subject to R-M and is propagated in *Escherichia coli* C, a strain devoid of any known R-M system, was chosen to study the role of methylated bases not related to R-M. A single methyl group has been found in the φX genome and it has been suggested that this group is essential in the final stages of phage maturation (Razin et al. 1975).

ROLE OF DNA METHYLATION IN RESTRICTION-MODIFICATION

φX is not restricted by the *Eco*K or *Eco*B R-M systems,[1] whereas the filamentous phages fd, f1, and M13 are restricted and undergo host-controlled modification in *Eco*B (but not *Eco*K) strains (Arber 1968). The host-controlled modification correlates with enzymatic methylation of the DNA of these phages, thereby protecting the phage DNA from the corresponding restriction endonuclease.

[1] Strains possessing the specificity of *E. coli* K12 and *E. coli* B, respectively.

The *Eco*B R-M System

The hypothesis that modification is based on specific methylation of sites on the DNA was first tested with phage fd. The fd·B[2] DNA molecule carries, in addition to the four methylated adenines encountered in unmodified fd·0 DNA, about two 6-methyladenine (m^6Ade) residues which are responsible for the B-specific modification (Arber 1968; Horiuchi et al., this volume). These methylation measurements suggested that two host specificity sites were present in the fd DNA and led to the assumption that mutations occurring within the sequence determining each site prevented recognition. The loss of one specificity site would therefore result in a reduction in restriction of unmodified phage and less methylation of phage DNA.

A simple procedure was applied to enrich for mutants of fd with reduced restriction in *Eco*B strains (Arber and Kühnlein 1967). Growth of the phage alternately on *Eco*0[3] and *Eco*B hosts for three or four double passages allowed the isolation of mutants that showed reduced restriction relative to wild-type fd·0. Repetition of the same procedure with a first-step mutant resulted in the isolation of a completely unrestricted phage. These findings were interpreted as reflecting a stepwise loss of B specificity sites by mutational changes in the fd DNA and confirmed the hypothesis that fd DNA carries two B specificity sites, as suggested by the methylation measurements. The two sites that confer DNA susceptibility to restriction by the B host were called SB_1 and SB_2 (see Fig. 1 in Horiuchi et al., this volume).

Methylation of fd DNA by the meth M·*Eco*B[4]-specific modification enzyme and the correlation between the restriction site and site of modification were studied further in vivo (Smith et al. 1972) and in vitro (Kühnlein and Arber 1972). The SB_1^0 SB_2^0 two-step mutant of fd grown on *Eco*B carried about two fewer m^6Ade residues than *Eco*B-modified wild-type phage DNA. This result suggested that the mutated specificity sites on the DNA of SB_1^0 SB_2^0 had not only lost their affinity for endo R·*Eco*B[5]-specific restriction, but also for meth M·*Eco*B-specific modification. This conclusion was verified by in vitro experiments (Kühnlein and Arber 1972). By use of a partially purified DNA modification enzyme from *E. coli* B, it was shown that the replicative-form (RF) DNA of a one-step mutant of fd (SB_1^0 SB_2) had lost two of the four methyl acceptor sites. The RF DNA of a two-step mutant (SB_1^0 SB_2^0), which is not subject to endo R·*Eco*B-specific restriction, was completely refractory to methylation by the modification enzyme.

The two B specificity sites have been located in relation to each other on the fd genome. In view of the fact that restriction and modification of fd

[2] fd·B, as well as fd·K, fd·0, etc., indicates the type of host-specific modification carried by a phage stock.

[3] *Eco*0 indicates strains with no detectable R-M (e.g., *E. coli* C).

[4] Meth M·*Eco*B, meth M·*Eco*K, etc., indicate specific modification methylases of the various strains.

[5] Endo R·*Eco*B, endo R·*Eco*K, etc., indicate restriction endonucleases of the various strains.

DNA occur in RF DNA rather than in viral single-stranded (SS) DNA (Kühnlein et al. 1969), unmodified fd RF DNA was used as substrate for cleavage in vitro with endo R·*Eco*B. The products of this cleavage were two fragments of unequal length (Boyer et al. 1971). Similar properties were found for the restriction-modification of the closely related bacteriophage f1. In contrast to fd and f1 which contain two B specificity sites that are susceptible to the *Eco*B R-M system, phage M13 was restricted less efficiently by endo R·*Eco*B. The degree of restriction of M13 on *Eco*B strains was close to that found for one-step mutants of fd. In addition, M13 yielded completely unrestricted variants in one mutational step. Arber and Kühnlein (1967) suggested, therefore, that M13 had only one B specificity site.

Because endo R·*Eco*B did not cleave modified DNA or the SB_1^0 SB_2^0 mutant DNA, it seemed reasonable to assume that the restriction enzyme recognized and cleaved the same B specificity sites on the phage DNA as were recognized by meth M·*Eco*B. However, it was found that this is not the case. Horiuchi and Zinder (1972) performed denaturation-renaturation experiments with an in vitro restriction product of f1 RF DNA that contained only one functional B specificity site and showed that the B specificity site was not the site of the cleavage and that the number of possible restriction sites was much larger than the number of B specificity sites. It was suggested by Horiuchi and Zinder (1972) that the B specificity sites of f1 RF DNA were recognition sites for endo R·*Eco*B as well as for meth M·*Eco*B, but were not cleavage sites. This means that a genetically defined restriction site may be an entry point on the DNA molecule for the restriction enzyme, which then progresses along the DNA molecule and cleaves at any one of a number of sites that may be quite remote from the original target. A potent ATPase activity associated with this reaction might be related to the movement of the enzyme (Horiuchi et al. 1974).

The number of methyl groups introduced per DNA molecule reflects the number of B specificity sites (Smith et al. 1972; Kuhnlein and Arber 1972; Hattman 1973; Vovis and Zinder 1975), a fact suggesting that B-specific methylation occurs at the B specificity sites. Subsequently, Horiuchi et al. (1975) correlated the genetic map of phage f1 with a cleavage map constructed by the use of restriction enzymes from various *Haemophilus* species. The locations of the B specificity sites relative to the sites of B-specific methylation in vitro were determined, and methylation was found to occur at or close to (within 200 base pairs at the most) the B specificity sites.

Since R·*Eco*B is both an endonuclease and a modification methylase (Haberman et al. 1972; Vovis et al. 1974) and methylation occurs at the B specificity sites, the state of the DNA at these sites will determine which enzymatic activity will be expressed. If the substrate DNA is an unmodified homoduplex, the enzyme acts as an endonuclease and cleaves the DNA rapidly. The modification of such a DNA occurs at a very slow rate. If the DNA is a heteroduplex consisting of one modified and one unmodified strand, the enzyme acts as a methylase and rapid methylation of the DNA occurs (Vovis et al. 1974). No endonucleolytic activity of the enzyme can be detected on such a molecule (Meselson and Yuan 1968; Vovis et al. 1973). If the DNA is a modified homoduplex, the enzyme acts neither as an endonuclease nor as a methylase. In this case the enzyme cannot even recognize the DNA (Yuan and Meselson 1970; Boyer et al. 1971; Horiuchi et al. 1974).

Methylation of Filamentous Phage DNA by *E. coli* K12 Methylases

Although bacteriophages fd, f1, and M13 are not restricted by endo R·*Eco*K, their DNA is still methylated by the *E. coli dam* gene product (a DNA adenine methylase) and by the *mec* gene product (a DNA cytosine methylase) (see Horiuchi et al., this volume). Four m^6Ade residues per DNA molecule were found in fd when the phage was grown on *E. coli* K12 (Hattman 1973). That this methylation does not reflect the activity of meth M·*Eco*K was demonstrated by infecting a mutant of *E. coli* K12 that was defective in the *dam* gene but normal in meth M·*Eco*K activity (Marinus and Morris 1975). Phage fd grown on the dam^- mutant was found to contain no m^6Ade. This result suggests that there are no base sequences in fd DNA that are recognized by endo R·*Eco*K or meth M·*Eco*K and that the fd DNA is methylated by the *dam* product.

The locations of the *dam* sites on f1 DNA have been determined recently (Vovis and Lacks 1977). The *E. coli dam* methylase was found to be responsible for making f1 DNA resistant to endo R·*Dpn*II[6] and sensitive to endo R·*Dpn*I.[6] The former enzyme is specific for unmethylated DNA and the latter for methylated DNA (Lacks and Greenberg 1975). Endo R·*Dpn*I cleaved f1 DNA from dam^+ cells at four sites, as did endo R·*Dpn*II with f1 DNA from dam^- cells. The restriction fragments were identical in both cases (69%, 19%, 7.6%, and 4.7% the size of the f1 genome). Experiments with hybrid duplexes of f1 DNA (methylated in only one strand) revealed that both enzymes can act only if both strands are homologous with respect to methylation at the cleavage site. These results suggest that the f1 DNA possesses four sites recognized by the *dam* methylase.

Methylation of Filamentous Phage DNA by Meth M·*Eco*RII

In addition to the R-M systems specified by the chromosomes of certain bacteria, some extrachromosomal elements (plasmids) harbored in bacteria control R-M systems of their own. One of these plasmids is the drug-resistance factor N-3.

[6] Endo R·*Dpn*I and endo R·*Dpn*II are restriction endonucleases from *Diplococcus pneumoniae* specific for methylated and unmethylated DNA, respectively.

Although fd, f1, and M13 are not subject to restriction by endo R·*Eco*RII, the restriction enzyme specified by N-3 (Arber 1966; Bickle and Arber 1969), their DNA is subject to methylation by meth M·*Eco*RII, the cytosine methylase specified by N-3 (Hattman 1973; Schlagman and Hattman 1974). fd DNA was found to contain one 5-methylcytosine (m^5Cyt) residue when grown on cells not harboring the N-3 factor. In cells harboring the N-3 factor the fd DNA "gained" one to two more m^5Cyt residues per fd DNA molecule. This methylation of the fd DNA and lack of restriction by endo R·*Eco*RII called for clarification of the function of the m^5Cyt residues in the fd DNA. It was clear that the methylation was carried out both by the *mec*-encoded DNA cytosine methylase of the host and the meth M·*Eco*RII of the plasmid. The question was whether the *mec* and the meth M·*Eco*RII DNA cytosine methylases methylated the same sequence of the fd DNA. The distribution of m^5Cyt in pyrimidine tracts derived from labeled viral DNA propagated in *mec*$^+$ and *mec*$^-$ N-3 was studied, and a similarity in the two specified methylation patterns was observed. For both DNAs the methylated sequence was Pu-C-m^5C-T-Pu (May and Hattman 1975a,b). These results suggested that the two enzymes methylated the same sites and that the differences in methylation levels observed in vivo might be due to differences in the respective rates of methylation or to incomplete methylation by the *mec*-gene product alone. Another possibility was that, in vivo, fd DNA was protected from N-3 restriction by the *mec*-encoded DNA cytosine methylase. That this was not the case was shown when fd was propagated on *mec*$^-$ cells and tested in cells harboring the N-3 factor. The phage was still not restricted in vivo by endo R·*Eco*RII (Hattman et al. 1973). In contrast to these experiments in vivo, restriction of fd RF DNA isolated from *mec*$^-$ and *mec*$^+$ cells showed that fd *mec*$^+$ RF DNA was not subject to cleavage by endo R·*Eco*RII, but fd *mec*$^-$ RF DNA was cleaved into two fragments, which corresponded to 85% and 15% of the fd DNA length, (Schlagman et al. 1976). The cleavage of fd *mec*$^-$ RF DNA at two sites by endo R·*Eco*II in vitro, together with the observation of two to three m^5Cyt residues per SS DNA molecule in mature fd (Hattman 1973; Hattman et al. 1973), suggested that the fd DNA contains at least two N-3 specificity sites. The discrepancy between the results obtained in vivo and in vitro leaves the mechanism by which fd *mec*$^-$ DNA escapes endo R·*Eco*RII restriction within the host cell unexplained.

ROLE OF DNA METHYLATION IN VIRUS MATURATION

The Pattern of φX DNA Methylation

The only methylated base in φX DNA is m^5Cyt; it appears in the phage DNA at a frequency of one m^5Cyt residue per molecule of SS DNA. Analysis of pyrimidine isopliths obtained from φX DNA labeled in vivo in the methyl group revealed that the single m^5Cyt residue was located in the phage DNA almost exclusively in the dinucleotide fraction next to thymine. This fact suggested that m^5Cyt appears in φX DNA in one of two sequences: Pu-T-m^5C-Pu or Pu-m^5C-T-Pu (Razin et al. 1970).

Experiments designed to elucidate the methylation of intracellular DNA molecules revealed that only the replicating intermediates in SS DNA synthesis were methylated in vivo (Razin et al. 1973). No methylation of RFI (covalently closed RF) or RFII (RF in which at least one stand is not continuous) DNA could be detected. Replicating intermediates, the immediate precursors of progeny SS DNA, are composed of a covalently closed complementary strand and a linear viral strand ranging in length from one to two viral genomes. In the course of SS DNA synthesis, nucleotides are added to the 3′ end while the 5′ end is packed into the coat protein to form the virion (Knippers et al. 1969). It is conceivable that the SS DNA tails of the replicating intermediates are the substrates for methylation, as this is the DNA that is packaged in the phage particles. That this may be the case was shown by several types of experiments; denatured methylated DNA was found to be susceptible to exonuclease I and the pattern of sedimentation of the denatured methylated DNA in a sucrose gradient at alkaline pH revealed that it was linear SS DNA almost double the length of the phage φX genome. These observations suggested that methylation occurred at the final stage of phage maturation. If the rolling-circle model (Gilbert and Dressler 1969) is accepted for the mode of φX DNA replication, termination of one round of replication determines the initiation of the next round of replication. Since the methyl label appears exclusively in the long SS DNA tails of the replicative forms, it is unlikely that methylation is involved in the initiation of SS DNA synthesis. Therefore, we proposed that methylation in the φX system serves as a signal for termination of a round of replication, an essential role in phage maturation (Razin et al. 1973).

Strong support for this concept was provided by experiments reported by Lee and Sinsheimer (1974). They prepared φX DNA labeled in vivo in the methyl group. This DNA was converted in vitro to RF by synthesis of the complementary strand and then cleaved into specific fragments with restriction enzymes. The restriction fragments were correlated with the genetic map to determine the location of the methyl group. The results showed that the single m^5Cyt residue was located in gene *H*, close to its junction with gene *A*. In view of the continuous mode of φX DNA synthesis (see Fig. 1), initiated from a unique position in gene *A* (close to the gene-*H*–gene-*A* junction) and proceeding clockwise around the genetic map (genes $A \rightarrow H$), it could be concluded that the methylated base was located close to a termination site. This too suggested an important role for the single methyl group in phage maturation. When nonmethylated viral DNA was synthesized in vitro with *E. coli* polymerase I and isolated complementary strands as templates, the RFII DNA produced was shown to accept one methyl group per molecule. This methyl group was found on the viral strand

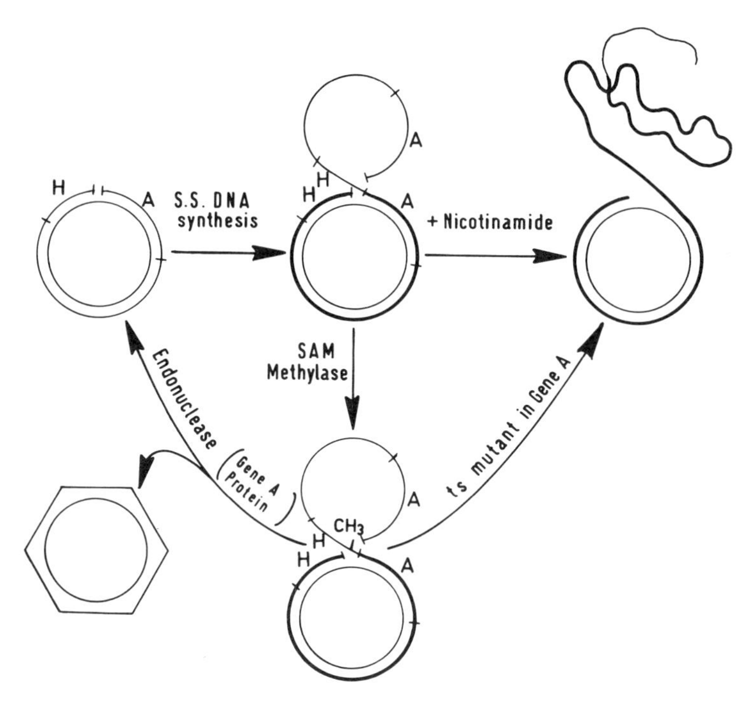

Figure 1 Hypothetical scheme for synthesis of φX viral SS DNA in normal infections, in infections in the presence of nicotinamide, and in infections with a gene-*A* mutant phage. *H* and *A* correspond to genes *H* and *A* on the genetic map of φX.

only. Viral strands isolated from the virus showed no methyl-accepting activity in vitro, whereas RF DNA isolated from phage-infected cells could be methylated in vitro. It is probable, therefore, that RF molecules are not methylated in vivo and that only viral strands are substrates for the methylation (Razin and Friedman 1975).

The Biological Role of φX DNA Methylation

In subsequent studies we looked for the biological significance of methylation in maturation of phage DNA. We attempted to impair the DNA methylation capacity of the cell by depleting the intracellular pool of the methyl donor, S-adenosylmethionine (SAM), or by inhibiting DNA methylase activity directly. Experiments designed to deplete SAM included: methionine starvation of a methionine auxotroph, inhibition of methionine synthesis by trimethoprim, inhibition of the synthesis of SAM by cycloleucine, and use of nicotinamide as a possible trap for methyl groups or as an activator of SAM splitting activity (Razin et al. 1975). Although none of these procedures resulted in a significant reduction in the intracellular level of SAM, it was found that nicotinamide acted as an inhibitor of DNA methylase by competing with SAM. Its effect on the methylation of DNA within the infected cells caused a parallel decrease in phage production. These effects of nicotinamide, together with its failure to affect protein synthesis and elongation of DNA chains, suggested that DNA methylation was an indispensable step in replication and prompted experiments to elucidate the stage in phage DNA replication that was critically dependent upon methylation (Friedman and Razin 1976). Since phage DNA methylation had been shown to occur late in the life cycle, at the stage of SS DNA synthesis (Razin et al. 1973), experiments were designed to study DNA methylation at that stage.

Pulse-label experiments revealed that initiation of SS DNA synthesis and also elongation of the nascent strand could take place in the presence of nicotinamide; however, termination appeared to be suppressed in the presence of the inhibitor. As a result, replicating intermediate molecules with abnormally long SS DNA tails accumulated, and the products of the normal circularization-excision process, progeny SS DNA rings, were not released (see Fig. 1). The abnormal replicating intermediate molecules have been visualized by electron microscopy (Friedman et al. 1977).

Since nicotinamide apparently interfered with DNA methylation by inhibiting DNA methylase activity competitively with respect to SAM, it was assumed that this inhibition would be reversible. In fact, when nicotinamide was washed from the cells, DNA methylation, excision of SS DNA rings, and production of mature virus resumed. These results suggested that when nicotinamide was used to prevent methylation of replicating intermediate molecules, the excision of the genome-length DNA essential for the maturation of the virus was impaired (see Fig. 1). Although the circularization-excision step is not yet well understood, it is likely that a specific endonuclease is involved.

The results described above strongly suggest that the methyl group in φX DNA, which is located close to the DNA replication termination site, serves as a signal for termination when it is recognized by the specific endonuclease that mediates excision of genome-length viral DNA. However, since RF molecules are not methylated, it appears that methylation is not a significant event during RF DNA replication.

Since the single methyl group plays a crucial role in phage maturation, we wished to know whether the phage DNA codes for a specific methylase or whether it relies instead on the DNA cytosine methylase known to be present in *E. coli* C strains. To answer this question we made use of a mutant of φX (isolated by D. Denhardt) which is capable of infecting *E. coli* B cells (Razin 1973). *E. coli* B is devoid of a DNA cytosine methylase and therefore provides an ideal tool for studying the genetic information directing cytosine-specific methylation in genomes that can be introduced into the cells. After infection of *E. coli* B cells with Denhardt's mutant, a DNA cytosine methylase activity was acquired by the cells. This was determined from the methylation patterns of the phage and host DNAs, as well as from the capacity of extracts of infected *E. coli* B cells to methylate DNA derived from uninfected cells. Experiments in vivo revealed that the phage DNA contained a single m^5Cyt residue per molecule, and m^5Cyt residues were found in the *E. coli* B host DNA in addition to the m^6Ade residues normally present in the uninfected *E. coli* B cells. In vitro, a partially purified enzyme from infected cells methylated DNA from uninfected cells but showed no activity with host DNA from infected cells. These results suggested that the phage controls a cytosine-specific DNA methylase.

All these data provide evidence for a function for the single m^5Cyt residue in φX DNA. The methyl group may serve as a recognition site for the gene-*A* product involved in excision of genome-length single strands during formation of phage particles. The recently developed in vitro system for φX SS DNA synthesis (Scott et al. 1977) may provide another tool for testing this point.

ACKNOWLEDGMENTS

I am grateful to S. Hattman and G. F. Vovis for providing me with their recent data. The helpful discussions with J. Mager during the preparation of the manuscript are deeply acknowledged. The work done in our laboratory was partially supported by the U.S.–Israel Binational Foundation grant No. 733 and by U.S. Public Health Service grant GM20483.

REFERENCES

Arber, W. 1966. Host specificity of DNA produced by *Escherichia coli*. 9. Host-controlled modification of bacteriophage fd. *J. Mol. Biol.* **20:** 483.

———. 1968. Host-controlled restriction and modification of bacteriophage fd. *Symp. Soc. Gen. Microbiol.* **18:** 295.

Arber, W. and U. Kühnlein. 1967. Mutationeller verlust B-spezifisher restriktion des bacteriophagen fd. *Pathol. Microbiol*. **30:** 946.

Bickle, T. and W. Arber. 1969. Host-controlled restriction and modification of filamentous I and F-specific bacteriophages. *Virology* **39:** 605.

Boyer, H., E. Scibienski, H. Slocum, and D. Roulland-Dussoix. 1971. The *in vitro* restriction of the replicative form of w.t. and mutant fd phage DNA. *Virology* **46:** 703.

Friedman, J. and A. Razin. 1976. Studies on the biological role of DNA methylation. II. Role of φX174 DNA methylation in the process of viral progeny DNA synthesis. *Nucleic Acids Res*. **3:** 2665.

Friedman, J., A. Friedmann, and A. Razin. 1977. Studies on the biological role of DNA methylation. III. Role of excision of one-genome long single-stranded φX174 DNA. *Nucleic Acids Res*. **4:** 3483.

Gilbert, W. and D. Dressler. 1969. DNA replication: The rolling circle model. *Cold Spring Harbor Symp. Quant. Biol.* **33:** 473.

Haberman, A., J. Heywood, and M. Meselson. 1972. DNA modification methylase activity of *Escherichia coli* restriction endonuclease K and P. *Proc. Natl. Acad. Sci.* **69:** 3138.

Hall, R. J. 1971. *The modified nucleosides in nucleic acids.* Columbia University Press, New York.

Hattman, S. 1973. Plasmid-controlled variation in the content of methylated bases in single-stranded DNA bacteriophages M13 and fd. *J. Mol. Biol.* **74:** 749.

Hattman, S., S. Schlagman, and L. Cousens. 1973. Isolation of a mutant of *Escherichia coli* defective in cytosine-specific deoxyribonucleic acid methylase activity and in partial protection of bacteriophage λ against restriction by cells containing the N-3 drug-resistance factor. *J. Bacteriol.* **115:** 1103.

Horiuchi, K. and N. D. Zinder. 1972. Cleavage of bacteriophage f1 DNA by the restriction enzyme of *Escherichia coli* B. *Proc. Natl. Acad. Sci.* **69:** 3220.

Horiuchi, K., G. F. Vovis, and N. D. Zinder. 1974. Effect of deoxyribonucleic acid length on the adenosine triphosphatase activity of *Escherichia coli* restriction endonuclease B. *J. Biol. Chem.* **249:** 543.

Horiuchi, K., G. F. Vovis, U. Enea, and N. D. Zinder. 1975. Cleavage map of bacteriophage f1: Location of the *Escherichia coli* B-specific modification sites. *J. Mol. Biol.* **95:** 147.

Knippers, R., A. Razin, R. Davis, and R. L. Sinsheimer. 1969. The process of infection with bacteriophage φX174. XXIX. *In vivo* studies on the synthesis of the single-stranded DNA of progeny φX174 bacteriophage. *J. Mol. Biol.* **45:** 237.

Kühnlein, U. and W. Arber. 1972. Host specificity of DNA produced by *Escherichia coli*. XV. The role of nucleotide methylation in *in vitro* B-specific modification. *J. Mol. Biol.* **63:** 9.

Kühnlein, U., S. Linn, and W. Arber. 1969. Host specificity of DNA produced by *Escherichia coli*. XI. *In vitro* modification of phage fd replicative form. *Proc. Natl. Acad. Sci.* **63:** 556.

Lacks, S. and B. Greenberg. 1975. A deoxyribonuclease of *Diplococcus pneumoniae* specific for methylated DNA. *J. Biol. Chem.* **250:** 4060.

Lee, A. S. and R. L. Sinsheimer. 1974. Location of the 5-methylcytosine group on the bacteriophage φX174 genome. *J. Virol.* **14:** 872.

Marinus, M. G. and N. R. Morris. 1975. Pleiotropic effects of a DNA adenine methylation mutation (dam-3) in *Escherichia coli* K12. *Mutat. Res.* **28:** 15.

May, M. S. and S. Hattman. 1975a. Deoxyribonucleic acid-cytosine methylation by host and plasmid-controlled enzymes. *J. Bacteriol.* **122:** 129.

———. 1975b. Analysis of bacteriophage deoxyribonucleic acid sequences methylated by host and R-factor-controlled enzymes. *J. Bacteriol.* **123:** 768.

Meselson, M. and R. Yuan. 1968. DNA restriction enzyme from *E. coli. Nature* **217:** 1110.

Razin, A. 1973. DNA methylase induced by bacteriophage φX174. *Proc. Natl. Acad. Sci.* **70:** 3773.

Razin, A. and J. Friedman. 1975. *In vitro* methylation of bacteriophage φX174 DNA. *FEBS Proc. Meet.* (Abstr.) 358.

Razin, A., D. Goren, and J. Friedman. 1975. Studies on the biological role of DNA methylation. Inhibition of methylation and maturation of the bacteriophage φX174 by nicotinamide. *Nucleic Acids Res.* **2:** 1967.

Razin, A., J. W. Sedat, and R. L. Sinsheimer. 1970. Structure of the DNA of bacteriophage φX174. VII. Methylation. *J. Mol. Biol.* **53:** 251.

———. 1973. *In vitro* methylation of replicating bacteriophage φX174 DNA. *J. Mol. Biol.* **78:** 417.

Schlagman, S. and S. Hattman. 1974. Mutants of the N-3 R-factor conditionally defective in *Hsp*II modification and deoxyribonucleic acid cytosine methylase activity. *J. Bacteriol.* **120:** 234.

Schlagman, S., S. Hattman, M. S. May, and L. Berger. 1976. *In vitro* methylation by *Escherichia coli* K-12 mec^+ deoxyribonucleic acid-cytosine methylase protects against *in vitro* cleavage by the RII restriction endonuclease (R·*Eco*RII). *J. Bacteriol.* **126:** 990.

Scott, J. F., S. Eisenberg, L. L. Bertsch, and A. Kornberg. 1977. A mechanism of duplex DNA replication revealed by enzymatic studies of phage φX174: Catalytic strand separation in advance of replication. *Proc. Natl. Acad. Sci.* **74:** 193.

Smith, J. D., W. Arber, and U. Kühnlein. 1972. Host specificity of DNA produced by *Escherichia coli.* XIV. The role of nucleotide methylation in *in vivo* B-specific modification. *J. Mol. Biol.* **63:** 1.

Vovis, G. F. and S. Lacks. 1977. Complementary action of restriction enzymes endo R·*Dpn*I and endo R·*Dpn*II on bacteriophage f1 DNA. *J. Mol. Biol.* **115:** 525.

Vovis, G. F. and N. D. Zinder. 1975. Modification of f1 DNA by a restriction endonuclease from *Escherichia coli* B. *J. Mol. Biol.* **95:** 557.

Vovis, G. F., K. Horiuchi, and N. D. Zinder. 1974. Kinetics of methylation of DNA by a restriction endonuclease from *Escherichia coli* B. *Proc. Natl. Acad. Sci.* **71:** 3810.

Vovis, G. F., K. Horiuchi, N. Hartman, and N. D. Zinder. 1973. Restriction endonuclease B and f1 heteroduplex DNA. *Nat. New Biol.* **246:** 13.

Yuan, R. and M. Meselson. 1970. A specific complex between restriction endonuclease and its DNA substrate. *Proc. Natl. Acad. Sci.* **65:** 357.

Single-stranded DNA Mycoplasmaviruses

Jack Maniloff,*† Jyotirmoy Das,* and Jan A. Nowak†
Departments of *Microbiology and of †Radiation Biology and Biophysics
University of Rochester Medical Center
Rochester, New York 14642

We have been studying the replication of group-1 mycoplasmaviruses. Mycoplasmas are the smallest free-living prokaryotes (see review by Maniloff and Morowitz 1972). The virus host is *Acholeplasma laidlawii*, a cell that has a genome of 1×10^9 daltons and, like other mycoplasmas, is bounded by only a 7-nm plasma membrane. The group-1-virus isolate that has been used in most studies, MVL51, is a bullet-shaped particle containing a molecule of circular single-stranded (SS) DNA of about 1.5×10^6 daltons. Virus infection is not lytic: infected cells continue to grow and extrude progeny viruses (Liss and Maniloff 1973a).

The biology of this virus and of the two groups of double-stranded DNA mycoplasmaviruses has been reviewed recently (Maniloff et al. 1977). We will consider here experimental data describing the molecular aspects of MVL51 structure and replication.

STRUCTURE

Morphology

Virions of MVL51 are unenveloped, bullet-shaped particles (Fig. 1) about 14 nm by 80 nm in size (literature reviewed by Maniloff et al. 1977). Optical analysis of electron micrographs of negatively stained virions has shown that the particles have helical symmetry (Bruce et al. 1972; Liss and Maniloff 1973a). The optical transforms have a near-meridional reflection at 4.8 nm, so the distance between helix turns is 4.8 nm. A model for MVL51 has been proposed by Bruce et al. (1972) based on the diffraction data. This model is a T=1 icosahedron with a diameter of 14 nm, cut across a twofold axis, with structure units added to form a helix. This is a two-start helix with 5.6 structure units per turn and a 20° pitch.

Proteins

Purified MVL51 preparations have been analyzed by SDS-polyacrylamide gel electrophoresis and found to contain four proteins with molecular weights of 70,000, 53,000, 30,000, and 19,000 (Maniloff and Das 1975). The approximate stoichiometric ratios of these proteins are 7:1:2:9.

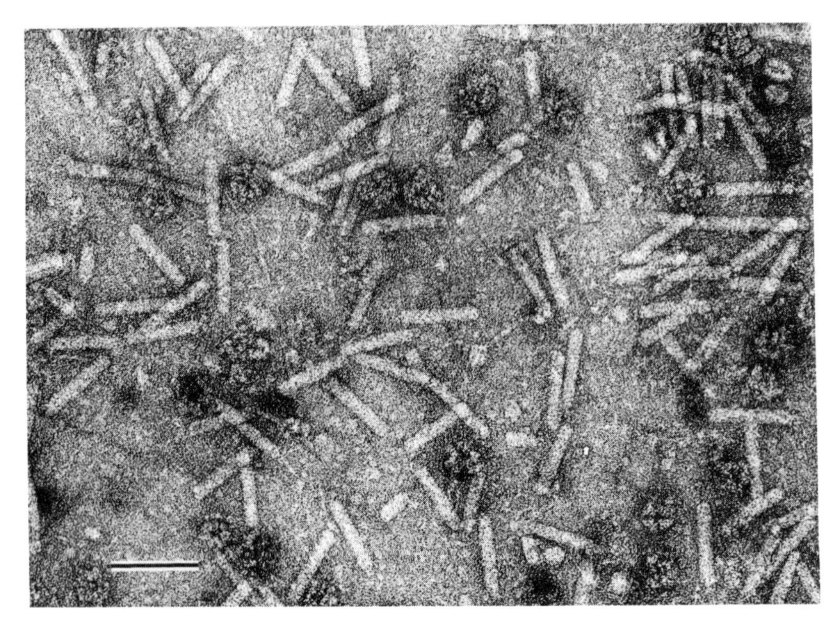

Figure 1 Electron micrograph of negatively stained MVL51 mycoplasmavirus. The virus was concentrated using polyethylene glycol 6000, purified by velocity sedimentation in a sucrose gradient, and negatively stained with uranyl formate. Bar = 0.1 μm.

DNA

The MVL51 genome is circular SS DNA. This was first shown by studies on the effects of various specific nucleases on viral DNA as assayed by transfection (Liss and Maniloff 1973b) and has recently been confirmed by electron microscope observations (Fig. 2a). Measurement of double-stranded MVL51 replicative-form (RF) DNA by electron microscopy has allowed a more accurate determination of the molecular weight (J. A. Nowak et al., in prep.) than was possible with previous sedimentation studies (Liss and Maniloff 1973b). The MVL51 RF DNA contour length is about 0.71 that of the M13 RF DNA used to calibrate these measurements. Taking a value of 6370 nucleotides for M13 DNA (Day and Berkowitz 1977), the MVL51 SS DNA contains about 4520 nucleotides and has a molecular weight (assuming 330 daltons/nucleotide) of about 1.5×10^6.

REPLICATION

DNA Replicative Intermediates

The first step of replication involves conversion of the parental viral SS DNA to double-stranded replicative forms (Das and Maniloff 1976a). As

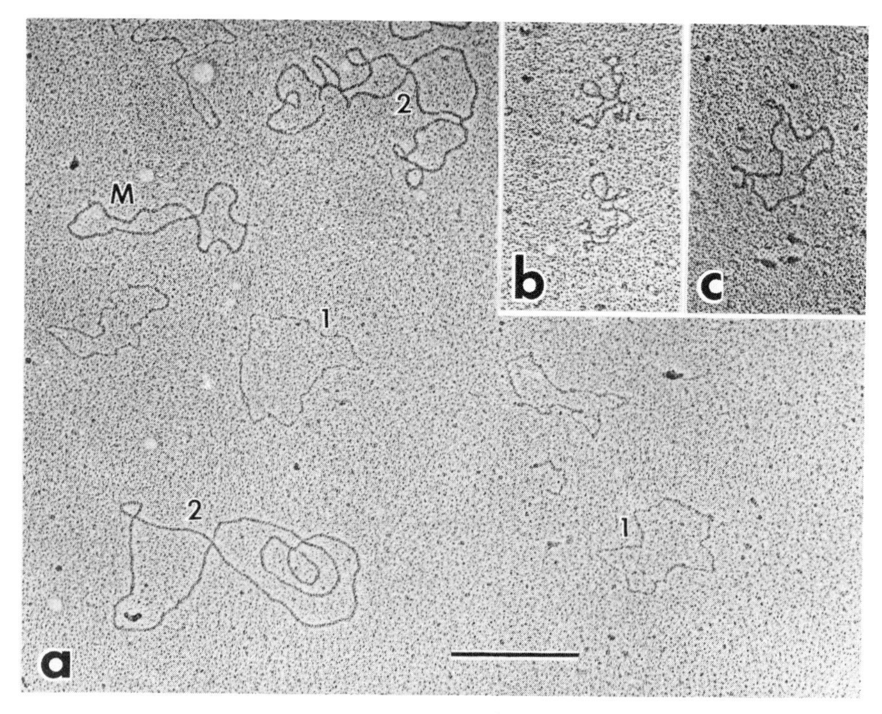

Figure 2 Electron micrographs of viral DNA prepared as described by Davis et al. (1971). (*a*) DNA from SDS-disrupted virions of MVL51 and MVL2 (a double-stranded DNA group-2 mycoplasmavirus) was mixed with purified coliphage M13 RF DNA. The following DNAs can be seen: 1, MVL51 viral circular SS DNA; 2, MVL2 viral DNA; M, M13 RF DNA. (*b*) MVL51 RFI DNA. (*c*) MVL51 RFII DNA. Bar = 0.5 μm.

this step is not affected by the pretreatment of cells with chloramphenicol, presumably it is carried out by preexisting host-cell enzymes (Das and Maniloff 1976c).

The replicative forms are RFI, a covalently closed, double-stranded, circular derivative of the viral DNA (Fig. 2b), and RFII, a relaxed form of RFI (Fig. 2c). These forms sediment with φX174 RFI and RFII in sucrose gradients (Fig. 3). The identification of MVL51 RFI and RFII has been confirmed by ethidium bromide-CsCl equilibrium gradient analysis (Das and Maniloff 1975) and by electron microscopy (Maniloff et al. 1977).

During the second stage of replication the parental RF DNA replicates to produce a pool of progeny RF molecules (Das and Maniloff 1975). This step requires at least one cellular function; Rep$^-$ cells, although capable of forming parental RF DNA, are blocked in further replication (Nowak et al. 1976).

In the final phase of replication RFII molecules serve as precursors for

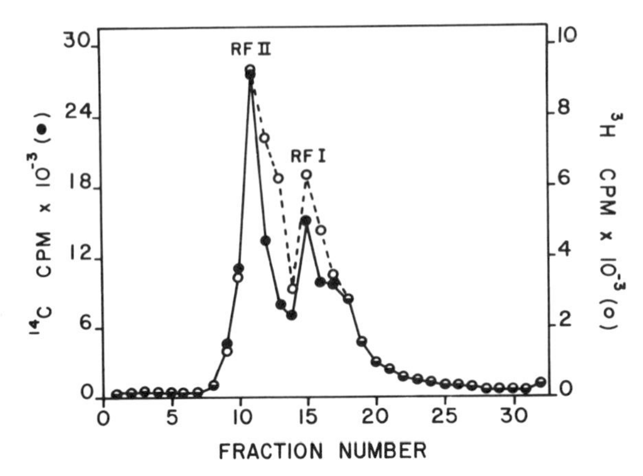

Figure 3 Velocity sedimentation of an MVL51-infected cell lysate. *A. laidlawii* cells were infected with MVL51. After 10 min of adsorption the cells were washed and resuspended in fresh medium containing 5 μCi [^{14}C]deoxythymidine per ml. At 60 min the cells were harvested, washed, mixed with ^{3}H-labeled φX RF DNA, and lysed with Sarkosyl. The lysate was analyzed by velocity sedimentation in a high-salt sucrose gradient, as described by Das and Maniloff (1975). The φX DNA was obtained from Bethesda Research Laboratories. Sedimentation was from left to right.

the synthesis of progeny viral, circular, single-stranded (SSI) DNA. Growth in nutrient-poor medium rather than in richer tryptose broth allows identification of an intermediate between nascent progeny viral SSI chromosomes and mature viruses (Das and Maniloff 1976b). This intermediate, SSII, is a protein-associated form of SSI DNA and is stable in 1 M NaCl. The conversion of SSI to SSII DNA is inhibited by chloramphenicol. The SSI form contains two proteins, which are electrophoretically identical to two of the four virion proteins (Maniloff and Das 1975). These are the 70,000- and 53,000-dalton virion proteins, and in SSII DNA they have an approximate stoichiometric ratio of 1.4:1.

Steps in DNA Replication

Parental viral DNA becomes associated with the host-cell membrane after infection, and this association is sensitive to pronase (Das and Maniloff 1976a). The association was shown by infecting cells with ^{32}P-labeled virus, lysing the cells by freezing and thawing, and sedimenting the lysate through a sucrose gradient over a CsCl shelf. Most of the parental label sedimented with the cell membrane as a fast-sedimenting complex. The remaining label was in slow-sedimenting material.

The amounts of fast- and slow-sedimenting parental viral DNA were measured as a function of multiplicity of infection (m.o.i.) (Das and Maniloff 1976a). Membrane-associated parental DNA was found to reach a saturation value, and from this value it was calculated that there are only two or three sites for membrane-associated parental DNA per cell.

Parental viral DNA is not transferred to progeny viruses. This conclusion is based on experiments in which cells were infected with ^{32}P-labeled MVL51. The amount of intracellular ^{32}P-labeled viral DNA remained constant, and no extracellular acid-insoluble ^{32}P could be detected (Das and Maniloff 1976a).

Several types of experiments have shown that synthesis of progeny RF DNA occurs by symmetric replication at sites on the cell membrane (Das and Maniloff 1976a): (1) During early infection, when most nascent DNA is RF, 80% of the nascent DNA is membrane-associated. (2) Equilibrium centrifugation of membrane-associated progeny RFII DNA in alkaline CsCl gradients showed nascent DNA in both the viral and the complementary strands, indicating symmetrical replication of the double-stranded RF DNA. (3) When infected cells were pulse-labeled with [^{3}H]deoxythymidine for short periods early in infection (when RF DNA replication predominates), the amount of labeled DNA in slow-sedimenting material relative to the amount in the fast-sedimenting material increased with increasing pulse length (Maniloff et al. 1977). This suggests that RF molecules are synthesized at the membrane and then released into the cytoplasm.

Later in infection, when progeny RF DNA replication continues and progeny SSI DNA synthesis increases, most nascent DNA is found as slow-sedimenting material. Analysis of slow-sedimenting progeny RFII DNA by equilibrium centrifugation in alkaline CsCl gradients showed nascent DNA only in the viral strand (Das and Maniloff 1976a), indicating that asymmetric replication is involved in the synthesis of progeny SSI DNA. These results do not rule out some asymmetric synthesis on the membrane.

Electron micrographs show clusters of extracellular progeny viruses at a few membrane sites, indicating that virus assembly and release occur at a limited number of sites per cell (Liss and Maniloff 1973a). Virus maturation and release involve interactions with the cell membrane which do not affect cell viability.

Synthesis of Virus-specific Proteins

In vivo synthesis of MVL51-specific proteins has been studied (Das and Maniloff 1978). Cells were irradiated with UV, to reduce synthesis of cell proteins, and divided into two parts. One part was infected with MVL51 and incubated with [^{3}H]leucine; the other part was not infected and was incubated with [^{14}C]leucine. After several infection times, equal volumes of both cultures were removed and mixed. Proteins were then extracted and analyzed by SDS-polyacrylamide gel electrophoresis. After electrophoresis, gels were sliced and each slice was assayed for ^3H and ^{14}C label. Virus-specific proteins could be identified by a ^3H:^{14}C ratio greater than unity in a particular gel slice. Six virus-specific proteins were resolved, with molecular weights of 70,000, 50,000, 30,000, 19,000, 14,000, and 10,000. The four largest proteins are probably the four virion proteins. The production of the 70,000-dalton protein remains about the same throughout infection, whereas the production of the other three structural proteins (at 50,000, 30,000, and 19,000 daltons) increases during infection. As for the two nonstructural proteins, the relative rate of synthesis of the 14,000-dalton protein is maximal when most viral DNA replication is in RF→RF synthesis, which suggests that this protein might have a role in progeny RF DNA synthesis. The relative amount of the 10,000-dalton protein increases late in infection, when most viral DNA synthesis is in SSI formation; this suggests that this protein might have a role in SSI DNA synthesis.

Replication in Rep$^-$ Cells

A. laidlawii cells with the Rep$^-$ phenotype are able to propagate double-stranded DNA mycoplasmaviruses but not single-stranded DNA mycoplasmaviruses. The block in MVL51 replication in Rep$^-$ cells is in the viral RF→RF step (Nowak et al. 1976). Studies of protein synthesis in infected Rep$^-$ cells, at an infection time when most viral replication should involve RF→RF synthesis, have shown that the amount of 70,000-dalton protein is the same as in wild-type cells (Das and Maniloff 1978). The amounts of the 50,000-, 30,000-, and 19,000-dalton structural proteins are smaller in Rep$^-$ cells than in wild-type cells, which suggests that RF DNA replication may be necessary to stimulate further synthesis of these three virion proteins. The amounts of the 10,000-dalton protein are small in both types of cells at this time in infection. However, the 14,000-dalton protein cannot be detected in Rep$^-$ cells. This suggests that the 14,000-dalton protein may be encoded by the cell and induced during viral replication.

Other Aspects of DNA Synthesis

The repair of UV-induced damage to MVL51 DNA has been studied by Das et al. (1977). As expected for a single-stranded DNA virus, it was found that MVL51 could not be host-cell-reactivated (a type of repair utilizing the cell's excision-repair system). However, MVL51 could be UV-reactivated (a type of repair presumably involving a UV-inducible cellular repair system).

Although *A. laidlawii* host strains can modify and restrict group-2 double-stranded DNA mycoplasmaviruses, no significant restriction or modification of MVL51 has been observed (Maniloff and Das 1975).

CONCLUSIONS

Two general types of single-stranded DNA bacteriophages have been described: icosahedral virions (e.g., φX) and filamentous virions (e.g., M13). Mycoplasmavirus MVL51 appears to represent another type of single-stranded DNA phage, with a genome size close to that of φX and a nonlytic mode of infection like that of filamentous phages. The bullet-shaped MVL51 morphology is unlike that of other known phages.

ACKNOWLEDGMENTS

Studies in this laboratory have been supported by U.S. Public Health Service grant AI-10605 from the National Institute of Allergy and Infectious Diseases. We thank Dr. Stanley Hattman for kindly providing the M13 RF DNA used in these studies.

REFERENCES

Bruce, J., R. N. Gourlay, R. Hull, and D. J. Garwes. 1972. Ultrastructure of *Mycoplasmatales* virus *laidlawii* 1. *J. Gen. Virol.* **16:** 215.

Das, J. and J. Maniloff. 1975. Replication of mycoplasmavirus MVL51. I. Replicative intermediates. *Biochem. Biophys. Res. Commun.* **66:** 599.

———. 1976a. Replication of mycoplasmavirus MVL51. Attachment of MVL51 parental DNA to host cell membrane. *Proc. Natl. Acad. Sci.* **75:** 1489.

———. 1976b. Replication of mycoplasmavirus MVL51. III. Identification of a progeny viral DNA-protein intermediate. *Microbios* **15:** 127.

———. 1976c. Replication of mycoplasmavirus MVL51. IV. Inhibition of viral synthesis by rifampin. *J. Virol.* **18:** 969.

———. 1978. Replication of mycoplasmavirus MVL51. V. *In vivo* synthesis of virus specific proteins. *Virology* (in press).

Das, J., J. A. Nowak, and J. Maniloff. 1977. Host cell and ultraviolet reactivation of ultraviolet irradiated mycoplasmaviruses. *J. Bacteriol.* **129:** 1424.

Davis, R. W., M. Simon, and N. Davidson. 1971. Electron microscope heteroduplex methods for mapping regions of base sequence homology in nucleic acids. *Methods Enzymol.* **21D:** 413.

Day, L. A. and S. A. Berkowitz. 1977. The number of nucleotides and the density and refractive index increments of fd virus DNA. *J. Mol. Biol.* **116:** 603.

Liss, A. and J. Maniloff. 1973a. Infection of *Acholeplasma laidlawii* by MVL51 virus. *Virology* **55:** 118.

———. 1973b. Characterization of *Mycoplasmatales* virus DNA. *Biochem. Biophys. Res. Commun.* **51:** 214.

Maniloff, J. and J. Das. 1975. Replication of mycoplasmaviruses. In *DNA synthesis and its regulation* (ed. M. Goulian et al.), p. 445. W. A. Benjamin, Reading, Massachusetts.

Maniloff, J. and H. J. Morowitz. 1972. Cell biology of the mycoplasmas. *Bacteriol. Rev.* **36:** 263.

Maniloff, J., J. Das, and J. R. Christensen. 1977. Viruses of mycoplasmas and spiroplasmas. *Adv. Virus Res.* **21:** 343.

Nowak, J. A., J. Das, and J. Maniloff. 1976. Characterization of an *Acholeplasma laidlawii* variant with a Rep⁻ phenotype. *J. Bacteriol.* **217:** 832.

Replication of the Viral DNA

The DNA Replication Cycle of the Isometric Phages

David Dressler, Dennis Hourcade, Kirston Koths, and John Sims
Department of Biochemistry and Molecular Biology
Harvard University
Cambridge, Massachusetts 02138

It is just twenty years since Robert Sinsheimer reported the results of his initial studies on a small and previously uncharacterized virus, φX174. It is one of the ironies of research that a virus initially so different and odd should gradually emerge as a system offering some of the deepest insights into a major area of molecular biology, DNA replication.

In contrast to the disruptive entrance into the host cell of larger viruses such as T4 and T7, bacteriophage φX and the related isometric phage G4 set to work quietly, not disturbing the host cell except for an eventual cessation of bacterial DNA synthesis (Tessman 1966; Stone 1970). Even in the midst of this seeming order, however, the virus diverts the host sufficiently to generate a thousandfold replication of itself—a process which is readily observed if one studies a mutant virus that does not lyse the host prematurely.

The host cell supplies almost all of the enzymatic machinery required for the viral DNA replication cycle. It is precisely for this reason that the single-stranded (SS) DNA phages are able to serve as a window into the DNA replication apparatus of *Escherichia coli*. The way in which the viral chromosome is processed during its replication reveals the reaction mechanisms of many host enzymes that normally play a role in bacterial DNA synthesis but which are commandeered by the virus after infection. In this article we will consider the various steps in the φX and G4 replication cycles and discuss the results in terms of DNA replication in general.

NEGATIVE-STRAND SYNTHESIS

A general problem in DNA synthesis concerns the mechanism by which DNA fragments are initiated, and this problem is brought into experimental focus by considering the first step in the life cycle of the isometric phages.

Conversion of Single-stranded Circles to Duplex Rings

For both φX and G4 the DNA replication cycle begins with the conversion of the infecting positive-strand circle to a duplex ring. In this event a negative strand is formed de novo using the infecting positive-strand circle as a template.

The conversion of the single-stranded circle to the duplex ring has a general relevance to DNA replication in that this synthesis shares with the DNA growing point the problem of initiating a polynucleotide strand de novo (cf. Fig. 1 A and B). In the case of the SS DNA phages, however, the situation is much more accessible to experimental analysis. Whereas the synthesis of DNA fragments in the growing points of the bacterial chromosome or of large viruses involves the formation of hundreds or even thousands of individual fragments, the small size of the viral DNA circle (5500 nucleotides) leads to the expectation that only one or at most a very few fragments are required. Thus, the conversion of SS DNA phage circles to duplex rings offers a model experimental system for studying the formation of DNA fragments.

One question that immediately presents itself concerns the nature of the enzymes involved in negative-strand synthesis. The synthesis of the negative strand occurs even if the infected cells are cultured in the presence of a high concentration of chloramphenicol, a condition under which virus-directed protein synthesis is completely inhibited (Tessman 1966). Thus, it can be concluded that all of the enzymes required for the formation of the negative strand are present in the host cell before infection. These proteins have been identified in part through studies in vivo (Dumas, this volume) and more completely by studies in which the conversion was carried out in vitro (McMacken et al.; Wickner; both this volume). The in vivo studies have shown that negative-strand synthesis involves a parasitization of host-cell DNA replication enzymes (rather than the use of a DNA repair pathway as was once thought). Mutations in several genes which render bacterial DNA

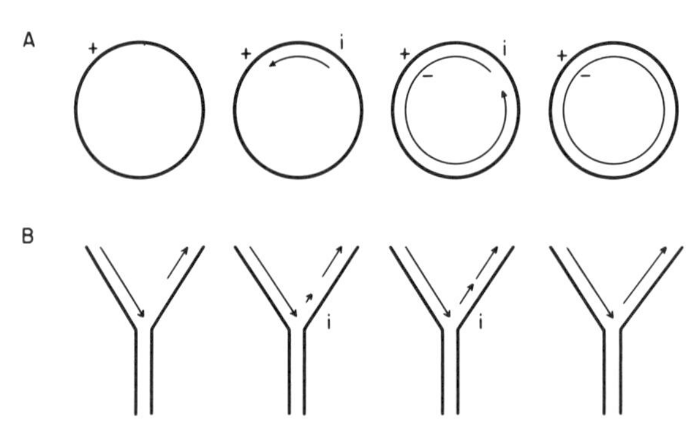

Figure 1 The first step in the DNA replication cycle of the isometric phages is the conversion of the infecting single-stranded circle to a duplex ring. This event involves the de novo initiation of a DNA strand (*A*). Analogous initiation events, repeated many times and in many places, are required in the bacterial DNA growing points (*B*).

replication labile at elevated temperature also block the synthesis of the viral negative strand. The relevant bacterial replication proteins are the products of genes *dnaB, dnaC, dnaE* (DNA polymerase III), *dnaG,* and *dnaZ*.

Catalysis of the reaction in vitro has provided a complete description of the proteins required (some of which have not yet been identified by mutations) and has led to an assignment of functions (Schekman et al. 1974; Wickner and Hurwitz 1974; Bouché et al. 1975, 1978; Wickner 1977; Rowen and Kornberg 1978a,b). For both G4 and φX the reaction can be divided into two steps. In the first step, a priming event prepares the template circle for replication. A short oligonucleotide is synthesized, either directly by the *dnaG* protein in the case of G4 or by a collection of enzymes (*dnaB*, *dnaC*, *dnaG*, and three additional protein factors) in the case of φX. The oligonucleotide primer remains associated with the template DNA after the priming protein leaves and in the second stage of the reaction is extended by DNA polymerase III holoenzyme to give a nearly full-length negative strand. These in vitro results form the experimental basis for the *primer hypothesis*, which in its most general form states that the initiation of DNA strands is accomplished by a set of specialized initiation polymerases that are distinct from the polymerases that elongate DNA chains (none of which can begin polynucleotide chains).

Aside from defining the proteins involved in fragment formation (negative-strand synthesis), there is another aspect to the initiation problem: Is the new strand begun at a specific site or randomly on the template DNA (in the case of G4 and φX the 5500-base positive-strand ring)? Essentially, one is asking whether the bacterial DNA replication enzymes that perform this synthesis initiate chains in response to specific recognition signals. G4 and φX give somewhat contrasting answers, and we will consider each case separately.

The G4 Negative Strand

To determine whether the G4 negative strand is initiated at a specific site, one needs to arrest its synthesis part way around the template positive-strand circle. The aim is to prevent the growing 3′ end of the negative strand from approaching the 5′ initiation point and being sealed to it by ligase, thereby erasing the initiation point. The synthesis of a full-length negative strand can be prevented by irradiating G4 phage with ultraviolet light prior to infection so that thymine dimers are formed in the positive-strand template DNA. The thymine dimers block the forward movement of the replication machinery (Rupp and Howard-Flanders 1968; Benbow et al. 1974; Masamune 1976; Caillet-Fauquet et al. 1977), and the cells then produce only partially duplex rings—molecules in which a negative strand has been initiated but not completed.

Two general results may be expected. If replication were to begin at a unique site, then roughly the same region of the G4 ring would be converted to the duplex state in all partially replicated molecules (Fig. 2A). On the other hand, if negative-strand synthesis were to begin at a different site on each chromosome, all G4 sequences would become duplex in one or another partially replicated molecule (Fig. 2B). Experimentally, one expects either that the partially duplex rings will release only a subset of the restriction fragments present in fully duplex rings or, alternatively, that all restriction fragments will be found in the collection of partially replicated rings.

With these alternatives in mind, G4 partially duplex rings were recovered from infected cells and treated with S1 endonuclease to remove that portion of the template DNA that was still single-stranded. The remaining double-helical regions were then analyzed with restriction enzymes for their degree of sequence heterogeneity.

Figure 3 shows the results obtained with the restriction enzyme *Hin*dII. While this enzyme cleaves G4 fully duplex rings into six specific fragments (Fig. 3A,B), cleavage of the duplex material derived from the partially replicated molecules yields only fragments B, C, D, and E, which are clustered together in one region of the G4 chromosome (Fig. 3C). Furthermore, bands B and D are prominent, whereas bands C and E are very faint.

These observations are readily explained if initiation of the negative strand were to begin in fragment A, with synthesis proceeding sequentially through areas B, D_1, E, C, and D_2 and then back through region A (as diagrammed in Fig. 3A). Because replication would be stopped by thymine dimers at various sites around the positive-strand template, fragment A would almost never be converted to a completely duplex state and thus would be missing from the restriction enzyme digest. Since replication would usually proceed as far as regions B and D_1, and sometimes go beyond into E,

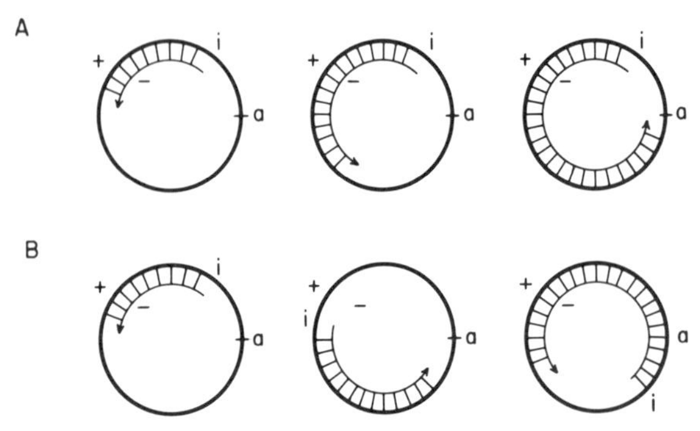

Figure 2 The set of partially replicated molecules expected if (*A*) initiation of the negative strand always occurs at a specific site on the positive-strand circle (with respect to a reference marker "a") or (*B*) if initiation occurs randomly on the template DNA.

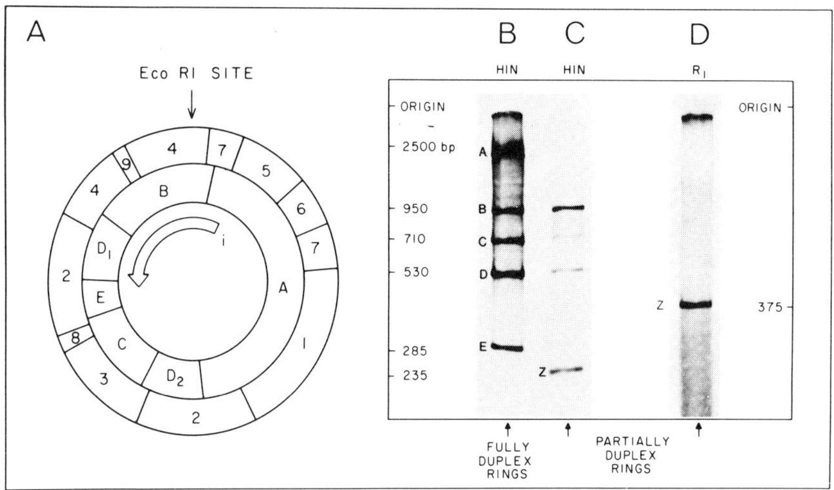

Figure 3 (*A*) A G4 duplex ring with the restriction enzyme sites for cleavage by *Hin*dII, *Eco*RI, and *Hae*III marked. (*B*) A polyacrylamide gel that displays, according to size, the set of fragments released from fully duplex rings by cleavage with *Hin*dII. (*C*) The fragments released by *Hin*dII cleavage of S1-treated partially duplex rings. Bands B and D are most heavily represented, which leads to the conclusion that negative-strand initiation occurs in this region of the G4 chromosome. Band Z (235 base pairs in size) is not found in fully duplex rings and is interpreted as representing the region of the G4 positive strand made duplex from the origin to the first *Hin*dII cut. (*D*) The fragment released from S1-treated partially duplex rings by treatment with *Eco*RI. This band (375 base pairs in size) is interpreted as representing the distance from the initation site to the single RI cut. (Data from Hourcade and Dressler 1978.)

C, and D_2, fragments B and D are recovered in greater quantity than E and C.

From the point of view of locating the negative-strand initiation site precisely, one result of the *Hin*dII analysis is particularly useful—the appearance of a new fragment not found in G4 fully duplex rings (Fig. 3C, band Z). This fragment cannot be a partial digestion product because it is smaller than all of the other G4 *Hin*dII fragments. It can be explained, however, as representing the sequence from the replication origin to the first restriction site along the duplex DNA (see Fig. 3A). The size of the new fragment places the initiation site on the positive-strand template at about 235 bases from the first *Hin*dII site—the cut between fragments A and B.

This interpretation of the *Hin*dII digest can be verified by further analysis with different restriction enzymes. A fully duplex G4 chromosome is cleaved only once by *Eco*RI (Fig. 3A). If replication indeed begins in fragment A 235 base pairs from the first *Hin*dII site, then RI cleavage of S1-treated partially replicated molecules is expected to produce a unique fragment representing the region from the origin to the RI cleavage site some 375 base pairs away. This fragment is in fact released by *Eco*RI, as seen in Figure 3D.

A final confirmation of the results derived from the studies with *Hin*dII and *Eco*RI was obtained by use of the restriction enzyme *Hae*III. When the partially replicated rings were digested with *Hae*III, only fragments 4 and 7 were often found converted to the duplex state (see Fig. 3A). An additional fragment about 100 base pairs long was also released from the partially duplex rings. This fragment is interpreted as representing the sequence from the origin to the first *Hae*III site.

In sum, by analysis with three different restriction enzymes it may be concluded that G4 positive strands are converted to double-stranded circles by initiation of the negative strand at a specific point 375 bases from the *Eco*RI cleavage site, 235 bases from a defined *Hin*dII site, and about 100 bases from a defined *Hae*III site (Hourcade and Dressler 1978; see also Martin and Godson 1977). Thus, the initiation of the G4 negative strand, an example of the initiation of a DNA fragment, is site-specific.

These in vivo data on G4 interlock with the results obtained in vitro by Zechel et al. (1975). Using a purified enzyme system (*dnaG* protein, DNA-binding protein, and DNA polymerase III holoenzyme) to convert G4 SS DNA circles to duplex rings, they have found that a small gap remains in the newly synthesized negative strand. The gap is expected to be bounded by the beginning (5′) and the end (3′) of the nascent negative strand, and analysis of its position has shown that the gap is located about 250 bases from the single *Eco*RI cleavage site. Thus, the in vivo and in vitro results reinforce one another and lead to the same conclusion: the G4 negative strand is initiated at a specific site.

Nucleotide Sequence of the G4 Negative-strand Origin

Knowing that the G4 negative strand is initiated at a unique site, one would like to know the nature of this regulatory DNA element. This site serves as the nucleic acid counterpart of the *dnaG* protein, providing a unique recognition element on the positive strand at which the replication enzymes can begin negative-strand synthesis.

The DNA sequence of the negative-strand initiation site has been determined by constructing a detailed restriction enzyme map of this area of the G4 chromosome and then sequencing the relevant restriction fragments (Sims and Dressler 1978; Fiddes et al. 1978). The relevant portion of the G4 chromosome is shown in Figure 4. In accord with in vivo data already discussed, the initiation site of the negative strand can be placed approximately 100 bases from the *Hae*III site (at the position marked by the horizontal bar).

Examination of the sequence immediately reveals one interesting result.

```
                                +150          +140  Hin f
         Positive Strand 5' ATGCCTACGA CACGTGACTC
         Negative Strand 3' TACGGATGCT GTGCACTGAG

 end of gene F                    Hae
+130          ↓          +110   III   +100        +90
 AATCATGACC TCGTAACGCA ACAAAGGCCG CCCCTCTACT GGTCAGATAC
 TTAGTACTGG AGCATTGCGT TGTTTCCGGC GGGGAGATGA CCAGTCTATG

+80        +70        +60        +50        Hin f
 CTGCCCAATG TGGGGCGGAC CGTGCCTACG GAGATACTCG AGTCTCCGAT
 GACGGGTTAC ACCCCGCCTG GCACGGATGC CTCTATGAGC TCAGAGGCTA

+30        +20        +10        +1 -1        -10        -20
 ACATGGACGG CGAAAGCCGC CGTCCCTACT GCAAAGCCAA AAGGACTAAC
 TGTACCTGCC GCTTTCGGCG GCAGGGATGA CGTTTCGGTT TTCCTGATTG
3' HO UACCUGCC GCUUUCGGCG GCAGGGAUGA ppp 5' primer
   Start of gene G
        -30        -40        -50        -60        -70
 ATATGTTCCA GAAATTCATT TCTAAGCACA ATGCTCCAAT TAACTCTACT
 TATACAAGGT CTTTAAGTAA AGATTCGTGT TACGAGGTTA ATTGAGATGA
```

Figure 4 The sequence of the G4 positive and negative strands in the intercistronic region containing the negative-strand initiation site. The sequence has been numbered so that +1 corresponds to the first nucleotide of the nascent negative strand. (Data from Sims and Dressler 1978.)

The initiation site lies in an intercistronic region between genes *F* and *G*. This is deduced by scanning the sequence for AUG start codons and UAG, UAA, and UGA stop codons and then observing that the triplets following the AUG at position −23 code for a protein essentially identical to the gene-*G* coat protein of the related phage φX (whose sequence is known from the work of Air et al. 1975, 1976; see Appendix I, this volume). Correspondingly, the triplets leading up to the UAA stop codon at DNA position +117 code for a protein nearly identical to the φX gene-*F* protein. The intercistronic region that contains the G4 negative-strand initiation site is 136 bases long. It can be drawn as having three regions of secondary structure which may form the recognition element for the *dnaG* priming protein (Fig. 5).

The most important part of the initiation-site sequence is the 28-base stretch 5′ ATGGACGGCGAAAGCCGCCGTCCCTACT 3′. This stretch is complementary to the sequence of the short oligonucleotide synthesized in vitro by the *dnaG* protein (Bouché et al. 1978). The presence of this sequence at the physiological initiation site verifies one aspect of the primer hypothesis: the oligonucleotide synthesized by the *dnaG* protein in vitro

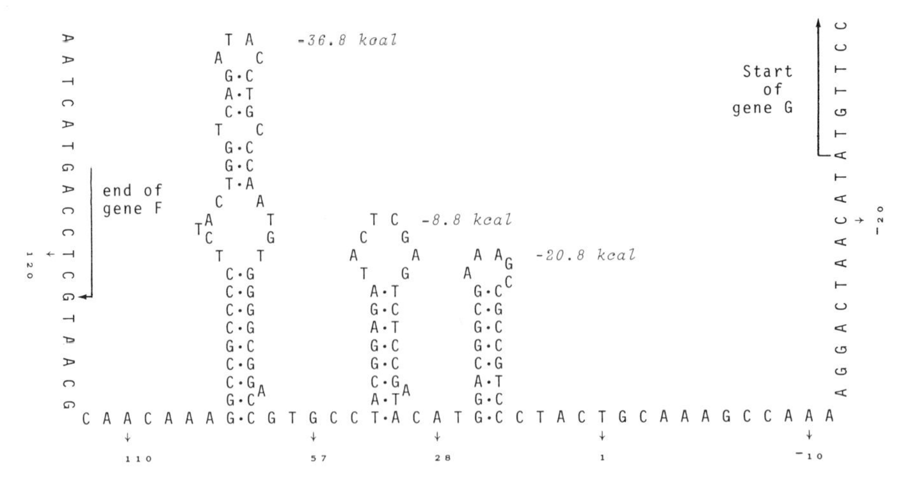

Figure 5 The G4 positive-strand template at the negative-strand initiation site. The diagram is drawn to show the potential regions of secondary structure that may form the recognition site for the *dnaG* priming protein. (Data from Sims and Dressler 1978.)

and used to prime the formation of the negative strand is encoded intact at the in vivo negative-strand initiation site. The 5′ end of the primer synthesized in vitro is found 104 bases from a *Hae*III cut; our experiments in vivo had located the start of negative-strand synthesis about 100 bases from this same *Hae*III cut.

Thus, for G4, the in vivo, in vitro, and DNA sequencing studies complement one another. Taken together, the data indicate exactly where polymerization starts, and thus the G4 origin is the first initiation site which makes use of the *E. coli* replication *dnaG* protein for which the start point of DNA synthesis is known at the nucleotide level.

The φX Negative Strand

Given a clear and simple result for negative-strand synthesis in G4, one might hope for a similar result with the related phage φX. However, the situation with φX is less clear. For instance, when the thymine-dimer experiments carried out with G4 are performed with φX, we find that the initiation of the negative strand is not strictly confined to a single site. After digestion of S1-treated partially duplex rings we do not find a simple subset of contiguous restriction fragments plus one new duplex piece (as is found for phage G4, Fig. 3). There may be a preferred initiation site near the boundary between φX genes *A* and *H*, but there appear to be at least one or two other starts in the adjacent 50% of the template ring, which contains genes *F*, *G*, and *H*. The region containing genes *A*, *B*, *C*, *D*, and *E* appears not to have a strong initiation site. Thus, the data provide no evidence of a single, unique site for the initiation of the φX negative strand, and the presence of multiple initiation sites makes the analysis difficult.

How can the apparent difference between G4 and φX be explained? Of course, if it is simply that there are more (but still defined) start signals in φX, there is really no problem. However, some investigators have reported numerous, virtually random starts in φX (Eisenberg and Denhardt 1974; McMacken et al. 1977). If this is in fact the case, negative-strand synthesis in φX is fundamentally different from that in G4. Such a difference could reflect the more complex enzymology involved in the conversion of the φX positive-strand circle to the duplex ring. Although initiation of negative-strand synthesis in both φX and G4 involves the *dnaG* protein, fragment formation in the case of φX (and probably in the bacterial DNA growing points) requires in addition the participation of the products of genes *dnaB* and *dnaC* and three additional protein factors. The difference between negative-strand initiation in G4 (simple site-specific) and that in φX (complex, apparently less specific) may then be understood in terms of enzymology. Still, the difference is somewhat puzzling. One usually associates an increasing number of proteins with increasing specificity, but here the introduction of additional proteins leads to less specificity for initiation. Perhaps the initiation enzymes that work on templates like φX select a specific entry site but then move along the template to a random place before initiating the negative strand (McMacken et al. 1977). A precedent for this possibility can be found in the behavior of the *E. coli* B restriction enzyme, which has genetically defined recognition sites but cleaves DNA randomly at distances up to 1000 nucleotides from the recognition site (Horiuchi and Zinder 1972). A rationale for randomized initiation sites for the φX negative strand could be that the replication system is deliberately beginning strands in various places so that errors introduced into the 5′ regions of the nascent strands will be distributed around the genome and not focused at particular sites.

In sum, the site specificity in negative-strand initiation observed with bacteriophage G4 cannot, at least now, be regarded as a general phenomenon. What then is its significance? The importance of the *dnaG* initiation site is that it demonstrates that it is possible in nature for a DNA strand to be initiated at a specific place by means of a relatively simple system consisting of only the *dnaG* priming protein and a recognition element formed by the template DNA itself. The defined protein-nucleic acid interaction leading to the G4 initiation event is valuable in that it provides a system that can be analyzed experimentally to determine what constitutes an origin of DNA synthesis.

FUNCTIONS OF THE DUPLEX RING

The synthesis of the negative strand converts the isometric phage chromosome to a double helix and thus sets the stage for the transcription of the phage genes by RNA polymerase. The nine or ten genes of the viral chromosome are then transcribed into messenger RNA by the host RNA polymerase. The polymerase recognizes several promoter sites and copies the newly synthesized negative strand (see Hayashi, this volume). No modification of the bacterial transcription machinery is induced; the turnoff of one set of genes and the activation of another through the synthesis of new polymerase specificity factors, or a new RNA polymerase, is not observed. Thus, the reason for the appearance of the viral proteins in unequal amounts is to be sought in the relative strengths with which the promoters attract the RNA polymerase, in the unequal efficiencies with which the RNA transcripts interact with the ribosomes to initiate protein chains, or in the dissimilar half-lives of the messengers. Thus, differential gene expression in the SS DNA phages is preprogrammed into the DNA as a static element.

In addition to their role in transcription, duplex rings serve two other purposes. One involves genetic recombination. Here again host enzymes (the RecA recombination system) are involved, and therefore the study of viral recombination sheds light on the recombination mechanism used by the host cell itself. As discussed by Warner and Tessman (this volume), recombination involves the fusion of two monomer rings, leading to the formation of a Holliday-type recombination intermediate. The intermediate structure has the appearance of a figure eight and has been observed by electron microscopy.

The last function of the viral duplex ring concerns its central role in further DNA replication, and this will be discussed in detail below.

SYNTHESIS OF POSITIVE-STRAND DNA

The initial stage of the phage life cycle, the formation of the first duplex ring, has never been especially controversial and is now understood in considerable detail. The events that come next, however, have been the subject of much spirited inquiry, as they represent a break with the traditional view in which daughter DNA strands are synthesized by de novo initiation.

The remainder of the DNA replication cycle consists of two events. First comes an early period in which double-stranded DNA circles are the net product of synthesis (Sinsheimer et al. 1962; Denhardt and Sinsheimer 1965; Lindqvist and Sinsheimer 1967). This is followed by a terminal period during which positive-strand DNA is asymmetrically produced to provide SS DNA for progeny virus particles (Dressler and Denhardt 1968; Knippers et al. 1968; Komano et al. 1968).

A proposal for the mechanism of this synthesis, the rolling-circle model, was made on the basis of data obtained when it became possible to isolate and study actively replicating φX DNA molecules—that is, molecules in the middle of a round of replication as opposed to unit-size duplex rings that were about to begin or had just completed a round of replication (Gilbert and Dressler 1969; Dressler 1970). Through gentler procedures for processing intracellular DNA, actively replicating DNA molecules were found as a continuum of structures sedimenting more rapidly than unit-size duplex rings (Fig. 6). The architecture of these partially replicated molecules was reconstructed from a knowledge of their constituent polynucleotide strands. Pulse-labeled intermediates recovered during the terminal stage of single-stranded circle synthesis were found to contain a radioactive positive strand of longer than genome length (Fig. 7A). The long strand was associated with a nonradioactive negative-strand template ring, which was released from the intermediate upon alkaline denaturation and rendered visible by virtue of its unique density in alkaline CsCl and its infectivity to spheroplasts (Fig. 7B). The 3′-OH end of the long positive strand was located on the negative-strand template ring and thus was in a position to be elongated directly by the DNA replication machinery (Gilbert and Dressler 1969; Knippers et al. 1969; Dressler 1970; see also Schröder and Kaerner 1972; Godson 1977).

From these data it was possible to extrapolate backward from the structure of the intermediate to determine the initiation events that would give rise to the rolling circle and also forward to the events that would allow the maturation of a progeny DNA molecule from the tail of the intermediate. Thus, by considering rolling circles with shorter and shorter tails and then no tail at all, one obtained the picture of a duplex ring with a specific nick in the positive strand at the origin of synthesis. Such a molecule would be derived from a covalently closed duplex ring acted upon by a "positive-strand nickase" encoded by the virus (Gilbert and Dressler 1969;

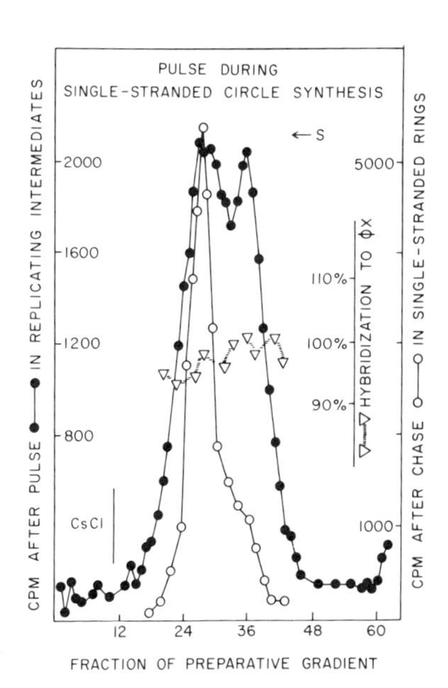

Figure 6 Isolation of φX rolling-circle intermediates. φX-infected cells were pulse-labeled during the period of circular SS DNA synthesis and the unfractionated cell lysate was centrifuged through a neutral velocity gradient. The radioactivity was recovered in a broad peak of structures (●——●) that sedimented more rapidly than relaxed duplex rings. Hybridization analysis (▽) showed that the label was present in φX DNA. During a chase the label flowed out of the replicating structures and into φX circular SS DNA (o——o). The polynucleotide strand substructure of the replicating intermediate is analyzed in Fig. 7. (Data from Dressler 1970.)

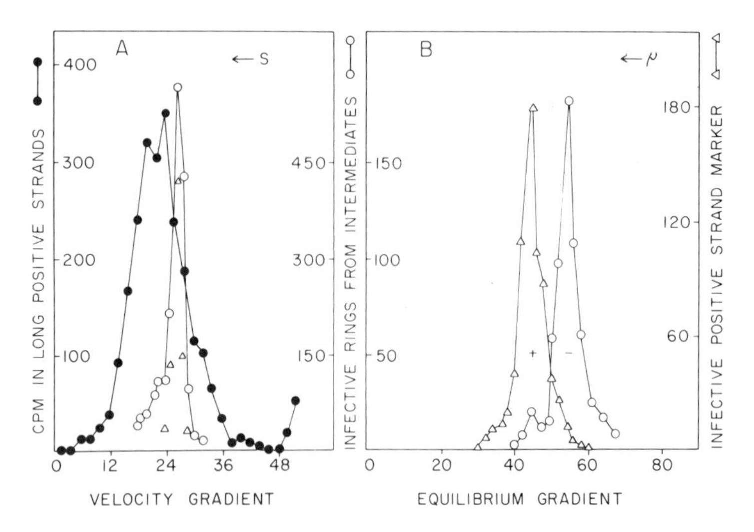

Figure 7 Strand analysis of φX replicating intermediates. Pulse-labeled φX replicating intermediates were recovered from the gradient shown in Fig. 6 and denatured into their component polynucleotide strands with alkali. The sedimentation profiles of the radioactive and the infective strands from the replicating intermediates are shown in *A*. In addition to the radioactive long positive strands (fractions 12–24), the denatured intermediates release an infective component which is responsible for the ability of fractions 25–28 to generate φX *am*3 particles upon incubation with *E. coli* spheroplasts. The infective component from the intermediate is identified as a single-stranded circle because it sediments with the same velocity as marker positive-strand circles (added at the time of centrifugation and identified in the spheroplast assay by their production of φX h4 phage).

(*B*) The infective single-stranded circles were recovered from fractions 25–28 of the gradient shown in *A* and then centrifuged to equilibrium in alkaline CsCl, which separates φX positive and negative strands. The rings corresponding in genotype to the replicating intermediates (*am*3) separated from the marker positive-strand circles (h4) and came to equilibrium at the density characteristic of φX negative-strand DNA (fractions 50–60). (Data from Dressler 1970.)

Dressler 1970). Hutchison and Sinsheimer (1966) had already found that one viral gene was essential for DNA replication beyond the synthesis of the first duplex ring, and studies of mutant phage by Francke and Ray (1972) showed that this gene was in fact responsible for producing nicked duplex rings in vivo. The nicking protein, the product of cistron *A*, has had a somewhat complicated history because of genetic experiments which suggested that it could not participate in complementation (see Baas and Jansz, this volume). Nevertheless, the nicking protein was detectable in extracts of φX-infected cells (Henry and Knippers 1974) and has been shown to be capable of introducing a discontinuity into φX positive strands at a specific site (Langeveld et al. 1978; Baas and Jansz, this volume). The φX gene-*A* nickase is able to function specifically on both φX and G4. The protein has been highly purified in several laboratories and is used in vitro to initiate synthesis upon duplex rings. In conjunction with DNA polymerase III holoenzyme, DNA-binding protein, and the *rep* unwinding protein, the gene-*A* protein leads to the production of a rolling-circle intermediate with a negative-strand template ring and an elongated positive strand (Eisenberg et al. 1977; Marians et al. 1977).[1]

Rolling-circle Intermediates during Single-stranded Circle Synthesis

The φX rolling circle is not a rare structure. In fact, there are about 35 per cell. This conclusion comes from a quantitative analysis of viral DNA forms as they develop during a life cycle that reflects as closely as possible the normal infection of individual cells by single virions (Koths and Dressler 1978). In this study, the host cells were grown in heavy-isotope medium, transferred to light medium, and infected with isotopically light phage. During the subsequent incubation period the bacterial DNA remained heavy, or shifted toward hybrid. In contrast, as they developed, the intracellular φX DNA forms had densities that were essentially light. Separation of the viral forms from host DNA could thus be accomplished by centrifugation of the total cell lysate to equilibrium in CsCl (Fig. 8).

This procedure concentrates and purifies the viral DNA, making it possible to use the electron microscope to determine both the relative and absolute numbers of φX forms as the virus life cycle proceeds. This may be accomplished by adding a known amount of an internal standard molecule to the sample and then spreading the mixture for electron microscopy. The internal standard chosen was a circular, supercoiled plasmid DNA that

[1] In the rolling-circle mechanism, the nick introduced by the gene-*A* protein both defines the origin for positive-strand synthesis and provides a 3′-OH primer for strand elongation. In our experiments we find that even for pulses as short as 5 sec, the great majority of label in the rolling circles is present in long positive strands. However, there is some label present in positive strands that are shorter than unit length (see, for instance, Fig. 7A). We can readily accept the existence of the short positive strands because, aside from the 3′-OH end of the positive strand which can be elongated directly, the possibility exists that *rep*-mediated unwinding of the double helix ahead of the growing point may expose a segment of negative-strand template that can be used for the de novo initiation of a positive-strand fragment. Such an event does not alter the essential character of the rolling-circle process in which the nick introduced by the gene-*A* protein defines the origin of replication and creates a 3′-OH terminus that can be elongated by DNA polymerase III. We view the possible formation of short positive strands as an additional option available to the cell because it contains enzymes capable of initiating DNA chains de novo. In contrast to this view, Okazaki and Denhardt (Machida et al. 1977; Denhardt 1975, 1977; Hours et al., this volume) have discussed the short strands as obligatory intermediates in positive-strand synthesis. But, before accepting this role for the short strands in the rolling-circle intermediates, it will be necessary to know that they are not due to (1) breakage of the DNA during isolation or (2) the misincorporation during synthesis of dUTP nucleotides followed by an excision and repair process that leaves the positive-strand transiently nicked (Tye et al. 1977).

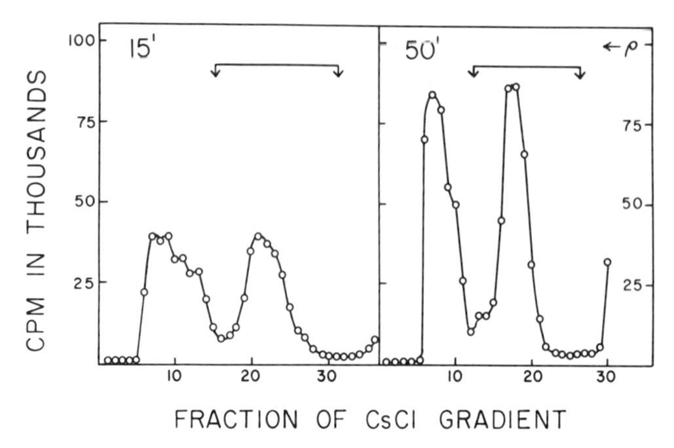

Figure 8 Isolation of intracellular φX DNA forms. Cells grown in heavy isotope medium were shifted to light medium and infected with φX at a multiplicity of 0.7. At various times after infection aliquots of the culture were pulse-labeled with tritiated thymidine and harvested. After centrifugation of the unfractionated cell lysates to equilibrium in CsCl, two peaks of radioactivity were separated: a viscous, denser peak containing bacterial DNA, and a less dense peak containing the viral DNA forms. The φX DNA forms were analyzed quantitatively by electron microscopy (see Figs. 9 and 10 and Table 1). (Data from Koths and Dressler 1978.)

could be readily distinguished from φX DNA forms in the electron microscope because of its larger size.

For a typical analysis (for instance of the viral DNA harvested at 50 min), 1/2500 of the CsCl-purified DNA sample was supplemented with 8×10^8 plasmid molecules and spread for electron microscopy. A random sampling of 1000 plasmid DNA molecules was then counted along with the number of each φX DNA form. When the number of φX rolling circles proved to be, for instance, 180 in the material harvested at 50 minutes, it could be concluded that the sample aliquot contained 18% as many rolling circles as plasmid molecules, or 1.4×10^8 rolling circles. Since the sample spread for the electron microscope represented 1/2500 of the material from the CsCl gradient and had been derived from 10^{10} infected cells, it could be calculated that the number of rolling-circle intermediates was $(1.4 \times 10^8) \times (2500)$ rolling circles divided by 10^{10} cells, or 35 rolling circles per cell.

The intracellular viral DNA forms recovered from cells at various times after infection were cataloged (Fig. 9 and Table 1), and calculations similar to the one described above were made to determine the number of φX DNA forms as they appeared with time after infection. The results are summarized in Figure 10. It is seen that the number of duplex rings increases during the first 25 minutes after infection, reaching a plateau level of about 60 per cell. The duplex rings are of two kinds, which occur in approximately equal numbers. One type is a double-stranded DNA circle (Fig. 9C) in which

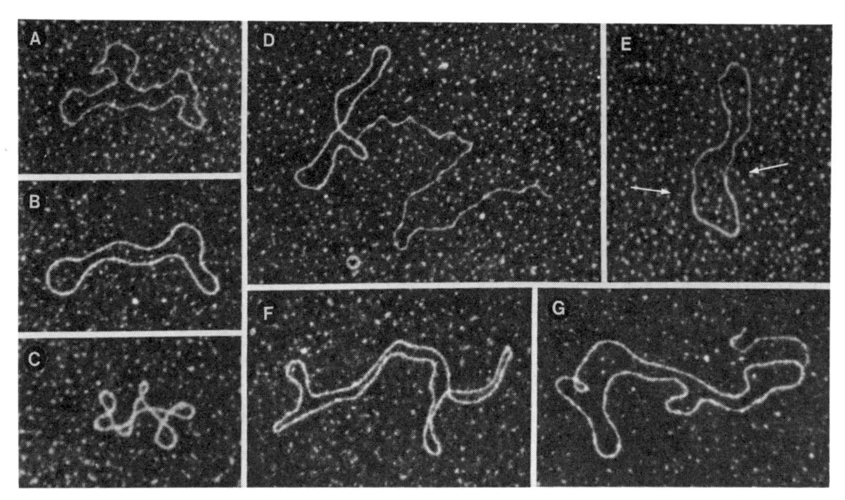

Figure 9 Electron microscopy of φX DNA forms. Intracellular φX DNA forms present at various stages in the virus life cycle were recovered from CsCl gradients (see the pooled fractions in Fig. 8) and spread for electron microscopy without further processing. The DNA was rendered visible simply by staining with uranyl acetate. (*A*) A single-stranded circle. (*B*) A relaxed duplex ring. (*C*) A supercoiled duplex ring. (*D*) A rolling-circle intermediate. (*E*) A partially duplex ring. (*F*) A dimer-size duplex ring. (*G*) A rolling-circle intermediate in which the template ring is a dimer circle. (Data from Koths and Dressler 1978.)

both strands are covalently closed (Vinograd and Lebowitz 1966) and superhelical twists have been introduced into the molecule by DNA gyrase (Gellert et al. 1976). The other form is a relaxed duplex ring whose open appearance is the result of either an internucleotide bond interruption (Burton and Sinsheimer 1965; Schekman et al. 1971; Johnson and Sinsheimer 1974) or a deficiency of superhelical twists (Fig. 9B).

During the latter part of the life cycle, single-stranded circles are produced at a linear rate of about 30 to 60 per minute per cell (at 30°C and 37°C, respectively), reaching a final level of several thousand. The rolling circles increase from a level of about 1 per cell at 8 minutes to about 35 per cell by 40 minutes.

The relatively large number of rolling-circle intermediates recovered per cell is encouraging from an experimental point of view; there is at this time no other system from which a uniform replicating intermediate can be isolated in such quantity. We have taken advantage of this numerical windfall to study the intermediates biochemically.

Late in the life cycle the rolling circles exist not as free DNA structures, but rather as complexes with proteins that are involved in phage maturation (Fujisawa and Hayashi 1976; K. Koths and D. Dressler, in prep.). If the intracellular φX DNA forms are isolated from the cell under conditions

Table 1 Number by type of intracellular φX DNA forms as they appear with time after infection

Time after infection (min)	Common forms				Partially duplex rings	Dimeric forms	Other	Total φX	Total internal standard
	single-stranded circles	duplex rings		rolling circles					
		supercoiled	relaxed						
15	15	266	407	73	10	17	2	790	1200
25	868	212	350	172	1	26	5	1634	1500
35	1898	137	194	185	1	17	2	2434	1000
50	3454	136	248	180	0	13	1	4032	1000
65	4746	92	201	159	1	18	0	5217	1000
Cells grown in 35 μg/ml chloramphenicol									
26	3	430	125	16	14	11	1	600	1417
26	5	438	121	16	10	8	2	600	1385
26	8	414	145	15	5	13	0	600	1452
Total for 26, 41, and 55 min	38	3697	995	100	55	111	4	5000	7792

Data from Koths and Dressler (1978).

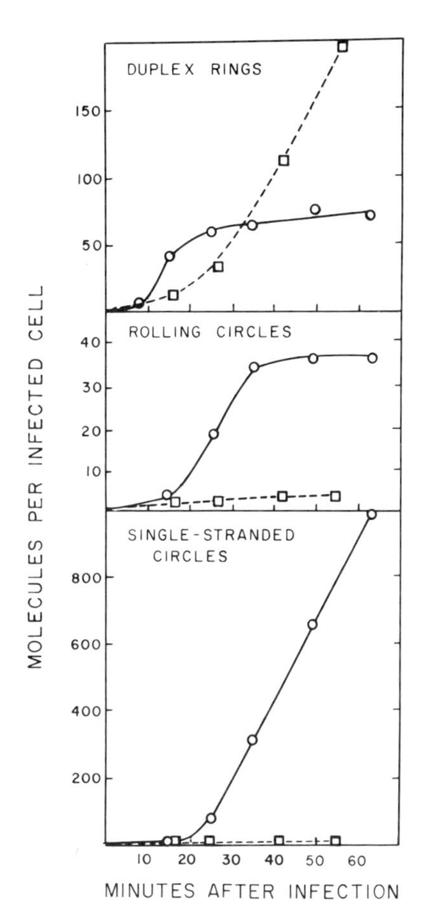

Figure 10 Numbers of the different φX DNA forms per cell as calculated from the data in Table 1. The 8-min point is from an identical experiment which gave identical time curves. Normal infection (o——o); infection in the presence of chloramphenicol at 35μg/ml (□——□).

whereby deproteinization is avoided, one is able to purify to homogeneity a protein-nucleic acid complex that contains a rolling-circle intermediate and an intact phage capsid (Fig. 11A).

Spreading the complex for the electron microscope at 4°C, one sees a duplex ring with an attached capsomeric structure (Fig. 11B, left). All of the molecules have this appearance. However, when the material is prepared for electron microscopy and briefly warmed to room temperature prior to spreading, the complex becomes partially disrupted. The complex is then seen to contain a rolling-circle intermediate with a tail of variable length to which a capsid is attached (Fig. 11B, right). Evidently, the displaced DNA is enclosed within the capsomeric structure during synthesis.

Spreading the complex for the electron microscope after incubation in 50% formamide at room temperature dislodges the capsid component completely. One obtains a population consisting entirely of rolling circles. Using these molecules one can show that the tails in the rolling-circle

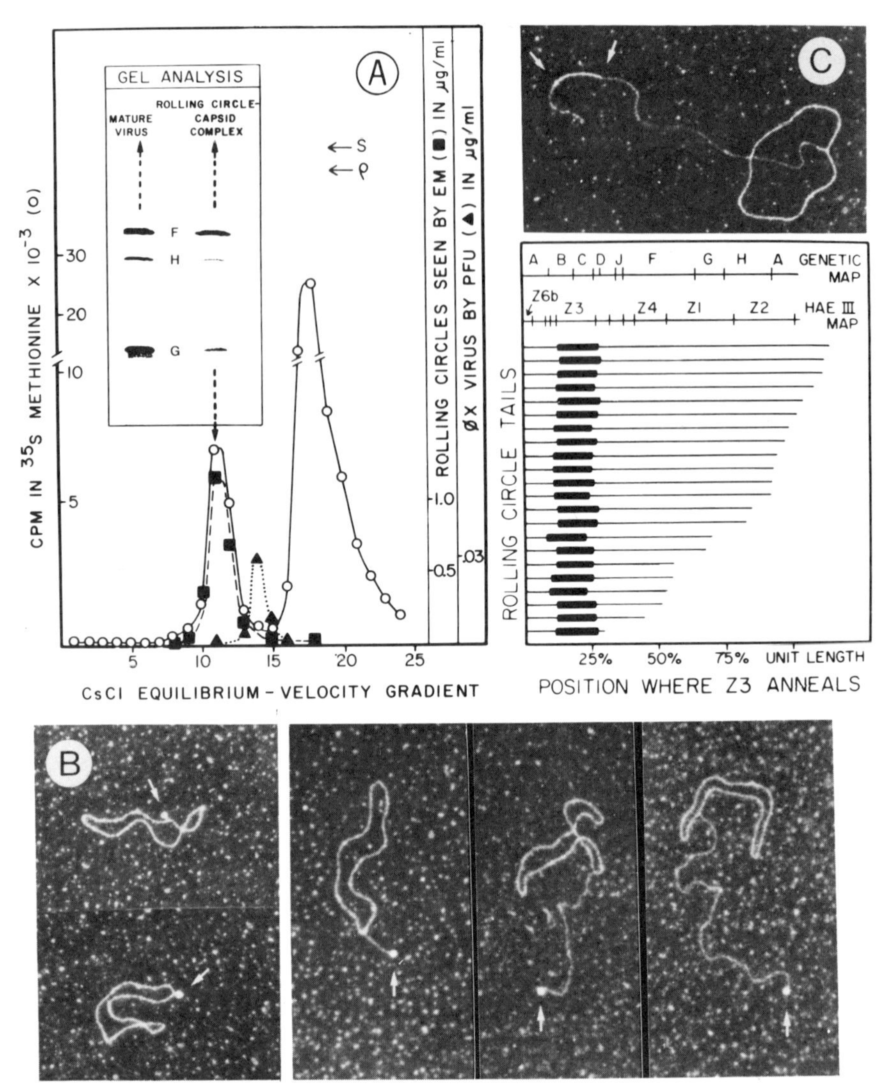

Figure 11 *(A)* Rolling circle-capsid complexes. φX-infected cells were grown in the presence of [^{35}S]methionine and harvested during the period of single-stranded circle synthesis. Rolling circle-capsid complexes were purified through two gradients: a neutral sucrose velocity gradient followed by centrifugation into a preformed CsC1 gradient. In CsC1 the complexes reached their buoyant density (*A*, fractions 10–13). The inset in *A* shows a gel analysis of the proteins present in the rolling circle-capsid complex; they are the major capsid proteins (the products of genes *F*, *G*, and *H*) and are present in approximately the same ratio as in mature virions.

(B) Electron micrographs of intact (left) and partially disrupted (right) rolling circle-capsid complexes. During the synthesis of circular SS DNA, the rolling-circle intermediate evidently displaces positive-strand DNA into a preformed capsid.

(C) The purified *Hae*III restriction fragment Z3 from φX duplex rings has been denatured and allowed to reanneal in the presence of rolling-circle intermediates. The position on the tail at which the negative strand of the fragment anneals serves to map the end of the tail.

intermediates have a unique end and that they have been formed by chain elongation from the place where the positive-strand nickase is known to cut. This was determined by purifying separately a defined fragment of φX DNA and then annealing it quantitatively to the tails of the rolling circles. The hybridization of the negative-strand fragment creates a duplex region on the tail which is visible in the electron microscope (Fig. 11C). The unique location of the duplex region on the tail serves to map the location of the end of the tail—that is, the origin-terminus of φX positive-strand synthesis. The origin occurs in the region of gene *A*, more specifically in the *Hae*III fragment Z6b (Fig. 11C). Thus, the initiation site for positive-strand synthesis, as mapped in the actively replicating intermediate, corresponds to the site of nicking by the gene-*A* protein (see Baas and Jansz, 1972; Baas and Jansz; Eisenberg et al.; both this volume).

The second component of the actively synthesizing complex, the associated phage capsid, can be seen by labeling with radioactive amino acids (Fig. 11A). Upon denaturation and polyacrylamide gel analysis, the major φX capsid proteins (the products of genes *F*, *G*, and *H*) are found to be present (Fig. 11A, inset). Slicing and counting of the gel shows that these proteins are present in approximately the same ratio in which they occur in the intact phage capsid. Furthermore, the complex releases one (or a very few) molecules of the gene-*A* nickase (m.w. 56,000); this protein appears to be attached covalently to the DNA of the intermediate as it will not enter the gel unless the disrupted complex is first digested with DNase. These results (K. Koths and D. Dressler, in prep.) confirm and extend the work of Fujisawa and Hayashi on φX assembly (see Hayashi, this volume).

SYNTHESIS OF DUPLEX RINGS

The data cited above demonstrate that during single-stranded circle synthesis φX positive-strand replication occurs in a rolling-circle intermediate. The simplest way to imagine the formation of duplex rings, and the method proposed in the rolling-circle model, is to assume that the two asymmetric replication mechanisms we have discussed (the de novo initiation of negative strands and the rolling-circle synthesis of positive strands) occur together. Thus, as positive-strand material is displaced from the circular part of the intermediate into a tail of increasing length, at some point a sequence would become available on the tail at which the replication enzymes could begin the synthesis of the negative strand. If this were to happen before the tail reached unit length and became detached, one would observe a rolling circle with a double-stranded tail. If the tail were first detached and circularized before the negative strand is initiated, then one would see negative-strand synthesis occurring in a partially duplex ring. Support for this doubly asymmetric mechanism for duplex ring synthesis comes from the quantitative electron microscopic data of Figure 10 and Table 1 (see also Baas and Jansz, this volume). The relevant data relate both to the properties of rolling circles and to the existence of partially duplex rings.

Rolling-circle Intermediates during Duplex Ring Synthesis

Rolling circles are observed throughout the φX replication cycle—not only late when single-stranded circles are the product of replication, but also earlier during the period of duplex ring synthesis (Table 1). Moreover, during duplex ring synthesis, as during single-stranded circle synthesis, the rolling circles have tails that are single-stranded. Only rarely is a rolling circle observed in which the tail has been converted to the duplex state (Fig. 12).

The lengths of the tails of the rolling circles vary from a few hundred nucleotides (the shortest length measurable in the electron microscope) to about unit length—5500 nucleotides (as in Fig. 11C). This result indicates that the maturation of DNA from the tail of the intermediate occurs rapidly, for instance, as soon as the proper nicking sequence has been exposed in the tail or by repeated nicking of the positive strand on the duplex ring at the origin of replication, permitting the separation of a unit viral genome from the intermediate each time the growing point approaches the origin.

We interpret these data to indicate that rolling-circle intermediates operate throughout the φX life cycle to produce positive-strand material which is rapidly detached from the tails to yield single-stranded rings.

Partially Duplex Rings

Are the positive-strand rings which appear to be synthesized by rolling circles early in the phage life cycle used to make duplex rings via the synthesis of a negative strand? If so, then during the period when the number of duplex rings is increasing one may hope to see single-stranded circles in which a complementary strand has begun to be laid down. We have in fact

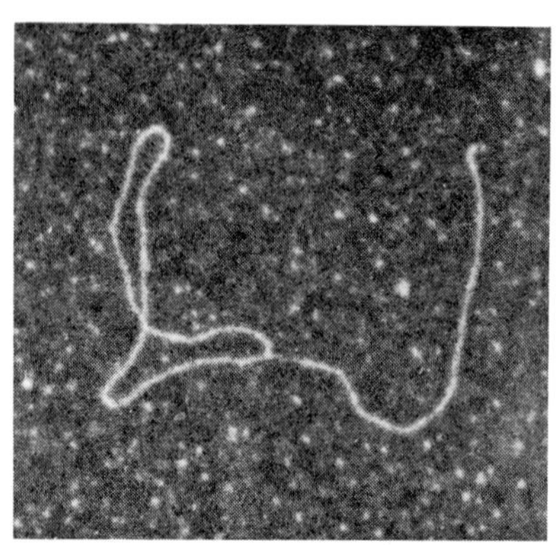

Figure 12 A rare φX rolling-circle intermediate with a double-stranded tail.

observed such molecules. An example is shown in Figure 9E. The part of the single-stranded circle that is duplex is marked off by arrows and has ranged from 10% to 90% in the 100 such molecules we have observed.

During the early stages of φX infection the number of single-stranded circles that appear to be in the process of conversion to a duplex state is low (about one to three per cell). Therefore, one must be concerned that the partially duplex rings observed at early times might just represent the conversion of late-infecting φX SS DNA rings to the duplex state—that is, the formation of the first φX double-stranded DNA circle. If this were so, these structures would represent only the initial stage of φX DNA replication and would have nothing to do with duplex ring synthesis in general. To obtain additional evidence relating to whether the partially duplex circles are involved in duplex ring synthesis we took advantage of the fact that in the presence of chloramphenicol accumulation of φX SS DNA rings does not occur late in the life cycle and duplex ring synthesis continues for several hours (Sinsheimer et al. 1962; Tessman 1966; see also Table 1 and Fig. 10). It is thus possible to harvest cells synthesizing duplex rings long after it would be reasonable to expect single-stranded circles to enter from late-infecting phage particles. Using chloramphenicol to obtain an extended period of φX duplex ring synthesis, we again observed both rolling-circle intermediates with single-stranded tails and partially duplex rings in roughly equal numbers (Table 1, lower half).

The observation during duplex ring synthesis of both rolling-circle intermediates with single-stranded tails and partially duplex circles would appear to provide the complementary pair of replicating intermediates needed for duplex ring synthesis.

On the basis of the data above, the synthesis of φX duplex rings appears to involve two complementary asymmetric events (diagrammed in Fig. 13): (1) the synthesis of positive-strand circles from a rolling-circle intermediate and (2) the conversion of positive-strand circles to duplex rings early in the phage life cycle.

This view is consistent with pulse-labeling studies in φX which show the incorporation of DNA precursors into longer-than-unit-length positive strands and shorter-than-unit-length negative strands during duplex ring synthesis (Schröder and Kaerner 1972; Dressler and Wolfson 1970).

POSITIVE AND NEGATIVE ASPECTS OF BEING SINGLE-STRANDED

Whereas early in the life cycle two types of asymmetric synthesis occur, accounting for the production of duplex rings, late in the virus life cycle only one of these asymmetric events occurs. At this time the positive-strand DNA produced by the rolling circles rapidly associates with the accumulating viral coat proteins and thus is prevented from becoming duplex (Fig. 11B). In this way a natural switchover from duplex ring synthesis to single-stranded circle synthesis occurs when a sufficient concentration of φX proteins has accumulated in the infected cell. The advantage gained from the evolution of such a system is that the SS DNA viruses can produce twice as many progeny from the same replication enzyme pools. The virus loses nothing in

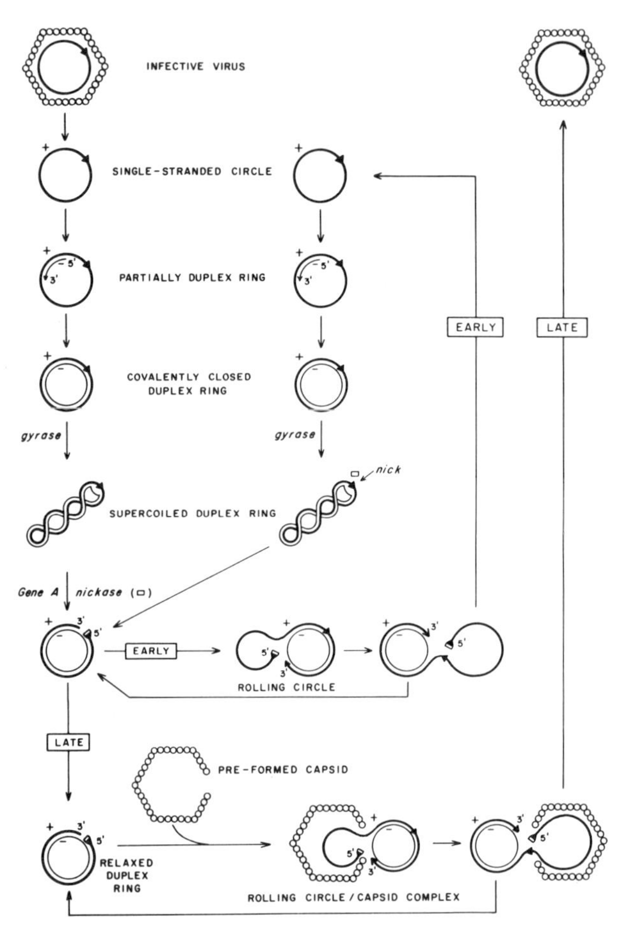

Figure 13 The φX DNA replication cycle, showing the involvement of the rolling-circle intermediate and the partially duplex ring. The triangle (▲) marks the site of nicking by the gene-*A* protein—the origin-terminus of positive-strand replication. The specific mechanism shown here for circularization of the progeny positive strand includes the suggestion by Eisenberg et al. (1977) that the φX cistron-*A* nicking protein (open box) remains associated with the 5′ terminus of the opened positive strand until a second nicking event generates a second 3′-OH end (one genome length away) which displaces the gene-*A* protein and closes the displaced SS DNA into a ring. During replication the end of the rolling-circle tail may not be free in solution, but rather be held back to the DNA growing point by a noncovalent interaction (disrupted by formamide) between the proteins on the tail and the replication enzyme complex.

having a single-stranded chromosome except for the ability to produce mature DNA capable of supporting those types of DNA repair that involve the excision of damaged bases and their replacement using a complementary DNA strand as template.

RATE OF DNA SYNTHESIS

By means of quantitative electron microscopy it has been possible to calculate the rate at which DNA synthesis is occurring in the rolling-circle intermediates. In experiments at 37°C one finds that the number of rolling-circle intermediates reaches a level of about 35 per cell by 40 minutes and remains nearly constant (Table 1 and Fig. 10). During this period there is a linear production of about 65 SS DNA circles (350,000 nucleotides of DNA) per minute per infected cell. Thus, the net rate of synthesis is about 350,000/35 or 10,000 nucleotides per minute per intermediate. This corresponds to a net polymerization rate of 150 to 200 nucleotides per second per rolling circle.

If each rolling circle is associated with a DNA polymerase system, then the number of DNA polymerase III molecules per cell must be at least 35. This figure is in agreement with the number of polymerase molecules per cell (10–40) that can be calculated from the data of Livingston et al. (1975) who monitored the recovery of active enzyme from a known number of cells.

On basis of these data it seems likely that the one effect of virus infection on host-cell metabolism, the eventual cessation of bacterial DNA synthesis, may be due to a titration of the cellular DNA polymerase III complexes so that they are no longer available for replication at the growing points of the host chromosome.

SUMMARY

The isometric phages are minute viruses with only minimal autonomy. They rely almost entirely on host-cell gene products for the replication of their DNA. These host-cell enzymes are the same ones that are involved in the replication of the bacterial chromosome. For this reason, studies of the structure of the intracellular viral DNA forms are valuable, as they provide descriptions of natural substrates upon which the DNA replication enzymes work.

We have presented an overview of the φX and G4 DNA replication cycles. In particular, we have discussed the results from our laboratory and from other laboratories concerning the naturally occurring viral DNA forms. The summary replication scheme shown in Figure 13 is based on DNA forms observed by electron microscopy and in pulse-labeling studies.

Data obtained both in vivo and in vitro provide strong support for the proposals of the rolling-circle model (1) that both the early synthesis of duplex rings and the later synthesis of single-stranded circles can be under stood in terms of one replication strategy and (2) that φX positive and negative strands are synthesized in two asymmetric events—the positive strand being elongated continuously after a 3′-OH primer group is created by the nicking of a covalently closed φX duplex ring and the negative strand being synthesized discontinuously by de novo initiation of a DNA fragment on a segment of an SS DNA template.

In analyzing the DNA replication cycle of the isometric phages we have noted two general aspects of DNA synthesis. One is the de novo initiation of DNA strands—exemplified by the synthesis of the G4 and φX negative strands. The other is the "nick, define an origin, and elongate" mechanism—exemplified by the rolling-circle synthesis of the positive strand. The former reaction occurs on one side of the DNA growing point where a daughter strand with a 3′ OH terminus does not exist and DNA fragments must be initiated de novo (Okazaki et al. 1969; Inman and Schnös 1971; Wolfson and Dressler 1972; Tye et al. 1977). The latter reaction is involved in replication systems that employ a rolling-circle intermediate—for example, the bacterial chromosome during mating, DNA synthesis in phages φX, G4, M13, P2, 186, PM2, and λ (late in the life cycle), and the special DNA synthesis that leads to the amplification of ribosomal RNA genes during frog oogenesis (see Ohki and Tomizawa 1969; Rupp and Ihler 1969; Ray 1969; Vapnek and Rupp 1971; Schnös and Inman 1971; Kiger and Sinsheimer 1969; Espejo et al. 1971; Hourcade et al. 1973; Rochaix et al. 1974; Godson 1977). The nick and elongate mechanism of the rolling-circle model may also be relevant both to the replication of the ends of eukaryotic chromosomes (Cavalier-Smith 1974) and to the replication of viruses such as adeno-associated virus (Strauss et al. 1976; Tattersall and Ward 1976).

REFERENCES

Air, G., F. Sanger, and A. Coulson. 1976. Nucleotide and amino acid sequences of gene G of φX174. *J. Mol. Biol.* **108:** 519.

Air, G., E. Blackburn, F. Sanger, and A. Coulson. 1975. The nucleotide and amino acid sequences of the N (5′) terminal region of gene G of bacteriophage φX174. *J. Mol. Biol.* **96:** 703.

Baas, P. and H. Jansz. 1972. φX174 replicative form DNA replication, origin and direction. *J. Mol. Biol.* **63**: 569.

Benbow, R., A. Zuccarelli, and R. Sinsheimer. 1974. A role for single-strand breaks in bacteriophage φX174 genetic recombination. *J. Mol. Biol.* **88**: 629.

Bouché, J.-P., L. Rowen, and A. Kornberg. 1978. The RNA primer synthesized by primase to initiate phage G4 DNA replication. *J. Biol. Chem.* **253**: 756.

Bouché, J.-P., K. Zechel, and A. Kornberg. 1975. *DnaG* gene product, a rifampicin-resistant RNA polymerase, initiates the conversion of a single-stranded coliphage DNA to its duplex replicative form. *J. Biol. Chem.* **250:** 5995.

Burton, A. and R. Sinsheimer. 1965. The process of infection with bacteriophage φX174. VII. Ultracentrifugal analysis of the replicative form. *J. Mol. Biol.* **14:** 327.

Caillet-Fauquet, P., M. Defais, and M. Radman. 1977. Molecular mechanisms of induced mutagenesis. Replication *in vivo* of bacteriophage φX174 single-stranded, ultraviolet light-irradiated DNA in intact and irradiated host cells. *J. Mol. Biol.* **117:** 95.

Cavalier-Smith, T. 1974. Palindromic base sequences and replication of eukaryote chromosome ends. *Nature* **250:** 467.

Denhardt, D. 1975. The single-stranded DNA phages. *CRC Crit. Rev. Microbiol.* **4:** 161.

———. 1977. The isometric single-stranded DNA phages. In *Comprehensive virology* (ed. H. Fraenkel-Conrat and R. Wagner), vol. 7, p. 1. Plenum Press, New York.

Denhardt, D. and R. Sinsheimer. 1965. The process of infection with bacteriophage φX174. IV. Replication of the viral DNA in a synchronized infection. *J. Mol. Biol.* **12:** 647.

Dressler, D. 1970. The rolling circle for φX DNA replication. II. Synthesis of single-stranded circles. *Proc. Natl. Acad. Sci.* **67:** 1934.

Dressler, D. and D. Denhardt. 1968. On the mechanism of φX single-stranded DNA synthesis. *Nature* **219:** 346.

Dressler, D. and J. Wolfson. 1970. Rolling circle for φX DNA replication. III. Synthesis of supercoiled duplex rings. *Proc. Natl. Acad. Sci.* **67:** 456.

Eisenberg, S. and D. Denhardt. 1974. Structure of nascent φX174 replicative form: Evidence for discontinuous replication. *Proc. Natl. Acad. Sci.* **71:** 984.

Eisenberg, S., J. Griffith, and A. Kornberg. 1977. φX174 cistron A protein is a multifunctional enzyme in DNA replication. *Proc. Natl. Acad. Sci.* **74:** 3198.

Espejo, R., E. Canelo, and R. Sinsheimer. 1971. Replication of bacteriophage PM2 deoxyribonucleic acid: A closed circular double-stranded molecule. *J. Mol. Biol.* **56:** 597.

Fiddes, J., B. Barrell, and N. Godson. 1978. Nucleotide sequences of the separate origins of synthesis of bacteriophage G4 viral and complementary DNA strands. *Proc. Natl. Acad. Sci.* **75:** 1081.

Francke, B. and D. Ray. 1972. *Cis*-limited action of the gene-A product of bacteriophage φX174 and the essential bacterial site. *Proc. Natl. Acad. Sci* **69:** 475.

Fujisawa, H. and M. Hayashi. 1976. Viral DNA-synthesizing intermediate complex isolated during assembly of bacteriophage φX174. *J. Virol.* **19:** 409.

Gellert, M., K. Mizuuchi, M. O'Dea, and H. Nash. 1976. DNA gyrase: An enzyme that introduces superhelical turns into DNA. *Proc. Natl. Acad. Sci.* **73:** 3872.

Gilbert, W. and D. Dressler. 1969. DNA replication: The rolling circle model. *Cold Spring Harbor Symp. Quant. Biol.* **33:** 473.

Godson, G. N. 1977. G4 DNA replication. II. Synthesis of viral progeny single-stranded DNA. *J. Mol. Biol.* **117:** 337.

Henry, T. and R. Knippers. 1974. Isolation and function of the gene A initiator of bacteriophage φX174, a highly specific endonuclease. *Proc. Natl. Acad. Sci.* **71:** 1549.

Horiuchi, K. and N. Zinder. 1972. Cleavage of bacteriophage f1 DNA by the restriction enzyme of *Escherichia coli* B. *Proc. Natl. Acad. Sci.* **69:** 3220.

Hourcade, D. and D. Dressler. 1978. The site specific initiation of a DNA fragment. *Proc. Natl. Acad. Sci.* (in press).

Hourcade, D., D. Dressler, and J. Wolfson. 1973. The amplification of ribosomal RNA genes involves a rolling circle intermediate. *Proc. Natl. Acad. Sci.* **70:** 2926.

Hutchison, C., III and R. Sinsheimer. 1966. The process of infection with bacteriophage φX174. X. Mutations in a φX lysis gene. *J. Mol. Biol.* **18:** 429.

Inman, R. and M. Schnös. 1971. Structure of branch points in replicating DNA: Presence of single-stranded connections in λ DNA branch points. *J. Mol. Biol.* **56:** 319.

Johnson, P. and R. Sinsheimer. 1974. Structure of an intermediate in the replication of bacteriophage φX174 DNA: The initiation site for DNA replication. *J. Mol. Biol.* **83:** 47.

Kiger, J. and R. Sinsheimer. 1969. Vegetative lambda DNA. V. Evidence concerning single-stranded elongation. *J. Mol. Biol.* **43:** 567.

Knippers, R., T. Komano, and R. Sinsheimer. 1968. The process of infection with bacteriophage φX174. XXI. Synthesis of progeny single-stranded DNA. *Proc. Natl. Acad. Sci.* **59:** 911.

Knippers, R., A. Razin, R. Davis, and R. Sinsheimer. 1969. The process of infection with bacteriophage φX174. XXIX. *In vivo* studies on the synthesis of the single-stranded DNA of progeny φX174 bacteriophage. *J. Mol. Biol.* **45:** 237.

Komano, T., R. Knippers, and R. Sinsheimer. 1968. The process of infection with bacteriophage φX174. XXII. Synthesis of progeny single-stranded DNA. *Proc. Natl. Acad. Sci.* **59:** 911.

Koths, K. and D. Dressler. 1978. Analysis of the φX DNA replication cycle by electron microscopy. *Proc. Natl. Acad. Sci.* **75:** 605.

Langeveld, S., A. van Mansfeld, P. Baas, H. Jansz, G. van Arkel, and P. Weisbeek. 1978. Nucleotide sequence of the origin of replication in bacteriophage φX174 RF DNA. *Nature* **271:** 417.

Lindqvist, B. and R. Sinsheimer. 1967. Process of infection with bacteriophage φX174. XIV. Studies on macromolecular synthesis during infection with a lysis-defective mutant. *J. Mol. Biol.* **28:** 87.

Livingston, D., D. Hinkle, and C. Richardson. 1975. Deoxyribonucleic acid polymerase III of *Escherichia coli*. *J. Biol. Chem.* **250:** 461.

Machida, Y., T. Okazaki, and R. Okazaki. 1977. Discontinuous replication of replicative form DNA from bacteriophage φX174. *Proc. Natl. Acad. Sci.* **74:** 2776.

Marians, K., J. Ikeda, S. Schlagman, and J. Hurwitz. 1977. Role of DNA gyrase in φX replicative-form replication *in vitro*. *Proc. Natl. Acad. Sci.* **74:** 1965.

Martin, D. and G. N. Godson. 1977. G4 DNA replication. I. Origin of synthesis of the viral and complementary DNA strands. *J. Mol. Biol.* **117:** 321.

Masamune, Y. 1976. Effect of ultraviolet irradiation of bacteriophage f1 DNA on its conversion to replicative form by extracts of *Escherichia coli*. *Mol. Gen. Genet.* **149:** 335.

McMacken, R., K. Ueda, and A. Kornberg. 1977. Migration of *Escherichia coli* *dna*B protein on the template DNA strand as a mechanism in initiating DNA replication. *Proc. Natl. Acad. Sci.* **74:** 4190.

Ohki, M. and J. Tomizawa. 1969. Asymmetric transfer of DNA strands in bacterial conjugation. *Cold Spring Harbor Symp. Quant. Biol.* **33:** 651.

Okazaki, R., T. Okazaki, K. Sakabe, K. Sugimoto, R. Kainuma, A. Sugino, and N. Iwatsuki. 1969. *In vivo* mechanism of DNA chain growth. *Cold Spring Harbor Symp. Quant. Biol.* **33:** 129.

Ray, D. 1969. Replication of bacteriophage M13. II. The role of replicative forms in single-strand synthesis. *J. Mol. Biol.* **43:** 631.

Rochaix, J. D., A. Bird, and A. Bakken. 1974. Ribosomal RNA gene amplification by rolling circles. *J. Mol. Biol.* **87:** 473.
Rowen, L. and A. Kornberg. 1978a. Primase, the *dna*G protein of *Escherichia coli*. *J. Biol. Chem.* **253:** 758.
———. 1978b. A ribo-deoxyribonucleotide primer synthesized by primase. *J. Biol. Chem.* **253:** 770.
Rupp, W. and P. Howard-Flanders. 1968. Discontinuities in the DNA synthesized in an excision-defective strain of *Escherichia coli* following ultraviolet irradiation. *J. Mol. Biol.* **31:** 291.
Rupp, W. and G. Ihler. 1969. Strand selection during bacterial mating. *Cold Spring Harbor Symp. Quant. Biol.* **33:** 647.
Schekman, R., A. Weiner, and A. Kornberg. 1974. Multienzyme systems of DNA replication. *Science* **186:** 987.
Schekman, R., M. Iwaya, K. Bromstrup, and D. Denhardt. 1971. The mechanism of replication of φX174 single-stranded DNA. II. An enzymic study of the structure of the replicative form II DNA. *J. Mol. Biol.* **57:** 177.
Schnös, M. and R. Inman. 1971. Starting point and direction of replication in P2 DNA. *J. Mol. Biol.* **55:** 31.
Schröder, C. and H.-C. Kaerner. 1972. Replication of bacteriophage φX174 replicative form *in vivo J. Mol. Biol.* **71:** 351.
Sims, J. and D. Dressler. 1978. The site specific initiation of a DNA fragment: DNA sequence of the bacteriophage G4 negative strand initiation site. *Proc. Natl. Acad. Sci.* (in press).
Sinsheimer, R., B. Starman, C. Nagler, and S. Guthrie. 1962. The process of infection with bacteriophage φX174. I. Evidence for a "replicative form." *J. Mol. Biol.* **4:** 142.
Stone, A. 1970. Protein synthesis and the arrest of bacterial DNA synthesis by bacteriophage φX174. *Virology* **42:** 171.
Strauss, S., E. Sebring, and J. Rose. 1976. Concatemers of alternating plus and minus strands are intermediates in adenovirus-associated virus DNA synthesis. *Proc. Natl. Acad. Sci.* **73:** 742.
Tattersall, P. and D. Ward. 1976. Rolling hairpin model for replication of parvovirus and linear chromosomal DNA. *Nature* **263:** 106.
Tessman, E. 1966. Mutants of bacteriophage S13 blocked in infectious DNA synthesis. *J. Mol. Biol.* **17:** 218.
Tye, B.-K., P.-O. Nyman, I. Lehman, S. Hochhauser, and B. Weiss. 1977. Transient accumulation of Okazaki fragments as a result of uracil incorporation into nascent DNA. *Proc. Natl. Acad. Sci.* **74:** 154.
Vapnek, D. and W. Rupp. 1971. Identification of individual DNA strands and their replication during conjugation in thermosensitive DNA mutants of *Escherichia coli*. *J. Mol. Biol.* **60:** 2682.
Vinograd, J. and J. Lebowitz. 1966. Physical and topological properties of circular DNA. *J. Gen. Physiol.* (Suppl. 6, Part 2) **49:** 103.
Wickner, S. 1977. DNA or RNA priming of bacteriophage G4 DNA synthesis by *Escherichia coli dna*G protein. *Proc. Natl. Acad. Sci.* **74:** 2815.
Wickner, S. and J. Hurwitz. 1974. Conversion of φX174 viral DNA to double-stranded form by purified *Escherichia coli* proteins. *Proc. Natl. Acad. Sci.* **71:** 4120.

Wolfson, J. and D. Dressler. 1972. Regions of single-stranded DNA in the growing points of replicating bacteriophage T7 chromosomes. *Proc. Natl. Acad. Sci.* **69:** 2682.

Zechel, K., J.-P. Bouché, and A. Kornberg. 1975. Replication of phage G4. A novel and simple system for the initiation of deoxyribonucleic acid synthesis. *J. Biol. Chem.* **250:** 4684.

Replication of φX174 RF DNA In Vivo

Pieter D. Baas and Hendrik S. Jansz
Laboratory for Physiological Chemistry
and Institute of Molecular Biology
State University of Utrecht
Vondellaan 24 A, Utrecht, The Netherlands

The first clue toward understanding the mode of φX174 DNA replication was the discovery by Sinsheimer et al. (1962) of an intracellular double-stranded form (replicative form [RF]) of bacteriophage φX DNA. Until that time it was a puzzle how a single-stranded DNA molecule could replicate. The mediation of a double-stranded form seemed to solve this problem, and a few years later Denhardt and Sinsheimer (1965a) and Stone (1967) were able to show that φX RF DNA replicates in a semiconservative manner.

Three stages of DNA synthesis can be distinguished during the φX life cycle (for reviews see Sinsheimer 1968; Denhardt 1975, 1977; Dressler et al., this volume). The first RF in the φX-infected *Escherichia coli* C host derives from the de novo synthesis of a complementary strand on the infecting viral single-stranded (SS) DNA (SS→RF), which is accomplished by preexisting host enzymes. This parental duplex RF DNA then replicates to form a pool of some 20 RF molecules (RF→RF) at 15 minutes after infection; this requires several host enzymes as well as one phage-encoded enzyme, the gene-*A* protein. In the final stage of the φX life cycle, SS DNA of the progeny virus is derived from RF DNA by asymmetric synthesis (RF→SS). This paper reviews the second stage of φX DNA replication (RF→RF) in vivo and is for the most part limited to labeling studies and to analyses of replicating intermediates by biochemical and electron microscopic methods. Other approaches, namely, the effect of blocking host-cell functions by mutation and studies of RF DNA replication in vitro, are dealt with elsewhere in this volume (see Dumas; Eisenberg et al.; Sumida-Yasumoto et al.).

Several studies during the 1960s suggested that two different modes of DNA synthesis were in operation in stages II and III. It was thought that the infecting (parental) viral DNA, after conversion to the parental RF molecule, was used repeatedly in a semiconservative way as a template for the synthesis of all progeny RF molecules in stage II, and that this mode of DNA synthesis was reversed in stage III, in which a few hundred progeny viral SS DNA molecules are synthesized by using the complementary strand of the parental RF as a template in a conservative way (Denhardt and Sinsheimer 1965c). This model was later modified so that stage-III progeny phage DNA is derived from the RF DNA pool. The viral strands of the RF molecules are displaced into coat proteins, whereas the remaining complementary strands serve as templates for the synthesis of new phage strands (Lindqvist and Sinsheimer 1968; Knippers et al. 1968; Komano et al. 1968).

More recent evidence indicates that after the formation of the parental RF molecule, the entire φX life cycle is governed by only one mode of DNA synthesis. This involves the repeated use of the complementary strands of parental and progeny RF molecules as templates for the formation of new viral DNA molecules, which are then either converted into RF molecules in stage II or packaged into viral coat proteins in stage III. This view was expressed first by Gilbert and Dressler (1969) in their rolling-circle model of φX DNA replication.

Departing from this viewpoint, it is attractive in dealing with the mechanism of φX RF DNA replication to discuss the syntheses of the viral and complementary strands separately, as they are brought about by two completely different mechanisms. There is evidence from in vitro studies that different proteins are required for the synthesis of the viral and complementary strands of φX RF DNA (Wickner; Eisenberg et al.; both this volume). Only during RF DNA replication are both types of φX DNA synthesis operational. Synthesis of the complementary strand during parental RF DNA formation can be considered as equivalent to the second phase in φX RF DNA replication, which follows the synthesis of the viral strand. The synthesis of complementary strands does not occur during progeny viral-strand synthesis because of the coupling of viral-strand synthesis with the process of phage maturation.

Generally, the biosynthesis of macromolecules can be divided into the stages of initiation, chain elongation, and termination. The replication of φX RF DNA is no exception to this rule, and the relevant subject matter will be discussed in the following order:

1. forms of double-stranded φX DNA in infected cells,
2. template selectivity during φX RF DNA replication,
3. initiation of φX RF DNA replication,
4. strand elongation during φX RF DNA replication,
5. termination of φX RF DNA replication,
6. summarizing scheme of φX RF DNA replication.

Forms of Double-stranded φX DNA in Infected Cells

Many studies of double-stranded φX DNA have made use of the fact that moderate concentrations of chloramphenicol (30–40 μg/ml) extend the period of φX RF DNA replication to several hours while inhibiting the synthesis of progeny SS DNA (Sinsheimer et al. 1962). In φX DNA preparations obtained from cells infected in the presence of chloramphenicol, two RF components are found which differ in sedimentation coefficient: a 21S major component, RFI (ca. 90%), and a 17S minor component, RFII (ca. 10%) (Burton and Sinsheimer 1965; Jansz and Pouwels 1965; Roth and Hayashi 1966; Jaenisch et al. 1966). RFI is a covalently closed, duplex DNA which is refractory to strand separation, whereas RFII can readily be converted to linear and circular SS DNA by denaturation. Therefore, RFII has long been considered a randomly nicked form of RFI.

However, Iwaya and Denhardt (1971) have shown that RFI and RFII are present in roughly comparable amounts throughout the latent period in normal infections without chloramphenicol. RFII is specifically lost at the phenol-water interphase during phenol extraction unless the preparation is pretreated with pronase. Schekman et al. (1971) have shown that RFII consists of a complex mixture of molecules containing a few (some 10%) RFII molecules with nicks bounded by a 3′-OH and a 5′-P residue. The majority of the RFII molecules contain one or more gaps, or short, single-strand tails, or both. Analysis of these structures has revealed some interesting aspects of φX RF DNA replication (see section on "Termination of φX RF DNA Replication").

Rush et al. (1967) discovered the presence of a small fraction (ca. 4%) of multimers (mostly circular dimers) in RF DNA preparations. Some of these multimeric forms were identified as catenanes, i.e., interlocked monomers (Benbow et al. 1972), and figure-eight molecules, i.e., molecules consisting of two monomer-length strands and one dimer-length circular strand (Gordon et al. 1970; Thompson et al. 1975). Their role in replication, if any, is obscure (Hofs et al. 1973). Several lines of evidence indicate that multimeric forms of φX RF DNA are intermediates in recombination (see Zinder; Warner and Tessman; both this volume).

RFI is the initial substrate as well as the final product of φX RF DNA replication. In this respect, some of the structural features of RFI may be significant. φX RFI is a DNA double helix composed of one viral and one complementary strand, each of which is covalently closed. In such a structure the two strands are interlocked, and, consequently, the two strands cannot separate from each other after denaturation. Exposure to strong alkaline conditions yields a denatured form of RFI that sediments at 53S at pH 12 and 40S at neutral pH in 1 M NaCl (Pouwels and Jansz 1964; Burton and Sinsheimer 1965; Jansz and Pouwels 1965; Pouwels et al. 1966, 1968). The two strands are not cross-linked, as the interlocking can be relieved by a single phosphodiester cleavage by DNase I (Pouwels and Jansz 1964; Jansz and Pouwels 1965; Roth and Hayashi 1966; Jaenisch et al. 1966; Siegel and Hayashi 1967; Jansz et al. 1968).

As isolated from infected cells, φX RFI contains superhelical turns at neutral pH, which give these molecules a highly twisted appearance in the electron microscope (Jaenisch et al. 1966). Superhelical φX RFI has a higher sedimentation coefficient (21S) than the extended circular form RFII (16–17S) (Burton and Sinsheimer 1965; Pouwels et al. 1966). The number of superhelical turns (τ) in covalently closed, duplex DNA is related to the number of turns in the double helix (β) as $\tau=\alpha-\beta$ (Bauer and Vinograd 1968); α is the topological winding number, i.e., the number of times that one strand winds around the other. In covalently closed DNA, α is invariant, and therefore a change in β must be compensated for by an equal change in τ. The superhelical turns (τ) in φX RFI are called negative turns because they are eliminated by treatments that reduce the number of turns in the double helix (β), e.g., by denaturation or dye intercalation. The elimination of superhelical turns from φX RFI causes a decrease in the sedimentation coefficient. The relation between the number of superhelical turns in covalently closed DNA and the sedimentation coefficient is not simple (Upholt et al. 1971; Woodward-Gutai and Lebowitz 1976).

The gradual unwinding of the double helix of φX RFI produced by increasing the pH (Pouwels et al. 1966) or ethidium bromide intercalation (Denhardt and Kato 1973) seems to cause an initial increase in the sedimentation coefficient that is followed by a drop in sedimentation coefficient to a minimum (16–17S at pH 11.4 in 1 M NaCl), the equivalence point at which the number of turns removed by denaturation equals the original number of turns in RFI.

From the molar binding ratio of ethidium per φX RFI molecule at equivalence (Wang 1969; Waring 1970), the superhelical density (number of superhelical turns per ten base pairs) can be computed. The superhelical density of φX RFI had initially been underestimated because of an underestimation of the degree of unwinding of the DNA double helix by intercalating agents. Taking the unwinding angle reported recently of 26° per intercalating molecule of ethidium (Wang 1974), one can calculate a superhelical density of approximately −0.08, which agrees reasonably well with estimates for SV40 DNA (−0.084) and PM2 DNA (−0.11) (Upholt 1977).

Although molecules like φX RFI that have high negative superhelical density have regions of interrupted secondary structure containing unpaired bases, these regions have not been found at specific sites and there is no evidence for the presence of cruciform structures (Bartok and Denhardt 1976). However, the fact that the φX-encoded gene-*A* protein nicks supertwisted φX RFI but not relaxed φX RFI (Marians et al. 1977) indicates that at least some structural feature of supertwisted RFI, probably a region of unpaired bases, is functionally significant in replication.

Template Selectivity during φX RF DNA Replication

The formation of the parental RF DNA is essentially complete about 3 minutes after synchronized infection (Denhardt and Sinsheimer 1965a). This is followed by progeny RF DNA synthesis, which continues at a constant rate—nearly two RF molecules per minute per cell—until about 15 minutes, at which time net RF DNA synthesis and host DNA synthesis cease. Then there is a sharp increase in the rate of viral DNA synthesis due to progeny SS DNA synthesis.

The linearity of φX RF DNA synthesis suggests that a limited number of RF molecules (probably one) participate in the formation of progeny RF DNA. In earlier work the favored opinion was that during the formation of all progeny RF DNA only the RF molecule that contained the parental viral strand replicated semiconservatively. The other progeny RF molecules were thought not to replicate. The experiments that seemed to support this view have been critically reviewed recently by Denhardt (1977), who came to the conclusion that in most cases the interpretation of the results was questionable. These experiments included ^{32}P- and ^{3}H-decay studies which suggested a unique role for the parental viral strand even after RF DNA replication had commenced (Denhardt and Sinsheimer 1965b; Datta and Poddar 1970). Also, density labeling studies (Denhardt and Sinsheimer 1965a; Stone 1967; Knippers and Sinsheimer 1968) in infected cells grown in dense medium and shifted at various times after infection to light medium indicated that the ^{32}P-labeled parental viral strand exchanged partners throughout the period of RF DNA replication. However, the possibility that the parental RF DNA did not replicate in the dense medium and that the observed shift in density of the parental RF DNA might represent the first round of RF DNA replication has not been excluded.

A special role for the parental RF molecule was also inferred from the observed specific association of the RF DNA containing ^{32}P-labeled viral DNA with a rapidly sedimenting membrane fraction (Knippers and Sinsheimer 1968; Salivar and Sinsheimer 1969). Pulse label flows from the membrane fraction into the slowly sedimenting progeny RF DNA (Knippers and Sinsheimer 1968). However, it should be realized that, apart from replicating parental RF DNA, other replicating intermediates may also become associated with the membrane fraction because of their single-stranded character or through their association with hydrophobic proteins. The low level of transfer of ^{32}P-labeled parental DNA to progeny SS DNA at low multiplicities of infection (Sinsheimer 1961; Kozinski 1961; Knippers et al. 1969b) had also been considered as support for the special role of the parental viral strand. However, more recent experiments indicate that part of the parental viral DNA is transferred conservatively to progeny phage under conditions of lysis inhibition (Schnegg and Hofschneider 1970; Iwaya and Denhardt 1973a) even at low multiplicities of infection.

In contrast to the earlier views, the following biochemical and genetic experiments indicate that the viral strand of the parental RF DNA is not used preferentially as a template for RF DNA replication. Matsubara et al. (1967) concluded from several density-shift experiments that, late in infection, progeny RF DNA also participates in the process of RF DNA replication. Iwaya and Denhardt (1973b) extended this observation by [^{3}H]-

thymidine pulse-labeling of *E. coli* cells infected with ^{2}H-^{15}N density-labeled phage at different times after infection. Their conclusion was that late in infection the RF molecule containing the parental viral strand was rarely used as a template for RF DNA replication. Schröder et al. (1973) came to the same conclusion. No experiments have been published which show directly that the RF molecule containing the parental complementary strand is replicated preferentially after the initial round of RF DNA replication has been completed. It appears, however, that the bulk of the RF molecules found late in infection derive their information from the complementary strand of the parental RF DNA.

Baas and Jansz (1972a) investigated the progeny phage produced after infection of spheroplasts with heteroduplex RF molecules containing different genetic markers in their viral and complementary strands. They showed in single-burst experiments that most spheroplasts produced progeny of only one genotype, which suggests that the heteroduplex RF DNA is repaired to a homoduplex prior to replication. The great majority of the phage obtained from spheroplasts that produced progeny of both genotypes had the genotype of the complementary strand. Assuming that all of the RF molecules have an equal chance to participate in the production of progeny SS DNA, this indicates that the majority of the RF molecules in the cell have the genotype of the complementary strand.

Similar conclusions can be drawn from the experiments of Merriam et al. (1971a,b). Besides studying the result of heteroduplex infection in spheroplasts, they also studied the result of heteroduplex infection in whole bacteria. Artificial heteroduplexes were constructed within the bacterium by means of the mutagenic properties of [5-^{3}H]deoxycytidine. The decay product of this labeled pyrimidine pairs as thymine, which efficiently effects the C→T transitions that increase the reversion rate of mutants susceptible to this transition. Their experiments showed clearly that it is the mutagenesis of the parental complementary strand, and not that of the parental viral strand, that affects the genotype of the progeny phage.

Initiation of φX RF DNA Replication

The first clue as to the origin and direction of φX RF DNA replication came from an observation made by Baas and Jansz (1972b). They analyzed the progeny phage from single-burst spheroplasts infected with heteroduplex RF DNA. A fraction of the spheroplasts produced a mixed burst of both phage genotypes. It was argued that to produce a mixed burst, the heteroduplex RF DNA must escape the repair system of the host by replication. Therefore, a heteroduplex region located near the origin of replication has a better chance of escaping the repair process than a heteroduplex region located near the terminus of RF DNA replication. Indeed, the percentages of mixed bursts varied considerably after infection of spheroplasts with heteroduplex RF molecules containing amber or temperature-sensitive mutations at different sites on the φX genome. As shown in Figure 1, a plot of the percentage of mixed bursts against the position of the mutation on the genetic map of φX yields a gradient with a sharp discontinuity between mutations *am*16 (gene *B*) and *ts*128 (gene *A*). These results suggest that φX RF DNA replication starts in gene *A* between *am*16 (high percentage of mixed bursts), which is located in the C-terminal *A*/*B* overlap, and the N-terminal mutation *ts*128 (low percentage of mixed bursts), and that RF DNA replication proceeds unidirectionally in the direction of the genes *A*, *B*, through *H*.

Weisbeek and van Arkel (1976) refined this study and investigated a number of different gene-*A* mutants that had been located on different restriction fragments by physical mapping using marker rescue (Weisbeek et al. 1976). They observed a discontinuity in the percentage of mixed bursts between *am*35 (low percentage of mixed bursts) and *am*18 (high percentage of mixed bursts). These two mutations were thought to be located close together in the C-terminal region of gene *A* on the *Hin*dII-5 cleavage fragment. However, other data have accumulated since then which argue against the location of the origin of φX RF DNA replication between these two mutations.

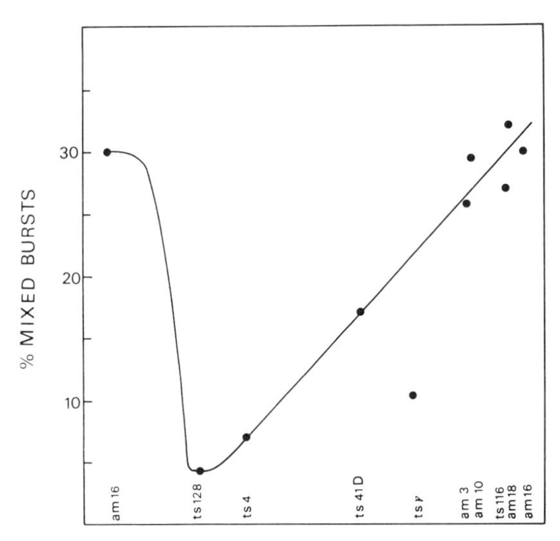

Figure 1 Localization of the origin of φX RF DNA replication between *am*16 and *ts*128 by heteroduplex mapping. (Modified from Baas and Jansz 1972b.) The percentage of spheroplasts producing a mixed burst after infection with heteroduplex RF DNA is plotted against the position of the mutation on the physical map of φX RF DNA (Weisbeek et al. 1976, 1977; Baas et al. 1976b; Smith et al. 1977).

P. J. Weisbeek (pers. comm.) has shown that *am*35 is a double mutant. One of the *am*35 mutations is in the same position as the *am*18 mutation (Smith et al. 1977). The position of the other mutation is not known, and this makes the significance of the results of heteroduplex assay doubtful in the case of *am*35. Also, the fact that according to recent marker-rescue (Weisbeek et al. 1977) and sequence data (Smith et al. 1977) *am*16 (which yields a high percentage of mixed bursts in the heteroduplex assay) maps upstream of *am*18 argues against a discontinuity in the frequency of mixed bursts between *am*35 and *am*18. Rather, this discontinuity should be expected to occur upstream of *am*16. However, additional experiments are needed to locate this discontinuity more accurately.

Other evidence, which will be discussed below, identifies the exact location of the origin of φX RF DNA replication in the region of gene-*A*. This evidence is based on the knowledge that has been collected over the years on the properties and mode of action of the gene-*A* protein, the only φX-encoded protein required for φX RF DNA replication (Tessman and Tessman, this volume).

Francke and Ray (1971, 1972) studied the action of the gene-*A* protein in *rep* cells. They made use of the discovery by Denhardt et al. (1967) of a class of *E. coli* mutants (*rep* mutants) in which parental RF DNA can be formed but further replication is blocked. The φX RF DNA that is formed is transcribed and translated, and all φX proteins, including the gene-*A* protein, are synthesized (Denhardt et al. 1972). These mutants are also unable to replicate phage P2 (Calendar et al. 1970).

In a series of elegant experiments Francke and Ray (1971, 1972) showed that some of the parental RF DNA in *rep*3 cells had a discontinuity specifically in the viral strand. This discontinuity was observed only if functional gene-*A* protein was present. In the presence of chloramphenicol at 100 μg/ml or after infections of *rep* cells under conditions of impaired gene-*A* function only RFI was found. Besides RFI and RFII, some RF molecules with viral strands somewhat longer than genome length were observed. This suggested that the viral strand was elongated at one end and displaced at the other end and that this process was soon halted because of the lack of the *rep* factor. Furthermore, the gene-*A* protein was shown to be *cis*-acting; in simultaneous infections of *rep* cells with phage with functional gene *A* and phage with mutated gene *A*, only the RF DNA derived from the phage with a functional gene *A* acquired a discontinuity in the viral strand. This confirmed and extended earlier observations of Tessman (1966) on the *cis*-acting properties of gene *A* in S13. In complementation experiments, gene-*A* mutations cannot be complemented by the wild-type gene in mixed infections, whereas mutations in other phage genes can.

On the basis of these results, Francke and Ray (1971, 1972) proposed a model (illustrated in Fig. 2) in which φX RF DNA replication is initiated on φX RFI after the introduction of a nick into the viral strand by the gene-*A*

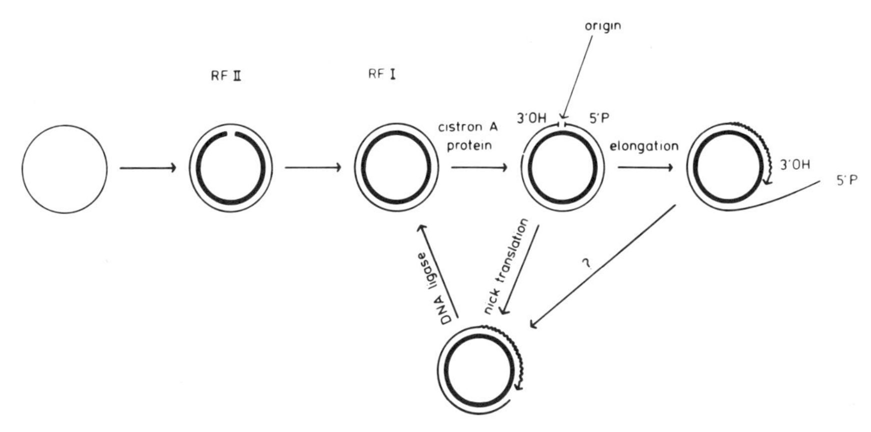

Figure 2 Model for φX RF DNA synthesis in *rep*3 cells according to Francke and Ray (1971) as modified by Baas et al. (1976a). (Reprinted, with permission, from Baas et al. 1976a.) After infection, a complementary strand (thick line) is synthesized using the parental DNA (thin line) as a template. The complementary strand is closed by ligase to yield an RFI molecule (made superhelical by gyrase) in which both strands are covalently closed. The gene-*A* protein then nicks the viral strand at the origin of φX RF DNA replication. In an attempt to start DNA replication, some nicked viral strands are elongated and form tailed molecules. But DNA synthesis stops because no functional *rep* gene product is present. The majority of the nicked molecules undergo nick translation, starting at the nick introduced by the gene-*A* protein. During this process the newly synthesized viral-strand DNA replaces the parental viral-strand DNA, but no net synthesis occurs. At any time during this nick-translation process the RFII molecules can be converted back to RFI molecules.

protein. According to this view, there is only one nick per RF molecule and this nick resides at a specific site—the origin of RF DNA replication. The 3′ OH of the nick serves as a primer for DNA polymerase, which elongates the viral strand and displaces the 5′ end as proposed in the rolling-circle model of DNA replication (Gilbert and Dressler 1969). For the displacement synthesis to continue, the *rep* protein (among other factors) is required.

Using this system, Baas et al. (1976a) have determined the position of the nick in the viral strand of φX RF DNA. Biochemical studies showed that in *rep* mutants, after the introduction of the nick, extensive degradation of the parental viral strand takes place, presumably by nick translation caused by DNA polymerase I (as shown in Fig. 2) (see also Bowman and Ray 1975). However, the original position of the nick could be determined with high precision. After infecting *rep* cells with ^{32}P-labeled φX in the presence of [^{3}H]thymidine, the labeled DNA was extracted and digested with *Hae*III restriction endonuclease. The ^{3}H/^{32}P ratios of the fragments were determined after separation by gel electrophoresis. A calculation from the different ^{3}H/^{32}P ratios of the various fragments placed the position of the original nick in restriction fragment Z6b approximately 100 nucleotides to the right of the Z2/Z6b junction on the *Hae*III cleavage map of φX RF DNA. Figure 3 shows these results.

Direct visualization of the position of the origin was obtained in the electron microscope. Replicating intermediates were isolated from *E. coli* H502 cells infected with φX in the presence of chloramphenicol at 40 μg/ml. Rolling-circle-type molecules, among others, were observed. After digestion with the restriction enzyme *Pst*I, which cleaves φX RF DNA only once, linear, branched molecules were produced. Examples are shown in Figure 4. Analysis of these molecules indicates that most of the tails originate from a position 20% away from the *Pst*I cleavage site. This is the distance between

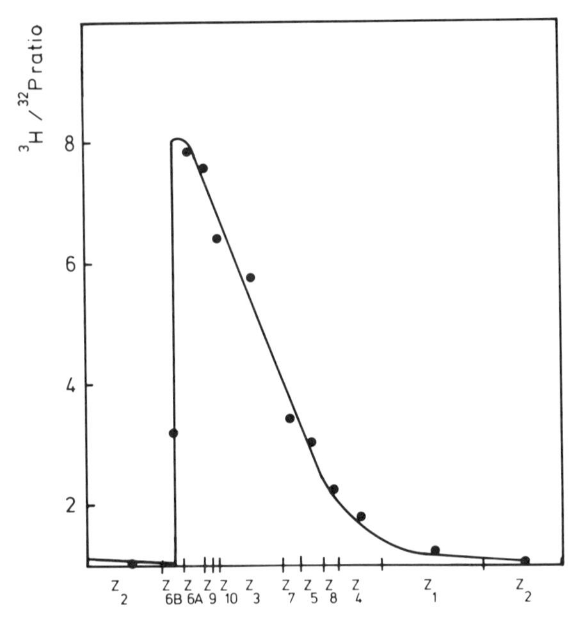

Figure 3 Position of the nick produced by the gene-*A* protein in φX RFI in vivo. (Reprinted, with permission, from Baas et al. 1976a.) *Rep* cells (*E. coli* C1704) infected with ^{32}P-labeled φX phage in the presence of [^{3}H]thymidine form parental RF DNA that contains ^{32}P label in the viral strand and ^{3}H label in the complementary strand. Gene-*A* protein nicks the viral strand of the parental RF DNA at a specific site. Nick translation occurs, gradually replacing the ^{32}P label in the viral strand with ^{3}H label. The original position of the gene-*A*-protein nick in φX RF DNA was determined by restriction enzyme analysis of nick-translated RFII. The ^{3}H/^{32}P ratio of the *Hae*III restriction fragments (normalized with respect to Z2 = 1.00) is plotted against the physical map of φX DNA. The ^{3}H/^{32}P gradient indicates that nick translation starts in fragment Z6b and proceeds clockwise around the physical map, i.e., in the direction of the genes *A*, *B*, through *H*. From the ^{3}H/^{32}P ratio in Z6b it was calculated that the original position of the nick was approximately 100 nucleotides from the Z2/Z6b junction.

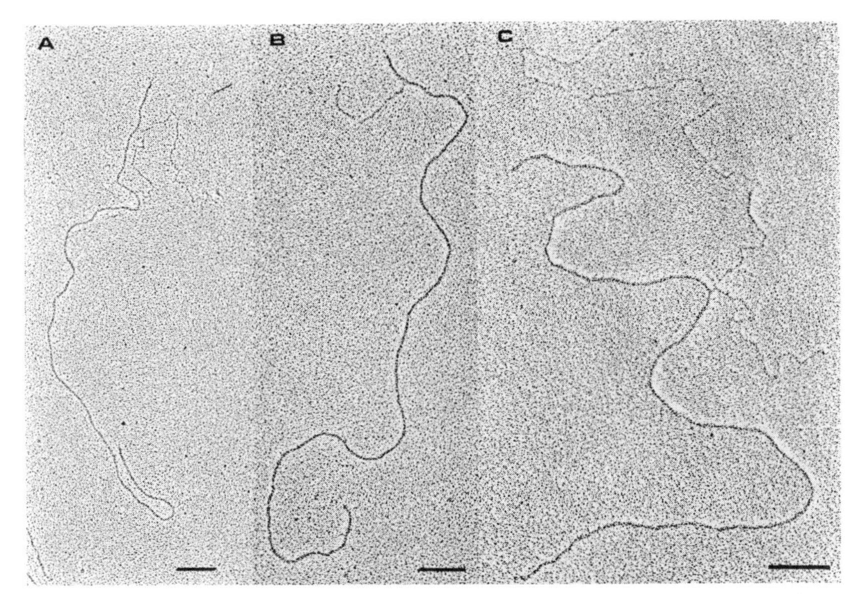

Figure 4 *Pst*I-cleaved rolling circles of φX DNA (W. Keegstra et al., unpubl.). The electron microscopic analysis consisted of 67 molecules, which have been subdivided into the three classes represented here. In *A* the 5′ end of the tail is completely branch-migrated. Tails of variable length originate from a position approximately 20% away from one end of the linear, double-stranded DNA. Average distance to the *Pst*I cleavage site: 20.3% ± 2.7%; molecules observed: 25. In *B* the 5′ end of the tail is in its original position. Tails of variable length originate from different positions on the linear, double-stranded DNA. The tails can be folded back along the molecule to a position approximately 20% away from the *Pst*I cleavage site. Average distance to the *Pst*I cleavage site: 20.0% ± 2.8%; molecules observed: 28. In *C* the 5′ end of the tail is partly branch-migrated. Two tails of variable length originate from the same point on different places on the linear, double-stranded DNA. One of the tails (the 5′ end) can be folded back along the molecule to a position approximately 20% away from the *Pst*I cleavage site. Average distance to the *Pst*I cleavage site: 17.9% ± 2.5%; molecules observed: 5. In the total analysis nine molecules gave other values, which were scattered around the molecule. Bar = 0.1 nm.

the position of the gene-*A* nick in the *Hae*III restriction fragment Z6b and the *Pst*I cleavage site. Additional data on the nature and position of the nick in φX RFI have come from studies of the interaction between the gene-*A* protein and φX RF DNA in vitro.

Linney and Hayashi (1973; Linney et al. 1972) showed that gene *A* codes for two proteins, the A protein with a molecular weight of about 60,000 and the A* protein with a molecular weight of about 35,000. That both proteins derive from the same gene has been shown by peptide mapping of the two proteins; with the exception of one tryptic peptide, both maps are the same.

From these results and the fact that nonsense mutations in the N-terminal part of gene *A* eliminate only the A protein, whereas nonsense mutations in the C-terminal part eliminate both the A and A* proteins, it was concluded that the start signal for the A* protein is located within gene *A* and that both proteins are read in the same reading frame.

The gene-*A* protein (A/A* protein mixture) has been partially purified by Linney et al. (1972) and Henry and Knippers (1974). The latter investigators showed that purified gene-*A* protein nicks φX RFI only in the viral strand. The nuclease activity was highly specific; RFI DNAs of f1, SV40, and PM2 were not nicked. The gene-*A* protein also acts on φX and M13 SS DNA (see also Singh and Ray 1975). Very recently Ikeda et al. (1976) and Eisenberg et al. (1976) have reported the isolation and characterization of highly purified gene-*A* protein.

The exact position of the gene-*A*-protein nick in φX RFI has been determined recently by Langeveld et al. (1978). Nicked RFI was digested with the restriction enzyme *Hae*III, and the fragments Z1 to Z10 were separated by gel electrophoresis. Only fragment Z6b was nicked, as shown by the fact that it gave rise to three single-stranded fragments upon denaturation and electrophoresis, one of fragment length (ca. 280 nucleotides) and two smaller fragments (ca. 180 and 100 nucleotides). Labeling of the 3′-OH-terminated nucleotide residue at the nick with terminal transferase followed by sequence analysis established the unique sequence: –TGCTCCCCCAACTTG$_{OH}$. This sequence corresponds to nucleotides 4283–4297 of the φX DNA sequence (Sanger et al. 1977). These results indicate that gene-*A* protein cleaves the 3′ sugar-phosphate bond of the guanylic acid residue at position 4297 in the viral strand of φX RFI DNA (this is position 4305 in the definitive sequence of φX, which is given in Appendix I). This agrees with the site of gene-*A* protein cleavage between nucleotides 4290–4330 (Eisenberg et al. 1977; see also Eisenberg et al., this volume).

The experimental evidence presented in this section identifies the gene-*A* protein as an initiator protein, in the sense of the replicon model proposed by Jacob et al. (1964). It is encoded by one of the genes of the viral replicon and it must be made before DNA replication can occur. Gene-*A* protein nicks the viral strand of φX RFI at a specific site, the origin of replication, thereby creating a 3′-OH-bounded strand which may serve as a primer for DNA polymerase. The subsequent elongation of the viral strand and the synthesis of the complementary strand will be discussed in the next section.

Comparison of the φX DNA sequence at the origin of the viral strand with the corresponding sequence of bacteriophage G4 DNA (Fiddes et al. 1978) shows that there is a conserved region of 30 nucleotides. This is the boxed sequence shown in Figure 5. Gene-*A* protein of φX nicks G4 RFI at the same site as in φX RFI (A. D. M. van Mansfeld and S. A. Langeveld, pers.

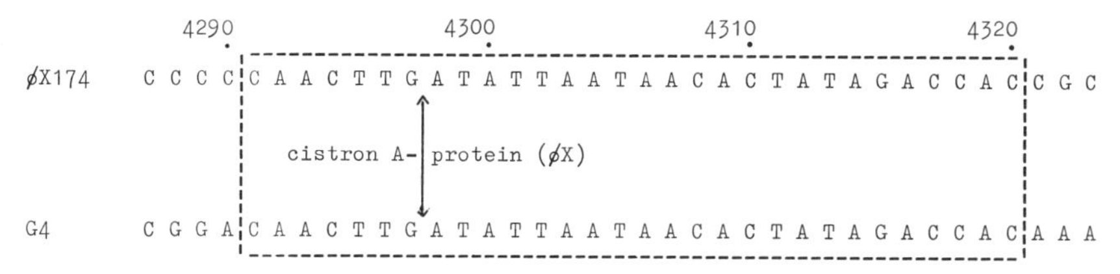

Figure 5 Position of the φX gene-*A*-protein nick in φX RFI and G4 RFI. Gene-*A* protein cleaves the phosphodiester bond in φX RFI (Langeveld et al. 1978) and G4 RFI (A. D. M. van Mansfeld and S. A. Langeveld, pers. comm.) at the position indicated, creating a 3′-OH G terminus in both cases. The φX DNA sequence and numbering are from Sanger et al. (1977); the G4 DNA sequence is from Fiddes et al., (1978).

comm.). The nucleotide sequence of the conserved region shows some characteristics that may be relevant to the mechanism of gene-*A*-protein action. The nick is located at the start of an A-T-rich region, which includes the palindrome TATTAATA, that is flanked on both sides by a G-C-rich sequence (not shown). Origin regions of circular DNA of SV40 (Subramanian et al. 1977), polyoma virus (Soeda et al. 1977), and ColE1 (Tomizawa et al. 1977) have also been found to contain A-T-rich sequences within G-C-rich tracts. Since the φX gene-*A* protein acts only on supertwisted RFI DNA and not on relaxed RFI DNA (Marians et al. 1977), it is suggested that the gene-*A* protein requires a base-unpaired duplex region, which is known to occur at A-T-rich sequences in superhelical DNA (Dasgupta et al. 1977), as well as a specific nucleotide recognition sequence. In this respect it is interesting to note that the nucleotide sequence around the gene-*A*-protein nick is not very unique in the φX DNA sequence (Sanger et al. 1977). Each of the five hexanucleotides ranging from ACTTGA to GATATT occurs more than once in the sequences of the viral and complementary strands. The nucleotide sequence ATCTTGATAAA at position 68 in the sequence shows a striking resemblance in sequence and base-pairing capacity to the gene-*A*-protein-nick region. However, none of these sequences in φX RFI DNA are acted upon by the gene-*A* protein, which indicates that the gene-*A*-protein recognition sequence has additional primary- or secondary-structure properties.

Alternative mechanisms of initiation of φX RF DNA replication have been considered on account of the presence of displacement loops in a fraction of the RFI molecules isolated from infected cells. These D loops were observed in the electron microscope (W. Keegstra, pers. comm.). Their average length is about 250 nucleotides. However, combined biochemical and electron microscopic analysis has shown that the D loops contain RNA as the displacing strand and that they are located at many different sites on the φX genome, including the known promoter sites.

Strand Elongation during φX RF DNA Replication

Our insight into the process of strand elongation during φX RF DNA replication comes from biochemical analysis of pulse-labeled replicating intermediates as well as from direct visualization of these intermediates in the electron microscope.

Detailed analysis of φX RF replicating intermediates, labeled by brief exposure to radioactive precursors (pulse-labeled), has been reported by Schröder et al. (Schröder and Kaerner 1972; Schröder et al. 1973) and by Fukuda and Sinsheimer (1976). Pulses shorter than the replication time of one RF molecule were used, and the pulse as a whole was incorporated symmetrically into the viral and complementary strands. This suggested that the pulse-labeled molecules were true intermediates in φX RF DNA replication. Under these conditions, negligible amounts of pulse label were detected in RFI. Both groups first separated the total pulse-labeled φX DNA into different fractions. Schröder et al. (1972, 1973) used sucrose gradient centrifugation for this separation, and Fukuda and Sinsheimer (1976) used propidium diiodide-CsCl equilibrium density gradient centrifugation. After the initial fractionation, length analysis by velocity sedimentation in alkaline sucrose gradients and strand analysis by buoyant-density centrifugation in alkaline CsCl gradients of the various fractions were carried out.

The most important conclusion from these studies was that replicating structures sedimenting more rapidly than RFI (Schröder and Kaerner 1972) or banding at a greater density than RFI (Fukuda and Sinsheimer 1976) contained pulse-labeled viral strands longer than genome length and extensive single-stranded regions. The ratio of pulse-labeled complementary strands, which were found as short fragments, to pulse-labeled viral strands was lowest in molecules with the most SS DNA. This is consistent with a rolling-circle mode of φX RF DNA replication, which involves the continuous elongation of the viral strand from one end displacing the other end as a linear single-stranded tail (Dressler and Wolfson 1970). The viral-strand tail remains largely single-stranded because the discontinuous synthesis of the complementary strand lags behind. Pulse-labeling experiments carried out at 41°C during the different stages of φX DNA replication in a host with a thermosensitive ligase support this conclusion (McFadden and Denhardt 1975). Replicating intermediates contain single-stranded regions in their viral strands, which suggests that synthesis of the viral strand precedes synthesis of the complementary strand.

The circularization of the displaced and partially double-stranded tail of the rolling circle is believed to yield an RFII molecule with multiple gaps in the complementary strand. The other partner of the rolling circle yields an RFII molecule with a closed complementary strand and a small gap in the viral strand at a specific place (see next section).

The various replicating intermediates predicted by a model in which the displacement synthesis of one strand precedes the synthesis of the other strand can be observed in the electron microscope. The following classes of molecules can be seen (Figs. 6 and 7; see also Koths and Dressler 1978):

1. Rolling-circle molecules with tails that are completely single-stranded (Fig. 6). The length of a tail only occasionally exceeds genome length. Branch migration of the tail is very often observed.
2. In a minor fraction of the rolling circles, complementary-strand synthesis is initiated on the tail (Fig. 6). Analysis of these molecules reveals no preferred sites for initiation of synthesis of complementary-strand DNA.
3. Circular, double-stranded molecules with single-stranded gaps (Fig. 7). This is the predominant class of replicating intermediates. Frequently, more than one gap per molecule is present. Restriction enzyme analysis showed that these gaps occur at many sites on the φX genome (see also next section).

Although the biochemical and electron microscopic studies of the replicating intermediates support the elongation mechanism predicted by the rolling-circle model, a few comments must be made. Viral strands longer

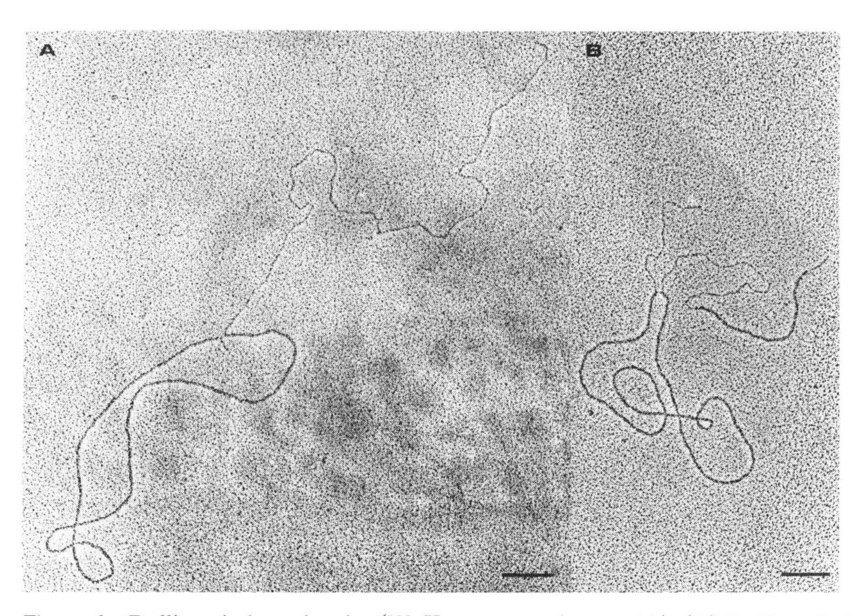

Figure 6 Rolling-circle molecules (W. Keegstra et al., unpubl.). (*A*) Rolling-circle molecule with a single-stranded tail. (*B*) Rolling-circle molecule with a branch-migrated tail on which complementary-strand synthesis has started. In one preparation 37 rolling-circle molecules were observed: 29 of these molecules had completely single-stranded tails, the other 8 had partially double-stranded tails. Analysis of 16 molecules with partially double-stranded tails before and after *Pst*I treatment did not reveal specific starting sites for synthesis of the complementary strand. Bar = 0.1 nm.

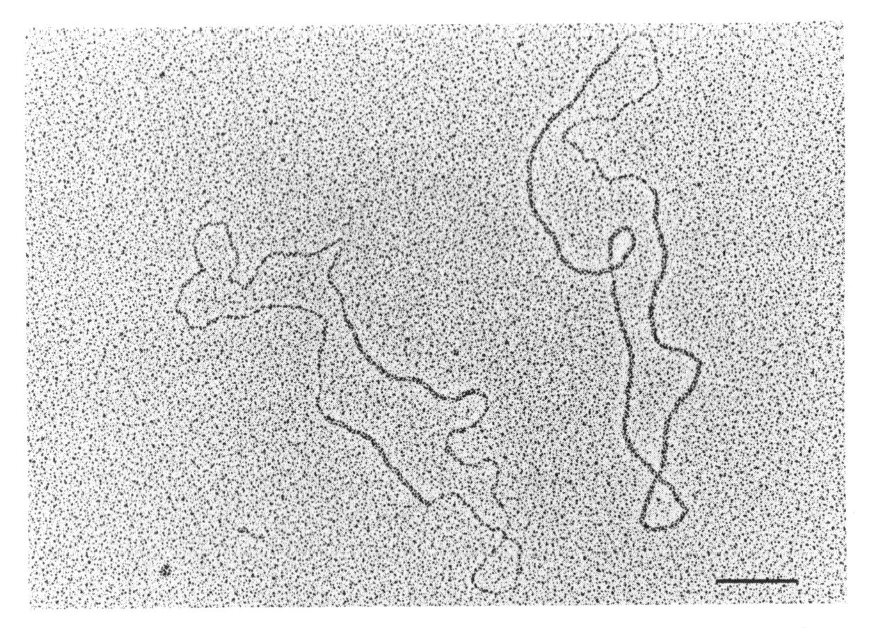

Figure 7 Gapped RFII molecules of φX DNA (W. Keegstra et al., unpubl.). The electron micrographs show partially double-stranded circles. One molecule has one single-stranded region and the other has two. In one preparation 73 gapped molecules were observed: 56 with one gap, 14 with two gaps, and 3 with three gaps. Analysis of 33 *Pst*I-cleaved gapped molecules did not indicate that the gaps were located at specific positions on the genome. Bar = 0.1 nm.

than unit length can easily be detected after separation of the different classes of replicating intermediates. However, it cannot rigorously be excluded that small fragments of the viral strand exist in vivo. Their joining may proceed extremely fast in vivo and it may be difficult to trap these intermediates. Our own results and those of others (Schekman et al. 1971; Kurosawa and Okasaki 1975) indicate that it is difficult in short pulses to stop efficiently viral-strand synthesis and ligase action. On the other hand, the existence of a small fraction of viral strands of less than unit length, as described by Yokohama et al. (1971) and McFadden and Denhardt (1975), during progeny SS DNA synthesis gives no definitive proof that the short strands are intermediates in synthesis, as these can also be the result of misincorporation of uracil into DNA (Tye et al. 1977). Machida et al. (1977) found pulse-labeled viral-strand fragments containing RNA at their 5′ ends during φX RF DNA replication. They have suggested that these RNA-primed DNA fragments are the precursors of the viral strands of greater than unit length (see also Hours et al., this volume).

Termination of φX RF DNA Replication

In the preceding section we have anticipated the termination events of φX RF DNA replication. "Gapped" RFII molecules (Fig. 7) can be considered as products of a round of rolling-circle replication. Molecules with small gaps were first discovered by Schekman et al. (1971). Their existence was deduced from the fact that a large proportion of RFII molecules can be converted by polynucleotide ligase to RFI molecules only with the aid of a DNA polymerase. φX RFII DNAs isolated from *polA* cells contain significantly more gapped molecules, which suggests a role for DNA polymerase I in the gap-filling process. Geider et al. (1972) and Schröder et al. (1973) have confirmed the existence of these gapped RFII molecules.

Eisenberg et al. (1974a,b, 1975) made a thorough study of these molecules. The gaps were filled in the presence of ^{32}P-labeled deoxynucleoside triphosphates and the incorporated label was analyzed by hybridization studies and restriction enzyme and pyrimidine tract analyses. The following results were obtained:

1. The average size of the gaps was 15 nucleotides. The gaps in the molecules shown in Figure 7 are of course much larger, and these are considered to be precursors of the molecules studied by Eisenberg et al. (1974a,b, 1975).
2. Under the conditions used, the gaps in the complementary strand were present at many, although not random, locations along the φX genome. There were molecules with one gap and molecules with several.
3. The gaps in the viral strand were restricted to a specific location in the *Hin*dIII-3 fragment, within gene *A*. A gap in the same region had also been found by Johnson and Sinsheimer (1974) in the viral strand of RFII DNA isolated from cells late in infection during the stage of progeny viral DNA synthesis. Although the gap was located in a specific region of the genome, the complexity of the pyrimidine tracts found in it indicated that the gap did not consist of a unique 15-nucleotide sequence. However, pyrimidine tract analysis showed a fourfold enrichment for the unique C_6T tract as compared with the whole viral strand. The C_6T tract is located in the *Hae*III restriction fragment Z6b within the *Hin*dIII-3 fragment, 13 nucleotides upstream from the position of the nick introduced into φX RFI by the gene-*A* protein.

The presence of multiply gapped RF molecules is consistent with the idea that complementary-strand synthesis is discontinuous and that it can start and terminate at many places on the φX genome. The origin of the gaps is not completely understood: they could arise from the presence of RNA primers made by the *dnaG* protein or they could originate because the synthesis of the complementary-strand pieces cannot reach the 5′ end of the adjacent piece, e.g., because of the presence of a DNA-binding, -priming, or unwinding protein, or because of some secondary-structure features in the viral strand.

The presence of only one gap in the viral strand, just before the position of the nick by the gene-*A* protein, is consistent with the idea that synthesis of the viral strand is continuous and unidirectional. This conclusion comes from analysis of unlabeled RFII molecules. Additional support for this conclusion was obtained by analysis of ^{3}H-pulse-labeled RFII molecules. The procedure outlined in the legend to Figure 8 yields RFII molecules that contain linear viral strands and circular complementary strands. The fact that 90% of the pulse label is found in the linear viral strands indicates that these RFII molecules have just finished their viral-strand synthesis, leaving a small gap or nick. Restriction enzyme analysis of these RFII molecules shows that the pulse label increases along the physical map of φX RF DNA, terminating (as shown in Fig. 8) in the restriction fragment Z6b. This indicates that the synthesis of the viral strand occurs in one direction and terminates close to the position at which it started.

Analysis of replicating intermediates of φX RF DNA in the electron microscope suggests the way in which the rolling circle separates into the two gapped daughter RFII molecules. DNA molecules containing two circles, at least one of which is completely double-stranded and the other is single-stranded or partially double-stranded (Fig. 9 A and B, respectively), were observed. The connection between the two molecules is about 20% away from the *Pst*I cleavage site (Fig. 10), i.e., at the position of the gene-*A*-protein nick in the *Hae*III fragment Z6b. These electron micrographs suggest that the tail of the rolling circle circularizes before the two daughter molecules separate. As circularization of the displaced strand was observed only in the termination phase, it may be that the association between the terminus and origin of the displaced strand requires that the origin be exposed in a single-stranded form. The presence of two, circular, double-stranded molecules connected by a piece of single-stranded DNA (Fig. 9C) supports this idea. Such a mechanism would assure that only strands of genome length are circularized.

We cannot exclude the possibility that in earlier stages of replication the end of the tail of the rolling circle has a loose connection in the origin region (Knippers et al. 1969a) or that the end of the tail travels along with the replication fork in a looped rolling circle, as has been observed in vitro (Eisenberg et al. 1977). If such a connection exists in vivo it must be less stable than that in the termination structures.

The last step in φX RF DNA replication is the conversion of the gapped RFII molecules into RFI molecules. Several investigators have noticed that this conversion is a rate-limiting step in the replication of φX RF DNA. Only after relatively long pulse times, exceeding 10 to 20 seconds, do significant amounts of radioactivity accumulate in RFI molecules. Also, several investigators have confirmed the observation made by Dressler and Wolfson

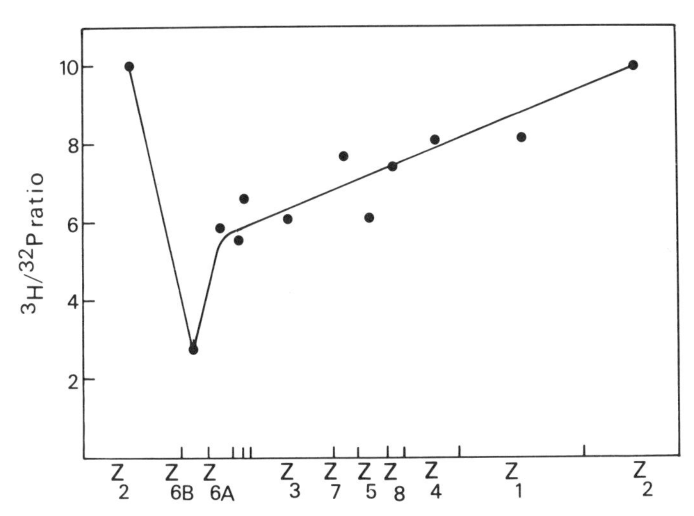

Figure 8 Position of the terminus of viral-strand synthesis in φX RFII DNA. After treatment with mitomycin C, *E. coli* H502 (*thy*$^-$, *uvrA*, *sup*$^-$) cells were infected with φX (m.o.i.=5) in the presence of chloramphenicol at 40 μg/ml at 30°C. Twenty minutes after infection the culture was pulse-labeled with [^{3}H]thymidine for 5 sec. The pulse was stopped by pouring the infected culture into a stainless-steel beaker containing an equal volume of crushed ice made with 0.05 M Tris, 0.005 M EDTA, 0.005 M KCN, and 0.05 M NaN_3, pH 8.0. This beaker was placed in a dry ice-acetone bath. After collection and washing, the cells were lysed in sucrose buffer with lysozyme, pronase, and SDS. *E. coli* DNA, together with SDS, was precipitated and the supernatant was extracted with phenol. After RNase treatment, total φX DNA was isolated in sucrose gradients and thereafter separated into different fractions by means of propidium diiodide-CsCl centrifugation. The resulting RFII fraction was passed through a BND column, and the double-stranded fraction, which eluted at 1 M salt, was collected. The pulse-labeled RFII DNA thus obtained contained a linear viral strand and a circular complementary strand, as shown by alkaline sucrose gradient and poly(UG)-CsCl centrifugations. Ninety percent of the label was found in the linear viral strands.

Pulse-labeled RFII was mixed with uniformly ^{32}P-labeled φX RF DNA and subjected to *Hae*III restriction enzyme analysis. The ^{3}H/^{32}P ratios of the *Hae*III fragments (normalized with respect to Z2 = 10.00) are plotted against the physical map of φX DNA. The ^{3}H/^{32}P gradient indicates that viral-strand synthesis proceeds clockwise around the physical map, i.e., in the direction of the genes *A*, *B*, through *H*, and terminates in fragment Z6b.

(1970) that these first pulse-labeled RFI molecules are preferentially labeled in their complementary strands. Restriction enzyme analysis of a mixture of uniformly ^{32}P-labeled φX RF DNA and ^{3}H-pulse-labeled RFI DNA, containing equal amounts of label in viral and complementary strands, showed a gradient with a minimum ^{3}H/^{32}P ratio in the *Hin*dII-3 fragment and a maximum ratio in the *Hin*dII-4 fragment (Godson 1974). It

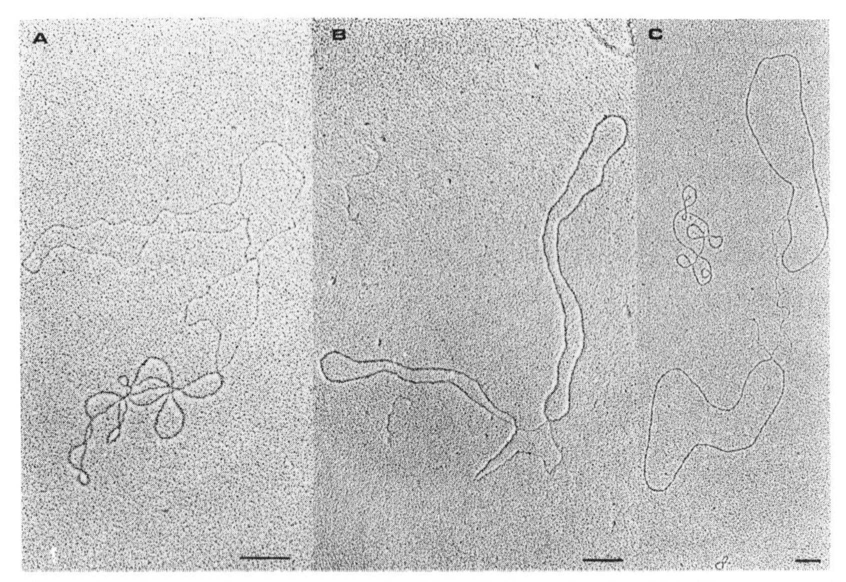

Figure 9 Termination structures of rolling-circle replication (W. Keegstra et al., unpubl.). The following termination structures were observed in the electron microscope. (*A*) A double-stranded circle connected to a single-stranded circle. (*B*) A double-stranded circle connected to a partially double-stranded circle. (*C*) Two double-stranded circles connected by a piece of single-stranded DNA. Bar = 0.1 nm.

is not clear whether the effect observed in this experiment represents the termination event of viral-strand synthesis or gap-filling in the complementary strand. Our own restriction analysis of pulse-labeled RFI DNA, labeled predominantly in the complementary strand, yielded a flat distribution, with the exception of restriction fragments Z6a and Z6b, which correspond to the *Hin*dII-3 region. These fragments contained significantly more pulse label and probably reflect filling in at the specific viral-strand gap.

Finally, the question of the introduction of superhelical turns into φX RFI DNA should be considered. Fukuda and Sinsheimer (1976) showed that in propidium diiodide-CsCl gradients a fraction of pulse-labeled φX RF DNA banded at a heavier position than uniformly labeled RFI. Biochemical analysis of this material revealed that this fraction contained two different classes of φX DNA: rolling-circle molecules and covalently closed, duplex circles, presumably with a low superhelical density.

We have confirmed this observation and have separated both classes of replicating φX DNA by BND-cellulose chromatography. The RFI molecules elute with 1 M NaCl and the rolling-circle-type molecules elute only after the addition of 2% caffeine. These RFI molecules band heterogeneously in propidium diiodide-CsCl gradients at a heavier position than the ^{14}C-labeled marker RFI molecules. They also sediment heterogeneously in

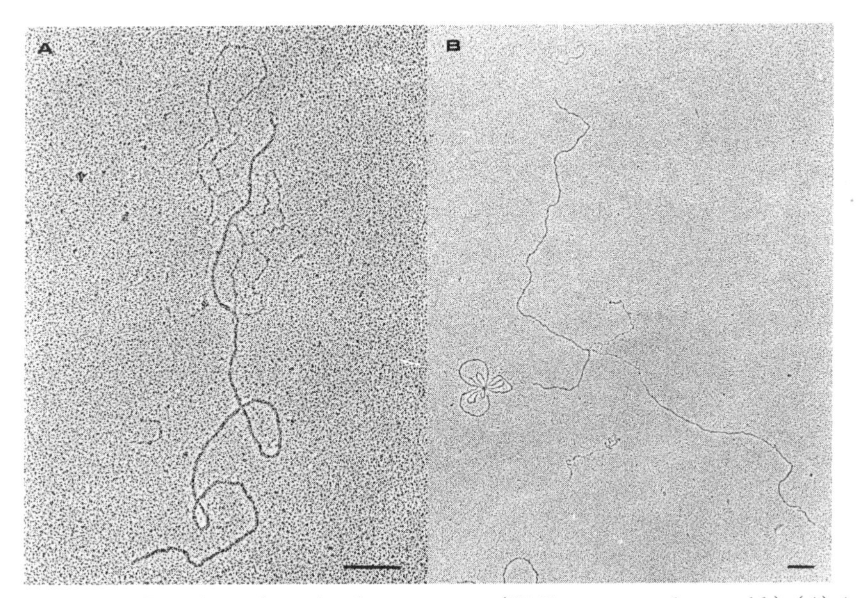

Figure 10 *Pst*I-cleaved termination structures (W. Keegstra et al., unpubl.). (*A*) A single-stranded circle connected to a double-stranded circle cleaved by *Pst*I. (*B*) A double-stranded circle connected to a partially double-stranded circle, both cleaved by *Pst*I. The connection site is approximately 20% away from the *Pst*I cleavage site. Bar = 0.1 nm.

neutral sucrose gradients between the positions of RFI and RFII. Figure 11 shows that in agarose gels in the absence of ethidium bromide they migrate as specific bands (at least 15 are visible) between the positions of RFI and RFII. In the presence of ethidium bromide they migrate as one band slightly faster than RFI. These results show clearly that this fraction is a heterogeneous population of RFI molecules with different superhelical densities. Presumably, the recently discovered gyrase (Gellert et al. 1976) introduces, turn by turn, the required number of negative superhelical twists into the covalently closed RFI DNA. Thus, the relaxed RFI molecules (also called RFIV) are converted to supertwisted RFI molecules, which, at least in vitro, are the preferred substrates for RNA polymerase (Hayashi and Hayashi 1971) and the gene-*A* protein (Ikeda et al. 1976).

SUMMARIZING SCHEME OF φX RF DNA REPLICATION IN VIVO

The results of biochemical and electron microscopic studies as discussed in the preceding sections suggest a mechanism of φX RF DNA replication (Fig. 12) that confirms some essential predictions made by the rolling-circle model (Gilbert and Dressler 1969).

Major advances have been made in our understanding of the mechanism

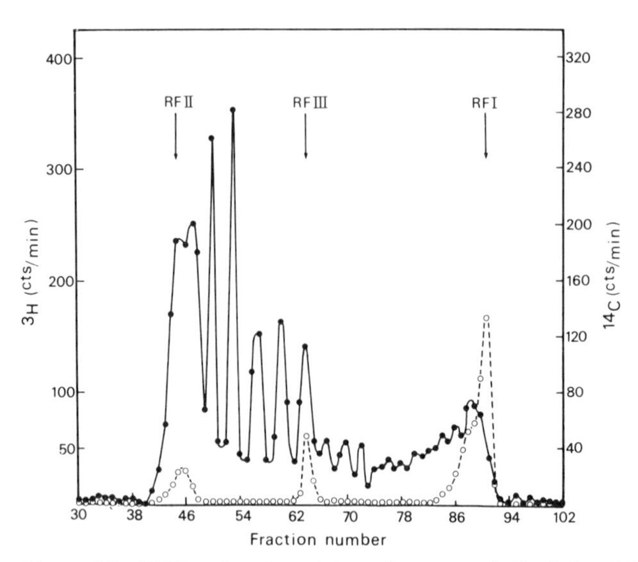

Figure 11 RFI molecules with various superhelical densities. Pulse-labeling (as outlined in the legend to Fig. 8) yielded labeled φX DNA that banded at the same or a heavier position than the RFI position in a propidium diiodide-CsCl gradient. This material was collected and subjected to BND-cellulose chromatography. The double-stranded fraction was mixed with ^{14}C-labeled RFI, RFII, and RFIII DNA and layered on a 1.4% agarose gel (length = 22 cm, diameter = 0.6 cm). Electrophoresis was carried out in 0.04 M Tris, 5 mM sodium acetate, and 1 mM EDTA, pH 7.8, for 20 hr at 1 mA/gel. After electrophoresis, the gel was sliced (0.18 cm/fraction), solubilized, and counted. (●——●) ^{3}H cpm, representing pulse-labeled φX DNA. (o——o) ^{14}C cpm, representing RFI, RFII, and RFIII DNA markers.

of initiation. Before initiation of φX RF DNA replication can take place, the SS DNA of the infecting phage must be converted into supertwisted RFI DNA and φX gene *A* must be transcribed and translated. The gene-*A* protein is a highly specific nuclease which introduces one nick specifically in the viral strand of the parental RF molecule in vivo (Francke and Ray 1971, 1972). This nick is located within the *Hae*III restriction fragment Z6b on the physical map of φX RF DNA at approximately 100 nucleotides from the Z2/Z6b junction (Baas et al. 1976a; Fig. 3); this position corresponds to the origin of tail growth of the rolling circle (Fig. 4; see also Koths and Dressler 1978). Langeveld et al. (1978) have shown that the gene-*A* protein cleaves specifically the phosphodiester bond in the viral strand of φX RFI in vitro between the G and A residues in the sequence ACTTGATATT, creating a 3′-OH G terminus at position 4305 (Appendix I). This position of the gene-*A* cleavage site agrees with the position between nucleotides 4290–4330 as determined by Eisenberg et al. (1977; see also Eisenberg et al., this volume).

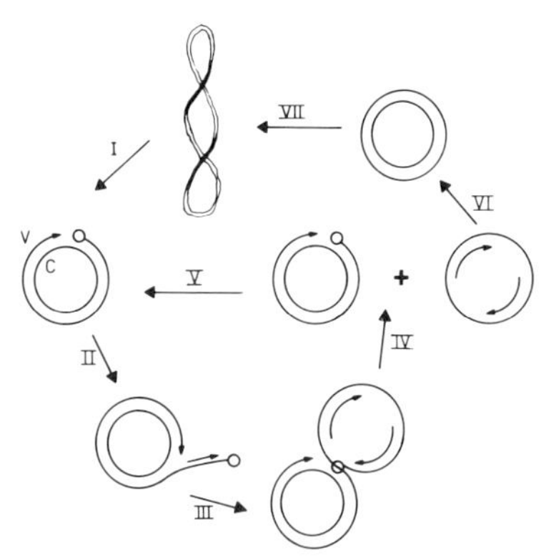

Figure 12 Summarizing scheme of φX RF DNA replication in vivo. (I) Initiation, or the introduction of a nick in the viral strand of RFI at the origin which creates a 3′-OH G terminus. (II) Strand elongation, or viral-strand synthesis starting at the origin according to the rolling-circle mechanism, and discontinuous synthesis of the new complementary strand on the displaced viral-strand tail. (III) Termination of a round of rolling-circle replication by circularization of the genome-length tail and (IV) separation of the two daughter RFII molecules, one containing a specific gap in the viral strand (which [V] continues to replicate) and one with multiple gaps in the complementary strand (which is converted via [VI] relaxed RFI into [VII] supertwisted RFI).

The minimum recognition sequence of the gene-*A* protein remains to be elucidated. It seems likely that local base-unpairing of the double helix is important for gene-*A*-protein action as supertwisted RFI is nicked but relaxed RFI is not (Marians et al. 1977).

Alternative mechanisms of initiation by RNA or protein priming have been suggested as part of the reciprocating strand model (Denhardt 1972, 1975, 1977). Indeed, RFI molecules with displacement loops containing short RNA strands have been found in φX-infected cells. However, the nonspecificity of their location on the φX RF molecule does not support their role in replication.

Most data indicate that synthesis of the viral strand takes place as proposed by the rolling-circle model. This involves continuous elongation of the old viral strand, starting at the 3′-OH end of the nick, and displacement of the 5′ end. The complementary strand serves as a template. Rolling circles have been identified in a fraction of pulse-labeled replicating intermediates by biochemical studies (Schröder and Kaerner 1972; Schröder et al. 1973; Fukuda and Sinsheimer 1976) and by electron microscopy (Fig. 6; see also Koths and Dressler 1978). They contain viral strands longer than genome length. Machida et al. (1977) have suggested that transient, RNA-primed DNA fragments are the precursors of these longer viral strands. The tails of the rolling circles are either completely single-stranded or partially double-stranded. The combined data indicate that φX RF DNA replication involves two different mechanisms of chain elongation. The synthesis of the viral strand starts at one position, the origin of replication, with displacement of a linear tail. In contrast, the complementary strand can start at many places on the displaced viral strand, and this synthesis lags behind that of the viral strand.

The circularization of the displaced single- or partially double-stranded tail of the rolling circle yields a circular viral strand or an RFII molecule with a closed viral strand and a complementary strand with multiple gaps. These RFII molecules are eventually converted to RFI. The other partner of the rolling circle yields an RFII molecule with a closed complementary strand and a discontinuity in the viral strand at a specific place containing the unique C_6T tract (Eisenberg et al. 1975), which is located 13 nucleotides upstream of the G residue at the origin of replication. The reason why viral-strand synthesis is interrupted before it reaches the origin of replication is not known. Plausible explanations involve the possibility that unwinding of the parental strands precedes the growing point of viral-strand synthesis in the termination phase.

The final step in φX RF DNA replication, the formation of the superhelical RFI molecule, is believed to occur by the closure of the gapped RFII molecule to form a relaxed RFI molecule, which is then rapidly converted by gyrase to superhelical RFI. This is suggested by the presence of pulse-labeled RFI molecules with increasing superhelical density in fractions of replicating intermediates (Fig. 11).

The daughter RF molecule with a gap in the viral strand and containing the closed parental complementary strand is the preferred template for further replication. This is suggested from heteroduplex infection experiments (Baas and Jansz 1972a; Merriam et al. 1971a,b), which indicate that the genetic information of the progeny phage is largely derived from the complementary strand of the parental RF DNA.

Finally, the distinction between rolling circles, Cairns structures (Knippers et al. 1969a), and looped rolling circles should be mentioned. Looped rolling circles, i.e., rolling circles in which the 5′ end of the tail travels along with the replication fork, were observed as intermediates during the synthesis of circular viral DNA from φX RF DNA in vitro (Eisenberg et al. 1977). We can not exclude the possibility that the end of the tail of the rolling circle remains associated, by a protein linkage, with the origin of the replication fork during RF DNA replication in vivo, and that this linkage is broken during isolation. The only looped structures that we have observed among many rolling circles seen in electron micrographs are those with a displaced strand that is of genome length or longer (Fig. 9). The two circles are connected to each other at the region of the replication origin (Fig. 10). These structures suggest that the 5′ end of the tail of the rolling circle becomes associated with the origin region at the end of a round of displacement synthesis, i.e., when the origin of the displaced strand is exposed in a single-stranded form. An attractive hypothesis, for which there is no evidence, is that this association is caused by a protein, e.g., the gene-*A* protein, which nicks the displaced strand at the origin and then joins the 3′ and 5′ ends of the displaced strand. This would terminate or interrupt viral-strand synthesis and yield circular viral strands of genome length.

REFERENCES

Baas, P. D. and H. S. Jansz. 1972a. Asymmetric information transfer during φX174 DNA replication. *J. Mol. Biol.* **63:** 557.

———. 1972b. φX174 replicative form DNA replication, origin and direction. *J. Mol. Biol.* **63:** 569.

Baas, P. D., H. S. Jansz, and R. L. Sinsheimer. 1976a. Bacteriophage φX174 DNA synthesis in a replication-deficient host: Determination of the origin of φX DNA replication. *J. Mol. Biol.* **102:** 633.

Baas, P. D., G. P. H. van Heusden, J. M. Vereijken, P. J. Weisbeek, and H. S. Jansz. 1976b. Cleavage map of bacteriophage φX174 RF DNA by restriction enzymes. *Nucleic Acids Res.* **8:** 1947.

Bartok, K. and D. T. Denhardt. 1976. Site of cleavage of superhelical φX174 replicative form DNA by the single-strand specific *N. crassa* nuclease. *J. Biol. Chem.* **251:** 530.

Bauer, W. and J. Vinograd. 1968. The interaction of closed circular DNA with intercalative dyes. I. The superhelix density of SV40 DNA in the presence and absence of dye. *J. Mol. Biol.* **33:** 141.

Benbow, R. M., M. Eisenberg, and R. L. Sinsheimer. 1972. Multiple length DNA molecules of bacteriophage φX174. *Nature* **237:** 141.

Bowman, K. L. and D. S. Ray. 1975. Degradation of the viral strand of φX174 parental replicative form DNA in a Rep⁻ host. *J. Virol.* **16:** 838.

Burton, A. and R. L. Sinsheimer. 1965. The process of infection with bacteriophage φX174. VII. Ultracentrifugal analysis of the replicative form. *J. Mol. Biol.* **14:** 327.

Calendar, R., B. Lindqvist, G. Sironi, and A. J. Clark. 1970. Characterization of Rep⁻ mutants and their interaction with P2 phage. *Virology* **40:** 72.

Dasgupta, S., D. P. Allison, C. E. Snyder, and S. Mitra. 1977. Base-unpaired regions in supercoiled replicative form DNA of coliphage M13. *J. Biol. Chem.* **252:** 5916.

Datta, B. and R. K. Poddar. 1970. Greater vulnerability of the infecting viral strand of replicative form DNA of bacteriophage φX174. *J. Virol.* **6:** 583.

Denhardt, D. T. 1972. A theory of DNA replication. *J. Theor. Biol.* **34:** 487.

———. 1975. The single-stranded DNA phages. *CRC Crit. Rev. Microbiol.* **4:** 161.

———. 1977. The isometric single-stranded DNA phages. In *Comprehensive virology* (ed. H. Fraenkel-Conrat and R. R. Wagner), vol. 7, p. 1. Plenum Press, New York.

Denhardt, D. T. and A. C. Kato. 1973. Comparison of the effect of ultraviolet radiation and ethidium bromide intercalation on the conformation of superhelical φX174 RF DNA. *J. Mol. Biol.* **77:** 479.

Denhardt, D. T. and R. L. Sinsheimer. 1965a. The process of infection with bacteriophage φX174. IV. Replication of the viral DNA in a synchronized infection. *J. Mol. Biol.* **12:** 647.

———. 1965b. The process of infection with bacteriophage φX174. V. Inactivation of the phage-bacterium complex by decay of ^{32}P incorporated in the infecting particle. *J. Mol. Biol.* **12:** 663.

———. 1965c. The process of infection with bacteriophage φX174. VI. Inactivation of infected complexes by ultraviolet irradiation. *J. Mol. Biol.* **12:** 674.

Denhardt, D. T., D. H. Dressler, and A. Hathaway. 1967. The abortive replication of φX174 DNA in a recombination deficient mutant of *E. coli. Proc. Natl. Acad. Sci.* **57:** 813.

Denhardt, D. T., M. Iwaya, and L. L. Larison. 1972. The *rep* mutation. II. Its effects on *E. coli* and on the replication of bacteriophage φX174. *Virology* **49:** 486.

Dressler, D. and J. Wolfson. 1970. The rolling circle for φX DNA replication. III. Synthesis of supercoiled duplex rings. *Proc. Natl. Acad. Sci.* **67:** 456.

Eisenberg, S. and D. T. Denhardt. 1974a. Structure of nascent φX174 replicative form: Evidence for discontinuous DNA replication. *Proc. Natl. Acad. Sci.* **71:** 1974.

———. 1974b. The mechanism of replication of φX174 single-stranded DNA. X. Distribution of the gaps in nascent RF DNA. *Biochem. Biophys. Res. Commun.* **61:** 532.

Eisenberg, S., J. Griffith, and A. Kornberg. 1977. φX174 cistron A protein is a multifunctional enzyme in DNA replication. *Proc. Natl. Acad. Sci.* **74:** 3198.

Eisenberg, S., J. F. Scott, and A. Kornberg. 1976. An enzyme system for replication of duplex circular DNA: The replicative form of phage φX174. *Proc. Natl. Acad. Sci.* **73:** 1594.

Eisenberg, S., B. Harbers, C. Hours, and D. T. Denhardt. 1975. The mechanism of replication of φX174 DNA. XII. Non-random location of gaps in nascent φX174 RF II DNA. *J. Mol. Biol.* **99:** 107.

Fiddes, J. C., B. G. Barrell, and G. N. Godson. 1978. Nucleotide sequences of the separate origins of synthesis of bacteriophage G4 viral and complementary strands. *Proc. Natl. Acad. Sci.* **75:** 1081.

Francke, B. and D. S. Ray. 1971. Formation of the parental replicative form DNA of bacteriophage φX174 and initial events in its replication. *J. Mol. Biol.* **61:** 565.

———. 1972. *Cis*-limited action of the gene *A* product of bacteriophage φX174 and the essential bacterial site. *Proc. Natl. Acad. Sci.* **69:** 475.

Fukuda, A. and R. L. Sinsheimer. 1976. Process of infection with bacteriophage φX174. XXXVIII. Replication of φX174 replicative form in vivo. *J. Virol.* **17:** 776.

Geider, K., H. Lechner, and H. Hoffmann-Berling. 1972. Nucleotide-permeable *E. coli* cells. V. Structure of newly synthesized φX174 RF DNA. *J. Mol. Biol.* **69:** 333.

Gellert, M., K. Mizuuchi, M. H. O'Dea, and H. A. Nash. 1976. DNA gyrase: An enzyme that introduces superhelical turns into DNA. *Proc. Natl. Acad. Sci.* **73:** 3872.

Gilbert, W. and D. Dressler. 1969. DNA replication: The rolling circle model. *Cold Spring Harbor Symp. Quant. Biol.* **33:** 473.

Godson, G. N. 1974. Origin and direction of φX174 double- and single-stranded DNA synthesis. *J. Mol. Biol.* **90:** 127.

Gordon, C. N., M. G. Rush, and R. C. Warner. 1970. Complex replicative form molecules of bacteriophage φX174 and S13 Su105. *J. Mol. Biol.* **47:** 495.

Hayashi, Y. and M. Hayashi. 1971. Template activities of the φX174 replicative allomorphic DNAs. *Biochemistry* **10:** 4212.

Henry, T. J. and R. Knippers. 1974. Isolation and function of the gene *A* initiator of bacteriophage φX174, a highly specific DNA endonuclease. *Proc. Natl. Acad. Sci.* **71:** 1549.

Hofs, E. B. H., G. A. van Arkel, P. D. Baas, D. J. Ellens, and H. S. Jansz. 1973. Mechanism of formation of bacteriophage φX174 circular and catenated dimer RF DNA. *Mol. Gen. Genet.* **126:** 37.

Ikeda, J., A. Yudelevich, and J. Hurwitz. 1976. Isolation and characterization of the protein coded by gene *A* of bacteriophage φX174 DNA. *Proc. Natl. Acad. Sci.* **73:** 2669.

Iwaya, M. and D. T. Denhardt. 1971. The mechanism of replication of φX174 single stranded DNA. II. The role of viral proteins. *J. Mol. Biol.* **57:** 159.

———. 1973a. Mechanism of replication of φX174. V. Dispersive and conservative transfer of parental DNA into progeny DNA. *J. Mol. Biol.* **73:** 291.

———. 1973b. Mechanism of replication of φX174 single stranded DNA. IV. The parental viral strand is not conserved in the replicating DNA structure. *J. Mol. Biol.* **73:** 279.

Jacob, F., S. Brenner, and F. Cuzin. 1964. On the regulation of DNA replication in bacteria. *Cold Spring Harbor Symp. Quant. Biol.* **28:** 329.

Jaenisch, R., P. H. Hofschneider, and A. Preuss. 1966. Uber infektiöse Substrukturen aus *Escherichia coli* Bakterophagen. VIII. On the tertiary structure and biological properties of φX174 replicative form. *J. Mol. Biol.* **21:** 501.

Jansz, H. S. and P. H. Pouwels. 1965. Structure of the replicative form of bacteriophage φX174. *Biochem. Biophys. Res. Commun.* **18:** 589.

Jansz, H. S., P. D. Baas, P. H. Pouwels, E. F. J. van Bruggen, and H. Oldenziel. 1968. Structure of the replicative form of bacteriophage φX174. V. Interconversions between twisted, extended and randomly coiled forms of cyclic DNA. *J. Mol. Biol.* **32:** 159.

Johnson, P. H. and R. L. Sinsheimer. 1974. Structure of an intermediate in the replication of bacteriophage φX174 DNA: The initiation site for DNA replication. *J. Mol. Biol.* **83:** 47.

Knippers, R. and R. L. Sinsheimer. 1968. Process of infection with bacteriophage φX174. XX. Attachment of the parental DNA of bacteriophage φX174 to a fast-sedimenting cell component. *J. Mol. Biol.* **34:** 17.

Knippers, R., T. Komano, and R. L. Sinsheimer. 1968. The process of infection with bacteriophage φX174. XXI. Replication and fate of the replicative form. *Proc. Natl. Acad. Sci.* **59:** 577.

Knippers, R., J. M. Whalley, and R. L. Sinsheimer. 1969a. The process of infection with bacteriophage φX174. XXX. Replication of double stranded DNA. *Proc. Natl. Acad. Sci.* **64:** 275.

Knippers, R., W. O. Salivar, J. E. Newbold, and R. L. Sinsheimer. 1969b. The process of infection with bacteriophage φX174. XXVI. Transfer of the parental DNA of bacteriophage φX174 into progeny bacteriophage particles. *J. Mol. Biol.* **39:** 641.

Komano, T., R. Knippers, and R. L. Sinsheimer. 1968. The process of infection with bacteriophage φX174. XXII. Synthesis of progeny single stranded DNA. *Proc. Natl. Acad. Sci.* **59:** 911.

Koths, K. and D. Dressler. 1978. An electron microscopic analysis of the φX DNA replication cycle. *Proc. Natl. Acad. Sci.* **75:** 605.

Kozinski, A. W. 1961. Uniform sensitivity to ^{32}P decay among progeny of ^{32}P-free phage φX174 grown on ^{32}P-labeled bacteria. *Virology* **13:** 377.

Kurosawa, Y. and R. Okazaki. 1975. Mechanism of DNA chain growth. XIII. Evidence for discontinuous replication of both strands of P2 phage DNA. *J. Mol. Biol.* **94:** 229.

Langeveld, S. A., A. D. M. van Mansfeld, P. D. Baas, H. S. Jansz, G. A. van Arkel, and P. J. Weisbeek. 1978. Nucleotide sequence of the origin of replication in bacteriophage φX174 RF DNA. *Nature* **271:** 417.

Lindqvist, B. H. and R. L. Sinsheimer. 1968. The process of infection with bacteriophage φX174. XVI. Synthesis of the replicative form and its relation to viral single-stranded DNA synthesis. *J. Mol. Biol.* **32:** 285.

Linney, E. and M. Hayashi. 1973. The two proteins of gene *A* of φX174. *Nat. New Biol.* **245:** 6.

Linney, E. A., M. N. Hayashi, and M. Hayashi. 1972. Gene *A* of φX174. I. Isolation and identification of the products. *Virology* **50:** 381.

Machida, Y., T. Okazaki, and R. Okazaki. 1977. Discontinuous replication of replicative form DNA from bacteriophage φX174. *Proc. Natl. Acad. Sci.* **74:** 2776.

Marians, K. J., J. Ikeda, S. Schlagman, and J. Hurwitz. 1977. Role of DNA gyrase in φX replicative form replication in vitro. *Proc. Natl. Acad. Sci.* **74:** 1965.

Matsubara, K., K. Shimada, and Y. Takagi. 1967. The replication process of single stranded DNA of bacteriophage φX174. III. Fate of parental phage DNA and replication of the RF. *J. Biochem.* **62:** 688.

McFadden, G. and D. T. Denhardt. 1975. The mechanism of replication of φX174 DNA. XII. Discontinuous synthesis of the complementary strand in an *E. coli* host with a temperature-sensitive polynucleotide ligase. *J. Mol. Biol.* **99:** 125.

Merriam, V., L. B. Dumas, and R. L. Sinsheimer. 1971a. Genetic expression in heterozygous replicative form molecules of φX174. *J. Virol.* **7:** 603.

Merriam, V., F. Funk, and R. L. Sinsheimer. 1971b. Genetic expression in whole cells of heterozygous replicative form molecules of φX174. *Mutat. Res.* **12:** 206.

Pouwels, P. H. and H. S. Jansz. 1964. Structure of the replicative form of bacteriophage φX174. *Biochim. Biophys. Acta* **91:** 177.

Pouwels, P. H., H. S. Jansz, J. van Rotterdam, and J. A. Cohen. 1966. Structure of the replicative form of bacteriophage φX174. Physicochemical studies. *Biochim. Biophys. Acta* **119:** 289.

Pouwels, P. H., C. M. Knijnenburg, J. van Rotterdam, and J. A. Cohen. 1968. Structure of the replicative form of bacteriophage φX174. VI. Studies on alkali-denatured double-stranded φX DNA. *J. Mol. Biol.* **32:** 169.

Roth, T. F. and M. Hayashi. 1966. Allomorphic forms of bacteriophage φX174 replicative DNA. *Science* **154:** 658.

Rush, M. G., A. K. Kleinschmidt, W. Hellman, and R. C. Warner. 1967. Multiple-length rings in preparations of φX174 replicative form. *Proc. Natl. Acad. Sci.* **58:** 1676.

Salivar, W. O. and R. L. Sinsheimer. 1969. Intracellular location and number of replicating parental DNA molecules of bacteriophages lambda and φX174. *J. Mol. Biol.* **41:** 39.

Sanger, F., G. M. Air, B. G. Barrell, N. L. Brown, A. R. Coulson, J. C. Fiddes, C. A. Hutchison III, P. M. Slocombe, and M. Smith. 1977. Nucleotide sequence of bacteriophage φX174 DNA. *Nature* **265:** 687.

Schekman, R. W., M. Iwaya, K. Bromstrup, and D. T. Denhardt. 1971. The mechanism of replication of φX174 single stranded DNA. III. An enzymic study of the structure of the replicative form II DNA. *J. Mol. Biol.* **57:** 177.

Schnegg, B. and P. H. Hofschneider. 1970. Transfer of parental φX174 DNA to progeny bacteriophage particles observed at low multiplicities of infection. *J. Mol. Biol.* **51:** 315.

Schröder, C. H. and H. C. Kaerner. 1972. Replication of bacteriophage φX174 replicative form DNA in vivo. *J. Mol. Biol.* **71:** 351.

Schröder, C. H., E. Erben, and H. C. Kaerner. 1973. A rolling circle model of the in vivo replication of bacteriophage φX174 replicative form DNA: Different fate of two types of progeny replicative form. *J. Mol. Biol.* **79:** 599.

Siegel, J. E. D. and M. Hayashi. 1967. Complementary strand infectivity in φX174 replicative form DNA. *J. Mol. Biol.* **27:** 443.

Singh, S. and D. S. Ray. 1975. A novel single-strand endonuclease specific for φX174 DNA. *Biochem. Biophys. Res. Commun.* **67:** 1429.

Sinsheimer, R. L. 1961. Replication of bacteriophage φX174. In *Proceedings of the R. A. Welch Foundation Conference on Chemical Research,* vol. V, p. 277.

Sinsheimer, R. L. 1968. Bacteriophage φX174 and related viruses. *Prog. Nucleic Acid Res. Mol. Biol.* **8:** 115.

Sinsheimer, R. L., B. Starman, C. Nagler, and S. Guthrie. 1962. The process of infection with bacteriophage φX174. I. Evidence for a "replicative form." *J. Mol. Biol.* **4:** 142.

Smith, M., N. L. Brown, G. M. Air, B. G. Barrell, A. R. Coulson, C. A. Hutchison III, and F. Sanger. 1977. DNA sequence at the C termini of the overlapping genes *A* and *B* in bacteriophage φX174. *Nature* **265:** 702.

Soeda, E., K. Miura, A. Nakaso, and G. Kimura. 1977. Nucleotide sequence around the replication origin of polyoma virus DNA. *FEBS Lett.* **79:** 383.

Stone, A. B. 1967. Some factors which influence the replication of the replicative form of bacteriophage φX174. *Biochem. Biophys. Res. Commun.* **26:** 247.

Subramanian, K. M., R. Dhar, and S. M. Weissman. 1977. Nucleotide sequence of a fragment of SV40 DNA that contains the origin of DNA replication and specifies the 5′ ends of "early" and "late" viral RNA. *J. Biol. Chem.* **252:** 355.

Tessman, E. S. 1966. Mutants of bacteriophage S13 blocked in infectious DNA synthesis. *J. Mol. Biol.* **17:** 218.

Thompson, B. J., C. Escarmis, B. Parker, W. C. Slater, J. Doniger, I. Tessman, and R. C. Warner. 1975. Figure 8 configuration of dimers of S13 and φX174 replicative form DNA. *J. Mol. Biol.* **91:** 408.

Tomizawa, J., H. Ohmori, and R. E. Bird. 1977. Origin of replication of colicin E1 plasmid DNA. *Proc. Natl. Acad. Sci.* **74:** 1865.

Tye, B., P. Nijman, I. R. Lehman, S. Hochhauser, and B. Weiss. 1977. Transient accumulation of Okasaki fragments as a result of uracil incorporation into nascent DNA. *Proc. Natl. Acad. Sci.* **74:** 154.

Upholt, W. B. 1977. Superhelix densities of circular DNAs. A generalized equation for their determination by the buoyant method. *Science* **195:** 891.

Upholt, W. B., H. B. Gray, and J. Vinograd. 1971. Sedimentation velocity behavior of closed circular SV40 DNA as a function of superhelix density, ionic strength, counterion and temperature. *J. Mol. Biol.* **61:** 21.

Wang, J. C. 1969. Degree of superhelicity of covalently closed cyclic DNAs from *E. coli*. *J. Mol. Biol.* **43:** 263.

———. 1974. The degree of unwinding of the DNA helix by ethidium. I. Titration of twisted PM2 DNA molecules in alkaline CsCl density gradients. *J. Mol. Biol.* **89:** 783.

Waring, M. 1970. Variation of the supercoils in closed circular DNA by binding of antibiotics and drugs: Evidence for molecular models involving intercalation. *J. Mol. Biol.* **54:** 247.

Weisbeek, P. J. and G. A. van Arkel. 1976. On the origin of bacteriophage φX174 replicative form DNA replication. *Virology* **72:** 72.

Weisbeek, P. J., W. E. Borrias, S. A. Langeveld, P. D. Baas, and G. A. van Arkel. 1977. Bacteriophage φX174: Gene *A* overlaps gene *B*. *Proc. Natl. Acad. Sci.* **74:** 2504.

Weisbeek, P. J., J. M. Vereijken, P. D. Baas, H. S. Jansz, and G. A. van Arkel. 1976. The genetic map of bacteriophage φX174 constructed with restriction enzyme fragments. *Virology* **72:** 61.

Woodworth-Gutai, M. and J. Lebowitz. 1976. Introduction of interrupted secondary structure in supercoiled DNA as a function of superhelix density. Consideration of hairpin structures in superhelical DNA. *J. Virol.* **18:** 195.

Yokohama, Y., T. Komano, and K. Onodera. 1971. The presence of short DNA chains in φX RF II in the late phase of infection. *Agric. Biol. Chem.* **34:** 1353.

Nascent Pieces of φX174 Viral-strand DNA Are Shorter Than Unit-length and Are Derived from All Regions of the Genome

Christian Hours, Margaret Matthes, Neil Miyamoto, and David T. Denhardt
Department of Biochemistry
McGill University
Montreal, Quebec, Canada H3G 1Y6

There is good evidence that the synthesis of the two strands of φX174 replicative-form (RF) DNA is asymmetric—i.e., that in each "round" of replication the viral strand is synthesized continuously and the complementary strand is synthesized discontinuously (Denhardt 1977). It is also commonly believed that once viral-strand synthesis is initiated, the DNA polymerase then simply moves around the complementary-strand ring indefinitely (Dressler 1970; Watson 1976). This was first promulgated as the rolling-circle model (Gilbert and Dressler 1969). According to this concept, the viral strand is elongated "endlessly" from the 3′-OH primer created by the insertion of a nick into the viral strand by the φX gene-*A* protein; the circular complementary strand serves repeatedly as the template, in the fashion of a "stamping machine." This simple mechanism produces, as an intermediate, a viral strand longer than unit length, and evidence for the existence of such strands under certain conditions has been reported (Dressler and Wolfson 1970; Eisenberg et al. 1977; Baas and Jansz, this volume).

However, little by little, data have been accumulating that bring this model into question. First, it was shown that a gap exists at a specific location in the viral strand of newly synthesized RF molecules produced during RF DNA replication (Eisenberg and Denhardt 1974; Eisenberg et al. 1975) and during progeny viral-strand DNA synthesis (Johnson and Sinsheimer 1974). Second, the *dnaG* protein of *Escherichia coli*, which is thought to produce primers for initiating DNA synthesis, is required for viral-strand synthesis as well as for complementary-strand synthesis (McFadden and Denhardt 1974). Third, it has been reported that on-going synthesis of the viral strand depends on the continued presence of active φX gene-*A* protein during progeny viral-strand synthesis (Fujisawa and Hayashi 1976; Tessman and Peterson 1977). This is inconsistent with the rolling-circle model as proposed originally and led us to state that the "rolling-circle model no longer rolls very far" because termination and a new initiation event are required at each round of replication rather than there being an endless "peeling off" of the viral strand (Denhardt and Hours 1978). Fourth, evidence for the existence of pieces of viral-strand DNA shorter than unit length, which suggests that the viral strand may be synthesized in a discontinuous fashion (Yokoyama et al. 1971; McFadden and Denhardt 1975), has been augmented recently (Machida et al. 1977). These findings have reopened the question as to how synthesis of viral-strand DNA is initiated. The data presented in this article were obtained using a new and unconventional technique for isolating nascent DNA pieces (Miyamoto and Denhardt 1977) and suggest that discontinuous synthesis of φX viral-strand DNA takes place over the entire genome.

EXPERIMENTAL PROCEDURES

The technique used was as follows: *E. coli* C in exponential growth was infected with φX *am*3 (lysis-defective) phage at a multiplicity of infection (m.o.i.) of 5. At the desired time the culture was pulse-labeled for approximately 3 seconds with [^{3}H]thymidine and poured into a one-half volume of a boiling aqueous solution containing 2% sodium dodecyl sulfate (SDS), 3% phenol, and 10 mM EDTA. Just as the mixture began to boil again it was removed from the heat and placed on ice. We believe this procedure stops instantly any enzymatic reactions that might affect replicating intermediates and that might operate to a significant extent when other stopping and lysing methods are used. This treatment disrupts the cell membrane, but not the cell wall, and allows small molecules ($< 10^7$ daltons) to be released into solution while high-molecular-weight molecules ($> 10^7$), particularly the bulk of the *E. coli* DNA, remain trapped inside the cell "ghost." A low-speed centrifugation removes these ghosts in a pellet with precipitated SDS.

The nucleic acids (and some proteins) were then precipitated out of the supernatant by isopropanol. The precipitate was dissolved in 1/20 of the initial culture volume, incubated briefly with dodecyl sulfate and pronase, and extracted twice with an equal volume of borate-saturated phenol-chloroform (1:1). The nucleic acids were again precipitated from the aqueous phase with isopropanol. The next step was to heat briefly (1 min at 100°C) the resuspended nucleic acids and separate the RNA from the DNA by chromatography on a nitrocellulose column as described by Miyamoto and Denhardt (1977). The sedimentation pattern of the pulse-labeled [^{3}H]DNA on an alkaline sucrose gradient is shown in Figure 1a. The molecular distribution is rather broad, ranging from very short molecules to roughly unit-length ones, with an apex around 8–9S. For further analysis, regions representing unit-length molecules and those of less than unit length were pooled separately as indicated.

We performed several controls to verify that the small DNA pieces were

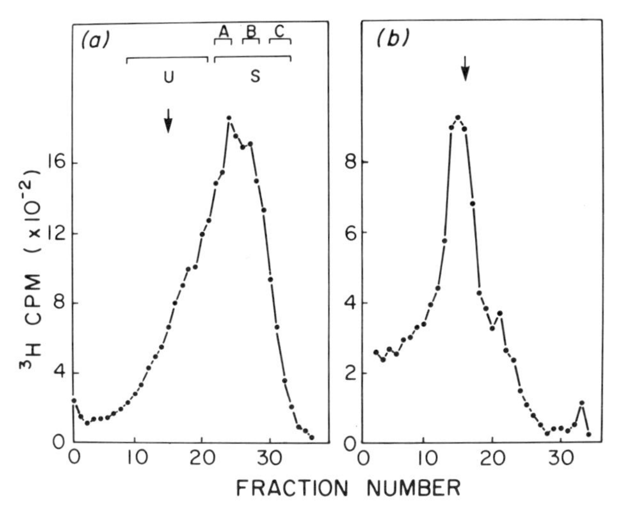

Figure 1 Velocity sedimentation in alkaline sucrose gradients of φX DNA pulse-labeled at 34°C during RF DNA replication and SS DNA synthesis. *E. coli* C cells (200 ml) were grown to 2–3 × 10^8 cells/ml in TKCaB (10 g Bacto-tryptone, 5 g KCl, 1 liter H_2O, 1 mM $CaCl_2$). At the appropriate time after infection with φX *am*3 at an m.o.i. of 5, cultures were labeled with [^{3}H]thymidine (5 μCi/ml) for 3 sec, and DNA synthesis was terminated in a boiling SDS-phenol solution (Miyamoto and Denhardt 1977). The viral-strand DNA was extracted and centrifuged in alkaline sucrose gradients (0.2 M NaOH, 0.8 M NaCl, 5–20% sucrose) for 16 hr at 34,000 rpm and 10°C in the SW40 rotor of a Beckman ultracentrifuge. Sedimentation is from right to left; arrows show the position of ^{32}P-labeled φX DNA. (*a*) Cells labeled at 10 min after infection; similar profiles were obtained with cells labeled at 6 min, 10 min in the presence of chloramphenicol (30 μg/ml), 20 min, or 35 min after infection. Brackets indicate the fractions pooled for further analysis. (*b*) Cells labeled at 10 min after infection and chased for 10 min with 100 μg/ml thymidine before stopping.

not produced by the isolation technique. First, just after addition of the culture to the boiling SDS-phenol solution, we introduced into the sample some ^{32}P-labeled φX single-stranded (SS) DNA as marker. Then the same extraction procedure was followed. Aliquots withdrawn after the first isopropanol precipitation, after the SDS-pronase digestion, and after the nitrocellulose column chromatography were analyzed on alkaline sucrose gradients: 89%, 84%, and 75%, respectively, of the molecules remained unit-length (circular or linear). As a second control we infected *E. coli* C with ^{32}P-labeled phage particles, stopped the culture 20 minutes after infection, and extracted the DNA as described above. In this case, after chromatography on the nitrocellulose column, the alkaline sucrose gradient profile revealed that 70% of the molecules were unit-length. The third control, and the most significant one, was a pulse-chase experiment. A culture was pulse-labeled with [^{3}H]thymidine 10 minutes after infection and then chased for 10 minutes with an excess of unlabeled thymidine. The alkaline sucrose gradient pattern of the [^{3}H]DNA extracted with the boiling SDS-phenol method is shown in Figure 1b: only 20% of the DNA was less than unit-length after the chase, whereas 60–70% of the DNA was less than unit-length immediately after the pulse. This evidence suggested that the short DNA molecules were replicating intermediates. The technique does not introduce more than about one break in 20,000 phosphodiester bonds (25% breakdown of a 5500-nucleotide chain).

RESULTS

Strand Polarity of the Nascent Pieces

Pooled fractions from the alkaline sucrose gradients (Fig. 1) were analyzed by hybridization to distinguish the viral DNA from the bacterial DNA, to measure the amount of φX-specific DNA, and to determine its polarity. φX viral SS DNA or heat-denatured φX RF DNA was dried onto nitrocellulose membrane filters. Hybridization with sonicated, labeled DNA was performed at 43°C for 36 hours in the presence of 2 × SSC in 50% formamide. ^{32}P-labeled RF DNA was added to the pulse-labeled [^{3}H]DNA as an internal marker in order to determine the hybridization efficiency. The results are shown in Table 1: columns *a* and *b* represent the hybridization of the [^{3}H]DNA from the "unit-length" region or the "shorter-than-unit-length" region of the alkaline sucrose gradient with φX RF DNA. We can see that by 10 minutes after infection most of the ^{3}H-label is incorporated into φX DNA (+ and −) sequences; *E. coli* DNA accounts for less than 20–30% of the labeled DNA. Columns *c* and *d* give the percentage of plus-strand DNA as estimated from the hybridization of a similar amount of ^{3}H-labeled DNA with φX viral SS DNA. This percentage is calculated from the difference in the amount of [^{3}H]DNA capable of annealing to filters carrying RF DNA or only viral-strand DNA. It is clear that during RF DNA replication roughly 50% of the ^{3}H-label is incorporated into viral-strand DNA sequences, and later, during progeny viral SS DNA synthesis, this percentage rises to 92–99%. The last column shows the calculated percentage of viral-strand DNA shorter than unit length obtained from the different preparations (a sample calculation is shown in the legend to Table 1). Except for the pulse-chase experiment, where the percentage dropped to 22%, the calculated values agree with the result of Machida et al. (1977) (60–70% of the pulse-labeled viral-strand DNA is shorter than unit-length).

Genome Location of the Nascent Pieces

To determine whether or not the short viral-strand DNA pieces arose only from the region of the origin of replication, we hybridized the pulse-labeled [^{3}H]DNA to φX DNA restriction enzyme fragments. After digestion with

Table 1 Determination of viral-strand DNA of less than unit length

Pulse time (min after infection)	Percentage of [³H]DNA that can anneal to φX RF DNA		Percentage of φX-specific [³H]DNA that cannot anneal to viral-strand DNA		Percentage of viral-strand DNA < unit length
	unit length (*a*)	<unit length (*b*)	unit length (*c*)	<unit length (*d*)	
6	32	31	57	64	58
10	83	72	61	56	65
10 (+ CAM)	85	84	52	69	65
10 (+ 10 min chase)	80	80	67	60	22
20	95	84	99	94	56
35	95	89	94	95	67

[³H]DNA from "unit-length" (U) or "shorter-than-unit-length" (S) pools from alkaline sucrose gradients (Fig. 1) was sonicated and hybridized to φX RF DNA and to φX viral SS DNA according to the technique described by Lane and Denhardt (1974). Uniformly ^{32}P-labeled RF DNA was added as an internal standard. Conditions of labeling are indicated in the first column: "10 min + CAM" represents DNA from a culture to which chloramphenicol (30 μg/ml) was added 3 min after infection and the culture was then labeled with [³H]thymidine at 10 min after infection; "10 min + chase" is as mentioned in the legend to Fig. 1b. Columns *a* and *b* give the percentage of [³H]DNA that can hybridize with φX RF DNA (the nonhybridizing DNA is *E. coli* DNA); columns *c* and *d* give the percentage of labeled φX DNA that cannot hybridize with φX viral SS DNA. The latter figures were derived by subtracting the amount of [³H]DNA capable of annealing to viral SS DNA from the amount capable of annealing to RF DNA. The last column shows the percentage of φX viral-strand DNA present as shorter-than-unit-length molecules.

The following is a sample calculation for DNA labeled at 10 min after infection: With an input of 4300 cpm of [³H]DNA from the U pool, 3560 cpm annealed to the RF DNA filter and 1390 cpm to the SS DNA filter (these values are corrected for the nonspecific background and for the efficiency of hybridization by means of the ^{32}P-labeled RF DNA internal standard; the cpm actually measured were 2670 cpm and 650 cpm). Thus, 2170 cpm (3560 − 1390), or 61% of the total 3560 cpm, was in the viral-strand DNA. With an input of 10,000 cpm of [³H]DNA from the S pool, 7200 cpm annealed to the RF DNA filter and 3170 cpm to the SS DNA filter. A calculation of the sort just described showed that 4030 cpm, or 56% of the total 7200 cpm, was incorporated into viral-strand DNA. Since the amount of [³H]DNA used for each hybridization represented equivalent portions of the original culture, it can be calculated that 65% (4030/4030 + 2170) of the total φX viral-strand DNA was present as shorter-than-unit-length molecules.

either *Hae*III or *Hha*I, the φX RF DNA restriction enzyme fragments were separated by electrophoresis on an agarose slab gel, denatured and neutralized in the gel, and "blotted" onto nitrocellulose filter paper (Southern 1975) for the subsequent hybridization. In this experiment the culture was pulse-labeled 35 minutes after infection so as to label almost exclusively viral-strand DNA. The shorter-than-unit-length region from the alkaline sucrose gradient was fractionated into three pools, labeled A, B, and C in Figure 1a, and the DNA from these pools was hybridized to the restriction fragments of φX RF DNA on the filters. The results for the DNA from pool C are shown in Figure 2 a and c. The unit-length DNA from the same

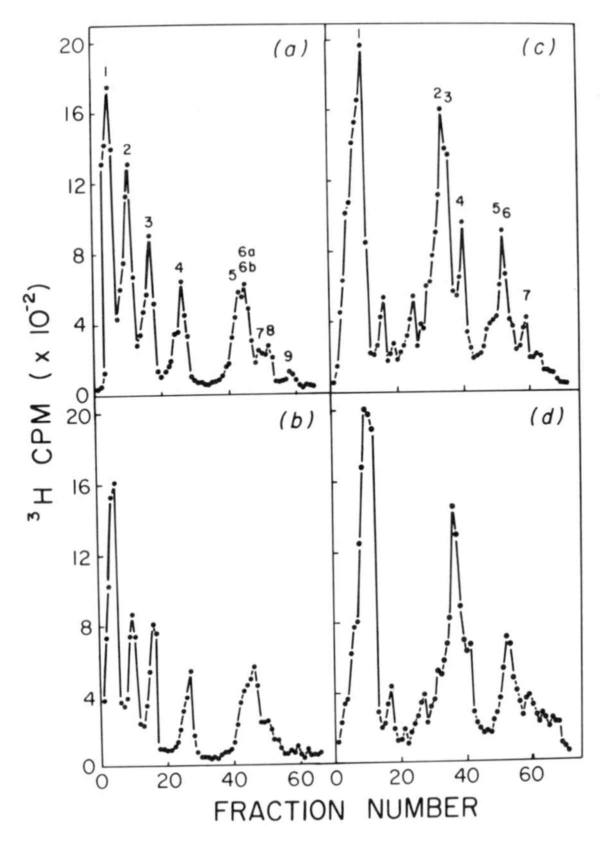

Figure 2 Distribution of the short DNA pieces on the genome. DNA from an infected culture that was pulse-labeled with [^{3}H]thymidine 35 min after infection was extracted and fractionated by alkaline velocity sedimentation into the "unit-length" and A, B, and C pools as described in the legend to Fig. 1. The [^{3}H]DNA was then hybridized to φX RF DNA restriction fragments that had been separated by electrophoresis on a 2% agarose slab gel and transferred to nitrocellulose filter strips as described by Southern (1975). Hybridization took place on the nitrocellulose strips under the conditions described by Denhardt (1966). When the annealing was completed, the filter strips were washed, cut into 2-mm slices, dried, and the radioactivity determined in a toluene-Omnifluor scintillant. The numbers in panels *a* and *c* indicate the positions of the individual restriction enzyme fragments. (*a, c*) [^{3}H]DNA from pool C hybridized to φX RF DNA restriction fragments generated by *Hae*III and *Hha*I, respectively. (*b, d*) [^{3}H]DNA from the "unit-length" pool hybridized to φX RF DNA restriction fragments generated by *Hae*III and *Hha*I, respectively.

preparation was hybridized to an identical set of restriction fragments; this DNA yielded the profiles shown in Figure 2 b and d. The frames on the left (Fig. 2a,b) and the right (Fig. 2c,d) show the annealing to *Hae*III and *Hha*I fragments, respectively. The similarity of the patterns in Figure 2 a and c to those in Figure 2 b and d, respectively, is obvious. These patterns resemble those obtained when uniformly labeled RF DNA is hybridized to the filters. We infer from this that the shortest pulse-labeled viral DNA pieces originate from many regions around the genome. Pools A and B yielded similar patterns (data not shown). The locations of the *Hae*III and *Hha*I fragments on the φX genome are given by Weisbeek and van Arkel (this volume); the *Hae*III-6b and *Hha*I-1 fragments contain the region of the origin (terminus).

DISCUSSION

We have verified the existence during the period of viral-strand DNA synthesis of viral-strand DNA molecules that are of less than unit length. These DNA pieces are replicating intermediates by virtue of the fact that they can be chased into mature DNA. Most significantly, they are randomly distributed around the φX genome. This latter point militates against two mechanisms that could give rise to short viral strands. These two mechanisms are based on the premise that viral-strand DNA synthesis is continuous during a round of replication.

The first mechanism is a modification of the rolling-circle model in which viral-strand DNA pieces arise from the "premature" action of the φX gene-*A* protein. If this protein should nick the nascent viral strand at the origin of replication, where the newly synthesized strand is covalently linked to the putative parental-strand primer before the completion of one round of replication, then a pulse-labeled short viral DNA piece would be generated. However, such a mechanism appears to be in conflict with the fact that the gene-*A* protein requires a superhelical structure (Marians et al. 1977) before it will act, and the rolling-circle intermediate is not superhelical.

The second model, which also assumes continuous viral-strand DNA synthesis but predicts that some viral-strand DNA will be shorter-than-unit-length and specific to the origin of replication, is the "reciprocating-strand model" (Denhardt 1975, 1977). It was speculated that a strand-displacement process catalyzed by the gene-*A* protein would lead to a certain number of newly synthesized viral strands of less than the unit length. Since both this model and the modified rolling-circle model require that the short viral-strand sequences originate predominantly from the origin of replication, they do not appear to be compatible with our results. As shown in Figure 2, the nascent viral-strand DNA pieces did not hybridize preferentially to the restriction enzyme fragments encompassing the origin.

There are at least three ways, none of which excludes the others, to explain the presence of viral strands shorter than unit length arising from all regions of the genome: (1) The hypothesis we favor is that the viral strand is synthesized by a discontinuous mechanism; in other words, there are multiple sites of initiation of viral-strand synthesis, as is the case for complementary-strand synthesis. Such sites could be specific or random. The presence of ribonucleotides at the 5′ end of nascent viral-strand DNA pieces, as reported recently by Machida et al. (1977), supports the idea of discontinuous synthesis. (2) We know that a replicating DNA molecule is the focus of intense enzymatic activity; the list of enzymes involved in DNA replication is extensive. Particularly relevant are the "nicking-closing" activities of ω protein (Wang 1971) and gyrase (Gellert et al. 1976). Such activities could account for the production of short DNA pieces even though the DNA strand might be synthesized in a continuous fashion. One might expect, however, that the parental strands would be more affected by these activities than the nascent growing chains, and we have shown that the parental viral strand remains relatively intact. (3) It is possible that the incorporation of uracil from dUTP accounts for the short DNA pieces (Tye et al. 1977). Unfortunately, no in vivo studies are yet available to determine the importance of this factor for φX.

We note in closing that Figure 1 reveals very little pulse-labeled viral-strand DNA that is longer than unit-length (we believe our extraction procedures would yield these molecules if they were there). This result raises the possibility that the strands of more than unit length observed when other methods are used to isolate nascent DNA (Gilbert and Dressler 1969; Dressler 1970) may be produced during isolation and may not be a true reflection of the in vivo replication process. Further work is necessary to resolve this situation. We are struck by the apparent facts that pieces of viral-strand DNA are joined together more rapidly to produce a unit-length strand than are pieces of complementary-strand DNA, and that the joining occurs by a process that does not seem to depend on the functioning of polynucleotide ligase (McFadden and Denhardt 1975).

ACKNOWLEDGMENTS

This research was supported by the Medical Research Council, the National Cancer Institute of Canada, and Le Ministère de l'Education du Québec.

REFERENCES

Denhardt, D. T. 1966. A membrane-filter technique for the detection of complementary DNA. *Biochem. Biophys. Res. Commun.* **23:** 641.

———. 1975. The single-stranded DNA phages. *Crit. Rev. Microbiol.* **4:** 161.

———. 1977. The isometric single-stranded DNA phages. In *Comprehensive virology* (ed. H. Fraenkel-Conrat and R. R. Wagner), vol. 7, p. 1. Plenum Press, New York.

Denhardt, D. T. and C. Hours. 1978. The present status of φX174 DNA replication *in vivo*. In *DNA synthesis: Present and future* (ed. M. Kohiyama and I. Molineux). Plenum Press, New York. (In press.)

Dressler, D. H. 1970. The rolling circle for φX DNA replication. II. Synthesis of single-stranded circles. *Proc. Natl. Acad. Sci.* **67:** 1934.

Dressler, D. H. and J. Wolfson. 1970. The rolling circle for φX DNA replication. III. Synthesis of supercoiled duplex rings. *Proc. Natl. Acad. Sci.* **67:** 456.

Eisenberg, S. and D. T. Denhardt. 1974. The mechanism of replication of φX174 single-stranded DNA. X. Distribution of the gaps in nascent RF DNA. *Biochem. Biophys. Res. Commun.* **61:** 532.

Eisenberg, S., J. Griffith, and A. Kornberg. 1977. φX174 cistron A protein is a multifunctional enzyme in DNA replication. *Proc. Natl. Acad. Sci.* **74:** 3198.

Eisenberg, S., B. Harbers, C. Hours, and D. T. Denhardt. 1975. The mechanism of replication of φX174 DNA. XII. Non-random location of gaps in nascent φX174 RF II DNA. *J. Mol. Biol.* **99:** 107.

Fujisawa, H. and M. Hayashi. 1976. Gene A product of φX174 is required for site-specific endonucleolytic cleavage during single-stranded DNA synthesis *in vivo*. *J. Virol.* **19:** 416.

Gellert, M., K. Mizuuchi, M. H. O'Dea, and H. A. Nash. 1976. DNA gyrase: An enzyme that introduces superhelical turns into DNA. *Proc. Natl. Acad. Sci.* **73:** 3872.

Gilbert, W. and D. H. Dressler. 1969. DNA replication: The rolling circle model. *Cold Spring Harbor Symp. Quant Biol.* **33:** 473.

Johnson, P. H. and R. L. Sinsheimer. 1974. Structure of an intermediate in the replication of bacteriophage φX174 DNA: The initiation site for DNA replication. *J. Mol. Biol.* **83:** 47.

Lane, H. E. D. and D. T. Denhardt. 1974. The *rep* mutation. III. Altered structure of the replicating *Escherichia coli* chromosome. *J. Bacteriol.* **120:** 805.

Machida, Y., T. Okazaki, and R. Okazaki. 1977. Discontinuous replication of replicative form DNA from bacteriophage φX174. *Proc. Natl. Acad. Sci.* **74:** 2776.

Marians, K. J., J. E. Ikeda, S. Schlagman, and J. Hurwitz. 1977. Role of DNA gyrase in φX174 replicative form replication *in vitro*. *Proc. Natl. Acad. Sci.* **74:** 1965.

McFadden, G. and D. T. Denhardt. 1974. Mechanism of replication of φX174 single-stranded DNA. IX. Requirement for the *Escherichia coli dna*G protein. *J. Virol.* **14:** 1070.

———. 1975. The mechanism of replication of φX174. XIII. Discontinuous synthesis of the complementary strand in an *Escherichia coli* host with a temperature-sensitive polynucleotide ligase. *J. Mol. Biol.* **99:** 125.

Miyamoto, C. and D. T. Denhardt. 1977. Evidence for the presence of ribonucleotides at the 5′ termini of some DNA molecules isolated from *Escherichia coli pol*A *ex*2. *J. Mol. Biol.* **116:** 681.

Southern, E. M. 1975. Detection of specific sequence among DNA fragments separated by gel electrophoresis. *J. Mol. Biol.* **98:** 503.

Tessman, E. S. and P. K. Peterson. 1977. Gene *A* protein of bacteriophage S13 is required for single-stranded DNA synthesis. *J. Virol.* **21:** 806.

Tye, B., P. Nyman, I. R. Lehman, S. Hochhauser, and B. Weiss. 1977. Transient accumulation of Okazaki fragments as a result of uracil incorporation into nascent DNA. *Proc. Natl. Acad. Sci.* **74:** 154.

Wang, J. C. 1971. Interaction between DNA and an *Escherichia coli* protein ω. *J. Mol. Biol.* **55:** 523.

Watson, J. C. 1976. *Molecular biology of the gene*, 3rd Ed., p. 240. W. A. Benjamin, Menlo Park, California.

Yokoyama, Y., T. Komano, and K. Onodera. 1971. The presence of short DNA chain in φX174 RF II in the late phase of infection. *Agric. Biol. Chem.* **35:** 1353.

Conversion of Phage Single-stranded DNA to Duplex DNA In Vitro

Sue Wickner
Laboratory of Molecular Biology
National Cancer Institute
National Institutes of Health
Bethesda, Maryland 20014

In recent years there has been much progress toward understanding the molecular mechanisms of DNA replication of small *Escherichia coli* phages. Three enzymatic pathways have been identified by which phage circular single-stranded (SS) DNA (fd [M13], St-1 [G4], and φX174) is converted to duplex DNA. Each of these pathways has been fractionated into many protein components, all of which are derived from the host. Some of the proteins required for the replication of phage SS DNA are also required for the replication of *E. coli* DNA. Some of the genes whose functions are required for *E. coli* DNA replication have been identified by the isolation of mutants conditionally lethal for DNA synthesis. Functions required for the initiation of *E. coli* DNA replication include *dnaA*, *dnaB*, *dnaC(D)*, *dnaI*, and *dnaP* gene products and RNA polymerase (Wechsler 1978). Those required for continuing a round of DNA replication include *dnaB*, *dnaC(D)*, *dnaG*, *dnaZ*, *polC*, *nalA*, and *cou* gene products (Wechsler 1978). By means of in vitro complementation assays it has been possible to isolate the proteins required for phage SS DNA replication that are also required for *E. coli* DNA replication. Extracts prepared from some *E. coli* mutants temperature-sensitive for DNA replication were found to be active for phage DNA synthesis at low temperature but not at high temperature; these included extracts of *dnaB*, *dnaC(D)*, *polC*, *dnaG*, and *dnaZ* mutants (Wickner et al. 1972; Schekman et al. 1972; Wickner and Hurwitz 1976). The corresponding wild-type proteins have been purified from wild-type cells by isolating the components that are able to restore DNA-synthesis activity in the mutant extracts at high temperature (Wright et al. 1973; Wickner et al. 1973a,b; Wickner and Hurwitz 1976). Presumably, the most heat-labile protein in the extract was the temperature-sensitive *dna* gene product, and the protein that restored activity was the product of the wild-type allele. Proof that the correct protein had been purified was obtained in each case by purifying the corresponding protein from the temperature-sensitive mutant and demonstrating that it was more thermolabile for activity in vitro than the wild-type protein. Other *E. coli* proteins required for the replication of phage SS DNA have been isolated on the basis of their essential role, in conjunction with purified *E. coli dna* gene products, in reconstituted DNA replication reactions.

Having purified proteins and defined DNA templates, it has been possible to reconstruct the three known phage circular SS DNA replication pathways in their entirety. Each of the pathways is distinct in its nucleotide and protein requirements and its DNA specificity (Table 1). These pathways have been divided into partial reactions: (a) initiation of DNA synthesis and (b) DNA elongation. Initiation of DNA synthesis by each of the pathways involves the synthesis of an oligonucleotide primer for DNA polymerases to elongate. The pathways differ in their mechanisms for synthesizing oligonucleotides. With fd and M13 DNA, RNA polymerase in conjunction with DNA-binding protein (Sigal et al. 1972) and several other *E. coli* proteins catalyzes the synthesis of a unique ribooligonucleotide. With St-1 and G4 DNA, *dnaG* protein catalyzes the synthesis of a unique deoxyribo-, ribo-, or mixed deoxyribo- and ribooligonucleotide only when the DNA is covered with DNA-binding protein. With φX DNA, *dnaG* protein also catalyzes the synthesis of an oligonucleotide, but only after a prepriming reaction involving *dnaB* and *dnaC(D)* proteins, DNA-binding protein, two or three other *E. coli* proteins, and ATP. These three pathways for the initiation of DNA synthesis are discussed below.

Table 1 Requirements for phage SS DNA synthesis with purified *E. coli* proteins

	fd, M13 DNA	G4, St-1 DNA	φX DNA
Specific DNA	+	+	+
ADP	−	+	−
ATP	+	−	+
UTP, CTP, GTP	+	−	−
dATP, dCTP, dGTP, dTTP	+	+	+
RNA polymerase	+	−	−
DNA-binding protein	+	+	+
RNase H, discriminatory proteins	+	−	−
dnaB protein	−	−	+
dnaC(*D*) protein	−	−	+
dnaG protein	−	+	+
Factors X, Y, Z; factors i and n	−	−	+
DNA polymerase III, *dnaZ* protein, DNA EFI, DNA EFIII; DNA polymerase III holoenzyme	+	+	+

Initiation of fd (M13) DNA Synthesis

Synthesis of fd (and M13) DNA has been accomplished in vitro with RNA polymerase, DNA-binding protein, several other *E. coli* proteins (RNase H and discriminatory proteins), DNA-elongation components, four ribonucleoside triphosphates (rNTPs), and four deoxynucleoside triphosphates (dNTPs) (Wickner and Kornberg 1974a; Schaller et al. 1976; Vicuna et al. 1977a,b; Gray et al. 1977). RNA polymerase functions in this DNA-synthesis reaction by catalyzing the synthesis of a ribooligonucleotide that provides a 3′-OH end for DNA elongation by DNA polymerase I alone or by DNA polymerase II or III in combination with *dnaZ* protein and elongation factors I and III (EFI and EFIII) (Fig. 1). The site of initiation of ribooligonucleotide synthesis is unique and is in a region of duplex structure (Schaller et al. 1976; Gray et al. 1977; K. Geider et al., pers. comm.). Presumably, when fd DNA is covered with DNA-binding protein and the other proteins required for primer synthesis, RNA polymerase is able to catalyze the synthesis of a specific ribooligonucleotide; however, it cannot synthesize a primer with G4, St-1, or φX DNA under similar conditions.

Initiation of G4 and St-1 DNA Synthesis

Synthesis of G4 and St-1 DNA has been accomplished in vitro with purified components (Schekman et al. 1974; Zechel et al. 1975; Bouché et al. 1975; Wickner 1977). The reaction requires *dnaG* protein, DNA-binding protein, ADP, four dNTPs, and DNA-elongation components (Table 2; Wickner 1977). The results summarized below suggest the following mechanism of initiation of G4 and St-1 DNA synthesis (Fig. 2). DNA-binding protein covers the DNA and facilitates the binding of *dnaG* protein to the DNA. An oligonucleotide is synthesized by the protein-DNA complex. The oligonucleotide is then elongated with DNA by one of the several DNA-elongation mechanisms.

The *dnaG* protein binds specifically to G4 and St-1 DNA. When *dnaG* protein is incubated with either of these DNAs and DNA-binding protein

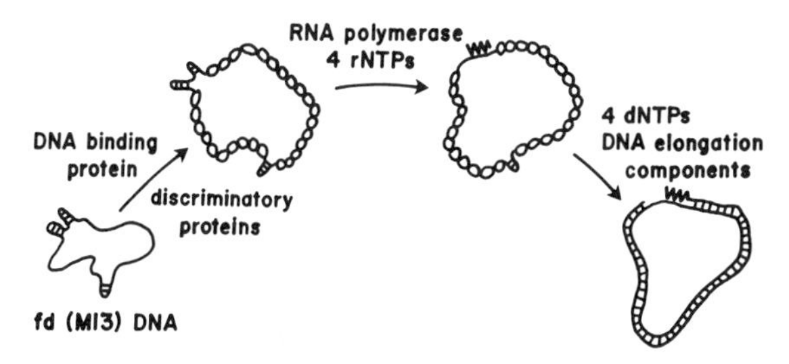

Figure 1 Mechanism of priming of fd (M13) DNA synthesis.

Table 2 Requirements for G4 and St-1 DNA synthesis

Additions	dTMP incorporated (pmoles/20 min)			
	G4 DNA	St-1 DNA	fd DNA	φX DNA
A. DNA polymerase III				
Complete	25.2	25.0	<0.1	<0.1
– *dnaG* protein	<0.1	<0.1		
– DNA-binding protein	0.3	0.3		
– DNA EFI	0.5	0.6		
– DNA EFIII	0.6	0.2		
– *dnaZ* protein	0.5	0.2		
– DNA polymerase III	0.4	0.4		
– ADP	1.9	5.0		
– ADP + other nucleotides[a]	1.7–3.1	–		
B. DNA polymerase II				
Complete	29.5	17.3		
– *dnaG* protein	0.3	0.2		
– DNA-binding protein	<0.1	0.4		
– DNA EFI	3.2	0.4		
– DNA EFIII	2.8	3.4		
– *dnaZ* protein	2.5	1.5		
– DNA polymerase II	0.4	0.4		
– ADP	1.9	7.5		
C. DNA polymerase I				
Complete	26.4	34.4	<0.1	0.5
– *dnaG* protein	<0.1	<0.1		
– DNA-binding protein	2.2	3.2		
– DNA polymerase I	<0.1	<0.1		
– ADP	0.7	9.0		

Reaction mixtures (30 μl) contained reaction buffer (1 mM dithiothreitol, 20 μg/ml rifampicin, 50 mM Tris-HCl, pH 7.5, 1 mg/ml bovine serum albumin, 7 mM $MgCl_2$), 50 μM each of dATP, dGTP, dCTP, and [^{3}H]dTTP (800 cpm/pmole), 12 μM ADP, 300 pmoles DNA, 0.8 μg DNA-binding protein, and 0.04 unit (U) of *dnaG* protein. Parts A and B contained 0.2 U *dnaZ* protein, 0.2 U DNA EFI, and 0.2 U DNA EFIII. Parts A, B, and C contained 0.3 U DNA polymerase III, 0.3 U DNA polymerase II, and 0.2 μg DNA polymerase I, respectively. After 15 min at 30°C, acid-insoluble radioactivity was measured (Wickner 1977).

[a]Nucleotides substituted for ADP at 20 μM were all four rNTPs, three rNDPs, four dNTPs, four dNDPs, AMP, and α,β-methylene ADP.

and then filtered through a Sepharose 6B gel, *dnaG* protein is found associated with the DNA (Fig. 3A,C). The binding of *dnaG* protein is dependent on the DNA being covered with DNA-binding protein (Fig. 3B) but is independent of the presence of nucleotides. With excess *dnaG* protein relative to G4 DNA, about one molecule of *dnaG* protein binds per DNA circle. The *dnaG* protein can be dissociated from the DNA by 1 M NaCl and recovered in an active form in the partially included volume after refiltration through Sepharose. In contrast to G4 and St-1 DNA, fd and φX DNA do not

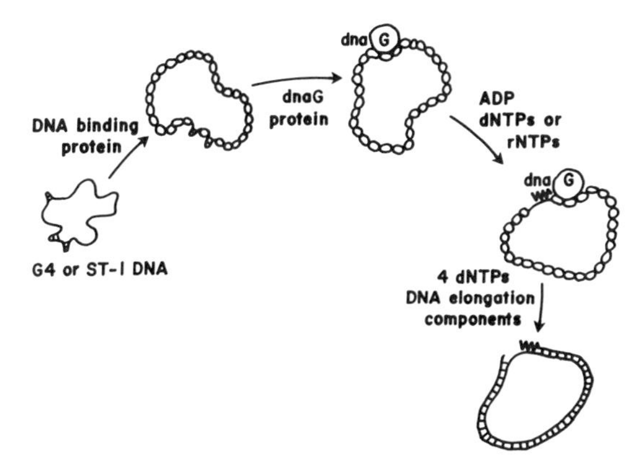

Figure 2 Mechanism of priming of G4 and St-1 DNA synthesis.

bind *dnaG* protein (Fig. 3D,E). Presumably, fd and φX DNA do not have binding sites that can be recognized by *dnaG* protein.

Synthesis of DNA by the isolated complex of G4 DNA, *dnaG* protein, and DNA-binding protein requires ADP, four dNTPs, and DNA-elongation components. This reaction has been separated into an ADP-dependent priming reaction (first reaction) and an ADP-independent DNA-elongation reaction (second reaction). ADP, dTTP, dGTP, *dnaG* protein, and DNA-binding protein are required in the first reaction for the incorporation of label from [α-^{32}P]dGTP or dTTP into a protein-DNA complex that can be isolated by filtration through Sepharose and in a second reaction for ADP-independent DNA synthesis by the complex (Fig. 4). After incubation of the isolated DNA-protein complex with four dNTPs and DNA-elongation components, the label from [α-^{32}P]dGTP or dTTP sediments in alkaline sucrose gradients at the position of nearly full-length linear G4 DNA (Fig. 5). Before DNA elongation the label from [α-^{32}P]dGTP or dTTP is contained in a unique oligonucleotide of about ten bases, as measured by acrylamide gel electrophoresis. Although dATP and dCTP are not required for primer synthesis, label from them is found in the oligonucleotide when they are included in the reaction mixture. UTP satisfies the dTTP requirement; dideoxyTTP and dTDP do not satisfy this requirement. GTP substitutes for dGTP; dGDP does not. The K_m for both the rNTPs and the dNTPs is about 10 μM, and there is no preferential utilization of rNTPs over dNTPs when both are present in the same reaction. ATP satisfies the ADP requirement; however, the K_m for ADP is 2 μM and that for ATP is 50 μM. The 5′ end of the oligonucleotide is a riboadenosine residue. The sequence of the oligonucleotide synthesized by *dnaG* protein in the presence of four rNTPs has been determined (McMacken et al., this volume) and is complementary to the sequence of the DNA at the origin of G4 DNA synthesis (Godson et al.; Dressler et al.; both this volume). The 5′ end of the oligonucleotide is

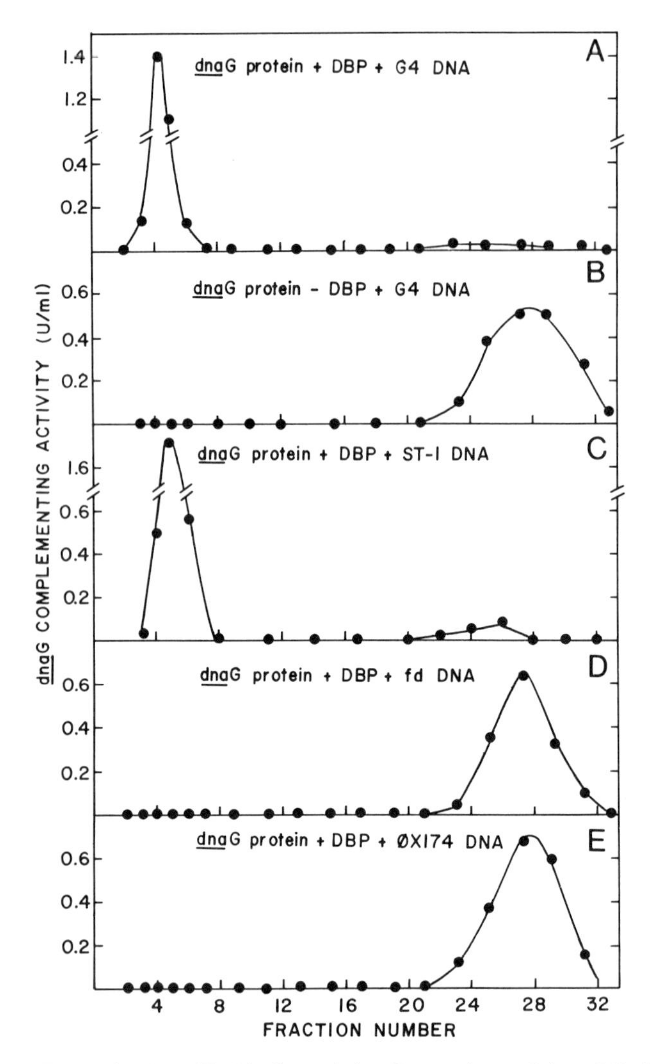

Figure 3 Specific binding of *dnaG* protein to G4 and St-1 DNA. (*A*) The first reaction (60 μl) contained reaction buffer (Table 2), 6 nmoles G4 DNA (as nucleotide), 8 μg DNA-binding protein, and 1.7 U *dnaG* protein. After 10 min at 30°C the mixture was applied at room temperature to a 22 × 0.5-cm. column of Sepharose 6B equilibrated with 50 mM Tris-HCl, pH 7.5, 1 mM dithiothreitol, 0.1 mg/ml bovine serum albumin, 5% glycerol, and 5 mM $MgCl_2$. The column was eluted with the same buffer, and 0.11-ml fractions were collected. The excluded volume (where DNA eluted) was fraction 4; the included volume was fraction 36. Fractions were assayed for *dnaG* activity. (*B*) The reaction mixture was as in *A* except that DNA-binding protein was omitted. (*C, D, E*) The reaction mixtures were as in *A* except that G4 DNA was omitted and 6 nmoles St-1, fd, and φX DNA were included in experiments *C*, *D*, and *E*, respectively (Wickner 1977).

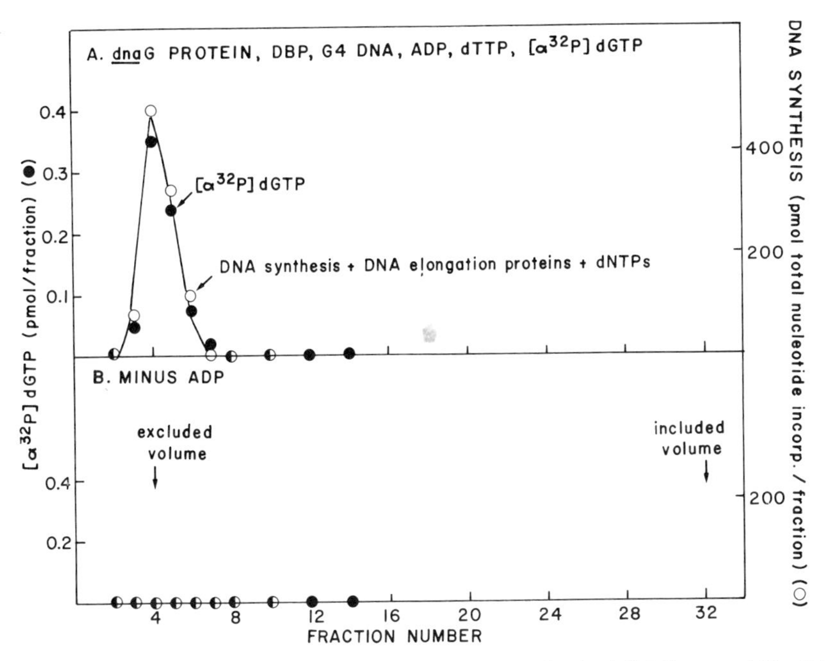

Figure 4 G4 DNA-dependent oligonucleotide synthesis. (*A*) Mixtures (60 μl) for the first reaction contained reaction buffer (Table 2), 6 nmoles G4 DNA, 2.1 U *dnaG* protein, 8.0 μg DNA-binding protein, and 16 μM ADP, dTTP, and [α-^{32}P]dGTP (50,000 cpm/pmole). After 15 min at 30°C mixtures were subjected to gel filtration as described in the legend to Fig. 3. The excluded volume fractions were measured for ^{32}P (●). They were also assayed for DNA synthesis in a second reaction (60 μl) containing 10–50 μl isolated protein-DNA complex, reaction buffer (Table 2), 50 μM each of dATP, dCTP, dGTP, and [^{3}H]dTTP (800 cpm/pmole), 0.4 U *dnaZ* protein, 0.4 U DNA EFI, 0.4 U DNA EFIII, and 0.6 U DNA polymerase III. After 10 min at 30°C acid-insoluble [^{3}H]dTMP was measured (o). The dTMP incorporation measured was the yield. (*B*) The experiment was the same as in *A* except that ADP was omitted from the first reaction (Wickner 1977).

unique, but the 3′-OH end varies depending on the nucleoside triphosphates present in the reaction mixture. It is not known whether rNTPs or dNTPs are the preferred substrates for *dnaG* protein in vivo. It may be possible to isolate G4 RFII (replicative-form II: circular, duplex DNA in which at least one strand is not continuous) synthesized in vivo and to determine whether the 5′ end of the complementary strand contains rNMPs or dNMPs.

With St-1 DNA, *dnaG* protein also catalyzes the synthesis of a unique oligonucleotide that can serve as a primer for DNA elongation. Incorporation of [α-^{32}P]dGTP or GTP into an oligonucleotide of about ten bases requires only ADP, ATP, dADP, or dATP. The oligonucleotide can be

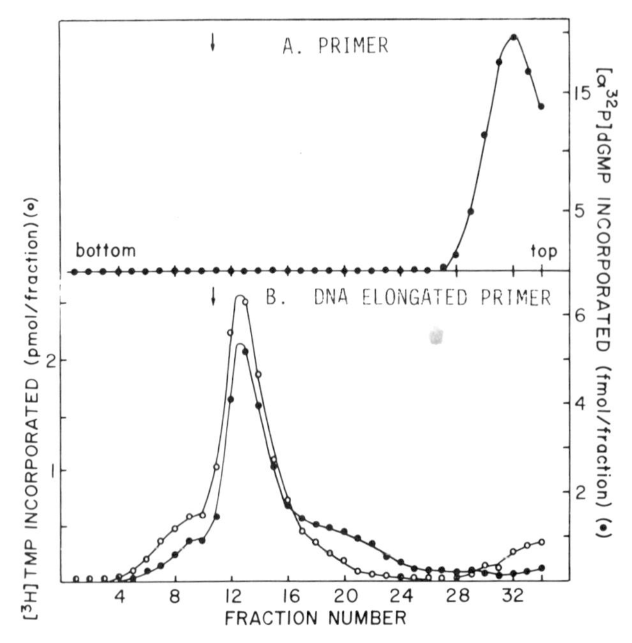

Figure 5 Alkaline sucrose sedimentation of the G4 primer and DNA-elongated primer. (*A*) Products of the priming reaction were prepared and isolated by gel filtration as described in the legend to Fig. 4. One portion of the isolated primer-DNA complex was adjusted to 20 mM EDTA, 0.8 M NaCl, and 0.2 M NaOH in 100 μl. The sample was sedimented in a 10–25% alkaline sucrose gradient containing 2 mM EDTA, 0.8 M NaCl, and 0.2 M NaOH for 6 hr in an SW 50.1 Spinco rotor at 50,000 rpm at 10°C. Fractions 1 to 34 were collected, and the radioactivity from [α-^{32}P]dGTP was measured directly (•). (*B*) Another portion of the isolated DNA complex was incubated in a second reaction (60 μl) (in the absence of ADP) with dATP, dCTP, dGTP, and [^{3}H]dTTP as described in the legend to Fig. 4 but with 0.6 μg T4 DNA polymerase. After 10 min at 30°C the reaction was adjusted to 20 mM EDTA, 0.8 M NaCl, and 0.2 M NaOH in 100 μl and sedimented as in *A*. Fractions 1 to 34 were collected, and the acid-insoluble radioactivity from [α-^{32}P]dGTP (•) and [^{3}H]dTTP (o) was measured. The arrow refers to the position of ^{14}C-labeled φX DNA. The primer was also found to be elongated by DNA polymerase I or by the combination of DNA polymerase III, *dnaZ* protein, DNA EFI, and DNA EFIII (Wickner 1977).

composed entirely of dNMP residues with a deoxyadenosine residue at the 5′ end. This is the first demonstration that DNA synthesis can be initiated by the synthesis of a deoxyribooligonucleotide primer. When fd or φX DNA is covered with DNA-binding protein, *dnaG* protein is unable to synthesize oligonucleotides with ADP, four dNTPs, and four rNTPs. Thus, the *dnaG*-protein-catalyzed priming reaction is template-specific. In addition, *dnaG* protein is unable to catalyze the incorporation of rNMPs or dNMPs at the

3′-OH end of RNA or DNA primers on an SS DNA template, or to catalyze oligonucleotide synthesis with synthetic deoxyribopolynucleotides or denatured *E. coli* DNA covered with DNA-binding protein.

That the *dnaG* protein is involved in oligonucleotide synthesis is shown by the fact that dGMP incorporation is thermolabile when thermolabile *dnaG* protein is used. In the experiments presented in Figure 6, wild-type and thermolabile *dnaG* proteins were heated and then incubated in first reactions with DNA-binding protein, G4 DNA, ADP, dTTP, and [α-^{32}P]dGTP. The products of this reaction were isolated by gel filtration and tested as substrates for DNA elongation in second reactions. Both [α-^{32}P]dGMP incorporation into primer in the first reaction and the ability of the isolated protein-DNA complex to support DNA elongation in the second reaction were more thermolabile when thermolabile *dnaG* protein was used than when wild-type *dnaG* protein was used. There is no evidence that *dnaG* protein is required for DNA elongation of the oligonucleotide primer in conjunction with DNA-elongation proteins.

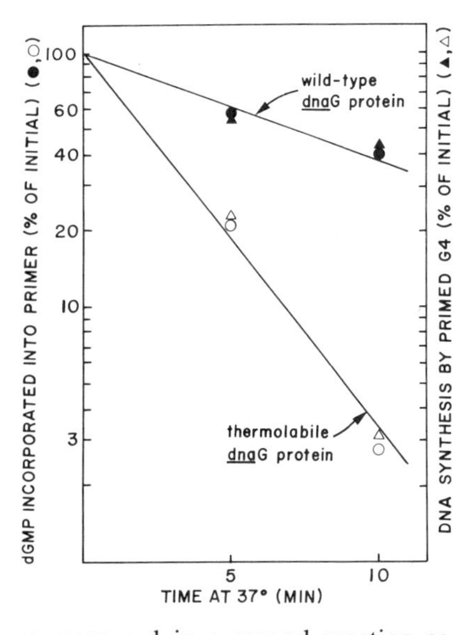

Figure 6 Thermolability of primer synthesis catalyzed by thermolabile *dnaG* protein. *DnaG* proteins (wild-type and thermolabile) were heated (diluted to 300 U/ml in 10% glycerol, 50 mM Tris-HCl, pH 8.0, 1 mM dithiothreitol, and 1 mg/ml bovine serum albumin) for 0, 5, or 10 min at 37°C. The *dnaG* proteins (10 μl) were then incubated in first reactions and subjected to gel filtration as described in the legend to Fig. 4. ^{32}P-radioactivity was measured in the DNA complexes isolated from reactions containing wild-type *dnaG* protein (●) and from reactions containing thermolabile *dnaG* protein (○). One hundred percent represented 1.02 and 0.80 pmoles of dGMP incorporated with wild-type and thermolabile *dnaG* protein-DNA complexes, respectively. ADP-independent dTMP incorporation was measured in a second reaction as described in the legend to Fig. 4, with DNA complexes formed using wild-type *dnaG* protein (▲) and thermolabile *dnaG* protein (△). One hundred percent represented 1040 and 712 pmoles of total nucleotide incorporated with wild-type and thermolabile *dnaG* protein-DNA complexes, respectively. When the first reaction contained a mixture of wild-type and thermolabile *dnaG* proteins heated 10 min at 37°C, the amounts of both dGMP incorporation into primer and ADP-independent DNA elongation were as expected (Wickner 1977).

Initiation of φX DNA Synthesis

Synthesis of φX DNA has been accomplished in vitro with purified *dnaB*, *dnaC(D)*, and *dnaG* proteins, DNA-binding protein, two or three other *E. coli* proteins defined only by their requirement in this reaction (referred to as replication factors X, Y, and Z [Wickner and Hurwitz 1974] or factors i and n [Schekman et al. 1975]), ATP, dNTPs, and DNA-elongation components (Table 3). Enzymatic activities of some of these proteins have been discovered. The *dnaB* protein is a ribonucleoside triphosphatase that is stimulated by SS DNA but not by double-stranded DNA or RNA; all four rNTPs, but not the dNTPs, are hydrolyzed (Wickner et al. 1974). The *dnaC(D)* protein inhibits the DNA-independent ATPase activity of *dnaB* protein and forms a physical complex with *dnaB* protein in the presence of ATP (Wickner and Hurwitz 1975a). The roles of the *dnaB* rNTPase and the *dnaB-dnaC(D)* protein complex in DNA synthesis are not yet known. No independent enzymatic activities have been found associated with *dnaC(D)* protein or replication factors X and Z. Replication factor Y is an SS DNA-dependent ATPase or dATPase (Wickner and Hurwitz 1975b). Under some assay conditions this ATPase activity is stimulated specifically by φX DNA but not by other phage SS DNAs whose replication does not require it.

The φX DNA-synthesis reaction, which is presented diagrammatically in Figure 7, has been divided into partial reactions (Wickner and Hurwitz

Table 3 Requirements for φX DNA synthesis

Additions	dTMP incorporated (pmoles/30 min)	
	φX DNA	fd DNA
Complete	23.4	<0.2
– *dnaB* protein	<0.2	
– *dnaC(D)* protein	<0.2	
– DNA-binding protein	2.2	
– replication factor X	0.4	
– replication factor Y	0.4	
– replication factor Z	0.9	
– *dnaG* protein	0.3	
– DNA polymerase III, *dnaZ* protein, DNA EFI, DNA EFIII	<0.2	
– ATP	<0.2	

Each reaction (25 μl) contained reaction buffer (Table 2), 1 mM ATP, 0.04 mM each of dATP, dCTP, dGTP, and [^{3}H]dTTP (800 cpm/pmole), 200 pmoles DNA, *dnaB* protein (0.05 U, 0.06 μg), *dnaC(D)* protein (0.03 U, 0.1 μg), DNA-binding protein (0.75 μg), factor X (0.05 U, 0.06 μg), factor Y (0.03 U), factor Z (0.05 U, 0.1 μg), *dnaG* protein (0.08 U, 0.06 μg), DNA polymerase III (0.2 U, 0.2 μg), *dnaZ* protein (0.1 U, 0.02 μg), DNA EFI (0.1 U, 0.02 μg), and DNA EFIII (0.1 U, 0.05 μg). After incubation at 30°C for 30 min acid-insoluble radioactivity was measured.

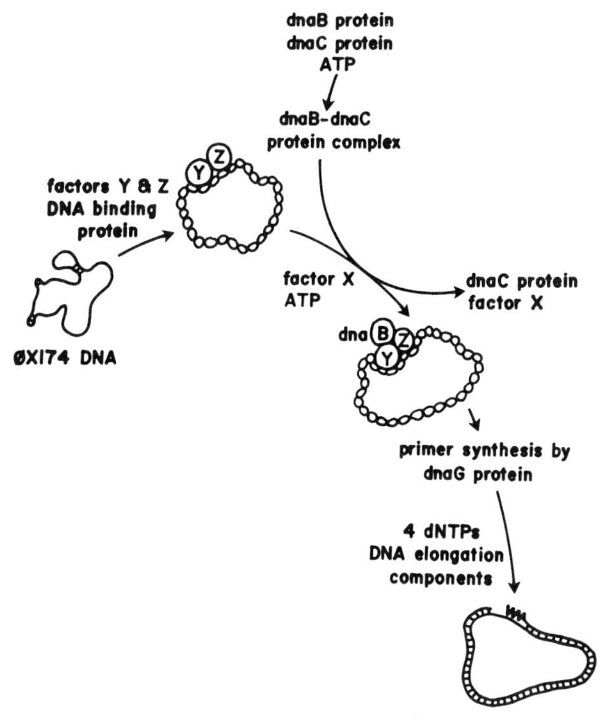

Figure 7 Mechanism of priming of φX phage DNA.

1974, 1975a,c; Weiner et al. 1976). The various intermediates have been isolated by agarose gel filtration. After the isolation of the protein-DNA complexes, their components were determined by dissociating the complexes with high salt and measuring the proteins present by complementation or reconstitution assays. When DNA-binding protein, replication factors Y and Z, and φX DNA are incubated and then filtered through Bio-Gel A-5m, all three proteins are associated with the DNA. These proteins bind, without specificity, to any SS DNA; the binding of any one of them is not dependent on the other two. Although φX DNA synthesis requires such high levels of DNA-binding protein so that the DNA is covered completely, it requires low levels of replication factors Y and Z, perhaps only one or several molecules per DNA circle. When DNA-binding protein, replication factors X, Y, and Z, *dnaB* and *dnaC(D)* proteins, ATP, and φX DNA are incubated and then filtered through agarose, *dnaB* protein is transferred from a protein complex with *dnaC(D)* protein to the protein-DNA complex containing DNA-binding protein and replication factors Y and Z. This protein-DNA complex contains *dnaB* protein, DNA-binding protein, and replication factors Y and Z; it does not contain *dnaC(D)* protein or factor X. If any one of the components of the first reaction is omitted, *dnaB* protein is not transferred to the φX DNA. This complex requires ATP for stabilization; it dissociates with a half-life of about 8 hours at 0°C. DNA synthesis by this isolated complex requires *dnaG* protein, dNTPs, and DNA-elongation components. It does not require, and it is not stimulated by, a second addition of *dnaC(D)* protein or replication factor X. When DNA-binding protein, replication factors X, Y, and Z, *dnaB*, *dnaC(D)*, and *dnaG* proteins, ATP, and φX DNA are incubated together and then filtered through agarose, the protein-DNA complex contains *dnaB* protein, DNA-binding protein, and replication factors Y and Z, but not *dnaG* protein. Furthermore, DNA synthesis by this complex still requires *dnaG* protein in addition to dNTPs and DNA-elongation components. Thus, although *dnaG* protein catalyzes the synthesis of a primer for DNA elongation, presumably by recognizing a site on the φX DNA created by the action of the proteins required for the prepriming reaction, it has not been possible to isolate such an intermediate containing *dnaG* protein. This complex mechanism by which φX DNA is primed is template-specific; under identical reaction conditions *dnaB* protein cannot be transferred to fd DNA. In sum, although enzymatic activities of a few of the proteins involved in φX prepriming are known and several DNA-protein intermediates have been isolated, the details of the molecular mechanism by which this reaction occurs are unknown.

DNA Elongation of Primed Single-stranded DNA

Once a primer is synthesized, DNA elongation can occur in vitro by any one of several mechanisms: (a) by DNA polymerase I alone, (b) by DNA polymerase II plus DNA-binding protein under some specific assay conditions (Sherman and Gefter 1976), or (c) by DNA polymerase II or III in combination with *dnaZ* protein, DNA EFI, and DNA EFIII under other specific conditions (Wickner and Hurwitz 1976; Wickner 1976). It is not clear whether phage SS DNA is converted to duplex DNA in vivo by one or more of these mechanisms. Since DNA polymerase III and *dnaZ* protein are required for *E. coli* DNA replication, the last mechanism probably shares some similarities with the in vivo mechanism of DNA replication and therefore has been studied extensively (W. Wickner et al. 1973; Wickner and Kornberg 1973; Hurwitz et al. 1973; Wickner and Kornberg 1974b; Hurwitz and Wickner 1974; Livingston et al. 1975; Livingston and Richardson 1975; Wickner and Hurwitz 1976; Wickner 1976; McHenry and Kornberg 1977).

The DNA-elongation reaction catalyzed by DNA polymerase III, *dnaZ* protein, DNA EFI, and DNA EFIII (Table 4) has been divided into partial reactions by the physical isolation of protein-protein and protein-DNA complexes preceding dNMP incorporation (Fig. 8).

DNA EFIII (65,000-dalton active protein) and *dnaZ* protein (125,000-

Table 4 Requirements for elongation of oligo(dT) primers on poly(dA) templates

Additions	dTMP incorporated (pmoles/20 min)
Complete	37.6
– DNA polymerase III	<0.2
– *dnaZ* protein	0.3
– DNA EFI	<0.2
– DNA EFIII	0.3
– oligo(dT)	<0.2
– ATP	<0.2
– ATP, + dATP	28.0
– ATP, + GTP, CTP, UTP, dTTP, dGTP, or dCTP	<0.2

Each reaction (30 μl) contained reaction buffer (Table 2), 0.1 mM ATP, 50 μM [^{3}H]dTTP (500 cpm/pmole), 1 nmole each of poly(dA) and oligo(dT), 0.1 U DNA EFI, 0.1 U DNA EFIII, 0.3 U *dnaZ* protein, and 0.3 U DNA polymerase III. After 20 min at 25°C acid-insoluble radioactivity was measured (Wickner 1976).

dalton active protein) can form a protein complex. This was shown by the coelution, following Sephadex gel chromatography, of *dnaZ* protein and DNA EFIII when these two proteins were mixed prior to filtration. Complex formation does not require DNA polymerase III, DNA EFI, ATP, or primed template. DNA polymerase III and DNA EFI have not been isolated as a larger complex with *dnaZ* protein and DNA EFIII.

Together, *dnaZ* protein and DNA EFIII catalyze the transfer of DNA EFI to a primed DNA template. When *dnaZ* protein, DNA EFI, DNA EFIII, primed template, and ATP or dATP are incubated and the reaction mixture subjected to Bio-Gel A-5m filtration, DNA EFI is recovered with the primed template; *dnaZ* protein, DNA EFIII, and ATP are not associated with the DNA. When ATP, primer, DNA template, DNA EFIII or *dnaZ* protein is omitted from the first reaction, DNA EFI is not transferred to the DNA. The function of ATP or dATP in this reaction is unknown. Incorporation of dTMP by the isolated complex requires DNA polymerase III and four dNTPs. When the primed template in the first reaction is poly(dA)-oligo(dT), dTMP incorporation by the complex of DNA EFI and DNA requires DNA polymerase III and dTTP; ATP or dATP is no longer required.

DNA polymerase III binds to the complex of primed template and DNA EFI, but not to primed template alone, and catalyzes DNA synthesis upon addition of dNTPs. It is not known whether DNA EFI is associated with DNA polymerase III during chain propagation. It is also not known whether, under these conditions, DNA-binding protein is required to melt secondary structure and allow synthesis of long DNA chains.

This mechanism of DNA elongation suggests that *dnaZ* protein and DNA

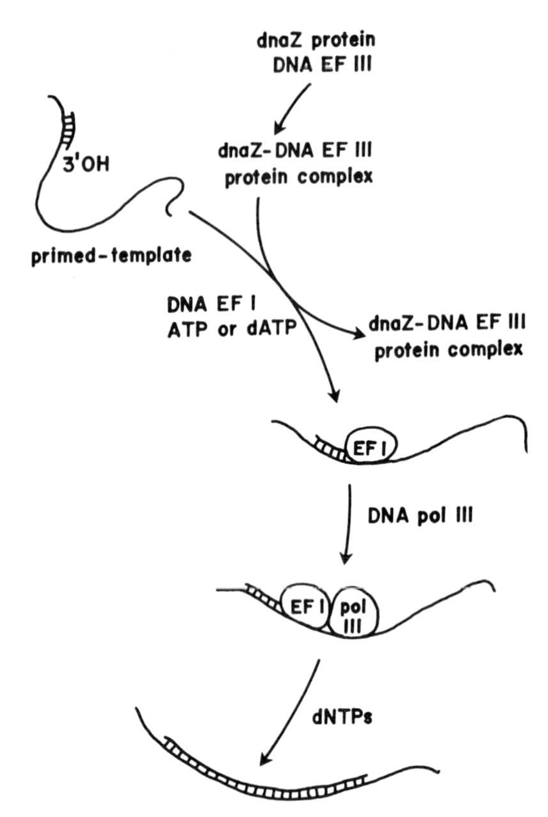

Figure 8 Mechanism of DNA elongation catalyzed by DNA polymerase III in combination with *dnaZ* protein, DNA EFI, and DNA EFIII.

EFIII must recognize DNA but that they are not associated with it tightly enough to be isolated as a complex. Recent experiments have shown that DNA-binding protein increases the affinity of *dnaZ* protein and DNA EFIII for SS DNA, such that complexes of these three proteins with any SS DNA can be isolated. Binding of *dnaZ* protein and DNA EFIII to DNA covered with DNA-binding protein occurs in the absence of ATP or dATP, primer, DNA EFI, and DNA polymerase III. The binding of DNA EFI to the complex still requires ATP or dATP and primer ends; the binding of DNA polymerase III still requires that DNA EFI be bound first.

The availability of *dnaZ* protein, DNA polymerase III, DNA EFI, and DNA EFIII in active forms and free from one another (Table 4) has facilitated the characterization of how elongation of primed SS DNA is initiated. Whether or not these four proteins exist in vivo as a protein complex is not known. This question could be answered by the isolation of temperature-resistant suppressors of *polC* cells that have alterations in *dnaZ* protein, DNA EFI, or DNA EFIII.

SUMMARY

In *E. coli*, both in vivo and in vitro, three pathways have been identified for initiating the conversion of phage circular SS DNA to duplex DNA. The enzyme pathway used by each phage is determined by the structure and sequence of the DNA, and each is replicated by only one of the three pathways. This diversity in mechanisms for the initiation of replication will probably be reflected in the discovery of multiple mechanisms for the regulation of DNA replication. In contrast to the initiation of phage DNA synthesis, DNA elongation is not DNA-specific; DNA elongation of any SS DNA (circular or linear, natural or synthetic) primed with DNA or RNA can be catalyzed by DNA polymerase I or by DNA polymerase II or III in combination with *dnaZ* protein, DNA EFI, and DNA EFIII.

REFERENCES

Bouché, J. P., K. Zechel, and A. Kornberg. 1975. *DnaG* gene product, a rifampicin-resistant RNA polymerase, initiates the conversion of a single-stranded coliphage DNA to its duplex replication form. *J. Biol. Chem.* **250:** 5995.

Gray, C., R. Sommer, C. Polke, E. Beck, and H. Schaller. 1977. Structure of the origin of DNA replication of bacteriophage fd. *Proc. Natl. Acad. Sci.* **75:** 50.

Hurwitz, J. and S. Wickner. 1974. Involvement of two protein factors and ATP in *in vitro* DNA synthesis catalyzed by DNA polymerase III of *Escherichia coli. Proc. Natl. Acad. Sci.* **71:** 6.

Hurwitz, J., S. Wickner, and M. Wright. 1973. Studies on *in vitro* DNA synthesis. II. Isolation of a protein which stimulates deoxynucleotide incorporation catalyzed by DNA polymerases of *Escherichia coli. Biochem. Biophys. Res. Commun.* **51:** 257.

Livingston, D. and C. C. Richardson. 1975. Deoxyribonucleic acid polymerase III of *Escherichia coli. J. Biol. Chem.* **250:** 470.

Livingston, D., D. Hinkle, and C. C. Richardson. 1975. Deoxyribonucleic acid polymerase III of *Escherichia coli. J. Biol. Chem.* **250:** 461.

McHenry, C. and A. Kornberg. 1977. DNA polymerase III holoenzyme of *Escherichia coli. J. Biol. Chem.* **252:** 6478.

Schaller, H., A. Uhlmann, and K. Geider. 1976. A DNA fragment from the origin of single-strand to double-strand DNA replication of bacteriophage fd. *Proc. Natl. Acad. Sci.* **73:** 49.

Schekman, R., A. Weiner, and A. Kornberg. 1974. Multiple systems of DNA replication. *Science* **186:** 987.

Schekman, R., J. H. Weiner, A. Weiner, and A. Kornberg. 1975. Ten proteins required for conversion of φX174 single-stranded DNA to duplex form *in vitro. J. Biol. Chem.* **250:** 5859.

Schekman, R., W. T. Wickner, O. Westergaard, D. Brutlag, K. Geider, L. L. Bertsch, and A. Kornberg. 1972. Initiation of DNA synthesis: Synthesis of φX174 replicative form requires RNA synthesis resistant to rifampicin. *Proc. Natl. Acad. Sci.* **69:** 2691.

Sherman, L. A. and M. L. Gefter. 1976. Studies on the mechanism of enzymatic DNA elongation by *Escherichia coli* DNA polymerase II. *J. Mol. Biol.* **103:** 61.

Sigal, N., H. Delius, T. Kornberg, M. Gefter, and B. Alberts. 1972. A DNA-unwinding protein isolated from *Escherichia coli*: Its interaction with DNA and with DNA polymerases. *Proc. Natl. Acad. Sci.* **69:** 3537.

Vicuna, R., J. E. Ikeda, and J. Hurwitz. 1977a. Selective inhibition of φX RFII compared with fd RFII DNA synthesis *in vitro*. *J. Biol. Chem.* **252:** 2534.

Vicuna, R., J. Hurwitz, S. Wallace, and M. Girard. 1977b. Selective inhibition of *in vitro* DNA synthesis dependent on φX174 compared with fd DNA. *J. Biol. Chem.* **252:** 2524.

Wechsler, J. A. 1978. The genetics of *E. coli* DNA replication. In *DNA synthesis: Present and future* (ed. M. Kohiyama and I. Molineux), p. 49. Plenum Press, New York.

Weiner, J. H., R. McMacken, and A. Kornberg. 1976. Isolation of an intermediate which precedes *dnaG* RNA polymerase participation in enzymatic replication of bacteriophage φX174 DNA. *Proc. Natl. Acad. Sci.* **73:** 752.

Wickner, R. B., M. Wright, S. Wickner, and J. Hurwitz. 1972. Conversion of φX174 and fd single-stranded DNA to replicative forms in extracts of *Escherichia coli*. *Proc. Natl. Acad. Sci.* **69:** 3233.

Wickner, S. 1976. Mechanism of DNA elongation catalyzed by *Escherichia coli* DNA polymerase III, *dnaZ* protein, and DNA elongation factors I and III. *Proc. Natl. Acad. Sci.* **73:** 3511.

———. 1977. DNA or RNA priming of bacteriophage G4 DNA synthesis by *Escherichia coli dnaG* protein. *Proc. Natl. Acad. Sci.* **74:** 2815.

Wickner, S. and J. Hurwitz. 1974. Conversion of φX174 viral DNA to double-stranded form by purified *Escherichia coli* proteins. *Proc. Natl. Acad. Sci.* **71:** 4120.

———. 1975a. Interaction of *Escherichia coli dnaB* and *dnaC(D)* gene products *in vitro*. *Proc. Natl. Acad. Sci.* **72:** 921.

———. 1975b. Association of φX174 DNA-dependent ATPase activity with an *Escherichia coli* protein, replication factor Y, required for *in vitro* synthesis of φX174 DNA. *Proc. Natl. Acad. Sci.* **72:** 3342.

———. 1975c. *In vitro* synthesis of DNA. In *DNA synthesis and its regulation* (ed. M. Goulian et al.), p. 227. W. A. Benjamin, Menlo Park, California.

———. 1976. Involvement of *Escherichia coli dnaZ* gene product in DNA elongation *in vitro*. *Proc. Natl. Acad. Sci.* **73:** 1053.

Wickner, S., M. Wright, and J. Hurwitz. 1973a. Studies on *in vitro* DNA synthesis. III. Purification of the *dnaG* gene product of *Escherichia coli*. *Proc. Natl. Acad. Sci.* **70:** 1613.

———. 1974. Association of DNA-dependent and independent ribonucleoside triphosphatase activities with *dnaB* gene product of *Escherichia coli*. *Proc. Natl. Acad. Sci.* **71:** 783.

Wickner, S., I. Berkower, M. Wright, and J. Hurwitz. 1973b. Studies on *in vitro* DNA synthesis: Purification of *dnaC* gene product containing *dnaD* activity from *Escherichia coli*. *Proc. Natl. Acad. Sci.* **70:** 2369.

Wickner, W. and A. Kornberg. 1973. DNA polymerase III star requires ATP to start synthesis on a primed DNA. *Proc. Natl. Acad. Sci.* **70:** 3679.

———. 1974a. A novel form of RNA polymerase from *Escherichia coli*. *Proc. Natl. Acad. Sci.* **71:** 4424.

———. 1974b. A holoenzyme form of deoxyribonucleic acid polymerase III. *J. Biol. Chem.* **249:** 6244.

Wickner, W., R. Schekman, K. Geider, and A. Kornberg. 1973. A new form of DNA polymerase III and a copolymerase replicate a long, single-stranded primer-template. *Proc. Natl. Acad. Sci.* **70:** 1764.

Wright, M., S. Wickner, and J. Hurwitz. 1973. Studies on *in vitro* DNA synthesis. V. Isolation of the *dnaB* gene product from *Escherichia coli*. *Proc. Natl. Acad. Sci.* **70:** 3120.

Zechel, K., J. P. Bouché, and A. Kornberg. 1975. Replication of phage G4. *J. Biol. Chem.* **250:** 4684.

Priming of DNA Synthesis on Viral Single-stranded DNA In Vitro

Roger McMacken,* Lee Rowen, Kunihiro Ueda,† and Arthur Kornberg
Department of Biochemistry
Stanford University School of Medicine
Stanford, California 94305

Among the many viral, bacterial, and animal DNA polymerases isolated thus far none can start a chain in vitro (Kornberg 1974). In addition to a template strand, each requires a 3′-OH-terminated chain (primer terminus). Because they lack such termini, the coliphage circular, single-stranded (SS) DNA chromosomes have proved to be ideal templates for studying the mechanisms used by *Escherichia coli* to initiate new DNA chains (Schekman et al. 1974; Wickner and Hurwitz 1975a).

Because RNA polymerases are known to initiate new RNA chains, it appeared that an RNA transcript might prime DNA synthesis (Brutlag et al. 1971), and rifampicin, a specific inhibitor of *E. coli* RNA polymerase, should therefore inhibit DNA synthesis. RNA polymerase was shown to be essential in the conversion of phage M13 SS DNA to the duplex replicative form (RF). Later it was demonstrated that soluble extracts from gently lysed *E. coli* cells support the conversion of various bacteriophage SS DNAs to their replicative forms (Wickner et al. 1972; Schekman et al. 1974). Fractionation and identification of the proteins involved in this apparently simple replication showed not only that synthesis of the complementary DNA chain is surprisingly complex, but also that there are at least three different DNA-strand initiation systems in *E. coli* (Schekman et al. 1974; Wickner and Hurwitz 1974; Schekman et al. 1975; Kornberg 1977). The chromosomes of the filamentous phage M13 and the isometric phages G4 and φX174 are each replicated in vitro by a different set of bacterial replication proteins, which differ primarily in the mechanism by which primer transcripts are generated (Table 1). In this article we shall focus on recent studies of the enzymology of primer synthesis on these viral-strand circles. For a more general review of RNA priming of DNA synthesis see Kornberg (1976) or McMacken et al. (1977b).

Present addresses: *Department of Biochemistry, School of Hygiene and Public Health, The Johns Hopkins University, 615 North Wolfe Street, Baltimore, Maryland 21205; †Department of Medical Chemistry, Kyoto University Faculty of Medicine, Yoshida, Sakyo-ku, Kyoto 606, Japan.

Table 1 Protein requirements for replication of viral SS DNAs

Stage	Template M13	G4	φX
Prepriming	none	none	DBP, *dnaB*, *dnaC*, proteins i and n
Priming	RNA polymerase DBP	primase DBP	primase *dnaB*, DBP
Elongation	holoenzyme DBP	holoenzyme DBP	holoenzyme DBP
Termination	DNA polymerase I DNA ligase	DNA polymerase I DNA ligase	DNA polymerase I DNA ligase

Replication of M13 DNA by Purified Proteins

Resolution and purification of the components needed for the conversion of SS DNA to RF DNA disclosed the requirements for the five proteins shown in Figure 1 (Geider and Kornberg 1974). DNA-binding protein (DBP) masks the single strand except at a single region, and primer synthesis can take place only in this position. In the absence of DNA polymerase I, the gap in RFII (circular, duplex DNA in which at least one strand is not continuous) is found at this "promoter" region (Geider and Kornberg 1974; Tabak et al. 1974).

The evidence for RNA priming of M13 DNA replication (Wickner et al. 1972; Westergaard et al. 1973), besides the effects of RNA polymerase inhibitors, is: (1) all four ribonucleoside triphosphates (rNTPs) are required; (2) conversion of SS DNA to RF DNA takes place in two stages (the

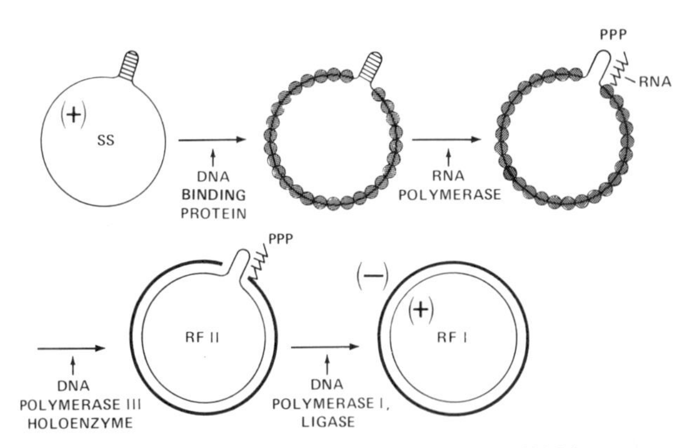

Figure 1 Scheme for conversion of M13 SS DNA to nonsuperhelical RFI (covalently closed, circular, duplex) DNA (RFIV).

initial stage of RNA synthesis produces a primed single strand that can be isolated and converted to RF DNA in a second stage in the absence of rNTPs and in the presence of rifampicin); (3) a phosphodiester linkage of a deoxyribonucleotide to a ribonucleotide in the isolated RF DNA is equivalent stoichiometrically to the number of RF molecules formed; and (4) an RNA fragment persists at the 5′ end of the newly synthesized complementary strand (this is inferred from the behavior of DNA polymerases and DNA ligases in filling and sealing the small gap in the RFII product). The size and nature of the primer transcript have been determined recently (Geider et al. 1978). The DNA sequence at the RNA-polymerase-binding site has been shown to have two "hairpin" structures (Ravetch et al. 1977; Gray et al. 1978; Schaller et al., this volume), each about 60 nucleotides long. The significance of the strongly duplex character of the SS DNA near the origin of M13 complementary-strand synthesis is not yet clear.

Factors present in cell extracts influence RNA polymerase to act upon DNAs from filamentous phages and discriminate against those from isometric phages. These include a subunit which appears to be associated with RNA polymerase (Wickner and Kornberg 1974) and additional factors (Vicuna et al. 1977a,b).

The mechanism of elongation of SS DNA phage chromosomes, primed and coated with DBP, appears to be universal, regardless of the template and primer (Table 1) (Schekman et al. 1974; Geider and Kornberg 1974; Wickner 1976; McHenry and Kornberg 1977; McMacken et al. 1977a). In all in vitro studies DNA polymerase III holoenzyme catalyzes covalent extension of the primer terminus into complementary-strand DNA. Although in this step DBP can be replaced by spermidine, binding protein is required to make DNA polymerase action maximally processive both for the rate and the extent of DNA synthesis (Geider and Kornberg 1974; Sherman and Gefter 1976).

Priming and Replication of G4 DNA In Vitro

In contrast to the case with M13 DNA, conversion of φX DNA to duplex RF DNA is unaffected by rifampicin both in vivo and in vitro (Wickner et al. 1972). Experimental evidence for rifampicin-resistant RNA priming of φX replication (Schekman et al. 1972) also shows that this initiating process involves a highly complex enzyme system (Table 1). However, when G4 DNA (Dumas; Godson; both this volume) is used in place of φX DNA, a markedly simpler rifampicin-resistant replication suffices. Only three proteins are required: DBP, primase (*dnaG* protein), and holoenzyme (Zechel et al. 1975) (Table 1). On the basis of the roles of the binding protein and the holoenzyme in M13 SS DNA replication, it was presumed that primase was responsible for the priming of G4 DNA needed for replication by holoenzyme. A specific origin for G4 SS-to-RF conversion was shown by the finding that the 5′ end of the gap in RFII occurred near the single site of cleavage by restriction endonuclease *Eco*RI (Zechel et al. 1975). This site has been shown in in vivo experiments to correspond to the physiological complementary-strand initiation site (Martin and Godson 1977; Hourcade and Dressler 1978; Godson et al., this volume).

With the purified proteins of G4 replication in hand, some significant features of primer synthesis were found. A unique RNA transcript results when primase acts on G4 DNA in the presence of DBP and the four rNTPs (Bouché et al. 1975). This RNA fragment is found at the sole site on the template strand used for initiation of the complementary strand; it is covalently linked to the 5′ portion of this chain after replication by holoenzyme. The sequence of the RNA primer covalently attached to the synthetic complementary strand (Fig. 2) (Bouché et al. 1978) contains a GC-rich duplex region. The complementary region in the SS DNA template left uncoated by DBP might serve as a recognition site for primase. Analysis of the covalent linkage formed between RNA and DNA when holoenzyme is incubated with G4 SS DNA (previously primed with RNA) shows that DNA synthesis can begin after primer residue 26, 27, 28, or 29. This variation in primer size may be due to premature termination of RNA synthesis or to removal of residues from the 3′ end by RNase. The sequence of the portion of the G4 chromosome containing the complementary-strand initiation site has been ascertained recently (Fiddes et al. 1978; Sims and Dressler 1978; Godson et al., this volume); it contains a sequence complementary to the primer sequence. The primer site is located about 400 residues from the *Eco*RI endonuclease cleavage site, in the intercistronic region between genes *F* and *G*. As anticipated from the more complex protein requirements for φX DNA replication, no such sequence is found in this DNA (Sanger et al. 1977).

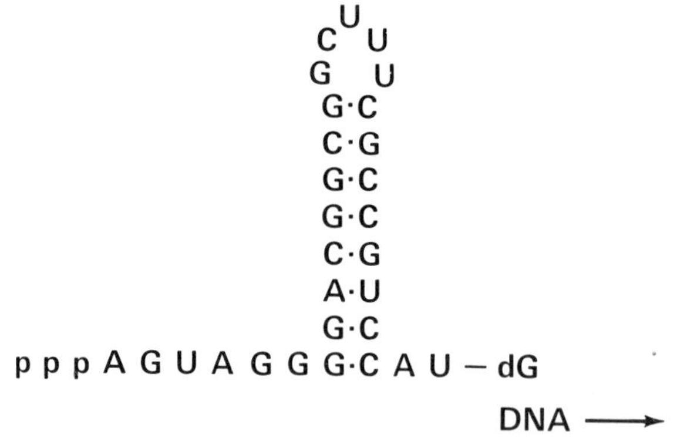

Figure 2 Sequence for G4 primer RNA.

Properties of Primase in Priming G4 DNA Synthesis

Using conversion of G4 viral-strand DNA to duplex RF DNA as an in vitro assay for primase, Rowen and Kornberg (1978a) purified this 60,000-dalton protein to near homogeneity. Three activities copurify with primase at each stage: (1) RNA synthesis on G4 DNA, (2) complementation of an extract from a temperature-sensitive *dnaG* mutant for replication of G4 DNA, and (3) priming of G4 DNA replication carried out by purified holoenzyme. Moreover, the role of primase in primer synthesis is corroborated by the thermolability of both nucleotide incorporation into primer (Wickner 1977) and DNA replication when primase from temperature-sensitive *dnaG* mutants is employed (Bouché et al. 1975).

In the presence of DBP, primase is remarkably specific for G4 DNA as a template for primer synthesis (Rowen and Kornberg 1978a; Wickner 1977). Other, unrelated, single- and double-stranded DNAs are inactive. This specificity may reside in a nucleotide sequence or secondary structure in G4 DNA (Bouché et al. 1978; Fiddes et al. 1978; Sims and Dressler 1978; Godson et al., this volume) which enables primase to bind only to this DBP-coated template (Wickner 1977; D. Bates and A. Kornberg, unpubl.). Primase action is not affected by rifampicin or streptolydigin, two potent inhibitors of *E. coli* RNA polymerase; primer synthesis is inhibited, however, by actinomycin D (Rowen and Kornberg 1978a).

Although ATP is the initiating nucleotide in primer synthesis on G4 DNA, several ATP analogs, such as adenosine 5′-tetraphosphate, ADP, adenosine 5′-*O*-(3-thiotriphosphate), and adenyl imidodiphosphate (Rowen and Kornberg 1978b), can be utilized by primase as substitutes. In addition, as discovered in studies of primase-dependent synthesis of primers for φX DNA replication (McMacken et al. 1977b), primase can incorporate deoxyribonucleoside triphosphates (dNTPs) as well as rNTPs (Kornberg 1977; Wickner 1977; Rowen and Kornberg 1978b). Primase adds either a ribonucleotide or a deoxyribonucleotide to the 3′ OH of a ribo- or deoxyribonucleotide residue of a primer terminus. Although dNTPs are incorporated, they inhibit rNTP incorporation profoundly even at low concentrations (5–50 μM), causing shorter transcripts to be synthesized (Rowen and Kornberg 1978b).

Synthesis of the complete 29-residue RNA transcript is not essential for effective priming of G4 replication, but synthesis of at least a dinucleotide is required (Rowen and Kornberg 1978b). The amount of primer actually synthesized in the complete system (i.e., coupled priming and replication) depends on the concentration of dNTPs (Rowen and Kornberg 1978b); at 20 μM rNTPs and 50 μM dNTPs only two to six residues of RNA are found in the replicated product, whereas at 5 μM dNTPs up to full-length RNA primer is synthesized before extension by holoenzyme occurs. These data suggest that primase transcribes G4 DNA until it reaches a termination site or is stopped prematurely by dNTPs at inhibitory levels. At any point after a dinucleotide has been formed, DNA polymerase may displace primase and extend the primer terminus.

Enzymology of φX DNA Replication

Conversion of φX DNA to its replicative form occurs by a rifampicin-resistant reaction both in vivo and in vitro (Wickner et al. 1972). Fractionation and identification of the required proteins indicate that at least nine proteins participate in the synthesis of the complementary strand, including each of those required for G4 replication and four additional ones: the *dnaB* and *dnaC* proteins and proteins i and n (Table 1) (Schekman et al. 1974, 1975; Wickner and Hurwitz 1974). Partially purified, the last four proteins slowly convert DBP-coated φX DNA to an activated nucleoprotein complex which is rapidly replicated upon the addition of primase and holoenzyme (Wickner and Hurwitz 1974; Weiner et al. 1976). We have termed this complex the φX replication intermediate (RI).

Because of its remarkable stability in the presence of ATP, the RI can be isolated readily (Weiner et al. 1976). Although the exact composition of the RI is not yet known, the evidence available suggests that proteins i and n are not essential components. Not only do these two proteins act catalytically during RI formation, but also the isolated RI is not affected by antibodies directed against them (Weiner et al. 1976; Ueda et al. 1978; McMacken and Kornberg 1978). On the other hand, the isolated RI contains a stoichiometric amount (approximately 150 molecules) of DBP (Weiner et al. 1976; J. Weiner and A. Kornberg, unpubl.). The *dnaB* protein of *E. coli* has been found to be a key constituent of the φX RI (McMacken et al. 1977b) on the basis of the following observations: (1) replicative activity of the isolated intermediate is sensitive to anti-*dnaB* gamma globulin (McMacken et al. 1977a; McMacken and Kornberg 1978); (2) DNA-dependent ATPase activity of *dnaB* protein (Wickner et al. 1974) is present in the isolated intermediate (McMacken et al. 1977a), and (3) labeled *dnaB* protein is incorporated into the complex (Ueda et al. 1978).

From the stoichiometry of *dnaB* protein used during RI formation and from the level of *dnaB* ATPase activity associated with the RI, it has been concluded that a single *dnaB* molecule is incorporated into the RI (Weiner et al. 1976; McMacken et al. 1977a). The precise functions of protein i, protein n, and *dnaC* protein in the transfer of *dnaB* protein to the DBP-coated φX DNA remain to be elucidated. It is known, however, that stoichiometric quantities of *dnaC* protein are required for the reaction, perhaps because *dnaB* protein may have to first form a complex with *dnaC* protein (Wickner and Hurwitz 1975b) before it is transferred to the φX DNA. Unfortunately, lack of an antibody directed against *dnaC* protein has prevented a rigorous analysis of *dnaC*-protein function in φX SS DNA replication.

A multienzyme system for priming replication of φX DNA in vitro (McMacken et al. 1977a,b; Kornberg 1977; McMacken and Kornberg 1978) consists of primase and each of the five proteins required for formation of the φX RI (Table 1), in addition to ATP, Mg^{++}, and the nucleoside triphosphates. Both rNTPs and dNTPs can serve as substrates; hybrid primers are formed when both are present (McMacken et al. 1977b; McMacken and Kornberg 1978). When the concentration of ribonucleotides is high in relation to that of deoxyribonucleotides, as in vivo, both the rate and the extent of incorporation are significantly higher for ribonucleotides than for deoxyribonucleotides. Both RNA and DNA primers appear to be initiated solely with ATP; neither GTP nor dATP can substitute as the initiating nucleotide (McMacken and Kornberg 1978). Because ATP is required during DNA-primer synthesis to stabilize the RI (containing template-bound *dnaB* protein and its ATPase activity), AMP residues are interspersed in the DNA transcripts (McMacken and Kornberg 1978). More than 98% of both the DNA and the RNA transcripts that are hybridized stably to the φX DNA template are extended by holoenzyme into complementary-strand DNA (McMacken et al. 1977a; McMacken and Kornberg 1978).

As judged from polyacrylamide gel electrophoresis in 7 M urea, RNA primers range from 14 to 50 nucleotides in length, whereas DNA primers have been reported to be substantially shorter, from 4 to 20 residues long (McMacken et al. 1977a; McMacken and Kornberg 1978). Primase will transcribe only those φX DNA circles that have been converted to the intermediate form (McMacken et al. 1977a; McMacken and Kornberg 1978). Using antibodies against the proteins required for primer synthesis, McMacken and Kornberg (1978) showed that, in addition to primase, both proteins known to be constituents of the RI (*dnaB* protein and DBP) also participate in primer synthesis (McMacken and Kornberg 1978).

The magnitude of RNA synthesis by the reconstituted system in the absence of coupled replication is far in excess of that required for priming complementary-strand synthesis. As many as 300 ribonucleotide residues (i.e., transcription of nearly 6% of the φX DNA template) or 80 deoxyribonucleotide residues can be incorporated per input circle (McMacken and Kornberg 1978). The complexity of the fingerprint pattern of the oligoribonucleotides produced after T1 RNase digestion of φX RNA primers indicates that transcripts are synthesized from multiple (probably random) sites on φX DNA (McMacken and Kornberg 1978). Furthermore, using [γ-^{32}P]ATP incorporation as a measure of chain initiation, it was discovered that several RNA primers (usually 5–8) are made on each circle (McMacken et al. 1977a). The single *dnaB* protein molecule bound in the RI was shown to participate in de novo initiation of each of these primer transcripts; antibodies directed against *dnaB* protein, when added to an ongoing RNA-priming reaction, blocked initiation of additional RNA chains immediately (McMacken et al. 1977a). From the sizes of the DNA chains synthesized when the gaps between the primers are subsequently filled by polymerase action, it appears that at high concentrations of primase the multiple primers are spaced at regular intervals, in the range of 70–500 nucleotides, along the template strand (McMacken et al. 1977a).

These data suggest (McMacken et al. 1977a) that *dnaB* protein, once fixed in the RI, migrates processively along the DBP-coated DNA strand, perhaps utilizing ATP hydrolysis to energize its movement, and that recognition by primase of bound *dnaB* protein or of an induced secondary structure in the template leads to initiation of a primer transcript at or near the locus. In the meantime, the *dnaB*-protein molecule, remaining on the DNA, apparently moves along the strand to a nearby site where synthesis of another primer is initiated by primase. The distance between neighboring primer transcripts appears to be an inverse function of the primase concentration (McMacken et al. 1977a). The polarity of movement is presumably in a direction opposite to the direction of growth of the primer, i.e., $5' \rightarrow 3'$ *on the single-stranded template*. Uncoupled from replication, this sequence is repeated many times over. With holoenzyme and dNTPs present, the first primer is rapidly elongated to a full-length, linear, complementary strand. These events are shown schematically in Figure 3.

Thus, *dnaB* protein acting on a φX DNA viral-strand circle appears to function as a "mobile promoter" for primer synthesis by primase to initiate

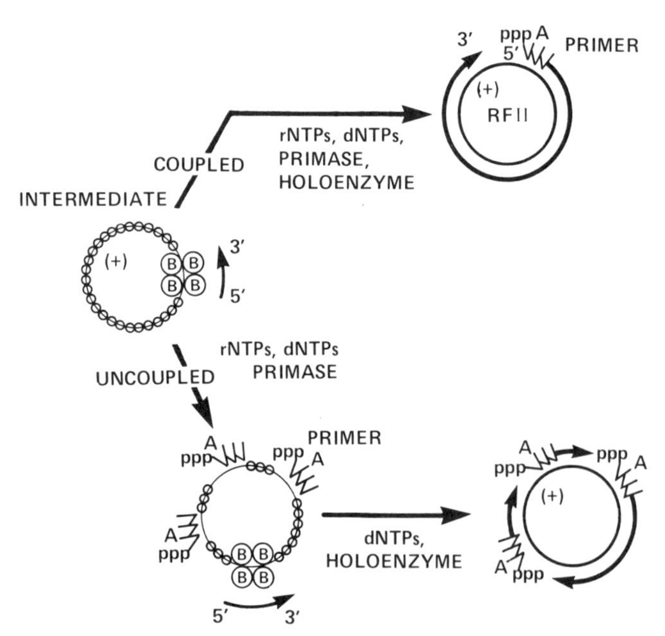

Figure 3 Scheme for *dnaB*-protein action as a "mobile promoter" in initiating DNA replication.

the growth of a DNA strand. Similarly, we have proposed that in the replication of the bacterial chromosome, the *dnaB* protein migrates processively along the newly exposed "lagging" strand, behind the replication fork, in the direction of fork movement (McMacken et al. 1977a) (Fig. 4). Primase recognizes *dnaB* protein (or a secondary structure which it generates in the template) to initiate primer synthesis for nascent (Okazaki) fragments of the *E. coli* chromosome. The *dnaB* protein-template complex would thus be a signal to primase that this SS DNA is engaged in replication rather than prepared for recombination, repair, or other metabolic processes involving SS DNA. Once bound, *dnaB* protein might travel with the replication fork without dissociating until the round of replication is completed. However, premature dissociation of the *dnaB* protein from the replicating DNA might be lethal if there were no means for rapidly restoring *dnaB* protein at or near the replication fork; *dnaC* protein and proteins i and n could participate in such a process.

The mechanism proposed for *dnaB*-protein action may apply to initiation of bidirectional replication at the chromosomal origin (Fig. 5) (McMacken et al. 1977a). A *dnaB*-protein molecule is bound on each strand at the

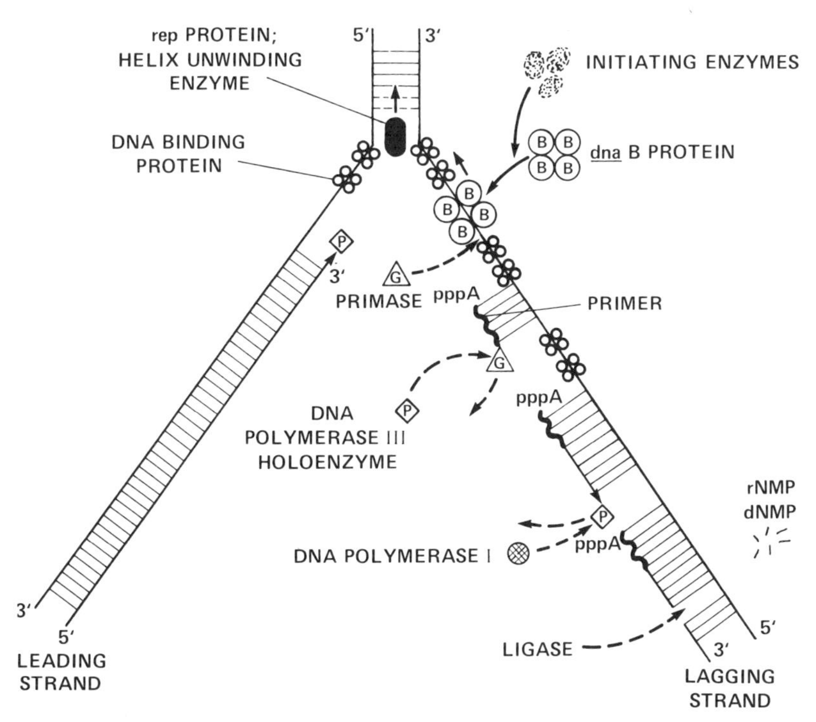

Figure 4 Scheme for *dnaB*-protein action at a replication fork in the *E. coli* chromosome.

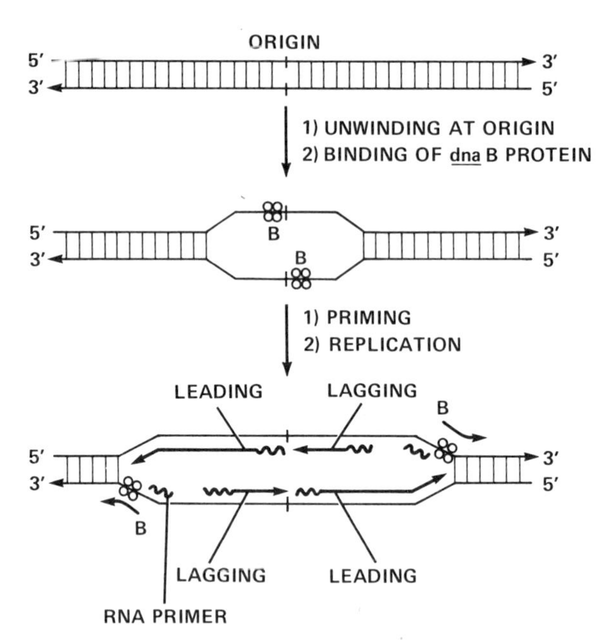

Figure 5 Scheme for *dnaB*-protein action in priming bidirectional replication at the chromosome origin (B represents *dnaB* protein).

replication origin and moves in the 5′ → 3′ direction. Thus, each *dnaB* molecule would direct priming of both the strand that is synthesized continuously (the single initiating event for the leading strand) and nascent fragments of the strand that is synthesized discontinuously (multiple events for the lagging strand).

SUMMARY

With conversion of phage SS DNA to duplex RF DNA in vitro used as an assay, a number of bacterial replication proteins that participate in initiation and synthesis of DNA chains have been identified (Table 1). After extensive purification of these proteins, three distinct systems for initiating DNA chains, required for replication of M13, G4, and φX DNA, have been reconstituted and studied. The same enzymatic mechanism for DNA chain elongation is employed by the three systems. DNA polymerase III holoenzyme extends a primer which is annealed to a template coated with binding protein. However, the three phages employ different systems for generating primers. Both RNA polymerase, in the case of M13, and primase, in the case of G4, appear to recognize and transcribe specific duplex (hairpin) regions in the viral DNA that have not been coated by binding protein. The φX replication system is unique in that a template-bound protein-recognition signal, *dnaB* protein, is required for primase to initiate primer synthesis. The movement of *dnaB* protein processively along the template strand provides fresh loci for primase action at almost any site. These findings on the *E. coli* proteins that are involved in the replication of coliphage DNA serve as a firm basis for understanding replication of the host chromosome.

ACKNOWLEDGMENTS

A portion of this investigation was completed at The Johns Hopkins University (by R. M.) and was supported by grant FR-05445 from the Biomedical Research Support Branch, Division of Research Facilities and Resources, National Institutes of Health. This work was supported in part by grants from the National Institutes of Health and the National Science Foundation.

REFERENCES

Bouché, J.-P., L. Rowen, and A. Kornberg. 1978. The RNA primer synthesized by primase to initiate phage G4 DNA replication. *J. Biol. Chem.* **253:** 765.

Bouché, J.-P., K. Zechel, and A. Kornberg. 1975. *DnaG* gene product, a rifampicin-resistant RNA polymerase, initiates the conversion of a single-stranded coliphage DNA to its duplex replicative form. *J. Biol. Chem.* **250:** 5995.

Brutlag, D., R. Schekman, and A. Kornberg. 1971. A possible role for RNA polymerase in the initiation of M13 DNA synthesis. *Proc. Natl. Acad. Sci.* **68:** 2826.

Fiddes, J. C., B. G. Barrell, and G. N. Godson. 1978. Nucleotide sequences of the separate origins of synthesis of bacteriophage G4 viral and complementary DNA strands. *Proc. Natl. Acad. Sci.* **75:** 1081.

Geider, K and A. Kornberg. 1974. Conversion of the M13 viral single strand to the double-stranded replicative forms by purified proteins. *J. Biol. Chem.* **249:** 3999.

Geider, K., E. Beck, and H. Schaller. 1978. An RNA transcribed from DNA at the origin of phage fd single strand to replicative form conversion. *Proc. Natl. Acad. Sci.* **75:** 645.

Gray, C. P., R. Sommer, C. Polke, E. Beck, and H. Schaller. 1978. Structure of the origin of DNA replication of bacteriophage fd. *Proc. Natl. Acad. Sci.* **75:** 50.

Hourcade, D. and D. Dressler. 1978. The site-specific initiation of a DNA fragment. *Proc. Natl. Acad. Sci.* **75:** 1652.

Kornberg, A. 1974. *DNA synthesis*. Freeman, San Francisco.

———. 1976. RNA priming of DNA replication. In *RNA polymerase* (ed. R. Losick and M. Chamberlin), p. 331. Cold Spring Harbor Laboratory, Cold Spring Harbor, New York.

———. 1977. Multiple stages in the enzymic replication of DNA. *Biochem. Soc. Trans.* **5:** 359.

Martin, D. M. and G. N. Godson. 1977. G4 DNA replication. I. Origin of synthesis of the viral and complementary DNA strands. *J. Mol. Biol.* **117:** 321.

McHenry, C. and A. Kornberg. 1977. DNA polymerase III holoenzyme of

Escherichia coli. Purification and resolution into subunits. *J. Biol. Chem.* **252:** 6478.

McMacken, R. and A. Kornberg. 1978. A multienzyme system for priming the replication of φX174 viral DNA. *J. Biol. Chem.* **253:** 3313.

McMacken, R., K. Ueda, and A. Kornberg. 1977a. Migration of *Escherichia coli dna*B protein on the template DNA strand as a mechanism in initiating DNA replication. *Proc. Natl. Acad. Sci.* **74:** 4190.

McMacken, R., J. -P. Bouché, S. L. Rowen, J. H. Weiner, K. Ueda, L. Thelander, C. McHenry, and A. Kornberg. 1977b. RNA priming of DNA replication. In *Nucleic acid-protein recognition* (ed. H. J. Vogel), p. 15. Academic Press, New York.

Ravetch, J. V., K. Horiuchi, and N. D. Zinder. 1977. Nucleotide sequences near the origin of replication of bacteriophage f1. *Proc. Natl. Acad. Sci.* **74:** 4219.

Rowen, L. and A. Kornberg. 1978a. Primase, the *dna*G protein: An enzyme which starts DNA chains. *J. Biol. Chem.* **253:** 758.

———. 1978b. A ribo-deoxyribonucleotide primer synthesized by primase. *J. Biol. Chem.* **253:** 770.

Sanger, F., G. M. Air, B. G. Barrell, N. L. Brown, A. R. Coulson, J. C. Fiddes, C. A. Hutchison III, P. M. Slocombe, and M. Smith. 1977. Nucleotide sequence of bacteriophage φX174 DNA. *Nature* **265:** 687.

Schekman, R., A. Weiner, and A. Kornberg. 1974. Multienzyme systems of DNA replication. *Science* **186:** 987.

Schekman, R., J. H. Weiner, A. Weiner, and A. Kornberg. 1975. Ten proteins required for conversion of φX174 single-stranded DNA to duplex form in vitro. *J. Biol. Chem.* **250:** 5859.

Schekman, R., W. Wickner, O. Westergaard, D. Brutlag, K. Geider, L. L. Bertsch, and A. Kornberg. 1972. Initiation of DNA synthesis: Synthesis of φX174 replicative form requires RNA synthesis resistant to rifampicin. *Proc. Natl. Acad. Sci.* **69:** 2691.

Sherman, L. A. and M. L. Gefter. 1976. Studies on the mechanism of enzymatic DNA elongation by *Escherichia coli* DNA polymerase II. *J. Mol. Biol.* **103:** 61.

Sims, J. and D. Dressler. 1978. The site-specific initiation of a DNA fragment: Nucleotide sequence of the bacteriophage G4 negative-strand initiation site. *Proc. Natl. Acad. Sci.* (in press).

Tabak, H. F., J. Griffith, K. Geider, H. Schaller, and A. Kornberg. 1974. Initiation of deoxyribonucleic acid synthesis. VII. A unique location of the gap in the M13 replicative duplex synthesized in vitro. *J. Biol. Chem.* **249:** 3049.

Ueda, K., R. McMacken, and A. Kornberg. 1978. *Dna*B protein of *Escherichia coli*: Purification and role in the replication of φX174 DNA. *J. Biol. Chem.* **253:** 261.

Vicuna, R., J.-E. Ikeda, and J. Hurwitz. 1977b. Selective inhibition of φX RFII compared with fd RFII DNA synthesis in vitro. II. Resolution of discrimination reaction into multiple steps. *J. Biol. Chem.* **252:** 2534.

Vicuna, R., J. Hurwitz, S. Wallace, and M. Girard. 1977a. Selective inhibition of in vitro DNA synthesis dependent on φX174 compared with fd DNA. I. Protein requirements for selective inhibition. *J. Biol. Chem.* **252:** 2524.

Weiner, J. H., R. McMacken, and A. Kornberg. 1976. Isolation of an intermediate which precedes *dna*G RNA polymerase participation in enzymatic replication of bacteriophage φX174 DNA. *Proc. Natl. Acad. Sci.* **73:** 752.

Westergaard, O., D. Brutlag, and A. Kornberg. 1973. Initiation of deoxyribonucleic acid synthesis. IV. Incorporation of the ribonucleic acid primer into the phage replicative form. *J. Biol. Chem.* **248:** 1361.

Wickner, S. 1976. Mechanism of DNA elongation catalyzed by *Escherichia coli* DNA polymerase III, *dna*Z protein, and DNA elongation factors I and III. *Proc. Natl. Acad. Sci.* **73:** 3511.

———. 1977. DNA or RNA priming of bacteriophage G4 DNA synthesis by *Escherichia coli dna*G protein. *Proc. Natl. Acad. Sci.* **74:** 2815.

Wickner, S. and J. Hurwitz. 1974. Conversion of φX174 viral DNA to double-stranded form by purified *Escherichia coli* proteins. *Proc. Natl. Acad. Sci.* **71:** 4120.

———. 1975a. In vitro synthesis of DNA. In *DNA synthesis and its regulation* (ed. M. Goulian et al.), p. 227. W. A. Benjamin, Menlo Park, California.

———. 1975b. Interaction of *Escherichia coli dna*B and *dna*C(D) gene products in vitro. *Proc. Natl. Acad. Sci.* **72:** 921.

Wickner, S., M. Wright, and J. Hurwitz. 1974. Association of DNA-dependent and independent ribonucleoside triphosphatase activities with *dna*B gene product of *Escherichia coli. Proc. Natl. Acad. Sci.* **71:** 783.

Wickner, W. and A. Kornberg. 1974. A novel form of RNA polymerase from *Escherichia coli. Proc. Natl. Acad. Sci.* **71:** 4425.

Wickner, W., D. Brutlag, R. Schekman, and A. Kornberg. 1972. RNA synthesis initiates in vitro conversion of M13 DNA to its replicative form. *Proc. Natl. Acad. Sci.* **69:** 965.

Zechel, K., J.-P. Bouché, and A. Kornberg. 1975. Replication of phage G4. A novel and simple system for the initiation of deoxyribonucleic acid synthesis. *J. Biol. Chem.* **250:** 4684.

An Enzyme System for Replicating the Duplex Replicative Form of φX174 DNA

Shlomo Eisenberg, John F. Scott, and Arthur Kornberg
Department of Biochemistry
Stanford University School of Medicine
Stanford, California 94305

Because the small coliphages M13, G4, and φX174 rely almost completely on host enzymes for development, their viral DNAs are model templates for investigating the role of these enzymes in the replication of the host chromosome.

The life cycle of these viruses includes three stages of DNA replication: (I) conversion of the viral single-stranded DNA circle to the duplex replicative form (SS → RF), (II) multiplication of the replicative form (RF → RF), and (III) synthesis of viral single strands using the complementary strand of the replicative form as template (RF → SS) (Denhardt 1975; Baas and Jansz; Ray; Dressler et al.; all this volume).

The enzymatic events in the SS → RF conversion, resolved and reconstituted, include three distinct mechanisms for initiating DNA chains (McMacken et al.; Wickner; both this volume). In this article we summarize the progress that has been made in resolving and reconstituting the second stage in the life cycle, the replication of RF DNA.

A Soluble Enzyme System for RF DNA Replication

A soluble enzyme fraction (fraction II, Table 1) prepared from uninfected *E. coli* cells was shown to support the synthesis of complementary-strand DNA on φX SS DNA templates. The result was the production of duplex RF DNA. Fraction II also supported DNA synthesis on RF DNA templates provided the original fraction II was supplemented with a fraction II prepared from φX-infected cells. The products of this reaction, RFI (superhelical, covalently closed, circular, duplex DNA) and RFII (circular, duplex DNA in which at least one strand is not continuous), contained newly synthesized viral (+) and complementary (−) strands; this was shown by the centrifugation analysis illustrated in Figure 1.

For the soluble extract to support replication of RFI, both the cistron-*A* protein (the product of φX gene *A*) and the *E. coli rep* protein were required (Eisenberg et al. 1976a). As shown in Tables 2 and 3, no replication of RF DNA took place if either cistron-*A* or *rep* protein was absent.

Table 1 A soluble enzyme system utilizing RFI as a template for replication

Template	Source of fraction II		DNA synthesis (pmoles)
	uninfected	infected	
None	+	−	<1
	−	+	<1
	+	+	1.5
SS DNA	+	−	100
	−	+	<1
RFI	+	−	14
	−	+	<1
	+	+	118

The reaction mixture in a 25-μl final volume contained 50 mM Tris-HCl (pH 7.5), 6% sucrose, 10 mM dithiothreitol (DDT), 0.1 mg/ml bovine serum albumin, 5 mM $MgCl_2$, 50 μM each dATP, dCTP, and dGTP, 18 μM [^{3}H]dTTP (sp. act. = 150–700 cpm/pmole), 800 μM ATP, 100 μM each CTP, UTP, and GTP, and 2 mM spermidine chloride; occasionally [α-^{32}P]dCTP was used in place of [^{3}H]dTTP. To this reaction mixture were added 40 μg of fraction II from uninfected *E. coli* H560, 1 μg of fraction II from φX *am*3-infected *E. coli* HF4720 cells, and 1 μg of φX RFI DNA. The reaction was carried out at 30°C for 20 min and stopped by the addition of 0.2 ml of 0.1 M sodium pyrophosphate and 1 ml of 10% trichloroacetic acid (TCA). The precipitate was collected on glass fiber filters (Whatman GF/C), washed three times with a 1 M HCl-0.2 M sodium pyrophosphate solution, dried, and counted in 5 ml of a toluene-based scintillation fluid in a Nuclear Chicago Liquid scintillation counter. DNA synthesis is expressed as total nucleotide incorporation.

Both proteins were shown to be dispensable in vivo for the synthesis of parental RF DNA but they were absolutely required for RF DNA multiplication (Tessman 1966; Denhardt et al. 1972). The dependence of RF DNA replication in vitro on φX cistron-*A* protein and *E. coli rep* protein provided assays for the purification of both enzymes (Eisenberg et al. 1976a).

Two Stages of RF DNA Multiplication: Synthesis of the Viral Strand, Stage II (+), and Synthesis of the Complementary Strand, Stage II (−)

In the presence of four proteins, namely, cistron-*A* protein, *rep* protein, DNA-binding protein (DBP), and DNA polymerase III holoenzyme, viral DNA circles are synthesized on φX RF DNA in quantities that far exceed the amount of input template (Eisenberg et al. 1976b; Scott et al. 1977). That only viral DNA is synthesized in this reaction was shown by competition annealing analyses and direct hybridization experiments (Eisenberg et al. 1976b). Sedimentation and electron microscopic analyses of the product of the reaction indicated that circular SS DNA was synthesized.

These viral DNA circles then served as templates for complementary-strand synthesis (stage II[−]), which required the same multiprotein system used for synthesizing the complementary strand in the original conversion of viral strands to RF DNA (SS → RF). This system includes *E. coli* DBP,

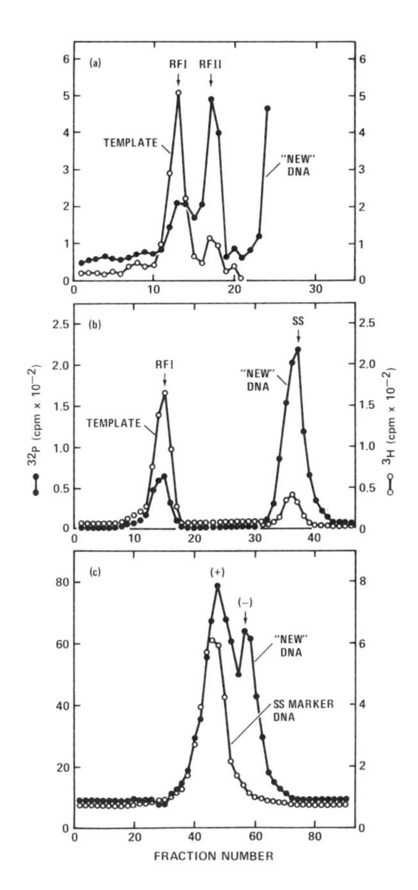

Figure 1 The products of RF DNA replication are RFI and RFII. The reaction mixture was as given in the notes to Table 1, with [^{3}H]thymidine-labeled φX RFI as template and [α-^{32}P]dCTP as the labeled deoxynucleoside triphosphate. (*a*) The product was purified by velocity sedimentation through a neutral sucrose gradient. (*b*) The RF peaks were pooled and concentrated with 2-propanol (Eisenberg et al. 1976a). The DNA was dissolved with 50 mM Tris-HCl (pH 7.5)-10 mM EDTA, denatured with alkali, and centrifuged on a linear, alkaline, 5–20% sucrose gradient (0.2 M NaOH, 0.8 M NaCl, 2 mM EDTA, 0.1% Sarkosyl). Centrifugation was performed in an SW50.1 rotor for 75 min at 50,000 rpm and 15°C in an L2-65B centrifuge. (*c*) RFII DNA was obtained in a reaction similar to that described above, the only exception being that unlabeled φX RFI template was used for banding on an alkaline CsCl gradient. The RFII DNA purified on neutral sucrose was concentrated, mixed with ^{3}H-labeled φX viral SS DNA marker, denatured with alkali, and analyzed on an equilibrium density gradient.

Table 2 Dependence on φX cistron-*A* protein for RF DNA replication

Soluble fraction	Amount (μg)	DNA synthesis (pmoles)
Uninfected fraction II only	40.0	14
+ fraction II infected with φX *am*3	0.5	51
+ fraction II infected with φX *am*3	1.0	114
+ fraction II (Cm-treated) infected with φX *am*3	1.0	13
+ fraction II infected with φX *am*50	1.0	13

Fraction II from uninfected *E. coli* H560 and fraction II from φX-infected HF4720 were each prepared as described elsewhere (Eisenberg et al. 1976a). φX *am*3 (cistron *E*, lysis-defective) and φX *am*50 (cistron *A*) were used for infections.

Chloramphenicol (Cm) was added to *E. coli* HF4720 (to a concentration of 200 μg/ml) 1 min before infection with φX *am*3 phage. Synthesis was measured as described in the notes to Table 1.

Table 3 Dependence on *E. coli* rep^+ for RF DNA replication

Fraction II		Complementing *rep*3^+ fraction	DNA synthesis (pmoles)
uninfected	infected		
rep	*rep*		0.3
rep^+	rep^+		24
rep	rep^+		11
rep	*rep*	fraction II, 0.4 μg	3.7
rep	*rep*	fraction II, 0.8 μg	7.0

Fraction II was prepared from uninfected and infected HF4704 *rep*3 and HF4704 rep^+ cells as described elsewhere (Eisenberg et al. 1976a). Complementing *rep*3^+ fraction II was prepared from HMS83. Synthesis was measured as described in the notes to Table 1, except that 10 μg DNA was used in each reaction.

proteins i and n, the protein products of genes *dnaB* and *dnaC*, primase (*dnaG* product), and holoenzyme (Eisenberg et al. 1976b).

Partial reconstitution of the replication of circular, duplex RF DNA with purified proteins has disclosed different requirements for the syntheses of the viral and complementary strands, which indicates that there is a different mechanism for the initiation and synthesis of the two strands. The asymmetry in the syntheses of the viral and complementary strands observed in vitro (Eisenberg et al. 1976b) is not inconsistent with in vivo observations (Dressler 1970; Dressler and Wolfson 1970; Schröder and Kaerner 1972; Eisenberg et al. 1975).

The discovery of the stage-II (+) reaction in which viral DNA circles were synthesized in quantities far exceeding the amount of input template was useful in a number of ways. (1) A simpler assay was available for the purification of cistron-*A* and *rep* proteins; with the stage-II(+) reaction as an assay, both proteins could be obtained more than 90% pure (Scott and Kornberg 1978; S. Eisenberg and A. Kornberg, unpubl.). (2) The several phases in DNA replication (initiation at the origin, elongation, and termination of a round of replication) could be studied readily. (3) The roles of the cistron-*A* and *rep* proteins in DNA replication could be examined.

Initiation of DNA Synthesis at the Origin of Replication

Cistron-*A* protein, which is required for RF DNA multiplication (stage II) and for stage-III SS DNA synthesis in vivo (Tessman 1966; Gilbert and Dressler 1969; Fujisawa and Hayashi 1976), acts specifically on the viral strand of RFI DNA (Francke and Ray 1971) to produce a nick. Purified preparations of cistron-*A* protein retained this nicking activity (Henry and Knippers 1974; Ikeda et al. 1976; Scott et al. 1977), which resulted in the formation of a cistron-*A* protein-RFII complex active in replication (Ikeda et al. 1976; Eisenberg et al. 1977). Figure 2A shows an electron micrograph of a purified complex. Of 100 DNA molecules observed, 69 had a single, globular, knoblike attachment, presumably cistron-*A* protein; 26 had no such attachment, 4 had 2, and 1 had 3. After treating a cistron-*A* protein-RFII complex with restriction endonuclease *Pst*I, which converts an RF circle to RFII (linear, duplex DNA of unit genome length) (Sanger et al. 1977), the knob on the cistron-*A* protein-RFIII complex was invariably found near one end of the linear DNA (Fig. 2B). The knob, measured from its center to each end of the molecule, was 19.82% (± 0.38) from the nearer end, or 1065 ± 20 nucleotides on the Sanger sequence map (Sanger et al. 1977; see also Appendix I, this volume) from residue 5375 at one end of the restriction cut. When a cistron-*A* protein-RFII complex was treated with *Hae*III restriction endonuclease, the resulting protein knob was associated

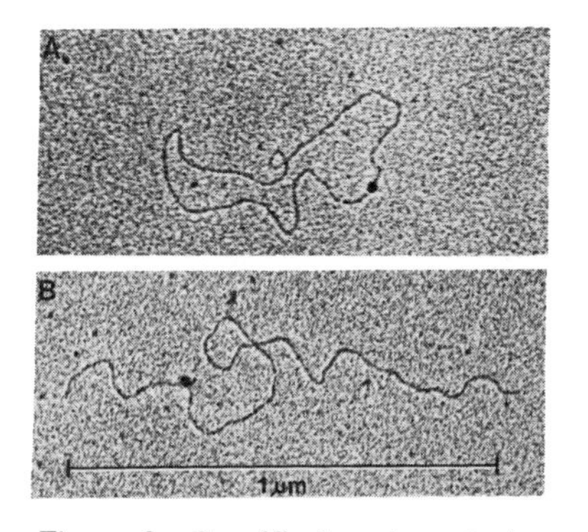

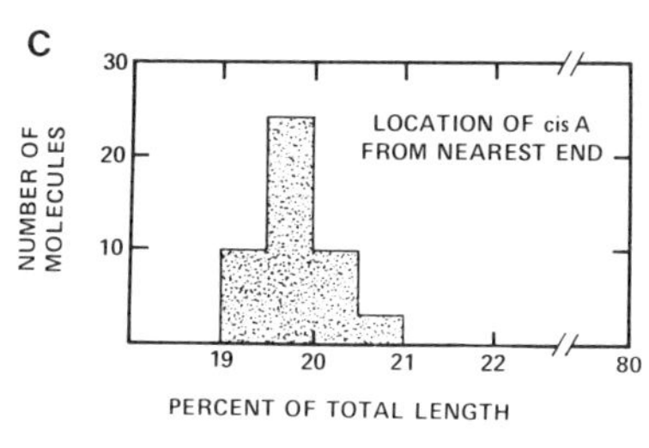

Figure 2 Specific location of cistron-*A* protein on φX RFII. The protein-DNA complex was prepared by incubating φX superhelical RFI with cistron-*A* protein. It was fixed with formaldehyde and glutaraldehyde and purified by velocity sedimentation through a neutral sucrose gradient. (*A*) Cistron-*A* protein-RFII complexes. Samples from pooled peak fractions containing RFII were mounted directly on grids for electron microscopic analysis (Eisenberg et al. 1977). (The distribution of protein-DNA complexes was similar when fixation by formaldehyde and glutaraldehyde was omitted.) Although the background of particulate material was variable, the frequency of unreacted DNA molecules with an attached knob was less than 1%. (*B*) A complex cleaved with *Pst*I nuclease. To locate the knob on the RF molecule, the cistron-*A* protein-RFII complex was diluted tenfold with 50 mM Tris-HCl (pH 8.1) and cleaved with *Pst*I restriction endonuclease in a mixture of 50 mM Tris-HCl (pH 8.1), 10 mM $MgCl_2$, 0.1 M NaCl, 0.1 μg DNA, and 6 units *Pst*I nuclease. The reaction was carried out at 30°C for 20 min and was stopped by addition of EDTA and NaCl to concentrations of 50 mM and 1 M, respectively. The DNA was mounted directly on EM grids for analysis. (*C*) The distance of the "knob" from the nearer of the two ends of the *Pst*I cut (*B*) was determined for 50 molecules (Sanger et al. 1977). The DNA length from the center of the knob to the nearer end of the molecule is expressed as a percentage of the total length of the molecule. The location of the knob, relative to the nearer end, is 19.82% (± 0.38), or 1065 ± 20 nucleotides based on the value of 5375 nucleotides for φX DNA (Sanger et al. 1977; see also Appendix I, this volume).

with DNA lengths of about 300 residues, which corresponds to the size of the Z6b fragments (S. Eisenberg and A. Kornberg, unpubl.).

The linear cistron-*A* protein-RFIII complex served as a substrate for replication upon addition of DBP, *rep* protein, holoenzyme, ATP, and deoxynucleoside triphosphates (as well as [α-^{32}P]dCTP). The strand containing the newly synthesized DNA sedimented at about 14S in alkaline sucrose (S. Eisenberg and A. Kornberg, unpubl.), a value corresponding to about 80% of the length of the φX genome. Thus, viral-strand synthesis was initiated by extension of the parental viral strand at the 3′-OH end that was made by the nicking action of the cistron-*A* protein.

Cleavage of the replicated RFIII by *Hae*III restriction endonucleases showed that the only region replicated was that from which the *Hae*III fragments 6a, 6b, 9, 10, and 3 and the *Hin*dII fragments 3, 5, and 8 were derived. This result, which is described in more detail in the legend to Figure 3, could have been obtained only if synthesis had been initiated in the region of the Z6b and R3 fragments and had then proceeded clockwise to the *Pst*I restriction site (Sanger et al. 1977). The amount of ^{32}P-label recovered in the unresolved Z6b and Z6a fragments was compared with that recovered in the Z3 fragment. On the basis of the number of deoxycytidylate (C) residues recovered and the total number of residues known to be in each of these fragments (Eisenberg et al. 1977), there is a deficiency of 27 ^{32}P-labeled C residues in the Z6b + Z6a fraction. From this it would appear that synthesis was initiated at 19–20% of a φX genome length from the *Pst*I restriction cut. A similar calculation, in which the R5 fragment was used as a reference, showed a deficiency of 26 ^{32}P-labeled C residues in the R3 fragment. The results of both calculations agree with the electron microscopic determination of the site where the knob, presumably cistron-*A* protein, is bound.

Thus, the site at which cistron-*A* protein nicks and binds is the origin of replication inferred from analyses in vivo (Baas and Jansz 1972; Godson 1974; Eisenberg et al. 1975; Baas et al. 1976; Koths and Dressler 1978; Baas and Jansz, this volume). Because the polarity of the complementary strand is 3′ → 5′ when the direction of replication is represented as clockwise (5′ → 3′), the strand synthesized must be viral DNA, and therefore cistron-*A* protein must nick the viral strand.

The Role of *rep* Protein in DNA Replication: Formation and Movement of a Replicating Fork

The replicating fork appears to move more slowly in *rep* mutants of *E. coli* than in the wild-type cell, and, as a consequence, more forks are formed (Lane and Denhardt 1975). Although the absence of *rep*-protein function does not appear to be lethal, its presence is required absolutely for RF DNA multiplication and SS DNA synthesis (Denhardt et al. 1972; Tessman and Peterson 1976).

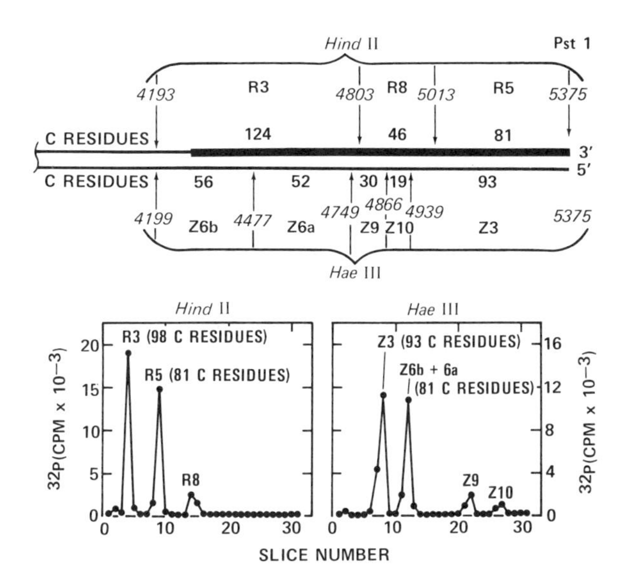

Figure 3 Location of the origin of replication by restriction endonuclease analysis. ^{32}P-labeled RFIII (prepared as described by Eisenberg et al. 1977) (150 pmoles) was treated with the *Hin*dII or the *Hae*III restriction enzymes in a 25-μl reaction mixture that contained 10 mM Tris-HCl (pH 8.1), 10 mM $MgCl_2$, 10 mM DTT, 25 mM NaCl, 40 μg/ml bovine serum albumin, 0.6 μg unlabeled φX RFI DNA marker, and 6 units of either nuclease. The reaction was carried out at 37°C for 3 hr, and the resulting fragments were electrophoresed on 10 × 10-cm agarose-acrylamide slab gels (Eisenberg et al. 1977). The products of each cleavage were identified by staining with ethidium bromide and autoradiography of the ^{32}P-labeled DNA. The radioactivity in the R3, R8, and R5 fragments of the *Hin*dII digest and in the Z6b + Z6a, Z9, Z10, and Z3 fragments of the *Hae*III digest was determined by slicing the gels into 3-mm slices and counting directly in a Nuclear Chicago gas flow counter. The numerical value for labeled cytidylate (C) residues in the R3 band was based on the known number of C residues (from the sequence map of Sanger et al. [1977; see also Appendix I, this volume]) and on the radioactivity recovered in the R5 fragment; similarly, Z3 served as a reference standard for Z6b + Z6a. The smaller fragments were not reliable standards because of variability in recovering them from the gel.

In the diagram, the *Hin*dII and *Hae*III cleavage sites near the right end of the *Pst*I cleavage maps of φX RFI are those of Sanger et al. (1977). The heavy line in the duplex DNA represents the stretch of newly synthesized ^{32}P-labeled viral DNA.

Three purified proteins—cistron-*A*, *rep*, and DBP—together catalyze the unwinding and the eventual complete separation of the two strands of superhelical RFI (Scott et al. 1977). Separation of the strands was demonstrated by sedimentation analysis in neutral sucrose gradients (Fig. 4). Only cistron-*A* protein and Mg^{++} were needed for nicking, but cistron-*A* protein, *rep* protein, DBP, Mg^{++}, and ATP were required for the separation of the strands.

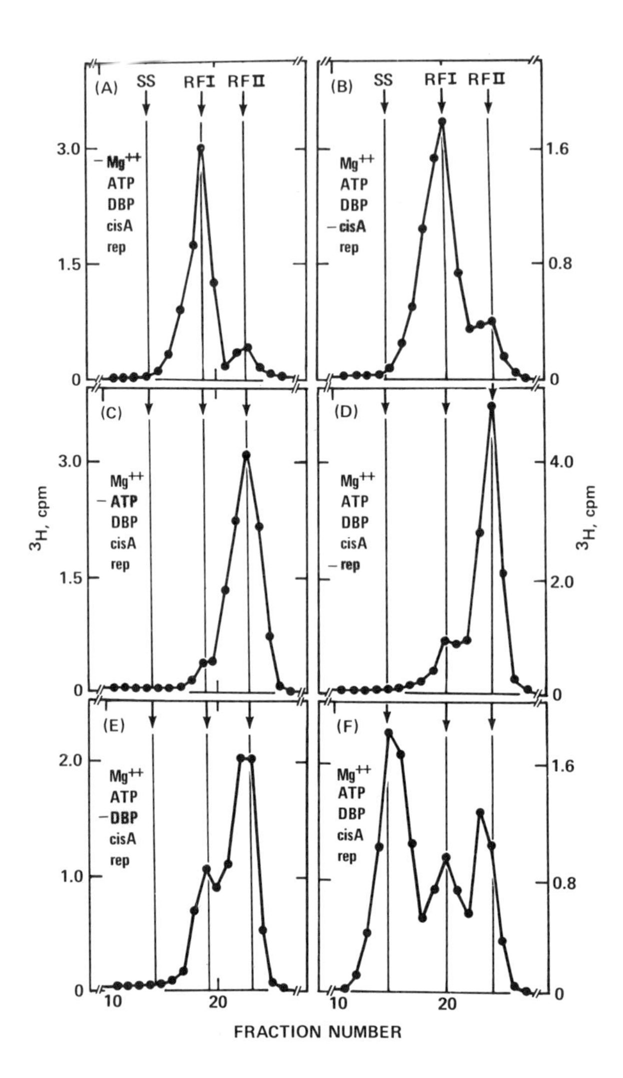

Figure 4 Requirements for the nicking and unwinding of superhelical RFI in the absence of replication. The complete reaction mixture was as described by Scott et al. (1977), except that holoenzyme and the four dNTPs were omitted in all cases. Reaction mixtures were incubated for 20 min, stopped by addition of EDTA to 0.1 M and SDS to 1%, and sedimented through 5–20% analytical neutral sucrose gradients as described previously (Eisenberg et al. 1976b). Sedimentations were performed in an SW 56 rotor at 50,000 rpm and 20°C for 2 hr in a Beckman centrifuge. The complete reaction mixture (*F*) contained cistron-*A* protein, *rep* protein, Mg^{++}, ATP, and DBP. The results shown in *A*–*E* were obtained with reaction mixtures from which Mg^{++} (*A*), cistron-*A* protein (*B*), ATP (*C*), *rep* protein (*D*), or DBP (*E*) had been omitted. (^3H cpm $\times$ 10^{-3})

Rep protein acts as an SS DNA-dependent ATPase (Kornberg et al. 1978). That hydrolysis of ATP to ADP by *rep* protein is coupled to SS DNA synthesis in a complete stage-II(+) reaction (Fig. 5) suggests that *rep* protein is indispensable in replication because its activity is required to supply the energy needed for helix unwinding.

Intermediate structures in the strand-separation reaction were observed when cistron-*A* protein-RFII complexes (isolated on neutral sucrose gradients) were incubated with DBP, *rep* protein, and ATP for 30 seconds at 30°C. Incubation was terminated when formaldehyde and glutaraldehyde were added to fix proteins complexed with DNA, and the DNA structures were then analyzed in the electron microscope (Fig. 6). All intermediates had a single-stranded loop (presumably the viral strand) attached to a duplex circle and a single-stranded stretch of complementary strand. Intermediates with free ends were not found. The ratio of the contour length of the displaced viral-strand loop to that of the single-stranded stretch of complementary strand was 1.0 ± 0.12 for 24 molecules measured. The products of the strand-separation reaction seen by electron microscopy were linear SS DNA (presumably viral strand) and circular SS DNA (presumably complementary strand), and these appeared in about equal numbers (186 and 213, respectively).

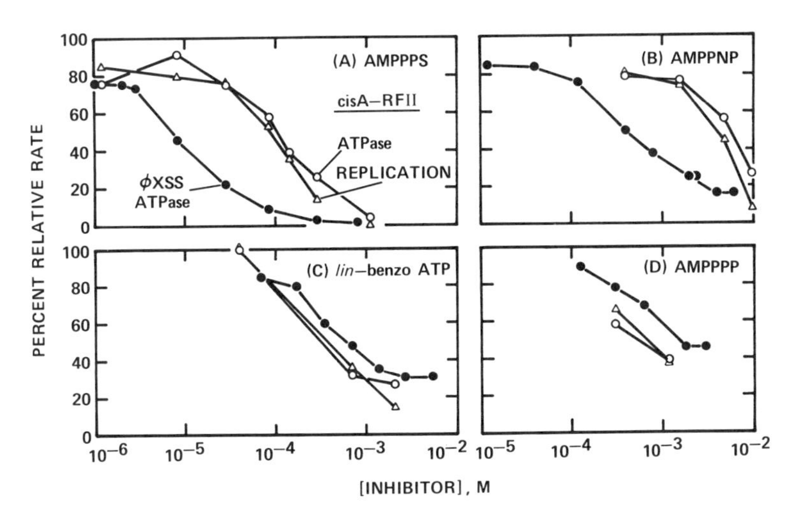

Figure 5 Inhibition of *rep* ATPase and replicative activities by analogs of ATP. Rates of *rep* activity with φX SS DNA (375 pmoles), expressed relative to those without inhibitor, were under ATPase assay conditions with 14 ng *rep* protein; rates with cistron-*A* protein-RFII complex (55 pmoles nucleotide) were under replication reaction conditions (Kornberg et al. 1978). Symbols in all panels are the same as those identified in *A*.

We infer from these results that *rep* protein interacts with cistron-*A* protein, presumably while the latter is bound to the 5′ end of the nick at the origin of replication, and that the energy from hydrolysis of ATP is used to displace the viral strand from the circle by movement of the 5′ end into the duplex fork.

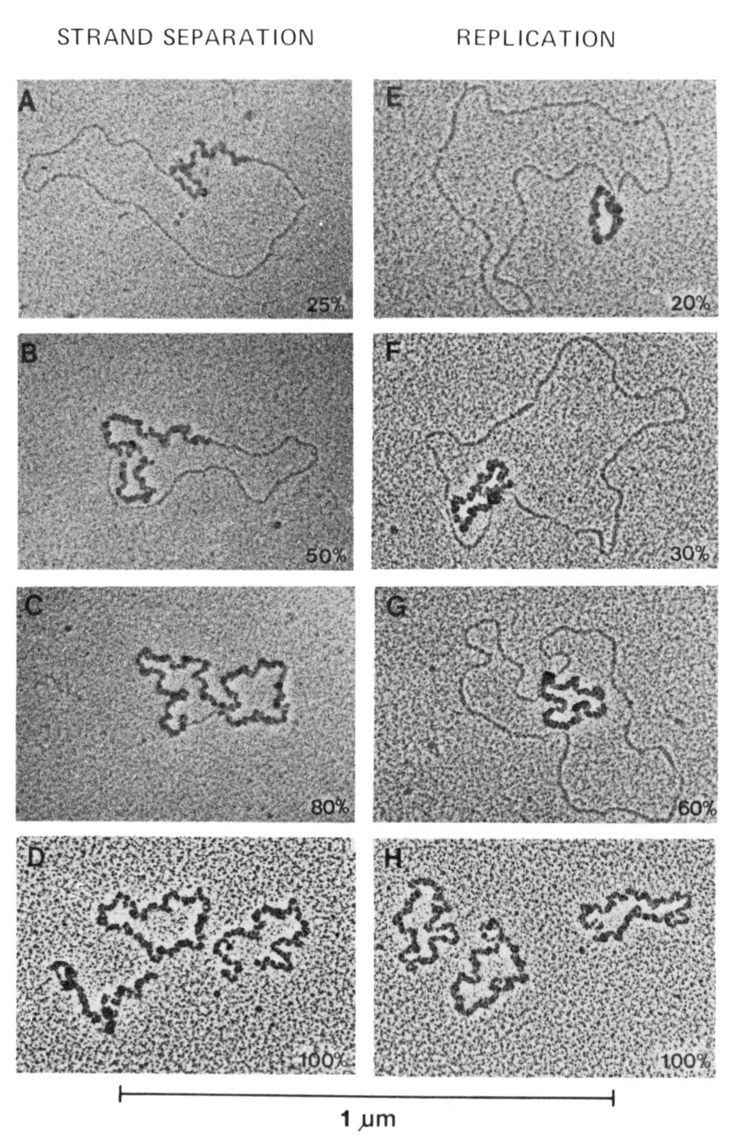

Figure 6 (See facing page for legend.)

The Role of Cistron-*A* Protein in RF DNA Replication

The cistron-*A* protein-RFII complex, with DBP, *rep* protein, and holoenzyme, supported a tenfold higher rate of replication than the standard reaction rate (S. Eisenberg and A. Kornberg, unpubl.). Repeated rounds of replication yielded an average of about 15 viral circles per cistron-*A* protein-RFII complex (Table 4). This synthesis was dependent upon DBP, *rep* protein, and holoenzyme (Table 4); no free cistron-*A* protein was added or believed to be present.

In another series of experiments (to be published elsewhere), φX circular SS DNA, coated with DBP, was cleaved by cistron-*A* protein. Upon subsequent removal of DBP and free cistron-*A* protein (by filtration through Bio-Gel A-1.5m) and incubation of the complex at 30°C in the presence of Mg^{++} but with no additional cistron-*A* protein, the proportion of circular molecules, measured by electron microscopy, increased from 13% to 45%. Thus, bound cistron-*A* protein acts catalytically in a continuous process of excision of unit lengths of DNA and ligation of the ends to form covalently closed viral circles.

The Replicating Intermediate Is a "Looped" Rolling Circle

Replicating intermediates were prepared by incubating a cistron-*A* protein-RFII complex with DBP, *rep* protein, holoenzyme, and [α-^{32}P]dCTP, dATP,

Figure 6 Intermediates in the RFII strand-separation and replication reactions. Strand separation (*A*, *B*, *C*, *D*): Cistron-*A* protein-RFII complexes were dialyzed against 50 mM Tris-HCl (pH 8.1), 50 mM NaCl, 4 mM DTT, and 0.1 mM EDTA for 60–90 min in collodion bags. Dialyzed complexes were mixed at 0°C with *rep* protein, DBP, and ATP, but holoenzyme and the dNTPs were omitted. After either 30 sec or 15 min at 30°C, the mixture was chilled in ice, treated with EDTA to 50 mM and fixed with formaldehyde and glutaraldehyde (Eisenberg et al. 1977). The protein-DNA complexes were purified through neutral sucrose gradients (in which they sedimented faster than φX RFII) and mounted directly for electron microscopic analysis (Eisenberg et al. 1977). *A, B,* and *C* are examples (30 sec) of progressive levels of strand separation of the RFII DNA. *D* shows the final product after 15 min. Replication (*E*, *F*, *G*, *H*): Replicating intermediates were obtained as described in the legend to Fig. 7. DNA from pooled fractions 12–15 was mounted directly on EM grids for analysis. The molecules in *E*, *F*, and *G* were from the pool of fractions 12–15. Intermediates with free tails were not observed. Panel *H* shows the final product of the replication reaction, the viral SS DNA circles, which were obtained as described in the notes to Table 4.

The thickness of the SS DNA is due to the coating of DBP; the contour length of the coated SS DNA is contracted by a factor of 2.56 (Ferguson and Davis 1978). The percentages shown in panels *A* through *F* describe the extent of strand separation and replication as determined by measuring the contour length of the displaced single-stranded loop (corrected for contraction by DBP).

Table 4 DNA synthesis starting with cistron-*A* protein-RFII complex as template

Additions and omissions	DNA synthesis (pmoles)
Complete	680
– DNA-binding protein	17
– *rep* protein	0
– holoenzyme	0

The cistron-*A* protein-RFII complex supports net synthesis of viral circular DNA. The complete reaction mixture in a 25-μl volume contained 50 mM Tris-HCl (pH 8.1), 10 mM $MgCl_2$, 4 mM DDT, 120 μg/ml bovine serum albumin, 800 μM ATP, 50 μM each dATP, dGTP, dCTP, and [α-^{32}P]-dTTP (124 cpm/pmole), 2000 units *rep* protein, 1.6 μg DBP, 20 units holoenzyme (G4 replication assay), and 96 pmoles cistron-*A* protein-RFII complex (containing 48 pmoles template nucleotide residues). The reaction was performed at 30°C for 15 min and was stopped by the addition of 0.2 ml of 0.1 M sodium pyrophosphate and 1 ml of 10% TCA. The precipitate was collected on glass fiber filters (Whatman GF/C), washed three times with a 1 M HCl-0.2 M sodium pyrophosphate solution, dried and counted in 5 ml of a toluene-based scintillation fluid in a liquid scintillation counter. DNA synthesis is expressed as total nucleotide incorporation.

dTTP, dGTP, and ATP for 30 seconds at 30°C. The complex was then fixed by the addition of formaldehyde and glutaraldehyde, and the labeled DNA was purified by sedimentation (Fig. 7). Most of the labeled replicating intermediates sedimented in a broad peak ahead of the RFII DNA. Fractions 12 to 15 (Fig. 7) were pooled and the complexes analyzed by electron microscopy (Fig. 6).

All replicating intermediates had a single-stranded, DBP-coated loop attached to a duplex circle (Fig. 6). The contour lengths of the single-stranded loops in Figure 6 show that 20%(E), 30%(F), and 60%(G) of the duplex circles had been replicated. (A factor of 2.56 was applied to correct

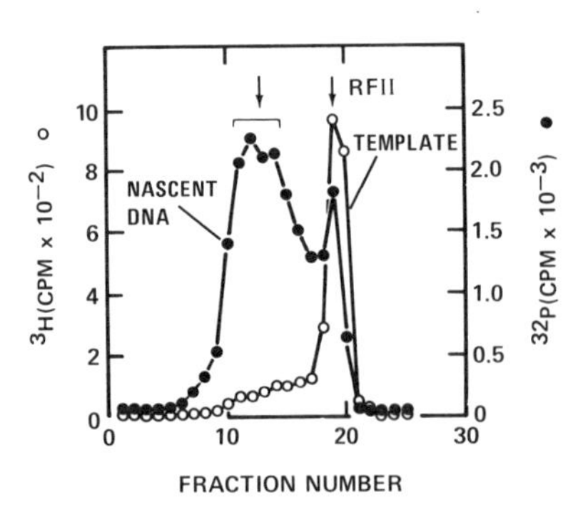

Figure 7 Purification of replicating intermediates by sedimentation through a neutral sucrose gradient. Replicating intermediates were synthesized in a 30-sec reaction as in the replication of the cistron-*A* protein-RFII complex. The reaction was terminated and fixed (see Fig. 6), layered on a 5–20% neutral sucrose gradient, and centrifuged at 50,000 rpm at 5°C for 2 hr. Sedimentation is from right to left.

for contraction of the SS DNA by DBP.) No intermediates with free ends were found. Rolling circles (Gilbert and Dressler 1969; Dressler 1970) with free single-stranded tails were found when the purified DNA was examined by electron microscopy on 40% formamide without prior cross-linking of protein to DNA (data not shown).

Thus, the 5′ end of the displaced strand, presumably with cistron-*A* protein attached (probably complexed with *rep* protein), is at the replicating fork and moves around the circular template.

DISCUSSION

The role of the cistron-*A* and *rep* proteins in DNA replication is described schematically in Figure 8. Cistron-*A* protein performs the following functions: (1) it nicks the viral strand at a specific sequence, the origin, to initiate a round of replication; (2) it participates, presumably by interacting with the

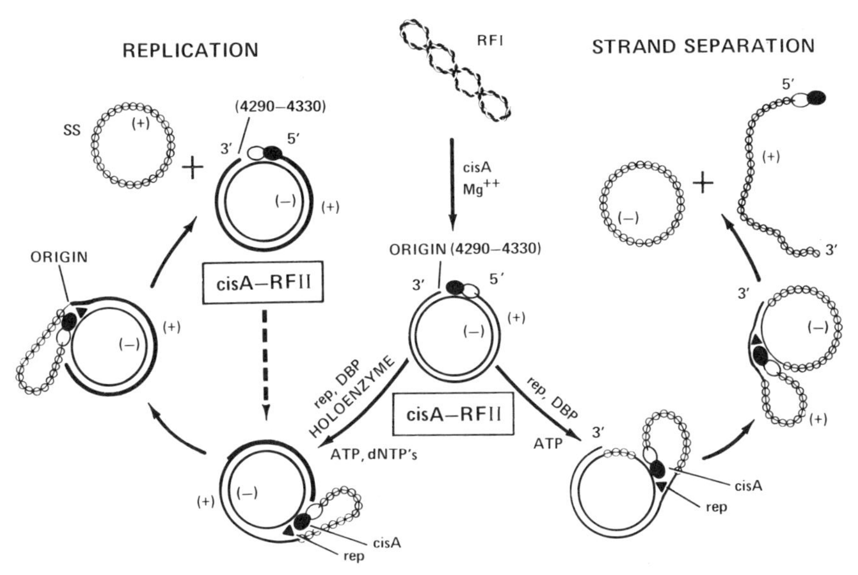

Figure 8 A scheme for cistron-*A* protein action illustrating its multiple functions. Cistron-*A* protein initiates a round of replication by nicking specifically the viral strand at the origin of replication. During strand separation, cistron-*A* protein, bound to the 5′ end, participates in the unwinding of the two strands. During replication, this protein, still bound to the 5′ end, travels around the complementary template, separating the two strands in advance of replication. After completing a full round, the free active site or subunit of cistron-*A* protein (●) nicks the regenerated origin, and the occupied active site (o) ligates the newly formed 3′ OH with the bound 5′ phosphoryl group. Thus, a single-stranded viral DNA circle is released, and the cistron-*A* protein-RFII complex, which can now enter another round of replication, is regenerated. We propose that in this catalytic action two active sites or subunits of cistron-*A* protein alternate in their covalent attachment to the 5′ end of the viral strand.

rep protein, in the unwinding of the two strands to create the replicating fork; (3) it excises a genome length of φX DNA after completion of a full round of replication, and (4) it ligates the excised DNA to form a covalently closed, circular molecule.

The unique nucleotide sequence that is exposed in superhelical RFI DNA[1] and nicked by cistron-*A* protein is located in the viral strand between nucleotides 4290 and 4330 on the φX DNA sequence map (Sanger et al. 1977; Appendix I, this volume). The nicking results in what we infer to be a covalent attachment of cistron-*A* protein to the 5′ end of the viral strand. This linkage, like that in the case of the untwisting enzymes (which carry out repeated nicking and ligating steps without an energy requirement [Wang 1971; Champoux 1976]), conserves the energy of the cleaved phosphodiester bond. To explain two successive nicking steps followed by ligation, we propose two active sites or subunits in the bound cistron-*A* protein, only one of which (open circle in Fig. 8) is bound covalently to the 5′ end in a given complex.

We suggest that *rep* protein recognizes the origin by its interaction with bound cistron-*A* protein. This is followed by unwinding of the two strands, which is sustained by ATP hydrolysis and DBP. Movement of the 5′ end into the duplex displaces the viral strand from its complementary circular template. The 3′-OH terminus at the origin serves as a primer and is extended by holoenzyme to regenerate the origin. When the cistron-*A* protein-*rep* protein complex traverses the full length of the template circle, a genome length of φX DNA is then excised by the unoccupied active site of the cistron-*A* protein (solid circle in Fig. 8) by displacing and nicking the regenerated origin in the viral strand. Concurrent with the nucleolytic scission carried out by the second cistron-*A* protein site (solid circle in Fig. 8), the first site (open circle) ligates the newly created 3′ OH with the bound 5′ end, using the conserved energy of the covalent protein-DNA bond. By this mechanism, which assumes that two active sites or subunits of the cistron-*A* protein are used alternately, the cistron-*A* protein-RFII complex is restored at the completion and release of each viral circle and the catalytic role of the complex is assured. The mechanism also provides for efficient use of cistron-*A* protein and its proper orientation for the continued synthesis of viral DNA on a single template.

Initiation of complementary-strand synthesis by host proteins (including *dnaB, dnaC*, and *dnaG* proteins, proteins i and n, holoenzyme, and other proteins) takes place either on the displaced viral loop of the replicating intermediate or on the released viral circle. Further studies in which the purified viral-strand-synthesizing system is coupled with that for the complementary strand are required to answer this question.

[1]*Relaxed*, covalently closed, circular, duplex φX DNA is not nicked by cistron-*A* protein (Henry and Knippers 1974; Ikeda et al. 1976; Marians et al. 1977).

A mechanism analogous to φX RF DNA replication may apply in the replication of other chromosomes. Initiation of synthesis of one strand by extension of a nick introduced by a cistron-*A*-like protein represents an origin of replication. (In chromosomes that replicate bidirectionally, nicking at the origin would involve both strands.) This "leading" strand, synthesized continuously, is analogous to the φX viral strand. The other, "lagging," strand, analogous to the φX complementary strand would be initiated repeatedly on the exposed parental viral-strand template.

Additional insights into the mechanism of DNA replication may emerge from further efforts to reconstitute the enzyme system for concerted, rapid, and efficient replication of RF DNA, as well as of the stages that precede and follow it.

ACKNOWLEDGMENTS

This work was supported in part by grants from the National Institutes of Health and the National Science Foundation. A grant from the Damon Runyon-W. A. Walter Winchell Cancer Fund provided postdoctoral fellowship support for S. E.

REFERENCES

Baas, P. D. and H. S. Jansz. 1972. φX174 replicative form DNA replication, origin and direction. *J. Mol. Biol.* **63:** 569.

Baas, P. D., H. S. Jansz, and R. L. Sinsheimer. 1976. φX174 DNA synthesis in a replication deficient host: Determination of the origin of φX DNA replication. *J. Mol. Biol.* **102:** 633.

Champoux, J. J. 1976. Evidence for an intermediate with a single-strand break in the reaction catalyzed by the DNA untwisting enzyme. *Proc. Natl. Acad. Sci.* **73:** 3488.

Denhardt, D. T. 1975. The single-stranded DNA phages. *CRC Crit. Rev. Microbiol.* **4:** 161.

Denhardt, D. T., M. Iwaya, and L. L. Larison. 1972. The *rep* mutation. II. Its effects on *E. coli* and on the replication of bacteriophage φX174. *Virology* **49:** 486.

Dressler, D. 1970. The rolling circle for φX DNA replication. III. Synthesis of single-stranded circles. *Proc. Natl. Acad. Sci.* **67:** 1934.

Dressler, D. and J. Wolfson. 1970. The rolling circle for φX DNA replication. III. Synthesis of supercoiled duplex rings. *Proc. Natl. Acad. Sci.* **67:** 456.

Eisenberg, S., J. F. Scott, and A. Kornberg. 1976. An enzyme system for replication of duplex circular DNA: The replicative form of phage φX174. *Proc. Natl. Acad. Sci.* **73:** 1594.

———. 1976b. Enzymatic replication of viral and complementary strands of duplex DNA of phage φX174 proceeds by separate mechanisms. *Proc. Natl. Acad. Sci.* **73:** 3151.

Eisenberg, S., J. Griffith, and A. Kornberg. 1977. φX174 cistron *A* protein is a multifunctional enzyme in DNA replication. *Proc. Natl. Acad. Sci.* **74:** 3198.

Eisenberg, S., B. Harbers, C. Hours, and D. T. Denhardt. 1975. The mechanism of replication of φX174 DNA. XII. Non-random location of gaps in nascent φX174 RFII DNA. *J. Mol. Biol.* **99:** 107.

Ferguson, J. and R. W. Davis. 1978. Quantitative electron microscopy of nucleic acids. In *Advanced techniques in biological electron microscopy II* (ed. J. K. Kochler), p. 123. Springer Verlag, New York.

Francke, B. and D. S. Ray. 1971. Formation of the parental replicative form DNA of bacteriophage φX174 and initial events in its replication. *J. Mol. Biol.* **61:** 565.

Fujisawa, H. and M. Hayashi. 1976. Gene A product of φX174 is required for site specific endonucleolytic cleavage during single-stranded DNA synthesis *in vivo. J. Virol.* **19:** 416.

Gilbert, W. and D. Dressler. 1969. DNA replication: The rolling circle model. *Cold Spring Harbor Symp. Quant. Biol.* **33:** 473.

Godson, G. N. 1974. Origin and direction of φX174 double- and single-stranded DNA synthesis. *J. Mol. Biol.* **90:** 127.

Henry, T. J. and R. Knippers. 1974. Isolation and function of the gene A initiator of bacteriophage φX174: A highly specific DNA endonuclease. *Proc. Natl. Acad. Sci.* **71:** 1549.

Ikeda, J., A. Yudelevich, and J. Hurwitz. 1976. Isolation and characterization of the protein coded by gene A of bacteriophage φX174 DNA. *Proc. Natl. Acad. Sci.* **731:** 2669.

Kornberg, A., J. F. Scott, and L. L. Bertsch. 1978. ATP utilization by *rep* protein in the catalytic separation of DNA strands at a replication fork. *J. Biol. Chem.* **253:** 3298.

Koths, K. and D. Dressler. 1978. Analysis of the φX DNA replication cycle by electron microscopy. *Proc. Natl. Acad. Sci.* **75:** 605.

Lane, H. E. D. and D. T. Denhardt. 1975. The *rep* mutation. IV. Slower movement of replication forks in *E. coli rep* strains. *J. Mol. Biol.* **97:** 99.

Marians, K. J., J. Ikeda, S. Schlagman, and J. Hurwitz. 1977. Role of DNA gyrase in φX replicative-form replication *in vitro. Proc. Natl. Acad. Sci.* **74:** 1965.

Sanger, F., G. M. Air, B. G. Barrell, N. L. Brown, A. R. Coulson, J. C. Fiddes, C. A. Hutchison III, P. M. Slocombe, and M. Smith. 1977. Nucleotide sequence of bacteriophage φX174 DNA. *Nature* **265:** 687.

Schröder, C. H. and H. C. Kaerner. 1972. Replication of bacteriophage φX174 replicative form DNA *in vivo. J. Mol. Biol.* **71:** 1351.

Scott, J. F. and A. Kornberg. 1978. Purification of the *rep* protein of *Escherichia coli. J. Biol. Chem.* **253:** 3292.

Scott, J. F., S. Eisenberg, L. L. Bertsch, and A. Kornberg. 1977. A mechanism of duplex DNA replication revealed by enzymatic studies of phage φX174: Catalytic strand separation in advance of replication. *Proc. Natl. Acad. Sci.* **174:** 193.

Tessman, E. S. 1966. Mutants of bacteriophage S13 blocked in infectious DNA synthesis. *J. Mol. Biol.* **17:** 218.

Tessman, E. S. and P. K. Peterson. 1976. Bacterial *rep*$^-$ mutations that block development of small DNA phages late in infection. *J. Virol.* **20:** 400.

Wang, J. C. 1971. Interaction between DNA and an *E. coli* protein. *J. Mol. Biol.* **55:** 523.

Studies on the In Vitro Synthesis of φX174 RFI DNA and Circular Single-stranded DNA

Chikako Sumida-Yasumoto, Joh-E Ikeda, Arturo Yudelevich,* Kenneth J. Marians, Samuel Schlagman, and Jerard Hurwitz
Department of Developmental Biology and Cancer
Division of Biological Sciences
Albert Einstein College of Medicine
Bronx, New York 10461

The replication of φX174 DNA in vivo has been studied extensively. Physically, the cycle can be divided into three steps: (1) the conversion of the entering viral circular single-stranded DNA to a circular duplex replicative form (SS → RF); (2) the duplication of the RF DNA to produce progeny RF molecules (RF → RF); and (3) the asymmetric synthesis of progeny viral SS DNA (RF → SS) (Sinsheimer 1968).

During the first stage, SS → RF, no new proteins other than those encoded in the host are required. This system has been reconstructed with purified proteins and has been shown to depend on a number of proteins known to be involved in *Escherichia coli* DNA replication (Wickner and Hurwitz 1975a; Schekman et al. 1975). These include the *E. coli dna* gene products *dnaB*, *dnaC*, *dnaG*, *dnaE* (DNA polymerase III), and *dnaZ* as well as a number of other proteins (*E. coli* DNA-binding protein, replication factors X, Y, and Z, and the DNA elongation factors I and III [EFI and EFIII]) that have not yet been defined genetically. It has been established that the proteins *dnaB*, *dnaC(D)*, DNA-binding protein, and replication factors X, Y, and Z generate a protein complex with φX SS DNA in the presence of ATP (Wickner and Hurwitz 1975b). The *dnaG* protein recognizes this complex and synthesizes an oligonucleotide which is used for priming of replication. The elongation of the primed DNA template requires the action of the DNA elongation system, which is comprised of *dnaZ*, DNA EFI, DNA EFIII, and DNA polymerase III (Wickner 1976).

The second step in the replication cycle, RF → RF, is dependent on only one viral gene product, the gene-*A* protein, both in vivo (Tessman 1966; Francke and Ray 1971) and in vitro (Henry and Knippers 1974; Ikeda et al. 1976; Eisenberg et al. 1976), to introduce a site-specific cleavage into φX RFI (superhelical, covalently closed, circular, duplex DNA). In vivo, in the

*Present address: Laboratoria de Bioquimica, Departamento de Ciencias Biologicas, Universidad Catolica de Chile, Casilla 114-D, Santiago, Chile.

absence of the φX gene-*A* protein, φX SS DNA is converted to RFI, which then is stable (Tessman 1966; Francke and Ray 1972). Most, and probably all, of the host proteins required for SS→ RF are required for RF→ RF also. In addition, Denhardt et al. (1967) identified a host gene (*rep*) that is essential for the replication, but not the formation, of the parental RF.

The final stage of φX DNA replication, the synthesis of progeny viral strands, occurs via a mechanism which is dependent on many of the proteins encoded in the φX genome, as well as a number of host proteins.

In this article we describe cell-free preparations capable of replicating φX RFI in both the RF→ RF and RF→ SS steps. A number of the proteins that play a role in these systems have been isolated and some of their properties will be presented.

Studies on RF → RF Replication

From in vivo studies (for a general review, see Denhardt 1977; Dumas, this volume) it was known that the synthesis of both progeny RF DNA and progeny φX DNA depends on the proteins encoded in the *E. coli dnaB dnaC(D)*, *dnaE*, *dnaG*, *dnaZ*, and *rep* gene loci, in addition to the φX gene-*A* protein. Cell-free preparations of uninfected *E. coli* cells that catalyzed the conversion of φX SS DNA to RFII (circular, duplex DNA in which at least one strand is not continuous) contained a number of these proteins. Extracts derived from φX-infected *E. coli* cells (fraction II) were used to supply additional proteins required for RF→ RF replication, among which was the φX gene-*A* protein. In the presence of these two fractions, externally added φX RFI readily supported DNA synthesis (Table 1). Similar observations have been reported by Eisenberg et al. (1976a). The reaction was dependent on the two protein fractions, ATP, Mg^{++}, and the four dNTPs (deoxynucleoside triphosphates of adenine, guanine, cytosine, and thymine). It was inhibited by *N*-ethylmaleimide, nalidixic acid, and novobiocin and was insensitive to rifampicin. Three other duplex DNA templates that were tested could not replace φX RFI in this system (Table 2).

Involvement of *E. coli* Proteins and the φX Gene-*A* Protein in RF → RF Replication

It has been established that φX RF DNA replication depends on a number of the *E. coli dna* gene products, the *rep* gene product, and the φX gene-*A* product. Utilizing extracts of *E. coli dna* temperature-sensitive (ts) mutants as a source of enzyme fractions, we have demonstrated the requirement for a number of these proteins in the in vitro RF→ RF system (Table 3). Mutants thermosensitive for *dnaC(D)* gene product were unable to support RF DNA replication at all because of the marked lability of the *dnaC(D)* protein. The addition of purified *dnaC(D)* protein readily restored RF DNA replication,

Table 1 Requirements for synthesis of φX progeny RF DNA in vitro

Additions	Incorporation of dTMP (pmoles/40 min)
Complete	374
− φX RFI	21
− ammonium sulfate fraction	6
− fraction II	6
− ATP; or − Mg^{++}; or − dATP, dGTP, dCTP	2–4
+ *N*- ethylmaleimide (4 mM)	3
+ rifampicin (10 μg/ml)	400
+ nalidixic acid (250 μg/ml)	41
+ novobiocin (240 μg/ml)	38

The conditions for the synthesis of DNA were as described by Sumida-Yasumoto et al. (1976). Standard reaction mixtures (0.05 ml) contained in order of addition: 20 mM Tris-HCl (pH 7.5), 10 mM $MgCl_2$, 4 mM dithiothreitol, 1 mM ATP, 1 mM NAD^+, 40 μM each dATP, dCTP, dGTP, and [α-^{32}P]dTTP (200−700 cpm/pmole), 100 μM each CTP, UTP, and GTP, and 1.8 nmoles φX RFI DNA. In this experiment the ammonium sulfate fraction (0.19 mg protein) was prepared from *E. coli* H560, and fraction II (23 μg protein) was prepared from *E. coli* H560 cells infected with φX *am* 3. Acid-insoluble radioactivity was measured.

whereas the addition of other purified *E. coli* gene products to the *dnaC* ts extract (*dnaB* or *dnaG*) did not. The requirements for *dnaB* and *dnaG* were observed only after heat treatment of their respective extracts, prepared from thermosensitive mutants, which inactivated the thermolabile protein. The addition of purified *dnaB* and *dnaG* proteins to their respective heat-inactivated extracts stimulated RF DNA replication approximately fivefold. For unknown reasons, the requirements for these components in the RF DNA replication system were substantially lower than for conversion of φX DNA to RFII (except for the *dnaC*[*D*] protein). The requirement for the *dnaE* gene product (DNA polymerase III) was marginal. Addition of

Table 2 Template specificity for RF → RF DNA replication

Template DNA (added nmoles)		Incorporation of dTMP (pmoles/40 min)
φX RFI	(1.8)	226
φX RFII	(0.9)	19
ColE1	(1.8)	8
PM2	(0.6)	18
Native T3	(0.8)	7
Omit DNA		

Reaction mixtures contained the amount of DNA indicated in parentheses (in nmoles of nucleotide residues), and the procedure was carried out as described by Sumida-Yasumoto et al. (1976).

Table 3 Involvement of *E. coli* and φX gene-*A* proteins in φX RF DNA synthesis

Source of *E. coli* extract			Incorporation of dTMP (pmoles/40 min)	
fraction II and the φX phage used for infection	ammonium sulfate fraction (uninfected)	Purified gene product added	before heat treatment	after heat treatment
dnaB ts (wt)	*dnaB* ts	none	106	41
dnaB ts (wt)	*dnaB* ts	*dnaB*	192	219
wt_1 (*am*3)	*dnaC* ts	none	28	—
wt_1 (*am*3)	*dnaC* ts	*dnaC*	568	—
dnaG ts (wt)	*dnaG* ts	none	461	57
dnaG ts (wt)	*dnaG* ts	*dnaG*	460	250
rep (wt)	*rep*	none	5	—
rep (wt)	rep^+	none	207	—
wt_2 (wt)	wt_2	none	121	—
wt_2 (H90)	wt_2	none	9	—
wt_2 (H90)	wt_2	φX gene *A*	120	—

The *E. coli* strains used corresponding to the various temperature-sensitive mutants described above were the following: *E. coli* BT1029 (*dnaB* ts), *E. coli* LD332 (*dnaC* ts), *E. coli* NY73 φX^s (*dnaG* ts), *E. coli* H560 (wild type [wt_1]), *E. coli* D92 (*rep*), *E. coli* 514 (wild type [wt_2]). The conditions for the isolation of the various extracts and additions were as described by Sumida-Yasumoto et al. (1976). H90 is a φX gene-*A* mutant.

purified DNA polymerase III to heat-treated extracts from two different *dnaE* ts mutants (*E. coli* BT1026 and BT1040) stimulated RF DNA replication at most only twofold. Similar difficulties were observed with such extracts in the conversion of φX DNA to RFII.

Protein fractions derived from *E. coli* strain D92 (*rep*) were inactive in RF DNA replication but were fully active in catalyzing φX DNA conversion to RFII. The combination of extracts derived from rep^+ strains and those from *rep* strains readily stimulated RF DNA replication. Protein fractions containing the φX gene-*A* protein were absolutely essential for any φX RFI-dependent dNMP incorporation. Infection of *E. coli* with φX strains that were A^- (φX H90 or φX N14) yielded extracts incapable of supporting RF DNA replication. The addition of purified φX gene-*A* protein to such reaction mixtures resulted in the synthesis of DNA.

Characterization of the φX Progeny RF DNA

The DNA synthesized in vitro was analyzed by zone sedimentation in neutral and alkaline sucrose. In neutral gradients, the product sedimented as expected for RFI and RFII, but the amount of each varied. Early in the reaction RFII predominated, whereas late in the reaction RFI predominated. Similar results were observed in alkaline sucrose gradient centrifugation. These observations suggest that the initial product of the reaction is φX RFII, which is then converted to RFI. The newly formed RFI possessed the same superhelical density as that of normal in vivo φX RFI, as measured by isopycnic banding in propidium diiodide-neutral CsCl.

The data in Figure 1 show that RFI replication occurred semiconservatively. For this experiment, DNA was synthesized in vitro with [α-^{32}P]dATP and 5-bromodeoxyuridine 5′-triphosphate (BrdUTP) in place of dTTP, and the product was subjected to neutral sucrose gradient and CsCl equilibrium gradient centrifugations (Fig. 1A). Approximately 20% and 80% of the BrdU-labeled DNA sedimented in neutral sucrose as RFI and RFII molecules, respectively. These products sedimented slightly faster than RFI and RFII containing only dTMP; in general, more RFII than RFI was observed in products of density-labeled DNA as compared with products formed with dTMP. When analyzed by neutral CsCl density centrifugation (Fig. 1B), over 65% of the newly synthesized [^{32}P]DNA and 20% of the

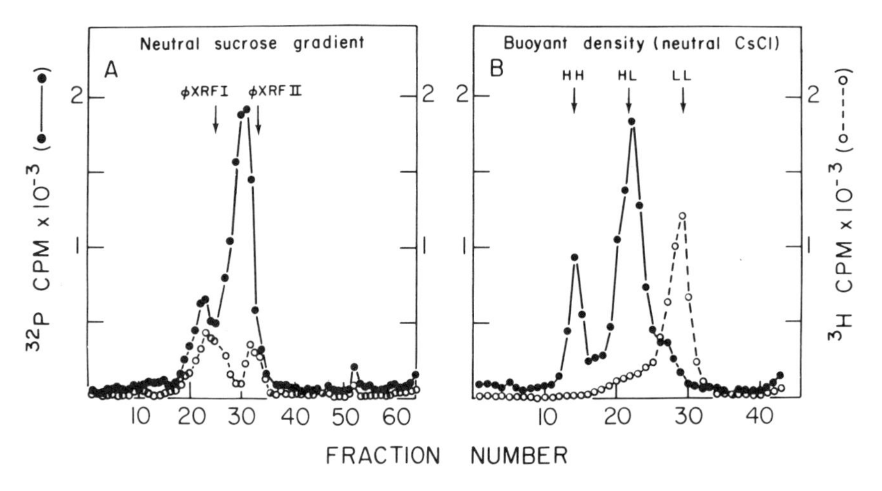

Figure 1 Neutral sucrose gradient and buoyant-density equilibrium centrifugation analyses of BrdU-labeled φX DNA synthesized in vitro. DNA was synthesized with [α-^{32}P]dATP (400 cpm/pmole), ^{3}H-labeled φX RFI (6 cpm/pmole), 40 μM BrdUTP (in place of dTTP), and fraction II (54 μg) prepared as described in the notes to Table 1. DNA was subjected (*A*) to velocity sedimentation in a neutral sucrose gradient and (*B*) to equilibrium buoyant-density centrifugation in neutral CsCl. (*A*) Arrows indicate the positions of reference ^{3}H-labeled φX RFI and RFII markers. (*B*) For neutral CsCl analysis, the volume was adjusted to 3 ml with 20 mM Tris-HCl (pH 7.5), 1 mM EDTA, and 10 mM NaCl to which 3.7 g of CsCl was added. The mixture was centrifuged for 64 hr at 35,000 rpm in an SW50.1 rotor at 20°C. Fractions were collected from the bottom of the tube, and acid-insoluble radioactivity was measured. Arrows indicate the positions expected for fully heavy (H·H), half-heavy (H·L), and completely light (L·L) φX RFI DNA.

template [^{3}H]DNA banded at the density of hybrid molecules (H·L). Approximately 20% of the ^{32}P-labeled product banded as expected for fully heavy (H·H) DNA molecules. In other experiments, in which products were subjected to isopycnic banding in propidium diiodide-neutral CsCl, both the RFI and RFII molecules were mixtures of H·H and H·L forms.

φX RFI-dependent Synthesis of φX Viral DNA In Vitro

The third step in the replication cycle of φX DNA in vivo has been shown to depend on a number of *E. coli* host proteins plus most of the phage-encoded proteins (Denhardt 1975; Dumas, this volume). Hayashi and coworkers (Fujisawa and Hayashi 1976; Mukai and Hayashi 1977) have shown that gently lysed *E. coli* cells infected with φX synthesize viral SS DNA specifically without externally added φX RFI DNA. They demonstrated that the incorporation of radioactive dTTP yields discrete DNA-protein complexes. No evidence for free circular SS DNA formation was obtained; all progeny viral-strand formation occurred via synthesis in protein-nucleic acid complexes (see Hayashi; Dressler et al.; both this volume).

In our studies, extracts freed of DNA by DEAE-cellulose treatment were prepared from *E. coli* cells infected with φX late in the infection cycle. The purpose of this was to accumulate large amounts of phage-encoded proteins. These fractions were supplemented with ammonium sulfate fractions from uninfected *E. coli* cells as well as with highly purified *dnaC* gene product. As in the case of RF→RF replication, the synthesis of SS DNA (more correctly the viral strand) depended on the presence of RFI, the two protein fractions, ATP, Mg^{++}, and the four dNTPs. It was inhibited by *N*-ethylmaleimide, nalidixic acid, and novobiocin but was unaffected by rifampicin (Table 4). The major difference between the fraction-II extracts used for RF→RF replication and those used for RF→SS synthesis was the time of infection; the former was isolated from cells infected with wild-type phage for 25 minutes, whereas the latter was isolated from cells infected with φX*am*3 for 50 minutes.

A comparison of the *E. coli* gene products required for RF DNA synthesis with those required for SS DNA synthesis is given in Table 5. To measure RF→RF replication we used ammonium sulfate fractions from uninfected *E. coli dna* mutants supplemented with purified *dnaC* (as indicated) and φX gene-*A* proteins. Where necessary, fractions were heated at nonpermissive temperatures and assayed in the presence or absence of purified gene products at 30°C. As shown, *dnaB*, *dnaC(D)*, *dnaG*, *dnaZ*, *rep*, and φX gene-*A* proteins were required for maximal RF DNA replication as well as for SS DNA synthesis. In reaction 13 in Table 5, fraction II derived from cells infected for 50 minutes with φX *am*H57 (gene-*F* mutant) added to uninfected ammonium sulfate receptors supported RF DNA replication but did not support SS DNA synthesis. When these extracts were mixed with

Table 4 Requirements for RF → SS DNA synthesis in vitro

Additions	Incorporation of dTMP (pmoles/40 min)
Complete	64
− φX RFI DNA	3
− receptor ammonium sulfate fraction	2
− fraction II	4
− ATP, or − Mg^{++}, or − dATP, dGTP, dCTP	2–4
+ *N*– ethylmaleimide (4 mM)	4
+ rifampicin (10 μg/ml)	67
+ nalidixic acid (240 μg/ml)	6
+ novobiocin (240 μg/ml)	16

Reaction mixtures (0.05 ml) contained in order of addition: 20 mM Tris-HCl (7.5), 10 mM $MgCl_2$, 4 mM dithiothreitol, 1 mM ATP, 0.2 mM NAD^+, 40 μM each dATP, dGTP, dCTP, and [α-^{32}P]dTTP (200–700 cpm/pmole), 0.1 mM each CTP, UTP, and GTP, 1 nmole ^{3}H-labeled RFI (1–6 cpm/pmole), 0.2 mg ammonium sulfate fraction from *E. coli* strain BT1029, 23 μg fraction II, and 0.1 unit (U) of *dnaC*. Reaction mixtures were incubated as indicated and acid-insoluble radioactivity measured. Reactions were stopped by addition of 10 mM (final) EDTA and 0.2% (final) SDS and heated at 65°C for 15 min followed by a 4-hr incubation at 37°C with autodigested proteinase K (1 mg/ml). The mixtures were dialyzed against 50 mM Tris-HCl (pH 8.0) and 5 mM EDTA for 12 hr. Viral-strand formation was measured after hybridization to poly(UG), followed by CsCl isopycnic centrifugation for 70 hr as described by Baas et al. (1976). The amount of viral-strand DNA that banded at the expected density is the value reported above. Under these conditions, $>$ 90% of the incorporated ^{32}P was recovered in this fraction. The reference markers, ^{14}C-labeled φX DNA and ^{3}H-labeled RFI and RFII, were run in parallel gradients.

extracts derived from cells infected for 50 minutes with φX *am*H90 (lacking the φX gene-*A* protein but containing the other φX-encoded proteins), RF → RF replication was not detected, whereas formation of SS DNA occurred. These results suggest that the temporal switch in RF DNA replication to SS DNA synthesis is due to viral proteins involved in viral DNA formation.

DNA Products Formed from RFI in the RF → RF and RF → SS Systems

As shown in Figure 2, the products formed in in vitro reactions primed with RFI depended on the φX-infected extract used. Fraction II prepared from cells infected for 25 minutes produced a mixture of φX RFII and RFI molecules, as determined by neutral sucrose gradient centrifugation. When the mixture of RFII and RFI molecules was denatured, annealed to poly-(UG), and subjected to analysis by isopycnic banding, three labeled products were detected (Fig. 2A); i.e., viral-strand DNA, complementary-strand DNA, and RFI. Viral and complementary strands coming from RFII were labeled equally. When RFI was converted to RFII by nicking with pancreatic DNase and analyzed as above, both strands were also found to be labeled

Table 5 Gene products required for RF DNA replication and SS DNA synthesis in vitro

Source of extract and φX phage used for fraction II	Incorporation of dTMP (pmoles/40 min)	
	RF DNA	SS DNA
1. BT1029 (*dnaB* ts) (*am*3)	43	21
2. + *dnaB* protein	210	68
3. LD332 (*dnaC* ts) (*am*3)	1	5
4. + *dnaC* protein	222	251
5. NY73 φX^s (*dnaG* ts) (*am*3)	55	35
6. + *dnaG* protein	218	109
7. AX729 (*dnaZ* ts)	4	11
8. + *dnaZ* protein	55	33
9. D92 (*rep*38) (*am*3)	3	4
10. + *rep* protein	248	102
11. BT1029 (*am*H90)	3	4
12. + φX gene-*A* protein	153	99
13. BT1029 (*am*H57)	117	0
14. + BT1029 (*am*H90)	0	118

Reaction mixtures (0.05 ml) were as described in the notes to Table 4 with the exception that the RF DNA system contained ammonium sulfate fractions from the *E. coli* strains indicated, 0.3 U of purified φX gene-*A* protein except in reaction 11, and 0.2 U of *dnaC* protein except in reaction 3. Fraction II was replaced by purified φX gene-*A* protein. In the case of the SS DNA system, we added the ammonium sulfate fractions used for the RF system supplemented with 0.2 U of *dnaC* protein. Fraction II (for SS DNA synthesis) was added as described in the notes to Table 4. This fraction was prepared from the necessary *E. coli* mutants infected with φX as indicated. In reaction 7, fraction II was the same as that used in reaction 1. In reactions 1, 2, 5, and 6, fraction II (for SS DNA synthesis) and uninfected ammonium sulfate fractions were heated as described previously (Sumida-Yasumoto et al. 1976). Fractions from LD332 and AX729 did not require heat treatment. Purified preparations of *dna* gene products (0.1–0.3 U) were added where indicated. In reactions 13 and 14, extracts were prepared from cells infected for 50 min with φX *am*H57 (a gene-*F* mutant).

equally. Thus, viral and complementary strands are labeled equally during RF DNA replication.

Products formed during a 20-minute incubation with fraction II isolated from *E. coli* cells infected for 50 minutes with φX *am*3 were shown by neutral sucrose gradient centrifugation to be a mixture consisting of 19% RFI molecules, 61% RFII molecules, and 20% SS DNA. The RFII molecules included a structure containing an extended, single-strand, viral tail (a rolling-circle intermediate [Dressler 1970]). Approximately 75%, 10%, and 15% of the total label incorporated was recovered in viral strands, in complementary strands, and in RFI molecules, respectively, after poly-(UG) annealing and CsCl isopycnic banding (Fig. 2B). The label present in RFI molecules was due to the selective synthesis of the viral strand.

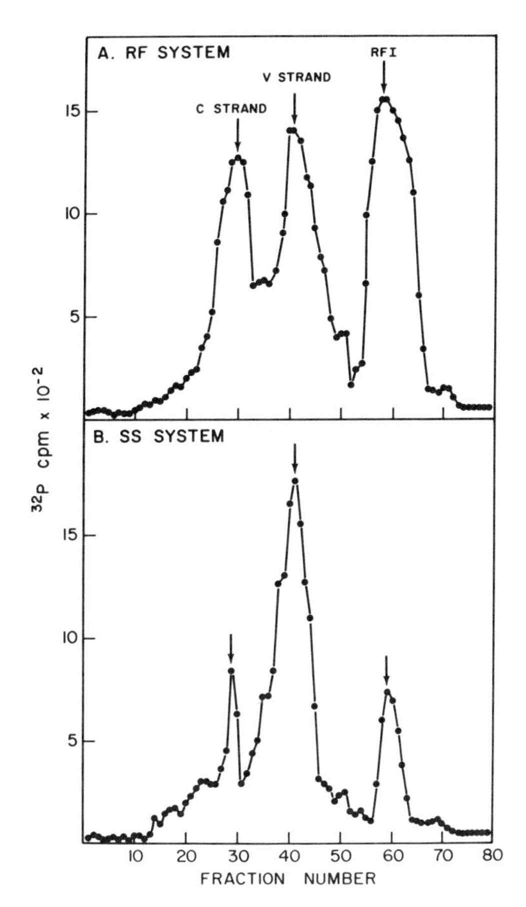

Figure 2 Poly(UG)/CsCl gradients of DNA products formed in (*A*) RF and (*B*) SS DNA synthesizing systems in vitro. Conditions were as described in the notes to Tables 1 and 4 except that reaction mixtures were incubated at 30°C for 20 min.

With the RF→ SS DNA system the following was observed: (a) the rate of synthesis was linear for 20 to 25 minutes and then plateaued abruptly; (b) the extent of synthesis varied—in some experiments net synthesis was achieved (Table 5); (c) prolonged incubation after viral DNA synthesis had stopped resulted in the formation of RFI and RFII; however, only the viral strands of both replicative forms were labeled; (d) no net formation of RFII or RFI occurred. These results suggest that fractions which catalyzed SS DNA synthesis did not catalyze RF DNA replication. Furthermore, RF DNA replication was blocked by the addition of fraction II, which catalyzed RF → SS DNA synthesis. Under these conditions, DNA synthesis was restricted to viral-strand formation.

Protein-DNA Complexes Formed in the RF → SS System

In vivo, φX viral DNA synthesis is coupled to φX maturation. Fujisawa and Hayashi (1976) and Dressler et al. (this volume) have studied maturation of

φX in vivo and demonstrated the formation of a protein-DNA complex having a sedimentation constant of 50. This 50S complex contained RFII (rolling-circle) molecules. We have detected similar complexes during in vitro RF → SS DNA formation. Under the conditions used, SS DNA synthesis stopped after 25 minutes, but the formation of different protein-nucleic acid complexes continued over a period of 140 minutes (Fig. 3). After incubation at 30°C for various times, reaction mixtures were subjected

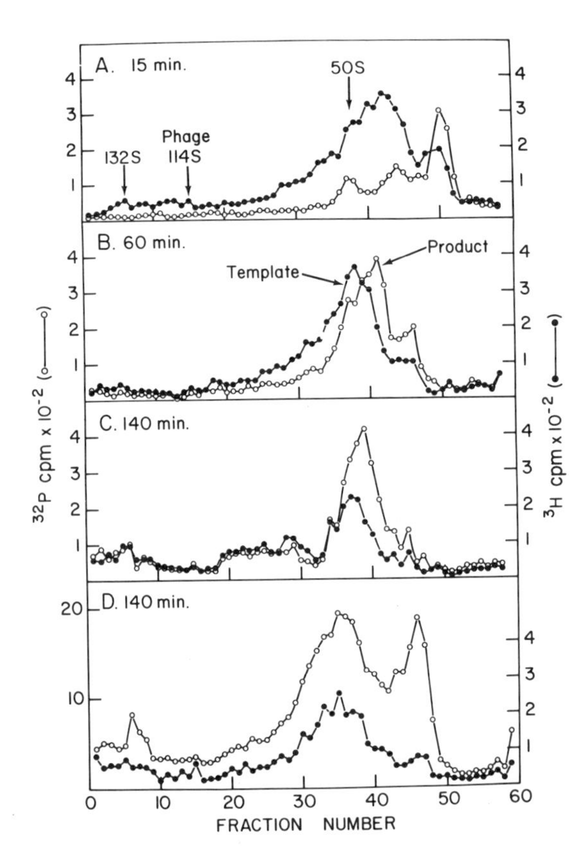

Figure 3 Neutral sucrose gradient analyses of DNA-protein complexes formed during in vitro SS DNA synthesis. Conditions were as described in the notes to Table 4, except that reactions were incubated for (*A*) 15 min, (*B*) 60 min, and (*C*, *D*) 140 min with 0.6 nmole of ^{3}H-labeled φX RFI. A different preparation of fraction II (25 μg protein) was used for SS DNA synthesis in the reaction shown in *D*. Reactions were stopped by addition of 20 mM (final) EDTA at 0°C, and the mixtures were centrifuged immediately through 10 ml of a 5–30% sucrose gradient in buffer containing 20 mM Tris-HCl (pH 7.5), 5 mM EDTA, and 0.1 M NaCl, with 0.5 ml of a 55% CsCl solution as a cushion at the bottom of the tube. Centrifugation was carried out in an SW 41 rotor (at 39,000 rpm) for 150 min at 4°C. ^{14}C-labeled φX *am*3 phage was run in a parallel gradient as a marker (114S). Fractions 1–4 represent the CsCl cushion.

to neutral (nondenaturing) sucrose gradients above a dense CsCl shelf. In these experiments both ^{3}H-labeled φX RFI template and newly labeled [^{32}P]DNA rapidly formed protein complexes that were distributed with time between a large 50S peak and material with a sedimentation constant of 132. In a separate experiment (Fig. 3D), a peak of ^{32}P-labeled product was detected in fraction 6 that corresponded to a complex of 132S; this fraction contained 3 × 10^8 plaque-forming units (PFU)/ml, whereas fraction 15 of the same gradient contained 10^8 PFU/ml. Infectious phage formation in vitro depended on the presence of φX RFI DNA, φX gene-*A* protein, φX gene-*F* protein, *rep* protein, and DNA synthesis. When fraction II (as used in Fig. 3D) was increased twofold, radioactivity from ^{3}H-labeled RFI DNA template was recovered in fractions 1–10; 10%, 20%, and 35% of the input ^{3}H-radioactivity was recovered in the 114–132S region after 20, 40, and 140 minutes of incubation, respectively. Kinetic studies suggest that 30–50S DNA complexes are formed first and are then converted to 60–70S complexes and eventually to 132S and 114S complexes. These large protein-DNA complexes (50S and greater) were not observed with the RF → RF replication system.

Structure of DNA in 50S and 60–70S Protein-DNA Complexes

After centrifugation (as in Fig. 3), material in the 50S region was pooled, deproteinized, and the DNA characterized by neutral sucrose gradient centrifugation (Fig. 4A). The DNA present in this complex was a mixture of RFI and RFII molecules; the latter sedimented faster than RFII and contained RFII structures with extended single-stranded tails (rolling circles or σ forms). The DNA in the 60–70S complex, analyzed in the same manner, contained a mixture of RFI, rolling circles (σ RFII), and circular and linear SS DNA (consisting of DNA of up to unit length). The nature of these DNA structures was also verified by electron microscopy (data not shown). After deproteinization, the DNA in the 50S complex was analyzed by alkaline sucrose gradient centrifugation (Fig. 4C); DNA products longer than unit length were detected. These results resemble those obtained in vivo (Gilbert and Dressler 1969; Knippers et al. 1969; Dressler 1970) and suggest that σ RFII (rolling-circle) molecules are intermediates in φX viral SS DNA formation (see also Dressler et al., this volume).

Overall Pathway of RF DNA Replication and SS DNA Synthesis In Vitro

At present, the complete details of the specific reactions involved in RF → RF replication and RF → SS DNA synthesis remain to be elucidated. However, a number of important observations regarding the reactions catalyzed by some of the proteins involved in DNA replication have been determined (these will be discussed further below). The overall pathway of

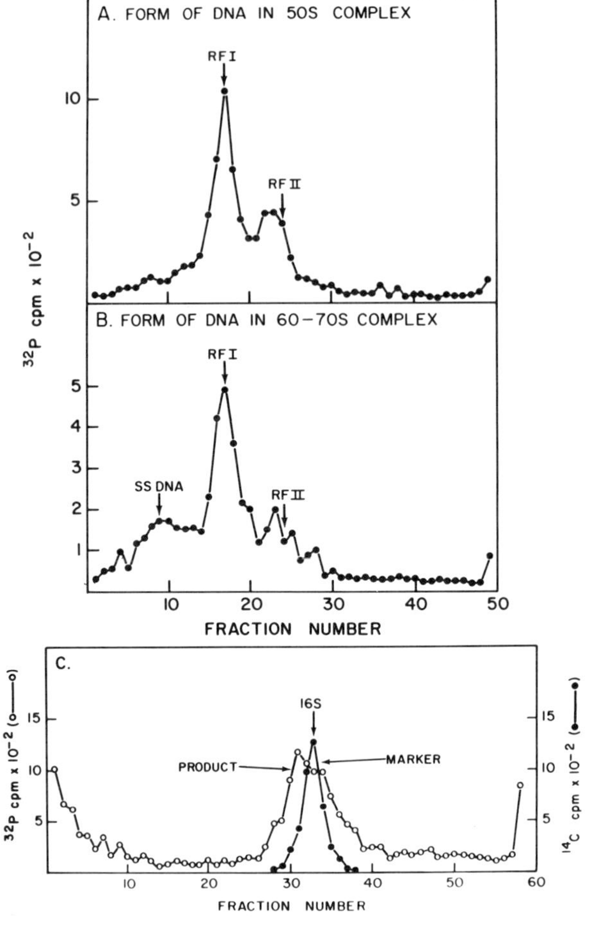

Figure 4 Sucrose gradient analysis of DNA from the 50S and 60–70S complexes. Conditions were as described in the notes to Table 4. Reactions were carried out for 40 min with the same fraction II (25 μg protein) used in Fig. 3D with [α-^{32}P]dTTP (2000 cpm/pmole) and 0.6 nmole of φX RFI; reactions were stopped by the addition of 20 mM (final) EDTA, and the mixtures were centifuged through a sucrose gradient as described in the legend to Fig. 3 (data not shown). Fractions 36–39 and 28–35, which corresponded to the same fractions indicated in Fig. 3, were pooled to obtain the 50S and 60–70S complexes, respectively. DNA was isolated from these complexes as described in the notes to Table 4. DNA samples were layered onto a 5-ml linear 5–20% neutral or alkaline sucrose gradient and sedimented as described previously (Sumida-Yasumoto et al. 1976), except that the alkaline sucrose gradient lacked NaCl and was run for 15 hr at 25,000 rpm at 4°C with ^{14}C-labeled φX DNA (16S) as a marker. Mixtures of ^{14}C-labeled DNA and ^{3}H-labeled φX RFI and RFII were run in parallel neutral sucrose gradients as markers. Neutral sucrose gradients of DNA isolated from 50S and 60–70S complexes are shown in *A* and *B*, respectively. An alkaline sucrose gradient of DNA from a 50S complex is presented in *C*. Sedimentation is from right to left.

the φX replication cycle is clearer, however, and Figure 5 summarizes the two replication modes considered here.

The proteins required for the reactions include (in part) the *E. coli* proteins *dnaB*, *dnaC(D)*, *dnaG*, *dnaZ*, *rep*, *nalA*, novobiocin-sensitive protein, *E. coli* DNA-binding protein, replication factors X, Y, and Z, DNA EFI and EFIII, and *dnaE* (DNA polymerase III). The requirements for the last six proteins are inferred from previous studies but have not been demonstrated. φX RFI (Fig. 5a) is attacked by the φX gene-*A* protein. This protein acts within cistron *A* of the viral strand (Baas and Jansz, this volume) and yields an RFII structure (Fig. 5a) containing the φX gene-*A* protein linked to the 5′ end. This intermediate can be replicated by fractions devoid of DNA and φX gene-*A*-protein activity. In the presence of the proteins listed above (as well as the φX gene-*A* protein—the only essential phage function—and other unknown host proteins), RFII is replicated via formation of a σ RFII (rolling-circle) structure containing double-stranded tails (such structures have been observed in the electron microscope). In these structures, viral and complementary strands are synthesized de novo as

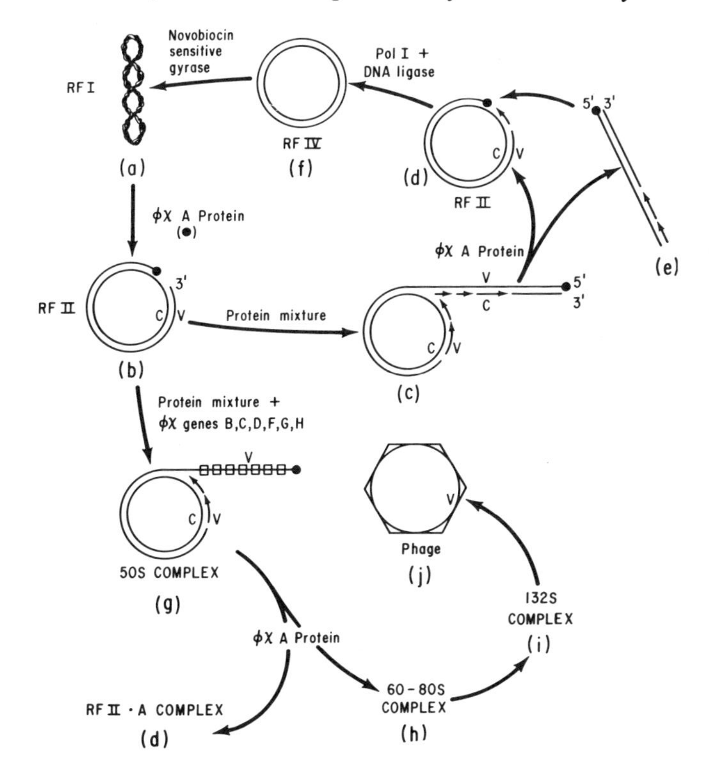

Figure 5 Postulated overall pathway of RF DNA replication and SS DNA synthesis in vitro (see text for details). The polarity of the viral strand is opposite to that chosen conventionally; v and c refer to the viral and complementary strands, respectively.

small pieces (Fig. 5c), which can then be chased into RFII and RFI (unpubl.). Synthesis continues until cistron *A* of the viral strand appears as a single-stranded region; this site can then be cleaved by the φX gene-*A* protein, probably generating the structure shown in Figure 5d and possibly that shown in Figure 5e. The latter structure is an RFIII-φX gene-*A* protein intermediate which could, upon circularization, yield the structure shown in Figure 5d. Studies by Eisenberg et al. (1977) suggest that the φX gene-*A* protein can ligate ends in an intramolecular reaction. RFIII molecules have been observed in the electron microscope (unpubl.), but their role in RF DNA replication is unknown. Conceivably, the φX gene-*A* protein itself could catalyze circularization of the RFIII structure and then be removed from the complex to yield RFII. RFII (Fig. 5d), after removal of the φX gene-*A* protein, can be converted in vitro by DNA ligase and DNA polymerase I to RFIV (relaxed RFI DNA; Fig. 5f), which can be converted to RFI by DNA gyrase (Gellert et al. 1976a; Sugino et al. 1977; Marians et al. 1977). The conversion of gapped φX RFII to RFIV is the rate-limiting step in RF → RF replication both in vivo and in vitro.

The replication of RFI is blocked by φX-encoded proteins essential for phage maturation. This is detected by a cessation of complementary-strand formation. In this way, RFI, upon conversion to a RFII-φX gene-*A* protein complex, can support only SS DNA synthesis. This pathway yields, in a complex with phage proteins, σ RFII (rolling-circle) molecules which contain single-stranded tails and in which only the viral strands are labeled (Fig. 5g). Viral-strand DNA synthesis occurs by formation of short, discontinuous fragments that can be rapidly chased into large fragments (unpubl.), as in the case of RF DNA replication. The 50S complex is attacked by the φX gene-*A* protein; this leads to the re-formation of a RFII-φX gene-*A* protein complex (Fig. 5d) and a new 60–80S complex (Fig. 5h). These complexes vary in size because of variations in the amounts and compositions of the viral proteins associated with the DNA. Eventually, these intermediates are converted to a 132S structure and then to phage (114S) (Fig. 5 i and j, respectively).

Characterization of the Proteins Involved in φX DNA Replication

Detailed studies from a number of laboratories continue to yield important clues as to the mechanism of DNA replication. The work carried out on the replication in vitro of different phage circular SS DNAs has revealed several mechanisms of initiation. There appear to be two mechanisms by which primer strands are formed on circular SS DNA. One involves DNA-dependent RNA polymerase (in fd and M13 DNA) (Brutlag et al. 1971; Schekman et al. 1974; Wickner et al. 1972), and the other involves the *dnaG* gene product (in φX, G4, and St-1) (Bouché et al. 1975; Wickner 1977). Both enzymes catalyze the formation of a small oligonucleotide primer under highly specific conditions. In the case of φX DNA, before the *dnaG* protein can catalyze the priming reaction, *dnaB*, *dnaC(D)*, *E. coli* DNA-binding protein, and replication factors X, Y, and Z must act on φX DNA in the presence of ATP to form a suitable complex which is then recognized by *dnaG*. In the case of G4 or St-1 DNA, *E. coli* DNA-binding protein is the only protein required for the action of the *dnaG* protein (Bouché et al. 1975; Wickner and Hurwitz 1975b). Once the primer is formed, however, a general mechanism of elongation of the primed DNA template occurs. This requires the action of the four proteins *dnaZ*, DNA EFI and EFIII, and DNA polymerase III (Wickner 1976).

The requirements for the replication of φX RFI include the proteins essential for SS → RF formation as well as additional proteins. The additional proteins that have been isolated and partially characterized so far include the φX gene-*A* protein, the *rep* protein, the *nal* protein, and the protein involved in part of the DNA gyrase activity. It is likely that these additional proteins are necessary for the utilization of duplex DNA structures, and for this reason their characterization is of particular importance if we are to understand more complicated DNA-replication systems. Attempts to reconstruct the RF → RF system with the 11 proteins characterized already (in the SS → RF system) plus the additional proteins involved in the RF → RF system have thus far been unsuccessful, which suggests that additional factors are needed.

Some of the properties of the proteins involved specifically in RF → RF replication have been studied in detail and the findings are presented below.

*φX Gene-*A *Protein*

As described above, the protein encoded in cistron *A* is essential for the initial attack on the φX RF DNA. It was purified initially by Henry and Knippers (1974) from extracts of *E. coli* cells infected with φX *am*3. The homogeneous preparation catalyzed a specific nicking of φX RFI to form an RFII in which the discontinuity resided in the viral strand. In our laboratory, we have purified the φX gene-*A* protein by means of the complementation assay in which φX RFI-dependent DNA synthesis requires the φX gene-*A* protein (Ikeda et al. 1976). This assay allowed the isolation of a homogeneous protein preparation of 59,000 daltons. The action of the purified enzyme was in keeping with the known physiological role of this protein. In accord with the results of Francke and Ray (1972) and Henry and Knippers (1974), the enzyme acted as an endonuclease which introduced a discontinuity in the viral strand in cistron *A* of RFI DNA. The requirements for action of the φX gene-*A* protein are precise. In the presence of Mg^{++} only superhelical φX RFI was attacked by the protein, whereas relaxed φX RFI (RFIV) was not (Fig. 6). No other superhelical DNA replaced φX RFI.

The product of the action of the φX gene-*A* protein was shown to be a

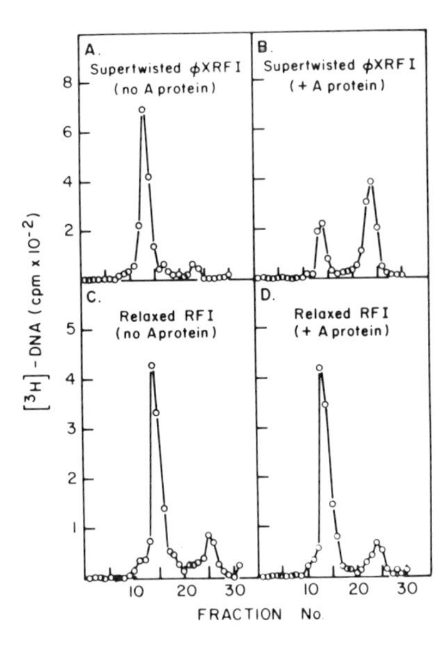

Figure 6 Endonuclease activity of φX gene-*A* protein on relaxed and supertwisted φX RFI DNA. Relaxed φX RFI (RFIV) was prepared by treating φX RFI with pancreatic DNase I, followed by phage T4 DNA ligase treatment. This RFI was isolated by centrifugation in propidium diiodide-CsCl gradients. The DNA had the density of completely relaxed DNA. Endonuclease assays were carried out with 430 pmoles of supertwisted φX RFI and relaxed φX RFI, which were incubated with 0.15 μg φX gene-*A* protein (step VI) for 30 min, and each reaction mixture was sedimented through a 5–20% alkaline sucrose gradient. (For additional details, see Ikeda et al. 1976.)

DNA-protein complex, φX RFII- φX gene-*A* protein. This complex was stable and could be isolated by sucrose gradient centrifugation (Fig. 7). The complex could be used in lieu of φX RFI DNA and the φX gene-*A* protein for RF → RF replication (Fig. 8). The nature of the complex has been examined. All of the information accumulated so far indicates that the φX gene-*A* protein is covalently linked to the 5′ end of the discontinuity. The isolated RFII- φX gene-*A* protein complex was attacked by exonuclease III but not by the 5′ → 3′ exonuclease component of DNA polymerase I either

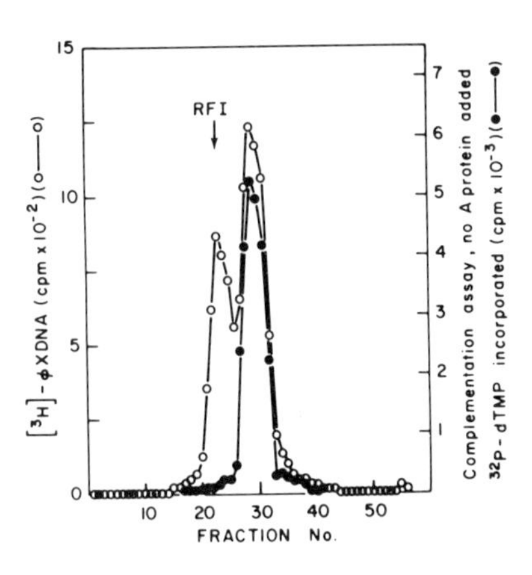

Figure 7 Isolation of the complex formed between RFI and φX gene-*A* protein (RF·A complex). ^{3}H-labeled φX RFI DNA (105 nmoles) was incubated with 7 μg φX gene-*A* protein (step VI) under standard conditions for endonuclease assay as described by Ikeda et al. (1976). After 20 min at 37°C, the reaction mixture was made 0.5 M in NaCl, layered onto a 10 ml 5–20% sucrose gradient in buffer A containing 0.5 M NaCl, and centrifuged in the SW 41 rotor for 14 hr at 30,000 rpm at 2°C. Gradients were fractionated, and the ^{3}H contents of the fractions were determined in 10-μl aliquots (o). Each fraction (10 μl) was also assayed for [^{32}P]dTMP incorporation in the complementation assay in the absence of added φX gene-*A* protein and template DNA (●).

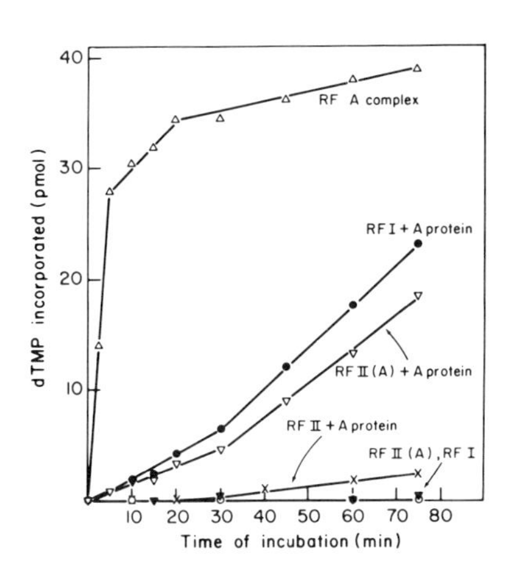

Figure 8 Rate of replication with RF-φX gene-*A* protein complex (RF·A) and other forms of φX DNA. Complementation assays were carried out in which φX RFI DNA was replaced with 340 pmoles of RF·A complex, and no φX gene-*A* protein was added (△); standard complementation assays contained 345 pmoles φX RFI and 0.4 μg φX gene-*A* protein (●). RF·A complex was treated with proteinase K (0.2 mg/ml) at 37°C for 20 min, extracted with phenol and ether, and the RFII(A) was isolated by propidium diiodide-CsCl centrifugation and then used as substrate (340 pmoles) in the complementation assay to which 0.04 μg φX gene-*A* protein was added (▽). RFII was isolated from φX *am*3-infected cells and was used as a substrate (350 pmoles) with 0.05 μg φX gene-*A* protein (x); 340 pmoles RFII(A) (▼) and 340 pmoles φX RFI (o) were used as templates with no φX gene-*A* protein added.

in the presence or absence of the four dNTPs; these results suggest the presence of a free 3′-OH end and a blocked 5′ terminus.

φX cistron *A* also codes for the φX A* protein (Linney and Hayashi 1973). This protein (33,500 daltons) has been purified to homogeneity and its properties have been studied (unpubl.). The φX A* protein does not substitute for the φX gene-*A* protein in any of the reactions catalyzed by the latter enzyme, nor does this protein affect the rate of φX RFII formation by the φX gene-*A* protein. Purified φX A* protein contains no detectable nuclease activity on duplex or single-stranded DNAs. The φX A* protein binds to duplex DNAs and inhibits the action of nucleases on these DNAs. Interestingly, in the presence of an excess of the φX A* protein, colicin E1 (Col E1), fd RFI, and PM2 DNAs were completely resistant to attack by a preparation of an *E. coli* nuclease which was a mixture of exonucleases (but free of endonuclease I). In the absence of the φX A* protein the DNAs are degraded to low-molecular-weight DNA fragments. With φX RFI DNA an excess of the φX A* protein also inhibits degradation of the DNA to low-molecular-weight fragments, but φX RFII and RFIII are formed. The site of breakage of the DNA is unknown.

DNA Gyrase Activity

The replication of duplex DNA, in contrast to the conversion of SS → RFII, was found to be inhibited by nalidixic acid and novobiocin, two known inhibitors of DNA replication. In our early studies on the replication of φX RFI DNA with limiting amounts of fraction II, we observed that φX RFIV

DNA (relaxed but covalently closed) was replicated, but only after a prolonged time lag. This contrasted with the rapid replication of φX RFI. Furthermore, it was noted that under these conditions the products formed during φX RFI DNA replication (initially φX RFII, which is then converted to φX RFI) varied in degree of superhelical density, but they eventually assumed the superhelical density characteristic of φX RFI formed in vivo. Both the finding that φX RFIV was not attacked by the φX gene-*A* protein and its sigmoidal replication rate suggested an active enzymatic process for introducing superhelicity. The system responsible for this reaction was discovered by Gellert et al. (1976a). These workers demonstrated that the conversion of ColE1 relaxed, closed, circular DNA (RFIV) to negatively superhelical DNA (RFI) required ATP and Mg^{++} and was inhibited by low concentrations of novobiocin. More recently, Sugino et al. (1977) showed that nalidixic acid also inhibits this system. We have made these same observations and have shown that the utilization of φX RFIV in φX gene-*A*-protein-mediated replication is completely dependent on DNA gyrase activity (Marians et al. 1977). The DNA gyrase reaction is a multiprotein system, at least two components of which have been resolved: the nalidixic acid protein, which has been purified to homogeneity both in our laboratory and in Cozzarelli's laboratory (Sugino et al. 1977), and the novobiocin target protein. There is possibly also a third component. We have repeatedly detected an additional protein fraction capable of stimulating the DNA gyrase reaction. At present, the mechanism operating in the generation of superhelical DNA is unknown. Recent studies indicate an important role for supertwisted, circular DNA structures in such biological reactions as integrative recombination of bacteriophage λ DNA (Mizuuchi and Nash 1976), replication of ColE1 (Gellert et al. 1976b) and φX DNA (Marians et al. 1977), and transcription (Hayashi and Hayashi 1971; Wang 1971).

The exact role of the DNA gyrase system in φX RFI DNA replication, however, remains to be elucidated. A number of observations remain to be explained. The effective inhibition of φX RFI DNA replication requires concentrations of novobiocin in the range of 200 μg/ml. In contrast, DNA gyrase activity is inhibited quantitatively by 1 μg/ml of the same drug. The role of DNA gyrase is evident when φX RFIV is used as the DNA substrate; however, RF $\rightarrow$ SS synthesis is also sensitive to novobiocin and nalidixic acid. It is possible that DNA gyrase plays some role in the mode of movement of the replicating fork in duplex DNA.

rep *Gene Function*

The *rep* protein has been purified to homogeneity in our laboratory. We have utilized for this purpose the φX RFI DNA replication complementation assay in which ammonium sulfate receptors derived from *rep* amber mutants of *E. coli*, supplemented with the φX gene-*A* protein or fraction II, served as the assay system. This procedure has resulted in the isolation of a 68,000-dalton protein which contains DNA-dependent ATPase activity. In contrast, *E. coli* extracts from a *rep* amber mutant carried through the same isolation procedure yielded fractions devoid of *rep*-gene complementing activity and free of the 68,000-dalton protein, but with the same DNA-dependent ATPase activity. Eisenberg et al. (1976a,b) have also isolated the *rep* protein. Others (Scott et al. 1977) have made use of an assay system in which the activity of the φX gene-*A* protein, *E. coli* DNA-binding protein, and a DNA elongation system (holoenzyme) containing DNA polymerase III is used to measure *rep*-protein activity. This assay resulted in the net synthesis of viral SS DNA. Scott et al. (1977) have shown that the isolated *rep* protein is a DNA-dependent ATPase and they have suggested that the strand-displacement reaction essential for the action of the holoenzyme-catalyzed dNMP incorporation is due to the combined action of the *rep* protein, *E. coli* DNA-binding protein, and the φX gene-*A* protein. In this system it would be essential that the *E. coli* DNA-binding protein bind to single-stranded DNA either in front or behind the moving replication site governed by the *rep* and φX gene-*A* proteins. This is an attractive model for the action of the *rep* protein. In our laboratory, attempts to detect in vitro φX RFI DNA-dependent incorporation of dNMP in the presence of highly purified φX gene-*A* protein, *E. coli* DNA-binding protein, *rep* protein, and the DNA elongation system (DNA polymerase III, *dnaZ*, and DNA EFI and EFIII) have been unsuccessful. It seems likely that the holoenzyme preparation used by Scott and coworkers may contain additional factors necessary for this reaction. Our studies on RF DNA replication with less highly purified fractions are in accord with in vivo observations, which indicate that a more complicated process is involved in the synthesis of viral and complementary strands (Denhardt 1977). A large number of different proteins that may play a role in moving replication protein complexes are now known, and a number of these proteins contain DNA-dependent ATPase activity. These include the *dnaB* protein (Wickner et al. 1974), the replication-factor-Y DNA-dependent ATPase (Wickner and Hurwitz 1975a), an *E. coli* ATP-dependent enzyme which "unwinds" duplex DNA (Abdel-Monem et al. 1977), the T4 44/62 protein complex (Alberts et al. 1975), the T7 gene-IV protein (Kolodner and Richardson 1977), and the *rep*-gene product (Scott et al. 1977). How these proteins are coupled to primer synthesis and DNA elongation reactions should be elucidated soon.

ACKNOWLEDGMENTS

This work was supported by grant NP-891 from the American Cancer Society, grants GM13344-12 and CA21622-01 from the National Institutes of Health, and grant PCM-76-04359 from the National Science Foundation. K. J. M. is a postdoctoral fellow of the American Cancer Society.

REFERENCES

Abdel-Monem, M., H. Lauppe, J. Kartenbeck, H. Dürwald, and H. Hoffmann-Berling. 1977. Enzymatic unwinding of DNA. III. Mode of action of *Escherichia coli* DNA unwinding enzyme. *J. Mol. Biol.* **110:** 667.

Alberts, B., C. F. Morris, D. Mace, N. Sinha, M. Bittner, and L. Moran. 1975. Reconstruction of the T4 bacteriophage DNA replication apparatus from purified components. In *ICN-UCLA Conference on DNA Synthesis and Its Regulation* (ed. M. Goulian et al.), p. 241. W. A. Benjamin, Menlo Park, California.

Baas, P. O., H. S. Jansz, and R. L. Sinsheimer. 1976. Bacteriophage φX174 DNA synthesis in a replication-deficient host: Determination of the origin of φX DNA replication. *J. Mol. Biol.* **102:** 633.

Bouché, J. P., K. Zechel, and A. Kornberg. 1975. *Dna* G gene product, a rifampicin-resistant RNA polymerase, initiates the conversion of a single-stranded coliphage DNA to its duplex replicative form. *J. Biol. Chem.* **250:** 5995.

Brutlag, D., R. Schekman, and A. Kornberg. 1971. A possible role for RNA polymerase in the initiation of M13 DNA synthesis. *Proc. Natl. Acad. Sci.* **68:** 2826.

Denhardt, D. T. 1975. The single-stranded DNA phages. *CRC Crit. Rev. Microbiol.* **4:** 161.

———. 1977. The isometric single-stranded DNA phages. In *Comprehensive virology* (ed. H. Fraenkel-Conrat and R. R. Wagner), vol. 7, p. 1. Plenum Press, New York.

Denhardt, D. T., D. H. Dressler, and A. Hathaway. 1967. The abortive replication of φX174 DNA in a recombination-deficient mutant of *E. coli. Proc. Natl. Acad. Sci.* **57:** 813.

Dressler, D. 1970. The rolling circle intermediate in φX DNA replication. III. Synthesis of single-stranded circles. *Proc. Natl. Acad. Sci.* **67:** 1932.

Eisenberg, S., J. Griffith, and A. Kornberg. 1977. φX174 cistron A protein is a multifunctional enzyme in DNA replication. *Proc. Natl. Acad. Sci.* **74:** 3198.

Eisenberg, S., J. F. Scott, and A. Kornberg. 1976a. An enzymatic system for replication of duplex circular DNA: The replicative form of phage φX174. *Proc. Natl. Acad. Sci.* **73:** 1594.

———. 1976b. Enzymatic replication of viral and complementary strands of duplex DNA of phage φX174 proceeds by separate mechanisms. *Proc. Natl. Acad. Sci.* **73:** 3151.

Francke, B. and D. S. Ray. 1971. Formation of the parental replicative form DNA of bacteriophage φX174 DNA and initial events in its replication. *J. Mol. Biol.* **61:** 565.

———. 1972. *Cis*-limited action of the gene A product of bacteriophage φX174 and the essential bacterial site. *Proc. Natl. Acad. Sci.* **69:** 475.

Fujisawa, H. and M. Hayashi. 1976. Viral DNA-synthesizing intermediate complex isolated during assembly of bacteriophage φX174. *J. Virol.* **19:** 409.

Gellert, M., K. Mizuuchi, M. H. O'Dea, and H. A. Nash. 1976a. DNA gyrase: An enzyme that introduces superhelical turns into DNA. *Proc. Natl. Acad. Sci.* **73:** 3872.

Gellert, M., M. H. O'Dea, T. Itoh, and J. Tomizawa. 1976b. Novobiocin and coumermycin inhibit DNA supercoiling catalyzed by DNA gyrase. *Proc. Natl. Acad. Sci.* **73:** 4474.

Gilbert, W. and D. Dressler. 1969. DNA replication: The rolling-circle model. *Cold Spring Harbor Symp. Quant. Biol.* **33:** 473.

Hayashi, Y. and M. Hayashi. 1971. Template activities of the φX174 replicative allomorphic deoxyribonucleic acids. *Biochemistry* **10:** 4212.

Henry, T. J. and R. Knippers. 1974. Isolation and function of the gene A initiator of bacteriophage φX174: A highly specific DNA endonuclease. *Proc. Natl. Acad. Sci.* **71:** 1549.

Ikeda, J.-E., A. Yudelevich, and J. Hurwitz. 1976. Isolation and characterization of the protein coded by gene A of bacteriophage φX174. *Proc. Natl. Acad. Sci.* **73:** 2669.

Knippers, R., A. Razin, R. Davis, and R. Sinsheimer. 1969. *In vivo* studies on the synthesis of the single-stranded DNA of progeny φX174 bacteriophage. *J. Mol. Biol.* **45:** 237.

Kolodner, R. and C. C. Richardson. 1977. Replication of duplex DNA by bacteriophage T7 DNA polymerase and gene 4 protein is accompanied by hydrolysis of nucleoside 5′-triphosphates. *Proc. Natl. Acad. Sci.* **74:** 1525.

Linney, E. A. and M. Hayashi. 1973. Two proteins of gene A of φX174. *Nat. New Biol.* **245:** 6.

Marians, K. J., J.-E. Ikeda, S. Schlagman, and J. Hurwitz. 1977. Role of DNA gyrase in φX replicative-form replication in vitro. *Proc. Natl. Acad. Sci.* **74:** 1965.

Mizuuchi, K. and H. A. Nash. 1976. Restriction assay for integrative recombination of bacteriophage λ DNA in vitro: Requirement for closed circular DNA substrate. *Proc. Natl. Acad. Sci.* **73:** 3524.

Mukai, R. and M. Hayashi. 1977. In vitro system that synthesizes circular viral DNA of bacteriophage φX174. *J. Virol.* **22:** 619.

Schekman, R., A. Weiner, and A. Kornberg. 1974. Multienzyme systems of DNA replication. *Science* **186:** 987.

Schekman, R., J. H. Weiner, A. Weiner, and A. Kornberg. 1975. Ten proteins required for conversion of φX174 single-stranded DNA to duplex form *in vitro*: Resolution and reconstitution. *J. Biol. Chem.* **250:** 5859.

Scott, J., S. Eisenberg, L. Bertsch, and A. Kornberg. 1977. A mechanism of duplex DNA replication revealed by enzymatic studies of phage φX174: Catalytic strand separation in advance of replication. *Proc. Natl. Acad. Sci.* **74:** 193.

Sinsheimer, R. L. 1968. Bacteriophage φX174 and related viruses. *Prog. Nucleic Acid Res. Mol. Biol.* **8:** 115.

Sugino, A., C. L. Peebles, K. N. Kreuzer, and N. R. Cozzarelli. 1977. Mechanism of action of nalidixic acid. Purification of *Escherichia coli nal* A gene product and its relationship to DNA gyrase and a novel nicking-closing enzyme. *Proc. Natl. Acad. Sci.* (in press).

Sumida-Yasumoto, C., A. Yudelevich, and J. Hurwitz. 1976. DNA synthesis *in vitro* dependent upon φX174 replicative form I DNA. *Proc. Natl. Acad. Sci.* **73:** 1887.

Tessman, E. S. 1966. Mutants of bacteriophage S13 blocked in infectious DNA synthesis. *J. Mol. Biol.* **17:** 218.

Wang. J. C. 1971. Interaction between DNA and an *Escherichia coli* protein ω. *J. Mol. Biol.* **55:** 523.

Wickner, R. B., M. Wright, S. Wickner, and J. Hurwitz. 1972. Conversion of φX174 and fd single-stranded DNA to replicative forms in extracts of *E. coli*. *Proc. Natl. Acad. Sci.* **69:** 3233.

Wickner, S. 1976. Mechanism of DNA elongation catalyzed by *Escherichia coli* DNA polymerase III, *dna*Z protein, and DNA elongation factors I and III. *Proc. Natl. Acad. Sci.* **73:** 3511.

———. 1977. DNA or RNA priming of bacteriophage G4 DNA synthesis by *Escherichia coli dna*G protein. *Proc. Natl. Acad. Sci.* **74:** 2815.

Wickner, S. and J. Hurwitz. 1975a. Association of φX174 DNA-dependent ATPase activity with an *E. coli* protein, replication factor Y, required for *in vitro* synthesis of φX174 DNA. *Proc. Natl. Acad. Sci.* **72:** 3342.

———. 1975b. *In vitro* synthesis of DNA. In *ICN-UCLA Conference on DNA Synthesis and Its Regulation* (ed. M. Goulian et al.), p. 227. W. A. Benjamin, Menlo Park, California.

Wickner, S., M. Wright, and J. Hurwitz. 1974. Association of DNA-dependent and independent ribonucleoside triphosphatase activities with *dna*B gene product of *Escherichia coli. Proc. Natl. Acad. Sci.* **71:** 783.

In Vivo Replication of Filamentous Phage DNA

Dan S. Ray
Molecular Biology Institute
and Department of Biology
University of California
Los Angeles, California 90024

Intracellular Forms of Filamentous Phage DNA

The first step in the replication of filamentous phage DNA is the conversion of the infecting viral DNA to a circular, duplex replicative form (RF). This RF molecule, termed the parental RF, replicates to produce a pool of progeny RF molecules. The accumulated RF molecules then serve both as templates for transcription of viral genes and as a source of progeny single-stranded (SS) DNA. The latter molecules are produced later in the life cycle by an asymmetric replication process in which the complementary strand of an RF molecule serves as a stable, circular template for the repeated displacement of viral single strands. The duplex replicative form occurs most often as a covalently closed, superhelical DNA (RFI). Infected cells also contain small, but significant, amounts of nicked circular RF molecules which contain one or more single-strand discontinuities (RFII) and relaxed circles in which both strands are covalently closed (RFIV). RFIII, a unit-length, linear RF, appears to occur only as an artifact.

Miniature forms of RF and SS DNA are observed in cells infected with phage preparations containing "miniphage." Such particles were first detected in phage preparations obtained after multiple passages beyond the original single-plaque isolation (Griffith and Kornberg 1974; Enea and Zinder 1975). These phage have extensive deletions of the genome and, consequently, do not contain any intact genes. Their replication and morphogenesis are dependent on gene functions provided by a helper phage. The miniphages contain only the region around the origin of replication and frequently one or more duplications of a portion of that region.

Double-length SS and RF species, as well as catenated RF DNA, have been detected in extremely small amounts (Jaenisch et al. 1969; Wheeler et al. 1974). The origin and significance of these rare forms are unknown. "Midiphage" that have somewhat more than a unit-length genome have also been observed and are described elsewhere in this volume (see Wheeler and Benzinger).

Kinetics and Regulation of Viral DNA Synthesis

The replication process of filamentous phage can be divided into three stages: the first stage, formation of the parental RF (phage → RF); the second stage, replication of the RF (RF → RF); and the third stage, asymmetric synthesis of viral single strands (RF → SS). The first stage of replication occurs so rapidly that no intermediates have yet been observed. The process does not require the synthesis of viral proteins and is mediated by a small number of host enzymes. Conversion of viral DNA to the parental replicative form (SS → RF) by purified enzymes has been studied extensively (R. B. Wickner et al. 1972; W. Wickner et al. 1972; Geider and Kornberg 1974) and likely reflects the essential elements of the phage → RF reaction. The latter reaction has not yet been achieved in a purified system. A model of the filamentous phage replication process is shown in Figure 1.

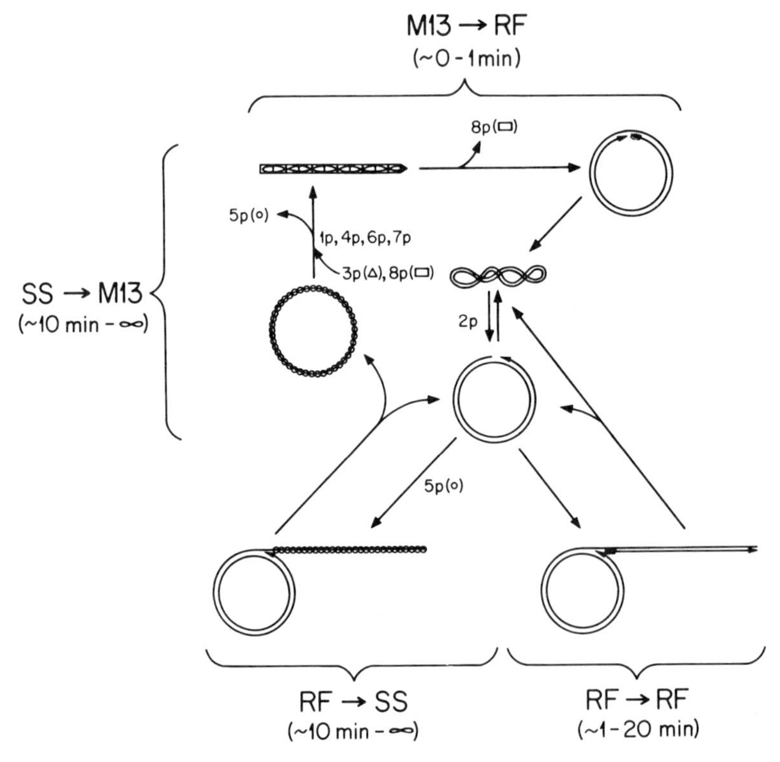

Figure 1 The replication cycle of filamentous phage. The three stages of replication of filamentous phage are indicated schematically along with an indication of the approximate duration of each stage and the requirements for viral gene products. (Outer circle) viral strand; (inner circle) complementary strand; (arrowhead) 3′-OH terminus; (~~~) RNA; (▭) gene-VIII protein, 8p; (o) gene-V protein, 5p; (Δ) gene-III protein, 3p.

During the first 10–15 minutes of infection the rate of DNA synthesis in infected cells is severalfold greater than in uninfected cells due to the synthesis of up to 100 progeny RF molecules per cell (Hohn et al. 1971). As expected for duplex DNA synthesis, radioactive labeling of RF molecules early in infection labels both strands equally. However, by 25 minutes after infection synthesis becomes totally asymmetric (Forsheit et al. 1971). Label is incorporated only into the viral strand of the RF during this stage of the replication process and flows with high efficiency into progeny single strands as these are formed. Prelabeled complementary strands remain covalently closed and retain their radioactivity, which suggests that the complementary strand serves as a stable template for the repeated synthesis of new viral strands. Preexisting viral strands are displaced as new viral strands are synthesized (Ray 1969). The single-strand pool that accumulates contains up to 200 viral single strands per cell (Ray et al. 1966).

Regulation of viral DNA synthesis is mediated by two viral proteins, the products of genes II and V (Pratt and Erdahl 1968). The gene-II protein is responsible for the asymmetry of the replication process. Its function is required for the accumulation of RFII molecules containing a single discontinuity in the viral strand (Fidanián and Ray 1972; Lin and Pratt 1972; Tseng and Marvin 1972b). The activity of gene II is essential both for RF DNA replication and for viral SS DNA synthesis. The shift from RF replication to the asymmetric synthesis of viral single strands is determined by the availability of the gene-V protein, a DNA-binding protein (Salstrom and Pratt 1971; Mazur and Model 1973; Mazur and Zinder 1975). Synthesis of progeny viral single strands increases in parallel with increased synthesis of the gene-V protein. In the absence of gene-V protein, viral DNA accumulates entirely as RF molecules. The functions of both of these regulatory proteins will be discussed in more detail later in this chapter.

The Phage → RF Reaction

The first stage of filamentous phage replication can be interrupted by rifampicin, an inhibitor of RNA polymerase, at a point following attachment of RNA polymerase but preceding DNA synthesis (Brutlag et al. 1971; Marco et al. 1974). This block to the formation of parental RF molecules must reflect inhibition of the synthesis of an RNA primer for the complementary strand since no such inhibition is observed in cells treated with inhibitors of protein synthesis. Furthermore, a requirement for all four ribonucleoside triphosphates (rNTPs) has been observed in vitro in the rifampicin-sensitive SS → RF reaction (Geider and Kornberg 1974). A noninfectious (eclipsed) phage particle accumulates in the presence of rifampicin, indicating that infection can proceed up to the point of attachment of RNA polymerase to the viral DNA at the unique origin of replication, while still preserving the phage structure. However, the sensitivity of the rifampicin-eclipsed phage to DNase indicates some alteration of the virion. The exposure of the DNA initiation sequence to the cytoplasm of the cell while the virion structure is maintained suggests that the RNA-polymerase-binding site is adjacent to or overlaps the binding site for the adsorption protein, the product of gene III. This hypothesis assumes that the viral DNA has a fixed and unique orientation in the virion. However, pyrimidine tract analysis of both whole phage and fragments from the adsorbing end of the virion show no difference, which implies a random orientation of the DNA in the virion (Tate and Petersen 1974). This apparent conflict might be resolved if the circular DNA were shown to be capable of a conveyor-belt type of movement within the virion at some point in the attachment process. Should this prove to be the case, the observed inhibition of phage attachment by either low temperature or cyanide poisoning (Marco et al. 1974) might reflect an energy requirement for the movement of the circular DNA within the filamentous virion.

An additional role for the gene-III protein (adsorption protein) has been proposed on the basis of its observed retention on the viral DNA even after conversion to the duplex replicative form (Jazwinski et al. 1973). This protein is thought to function as a "pilot" that might guide the DNA to some appropriate host system required for replication of the viral genome. This hypothesis is weakened, however, by the observation (Pratt et al. 1969) that viral DNA from polyphage particles lacking the adsorption protein is infectious to spheroplasts even though the DNA totally lacks the adsorption protein.

The mechanism by which the complementary strand is synthesized in the phage → RF reaction in vivo can be distinguished from that occurring during RF → RF replication by its resistance to nalidixic acid (Fidanián and Ray 1974). The complementary strand of parental RF DNA formed in the presence of this drug is fully functional, as judged by its ability to direct the synthesis of the gene-II protein. The basis for this sharp distinction is unclear.

RF → RF Replication

Replication of parental RF DNA to form a pool of RF molecules requires the function of a single viral gene product, that of gene II (Pratt and Erdahl 1968). Infection in the presence of chloramphenicol, an inhibitor of protein synthesis, or infection of a nonpermissive host cell with amber mutants defective in gene II leads to the formation of the parental RF, but further replication is inhibited (Fidanián and Ray 1972). In addition, there is a striking reduction in the ratio of RFII to RFI under these conditions, which suggests an involvement of gene II in the formation of the open, circular RFII. In contrast, RF DNA replication is also blocked after formation of the parental RF DNA in cells infected in the presence of nalidixic acid, an inhibitor of DNA replication, or in infection of *Escherichia coli rep*3 cells, which are defective in small-phage replication beyond the formation of the parental RF; but in both these cases gene expression can occur and normal amounts of RFII molecules are formed. Analysis of the RFII molecules formed under either of these conditions shows that the newly synthesized complementary strand is a covalently closed ring, whereas the parental viral strand contains a single discontinuity (Fidanián and Ray 1972). These results indicate clearly a requirement for expression of gene II for the formation of RFII and for the subsequent replication of RF DNA. The gene-II protein appears most likely to be a highly specific nuclease required for the initiation of a round of replication. The corresponding gene-*A* protein of φX174 plays a similar role in φX replication (Francke and Ray 1972) and has been shown to be a sequence-specific endonuclease.

Except for the gene-II protein, all of the functions required for RF DNA replication are provided by the host (Dumas, this volume). These include RNA polymerase, DNA polymerases I and III, ligase, a DNA-unwinding enzyme (the *rep* protein), gyrase, the *dnaB* protein, and the *dnaG* primase. Except for RNA polymerase, these enzymes are also involved in the replication of φX RF DNA (Wickner; McMacken et al.; Sumida-Yasumoto et al.; all this volume).

RF DNA replication occurs by a mechanism in which the two strands are nonequivalent. An asymmetry is introduced initially by the action of the gene-II protein which leads to the formation of RF molecules that have a single discontinuity in the viral strand and an intact circular complementary strand. Subsequent replication of the RF DNA involves rapidly sedimenting replicative intermediates (RI) that have a partially single-stranded structure (Tseng and Marvin 1972a). Alkaline denaturation of such intermediates yields viral strands of greater than unit length and both viral- and complementary-type strands of shorter than unit length (Forsheit and Ray 1971; Tseng and Marvin 1972a). These results are consistent with a rolling-circle mechanism of replication (Gilbert and Dressler 1969) in which the open viral strand of RFII is continuously displaced as a new viral strand is synthesized. This model predicts that viral strands of greater than unit length would arise by covalent attachment between the preexisting viral strand and the nascent viral strand. The occasional nicking of the RI at this junction would produce nascent viral-strand fragments of less than unit length.

The products of RF DNA replication are RFII molecules. Their conversion to superhelical RFI involves a relaxed, but covalently closed, intermediate termed RFIV (K. Horiuchi and N. D. Zinder, pers. comm.). Conversion of RFIV to RFI appears to be meditated by the host gyrase, since the conversion is inhibited by coumermycin, a specific inhibitor of gyrase. In the presence of this drug, RFIV molecules accumulate.

Initiation of the complementary strand most likely occurs by an RNA-polymerase-mediated priming, since complementary-strand synthesis is rapidly inhibited by rifampicin but not by chloramphenicol (Brutlag et al. 1971; Fidanián and Ray 1974). It has already been mentioned that there is some difference between complementary-strand synthesis during phage → RF synthesis and that during RF → RF replication, as indicated by the differential effects of nalidixic acid on complementary-strand synthesis during these two stages of replication. Synthesis of the complementary strand during RF → RF replication is much more sensitive to nalidixic acid than is synthesis of the viral strand, yet complementary-strand synthesis is resistant to inhibition by this drug during parental RF DNA formation.

While it is tempting to speculate that viral-strand synthesis is primed by the 3′ terminus of the initial RFII molecule (Fidanián and Ray 1972), there is as yet no evidence to support this hypothesis. Furthermore, there is a clear requirement for the *dnaG* protein in RF DNA replication (Ray et al. 1975; Dasgupta and Mitra 1976). Since this protein has been found to be involved in the synthesis of a primer for DNA chain initiation in the G4 and φX replication systems (McMacken et al.; Wickner; both this volume), presumably it functions in a similar capacity in filamentous phage replication. In the case of G4 and φX, however, the *dnaG* protein primes the synthesis of the complementary strand during SS → RF synthesis. This is clearly not the case for the filamentous phages as the *dnaG* protein is required for RF → RF replication but not for SS → RF synthesis. In addition, as discussed above, the complementary strand of filamentous phage RF DNA appears to be primed by RNA polymerase. Furthermore, experiments with both *dnaG* and *dnaB* mutants indicate that synthesis of both strands is inhibited equally in these mutants at a nonpermissive temperature (Ray et al. 1975). These results indicate that the *dnaB* and *dnaG* functions are not required exclusively for complementary-strand synthesis. Like *dnaG*, the *dnaB* function is required only during duplex DNA replication (Olsen et al. 1972). It seems likely, therefore, that both the *dnaG* and *dnaB* proteins may be involved in some way in the initiation of viral-strand synthesis. However, the possibility of an indirect effect of these mutations cannot be excluded entirely.

Another gene function possibly involved in RF DNA replication is that of *dnaA* (Mitra and Stallions 1976). However, the observation that the *dnaA* function is no longer required in Hfr strains in which the *dnaA* mutation is suppressed as a result of integrative suppression suggests the possibility that the requirement may be for host replication rather than for the *dnaA* product per se. The determination of the roles of specific host proteins in RF DNA replication will require the development of RF DNA replication systems similar to those already developed for φX as described elsewhere in this volume (Eisenberg et al.; Sumida-Yasumoto et al.).

The replicative origins for both the viral and the complementary strand are located within the intergenic region between genes II and IV (Tabak et al. 1974; Horiuchi and Zinder 1976; Suggs and Ray 1977). Both strands initiate within this region but are synthesized in opposite directions (Fig. 2).

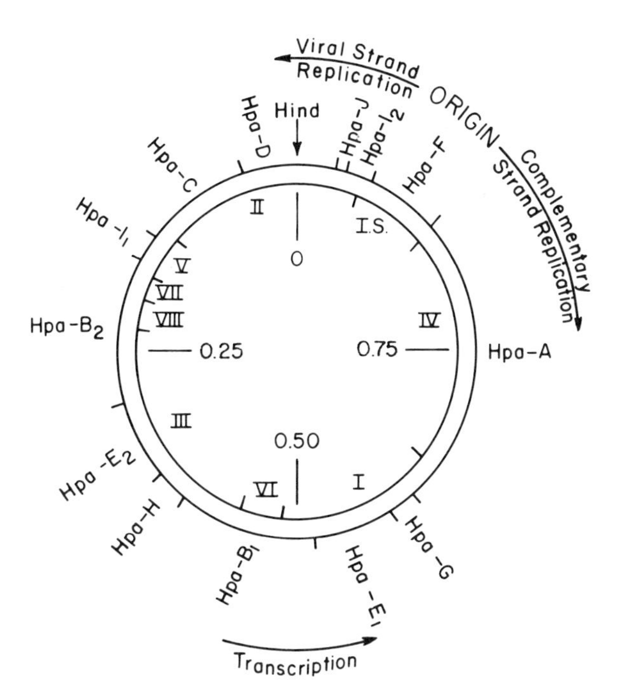

Figure 2 Origin and direction of synthesis of the two strands of M13. Both the viral and the complementary strand initiate within the intergenic space (I.S.) located between genes II and IV but they are synthesized in opposite directions. The *Hpa*II cleavage map and its alignment with the genetic map and with other restriction maps are described by Horiuchi et al., Konings and Schoenmakers, and Schaller et al. (all this volume).

If complementary-strand synthesis requires a single-stranded template, as in the SS → RF reaction, it may be necessary for an entire unit length of viral DNA to be synthesized prior to initiation of the complementary strand. In this case, the viral strand could possibly be circularized shortly after initiation of the complementary strand, thus leading to RI containing a circular viral strand and a nascent complementary strand (Staudenbauer et al., this volume).

RF DNA replication occurs in association with cellular membranes (Forsheit and Ray 1971; Staudenbauer and Hofschneider 1971; Kluge et al. 1971). Both parental labeled RF and pulse-labeled RF DNAs can be found in membrane fractions of gently lysed cells. The fraction of pulse-labeled RF DNA associated with the membrane is inversely proportional to the pulse length, which suggests that the RF molecules are synthesized on the membrane and then released. Gentle lysis of cells infected with an M13 *am*5 mutant results in the observation in the electron microscope of circular RF molecules attached to the cell envelope (Griffith and Kornberg 1972).

Under the less gentle conditions of osmotic disruption of the infected cells, the circular RF molecules are found separated from the cell envelope.

RF → SS Synthesis

The transition from duplex DNA replication to asymmetric viral-strand synthesis is regulated by the accumulation of a DNA-binding protein, the product of gene V (Salstrom and Pratt 1971; Mazur and Model 1973; Mazur and Zinder 1975). Mutants defective in gene V undergo normal RF DNA replication but fail to switch over to SS DNA synthesis. This protein has a very high affinity for SS DNA (Alberts et al. 1972; Oey and Knippers 1972) and is thought to affect the transition to SS DNA synthesis by binding to the nascent viral strands and thereby inhibiting initiation of complementary-strand synthesis. Although there is no direct evidence as yet for the association of gene-V protein with RI, such an interaction would be expected on the basis of the known properties of the gene-V protein and those of the RI.

Progeny single strands accumulate late in the infection process as DNA-protein complexes. By 75 minutes after infection the host cell contains up to 200 progeny single strands (Ray et al. 1966), each of which is coated with approximately 1300 gene-V proteins (Webster and Cashman 1973; Pratt et al. 1974). These complexes have been isolated from infected cells and shown to be rod-shaped structures 1.1 μm in length and 16 nm in width. Essentially all of the SS DNA and at least one-half to two-thirds of the gene-V protein are contained in such complexes. Neither of the known capsid proteins appears to be associated with these structures. During morphogenesis, the gene-V protein is displaced from the viral DNA at the cell surface as the filamentous virion is extruded out into the medium. These gene-V proteins can then be recycled, associating with newly formed viral single strands (Pratt et al. 1974). Recycling of the gene-V protein is an active process requiring viral morphogenesis. Inhibition of phage production immediately blocks the release of gene-V protein from the complex (Mazur and Zinder 1975).

Direct evidence for the association of gene-V protein with viral single strands in vivo has been obtained by UV cross-linking of the gene-V protein to viral DNA late in the infection process (Lica and Ray 1977). These complexes are readily isolated from detergent-treated lysates by CsCl equilibrium centrifugation. Only a single tryptic peptide remains associated with the viral DNA after trypsin treatment of cross-linked complexes, indicating a specific association of the gene-V protein with the DNA. The cross-link is located close to the carboxyl terminus, between residues 70 and 77.

The gene-V protein serves to maintain progeny viral strands in the single-stranded form. Cells infected with a gene-V temperature-sensitive mutant rapidly convert the pool of viral DNA to a double-stranded form upon a shift to the nonpermissive temperature (Salstrom and Pratt 1971). The ability of the gene-V protein to prevent complementary-strand synthesis on the progeny viral strands suggests that it could also function in that capacity on RI. Binding of the gene-V protein to nascent viral strands might also provide protection against exonucleolytic degradation of the viral strand prior to circularization (Oey and Knippers 1972).

Viral-strand synthesis normally follows the accumulation of a pool of progeny RF molecules. However, if a pool of gene-V protein is allowed to accumulate in the absence of RF DNA replication, SS DNA synthesis begins immediately upon release of the block to DNA synthesis (Mazur and Model 1973). Under such conditions the parental viral strand is rapidly transferred to the pool of progeny SS DNA. The accumulation of excess gene-V protein has also been shown to allow chloramphenicol-resistant SS DNA synthesis. In a normal infection the addition of chloramphenicol leads to a rapid switch back to duplex DNA replication (Ray 1970). It therefore appears that viral-strand synthesis is regulated entirely by the availability of gene-V protein.

Whether or not the gene-V protein may also play a positive role in SS DNA synthesis is unclear. It has been observed that SS DNA synthesis is temperature-sensitive in a *dnaB* mutant infected with a gene-V ts mutant (Staudenbauer and Hofschneider 1973). Since the *dnaB* function is required for RF DNA replication but not for SS DNA synthesis (Olsen et al. 1972), it was concluded that the gene-V protein plays a positive role in viral-strand synthesis. Whether or not the altered gene-V protein might be inhibitory to SS DNA synthesis at the nonpermissive temperature is unknown. Other experiments suggest that a distinction must be made between asymmetric viral-strand synthesis and SS DNA accumulation. It was found that both strands of RF molecules are synthesized equally in a nonpermissive host infected with a gene-V amber mutant, yet upon addition of either rifampicin or nalidixic acid, synthesis became entirely asymmetric but no SS DNA was formed (Fidanián and Ray 1974). This asymmetric synthesis of viral strands without the concomitant accumulation of SS DNA suggests that the function of the gene-V protein is to protect the viral strand from degradation during SS DNA synthesis. However, these experiments do not exclude the possibility that gene-V protein might be required not only for asymmetric viral-strand synthesis but also for stable accumulation of viral SS DNA, but at different levels. A low level of gene-V protein might be sufficient at a replication fork, whereas a greater quantity would be needed to coat the nascent viral strand completely. Possibly a low level of suppression in the nonpermissive host might provide sufficient gene-V protein to allow some asymmetric synthesis but not enough to allow SS DNA accumulation. Resolution of this question will be accomplished most easily in an in vitro system with purified proteins.

The intermediates in SS DNA synthesis contain a closed, circular complementary strand and an elongated viral strand (Ray 1969; Forsheit et al. 1971; Staudenbauer and Hofschneider 1972a; Kluge 1974). Electron microscopic analysis of these RI shows open, circular, duplex rings with single-stranded tails (Allison et al. 1977; Ray 1977) of the type proposed by Gilbert and Dressler (1969). The viral strand initiates within the intergenic region between genes II and IV and is synthesized unidirectionally in a counterclockwise direction on the standard physical and genetic maps (Horiuchi and Zinder 1976; Suggs and Ray 1977; Allison et al. 1977). The origin-terminus of the viral strand has been determined by three independent methods: (1) location of the site of the discontinuity in the late life cycle of RFII molecules, (2) determination of the temporal order of synthesis of each segment of the viral strand of the product RFI, and (3) electron microscopic mapping of the 5′ end of RI. One of the products of a round of viral-strand synthesis is an RFII molecule containing a single discontinuity. The site of this discontinuity has been located by carrying out limited repair synthesis with polymerase I and labeled triphosphates and determining the site of the repair label by restriction analysis. The discontinuity has been located near the center of the intergenic space. In the second method, supercoils formed during SS DNA synthesis have been analyzed in a Dintzis type of terminal-labeling experiment. These results have established the direction of synthesis of the viral strand and place the origin-terminus of the viral strand in the same region as the discontinuity in the product RFII. Partial denaturation mapping of rolling-circle intermediates likewise places the viral-strand origin at this same site on the genome (Allison et al. 1977).

Entry of an RFI molecule into the replication process requires the introduction of the gene-II-specific discontinuity into the viral strand of the RF (Fidanián and Ray 1972). As discussed above, the role of the discontinuity in RFII molecules is uncertain. Although it has generally been assumed to serve as a primer for viral-strand synthesis, it could also serve some other function, such as to permit unwinding of the circular DNA and propagation of a viral strand initiated at another site. The appealing simplicity of the rolling-circle model (Gilbert and Dressler 1969) and the discovery that single-stranded phages specify functions required for the accumulation of specifically nicked RF molecules (Francke and Ray 1971; Fidanián and Ray 1972) have perhaps inhibited consideration of de novo initiation of the viral strand. However, there are now several observations that we think require such consideration.

First of all, there is the unexplained sensitivity of viral-strand synthesis to rifampicin (Staudenbauer and Hofschneider 1972a,b; Mitra 1972; Fidanián and Ray 1974). This observation alone is not entirely compelling, since viral-strand synthesis is much less sensitive to rifampicin than is complementary-strand synthesis and since at least two rounds of rifampicin-resistant viral-strand synthesis can occur under conditions that allow accumulation of excess gene-V protein prior to addition of the drug (Fidanián and Ray

1974). Yet, rifampicin inhibits viral-strand synthesis much more rapidly than does chloramphenicol, which suggests that RNA synthesis plays some role other than just for the continued expression of gene V (Staudenbauer and Hofschneider 1972a,b). Second, a mutation affecting only the 5′ → 3′ exonuclease activity of DNA polymerase I (Konrad and Lehman 1974) inhibits completely the sealing of the discontinuity in RFII molecules formed during SS DNA synthesis (Chen and Ray 1976; T.-C. Chen and D. S. Ray, in prep.). The lack of closure of the viral strands of these molecules in the absence of a functional DNA polymerase I 5′ → 3′ exonuclease suggests that a primer that would normally be removed by this activity still remains at the 5′ terminus and inhibits closure of the RFII. This interpretation is supported by the observation of greatly increased amounts of RNA-primed Okazaki fragments in this mutant (Kurosawa et al. 1975). Finally, there is the observed requirement in vivo for both *dnaB* and *dnaG* functions for RF → RF replication but not for RF → SS synthesis (Olsen et al. 1972; Ray et al. 1975). The very high sensitivity of complementary-strand synthesis to rifampicin suggests that the complementary strand may be primed by RNA polymerase just as in the SS → RF reaction in vitro. In view of the evidence implicating *dnaB* and *dnaG* functions in DNA chain initiation (cf. McMacken et al.; Wickner; both this volume), the requirements for *dnaB* and *dnaG* functions during RF DNA replication may reflect an involvement of these functions in initiating viral-strand synthesis at this stage of the replication cycle.

The possibility of a positive role for gene-V protein and the rifampicin sensitivity of SS DNA synthesis have led to some speculation that gene-V protein and RNA polymerase might play a role in the initiation of the viral strand as gene-V protein accumulates. However, there is no direct evidence in support of this hypothesis.

SUMMARY

Replication of filamentous phage DNA involves duplex RI in which the two strands are replicated by different mechanisms. Asymmetry is introduced into the replication process by the phage gene-II protein, which is required for the formation of RFII molecules specifically nicked in the viral strand. Synthesis of viral single strands also requires the function of an additional phage protein, the gene-V protein, a DNA-binding protein that prevents synthesis of the complementary strand late in infection. Initiation of the viral strand occurs in the intergenic space between genes II and IV at a site close to the complementary-strand origin.

During both phage → RF synthesis and RF → RF replication, synthesis of the complementary strand is initiated by a rifampicin-sensitive mechanism at a single, specific site within the intergenic space. However, complementary-strand synthesis during the phage → RF reaction is catalyzed by a nalidixic acid-resistant mechanism, whereas that occurring during RF → RF replication is sensitive to this drug.

Gene-II-specific nicking of the viral strand is required for viral-strand synthesis during both RF → RF replication and RF → SS synthesis, but the role of the discontinuity remains obscure. The simplest hypothesis is that the 3′ terminus created by nicking of the viral strand serves as a primer for direct elongation of the viral strand. Several lines of evidence now suggest that consideration must also be given to the possibility of de novo initiation of the viral strand. Purification of the proteins required for RF DNA replication and for SS DNA synthesis and the reconstitution in vitro of these stages of the filamentous phage replication cycle should provide us with considerably greater insight into the precise biochemical mechanisms regulating the initiation and elongation of both strands of the viral genome.

ACKNOWLEDGMENTS

Research in the author's laboratory was supported by grants from the National Institutes of Health (AI 10752) and the National Science Foundation (PCM 76-02709).

REFERENCES

Alberts, B., L. Frey, and H. Delius. 1972. Isolation and characterization of gene 5 protein of filamentous bacterial viruses. *J. Mol. Biol.* **68:** 139.

Allison, D. P., A. T. Ganesan, A. C. Olson, C. M. Snyder, and S. Mitra. 1977. Electron microscopic studies of bacteriophage M13 DNA replication. *J. Virol.* **24:** 673.

Brutlag, D., R. Schekman, and A. Kornberg. 1971. A possible role for RNA polymerase in the initiation of M13 DNA synthesis. *Proc. Natl. Acad. Sci.* **68:** 2826.

Chen, T.-C. and D. S. Ray. 1976. Replication of bacteriophage M13. X. M13 replication in a mutant of *Escherichia coli* defective in the 5′ → 3′ exonuclease associated with DNA polymerase I. *J. Mol. Biol.* **106:** 589.

Dasgupta, S. and S. Mitra. 1976. The role of *Escherichia coli dna*G function in coliphage M13 DNA synthesis. *Eur. J. Biochem.* **67:** 47.

Enea, V. and N. D. Zinder. 1975. A deletion mutant of bacteriophage f1 containing no intact cistrons. *Virology* **68:** 105.

Fidanián, H. M. and D. S. Ray. 1972. Replication of bacteriophage M13. VII. Requirement of the gene 2 protein for the accumulation of a specific RFII species. *J. Mol. Biol.* **72:** 51.

———. 1974. Replication of bacteriophage M13. VIII. Differential effects of rifampicin and nalidixic acid on the synthesis of the two strands of M13 duplex DNA. *J. Mol. Biol.* **83:** 63.

Forsheit, A. B. and D. S. Ray. 1971. Replication of bacteriophage M13. VI. Attachment of M13 DNA to a fast-sedimenting host cell component. *Virology* **43:** 647.

Forsheit, A. B., D. S. Ray, and L. Lica. 1971. Replication of bacteriophage M13. V. Single-strand synthesis during M13 infection. *J. Mol. Biol.* **57:** 117.

Francke, B. and D. S. Ray. 1971. Fate of parental φX174 DNA upon infection of starved thymine-requiring cells. *Virology* **44:** 168.

———. 1972. Ultraviolet-induced cross-links in the deoxyribonucleic acid of single-stranded deoxyribonucleic acid viruses as a probe of deoxyribonucleic acid packaging. *J. Virol.* **9:** 1027.

Geider, K. and A. Kornberg. 1974. Initiation of DNA synthesis. VIII. Conversion of the M13 viral single-strand to the double-stranded replicative forms by purified proteins. *J. Biol. Chem.* **249:** 3999.

Gilbert, W. and D. Dressler. 1969. DNA replication: The rolling circle model. *Cold Spring Harbor Symp. Quant. Biol.* **33:** 473.

Griffith, J. and A. Kornberg. 1972. DNA-membrane associations in the development of a filamentous bacteriophage, M13. In *Membrane research* (ed. C. F. Fox), p. 281. Academic Press, New York.

———. 1974. Mini M13 bacteriophage: Circular fragments of M13 DNA are replicated and packaged during normal infections. *Virology* **59:** 139.

Hohn, B., H. Lechner, and D. A. Marvin. 1971. Filamentous bacterial viruses. I. DNA synthesis during the early stages of infection with fd. *J. Mol. Biol.* **56:** 143.

Horiuchi, K. and N. D. Zinder. 1976. Origin and direction of synthesis of bacteriophage f1 DNA. *Proc. Natl. Acad. Sci.* **73:** 2341.

Jaenisch, R., P. H. Hofschneider, and A. Preuss. 1969. Isolation of circular DNA by zonal centrifugation: Separation of normal length, double length and catenated M13 replicative form DNA and of host specific "episomal" DNA. *Biochim. Biophys. Acta* **190:** 88.

Jazwinski, S. M., R. Marco, and A. Kornberg. 1973. A coat protein of the bacteriophage M13 virion participates in membrane-oriented synthesis of DNA. *Proc. Natl. Acad. Sci.* **70:** 205.

Kluge, F. 1974. Replicative intermediates in bacteriophage M13: Single-stranded DNA synthesis. *Hoppe-Seyler's Z. Physiol. Chem.* **355:** 410.

Kluge, F., W. L. Staudenbauer, and P. H. Hofschneider. 1971. Replication of bacteriophage M13: Detachment of the parental DNA from the host membrane and transfer to progeny phages. *Eur. J. Biochem.* **22:** 350.

Konrad, E. B. and I. R. Lehman. 1974. A conditional lethal mutant of *Escherichia coli* K12 defective in the $5' \rightarrow 3'$ exonuclease associated with DNA polymerase I. *Proc. Natl. Acad. Sci.* **71:** 2048.

Kurosawa, Y., T. Ogawa, S. Hirose, T. Okazaki, and R. Okazaki. 1975. Mechanism of DNA chain growth. XV. RNA-linked nascent DNA pieces in *Escherichia coli* strains assayed with spleen exonuclease. *J. Mol. Biol.* **96:** 653.

Lica, L. and D. S. Ray. 1977. Replication of bacteriophage M13. XII. *In vivo* crosslinking of a phage-specific DNA binding protein to the single-stranded DNA of bacteriophage M13 by ultraviolet light. *J. Mol. Biol.* **115:** 45.

Lin, N. and D. Pratt. 1972. Role of bacteriophage M13 gene 2 in viral DNA replication. *J. Mol. Biol.* **72:** 37.

Marco, R., S. M. Jazwinski, and A. Kornberg. 1974. Binding, eclipse and penetration of the filamentous bacteriophage M13 in intact and disrupted cells. *Virology* **62:** 209.

Mazur, B. J. and P. Model. 1973. Regulation of f1 single-stranded DNA synthesis by a DNA binding protein. *J. Mol. Biol.* **78:** 285.
Mazur, B. J. and N. D. Zinder. 1975. The role of gene V protein in f1 single-strand synthesis. *Virology* **68:** 49.
Mitra, S. 1972. Inhibition of M13 phage synthesis by rifampicin in some rifampicin-resistant *Escherichia coli* mutants. *Virology* **50:** 422.
Mitra, S. and D. R. Stallions. 1976. The role of *Escherichia coli dna*A gene and its integrative suppression in M13 coliphage DNA synthesis. *Eur. J. Biochem.* **67:** 37.
Oey, J. L. and R. Knippers. 1972. Properties of the isolated gene 5 protein of bacteriophage fd. *J. Mol. Biol.* **68:** 125.
Olsen, W. L., W. L. Staudenbauer, and P. H. Hofschneider. 1972. Replication of bacteriophage M13: Specificity of the *Escherichia coli dna*B function for replication of double-stranded M13 DNA. *Proc. Natl. Acad. Sci.* **69:** 2570.
Pratt, D. and W. S. Erdahl. 1968. Genetic control of bacteriophage M13 DNA synthesis. *J. Mol. Biol.* **37:** 181.
Pratt, D., P. Laws, and J. Griffith. 1974. Complex of phage M13 single-stranded DNA and gene 5 protein. *J. Mol. Biol.* **82:** 425.
Pratt, D., H. Tzagoloff, and J. Beaudoin. 1969. Conditional lethal mutants of the small filamentous coliphage M13. II. Two genes for coat proteins. *Virology* **39:** 42.
Ray, D. S. 1969. Replication of bacteriophage M13. II. The role of replicative forms in single-strand synthesis. *J. Mol. Biol.* **43:** 631.
———. 1970. Replication of bacteriophage M13. VI. Synthesis of M13-specific DNA in the presence of chloramphenicol. *J. Mol. Biol.* **53:** 239.
———. 1977. Replication of filamentous bacteriophages. In *Comprehensive virology* (ed. H. Fraenkel-Conrat and R. R. Wagner), vol. 7, p. 105. Plenum Press, New York.
Ray, D. S., H. P. Bscheider, and P. H. Hofschneider. 1966. Replication of the single-stranded DNA of the male-specific bacteriophage M13: Isolation of intracellular forms of phage-specific DNA. *J. Mol. Biol.* **21:** 473.
Ray, D. S., J. Dueber, and S. Suggs. 1975. Replication of bacteriophage M13. IX. Requirement of the *Escherichia coli dna*G function for M13 duplex DNA replication. *J. Virol.* **16:** 348.
Salstrom, J. S. and D. Pratt. 1971. Role of coliphage M13 gene 5 in single-stranded DNA production. *J. Mol. Biol.* **61:** 489.
Staudenbauer, W. L. and P. H. Hofschneider. 1971. Membrane attachment of replicating parental DNA molecules of bacteriophage M13. *Biochem. Biophys. Res. Commun.* **42:** 1035.
———. 1972a. Replication of bacteriophage M13: Inhibition of single-strand DNA synthesis in an *Escherichia coli* mutant thermosensitive in chromosomal DNA replication. *Eur. J. Biochem.* **30:** 403.
———. 1972b. Replication of bacteriophage M13: Inhibition of single-strand DNA synthesis by rifampicin. *Proc. Natl. Acad. Sci.* **69:** 1634.
———. 1973. Replication of bacteriophage M13: Positive role of gene 5 protein in single-strand DNA synthesis. *Eur. J. Biochem.* **34:** 569.
Suggs, S. and D. S. Ray. 1977. Replication of bacteriophage M13. XI. Localization of the origin for M13 single-strand synthesis. *J. Mol. Biol.* **110:** 147.
Tabak, H. F., J. Griffith, K. Geider, H. Schaller, and A. Kornberg. 1974. Initiation of

deoxyribonucleic acid synthesis. VII. A unique location of the gap in the M13 replicative duplex synthesized *in vitro*. *J. Biol. Chem.* **249:** 3049.

Tate, W. P. and G. B. Petersen. 1974. Structure of the filamentous bacteriophages: Orientation of the DNA molecule within the phage particle. *Virology* **62:** 17.

Tseng, B. Y. and D. A. Marvin. 1972a. Filamentous bacterial viruses. V. Asymmetric replication of fd duplex deoxyribonucleic acid. *Virology* **10:** 371.

———. 1972b. Filamentous bacterial viruses. VI. Role of fd gene 2 in deoxyribonucleic acid replication. *Virology* **10:** 384.

Webster, R. E. and J. S. Cashman. 1973. Abortive infection of *Escherichia coli* with the bacteriophage f1: Cytoplasmic membrane proteins and the f1 DNA gene 5 protein complex. *Virology* **55:** 20.

Wheeler, F. C., R. H. Benzinger, and H. Bujard. 1974. Double-length, circular, single-stranded DNA from filamentous phage. *J. Virol.* **14:** 620.

Wickner, R. B., M. Wright, S. Wickner, and J. Hurwitz. 1972. Conversion of φX174 and fd single-stranded DNA to replicative forms in extracts of *Escherichia coli*. *Proc. Natl. Acad. Sci.* **69:** 3233.

Wickner, W., D. Brutlag, R. Schekman, and A. Kornberg. 1972. RNA synthesis initiates *in vitro* conversion of M13 DNA to its replicative form. *Proc. Natl. Acad. Sci.* **69:** 965.

Requirements for Host Gene Products in Replication of Single-stranded Phage DNA In Vivo

Lawrence B. Dumas
Department of Biochemistry and Molecular Biology
Northwestern University
Evanston, Illinois 60201

The identification of those proteins required for host-cell DNA synthesis that are essential for replication of the DNA of the small, single-stranded DNA phages has closely paralleled the genetic dissection of the DNA-synthesis apparatus of the host cell itself. Conditionally lethal mutants of the host cell marked in those genes whose products are required for host DNA synthesis have been used to determine which bacterial proteins are also essential for synthesis of viral DNA. The genes identified by temperature-sensitive mutations are designated *dna* genes. To a lesser extent, highly specific drug inhibitors of host proteins have been used in similar analyses.

Those genes pertinent to this discussion, as well as the role played by each gene's product in DNA replication in *Escherichia coli*, are listed in Table 1.

Table 1 Genes involved in *E. coli* DNA synthesis

Gene	Role in DNA synthesis	Reference[a]
dnaA	initiation at origin	7, 8, 21
dnaB	fragment initiation	7, 21, 22, 24
dnaC	initiation at origin	2, 19, 21
dnaE (*polC*)	chain elongation	5, 17, 21
dnaG	fragment initiation	14, 21, 23, 26
dnaI	initiation at origin	1
dnaP	initiation at origin	20
dnaZ	chain elongation	4, 25
polA	gap filling and primer excision	10, 11, 18
lig	fragment ligation	6, 9
rpoB (*rif*)	initiation at origin	15, 16
rep	fork migration	3, 12, 13

[a] References: 1. Beyersmann et al. (1974); 2. Carl (1970); 3. Denhardt et al. (1967); 4. Filip et al. (1974); 5. Gefter et al. (1971); 6. Gottesman et al. (1973); 7. Hirota et al. (1969); 8. Hirota et al. (1970); 9. Konrad et al. (1973); 10. Konrad and Lehman (1974); 11. Kuempel and Veomett (1970); 12. Lane and Denhardt (1974); 13. Lane and Denhardt (1975); 14. Lark (1972a); 15. Lark (1972b); 16. Messer (1972); 17. Nüsslein et al. (1971); 18. Okazaki et al. (1971); 19. Schubach et al. (1973); 20. Wada and Yura (1974); 21. Wechsler and Gross (1971); 22. Weiner et al. (1976); 23. Wickner (1977); 24. Wickner and Hurwitz (1974); 25. Wickner and Hurwitz (1976); 26. Zechel and Kornberg (1975).

Some of these gene products play crucial roles in the initiation of DNA synthesis, either at the origin of replication or in the initiation of the synthesis of Okazaki fragments. Others play important roles in chain elongation and fork migration or in termination reactions, including gap filling, primer excision, and fragment ligation.

The specific inhibitors of *E. coli* DNA synthesis that are pertinent to this discussion include rifampicin, nalidixic acid, novobiocin, and coumermycin. Rifampicin inhibits DNA-dependent RNA polymerase, which plays an essential role in the initiation of DNA synthesis at the origin of replication (Table 1). Novobiocin and coumermycin hinder DNA synthesis (Smith and Davis 1967; Ryan 1976) by inhibiting specifically DNA gyrase (Gellert et al. 1976). Nalidixic acid inhibits DNA synthesis (Goss et al. 1965) by interfering with the nicking-closing activities of the *nalA* gene product, which appears to be a component of DNA gyrase (A. Sugino et al., in prep.).

Because small-phage DNA codes for only a few proteins, it could be expected that these phages would take advantage of most or all of the DNA replication machinery of the host cell, and this is indeed the case.

Synthesis of Infectious Progeny Phage

It is useful to consider first which bacterial gene products are necessary for the normal synthesis of infectious progeny phage, i. e., the complete cycle of replication. If any one of the three stages of viral DNA synthesis is defective in a mutant host, the rate of formation of infectious phage particles should be reduced. This analysis thus indicates which bacterial mutants to investigate further and also provides a means for comparing phages on the basis of the extent to which they make use of host DNA-synthesis proteins.

The results of such analyses performed in many laboratories are summarized in Table 2. It is immediately evident that the isometric phages tested thus far fall into two groups with respect to their requirements for host DNA-synthesis proteins. Group-I phages, which include φX174, S13, φA, and G4, require the products of the host *dnaB, dnaC, dnaE, dnaG, dnaZ, rep*, and *lig* genes for normal phage synthesis; the products of *dnaA, dnaI*, and *dnaP* apparently are not required. Group-II phages, which include St-1, φXtB,[1] φK, and α3, require the products of the host *dnaE, dnaG, dnaZ*, and *lig* genes but apparently not those of genes *dnaA, dnaB, dnaC, dnaI*, and *dnaP*. The major difference between the groups is that group-II phages appear not to require the products of genes *dnaB* and *dnaC*, whereas group-I phages do. It is interesting that the phages in these two groups can also be distinguished on the basis of such properties as host range and serological reactivity (Bradley 1970; Taketo 1977b). Group-I phages fall into Bradley's group B (except for G4, which was not tested), whereas

[1]Evidence that φXtB is St-1 has been published recently by Matthes and Denhardt (1978).

group-II phages fall into Bradley's groups A and C. Taketo (1977b) has suggested recently that Bradley's groups A and C no longer need to be classified as two distinct groups. Thus, the isometric phages not only have distinct host ranges and serological properties, but they also have distinct requirements for specific host DNA-synthesis proteins.

Neither group-I nor group-II isometric phages require the products of host genes *dnaA, dnaI*, and *dnaP* for DNA synthesis. These proteins appear to be needed by the host for the initiation of DNA synthesis at the origin of replication (Table 1). That they are not required by these phages is perhaps because in each case the phage DNA codes for a substitute initiation protein, analogous to the gene-*A* protein of φX (Henry and Knippers 1974; Sumida-Yasumoto et al. 1976; Eisenberg et al. 1976). However, group-I phages do require one host protein that is essential for the initiation of host DNA synthesis at the origin of replication, namely, the *dnaC* protein. Therefore, initiation of DNA synthesis on group-I phages may be more like that on the host chromosome than that on group-II phages.

The filamentous phages have requirements for host DNA-synthesis proteins similar to those of the group-I isometric phages; they require the products of host genes *dnaB, dnaC, dnaE, dnaG, dnaZ, rep*, and *polA*. In addition, the products of genes *dnaA* and *rpoB* also appear to be essential for DNA synthesis on the filamentous phages, whereas none of the isometric phages tested required these two functions directly for DNA synthesis. Thus, the filamentous phages are closer to the host in terms of the initiation factors required for DNA synthesis. This suggests that the control of DNA replication in these two phage types is different.

Synthesis of Parental Replicative-form (RF) DNA

The host-protein requirements for phage parental RF DNA synthesis in vivo have been studied extensively in group-I isometric phage and filamentous phage infections. Group-II isometric phage DNA templates have only been used for in vitro studies. The results of the in vivo analyses are summarized in Table 3. The host *dnaG* and *dnaZ* proteins are required by G4 for parental RF DNA synthesis, most likely for initiation and chain elongation, respectively. G4 parental RF DNA synthesis is normal in *dnaA, dnaB, dnaC, dnaE*, and *rep* mutants of *E. coli*. Thus, not all of the host gene products required for normal progeny phage production (Table 2) are required by G4 for the first stage of DNA synthesis. Specifically, the *dnaB, dnaC, dnaE*, and *rep* proteins must be called upon at a later stage of DNA synthesis.

The requirements for parental RF DNA synthesis on the other group-I isometric phages, particularly φX, are more numerous. In addition to those proteins required for G4 parental RF DNA synthesis, φX also requires the *dnaB* and *dnaC* proteins. The latter protein is needed only in the presence of rifampicin, which presumably inhibits the host DNA-dependent RNA

Table 2 Bacterial gene products required for normal phage growth

Host gene product	Isometric phages[a]								Filamentous phage
	group I				group II				
	φX	S13	φA	G4	St-1	φXtB[b]	φK	α3	M13
dnaA	– (12, 22)	– (27)	– (22)		– (2)	– (25)	– (24)		+ (14, 15)
dnaB	+ (8, 21, 22)	+ (27)	+ (22)	+ (6)	– (2)	– (1, 25)	– (24)	– (27)	+ (14, 17)
dnaC	+ (11, 22)	+ (27)	+ (22)	+ (6)	– (2)	– (10, 25)	– (24)	– (27)	+ (14)
dnaE	+ (7, 22)	+ (27)	+ (22)	+ (6)	+ (2)	+ (25)	+ (24)	+ (26)	+ (14, 20)
dnaG	+ (13, 22)	+ (27)	+ (22)	+ (6, 26)		+ (25)	+ (24)	+ (26)	+ (14, 16, 18)
dnaI	– (23)	– (27)	– (23)			– (25)	– (24)		

dnaP	– (23)	– (27)	– (23)		– (25)	– (24)		
dnaZ	+ (23, 28)	+ (27)	+ (23)	+ (26)	+ (25)	+ (24)	+ (26)	+ (28)
rep	+ (5)	+ (27)	+ (26)	+ (26)	+ (25)	+ (24)	+ (26)	+ (5)
lig	+ (23)	+ (26)	+ (23)		+ (25)	+ (24)	+ (26)	
polA								+ (4)
rpoB								+ (3, 9, 19)

[a]–, Not required; +, required. The references, given in parentheses, are as follows: 1. Bone and Dowell (1973); 2. Bowes (1974); 3. Brutlag et al. (1971); 4. Chen and Ray (1976); 5. Denhardt et al. (1967); 6. Derstine et al. (1976); 7. Dumas and Miller (1973); 8. Dumas and Miller (1974); 9. Fidanián and Ray (1974); 10. Haworth et al. (1975); 11. Kranias and Dumas (1974); 12. Mannheimer and Truffaut (1974); 13. McFadden and Denhardt (1974); 14. Mitra and Stallions (1973); 15. Mitra and Stallions (1976); 16. Dasgupta and Mitra (1976); 17. Olsen et al. (1972); 18. Ray et al.(1975); 19. Staudenbauer and Hofschneider (1972b); 20. Staudenbauer et al. (1973); 21. Steinberg and Denhardt (1968); 22. Taketo (1973); 23. Taketo (1975); 24. Taketo (1976a); 25. Taketo (1976b); 26. Taketo (1977a); 27. Taketo (1977b); 28. Truitt and Walker (1974).

[b]φXtB is probably St-1 (Matthes and Denhardt 1978).

Table 3 Bacterial gene products required for parental RF DNA synthesis in vivo

Phage	Gene products required[a]	Gene products not required[a]
φX	*dnaB* (7, 20)	*dnaE* (4, 6, 10)
	dnaC (or *rpoB*) (8, 12)	*rep* (3)
	dnaG (13, 22)	
	dnaZ (11)	
	lig (14)	
G4	*dnaG* (21)	*dnaA* (5, 21)
	dnaZ (21)	*dnaB* (5, 21)
		dnaC (5, 21)
		dnaE (5, 21)
		rep (21)
M13	*dnaE* (or *polA*) (19)	*dna*A (1, 15)
	dnaZ (11)	*dnaB* (17)
	rpoB (2)	*dnaG* (16, 18)
		rep (9)

[a] The references, given in parentheses, are as follows: 1. Bouvier and Zinder (1974); 2. Brutlag et al. (1971); 3. Denhardt et al. (1967); 4. Denhardt et al. (1973); 5. Derstine et al. (1976); 6. Dumas and Miller (1973); 7. Dumas and Miller (1974); 8. Dumas et al. (1975); 9. Fidanián and Ray (1972); 10. Greenlee (1973); 11. Haldenwang and Walker (1976); 12. Kranias and Dumas (1974); 13. McFadden and Denhardt (1974); 14. McFadden and Denhardt (1975); 15. Mitra and Stallions (1976); 16. Dasgupta and Mitra (1976); 17. Olsen et al. (1972); 18. Ray et al. (1975); 19. Staudenbauer (1974); 20. Steinberg and Denhardt (1968); 21. Taketo (1977a); 22. Truffaut and Mannheimer (1975).

polymerase (Dumas et al. 1975). Apparently, the presence of DNA-dependent RNA polymerase can obviate the requirement for the *dnaC* protein at this stage of φX DNA synthesis.

A multiple mutant of φX, φXahb, which apparently is altered in a phage capsid protein, is able to synthesize parental RF DNA in at least one temperature-sensitive *dnaB* mutant host (Vito and Dowell 1976). φX parental RF DNA synthesis is defective in the same host (Vito et al. 1975). Thus, alterations in the phage DNA, and perhaps in phage proteins, affect either the host factors required for phage DNA synthesis or the ability of the phage to make use of the altered host *dnaB* protein. Interestingly, φXahb parental RF DNA synthesis is rifampicin-sensitive in this *dnaB* mutant at the nonpermissive temperature but not at the permissive temperature. This suggests an obviating role for DNA-dependent RNA polymerase in this system as well.

This comparison of two group-I isometric phages on the basis of their requirements for host proteins for parental RF DNA synthesis demonstrates that even though they both require the same proteins overall for the production of progeny phage, they differ in the proteins required for the first stage of DNA replication. Since they differ in those proteins presumed to be required for the initiation of complementary-strand DNA synthesis rather than in those required for chain elongation, it appears that initiation is template-specific.

The product of the *E. coli dnaE* gene, DNA polymerase III, apparently is not essential for chain elongation in parental RF DNA synthesis on either phage DNA template. However, N. Truffaut and I. Mannheimer (unpubl.) have demonstrated temperature-sensitive φX parental RF DNA synthesis in a *polA polB dnaE*^{ts} mutant host, which indicates that DNA polymerase III may be essential when the other two bacterial DNA polymerases are defective. (It should be noted that Denhardt et al. [1973] obtained the opposite result.) Evidently, any one of the three host-cell DNA polymerases can catalyze chain elongation at this stage of viral DNA synthesis.

Although parental RF DNA synthesis on group-II isometric phages has not been studied in vivo, we can tentatively conclude that the *dnaB* and *dnaC* proteins are not essential at this stage. Since phage production is normal in host cells lacking either of these functions (Table 2), it is unlikely that they play an essential role in parental RF DNA synthesis. This is corroborated by studies in vitro which demonstrated a requirement for the *dnaG* protein, but not for the *dnaB* and *dnaC* proteins, for the initiation of complementary-strand synthesis on phage St-1 and φXtB DNA templates (Wickner and Hurwitz 1975b). Thus, the group-I phage G4 is more similar to group-II phages than to other group-I phages in the bacterial proteins required for the initiation of parental RF DNA synthesis.

The filamentous phages differ markedly from the isometric phages in their requirements for parental RF DNA synthesis. Such synthesis on filamentous phages is sensitive to rifampicin (as indicated in Table 3 by the requirement for the *rpoB* protein), and the host DNA-dependent RNA polymerase is essential for chain initiation in this system (Brutlag et al. 1971). The filamentous phages do not use those host DNA-synthesis proteins essential for chain initiation on isometric phage DNA. The products of genes *dnaB* and *dnaG*, for example, do not play an essential role even in the presence of rifampicin. This corroborates the suggestion that initiation of parental RF DNA synthesis is template-specific.

Parental RF DNA synthesis of the filamentous phages, like that of the isometric phages, is not dependent absolutely on DNA polymerase III, the product of the *dnaE* gene. Unlike the isometric phages, however, the filamentous phages exhibit defective parental RF DNA synthesis in a *polA dnaE* double mutant (Staudenbauer 1974). DNA polymerase I (the product of the *polA* gene) can possibly substitute for DNA polymerase III in chain elongation on these templates, but one or the other is essential. The product of the *dnaZ* gene is essential for parental RF DNA synthesis on both phage types.

Replication of RF DNA

The requirements for host DNA-synthesis proteins in the second stage of phage DNA replication have been examined in φX-, G4-, and M13-infected cells. A summary of the known requirements is given in Table 4. For the group-I isometric phage φX, the requirements for RF DNA replication are essentially the same as those for normal synthesis of infectious phage (compare Table 4 with Table 2). The product of the *dnaH* gene (Sakai et al. 1974) has been omitted because it apparently is not involved directly in φX DNA synthesis (Derstine and Dumas 1976). Since the products of the host genes *dnaA, dnaI*, and *dnaP* are not essential for the complete cycle of φX DNA replication (see Table 2), we can assume that these factors are not essential for RF DNA replication.

G4, the other group-I isometric phage that has been examined, appears to require the same host DNA-synthesis proteins as φX for RF DNA replication. In this case it has been demonstrated that the product of the *dnaA* gene is not essential for RF DNA replication. Although the effects of mutations in the host *dnaZ, polA*, and *lig* genes on G4 RF DNA replication have not yet been reported, when they are they probably will suggest essential roles for these gene products similar to their roles in φX RF DNA replication.

There have been no reports as yet of direct examination of the group-II isometric phages for their requirements for RF DNA replication. However, since the *dnaB* and *dnaC* proteins apparently are not required for the complete cycle of DNA replication, we can assume that they are not essential for RF DNA replication. The products of host genes *dnaE, dnaG, dnaZ, rep*, and *lig* probably are required, as they are for the complete cycle of replication (Table 2).

The filamentous phages, like the group-I isometric phages, require the products of the host *dnaB, dnaE, dnaG, dnaZ*, and *rep* genes for RF DNA replication. They presumably require the *dnaC* protein also, as it has been reported that this gene product is needed for the complete cycle of replication (Mitra and Stallions 1973) (Table 2). In addition, these phages require the *dnaA* protein and DNA-dependent RNA polymerase (*rpoB*) for RF DNA replication. The RNA polymerase apparently plays a specific role in the synthesis of the new complementary strand (Fidanián and Ray 1974), perhaps in a manner analogous to its role in the initiation of the synthesis of the complementary strand on the viral strand during the synthesis of parental RF DNA (Brutlag et al. 1971). The role of the host *dnaA* protein is unknown.

Nalidixic acid has an inhibitory effect like that of rifampicin on filamentous phage RF DNA replication in that it inhibits specifically the synthesis of

Table 4 Bacterial gene products required for RF DNA replication in vivo

Phage	Gene products required[a]	Gene products not required[a]
φX	*dnaB* (8, 22)	*polA* (9)
	dnaC (14)	
	dnaE (5, 7, 12)	
	dnaG (15, 25)	
	dnaZ (13)	
	rep (3, 4)	
	lig (16)	
G4	*dnaB* (6)	*dnaA* (6)
	dnaC (6)	
	dnaE (6)	
	dnaG (6)	
	rep (20)	
M13	*dnaA* (17)	
	dnaB (19)	
	dnaE (23, 24)	
	dnaG (18, 21)	
	dnaZ (13)	
	rep (2, 10)	
	rpoB (1, 11)	

[a] The references, given in parentheses, are as follows: 1. Brutlag et al. (1971); 2. Denhardt (1975); 3. Denhardt et al. (1967); 4. Denhardt et al. (1972); 5. Denhardt et al. (1973); 6. Derstine et al. (1976); 7. Dumas and Miller (1973); 8. Dumas and Miller (1974); 9. Eisenberg and Denhardt (1974); 10. Fidanián and Ray (1972); 11. Fidanián and Ray (1974); 12. Greenlee (1973); 13. Haldenwang and Walker (1977); 14. Kranias and Dumas (1974); 15. McFadden and Denhardt (1974); 16. McFadden and Denhardt (1975); 17. Mitra and Stallions (1976); 18. Dasgupta and Mitra (1976); 19. Olsen et al. (1972); 20. Ray and Dueber (1975); 21. Ray et al. (1975); 22. Steinberg and Denhardt (1968); 23. Staudenbauer (1974); 24. Staudenbauer et al. (1973); 25. Truffaut and Mannheimer (1975).

the new complementary strand (Fidanián and Ray 1974). The role of the nalidixic acid-sensitive factor in this synthesis is obscure. Novobiocin is also inhibitory at this stage, causing accumulation of relaxed, circular, double-stranded RF molecules (RFIV) rather than normal supercoiled DNA (RFI) (K. Horiuchi, unpubl.). This implies that the host-cell DNA gyrase plays an essential role in the initiation or termination of RF DNA replication.

Single-stranded (SS) DNA Synthesis

Since only one strand of the RF DNA is copied during the last stage of phage DNA replication, we might expect that there would be fewer host DNA-synthesis proteins required for this stage than for RF DNA replication. This appears to be so for the two phages that have been studied in some detail (Table 5). The group-I isometric phage φX requires the products of the host *dnaE, dnaG, dnaZ, rep*, and *lig* genes for SS DNA synthesis. The *dnaB* and *dnaC* proteins essential for RF DNA replication are not required at this stage. In vitro studies imply that these two host proteins are essential in RF DNA replication for the initiation of the new complementary strand but not for the initiation of the new viral strand (Eisenberg et al. 1976). Since complementary-strand DNA is not made during the third stage of DNA synthesis, it is not surprising that these gene products are dispensable.

What is surprising is that the host *dnaG* protein is required. The *dnaG* protein is a chain-initiation polymerase for DNA synthesis (Zechel and Kornberg 1975; Wickner 1977). There is no apparent reason for its direct participation in de novo chain initiation during this stage of φX DNA synthesis. In its simplest form, the rolling-circle model for this stage of viral

Table 5 Bacterial gene products required for SS DNA synthesis in vivo

Phage	Gene products required[a]	Gene products not required[a]
φX	*dnaE* (3, 5)	*dnaB* (4, 5)
	dnaG (10)	*dnaC* (5, 9)
	dnaZ (7)	
	rep (19)	
	lig (8, 11, 14, 15)	
M13 (f1)	*dnaA* (1)	*dnaB* (13, 17)
	dnaE (16)	*dnaG* (12)
	dnaZ (7)	
	rpoB (6, 18)	
	polA (2)	

[a] The references, given in parentheses, are as follows: 1. Bouvier and Zinder (1974); 2. Chen and Ray (1976); 3. Dumas and Miller (1973); 4. Dumas and Miller (1974); 5. Dumas and Miller (1976); 6. Fidanián and Ray (1974); 7. Haldenwang and Walker (1977); 8. Iwaya et al. (1973); 9. Kranias and Dumas (1974); 10. McFadden and Denhardt (1974); 11. Miller and Sinsheimer (1974); 12. Dasgupta and Mitra (1976); 13. Olsen et al. (1972); 14. Schekman and Ray (1971); 15. Schekman et al. (1971); 16. Staudenbauer (1974); 17. Staudenbauer and Hofschneider (1972a); 18. Staudenbauer and Hofschneider (1972b); 19. Tessman and Peterson (1976).

DNA synthesis requires only an endonuclease (the gene-*A* protein) to cut the old viral DNA strand on the template RF molecule to create a 3′ end for chain elongation. If the *dnaG* protein does in fact participate directly in the synthesis of φX SS DNA, then chain initiation must be more complex than the basic model predicts.

Again, although it has not been shown directly that the *dnaA, dnaI*, and *dnaP* proteins are dispensable during φX SS DNA synthesis, we can assume that they are since they are not essential for the normal synthesis of infectious progeny phage.

The requirements for SS DNA synthesis have not been reported for the group-II isometric phages. However, it is anticipated that they will closely resemble those of φX.

The requirements established thus far for the filamentous phage M13 (and for f1 and fd) are similar to those for φX with respect to host DNA-elongation factors (the products of genes *dnaE* and *dnaZ*). Phage M13 SS DNA synthesis, like that of φX, requires a chain-initiation catalyst (the product of *rpoB* for M13 and the product of *dnaG* for φX). In addition, M13 requires the product of the *polA* gene, specifically the 5′ exonuclease activity associated with this enzyme. This implies that primer excision is essential for this stage as well. Thus, as with φX, SS DNA synthesis on M13 appears to be more complex than the basic rolling-circle model predicts.

SS DNA synthesis on the filamentous phages, like RF DNA replication, seems to require the *dnaA* protein; most probably it is required at the end of this stage since conversion of linear SS DNA molecules to circles has been shown to be defective in one *dnaA* mutant host (Bouvier and Zinder 1974). However, Mitra and Stallions (1976) failed to observe this defect in another *dnaA* mutant strain. Why these two observations differ cannot yet be explained.

CONCLUSIONS

The isometric and filamentous phages all use the same *E. coli* DNA-synthesis proteins for their chain-elongation (the *dnaE, dnaZ*, and *rep* proteins) and DNA-synthesis-termination reactions (the *polA* and *lig* proteins) that are needed for normal chain-elongation and termination reactions on the host-cell chromosome. Differences are evident, however, in their requirements for host DNA-initiation proteins, both those required for initiation at the origin and those required for initiation of synthesis of Okazaki fragments. None of the phages require all of the host DNA-initiation proteins. Each phage codes for its own DNA-initiation protein, and each group of phages uses the initiation proteins provided by the host to a different extent.

The group-II isometric phages are least dependent on the host DNA-initiation proteins. They require only the *dnaG* protein for chain initiation, whereas the group-I isometric phages require the *dnaB* and *dnaC* proteins in addition to the *dnaG* protein. In phage G4, all three proteins are required for RF DNA replication, yet the *dnaG* protein alone is sufficient for chain initiation during parental RF DNA synthesis. On the other hand, φX requires the *dnaB, dnaC*, and *dnaG* proteins for chain initiation during both parental RF DNA synthesis and RF DNA replication.

The roles of the *dnaB, dnaC*, and *dnaG* proteins in φX DNA synthesis have been studied extensively both in vitro and in vivo. During parental RF DNA synthesis the *dnaB* and *dnaC* proteins interact with the template DNA prior to chain initiation catalyzed by the *dnaG* protein (Wickner and Hurwitz 1974, 1975a; Weiner et al. 1976). It is not yet clear why with φX DNA the *dnaG* protein cannot act alone to catalyze chain initiation on the template DNA strand, as it does with G4 DNA and the group-II phage DNAs.

In vitro experiments have shown that the host DNA-initiation proteins *dnaB, dnaC*, and *dnaG* are required during φX RF DNA replication to initiate synthesis of the new complementary strand on the viral-strand template (Eisenberg et al. 1976), as is the case during parental RF DNA synthesis. These proteins are not needed for the initiation of synthesis of the new viral strand on the complementary strand of the RF DNA molecule (Eisenberg et al. 1976). These requirements are consistent with those that would be expected for a circular, double-stranded DNA molecule that replicates according to the rolling-circle model. The initiation of the synthesis of the new viral strand on the complementary strand requires a specific phage-encoded endonuclease (the gene-*A* protein) to cleave the old viral strand, thereby providing a primer 3′ end for the chain-elongation DNA polymerase (Francke and Ray 1971; Henry and Knippers 1974; Eisenberg et al. 1976; Ikeda et al. 1976). Initiation of complementary-strand DNA synthesis, however, requires de novo primer synthesis.

An in vivo experiment designed to detect the expected difference in the requirements of the template strands for the *dnaG* chain-initiation polymerase during φX RF DNA replication failed to show a clean asymmetry (McFadden and Denhardt 1974). The explanation for this could be the difficulty of performing such an experiment in the complex in vivo system or that chain initiation during φX RF DNA replication in vivo is more complex than the rolling-circle model predicts.

The same model could explain the protein requirements for initiation of G4 RF DNA replication. Although the *dnaG* protein is sufficient for the initiation of synthesis of the new complementary strand during parental RF DNA synthesis, the *dnaB* and *dnaC* proteins may be necessary for the synthesis of the complementary strand on the viral-strand template during RF DNA replication. As with φX, initiation of synthesis of the new viral strand on the complementary-strand template may require a specific phage-encoded endonuclease to cleave the old viral strand, thereby providing a primer 3′ end for the chain-elongation polymerase.

The protein requirements for initiation of filamentous phage RF DNA replication are more complex than those for initiation of isometric phage RF DNA replication. The products of the host *dnaA, dnaB, dnaC, dnaG*, and *rpoB* genes are all required. Yet, of these, only the product of *rpoB* plays an essential role in the initiation of parental RF DNA synthesis. It has been thought that filamentous phage RF DNA replicates via essentially the same mechanism as φX RF DNA. That is, a nick occurs in the viral strand and this initiates formation of a rolling-circle replicating intermediate (Tseng and Marvin 1972; Ray, this volume). The *dnaB, dnaC,* and *dnaG* proteins may be required for de novo initiation of the synthesis of a new viral strand on the complementary strand of the RF molecule, even though the classical rolling-circle model for replication requires no such de novo initiation. Alternatively, these proteins might be required for initiating synthesis of a new complementary strand on the viral strand in a manner unique to RF DNA replication; this would imply different mechanisms for complementary-strand synthesis during RF DNA replication and parental RF DNA synthesis.

In an attempt to determine which of these two possibilities was more likely, M. L. Bayne and L. B. Dumas (in prep.) isolated circular complementary strands from filamentous phage f1 RF DNA and examined chain initiation on this template in vitro. They observed no initiation catalyzed by the *dnaB, dnaC,* and *dnaG* proteins on this template. This observation is consistent with the rolling-circle model, which predicts that chain initiation on the complementary strand of the RF molecule in vivo should not need de novo primer synthesis, but rather should need only an endonuclease to provide a primer 3′ end on the old viral strand. However, this leaves us without a satisfactory explanation for the role of these three proteins in RF DNA replication.

The product of the host *rpoB* gene, as part of the host DNA-dependent RNA polymerase, plays a specific role in the initiation of synthesis of the new complementary strand during filamentous phage RF DNA replication (Fidanián and Ray 1974), as it does during parental RF DNA synthesis. There is not yet any indication as to the role of the *dnaA* protein.

It is of particular interest that the group-II isometric phages do not seem to require the host *dnaB* and *dnaC* proteins at all, whereas the other isometric phages and the filamentous phages do require them at least for RF DNA replication. It is not known whether other host- or phage-encoded proteins substitute for the *dnaB* and *dnaC* protein activities, or whether DNA molecules of this phage group have unique properties that obviate the need for the initiation activities of these two host proteins.

The filamentous phages apparently require more host gene products for the initiation of DNA synthesis than do the isometric phages. Filamentous phages do not inhibit host DNA replication or kill the cell. It seems reasonable to assume, therefore, that filamentous phage DNA replication is closely regulated by the host DNA-initiation system. The isometric phages, on the other hand, inhibit host DNA synthesis (Lindqvist and Sinsheimer 1967) and kill the cell. It obviously is to the advantage of these phages that they be more free from the host-controlled DNA-initiation system.

Finally, it seems clear from these studies that the small, single-stranded DNA phages have provided us with considerable information to help us to understand how a simple infecting virus genome uses the host cell's DNA replication apparatus to replicate its own DNA and how different phage genomes are able to use different parts of this replication apparatus. Further studies, especially those in vitro, should help us to understand the basis for the different requirements for host DNA-initiation proteins exhibited by different phages. In so doing, such studies should provide us with a detailed picture of the unique control of the replication of the DNA of these viruses.

ACKNOWLEDGMENTS

Work in the author's laboratory was supported by a Research Career Development Award (AI-70,632) and a Research Grant (AI-09882) from the National Institute of Allergy and Infectious Diseases. This chapter was completed in August, 1977.

REFERENCES

Beyersmann, D., W. Messer, and M. Schlict. 1974. Mutants of *Escherichia coli* B/r defective in deoxyribonucleic acid initiation: *DnaI*, a new gene for replication. *J. Bacteriol.* **118:** 783.

Bone, D. R. and C. E. Dowell. 1973. A mutant of bacteriophage φX174 which infects *E. coli* K12 strains. II. Replication of φXtB in ts DNA strains. *Virology* **52:** 330.

Bouvier, F. and N. D. Zinder. 1974. Effects of the *dnaA* thermosensitive mutation of *Escherichia coli* on bacteriophage f1 growth and DNA synthesis. *Virology* **60:** 139.

Bowes, J. M. 1974. Replication of bacteriophage St-1 in *Escherichia coli* strains temperature-sensitive in DNA synthesis. *J. Virol.* **13:** 1400.

Bradley, D. E. 1970. A comparative study of some properties of the φX174 type bacteriophages. *Can. J. Microbiol.* **16:** 965.

Brutlag, D., R. Schekman, and A. Kornberg. 1971. A possible role for RNA polymerase in the initiation of M13 DNA synthesis. *Proc. Natl. Acad. Sci.* **68:** 2826.

Carl, P. L. 1970. *Escherichia coli* mutants with temperature-sensitive synthesis of DNA. *Mol. Gen. Genet.* **109:** 107.

Chen, T.-C. and D. S. Ray. 1976. Replication of bacteriophage M13. X. M13 replication in a mutant of *Escherichia coli* defective in the 5′ → 3′ exonuclease associated with DNA polymerase I. *J. Mol. Biol.* **106:** 589.

Dasgupta, S. and S. Mitra. 1976. The role of *Escherichia coli dnaG* function in coliphage M13 DNA synthesis. *Eur. J. Biochem.* **67:** 47.

Denhardt, D. T. 1975. The single-stranded DNA phages. *CRC Crit. Rev. Microbiol.* **4:** 161.

Denhardt, D. T., D. H. Dressler, and A. Hathaway. 1967. The abortive replication of φX174 DNA in a recombination-deficient mutant of *Escherichia coli. Proc. Natl. Acad. Sci.* **57:** 813.

Denhardt, D. T., M. Iwaya, and L. L. Larison. 1972. The *rep* mutation. II. Its effect on *E. coli* and on the replication of bacteriophage φX174. *Virology* **49:** 486.

Denhardt, D. T., M. Iwaya, G. McFadden, and G. Schochetman. 1973. The mechanism of replication of φX174 single-stranded DNA. VI. Requirements for ribonucleoside triphosphates and DNA polymerase III. *Can. J. Biochem.* **51:** 1588.

Derstine, P. L. and L. B. Dumas. 1976. Deoxyribonucleic acid synthesis in a temperature-sensitive *Escherichia coli dnaH* mutant, strain HF4704S. *J. Bacteriol.* **128:** 801.

Derstine, P. L., L. B. Dumas, and C. A. Miller. 1976. Bacteriophage G4 DNA synthesis in temperature-sensitive *dna* mutants of *Escherichia coli. J. Virol.* **19:** 915.

Dumas, L. B. and C. A. Miller. 1973. Replication of bacteriophage φX174 DNA in a temperature-sensitive *dnaE* mutant of *Escherichia coli* C. *J. Virol.* **11:** 848.

———. 1974. Inhibition of bacteriophage φX174 DNA replication in *dnaB* mutants of *Escherichia coli* C. *J. Virol.* **14:** 1369.

———. 1976. Bacteriophage φX174 single-stranded viral DNA synthesis in temperature-sensitive *dnaB* and *dnaC* mutants of *Escherichia coli. J. Virol.* **18:** 426.

Dumas, L. B., C. A. Miller, and M. L. Bayne. 1975. Rifampicin inhibition of bacteriophage φX174 parental replicative form DNA synthesis in an *Escherichia coli dnaC* mutant. *J. Virol.* **16:** 575.

Eisenberg, S. and D. T. Denhardt. 1974. Structure of nascent φX174 replicative form: Evidence for discontinuous DNA replication. *Proc. Natl. Acad. Sci.* **71:** 984.

Eisenberg, S., J. F. Scott, and A. Kornberg. 1976. Enzymatic replication of viral and complementary strands of duplex DNA of phage φX174 proceeds by separate mechanisms. *Proc. Natl. Acad. Sci.* **73:** 3151.

Fidanián, H. M. and D. S. Ray. 1972. Replication of bacteriophage M13. VII. Requirement of the gene 2 protein for the accumulation of a specific RFII species. *J. Mol. Biol.* **72:** 51.

———. 1974. Replication of bacteriophage M13. VIII. Different effects of rifampicin and nalidixic acid on the synthesis of the two strands of M13 duplex DNA. *J. Mol. Biol.* **83:** 63.

Filip, C. C., J. S. Allen, R. A. Gustafson, R. G. Allen, and J. R. Walker. 1974. Bacterial cell division regulation: Characterization of the *dnaH* locus of *Escherichia coli. J. Bacteriol.* **119:** 443.

Francke, B. and D. S. Ray. 1971. Formation of the parental replicative form DNA of bacteriophage φX174 and initial events in its replication. *J. Mol. Biol.* **61:** 565.

Gefter, M. L., Y. Hirota, T. Kornberg, J. A. Wechsler, and C. Barnoux. 1971. Analysis of DNA polymerases II and III in mutants of *Escherichia coli* thermosensitive for DNA synthesis. *Proc. Natl. Acad. Sci.* **68:** 3150.

Gellert, M., M. H. O'Dea, T. Itoh, and J. Tomizawa. 1976. Novobiocin and coumermycin inhibit DNA supercoiling by DNA gyrase. *Proc. Natl. Acad. Sci.* **73:** 4474.

Goss, W. A., W. H. Deitz, and T. M. Cook. 1965. Mechanism of action of nalidixic acid on *Escherichia coli.* II. Inhibition of deoxyribonucleic acid synthesis. *J. Bacteriol.* **89:** 1068.

Gottesman, M. M., M. L. Hicks, and M. Gellert. 1973. Genetics and function of DNA ligase in *Escherichia coli. J. Mol. Biol.* **77:** 531.

Greenlee, L. L. 1973. Replication of bacteriophage φX174 in a mutant of *Escherichia coli* defective in the *dnaE* gene. *Proc. Natl. Acad. Sci.* **70:** 1757.

Haldenwang, W. G. and J. R. Walker. 1976. Parental RF formation of phages φX174 and M13 requires the *dnaZ* gene product of *Escherichia coli. Biochem. Biophys. Res. Commun.* **70:** 932.

———. 1977. Inhibition of bacteriophage M13 and φX174 duplex DNA replication and single-strand synthesis in temperature-sensitive *dnaZ* mutants of *Escherichia coli. J. Virol.* **22:** 23.

Haworth, S. R., C. F. Gilgun, and C. E. Dowell. 1975. Growth studies of three φX174 mutants in ts DNA mutants of *Escherichia coli. J. Virol.* **15:** 720.

Henry, T. J. and R. Knippers. 1974. Isolation and function of the gene A initiator of bacteriophage φX174, a highly specific DNA endonuclease. *Proc. Natl. Acad. Sci.* **71:** 1549.

Hirota, Y., J. Mordoh, and F. Jacob. 1970. On the process of cell division in *Escherichia coli.* III. Thermosensitive mutants of *Escherichia coli* altered in the process of DNA initiation. *J. Mol. Biol.* **53:** 369.

Hirota, Y., A. Ryter, and F. Jacob. 1969. Thermosensitive mutants of *E. coli* affected in the processes of DNA synthesis and cellular division. *Cold Spring Harbor Symp. Quant. Biol.* **33:** 677.

Ikeda, J. E., A. Yudelevich, and J. Hurwitz. 1976. Isolation and characterization of the protein coded by the gene A of bacteriophage φX174. *Proc. Natl. Acad. Sci.* **73:** 2669.

Iwaya, M., S. Eisenberg, K. Bartok, and D. T. Denhardt. 1973. Mechanism of replication of single-stranded φX174 DNA. VII. Circularization of the progeny viral strand. *J. Virol.* **12:** 808.

Konrad, E. B. and I. R. Lehman. 1974. A conditional lethal mutant of *Escherichia coli* K12 defective in the 5′→3′ exonuclease associated with DNA polymerase I. *Proc. Natl. Acad. Sci.* **71:** 2048.

Konrad, E. B., P. Modrich, and I. R. Lehman. 1973. Genetic and enzymatic characterization of a conditional lethal mutant of *Escherichia coli* K12 with a temperature-sensitive DNA ligase. *J. Mol. Biol.* **77:** 519.

Kuempel, P. L. and G. E. Veomett. 1970. A possible function of DNA polymerase in chromosome replication. *Biochem. Biophys. Res. Commun.* **41:** 973.

Kranias, E. G. and L. B. Dumas. 1974. Replication of bacteriophage φX174 DNA in a temperature-sensitive *dnaC* mutant of *Escherichia coli* C. *J. Virol.* **13:** 146.

Lane, H. E. D. and D. T. Denhardt. 1974. The *rep* mutation. III. Altered structure of the replicating *E. coli* chromosome. *J. Bacteriol.* **120:** 805.

———. 1975. The *rep* mutation. IV. Slower movement of replication forks in *E. coli rep* strains. *J. Mol. Biol.* **97:** 99.

Lark, K. G. 1972a. Genetic control over the initiation of the synthesis of short deoxyribonucleotide chains in *E. coli. Nat. New Biol.* **240:** 237.

———. 1972b. Evidence for the direct involvement of RNA in the initiation of DNA replication in *Escherichia coli* $15T^{-}$. *J. Mol. Biol.* **64:** 47.

Lindqvist, B. H. and R. L. Sinsheimer. 1967. Process of infection with bacteriophage φX174. XIV. Studies on macromolecular synthesis during infection with a lysis-defective mutant. *J. Mol. Biol.* **28:** 87.

Mannheimer, I. and N. Truffaut. 1974. Growth of bacteriophage φX174 in *Escherichia coli* strains carrying temperature-sensitive mutations for DNA initiation. *Mol. Gen. Genet.* **130:** 21.

Matthes, M. and D. T. Denhardt. 1978. Comparison of restriction endonuclease *Hae*III fragments of φX174, φXtB, and St-1: Evidence that φXtB is St-1. *Virology* **85:** 626.

McFadden, G. and D. T. Denhardt. 1974. Mechanism of replication of φX174 single-stranded DNA. IX. Requirement for the *Escherichia coli dnaG* protein. *J. Virol.* **14:** 1070.

———. 1975. The mechanism of replication of φX174 DNA. XIII. Discontinuous synthesis of the complementary strand in an *Escherichia coli* host with a temperature-sensitive polynucleotide ligase. *J. Mol. Biol.* **99:** 125.

Messer, W. 1972. Initiation of deoxyribonucleic acid replication in *Escherichia coli* B/r: Chronology of events and transcriptional control of initiation. *J. Bacteriol.* **112:** 7.

Miller, L. K. and R. L. Sinsheimer. 1974. Nature of φX174 linear DNA from a ligase-defective host. *J. Virol.* **14:** 1503.

Mitra, S. and D. R. Stallions. 1973. Role of *dna* genes of *Escherichia coli* in M13 phage replication. *Virology* **52:** 417.

———. 1976. The role of *Escherichia coli dnaA* gene and its integrative suppression in M13 coliphage DNA synthesis. *Eur. J. Biochem.* **67:** 37.

Nüsslein, V., B. Otto, F. Bonhoeffer, and H. Schaller. 1971. Function of DNA polymerase III in DNA replication. *Nature* **234:** 285.

Okazaki, R., M. Arisawa, and A. Sugino. 1971. Slow joining of newly replicated DNA chains in DNA polymerase I deficient *Escherichia coli* mutants. *Proc. Natl. Acad. Sci.* **68:** 2954.

Olsen, W. L., W. L. Staudenbauer, and P. H. Hofschneider. 1972. Replication of bacteriophage M13: Specificity of the *E. coli dnaB* function for replication of double-stranded M13 DNA. *Proc. Natl. Acad. Sci.* **69:** 2570.

Ray, D. S. and J. Dueber. 1975. Structure and replication of replicative forms of the φX-related bacteriophage G4. In *DNA synthesis and its regulation* (ed. M. Goulian et al.), p. 370. W. A. Benjamin, Menlo Park, California.

Ray, D. S., J. Dueber, and S. Suggs. 1975. Replication of bacteriophage M13. IX. Requirement of the *Escherichia coli dnaG* function for M13 duplex DNA replication. *J. Virol.* **16:** 348.

Ryan, M. J. 1976. Coumermycin A_1: A preferential inhibitor of replicative DNA synthesis in *Escherichia coli*. I. *In vivo* characterization. *Biochemistry* **15:** 3769.

Sakai, H., S. Hashimoto, and T. Komano. 1974. Replication of deoxyribonucleic acid in *Escherichia coli* C mutants temperature-sensitive in the initiation of chromosome replication. *J. Bacteriol.* **119:** 811.

Schekman, R. W. and D. S. Ray. 1971. Polynucleotide ligase and φX174 single-strand synthesis. *Nat. New Biol.* **231:** 170.

Schekman, R. W., M. Iwaya, K. Bromstrup, and D. T. Denhardt. 1971. The mechanism of replication of φX174 single-stranded DNA. III. An enzymatic study of the structure of the replicative form II DNA. *J. Mol. Biol.* **57:** 177.

Schubach, W. H., J. D. Whitmer, and C. I. Davern. 1973. Genetic control of DNA initiation in *Escherichia coli*. *J. Mol. Biol.* **74:** 205.
Smith, D. H. and B. D. Davis. 1967. Mode of action of novobiocin in *Escherichia coli*. *J. Bacteriol.* **93:** 71.
Staudenbauer, W. L. 1974. Involvement of DNA polymerases I and III in the replication of bacteriophage M13. *Eur. J. Biochem.* **49:** 249.
Staudenbauer, W. L. and P. H. Hofschneider. 1972a. Replication of bacteriophage M13. Mechanism of single-stranded DNA synthesis in an *Escherichia coli* mutant thermosensitive in chromosomal DNA replication. *Eur. J. Biochem.* **30:** 403.
———. 1972b. Replication of bacteriophage M13: Inhibition of single-stranded DNA synthesis by rifampicin. *Proc. Natl. Acad. Sci.* **69:** 1634.
Staudenbauer, W. L., W. L. Olsen, and P. H. Hofschneider. 1973. Analysis of bacteriophage M13 DNA replication in an *Escherichia coli* mutant thermosensitive in DNA polymerase III. *Eur. J. Biochem.* **32:** 247.
Steinberg, R. A. and D. T. Denhardt. 1968. Inhibition of synthesis of φX174 DNA in a mutant host thermosensitive for DNA synthesis. *J. Mol. Biol.* **37:** 525.
Sumida-Yasumoto, C., A. Yudelevich, and J. Hurwitz. 1976. DNA synthesis *in vitro* dependent upon φX174 replicative form I DNA. *Proc. Natl. Acad. Sci.* **73:** 1887.
Taketo, A. 1973. Sensitivity of *Escherichia coli* to viral nucleic acid. VI. Capacity of *dna* mutants and DNA polymerase-less mutants for multiplication of φA and φX174. *Mol. Gen. Genet.* **122:** 15.
———. 1975. Replication of φA and φX174 in *Escherichia coli* mutants thermosensitive in DNA synthesis. *Mol. Gen. Genet.* **139:** 285.
———. 1976a. Host genes involved in the replication of single-stranded DNA phage φK. *Mol. Gen. Genet.* **148:** 25.
———. 1976b. Host factor requirements and some properties of φXtB. An evolutionary aspect of φX-type phages. *Mol. Gen. Genet.* **148:** 139.
———. 1977a. Conversion of bacteriophage G4 single-stranded viral DNA to double-stranded replicative form in *dna* mutants of *Escherichia coli*. *Biochim. Biophys. Acta* **476:** 149.
———. 1977b. Effect of *dna* mutations on replication of S13, φR, and φKh-1. *J. Gen. Microbiol.* **23:** 29.
Tessman, E. S. and P. K. Peterson. 1976. Bacterial *rep*⁻ mutations that block development of small DNA bacteriophages late in infection. *J. Virol.* **20:** 400.
Truffaut, N. and I. Mannheimer. 1975. Development of bacteriophage Ha_2, a φX174 derivative, in *Escherichia coli* strains carrying a thermosensitive mutation in the *dnaG* gene. *Biochimie* **57:** 905.
Truitt, C. L. and J. R. Walker. 1974. Growth of phages λ, φX174, and M13 requires the *dnaZ* (previously *dnaH*) gene product of *Escherichia coli*. *Biochem. Biophys. Res. Commun.* **61:** 1036.
Tseng, B. Y. and D. A. Marvin. 1972. Filamentous bacterial viruses. V. Asymmetric replication of fd duplex deoxyribonucleic acid. *J. Virol.* **10:** 371.
Vito, C. C. and C. E. Dowell. 1976. Novel replicative properties of a capsid mutant of bacteriophage φX174. *J. Virol.* **18:** 942.
Vito, C. C., S. B. Primrose, and C. E. Dowell. 1975. The growth of a capsid mutant of bacteriophage φX174 in a temperature-sensitive strain of *Escherichia coli*. *J. Virol.* **15:** 281.
Wada, C. and T. Yura. 1974. Phenethylalcohol resistance in *Escherichia coli*. III. A

temperature-sensitive mutation (*dnaP*) affecting DNA replication. *Genetics* **77:** 199.

Wechsler, J. A. and J. D. Gross. 1971. *Escherichia coli* mutants temperature-sensitive for DNA synthesis. *Mol. Gen. Genet.* **113:** 273.

Weiner, J. H., R. McMacken, and A. Kornberg. 1976. Isolation of an intermediate which precedes *dnaG* RNA polymerase participation in enzymatic replication of bacteriophage φX174 DNA. *Proc. Natl. Acad. Sci.* **73:** 752.

Wickner, S. 1977. DNA or RNA priming of bacteriophage G4 DNA synthesis by *Escherichia coli dnaG* protein. *Proc. Natl. Acad. Sci.* **74:** 2815.

Wickner, S. and J. Hurwitz. 1974. Conversion of φX174 viral DNA to double-stranded form by purified *Escherichia coli* proteins. *Proc. Natl. Acad. Sci.* **71:** 4120.

———. 1975a. Interaction of *Escherichia coli dnaB dnaC(D)* gene products *in vitro*. *Proc. Natl. Acad. Sci.* **72:** 921.

———. 1975b. Association of φX174 DNA-dependent ATPase activity with an *Escherichia coli* protein, replication factor Y, required for *in vitro* synthesis of φX174 DNA. *Proc. Natl. Acad. Sci.* **72:** 3342.

———. 1976. Involvement of the *Escherichia coli dnaZ* gene product in elongation *in vitro*. *Proc. Natl. Acad. Sci.* **73:** 1053.

Zechel, K. and A. Kornberg. 1975. *DnaG* gene product, a rifampicin-resistant RNA polymerase, initiates conversion of single-stranded coliphage DNA to its duplex replicative form. *J. Biol. Chem.* **250:** 5995.

Host-factor Requirements and Some Properties of α3 and Related Phages

Akira Taketo and Ken-Ichi Kodaira
Department of Biochemistry
School of Medicine
Kanazawa University
Kanazawa, Ishikawa 920, Japan

More than 20 strains of circular single-stranded DNA phages (Microviridae) have been reported. Despite their morphological similarity, certain differences in their biological properties have been observed. On the basis of host range and antigenicity, Bradley (1970) classified circular single-stranded DNA phages into three groups: A, B, and C. Although considerable data are available for phages in groups B (e.g., φX174 and S13) and C (e.g., St-1 and φK), little is known about the replication properties of group-A phages. Moreover, the relationships among the three groups and several other strains isolated recently (such as U3, G4, G13, and G14) remain to be elucidated. In this article the properties of α3, a representative of the group-A phages, are compared with those of other microvirid phages, including U3, G4, and G14.

Properties of α3

The host range of α3 is usually limited to *Escherichia coli* C and B strains (Bradley 1970). Like φX, α3 can infect *E. coli* 15T$^-$, CR63·1, C600·1, and W2252 S-14 (Taketo 1977a). In contrast to φX and φA, α3 plates well on *E. coli* C/5, C/15, and AX729, but not on C/62. On the other hand, G4 forms plaques on C/62 but not on C/5, whereas G13 and G14 cannot infect C/15 (Godson 1974). In Table 1 the plating efficiency of α3 on newly isolated lipopolysaccharide (LPS) mutants of *E. coli* C (Feige and Stirm 1976) and B (Prehm et al. 1975) strains is compared with the plating efficiencies of several microvirid phages. Phages α3, G13, and G14 were unable to infect *E. coli* C71 lacking branched glucose residues. However, the role of this glucose residue is uncertain because strain C23·1, which is deficient in branched glucose as well as in galactose residues, was sensitive to these phages. The terminal galactose residue of *E. coli* C LPS is essential for G4 and φA, as it is for φX (Feige and Stirm 1976). Strain B and its LPS mutant BB5 (deficient in terminal glucose) are sensitive, whereas BB2 deficient in terminal and subterminal glucose) is resistant to α3, G13, and G14, which indicates that subterminal glucose is involved in the viral

Table 1 Efficiency of plating of $\alpha3$ and related phages

Strain	$\alpha3$	φKh-1	G4	G13	G14
C	2.2×10^{10}	2.0×10^{10}	2.0×10^{10}	2.1×10^{10}	2.2×10^{10}
C61	6.9×10^{9}	10^{1}	1.6×10^{10}	1.6×10^{10}	1.5×10^{10}
C71	$< 10^{1}$	$< 10^{1}$	2.9×10^{10}	$< 10^{1}$	3.2×10^{4}
C23	2.0×10^{10}	6.4×10^{10}	$< 10^{1}$	5.5×10^{9}	5.2×10^{8}
C23·1	1.3×10^{10}	6.4×10^{10}	$< 10^{1}$	1.4×10^{10}	5.2×10^{8}
BB	2.3×10^{8}	3.5×10^{10}	$< 10^{1}$	2.5×10^{8}	2.8×10^{6}
BB5	4.5×10^{8}	3.4×10^{9}	$< 10^{1}$	3.1×10^{9}	4.8×10^{9}
BB1	$< 10^{1}$	7.3×10^{7}	$< 10^{1}$	$< 10^{1}$	$< 10^{1}$
BB7	$< 10^{1}$	1.2×10^{10}	$< 10^{1}$	$< 10^{1}$	$< 10^{1}$

adsorption. The host range of φKh-1 (Taketo 1977a) is complicated, and the correlation between phage sensitivity and LPS structure is rather obscure. The phage-resistant mutants listed above can be transfected by the viral DNA upon treatment of the recipient cells with Ca^{++} or Ba^{++} (Taketo 1977b).

As reported by Bradley (1970), $\alpha3$ is not affected by anti-φX serum (Fig. 1a) but is slowly inactivated by antiserum directed against group-C phage St-1 (Fig. 1b). Group-C phage φK is neutralized by anti-$\alpha3$ serum at a

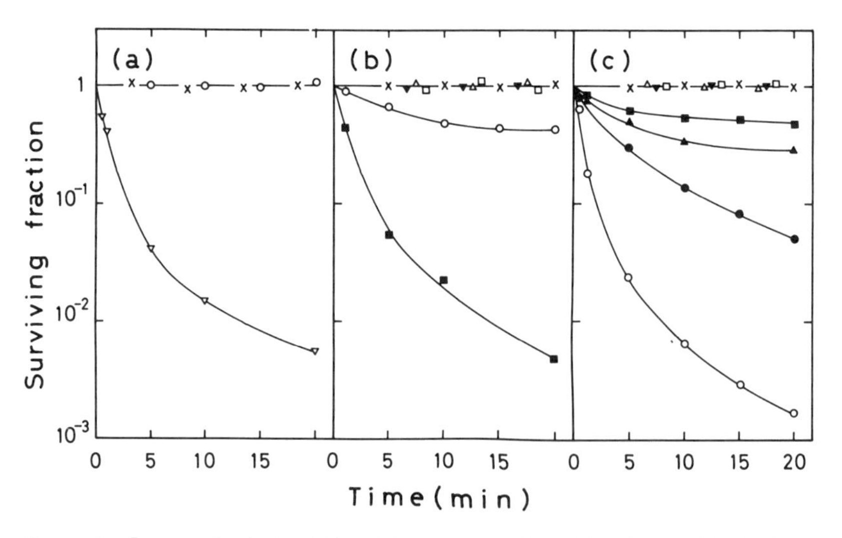

Figure 1 Immunological relationships among microvirid phages. Each phage was incubated at 37°C with antiserum against (*a*) φX174, (*b*) St-1, or (*c*) $\alpha3$, and the change in infectivity was followed. (o) $\alpha3$; (△) G4; (□) G14; (▽) φX174; (●) φK; (▲) φXtB; (■) St-1; (▼) G13; (x) U3.

considerable rate (Fig. 1c). St-1 and φXtB are partially sensitive, whereas group-B phages, such as φA and G6, are insensitive to antiserum against $\alpha 3$. In contrast to group-C phages, U3, G4, G13, and G14 are resistant to antisera directed against $\alpha 3$ or St-1. Like G4 and G14 (Godson 1974), U3 is not inactivated by anti-φX serum. These results indicate that group-A and group-C phages are related immunologically, whereas U3, G4, and G14 are distinct from all three groups.

The sedimentation rates for $\alpha 3$ phage particles, viral single-stranded (SS) DNA, and replicative-form (RF) DNA in sucrose gradients are only slightly higher than those for φX. Unlike with U3, G4, G13, and G14, multiplication of $\alpha 3$ proceeds at 42–43°C but not below 25°C.

Host functions involved in the growth of $\alpha 3$ have been investigated using various replication mutants of *E. coli*. Growth of virus was not particularly thermosensitive in *dnaA*, *dnaB*, *dnaC(D)*, *dnaH*, *dnaI*, and *dnaP* mutants. On the other hand, multiplication of $\alpha 3$ was restricted at 43°C in *dnaE*, *dnaF(nrdA)*, *dnaG*, and *dnaZ* cells, indicating requirements for DNA polymerase III, ribonucleotide reductase, *dnaG* protein, and DNA elongation factor II. With regard to *dna*-factor requirements, $\alpha 3$ is close to group-C phages (Taketo 1976a,b) and distant from group-B strains (Taketo 1973, 1975). At 43°C the yield of $\alpha 3$ was decreased in $polA^{ts}$ mutants and reduced markedly in a temperature-sensitive *lig*7 strain. In addition, phage growth was abortive even at 33–37°C in D43 *rep* mutants treated with Ca^{++} and then exposed to $\alpha 3$ SS or RF DNA. Multiplication of $\alpha 3$ was considerably thermosensitive in BW2001 *xth* mutants. Whether this effect is due to deficiency in exonuclease III and/or endonuclease II is unknown.

Recent experiments in vivo have shown that formation of $\alpha 3$ parental RF DNA requires *dnaG* and *dnaZ* gene products, whereas replication of $\alpha 3$ progeny RF DNA depends on *dnaE*, *dnaG*, *dnaZ*, and *rep* functions. Synthesis of RF DNA proceeds in the presence of 30 μg/ml chloramphenicol or 200 μg/ml rifampicin. In addition, synthesis of the parental RF DNA is resistant, but replication of progeny RF DNA is sensitive, to 50 μg/ml mitomycin C or nalidixic acid or 200 μg/ml novobiocin.

Host-factor Requirements of U3

As a member of the group-C phages, U3 is K12-specific (Watson and Paigen 1971), but growth of U3 was restricted at 42–43°C even in dna^+ hosts. Replication of the progeny RF DNA was markedly affected at the high temperature, as determined by [^{3}H]thymidine labeling. However, parental RF DNA was formed in dna^+ bacteria at 43°C. In PC7 *dnaC(D)* and PC3 *dnaG* cells no RF DNA was detected at the restrictive temperature (Fig. 2). In contrast, synthesis of RF DNA occurred at 43°C in FA21 *dnaB*,

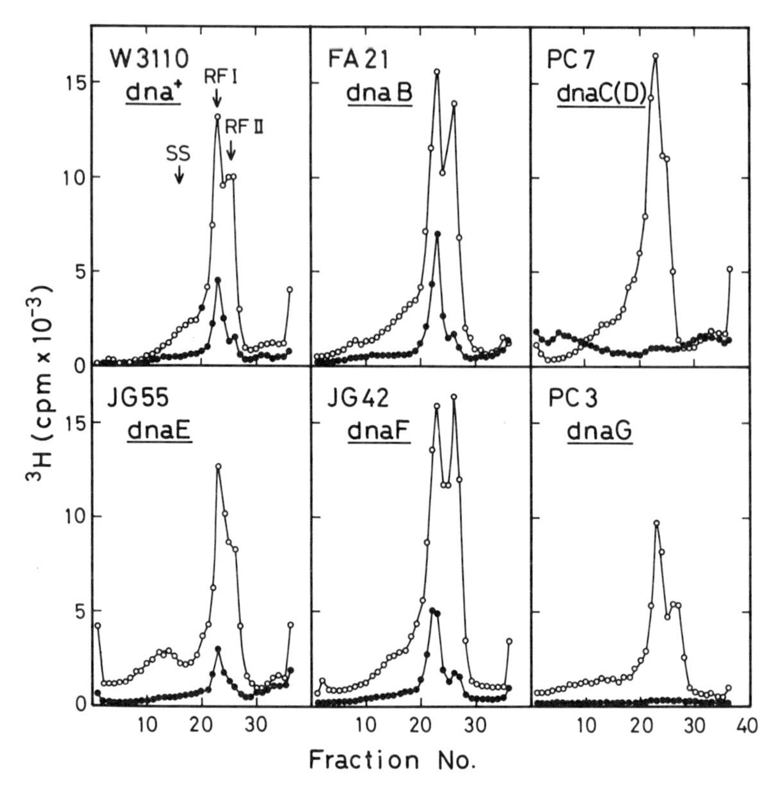

Figure 2 Synthesis of U3 DNA in *dna* mutants of *E. coli.* Cells grown exponentially at 30°C in Tris-casamino acids-glucose (TCG) medium supplemented with 2 μg/ml thymine (Kodaira and Taketo 1977) were divided into two portions. One portion was shifted up to 43°C and the other was kept at 30°C. At 10 min after the shift, 1/100 vol 1 M $CaCl_2$ and U3 (input multiplicity = 10) were added. After 20 sec, [*methyl*-^{3}H]thymidine was added to 10 μCi/ml and labeling was continued for 10 min. Isotope incorporation was terminated, the cells lysed, and viral DNA sedimented through neutral sucrose gradients as described previously (Kodaira and Taketo 1977). The direction of sedimentation is from right to left. (o) 30°C; (•) 43°C.

JG55 *dnaE*, JG42 *dnaF(nrdA)*, BT4113 *polA*^{ts}, and KS268 *lig*7 (ts) mutants.

Host-factor Dependence of G4

Synthesis of G4 parental RF DNA, like that of *a*3, is dependent on only the *dnaG* and *dnaZ* functions (Kodaira and Taketo 1977). On the other hand, *dnaB*, *dnaC(D)*, *dnaE*, *dnaG*, *dnaZ*, and *rep* gene products are essential for replication of G4 progeny RF DNA. Thus, with regard to the replicative system for progeny RF DNA, G4 resembles φX and G4 *tr*1 (Derstine et al. 1976), rather than α3.

Replicative Properties of G14

For G14, assignment of host factors was complicated by the unusual thermosensitivity of viral growth (Taketo 1977c). Conversion of infecting G14 SS DNA (labeled with [^{14}C]thymine) into RF DNA was affected at 43°C (Fig. 3a) and the rate of progeny RF DNA synthesis (as measured by pulse-labeling for 1 min with [^{3}H]thymidine) was negligible (Fig. 3b). By increasing the labeling time to 10 minutes, G14 RF DNA synthesis is detectable at 43°C in *dna*$^+$ and *dnaA* hosts (Fig. 3c,d). Even under these conditions, however, no RF DNA was detected in *dnaB*, *dnaC(D)*, and *dnaG* mutants. In addition, the amount of RF DNA synthesized in *dnaE* and *dnaZ* strains was insignificant. Thus, G14 resembles φX, rather than G4 and α3, with regard to host-factor requirements for parental RF DNA synthesis.

DISCUSSION

Regardless of their different host ranges, group-A phage α3 and some group-C phages (e.g., St-1 and φK) closely resemble each other in their host-factor requirements. Moreover, the two groups are to a certain degree immunologically related. It seems reasonable, therefore, to combine groups A and C into one large group. On the other hand, G14, like α3, is able to replicate in *E. coli* C and B strains but, unlike α3, is insensitive to anti-α3 serum and depends on *dnaB* and *dnaC(D)* functions. Although U3 and group-C phages are K12-specific, synthesis of U3 parental RF DNA is dependent on at least the *dnaC(D)* gene product. Thus, with regard to requirements for parental RF DNA synthesis, G4 is similar to α3 and different from group-B phage φX. However, G4 resembles φX in host range and with regard to the replicative system for progeny RF DNA. Immunologically, U3, G4, and G14 are not related to microvirid phages belonging to group A, B, or C, and further classification of these three phages is difficult at present. Although the permissive growth temperature for these strains (and G13 and φC as well) is relatively low, a direct relationship does not exist between growth-temperature range and host-factor requirements or host range.

Microvirid phages thus exhibit considerable divergence in biological properties. Recently, overlapping genes have been discovered in φX DNA (Barrell et al. 1976). Even point mutations, were they to occur in overlapping genes, might alter more than a single phenotype and thus facilitate such evolutional diversity.

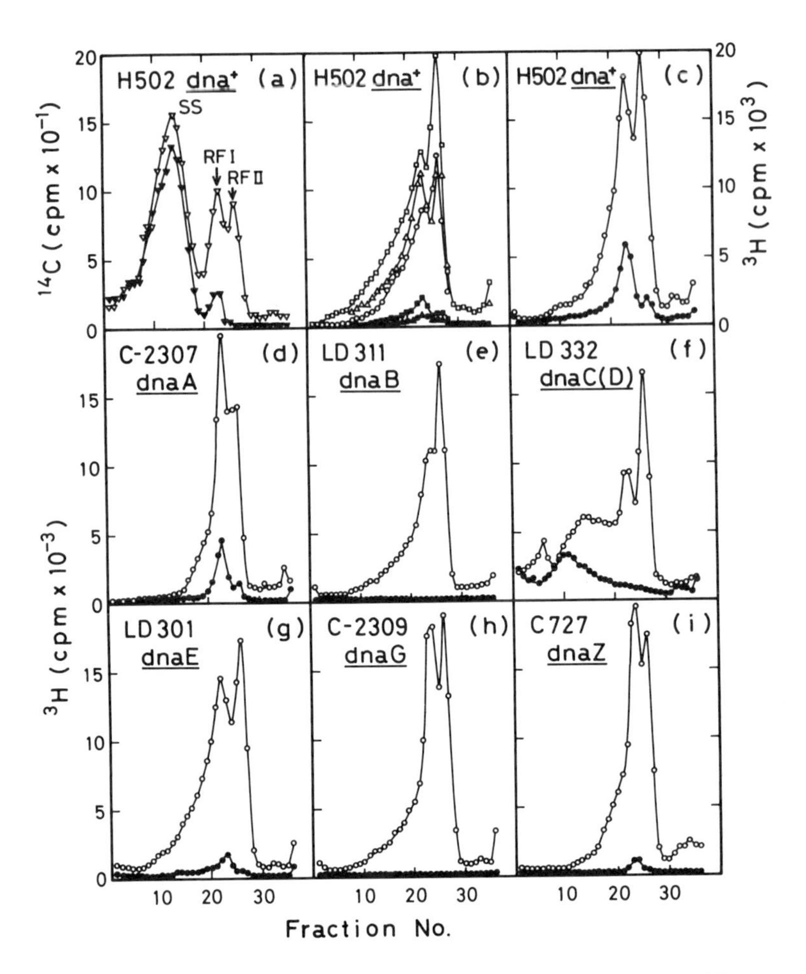

Figure 3 Synthesis of G14 DNA in *dna* mutants of *E. coli.* (*a*) Exponentially growing cells of H502 dna^+ in TCG medium supplemented with 2 μg/ml thymine were divided into two portions. One portion was shifted up to 43°C and the other retained at 30°C. After 10 min, G14 labeled with [^{14}C]thymine (1.3 × 10^{-6} cpm/plaque-forming unit) was added to a multiplicity of 5, together with 1/100 vol 1 M $CaCl_2$. At 10 min after infection the culture was poured into ice-cold 75% ethanol in Tris-EDTA, the cells lysed, and DNA extracted and analyzed as described previously (Kodaira and Taketo 1977). (▽) 30°C; (▼) 43°C. (*b*) A log-phase culture of H502 dna^+ in TCG medium containing 2 μg/ml thymine was divided into 5-ml portions and incubated at 43°C or 30°C. After 10 min, 0.05 ml 1 M $CaCl_2$ and G14 (input multiplicity = 10) were added to each culture. At 4, 9, or 14 min postinfection the cultures were pulse-labeled for 1 min with [^{3}H]thymidine (added to 10 μCi/ml). The incorporation was terminated, the cells lysed, and the DNA analyzed by sucrose gradient centrifugation. Open symbols, 30°C; closed symbols, 43°C; (△ and ▲) pulse-labeled at 4 min; (□ and ■) pulse-labeled at 9 min; (o and ●) pulse-labeled at 14 min. (*c–i*) Synthesis of G14 DNA in *dna* mutants was followed by means of [^{3}H]thymidine-labeling, as described for U3 (Fig. 2). (o) 30°C; (●) 43°C.

REFERENCES

Barrell, B. G., G. M. Air, and C. A. Hutchison III. 1976. Overlapping genes in bacteriophage φX174. *Nature* **264:** 34.

Bradley, D. E. 1970. A comparative study of some properties of the φX174 type bacteriophages. *Can. J. Microbiol.* **16:** 965.

Derstine, P. L., L. B. Dumas, and C. A. Miller. 1976. Bacteriophage G4 DNA synthesis in temperature-sensitive *dna* mutants of *Escherichia coli. J. Virol.* **19:** 915.

Feige, U. and S. Stirm. 1976. On the structure of the *Escherichia coli* C cell wall lipopolysaccharide core and on its φX174 receptor region. *Biochem. Biophys. Res. Commun.* **71:** 566.

Godson, G. N. 1974. Evolution of φX174. Isolation of four new φX-like phages and comparison with φX174. *Virology* **58:** 272.

Kodaira, K. and A. Taketo. 1977. Conversion of bacteriophage G4 single-stranded viral DNA to double-stranded replicative form in *dna* mutants of *Escherichia coli. Biochim. Biophys. Acta* **476:** 149.

Prehm, P., S. Stirm, B. Jann, and K. Jann. 1975. Cell-wall lipopolysaccharide from *E. coli* B. *Eur. J. Biochem.* **56:** 41.

Taketo, A. 1973. Sensitivity of *Escherichia coli* to viral nucleic acid. VI. Capacity of *dna* mutants and DNA polymerase-less mutants for multiplication of φA and φX174. *Mol. Gen. Genet.* **122:** 15.

———. 1975. Replication of φA and φX174 in *Escherichia coli* mutants thermosensitive in DNA synthesis. *Mol. Gen. Genet.* **139:** 285.

———. 1976a. Host genes involved in the replication of single-stranded DNA phage φK. *Mol. Gen. Genet.* **148:** 25.

———. 1976b. Host factor requirements and some properties of φXtB. *Mol. Gen. Genet.* **148:** 139.

———. 1977a. Effect of *dna* mutations on replication of S13, φR, and φKh-1. *J. Gen. Appl. Microbiol.* **23:** 29.

———. 1977b. Sensitivity of *Escherichia coli* to viral nucleic acid. XII. Ca^{2+}- or Ba^{2+}-facilitated transfection of cell envelope mutants. *Z. Naturforsch.* **32c:** 429.

———. 1977c. Sensitivity of *Escherichia coli* to viral nucleic acid. XIII. Replicative properties of φX174-type phages. *J. Gen. Appl. Microbiol.* **23:** 85.

Watson, G. and K. Paigen. 1971. Isolation and characterization of an *Escherichia coli* bacteriophage requiring cell wall galactose. *J. Virol.* **8:** 669.

Replication of M13 Duplex DNA In Vitro

Walter L. Staudenbauer, Barbara E. Kessler-Liebscher, Peter K. Schneck, Bob van Dorp, and Peter Hans Hofschneider
Max-Planck-Institut für Biochemie
8033 Martinsried, Federal Republic of Germany

Replication of M13 DNA proceeds through three consecutive stages: (I) conversion of the viral single-stranded DNA to the double-stranded replicative form (SS→RF), (II) multiplication of replicative forms (RF→RF), and (III) synthesis of progeny single strands (RF→SS). Stages II and III employ M13 duplex DNA as template and depend on the phage-encoded gene-II protein. The switch from double-strand to single-strand synthesis is brought about by the accumulation of M13 gene-V protein (see review by Ray 1977).

For a biochemical characterization of the enzymatic reactions involved in M13 DNA replication, appropriate in vitro systems are needed. The term "in vitro," as understood here, refers to a situation where the permeability barrier of the cell has been destroyed and DNA synthesis has become dependent on exogenous nucleotides both as substrates and as energy supply. It is useful to make a distinction between cellular and cell-free in vitro systems. Cellular in vitro systems (nucleotide-permeable cells) still retain the in vivo concentration and spatial arrangement of the components of the replication apparatus. Cell-free systems are obtained after complete disruption of the cells and removal of cellular debris by high-speed centrifugation. Cellular systems may resemble the living state more closely, but they have the disadvantage that they are in general not permeable to proteins. On the other hand, cell-free systems can be fractionated and complemented with macromolecules but they are prone to experimental artifacts due to the disruption of the cell.

As far as M13 DNA replication is concerned, only the first stage, the SS→RF conversion, has been reconstituted with purified components (Geider and Kornberg 1974). We have therefore embarked upon developing in vitro systems that can carry out the further replication of M13 duplex DNA. To study specifically RF→RF replication, *Escherichia coli* cells infected with the M13 *am*5 mutant (amber in gene V) were used as starting material. For the work described in this review, three types of in vitro systems were employed: a plasmolysed cellular system obtained by treating infected cells with 2 M sucrose (Staudenbauer 1974), a highly concentrated cell-free system prepared by concentrating crude extracts on a cellophane disk and then transferring the disk onto a drop of incubation mixture (Schneck et al. 1978), and a soluble cell-free system consisting of crude extract supplemented with DNA-binding protein(s) (B. van Dorp et al., in prep.).

All three systems are capable of incorporating deoxyribonucleoside triphosphates into endogenous M13 RF DNA by a semiconservative process. The incorporation is dependent on ATP and $MgCl_2$ and stimulated by the addition of KCl. Occasionally, DNA synthesis is enhanced upon addition of all four ribonucleoside triphosphates. Before describing some results obtained with these systems, we would like to emphasize that even gentle methods of disrupting the cell membrane may damage the replication apparatus and thus lead to artifactual reactions.

Structure of Replicative Intermediates

The classical approach for establishing a biochemical pathway consists of characterizing the reaction intermediates and ordering them into a logical sequence. For the purpose of isolating replicating DNA molecules, nucleotide-permeable cells offer several experimental advantages. Elimination of the permeability barrier and control of substrate concentration greatly facilitate pulse-labeling. Moreover, the ability to change reactants quickly makes possible short-time pulse-chase experiments that are not feasible in vivo. We therefore employed the plasmolysed cell system for analyzing the structure of replicative intermediates in M13 RF→RF replication (Kessler-Liebscher et al. 1978). In short, cells were pulsed with $[^3H]$dTTP, and the phage DNA was selectively extracted by a modified Hirt procedure and fractionated by BND-cellulose chromatography or dye-buoyant density centrifugation.

Electron microscopy of replicative intermediates revealed the two classes of molecules shown in Figure 1: circles with single-stranded tails (σ [sigma] structures) and partially double-stranded circles (o [omicron] structures). In addition, duplex rings with two single-stranded tails were observed, which probably arise from single-tailed molecules by branch migration. It was found that replicative intermediates in viral-strand synthesis (RI_v) that contain labeled viral strands of more than unit genome length exhibit restricted binding of intercalating dyes. Unwinding of the helix in these molecules is probably blocked by noncovalent interaction between the 3′- and 5′-terminal regions of the elongated viral strand (Espejo and Sinsheimer 1976).

In addition to elongated viral strands, pulse-labeled viral-strand DNA shorter than unit genome length was also detected. This observation does not necessarily imply a discontinuous mode of viral-strand synthesis. These DNA fragments might result from excision repair at sites of deoxyuridine

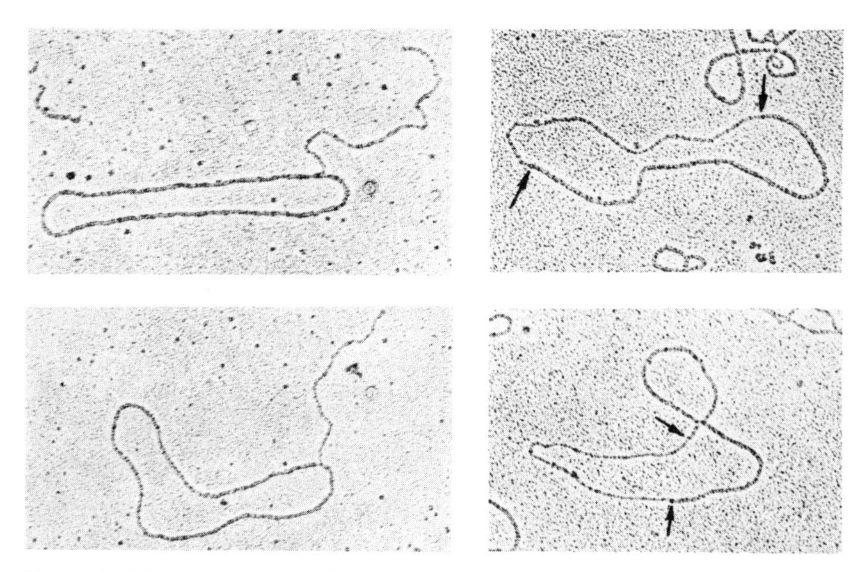

Figure 1 Electron micrographs of intermediate of M13 RF DNA replication in plasmolysed cells. The arrows indicate single-stranded regions in partially double-stranded molecules.

misincorporation (Tye et al. 1977). Moreover, site-specific nicking at the origin/terminus obviously is required for termination of viral-strand synthesis.

It should be noted that the tails of the σ structures as observed by electron microscopy are always single-stranded and never longer than one genome. We infer from this observation that complementary-strand synthesis takes place only after the release of the displaced viral single strand from the replicative intermediate (RI) upon termination of one round of viral-strand synthesis. The partially double-stranded o structures may represent intermediates in complementary-strand synthesis (RI_c). However, the data presented here do not preclude the possibility that such structures might be degradation products produced as a consequence of damage to the cell membrane. A hypothetical scheme of M13 RF→RF replication is presented in Figure 2. It is proposed that viral and complementary strands are synthesized separately in two consecutive replication cycles. Thus, M13 duplex DNA replication proceeds by a two-cycle mechanism (RF→SS→RF).

Role of DNA-binding Proteins

The gene-V protein of filamentous phages has been characterized as a DNA-binding protein (DBP) which binds cooperatively to single-stranded DNA (Oey and Knippers 1972; Alberts et al. 1972). It is assumed that the

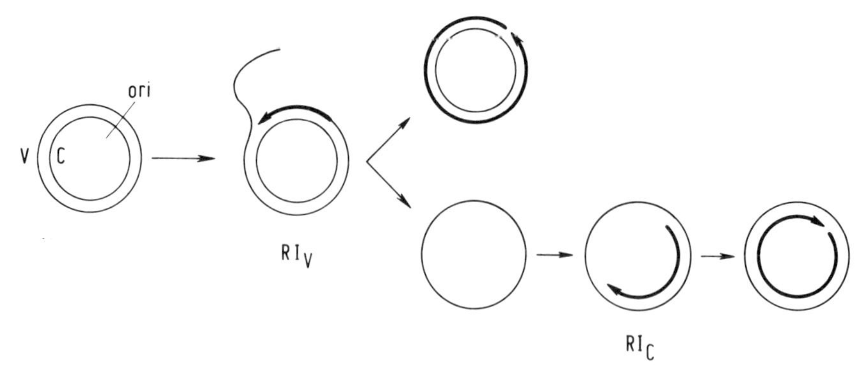

Figure 2 Two-cycle model for the replication of M13 RF DNA.

gene-V protein causes a switch from RF→RF replication to RF→SS synthesis by repressing complementary-strand synthesis on progeny viral single strands (Salstrom and Pratt 1971; Mazur and Model 1973; Mazur and Zinder 1975). Circumstantial in vivo evidence has also been presented which suggests that this protein might exert a positive effect on the production of progeny single strands (Staudenbauer and Hofschneider 1973).

To obtain further information on the role of gene-V protein, we investigated its effect on RF→RF replication in a cellophane-disk system prepared from M13 *am*5-infected cells. Phage DNA synthesized with or without addition of exogenous gene-V protein was density-labeled with bromodeoxyuridine triphosphate (BrdUTP) and analyzed by equilibrium centrifugation in a CsCl gradient as shown in Figure 3. This technique allows for the separation of the two types of hybrid DNA molecules according to differences in the thymine content of the viral and complementary strands (Kessler-Liebscher and Staudenbauer 1976). In the control experiment (Fig. 3A), peaks of radioactivity at densities 1.745, 1.76, and 1.81 are observed, which correspond, respectively, to light-heavy (complementary strand labeled with bromodeoxyuridine), heavy-light (viral strand labeled), and heavy-heavy (both strands labeled) RF DNA. It can be seen from the gradient shown in Figure 3B that addition of gene-V protein interferes specifically with the formation of light-heavy RF DNA containing a newly synthesized complementary strand.

Also, a DBP isolated from uninfected *E. coli* cells (Sigal et al. 1972) has been implicated in several aspects of bacterial DNA replication (Molineux et al. 1974; Weiner et al. 1975). It was therefore interesting to compare the effects of exogenous DBP with those of exogenous gene-V protein when added to the cellophane-disk system. As shown in Figure 3C, addition of DBP stimulates the synthesis of both types of hybrid RF molecules (note the difference in scale) and markedly increases the relative amount of heavy-heavy RF DNA. This intimates that DBP is involved not only in

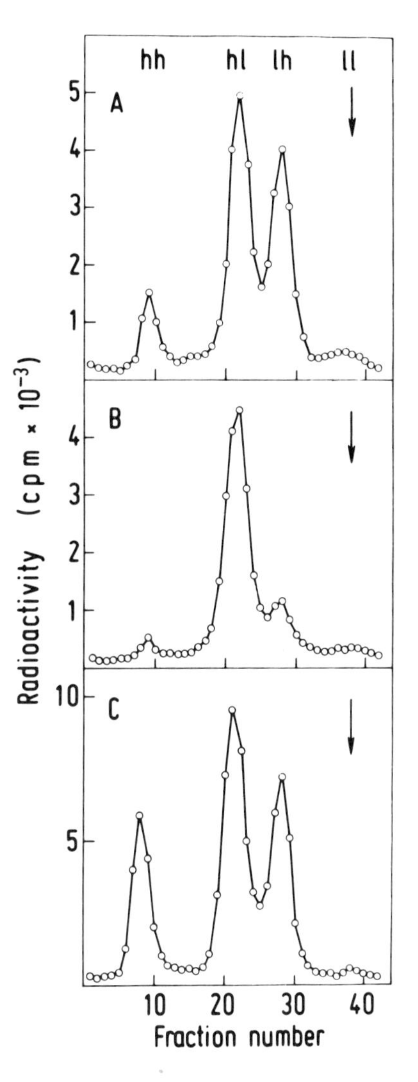

Figure 3 Isopycnic analysis of BrdU-labeled M13 RF DNA. An extract (3 μl) from M13 *am*5-infected cells was concentrated on a cellophane disk (*A*) without or (*B*) after addition of gene-V protein and (*C*) DNA-binding protein as described by Schneck et al. (1978). The disks were incubated on 25-μl drops of incubation mixture at 30°C for 30 min. Incubation mixtures contained 20 mM HEPES (pH 8.0), 0.1 M KCl, 10 mM $MgCl_2$, 2 mM ATP, 0.1 mM NAD, 0.05 mM each of CTP, GTP, and UTP, 0.5 mM of BrdUTP, 0.025 mM each of dATP, dCTP, and [^{3}H]dGTP. Density-labeled DNA was extracted by incubating the disks with 0.5% Sarkosyl and proteinase K. Equilibrium centrifugation in neutral CsCl was performed as described by Kessler-Liebscher and Staudenbauer (1976). The arrow indicates the position of completely light RF DNA.

complementary-strand synthesis (Geider and Kornberg 1974) but also in viral-strand synthesis. Similar observations have been made for the replication of φX174 RF DNA in vitro (Eisenberg et al. 1977; Scott et al. 1977). The finding that a significant portion of the reaction product is fully synthetic RF DNA provides strong evidence that the newly synthesized DNA can function as a template for further replication. It also indicates that the cell-free system has retained the capacity for reinitiation, as has been shown for the plasmolysed cell system (Kessler-Liebscher and Staudenbauer 1976).

The stimulatory effect of DBP on RF DNA replication in the cellophane-disk system prompted us to test the effect of DBPs on M13 DNA synthesis in a "soluble" system prepared from crude extracts (Fig. 4). It was found that the addition of these proteins stimulated markedly the low incorporation normally observed in unconcentrated extracts from M13 *am*5-infected cells. The DNA synthesis obtained with optimal amounts of DBP is about twice as high as that stimulated by gene-V protein. This is in accord with the assumption that DBP is involved in the synthesis of both the viral and the complementary strands, whereas gene-V protein selectively enhances viral-strand synthesis. It should be noted that the replicative capacity (expressed as nucleotide incorporation per mg of total protein) of the soluble in vitro system supplemented with DBP is still an order of magnitude lower than that of the cellophane-disk system. Thus, the concentration of other replication proteins might also be rate-limiting in the soluble system.

Effect of Inhibitors

Specific inhibitors have been extremely useful in identifying the enzymes involved in complex biochemical pathways. We have therefore investigated the effects of several commonly employed inhibitors of nucleic synthesis on the various stages of M13 DNA replication in cell-free extracts. The results are summarized in Table 1. As far as RF→RF replication is concerned, it should be kept in mind that a selective inhibition of viral-strand synthesis will also prevent complementary-strand synthesis because of the intrinsic asymmetry of the replication process.

A direct involvement of RNA polymerase in RF→RF replication is indicated by the inhibitory effect of rifampicin. Analysis of the reaction product provides evidence for a preferential inhibition of complementary-strand synthesis (Schneck et al. 1978). This finding supports the model shown in Figure 2, which suggests that during RF→RF replication complementary strands are initiated by an RNA priming mechanism identical to

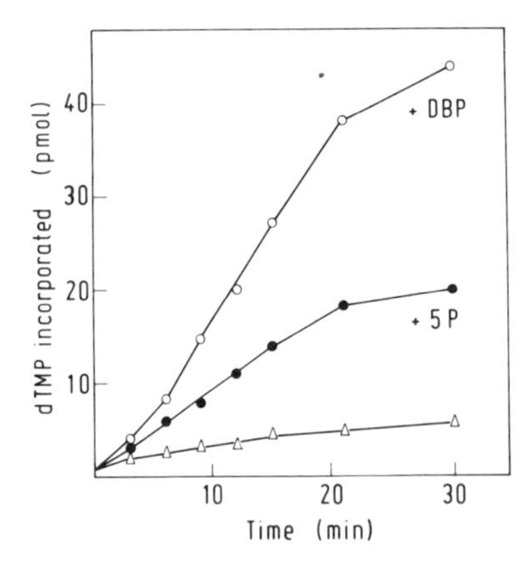

Figure 4 Stimulation of DNA synthesis in soluble extracts by DNA-binding proteins. Incubation mixtures (150 μl) containing 100 μl crude extract from M13 *am*5-infected cells were incubated at 30°C without and with addition of gene-V protein (100 μg/ml) or DBP (70 μg/ml). Components of the reaction mixture are as described in the legend to Fig. 3, except that BrdUTP was replaced with 0.025 mM [^{3}H]dTTP. At the times specified, aliquots (20 μl) were removed and assayed for acid-insoluble radioactivity.

Table 1 Effect of inhibitors on the different stages of M13 DNA replication in cell-free extracts.

	[^{3}H]dTMP incorporation (%)		
Addition	SS → RF	RF → RF	RF → SS
None	100	100	100
+ Rifampicin (5 μg/ml)	3	36	97
+ Streptolydigin (50 μg/ml)	85	22	39
+ AraCTP (0.5 mM)	3	3	4
+ Nalidixic acid (50 μg/ml)	102	25	27
+ Oxolinic acid (50 μg/ml)	97	12	11
+ Novobiocin (10 μg/ml)	100	23	24
+ Coumermycin (10 μg/ml)	97	7	13

Soluble extracts were prepared by the procedure of Wickner et al. (1972). Conversion of SS to RF DNA was assayed with extracts from uninfected cells upon addition of exogenous M13 SS DNA. RF → RF replication and RF → SS synthesis were assayed with extracts from M13 *am*5-infected cells supplemented with DBP (70 μg/ml) or gene-V protein (100 μg/ml), respectively. Reaction mixtures (40 μl) were as described in the legend to Fig. 4, and the incubations were carried out for 30 min at 30°C. One-hundred percent [^{3}H]dTMP incorporation corresponds to 50.4 pmoles (SS → RF), 117.9 pmoles (RF → RF), and 54.4 pmoles (RF → SS).

that involved in the SS→RF conversion (Wickner et al. 1972). Streptolydigin, which interferes with the elongation of RNA chains by RNA polymerase (Cassani et al. 1971), has an inhibitory effect on RF→RF replication as well as on RF→SS synthesis. This result intimates that blocking of transcription not only prevents complementary-strand synthesis but may also affect the synthesis of viral strands (Staudenbauer and Hofschneider 1972; Fidanián and Ray 1974). The relative insensitivity of complementary-strand synthesis to streptolydigin is due to the fact that high concentrations (600 μg/ml) of this drug are necessary to inhibit RNA synthesis on a single-stranded template (Schekman et al. 1972).

All three stages of M13 DNA replication exhibit similar sensitivities to arabinosyl nucleoside triphosphates. These substrate analogs have been shown to affect replicative DNA synthesis by interfering with DNA polymerase III function (Staudenbauer 1976). Therefore, this enzyme is most likely to be responsible for polymerization of deoxynucleotides during viral-strand, as well as complementary-strand, synthesis. However, participation of the other DNA polymerases in M13 DNA synthesis is by no means excluded.

Nalidixic acid and the structurally related oxolinic acid inhibit primarily viral-strand synthesis and have no inhibitory effect in the SS→RF system. This also applies to novobiocin and the related drug coumermycin. The evidence available suggests that these drugs interfere with the unwinding of double-stranded DNA (A. Sugino et al., pers. comm.). Novobiocin and coumermycin block DNA gyrase, an enzyme which introduces negative superhelical turns into covalently closed, circular DNA (Gellert et al. 1976).

Since most of the phage DNA present in the cell-free extract is already in the form of supercoiled RFI DNA (superhelical, covalently closed, circular duplex), it appears that DNA gyrase is not only essential for supertwisting of RFIV DNA (nonsuperhelical, covalently closed, circular duplex) (Marians et al. 1977) but may also be involved in the initiation and/or elongation of the viral strand.

CONCLUSIONS

Our studies on M13 DNA replication in vitro lead to the conclusion that there are basically only two reactions involved: viral-strand synthesis on double-stranded RF templates and complementary-strand synthesis on free viral single strands. In the absence of gene-V protein, these two reactions seem to proceed with comparable rates, and the newly produced viral strands are quickly converted to duplex DNA.

Both viral-strand and complementary-strand synthesis employ DNA polymerase III as the main polymerizing activity. The detection of replicative intermediates with elongated viral strands of more than unit genome length suggests that viral-strand synthesis is initiated by endonucleolytic cleavage of the parental viral strand. This initiation event apparently requires the M13 gene-II protein (Fidanián and Ray 1972). On the other hand, complementary-strand synthesis is primed by an RNA chain synthesized by RNA polymerase. Both viral-strand and complementary-strand synthesis require high concentrations of DBP. Gene-V protein stimulates viral-strand synthesis (possibly by promoting the recycling of DBP at the replication fork) but blocks complementary-strand synthesis. Adding this protein to a cell-free replication system results in a switch from duplex DNA replication to RF→SS synthesis.

The conclusion that a separate mechanism of RF→RF replication does not exist is in accord with the in vivo data of Horiuchi and Zinder (1976). It is apparently at variance with the streptolydigin sensitivity of viral-strand synthesis (Table 1) and with experiments showing that thermal inactivation of the *dnaB* or the *dnaG* protein affects specifically RF→RF replication (Staudenbauer et al. 1974; Ray et al. 1975; Dasgupta and Mitra 1976). However, thermosensitivity of M13 duplex DNA replication in these *dna* mutants does not necessarily imply a direct involvement of the corresponding gene products in RF DNA replication. Thus, it may be argued that a shift to nonpermissive conditions affects M13 duplex DNA synthesis by causing a depletion of the intracellular pool of DBP (Weiner et al. 1975). As far as RF→SS synthesis is concerned, a deficiency in DBP might be compensated for by accumulation of gene-V protein (Staudenbauer and Hofschneider 1973). Since cell-free systems capable of sustaining all three stages of M13 DNA replication are now available, in vitro complementation experiments with mutant extracts can be carried out to test this hypothesis.

ACKNOWLEDGMENT

This work was supported by a grant from the Deutsche Forschungsgemeinschaft.

REFERENCES

Alberts, B., L. Frey, and H. Delius. 1972. Isolation and characterization of gene 5 protein of filamentous bacterial viruses. *J. Mol. Biol.* **68:** 139.

Cassani, G., R. R. Burgess, H. M. Goodman, and L. Gold. 1971. Inhibition of RNA polymerase by streptolydigin. *Nat. New Biol.* **230:** 197.

Dasgupta, S. and S. Mitra. 1976. The role of *Escherichia coli dna*G function in coliphage M13 DNA synthesis. *Eur. J. Biochem.* **67:** 47.

Eisenberg, S., J. Griffith, and A. Kornberg. 1977. φX174 cistron A protein is a multifunctional enzyme in DNA replication. *Proc. Natl. Acad. Sci.* **74:** 3198.

Espejo, R. T. and R. L. Sinsheimer. 1976. The process of infection with bacteriophage φX174. XXXIX. The structure of a DNA form with restricted binding of intercalating dyes observed during synthesis of φX single-stranded DNA. *J. Mol. Biol.* **102:** 723.

Fidanián, H. M. and D. S. Ray. 1972. Replication of bacteriophage M13. VII. Requirement of the gene 2 protein for the accumulation of a specific RFII species. *J. Mol. Biol.* **72:** 51.

———. 1974. Replication of bacteriophage M13. VIII. Differential effects of rifampicin and nalidixic acid on the synthesis of the two strands of M13 duplex DNA. *J. Mol. Biol.* **83:** 63.

Geider, K. and A. Kornberg. 1974. Initiation of DNA synthesis. VIII. Conversion of the M13 viral single strand to the double-stranded replicative form by purified proteins. *J. Biol. Chem.* **249:** 3999.

Gellert, M., M. H. O'Dea, T. Itoh, and J. Tomizawa. 1976. Novobiocin and coumermycin inhibit DNA supercoiling catalyzed by DNA gyrase. *Proc. Natl. Acad. Sci.* **73:** 4474.

Horiuchi, K. and N. D. Zinder. 1976. Origin and direction of synthesis of bacteriophage f1 DNA. *Proc. Natl. Acad. Sci.* **73:** 2341.

Kessler-Liebscher, B. E. and W. L. Staudenbauer. 1976. Replication of M13 DNA in plasmolysed *Escherichia coli* cells. Formation of fully synthetic duplex DNA. *Eur. J. Biochem.* **70:** 523.

Kessler-Liebscher, B. E., W. L. Staudenbauer, and P. H. Hofschneider. 1978. Replication of M13 DNA in plasmolysed *Escherichia coli* cells. Structure of a replicative intermediate with restricted binding of intercalating dyes. *Eur. J. Biochem.* **84:** 233.

Marians, K. J., J. Ikeda, S. Schlagman, and J. Hurwitz. 1977. Role of DNA gyrase in φX replicative-form replication in vitro. *Proc. Natl. Acad. Sci.* **74:** 1965.

Mazur, B. J. and P. Model. 1973. Regulation of f1 single-stranded DNA synthesis by a DNA binding protein. *J. Mol. Biol.* **78:** 285.

Mazur, B. J. and N. D. Zinder. 1975. The role of gene V protein in f1 single-strand synthesis. *Virology* **68:** 49.

Molineux, I. J., S. Friedman, and M. L. Gefter. 1974. Purification and properties of the *Escherichia coli* deoxyribonucleic acid-unwinding protein. Effects on deoxyribonucleic acid synthesis in vitro. *J. Biol. Chem.* **249:** 6090.

Oey, J. L. and R. Knippers. 1972. Properties of the isolated gene 5 protein of bacteriophage fd. *J. Mol. Biol.* **68:** 125.

Ray, D. S. 1977. Replication of filamentous bacteriophages. In *Comprehensive virology* (ed. H. Fraenkel-Conrat and R. R. Wagner), vol. 7, p. 105. Plenum Press, New York.

Ray, D. S., J. Dueber, and S. Suggs. 1975. Replication of bacteriophage M13. IX. Requirement of the *Escherichia coli dna*G function for M13 duplex DNA replication. *J. Virol.* **16:** 348.

Salstrom, J. S. and D. Pratt. 1971. Role of coliphage M13 gene 5 in single-stranded DNA production. *J. Mol. Biol.* **61:** 489.

Schekman, R., W. Wickner, O. Westergaard, D. Brutlag, K. Geider, L. L. Bertsch, and A. Kornberg. 1972. Initiation of DNA synthesis: Synthesis of φX174 replicative form requires RNA synthesis resistant to rifampicin. *Proc. Natl. Acad. Sci.* **69:** 2691.

Schneck, P. K., B. van Dorp, W. L. Staudenbauer, and P. H. Hofschneider. 1978a. A cell-free in vitro system for the replication of bacteriophage M13 duplex DNA. *Nucleic Acids Res.* (in press).

Scott, J. F., S. Eisenberg, L. L. Bertsch, and A. Kornberg. 1977. A mechanism of duplex DNA replication revealed by enzymatic studies of phage φX174: Catalytic strand separation in advance of replication. *Proc. Natl. Acad. Sci.* **74:** 193.

Sigal, N., H. Delius, T. Kornberg, M. L. Gefter, and B. Alberts. 1972. A DNA-unwinding protein isolated from *Escherichia coli*: Its interaction with DNA and with DNA polymerases. *Proc. Natl. Acad. Sci.* **69:** 3537.

Staudenbauer, W. L. 1974. Replication of the double-stranded replicative form DNA of bacteriophage M13 in plasmolysed *Escherichia coli* cells. *Eur. J. Biochem.* **47:** 353.

———. 1976. Replication of small plasmids in extracts of *Escherichia coli*: Requirement for both DNA polymerases I and III. *Mol. Gen. Genet.* **149:** 151.

Staudenbauer, W. L. and P. H. Hofschneider. 1972. Replication of bacteriophage M13. Inhibition of single-strand synthesis by rifampicin. *Proc. Natl. Acad. Sci.* **69:** 1634.

———. 1973. Replication of bacteriophage M13. Positive role of gene 5 protein in single-strand DNA synthesis. *Eur. J. Biochem.* **34:** 569.

Staudenbauer, W. L., W. L. Olsen, and P. H. Hofschneider. 1974. Alternate pathways of bacteriophage M13 DNA replication. In *Mechanism and regulation of DNA replication* (ed. A. R. Kolber and M. Kohiyama), p. 117. Plenum Press, New York.

Tye, B., P. Nyman, I. R. Lehman, S. Hochhauser, and B. Weiss. 1977. Transient accumulation of Okazaki fragments as a result of uracil incorporation into nascent DNA. *Proc. Natl. Acad. Sci.* **74:** 154.

Weiner, J. H., L. L. Bertsch, and A. Kornberg. 1975. The deoxyribonucleic acid unwinding protein of *Escherichia coli:* Properties and functions in replication. *J. Biol. Chem.* **250:** 1972.

Wickner, W., D. Brutlag, R. Schekman, and A. Kornberg. 1972. RNA synthesis initiates in vitro conversion of M13 DNA to its replicative form. *Proc. Natl. Acad. Sci.* **69:** 965.

Control of DNA Structure by Proteins

Klaus Geider, Verena Berthold, Mahmoud Abdel-Monem, and Hartmut Hoffmann-Berling
Max-Planck-Institut für Medizinische Forschung
6900 Heidelberg, Federal Republic of Germany

DNA-binding proteins (Alberts and Frey 1970) play an important role in the regulation of nucleic acid metabolism in cells and in the life cycles of bacteriophages and animal viruses. They can influence the structure of nucleic acids and can affect enzymatic activities either directly by associating with other proteins or indirectly by binding to DNA. Proteins showing cooperative binding to single-stranded (SS) DNA denature the double helix in low salt by binding to frayed ends of the DNA. In addition, the double-stranded state of DNA is under the control of enzymes. In *Escherichia coli*, these include a nicking-closing enzyme (Wang 1971), DNA gyrase (Gellert et al. 1976), and ATP-dependent enzymes capable of separating the two strands of the DNA double helix. Three DNA-dependent ATPases that unwind duplex DNA, referred to here as DNA helicases (previously known as DNA-unwinding enzymes [Abdel-Monem et al. 1976]), will be discussed in this article. Two of these enzymes were purified from *E. coli* and the third was purified from cells infected with T4 bacteriophage. Still another DNA-dependent ATPase that unwinds DNA is the product of the *rep* gene of *E. coli*. Mutations in *rep* are not lethal for the cell but block replication of double-stranded replicative form (RF) DNA of small bacteriophages such as φX174 (Denhardt et al. 1972).

DNA-BINDING PROTEINS

The role of phage DNA-binding proteins in the cell was deduced from genetic defects. The absence of phage fd DNA-binding protein, which is encoded by gene V, leads to an accumulation of RF DNA (Salstrom and Pratt 1971) because the protein prevents the conversion of phage SS DNA to the double-stranded form (Geider and Kornberg 1974). Phage T4 DNA-binding protein, which is encoded by gene 32, is required in replication and recombination (Alberts and Frey 1970). For the classical *E. coli* DNA-binding protein (Sigal et al. 1972), referred to here as "DNA-binding protein I," no mutants are known. The results from studies in vitro suggest that this protein functions in DNA replication by helping in the recognition of the initiation site (Schaller et al. 1976) and also by stimulating DNA polymerase III holoenzyme progress (Geider and Kornberg 1974). The double-stranded DNA-binding proteins that have been identified may have a role in transcription (Rouvière-Yaniv and Gros 1975). Table 1 summarizes some properties of the DNA-binding proteins.

E. coli DNA-binding Protein II

We have isolated a low-molecular-weight DNA-binding protein from *E. coli* (Berthold and Geider 1976). This protein, previously called DNA-binding protein HD but now called DNA-binding protein II, is probably identical with DNA-binding protein HU (Rouvière-Yaniv and Gros 1975). It is not denatured by heat and it binds to double-stranded DNA. It also binds to SS DNA and to RNA. A cell can contain at least 10,000 monomers of the protein (Berthold and Geider 1976).

Physical Properties of DNA-binding Protein II

Amino acid analysis reveals 85 residues. The molecular weight of the monomer is 9000, which is in agreement with its migration in SDS-polyacrylamide gels. In the native state, the protein exists as a tetramer, as shown by cross-linking with dimethylsuberimidate. A cluster of lysine and arginine residues in its amino acid sequence (A. Tsugita, pers. comm.) may cause histonelike behavior, although the isoelectric point is around 7.5. Met-Asn-Lys is found at the N terminus of this protein as well as at the end of other DNA-binding proteins.

Binding to Nucleic Acids

Complexes of DNA-binding protein II with phage fd SS DNA, T7 double-stranded DNA, or phage fr RNA show an increase in sedimentation rate compared with free nucleic acids. The gradual shift in the sedimentation rate of the DNAs in the presence of increasing amounts of the protein indicates noncooperative binding. This was deduced for RNA, also, in a filter-binding assay. Competition experiments showed that DNA-binding protein II has a preference for SS DNA, but when the single-stranded parts of double-stranded DNA molecules were complexed with DNA-binding protein I,

Table 1 Properties of DNA-binding proteins

	Phage fd DNA-binding protein	*Eco* DNA-binding protein I	*Eco* DNA-binding protein II
Binding to SS DNA	+	+	+
Cooperative binding to SS DNA	+	+	−
Binding to double-stranded DNA	−	−	+
Binding to RNA	−	±	+
Inhibition of DNA synthesis	+	−	+
Inhibition of transfection	−	+	−

protein II bound readily to double-stranded DNA (Geider 1978). We assume, therefore, that, in the cell, DNA-binding protein I is bound to the SS DNA region near the replication fork, whereas protein II is part of the structure containing the double-stranded portions of the chromosome (Varshavsky et al. 1977). Furthermore, DNA-binding protein II may increase the stability of RNA. Preparations of washed ribosomes are enriched for this protein (V. Berthold, unpubl.). Suryanarayana and Subramanian (1978) have found the protein attached to the free 30S ribosomal subunit. At low pH the protein can be separated into two components which show microheterogeneity in their amino acid sequences.

Exchange of Proteins in Complexes with SS DNA

Phage fd SS DNA requires DNA-binding protein I (Sigal et al. 1972; Weiner et al. 1975) for its conversion to double-stranded RF DNA (Geider and Kornberg 1974). This step in DNA replication can be inhibited by the addition of DNA-binding protein II (Berthold and Geider 1976) or fd DNA-binding protein (Geider and Kornberg 1974). When one of these proteins is added to the ring-shaped complexes of fd DNA and DNA-binding protein I, complexes having the structure dictated by the added protein are formed, thus indicating that the original protein is displaced (Zentgraf et al. 1977). Transfection experiments with fd DNA indicated that fd DNA-binding protein can displace DNA-binding protein I completely, whereas DNA-binding protein II can substitute for only some of the DNA-binding protein I molecules (Uhlmann and Geider 1977). Experiments with ^{125}I-labeled DNA-binding protein I have confirmed these results. When complexes between fd DNA and the radioactive protein were exposed to fd DNA-binding protein, label comigrating with the DNA was displaced completely (Geider 1978). DNA-binding protein II can displace not more than half of DNA-binding protein I, regardless of which protein is added first. Each of the three DNA-binding proteins protects SS DNA from degradation by DNase I. The effect is less pronounced for protein II than for the other two binding proteins. A mixed complex of *E. coli* DNA-binding protein I and II is more protected than a complex with either protein alone.

DNA HELICASES

The first DNA helicase isolated from *E. coli* (DNA helicase I) is a large, fibrous molecule which forms aggregates. As seen in the electron microscope, these are fibrils (Abdel-Monem and Hoffmann-Berling 1976; Abdel-Monem et al. 1977c). The other two DNA helicases, DNA helicase II of *E. coli* (Abdel-Monem et al. 1977a,b) and the DNA helicase of T4-infected cells (H. Krell et al., in prep.), are globular proteins of moderate size (Table 2). It is believed that DNA helicase II is identical with the DNA-dependent ATPase described previously by Richet and Kohiyama (1976) but not with the *rep* protein (Takahashi et al., this volume). When DNA helicase II (75,000 daltons) and *rep* protein (70,000 daltons [Takahashi et al. 1977] or 65,000 daltons [Scott and Kornberg 1978]) are run in the same electrophoretic gel, they form separate bands (S. Takahashi, pers. comm.). The DNA helicase isolated from T4-infected cells is thought to be the same protein as the DNA-dependent ATPase encoded by gene *dda* of the phage (Ebisuzaki et al. 1972; Purkey and Ebisuzaki 1977). Deletion of this gene is not lethal for the virus (Homyk and Weil 1974).

Characteristics of the Three DNA Helicases

In an attempt to characterize the three DNA helicases, we have studied the unwinding of various DNAs and the requirements of the enzymes in DNA binding and their mode of action. Partially double-stranded DNA molecules synthesized in vitro on bacteriophage fd circular SS DNA were used for the unwinding studies. Incubation of such partial duplexes with any of the three DNA helicases in the presence of ATP results in the separation of the complementary DNA chains. The unwinding of the DNA can be measured on the basis of its increased sensitivity to degradation by a single-strand-specific endonuclease or by its changed sedimentation coefficient. Sedimentation analysis also showed that there is no breakdown of the DNA by the helicases. At saturating levels of DNA helicase I or T4 DNA helicase, the unwinding of partial duplexes of fd DNA occurs within 3 minutes; with DNA helicase II this takes 3 to 5 minutes.

The fraction of the DNA molecules denatured depends on the amount of enzyme added; when the ratio of enzyme to DNA is low, the unwinding of the DNA levels off prematurely. On the basis of this effect it has been estimated that 60 to 100 molecules of DNA helicase I are required to unwind one molecule of partially duplex fd DNA, whereas 600 to 1000 molecules of DNA helicase II and 3000 molecules of T4 DNA helicase are needed to achieve the same result.

Table 2 DNA helicases

Designation	Gene	Molecular weight	Molecular shape	V_{max} (moles ATP/mole enzyme/min; 35°C)	Copies/cell
DNA helicase I		180,000	fibrous	10,000	500–700[a]
DNA helicase II		75,000	globular	4000	~4000[b]
T4 DNA helicase	*dda*	56,000	globular	14,000	~200[b]

[a] Estimate based on protein gel electrophoresis of a crude cell extract.
[b] Estimate based on determinations of ATPase activity in cell fractions.

In more recent studies on DNA unwinding (B. Kuhn et al., unpubl.) λ bacteriophage DNA has been used. This DNA has 50,000 base pairs and is eight times longer than fd DNA. For reasons which will be discussed later, 0.4% of the DNA was digested with exonuclease III prior to incubation with the helicases. About twice as much DNA helicase I is needed to unwind one λ DNA molecule as is needed to unwind one fd DNA molecule. For the other two DNA helicases the difference is approximately tenfold; unwinding of one λ duplex requires ten times as many DNA helicase II or T4 DNA helicase molecules. It appears, therefore, that the amount of DNA helicase II or T4 DNA helicase required for unwinding is proportional to the length of the DNA molecule, whereas the amount of DNA helicase I needed is not. DNA helicase I and T4 DNA helicase unwind 80% of the λ DNA molecules, whereas DNA helicase II unwinds not more than 50%.

The three enzymes dephosphorylate ATP at the γ position. This DNA-stimulated hydrolysis of ATP can be inhibited by SH blocking agents such as *p*-chloromercuribenzoate or by the substitution of an analog such as the β, γ-imido derivative of ATP (AMPPNP) for ATP. In each case, loss of the ATPase activity is correlated with loss of the ability to unwind DNA. Studies in which SS DNA stimulated the ATPase activity of the three enzymes, whereas double-stranded DNA did not, suggest that duplex DNA is not recognized. Since ATP is hydrolyzed in a reaction between the enzyme and SS DNA which does not involve unwinding, it is difficult to decide whether there is stoichiometry of ATP utilization during the unwinding of a double strand.

Studies on the binding of enzyme-DNA complexes to nitrocellulose filters revealed that the enzymes do not bind to double-stranded DNA. It was also found that fully replicated fd RF DNA is not active in unwinding assays but that unwinding occurs when nicked circular fd DNA duplexes (RFII) are treated with exonuclease III or λ exonuclease before being incubated with a DNA helicase. To estimate the length of the single-stranded region required for the initiation of DNA unwinding, λ DNA was degraded with exonuclease III to various extents. It was found that DNA helicase I requires degradation of at least 0.4% of the DNA (i.e. about 200 nucleotides from each end). Since DNA helicase II and T4 DNA helicase can unwind 30% of untreated λ DNA molecules, the 12 unpaired nucleotides present at each 5′ terminus of λ DNA are sufficient to initiate unwinding. After treatment of the λ DNA with exonuclease III more than 80% of the DNA molecules may be unwound by T4 DNA helicase and 50% by DNA helicase II.

Interestingly enough, λ exonuclease-eroded fd RFIII (linear duplexes obtained by treating RFII DNA with the single-strand-specific nuclease S1) was not unwound by any of the enzymes, in contrast to exonuclease III-eroded RFIII and λ exonuclease-eroded RFII. The λ exonuclease-treated RFIII was not unwound even when the treated material was incubated with alkaline phosphatase to remove 5′-terminal phosphate groups which conceivably could inhibit the unwinding. Since λ exonuclease degrades a DNA chain in the 5′ to 3′ direction, whereas exonuclease III degrades in the 3′ to 5′ direction, it appears that the three DNA helicases unwind the duplex adjacent to the 3′ end of the single-stranded region where binding of the enzymes occurs, but not the duplex connected to the 5′ end. In other words, when starting on DNA with a gap, unwinding should be unidirectional. The validity of this prediction is presently under study (B. Kuhn et al., unpubl.).

Studies on enzyme-DNA complexes have shown that in the presence of ATP, DNA helicase I remains associated with the DNA for longer than 20 minutes on the average, whereas DNA helicase II and T4 DNA helicase dissociate within 2 minutes. To characterize the complex further, an excess of SS DNA was added to the partially double-stranded DNA substrate to trap free enzyme molecules in the assay system. In the case of DNA helicase I, SS DNA inhibited unwinding when added to the enzyme before the duplex DNA and not when added afterwards, which suggests that DNA unwinding, once initiated, proceeds regardless of whether or not enzyme molecules are permitted to adsorb to the enzyme-DNA complex. In contrast, the action of DNA helicase II or T4 DNA helicase was inhibited regardless of whether the SS DNA was added before or after the double-stranded DNA, which indicates that DNA unwinding depends on the continued uptake of enzyme molecules into the complex.

It seems, therefore, that there are different ways of coupling DNA unwinding to ATP hydrolysis. Since complexes formed by DNA helicase I are comparatively stable and since there is no uptake of enzyme molecules into the complex during unwinding, this enzyme presumably acts in a processive manner (with ATP providing the energy required for migration of the protein along the DNA). Consistent with this view is the fact that the amount of enzyme needed to unwind does not increase proportionally with DNA length. In view of their tendency to aggregate, it is likely that DNA helicase I molecules can form a multimolecular unit on the DNA and in this way unwind the duplex catalytically.

In contrast to DNA helicase I, DNA helicase II and T4 DNA helicase appear to act in a nonprocessive (distributive) manner. Since stoichiometric amounts of these enzymes are needed to unwind DNA, these two helicases behave rather like DNA-binding proteins—apart from the fact that their action depends on ATP hydrolysis. It should be noted that the amount of DNA helicase II or T4 DNA helicase extracted per cell does not suffice to unwind a long stretch of DNA (Table 2). Attempts to enhance the unwinding by these enzymes by the presence of DNA-binding protein I or T4 gene-32 protein were unsuccessful. It is doubtful, therefore, that these enzymes cooperate with DNA-binding proteins as does the *rep* protein (Scott et al. 1977).

GENERAL COMMENTS

DNA-binding protein II is similar to DNA-binding proteins studied in other laboratories (Gosh and Echols 1972; Crepin et al. 1975; Rouvière-Yaniv and Gros 1975). These other proteins may play a role in the organization of DNA in cells, as do histones in chromatin (Varshavsky et al. 1977). In addition, these proteins may assist in regulating nucleic acid metabolism by enhancing transcription at specific sites (Crepin et al. 1975) or by slowing down DNA replication (Berthold and Geider 1976). Furthermore, they may protect DNA against degradation or influence translation by binding to ribosomes.

DNA helicases, on the other hand, may be regarded as chemo-mechanical agents for the control of DNA structure and as such may be classified together with *rep* protein, DNA gyrase, and also *recBC* nuclease. In fact, the possibility of a DNA-dependent ATPase unwinding the DNA duplex was first proposed for *recBC*-type nucleases (Friedman and Smith 1973; MacKay and Linn 1974). Recently, MacKay and Linn (1976) have demonstrated that the nuclease activity of *recBC* is blocked in the presence of DNA binding-protein I and the enzyme actively unwinds the DNA duplex.

So far, we have not succeeded in assigning biological functions to any of the helicases. In particular, since no mutants are available, it is not clear whether one or the other of the two *E. coli* DNA helicases participates in DNA replication.

REFERENCES

Abdel-Monem, M. and H. Hoffmann-Berling. 1976. Enzymatic unwinding of DNA. Purification and characterization of a DNA dependent ATPase from *E. coli. Eur. J. Biochem.* **65:** 431.

Abdel-Monem, M., M. C. Chanal, and H. Hoffmann-Berling. 1977a. DNA unwinding enzyme II of *E. coli*. Characterization of the ATPase activity. *Eur. J. Biochem.* **79:** 33.

Abdel-Monem, M., H. Dürwald, and H. Hoffmann-Berling. 1976. Enzymatic unwinding of DNA. Chain separation by an ATP dependent DNA unwinding enzyme. *Eur. J. Biochem.* **65:** 441.

———. 1977b. DNA unwinding enzyme II of *E. coli.* Characterization of the DNA unwinding activity. *Eur. J. Biochem.* **79:** 39.

Abdel-Monem, M., H. F. Lauppe, J. Kartenbeck, H. Dürwald, and H. Hoffmann-Berling. 1977c. Enzymatic unwinding of DNA. Mode of action of *E. coli* DNA unwinding enzyme. *J. Mol. Biol.* **110:** 667.

Alberts, B. M. and L. Frey. 1970. T4 bacteriophage gene 32. A structural protein in the replication and recombination of DNA. *Nature.* **227:** 1313.

Berthold, V. and K. Geider. 1976. Interaction of DNA with DNA binding proteins. The characterization of protein HD from *Escherichia coli* and its nucleic acid complexes. *Eur. J. Biochem.* **71:** 443.

Crepin, M., R. Cukier-Kahn, and F. Gros. 1975. Effect of a low-molecular-weight DNA-binding protein, H_1 factor, on the *in vitro* transcription of the lactose operon of *Escherichia coli. Proc. Natl. Acad. Sci.* **72:** 333.
Denhardt, D. T., M. Iwaya, and L. L. Larison. 1972. The *rep* mutation. Its effect on *E. coli* and the replication of bacteriophage φX174. *Virology* **49:** 486.
Ebisuzaki, K., M. T. Behme, C. Senior, D. Shannon, and D. Dunn. 1972. An alternative approach to the study of new enzymatic reactions involving DNA. *Proc. Natl. Acad. Sci.* **69:** 515.
Friedman, E. A. and H. O. Smith. 1973. Production of possible recombination intermediates by an ATP dependent DNAase. *Nat. New Biol.* **241**: 54.
Geider, K. 1978. Interaction of DNA with DNA-binding proteins. *Eur. J. Biochem.* **87:** 617.
Geider, K. and A. Kornberg. 1974. Conversion of the M13 viral single strand to the double-stranded replicative form by purified proteins. *J. Biol. Chem.* **249**: 3999.
Gellert, M., K. Mizuuchi, M. H. O'Dea, and H. A. Nash. 1976. DNA gyrase: An enzyme that introduces superhelical turns into DNA. *Proc. Natl. Acad. Sci.* **73:** 3872.
Gosh, S. and H. Echols. 1972. Purification and properties of D protein: A transcription factor of *Escherichia coli. Proc. Natl. Acad. Sci.* **69:** 3660.
Homyk, T. and J. Weil. 1974. Deletion analysis of two nonessential regions of the T4 genome. *Virology* **61:** 505.
MacKay, V. and S. Linn. 1974. The mechanism of degradation of duplex DNA by the *rec*BC enzyme of *E. coli* K12. *J. Biol. Chem.* **249:** 4286.
———. 1976. Selective inhibition of the DNAase activity of the *rec*BC enzyme by the DNA binding protein from *E. coli. J. Biol. Chem.* **251:** 3716.
Purkey, R. M. and K. Ebisuzaki. 1977. Purification and properties of a DNA-dependent ATPase induced by bacteriophage T4. *Eur. J. Biochem.* **75:** 303.
Richet, E. and M. Kohiyama. 1976. Purification and characterization of a DNA dependent ATPase from *Escherichia coli. J. Biol. Chem.* **251:** 808.
Rouvière-Yaniv, J. and F. Gros. 1975. Characterization of a novel, low-molecular-weight DNA-binding protein from *Escherichia coli*. *Proc. Natl. Acad. Sci.* **72:** 3428.
Salstrom, J. S. and D. Pratt. 1971. Role of coliphage M13 gene 5 in single-stranded DNA production. *J. Mol. Biol.* **61:** 489.
Schaller, H., A. Uhlmann, and K. Geider. 1976. A DNA fragment from the origin of single strand to double strand DNA replication of bacteriophage fd. *Proc. Natl. Acad. Sci.* **73:** 49.
Scott, J. F. and A. Kornberg. 1978. Purification of the *rep* protein of *E. coli*: An ATPase which separates duplex DNA strands in advance of replication. *J. Biol. Chem.* **253:** 3292.
Scott, J. F., S. Eisenberg, L. L. Bertsch, and A. Kornberg. 1977. Mechanism of duplex DNA replication revealed by enzymatic studies of phage φX174. Catalytic strand separation in advance of replication. *Proc. Natl. Acad. Sci.* **74:** 193.
Sigal, N., H. Delius, T. Kornberg, M. L. Gefter, and B. Alberts. 1972. A DNA-unwinding protein isolated from *Escherichia coli*: Its interaction with DNA and with DNA polymerases. *Proc. Natl. Acad. Sci.* **69:** 3537.
Suryanarayana, T. and A.-R. Subramanian. 1978. Specific association of two

homologous DNA-binding proteins to the native 30S ribosomal subunits of *Escherichia coli. Biochim. Biophys. Acta* (in press).

Takahashi, S., C. Hours, and D. T. Denhardt. 1977. Specialized transduction of the *rep* gene of *Escherichia coli. FEMS Microbiol. Lett.* **2:** 279.

Uhlmann, A. and K. Geider. 1977. Interaction of DNA with DNA binding proteins. III. Infectivity of protein complexed phage fd DNA in *E. coli* spheroplasts. *Biochim. Biophys. Acta* **474:** 639.

Varshavsky, A. J., S. A. Nedospasov, V. V. Bakayev, T. G. Bakayeva, and G. P. Georgiev. 1977. Histone-like proteins in the purified *E. coli* deoxyribonucleoprotein. *Nucleic Acids Res.* **4:** 2725.

Wang, J. C. 1971. Interaction between DNA and an *E. coli* protein ω. *J. Mol. Biol.* **55:** 523.

Weiner, J. H., L. L. Bertsch, and A. Kornberg. 1975. The DNA unwinding protein of *E. coli*. Properties and function in replication. *J. Biol. Chem.* **250:** 1972.

Zentgraf, H., V. Berthold, and K. Geider. 1977. Interaction of DNA with DNA binding proteins. II. Displacement of *E. coli* DNA unwinding protein and the condensed structure of DNA complexed with protein HD. *Biochim. Biophys. Acta* **474:** 629.

Isolation and Partial Characterization of Gene-II Protein from fd-infected Cells

Thomas F. Meyer and Klaus Geider
Abteilung Molekulare Biologie
Max-Planck-Institut für Medizinische Forschung
6900 Heidelberg, Federal Republic of Germany

Propagation of the replicative-form (RF) DNA of phage fd is initiated by the action of phage-encoded gene-II protein. Defects in gene II prevent the formation of RFII molecules (circular duplexes in which at least one strand is not continuous) nicked specifically in the viral strand and thereby cause accumulation of RFI DNA (superhelical, covalently closed, circular duplexes) (Lin and Pratt 1972; Fidanián and Ray 1972). The protein isolated from membrane fractions of phage-infected cells migrates on SDS gels with an apparent molecular weight of 40,000 (Lin and Pratt 1974); when synthesized in vitro in a protein-synthesizing system it has a molecular weight of 40,000 (Model and Zinder 1974) or 46,000 (Konings et al. 1975). All previous attempts to demonstrate enzymatic activity of gene-II protein in vitro have failed. We report here the isolation of a gene-II protein that stimulates fd RFI replication in vitro and cleaves specifically fd RFI DNA.

Isolation of Gene-II Protein

Gene-II protein was isolated from *Escherichia coli* cells infected for 30 minutes with an fd*am*5 (gene V) mutant (multiplicity of infection [m.o.i.] = 5). Its activity was checked by incorporation of [α-^{32}P]dATP into fd RFI DNA, which was supplemented with partially purified extracts from uninfected cells, analogous to a system developed for the replication of φX174 RFI DNA (Eisenberg et al. 1976; Sumida-Yasumoto et al. 1976).

The infected cells were broken in an X-press, and the high-speed supernatant was precipitated with 0.25 g ammonium sulfate per ml. The pellet was dissolved in a small volume, and the protein was run through an Ultrogel AcA 44 column. The active protein fractions were applied to DEAE-cellulose and eluted with salt. The fractions with gene-II-protein activity were then applied to phosphocellulose. Gene-II protein eluted at 250 mM NaCl.

Coding of the Isolated Protein by fd Gene II

The synthesis of fd RF DNA by various ammonium sulfate-precipitated fractions of H560 cells is given in Table 1. Background synthesis was observed in extracts of uninfected cells and in extracts of cells infected with fd *am*11 (gene II) mutants or wild type. Chloramphenicol applied after infection stimulated the activity of gene-II protein in extracts of wild-type-infected cells but not in extracts of cells infected with *am*11 (gene II). Extracts of fd *am*5-infected cells, however, had high gene-II-protein activity. In a protein-synthesizing system gene-V protein decreases the synthesis of gene-II protein, and in infected cells this protein apparently also prevents the accumulation of gene-II protein. Chloramphenicol diminishes gene-V-protein synthesis, but it does not prevent the production of some gene-II protein.

Molecular Weight of the Isolated Gene-II Protein

Gel filtration of the partially purified protein was performed on Ultrogel AcA 54 in 0.1 M and 1 M NaCl. Bovine serum albumin (67,000 daltons), ovalbumin (45,000 daltons), and cytochrome *c* (12,500 daltons) were used as markers. The peak of gene-II protein activity comigrated with that of ovalbumin. This indicates a native molecular weight of 45,000 for gene-II protein.

On SDS gels, the isolated gene-II protein electrophoresed as a close doublet at 45,000 daltons. These bands comigrated with the band for gene-II protein prepared from membrane fractions (Lin and Pratt 1974) and with that for the protein encoded by gene II of fd RF DNA in an in vitro protein-synthesizing system (T. F. Meyer et al., unpubl.).

The gene-II protein band from membrane fractions was carefully compared with the doublet of the isolated active protein. Only the slower band

Table 1 Synthesis of fd RFI DNA in extracts of infected and noninfected H560 cells

Fraction II	Infection	DNA synthesized (pmoles)
H560	none	4
H560	fd wt	5
H560	fd wt + chloramphenicol[a]	9
H560	fd *am*11 (gene II)	3
H560	fd *am*11 (gene II) + chloramphenicol[a]	4
H560	fd *am*5 (gene V)	41

The extracts were treated with ammonium sulfate (Eisenberg et al. 1976), and 4 μg protein was incubated with 250 ng fd RFI DNA and ribodeoxynucleoside triphosphates for 30 min at 30°C.

[a] Chloramphenicol (30 μg/ml of medium) was added 10 min after infection.

migrated exactly with the "membrane-bound" gene-II protein band. The slower band represented about one-third of the total protein present.

Effect on RF DNA Replication

The isolated gene-II protein stimulated DNA synthesis specifically with fd RFI but not with φX RFI (Table 2). The system was complemented with partially purified extracts from H560 cells. Extracts from H517 cells, which are defective in small-phage RF DNA replication (*rep*$^-$), did not support fd RF DNA replication after addition of gene-II protein.

Nicking Activity of Gene-II Protein

As expected from the results on RF DNA replication, the gene-II protein preparation cleaved only fd RFI and not φX RFI (data not shown). The cleavage reaction required only fd RFI, which had to be supercoiled, and the addition of Mg^{++}. The extent of the reaction was very dependent on the amount of gene-II protein added to fd RFI.

DISCUSSION

The gene-II protein from fd was isolated to approximately 30% purity. The very labile protein had a molecular weight of 45,000 under native and denaturing conditions. In a crude extract it cleaved specifically supercoiled fd RFI DNA and stimulated specifically replication of fd RF DNA.

Larger quantities of gene-II protein were found in cells infected with amber mutants in gene V than in wild-type-infected cells. Gene-V protein might interfere actively with the synthesis of gene-II protein in the cell via control of transcription or translation. An additional effect could be due to an accumulation of RFI molecules in cells infected with gene-V mutants.

The gene-II protein is a part of a protein system which replicates fd RFI to

Table 2 Specificity of gene-II protein for the RF DNA substrate

Template	Fraction II (noninfected)	Gene-II protein (fraction V)	DNA synthesized (pmoles)
fd RFI	H560	−	1.8
fd RFI	H560	+	37
fd RFI	H517 (*rep*3)	+	3.2
φX RFI	H560	−	2.7
φX RFI	H560	+	2.2

The extracts were incubated for 20 min at 37°C with 250 ng fd RFI DNA and purified gene-II protein as indicated.

RF or viral single-stranded DNA. The product is fully hybrid RF DNA. DNA-binding protein I and DNA polymerase III holoenzyme are other stimulatory factors for the replication of fd RFI (K. Geider and T. F. Meyer, unpubl.).

REFERENCES

Eisenberg, S., J. F. Scott, and A. Kornberg. 1976. An enzyme system for replication of duplex circular DNA: The replicative form of phage φX174. *Proc. Natl. Acad. Sci.* **73:** 1594.

Fidanián, H. M. and D. S. Ray. 1972. Replication of bacteriophage M13. VII. Requirement of the gene 2 protein for the accumulation of a specific RFII species. *J. Mol. Biol.* **72:** 51.

Konings, R. N. H., T. Hulsebos, and C. A. van den Hondel. 1975. Identification and characterization of the in vitro synthesized gene products of bacteriophage M13. *J. Virol.* **15:** 570.

Lin, N. S.-C. and D. Pratt. 1972. Role of bacteriophage M13 gene 2 in viral DNA replication. *J. Mol. Biol.* **72:** 37.

———. 1974. Bacteriophage M13 gene 2 protein: Increasing its yield in infected cells, and identification and localization. *Virology* **61:** 334.

Model, P. and N. D. Zinder. 1974. In vitro synthesis of bacteriophage f1 proteins. *J. Mol. Biol.* **83:** 231.

Sumida-Yasumoto, C., A. Yudelevich, and J. Hurwitz. 1976. DNA synthesis in vitro dependent upon φX174 replicative form I DNA. *Proc. Natl. Acad. Sci.* **73:** 1887.

The *Escherichia coli rep* Gene

Seishi Takahashi, Christian Hours, Makoto Iwaya,*
H. E. David Lane,† and David T. Denhardt
Department of Biochemistry
McGill University
Montreal, Quebec, Canada H3G 1Y6

The *Escherichia coli rep* mutant was originally isolated from cells surviving infection with bacteriophage φX174 on the basis of its ability to adsorb φX phage but not to support phage reproduction (Denhardt et al. 1967). This was one of the first deliberate isolations of an *E. coli* mutant that had lost the capacity to support replication of a virus for reasons other than the loss of an adsorption site. Only one of the original three isolates behaved well; it was designated *rep*3 and has been used in much of the subsequent research. The mutant grows well and is not conditionally lethal. It supports the conversion of viral single-stranded (SS) DNA to the double-stranded replicative form (RF), but replication of the RF does not occur. Also, the mutant is altered slightly in its ability to perform cellular recombination and repair activities (Denhardt et al. 1967, 1972; Calendar et al. 1970). The *rep* gene maps at 83 minutes, between *ilv* and *metE*, and the mutant allele is recessive to the wild-type allele (Calendar et al. 1970). The mutation transduces with *ilv* at a high frequency, 70–80%, and thus can be easily transferred to other strains. The origin of replication of the *E. coli* chromosome is located in the same region as the *rep* gene.

Other genes that map in the *ilv* region are *bfm, rho, cya, uvrD*, and *dnaP* (Bachmann et al. 1976). *E. coli bfm* mutants were isolated as survivors of infection by bacteriophage BF23 that had lost the capacity to support BF23 DNA replication; the mutation in these cells was cotransducible with *ilv* at a frequency of 80% (Shinozawa 1973). However, it is not known to what other phages BF23 is most closely related. The *rho* gene produces a product that acts at certain sites in DNA to terminate transcription. The protein has a molecular weight of about 45,000 daltons in SDS-polyacrylamide gels (Das et al. 1976) and is clearly different from the 70,000-dalton *rep* protein (see below). The phenotype of *cya* mutants does not suggest any obvious relation to *rep* (*cya* codes for adenyl cyclase), and indeed a *cya* mutant was observed to support phage P2 replication (S. Takahashi, unpubl.). There is evidence that *uvrD*, *uvrE*, *recL*, and *mutU*4 are all mutations in the same gene (Seigel 1973; Kushner et al. 1978), which seems to be different from the *rep* gene.

Present addresses: *Department of Microbiology, University of Massachusetts Medical School, Worcester, Massachusetts 01650; †Department of Cell Biology, University of Auckland, Private Bag, Auckland, New Zealand.

UvrD and *uvrE* have been implicated in mismatch repair (Nevers and Spatz 1975). The *dnaP* gene also seems to be distinct from the *rep* gene since λ *prep*$^+$ phage do not suppress the temperature-sensitive phenotype of the *dnaP* mutant (S. Takahashi, unpubl.).

rep Mutants Do Not Replicate Certain Phage DNAs

The *rep*3 mutant does not support the replication of any of the single-stranded DNA phages, either filamentous or isometric, and it also does not support the replication of the temperate noninducible phages P2 and 186, which have a double-stranded, linear DNA genome. The P2-dependent phage P4 will replicate in *rep*3, however (Lindqvist and Six 1971). No φX or P2 mutants have been found that are able to grow on *E. coli rep*3.

A number of additional *rep* mutants have been isolated, in some cases through infections with φX plus a collection of φX extended-host-range mutants or φX plus P2 to improve the discrimination against mutants that do not support φX replication because of an inability to adsorb the phage. Among these additional *rep* mutants several have been found that can support the replication of phage P2 (Iwaya 1971; Tessman and Peterson 1976).

Iwaya (1971) found that his new *rep* mutants that could support P2 phage production were also capable of supporting φX RF DNA replication to a limited extent. In these *rep* mutants (tentatively designated *repB*) a defective, noninfectious phage particle containing unit-length DNA was formed. As illustrated in Figure 1, these particles banded in a CsCl equilibrium density gradient at a less dense position than mature φX particles. The defective particles made in the *repB* mutant were not comparable to φX 70S particles (Denhardt 1977) as they were more dense than 70S particles and sedimented at almost the same rate as mature virions. These defective particles, which represented less than 10% of the particles made in a normal infection, contained unit-length DNA, and at least some of this DNA appeared to be circular (Iwaya 1971). The fact that *repB* mutants are complemented by λ*prep*$^+$ (S. Takahashi, unpubl.) suggests that the *repB* mutation represents a second type of mutation in the *rep* gene. These results imply that the protein product of the *rep* gene plays a role in SS DNA synthesis and/or phage morphogenesis.

Independently, Tessman and Peterson (1976) discovered a class of late-acting *rep* mutants, which they designated *groL* or *repL* (for late block). Like *repB* mutants, the *repL* mutants supported RF DNA replication to a near-normal extent and allowed phage P2 production. Also, M13 was able to replicate on *repL* strains, though apparently quite poorly. Most interestingly, it was observed that φX (also S13 and the G phages) could mutate to a form that could grow on the *repL* mutants. These spontaneous phage mutants that can overcome the *repL* block are called *ogr* mutants, and their

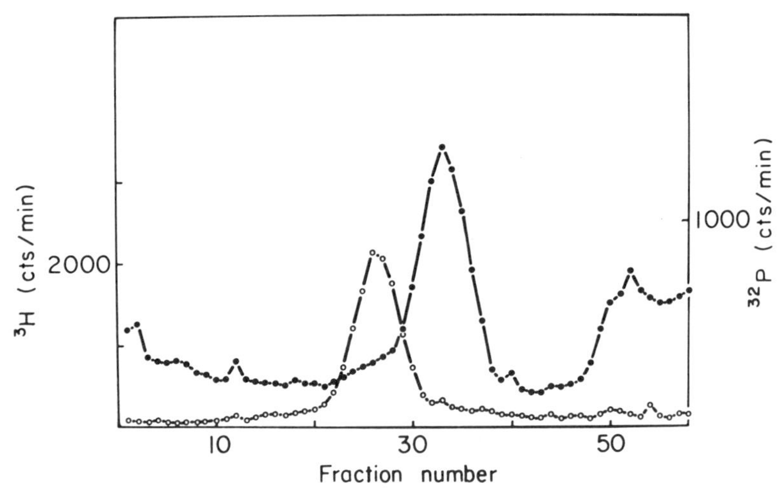

Figure 1 Neutral CsCl equilibrium density gradients of the defective particle made in *repB* cells. *E. coli* HF4704 *repB* was grown to about 2 × 10^8 cells/ml in 10 ml mT3XD at 37°C, treated with mitomycin C (50 μg/ml) for 10 min without aeration, and then infected with φX *am*3 at a multiplicity of about 5. [^{3}H]Thymidine (50 μCi at each time) was added to the culture at 0 min and 30 min after infection, and the incubation was continued at 30°C until 120 min after infection. The cells were washed with borate-EDTA and Tris-EDTA, suspended in 1 ml of 0.05 M Tris-HCl, pH 8, and lysed with lysozyme and EDTA at 0°C. This lysate was centrifuged to equilibrium in a neutral CsCl density gradient with infectious ^{32}P-labeled φX phage. The graph shows the distribution of ^{3}H (●—●) and ^{32}P(o—o) in this CsCl equilibrium density gradient. Density increases towards the left. (See Iwaya 1971, for further details.)

mode of origin showed them to be multiple mutants. Some of the *ogr* mutants were found to be mutated in gene *F*, which suggests that an interaction must occur between the *rep* protein and the major capsid protein. The *ogr* mutations were found to be *cis*-acting in the *repL* host. It was also observed that certain gene-*A* mutants could not mutate to the *ogr* state. Since phage mutants that can overcome the *repL* block are *cis*-acting (a gene-*A* characteristic) as well as mutated in gene *F*, it has been inferred by E. S. Tessman and P. K. Peterson (pers. comm.) that *ogr* mutants are *A*-*F* double mutants. It was suggested by Tessman and Peterson (1976) that a ternary complex consisting of the protein products of genes *A*, *rep*, and *F* is required for SS DNA synthesis and/or phage morphogenesis.

Reduced Replication-fork Velocity in *rep* Mutants

Although the comportment of the *rep* mutant was subtly different from that of the parent strain, the reason for this remained a mystery until the unexpected discovery by H. E. D. Lane that there was an increase in the ratio of DNA sequences near the origin of replication to DNA sequences near the terminus of replication in the bacterial DNA of the *rep* mutant. This discovery was a spin-off of an experiment to determine whether or not the *rep* mutant replicated its DNA bidirectionally; as shown in Figure 2, it does. There was an approximate twofold increase in the ratio of origin sequences (near *metA*) to sequences around the terminus of replication (near *trp*), which indicated that there were more replication forks per chromosome in the *rep* mutant than there were in wild-type cells under the same conditions (Lane and Denhardt 1974).

Because it was known that the overall rate of DNA synthesis per cell was the same for isogenic *rep* and *rep*$^+$ cells, it seemed likely that each of the replication forks operated at a reduced rate, and good evidence was obtained that this was the case (Lane and Denhardt 1975). Thus, the evidence is compelling that when *rep* function is defective the rate at which an individual replication fork moves in *E. coli* is halved (approximately) and, to compensate for this, the cell doubles the number of functional replication forks.

The *rep* Protein

The fact that the *rep* gene mapped close to *ilv* led us to test the *ilv*-transducing λ phage isolated by Shimada et al. (1973) for the presence of the *rep* gene. One class of *ilv*-transducing phage carrying the *rep* gene was identified by their ability to suppress the Rep$^-$ phenotype, both during lytic infection and in the lysogenic state. Figure 3 shows the complementation by λ*prep*$^+$ for both φX and P2 during lytic infection.

Because expression of the *rep* gene is not inhibited by the λ repressor in a λ

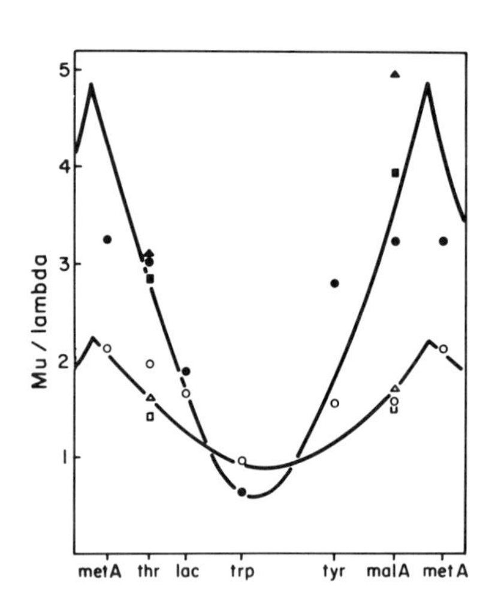

Figure 2 Relative proportions of the DNA from different regions of the *E. coli* chromosome in nonsynchronized populations. Bacteriophage Mu was integrated at one of the various chromosomal locations indicated along the abscissa. Bacteriophage λ was integrated at its normal attachment site in all cases. Each point represents the ratio of Mu DNA sequences to λ DNA sequences for an *E. coli* culture containing Mu integrated at the site indicated. The amounts of Mu and λ DNA were quantified by DNA-DNA hybridization. Further details are given by Lane and Denhardt (1974, 1975). The open symbols represent *rep*$^+$ and the solid symbols represent an isogenic *rep* mutant.

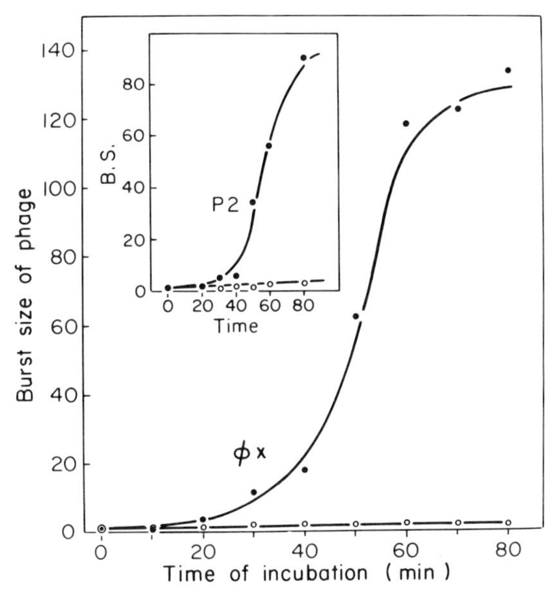

Figure 3 Provision of *rep* function in vivo by $\lambda prep^+$. D43(λ) was grown to a cell density of 2 $\times 10^8$/ml and then infected simultaneously with either $\lambda prep^+$ or $\lambda prep^-$ at a multiplicity of infection of 5 in all cases. Adsorption was allowed to take place for 15 min at 0°C and 10 min at 37°C, and then unadsorbed phage were removed by centrifugation. The cell pellet was resuspended in fresh, prewarmed medium at time 0. Portions of the culture were taken at intervals and lysed with $CHCl_3$ (for P2) or lysozyme (for φX) and then assayed for φX or P2 on *E. coli* HF4704(λ). (●) Cells infected with $\lambda prep^+$; (o) cells infected with $\lambda prep^-$ (a deletion mutant of $\lambda prep^+$ lacking *rep* function).

lysogen (S. Takahashi, unpubl.), one can radiochemically label the *rep* protein with a fair degree of specificity. Host protein synthesis is blocked by irradiation with ultraviolet light, and most lambda protein synthesis is suppressed by the repressor. When the proteins encoded by $\lambda prep^+$ and $\lambda prep^-$ were compared, a prominent band at about 70,000 daltons appeared to correlate with the presence or absence of the *rep* function (Takahashi et al. 1977). This protein was purified and found to possess *rep* activity in an in vitro complementation assay (S. Takahashi et al., in prep.). A protein of similar molecular weight has also been identified as the *rep* protein by Scott et al. (1977) and by Sumida-Yasumoto et al. (this volume). Deletion mapping of the $\lambda prep^+$ phage indicated that the *rep* gene mapped very close to the *ilv* locus, since the smallest deletion extending to the left from the attachment site that was isolated had lost the *rep* function.

The purified *rep* protein has an SS DNA-dependent ATPase activity and appears to be able to unwind duplex DNA by means of the energy of hydrolysis of ATP (Eisenberg et al. 1977). The data shown in Figure 4

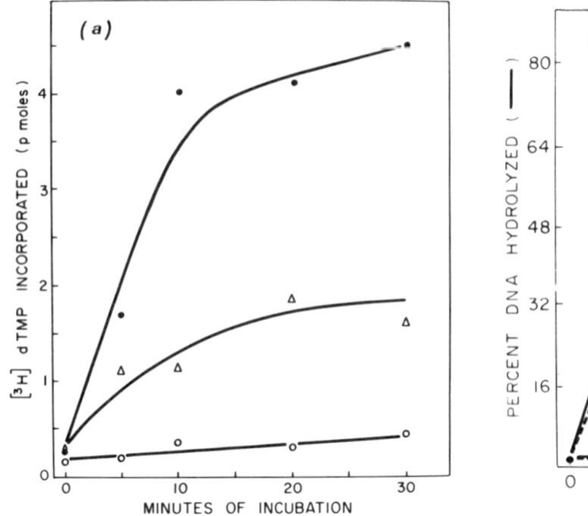

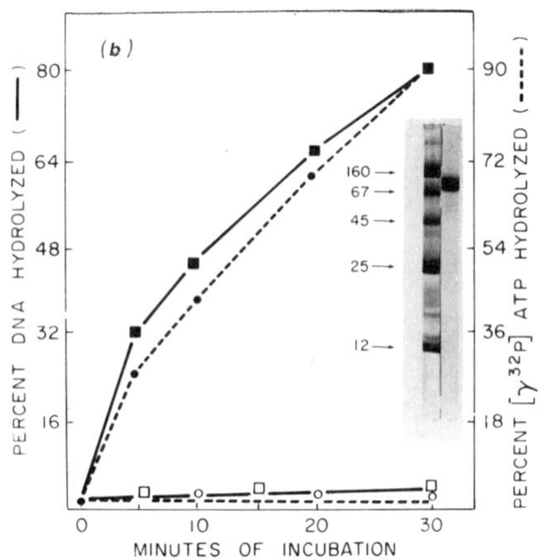

Figure 4 Properties of the *rep* protein. *E. coli* lysogenic for λ*prep*$^{+}$*C*1857 was grown to 3 × 10^8 cells/ml and then heat-induced at 42°C for 15 min. After vigorous aeration at 39°C for an additional 15 min the culture was cooled to 0°C and centrifuged; the cell pellet was stored at −20°C. The *rep*-gene protein was purified by ammonium sulfate fractionation, DEAE chromatography, affinity chromatography on DNA-cellulose, and gel filtration. The details will be published elsewhere.

(*a*) Stimulation of DNA synthesis in vitro by *rep* protein on an RFII DNA template. (●—●) Complete system; (△—△) minus RF DNA template; (o—o) minus *rep* protein. The details of this assay have been described by Takahashi et al. (1977).

(*b*) Presence of ATPase activity and apparent duplex-DNA-unwinding activities in a preparation of *rep* protein. The unwinding activity (———) was assayed as follows: ^{3}H-labeled φX RFII DNA (1.2 μg, 15,000 cpm) was preincubated at 37°C for 10 min in 80 μl of a solution containing 0.1 M Tris-HCl, pH 8.0, 0.05 M NaCl, 0.01 M $MgCl_2$, 1 mM ATP, and 8 μl of purified *rep* protein. Then 5 μl of a preparation of *N. crassa* single-strand-specific nuclease was added (time = 0 min). At various intervals, 10-μl aliquots were withdrawn, precipitated with trichloroacetic acid (TCA), and collected onto glass-fiber filters. After several washes with ice-cold 5% TCA and then ethanol, the filters were dried and the radioactivity determined. The percentage of ^{3}H-labeled RF DNA rendered acid-soluble by incubation first with *rep*-gene protein and then with the nucleases was determined (■). As a control, ^{3}H-labeled RFII DNA was incubated under the same conditions but in the absence of *rep* protein (□). The *rep* protein alone had no nuclease activity. The ATPase (- - -) was assayed as described by Abdel-Monem and Hoffmann-Berling (1976). Denatured calf thymus DNA was used. Solid symbols are the complete system; open symbols are the complete system minus *rep* protein. The inset shows an SDS-polyacrylamide gel of purified *rep* protein (stained with Coomassie blue) electrophoresed in parallel with the molecular weight markers cytochrome *c* (12,400 daltons), chymotrypsinogen (25,000 daltons), ovalbumin (45,000 daltons), bovine serum albumin (67,000 daltons), and human gamma globulin (160,000 daltons). The gel is deliberately overloaded to illustrate the absence of contaminating polypeptides; the *rep* protein reproducibly migrates just slightly more slowly than bovine albumin.

indicate that an essentially pure preparation of *rep* protein not only has an SS DNA-dependent ATPase activity, but also is capable of stimulating the action of the *Neurospora crassa* single-strand-specific nucleases on duplex DNA, presumably as the result of an active strand-separating process (Geider et al., this volume). We have not excluded the possibility that a contaminating activity (e.g., the *recBC* nuclease) is responsible for one or more of these activities; however, the ATP-dependent duplex-DNA-unwinding activity correlates nicely with the physiological defect exhibited by the mutant. It is a reasonable inference that the *rep* protein normally functions, using the energy of hydrolysis of ATP, to facilitate unwinding of the parental duplex during *E. coli* DNA replication, and that when the *rep* protein is defective, the two strands of the duplex are separated less efficiently and the rate of movement of the growing forks is retarded (Cairns and Denhardt 1968). It is interesting that there is a feedback process that results in an increase in the number of growing forks such that the same overall rate of DNA synthesis per cell is maintained. Presumably one or more additional functions become of greater significance to the cell in the absence of the *rep* function, and it might be revealing to isolate conditionally lethal mutants from a *rep* strain to see if a new class of "*dna*" mutants can be identified.

ACKNOWLEDGMENTS

The research described in this paper was supported by the National Cancer Institute of Canada and the Medical Research Council of Canada. We thank E. Tessman for contributions to the manuscript.

REFERENCES

Abdel-Monem, M. and H. Hoffmann-Berling. 1976. Enzymic unwinding of DNA. 1. Purification and characterization of a DNA-dependent ATPase from *Escherichia coli*. *Eur. J. Biochem.* **65:** 431.

Bachmann, B. J., K. B. Low, and A. L. Taylor. 1976. Recalibrated linkage map of *Escherichia coli* K-12. *Bacteriol. Rev.* **40:** 116.

Cairns, J. and D. T. Denhardt. 1968. Effect of cyanide and carbon monoxide on replication of bacterial DNA in vivo. *J. Mol. Biol.* **36:** 335.

Calendar, R., B. Lindqvist, G. Sironi, and A. J. Clark. 1970. Characterization of *rep*$^-$ mutants and their interaction with P2 phage. *Virology* **40:** 72.

Das, A., D. Court, and S. Adhya. 1976. Isolation and characterization of conditional lethal mutants of *Escherichia coli* defective in transcription termination factor *rho*. *Proc. Natl. Acad. Sci.* **73:** 1959.

Denhardt, D. T. 1977. The isometric single-stranded DNA phages. In *Comprehensive virology* (ed. H. Fraenkel-Conrat and R. R. Wagner), vol. 7, p.1. Plenum Press, New York.

Denhardt, D. T., D. H. Dressler, and A. Hathaway. 1967. The abortive replication of

φX174 DNA in a recombination-deficient mutant of *Escherichia coli*. *Proc. Natl. Acad. Sci.* **57:** 813.
Denhardt, D. T., M. Iwaya, and L. L. Larison. 1972. The *rep* mutation. II. Its effect on *Escherichia coli* and on the replication of bacteriophage φX174. *Virology* **49:** 486.
Eisenberg, S., J. Griffith, and A. Kornberg. 1977. φX174 cistron A protein is a multifunctional enzyme in DNA replication. *Proc. Natl. Acad. Sci.* **74:** 3198.
Iwaya, M. 1971. "On the mechanism of φX174 DNA replication." Ph.D. thesis, Harvard University, Cambridge, Massachusetts.
Kushner, S. R., J. Shepherd, G. Edwards, and V. F. Maples. 1978. *Uvr*D, *uvr*E, and *rec*L represent a single gene. *J. Supramol. Struct.* (Suppl.) **2:** 59.
Lane, H. E. D. and D. T. Denhardt. 1974. The *rep* mutation. III. Altered structure of the replicating *Escherichia coli* chromosome. *J. Bacteriol.* **120:** 805.
———. 1975. The *rep* mutation. IV. Slower movement of replication forks in *Escherichia coli rep* strains. *J. Mol. Biol.* **97:** 99.
Lindqvist, B. H. and E. W. Six. 1971. Replication of bacteriophage P4 DNA in a nonlysogenic host. *Virology* **43:** 1.
Nevers, P. and H.-C. Spatz. 1975. *Escherichia coli* mutants *uvr*D and *uvr*E deficient in gene conversion of λ heteroduplexes. *Mol. Gen. Genet.* **139:** 233.
Scott, J. F., S. Eisenberg, L. L. Bertsch, and A. Kornberg. 1977. A mechanism of duplex DNA replication revealed by enzymatic studies of phage φX174: Catalytic strand separation in advance of replication. *Proc. Natl. Acad. Sci.* **74:** 193.
Shimada, K., R. A. Weisberg, and M. E. Gottesman. 1973. Prophage lambda at unusual chromosomal locations. II. Mutations induced by bacteriophage lambda in *Escherichia coli* K12. *J. Mol. Biol.* **80:** 297.
Shinozawa, T. 1973. A mutant of *Escherichia coli* K12 unable to support the multiplication of bacteriophage BF23. *Virology* **54:** 427.
Siegel, E. C. 1973. Ultraviolet sensitive mutator *mut*U4 of *Escherichia coli* inviable with *pol*A. *J. Bacteriol.* **113:** 161.
Takahashi, S., C. Hours, and D. T. Denhardt. 1977. Specialized transduction of the *rep* gene of *Escherichia coli*. *FEMS Microbiol. Lett.* **2:** 279.
Tessman, E. S. and P. K. Peterson. 1976. Bacterial rep^- mutations that block development of small DNA bacteriophages late in infection. *J. Virol.* **20:** 400.

Recombination, Repair, and Genetic Engineering

Genetic Recombination

Norton D. Zinder
The Rockefeller University
New York, New York 10021

The small single-stranded DNA phages, such as φX174 (icosahedral phage, IP) or f1 (filamentous phage, FP), were appropriate materials with which to attempt the study of the mechanism of genetic recombination. The small size of their DNA should make intermediate forms in recombination readily separable from host DNA. Their few genes should allow markers to be distributed over the whole genome. Even their single-stranded nature should not create problems since immediately upon entry into their host cell the DNA becomes double-stranded (Denhardt 1975). The major difficulties are that their DNAs are ring molecules and that breakage of the rings renders them nonviable. In addition, for rings to recombine and yield a monomer at least two events may be required.

Still, a review of segregation data from single-burst genetic crosses of these phages is extremely interesting. From any single event one tends to recover one genetically intact parent (even when it is marked with mutations scattered about the ring) and a recombinant for a selected pair of genes, most of which recombinants contain the markers of the other parent. This somewhat remarkable finding must be telling us something about the mechanism of genetic recombination. We will refer to this hereafter as the "one parent-one recombinant" yield.

Genes and Marker Mutations

The first studies of genetic recombination in these small phages were performed by Zahler (1958). At the time, the only fact known about these phages was that they were small. He had only a few plaque-morphology mutants and concluded from his experiments that there was no detectable genetic recombination in IP. Around the time of the discovery of the single-stranded nature of the IP, the use of conditionally lethal mutants came into vogue. With the availability of temperature-sensitive and nonsense mutants, more precise studies became possible. Such mutants can be used to define all essential virus-encoded functions without knowledge of what they are. For the IP and FP, nine (Denhardt 1975) and eight genes, respectively, have been defined by complementation analysis (Pratt et al. 1966). For the IP, two-factor crosses can be done, and these yield a circular map more or less proportional to the distance between markers. The total map length is only 2×10^{-3} map units (Tessman 1966). There is a considerable anomaly in that crosses involving gene *A* show an excessively high frequency of recombination (Benbow et al. 1971). This anomaly disappears in *recA* hosts, where the total map length becomes about 5×10^{-4}. Not surprisingly, three-factor crosses show very high negative interference, with the segregation of an unselected genetic marker showing a maximal ratio of 7:1 among the selected recombinant progeny. Such crosses give a relatively additive map, with the segregation ratio of the unselected marker going through a maximum as the unselected marker proceeds in position around the ring (Baker and Tessman 1967). This shows that the recombining entity is indeed circular and not opened at a specific point. The mapping of the FP is much more complicated. The assembly mechanisms for these phages are such that diploid particles, containing two circular genomes, appear frequently (Salivar et al. 1967; Scott and Zinder 1967). In crosses between amber mutants, about 2% of the progeny are heterozygous diploids—containing both parental genomes—capable of forming plaques by continuous complementation. These pseudo wild-type progeny are 10–100 times as common as true recombinants and obscure the results of pairwise crosses.

The FP map was therefore constructed in a series of four-factor crosses in which only unique recombinants, those whose genotype could have occurred only by recombination, were scored. The crosses involved two selective markers which varied in position for each cross and two unselected markers at fixed points. In a cross $am_1 am_2^+ B_1 B_2^+ \times am_1^+ am_2 B_1^+ B_2$ (selecting for am^+), progeny with genotypes $am^+ B_1^+ B_2^+$ can only arise through recombination. The unselected markers in these crosses were about 20% of the genome away from each other and thus define a greater and a lesser arc on the circular genome. If both of the selected markers are within the same arc (B-*am*-*am*-B), recombination between them gives rise primarily to only one of the two possible unique recombinants, either $am^+ B_1^+ B_2^+$ or $am^+ B_1 B_2$, depending on the order of am_1 and am_2 with respect to B_1 and B_2. On the other hand, with one of the selected markers in the greater arc and the other within the lesser arc, both unique recombinants can be formed. The production of both types of unique recombinants indicates that the markers are in this staggered (*am*-B-*am*-B) configuration. A map was constructed by doing as many pairwise selective crosses as possible, with the same unselected markers scored in each. Ambiguities arise when a selected marker and an unselected marker are quite close; even if the markers are staggered, linkage between markers results in an excess of one or the other of the unique recombinants. The one error in the original genetic map of f1 was the misplacement of genes V and VII, two small genes very close to the unselected marker (Lyons and Zinder 1972; Vovis et al. 1975).

The finer details of the genetic maps of IP and FP have been developed with physical mapping by rescue experiments with genetically marked fragments produced by restriction enzymes and nucleotide sequencing (Hutchison and Edgell 1971; Horiuchi et al. 1975). (See Horiuchi et al. and

Schaller et al. [both this volume] for an updated sequence map, and Appendices I and II for φX and G4 sequences.)

Genetic Analysis of the Mechanism of Recombination

The majority of phage crosses present a problem in population biology. Phage DNA molecules are increasing in number, and there are numerous rounds of mating between progeny molecules. With the small phages, however, the recombination frequency is so low that further recombination of the recombinants is essentially ruled out. In fact, for both IP and FP the evidence supports the idea that the major source of recombinants derives from the mating of input genomes. It has been shown for IP that this mating of input particles is dependent on a functioning host *recA* gene, whereas the secondary recombination (about 1/20 of the primary amount) is independent of either *recA* or *recB* (Baker et al. 1971). Similarly, 95% of the recombination in FP is *recA*-dependent (Lyons and Zinder 1972). This ability of input FP to recombine, coupled with the continued production of phage from infected cells (reduced sampling problems), made the FP quite convenient material with which to analyze segregation data in detail. The general similarities between the two kinds of phage leads one to believe that what is true for one is true for the other. Both single-stranded phages enter host cells and become double-stranded replicative forms (RF) through the action of host enzymes. The first phage gene to function is gene *A* for IP and gene II for FP. These genes have a similar function: to nick the positive strand of the RF at a specific site so as to allow further replication to proceed. With the IP, gene *A* is *cis*-dominant and can affect only the specific genome that produces it. However, in FP, gene II complements all genes well and symmetrically. Thus, crosses between two gene-II mutants must await genetic recombination before further replication can occur. In addition, all elements that emerge from such crosses can be rescued by the selected wild-type recombinants. Since the recombination frequency is low, the phage particles that emerge from a single cell (which can be ascertained by plating infective centers before progeny are produced or in single-burst experiments) should be a reflection of all of the progeny that emerge from a single mating event (Boon and Zinder 1971).

In most of the experiments to be discussed below the progeny released by single infected cells were examined under single-burst conditions. Nonpermissive cells, infected at low multiplicity with gene-II amber mutants, were freed from unadsorbed phage, diluted, and distributed into many tubes so that a given tube was unlikely to contain more than one recombinant-producing cell. Since f1 is released by infected cells for many generations and not in a lytic burst, the cells were killed and phage production stopped at an arbitrary time after infection. The progeny contained in a single burst represent the products of individual recombination events between chromosomes. Since many of the experiments involve the *Escherichia coli* B restriction-modification system, we describe it in some detail below.

Bacteriophage f1 is sensitive to restriction and modification by the *E. coli* B restriction endonuclease. Phage grown on *E. coli* K form plaques with an efficiency of 7×10^{-4} on *E. coli* B, while phage grown on B form plaques equally well on B or K (Arber 1966). The sensitivity to B restriction is determined at two genetic loci, SB_1 and SB_2, linked, respectively, to gene V and to the N-terminal region of gene II (Arber and Kühnlein 1967; Boon and Zinder 1971; Horiuchi et al. 1975). Mutation at either site reduces sensitivity to B restriction 30-fold; mutation at both sites renders phage immune. Wild-type phage grown in B hosts become insensitive to further B restriction by virtue of a host-induced modification: the methylation of specific adenine residues within or near the SB sites renders their DNA resistant to cleavage by the B restriction endonuclease. Unmodified DNA is rapidly degraded upon entering the restricting host, first by specific action of the restriction enzyme and then by the action of other nucleases (Arber and Linn 1969).

In vitro, the isolated B restriction enzyme uses appropriate double-stranded DNA as a substrate for two reactions—double-strand cleavage and methylation (Horiuchi and Zinder 1972; Vovis et al. 1974). Surprisingly, although the enzyme must recognize specific sites, which can be obscured by mutation or methylation, cleavage occurs almost at random (Horiuchi and Zinder 1972; Hartman and Zinder 1974a).

For the analysis of segregation data obtained in B hosts it has been important to determine how the B restriction-modification system responds to heteroduplex DNA molecules. Heteroduplexes containing one methylated and one nonmethylated strand (SB^m/SB^+) are not cleaved but instead are quickly methylated in the second strand (Vovis et al. 1973, 1974). This reaction must be important to the host, which without it might become sensitive to its own restriction system during rapid semiconservative replication of DNA. Heteroduplexes containing one mutant and one wild-type strand (SB^0/SB^+) are neither cleaved nor modified (Vovis et al. 1973; Enea et al. 1975; Vovis and Zinder 1975). Upon replication and segregation, these give rise to two homoduplexes, one resistant to restriction and one sensitive.

To determine whether genetic recombination in f1 involves breakage and joining of DNA molecules (Meselson and Weigle 1961) or whether recombinant molecules contain only newly synthesized DNA, Boon and Zinder (1970) studied the transfer of DNA from infecting mutant phage into recombinant progeny, using B-specific modification as a label (Ihler and Meselson 1963). Arber (1966) had already demonstrated that during wild-type infection DNA from infecting phage reemerged in progeny particles. Boon and Zinder (1970) infected *E. coli* K with phage that had been modified by growth on B and found that one-fifth of the infected cells released at least one phage retaining its B-induced modifications. When B-modified mutants of gene II were crossed in K, equally efficient transfer occurred: one-fifth of the cells producing recombinant progeny yielded a recombinant that "remembered" its earlier growth on B. Two conclusions were drawn from this result: (1) that recombination can occur between input chromosomes without expression of gene II and thus without previous phage replication, and (2) that recombination involves some kind of breakage and joining. In addition, the result suggested that the exchange of DNA between recombining molecules must involve not just one strand but both strands of a double helix. The messenger RNA of f1 has the same polarity as the viral single strand and is transcribed from the complementary strand of the RF molecule (Sugiura et al. 1969). In the cross between two gene-II mutants, recombination must have produced a wild-type complementary strand, or else no functional gene-II message would have been transcribed and no progeny would have been formed. But the emergence of parental DNA (viral strand) in the wild-type progeny suggests that the breakage and joining has involved the viral strand as well. Therefore, either double-strand exchange has occurred or some special mechanism such as gene conversion has been involved. The evidence concerning gene conversion in these phages will be discussed in a later section.

A set of crosses between amber mutants of gene II, with varied combinations of wild-type and mutant SB sites as unselected markers, provided evidence that recombination in f1 did not occur through reciprocal double-strand exchange (Boon and Zinder 1971). The relative order of the markers is shown in Figure 1. About 300 nucleotides separate each marker from the next, except for *am*124, which is very close to SB_2^0. The phage progeny released in single bursts were analyzed and the bursts classified according to their contents. In the majority of the bursts one parental genotype emerged intact, along with one recombinant genotype. Most often the recombinant bore the unselected markers of the missing parent, but a significant fraction of these bursts contained parents and recombinants which shared unselected markers. These results resemble the results of yeast crosses in which gene conversion along with some asymmetric coconversion has occurred (Fogel

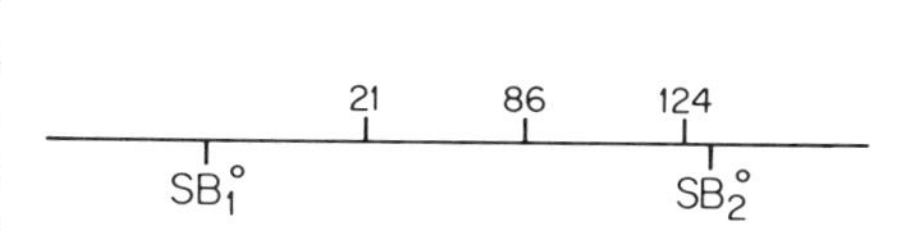

Figure 1 Map of the markers used in the f1 crosses: *am*21, *am*86, and *am*124 are amber mutations of gene II. They are approximately 300 nucleotides apart; *am*124 is closest to the amino terminus. All of the crosses, unless otherwise noted, are between *am*86 and *am*21. The SB markers define the two sites that determine the sensitivity of f1 to endonuclease R·*Eco* B.

and Mortimer 1969). Next to this class of mixed bursts, bursts containing a single recombinant genotype were observed most frequently. No reciprocal recombinants were found in the contents of a single burst. Control experiments established that all of the possible genotypes (except for the double amber mutant, which was not available to test) could grow well and be rescued by the wild-type recombinants, so failure to observe a particular genotype did not result from any selective bias, nor is there a limit to the number of genotypes that can be released from a properly infected cell.

Although these data were consistent with a number of models for genetic recombination, the conservation of the parent and the failure of reciprocal recombinants to appear could not be explained by reciprocal breakage and reunion. The simplest mechanisms to consider, given the asymmetric products of recombination, were those involving asymmetric single-strand exchange and heteroduplex formation. Meselson and Radding (1975) showed that such mechanisms could, with some modifications, generate reciprocal recombinants as well.

At this point it was still unclear whether the recombination event generated two double helices (e.g., one parental and one recombinant chromosome), as would be expected from certain breakage and copying models (Boon and Zinder 1971), or a single, segregating, heteroduplex chromosome. The effects of B restriction on segregation were examined in an attempt to resolve the two possibilities.

In the first set of these restriction crosses (Hartman and Zinder 1974a), one of the parents had been grown in *E. coli* B and was immune to B restriction, whereas the other was sensitive to restriction. Identical crosses were performed in B and K hosts, selecting for recombination between two gene-II mutations. The results of crosses in K were the same as before—most bursts contained one parent and one recombinant with the unselected markers of the missing parent. The crosses performed in B gave, in essence, half the results obtained in K: the major class of bursts contained one parent and one recombinant with the unselected markers of the missing parent, but neither the restriction-sensitive parent nor the recombinant expected with it appeared in any of the bursts. The frequencies of recombination in B hosts were not significantly lower than those in K, evidence that a double-strand break in one of the parents does not interfere with recombination. Considering the expected brief survival of the sensitive parent in the restricting host (Arber and Linn 1969), the recombination frequencies observed in restriction crosses were very high, which suggested that B restriction promotes recombination; recombination frequency may be limited by the requirement for a double-strand cleavage. Experiments in *recA* hosts, however, demonstrate that B restriction cannot replace the *recA* function (Hartman 1973). The frequency of recombination in B hosts is drastically reduced if both parents are sensitive to restriction (Hartman

1973). These results suggest that one participating chromosome in a recombination event must remain circular and are consistent with asymmetric donor-receptor models of recombination. Linkage values measured in K hosts were not changed in B, which confirmed in vivo the conclusion drawn from in vitro experiments (Horiuchi and Zinder 1972) that the B enzyme cleaves DNA at random sites.

In a second set of crosses (Hartman and Zinder 1974b) one of the parents was sensitive to restriction and the other was rendered immune by mutation at both SB sites (SB=0). Neither parent had been grown in B. Again, the progeny contained the immune parent and one recombinant, but this time the recombinant was mutant at both SB sites (SB=0)—it resembled the parent with which it had emerged. The recombinant progeny that emerged without a parent were also all SB=0.

When these crosses were repeated with a gene-II mutant (*am*124) closely linked to the SB_2^0 site (see map), the recombination frequency in B hosts dropped 20-fold. All the recombinants were again SB=0 (Enea and Zinder 1975).

The simplest way to explain all of the data is to suppose that during recombination one of the parental chromosomes donates a considerable portion of one strand to the other parent, displacing the homologous strand and creating a heteroduplex molecule. The donor parent is lost because recombination requires a double-strand cleavage of the donor, like the cleavage that occurs during B restriction. In the first set of restriction crosses described, in which one parent is immune to restriction by virtue of its modified SB sites, the sensitive parent is cleaved but still acts as donor. A heteroduplex molecule is formed which contains a heteroduplex SB site unmodified in the recombinant strand but modified in the other (SB^m/SB^+). Such sites are quickly recognized and modified by the B enzyme so that when the heteroduplex segregates during replication, both the recombinant and parental strands are immune to restriction and give rise to viable progeny. If the initial heteroduplex region is long enough to extend from the selected am^+ markers to the SB site, the parental site (SB^m/SB^m) remains and segregation has the same result.

When one of the parents is protected by mutation rather than by modification of the SB site, the same initial events occur, leading to the formation of a heteroduplex with either SB^0/SB^0 or SB^0/SB^+ structure depending on the length of the single strand donated. Neither of these structures is recognized by the B enzyme and neither restriction nor modification occurs (Vovis and Zinder 1975). In the first case, segregation gives rise to viable recombinant and parental progeny, which share the SB^0 marker and are genetically insensitive to restriction. But segregation of the second structure, which contains an unmodified SB^+ site in the recombinant strand, produces a recombinant chromosome sensitive to restriction and no progeny phage emerge. When the selected markers are far enough from the SB site, most heteroduplex chromosomes are still SB^0/SB^0 and the observed recombination frequency is high. But when the selected marker is quite close to the SB site (as in the crosses involving *am*124), most of the heteroduplexes are of the SB^0/SB^+ type and most recombinants are destroyed upon replication and segregation.

The model of recombination described here explains all the segregation data obtained in f1 crosses if the frequency of gene conversion is not high; parents and recombinants emerging from a single cell would bear identical unselected markers if gene conversion or base correction were frequent. I will deal first with the limited segregation data available for the IP and then with the evidence concerning gene conversion or base correction before returning to the definitive experiments with f1.

Benbow et al. (1974) have reported data obtained from nine recombinant single bursts of φX under conditions that permitted input molecules to replicate before recombination. In each of the bursts the majority of the progeny were one parent and one recombinant. The other parent and the reciprocal recombinant did not appear. Doniger et al. (1973) isolated "dimer" DNA molecules produced during a cross between phage S13 mutants and allowed them to segregate in single transfected spheroplasts. Such structures, which result from reciprocal recombination events, would be expected to give rise to reciprocal recombinant progeny. Remarkably, in most cases these dimers gave one parent and one recombinant. The authors suggest that the nonreciprocal recombinant arises through branch migration within a figure-eight heteroduplex which is converted to the dimer. Base correction and replication follow in such a way that from the dimer only one of the monomers is conserved. The parent is produced by breakdown of the dimer near the initial point of recombination. The first speculation is reasonable, but I find the latter somewhat gratuitous. First, I doubt that dimers break down by recombination; second, that they would remember where they had crossed over is most difficult to understand. At best, it could be that Doniger et al. (1973) isolated figure-eight dimers that broke down by replication, and that, subsequently, one of the monomers was destroyed or failed to replicate. Still, we see that the production of one parent and one recombinant is not unique to the FP.

Gene Conversion

The basic experiment was first performed by Baas and Jansz (1972a), who constructed heteroduplex molecules in vitro and transfected spheroplasts. In their experiments, the viral strand was an intact ring and the complementary strand contained a single, random break. Segregation of a single marker was followed: the vast majority of single cells produced phage of only one genotype, derived with equal frequency from the viral or complementary strand. The conclusion was that gene conversion must occur with very high frequency. Baas and Jansz (1972b) also observed some mixed bursts; the frequency with which these occurred was proportional to the distance, in one direction along the circular map, between the marker being studied and gene *A*. They concluded that replication, beginning in gene *A*, competes with repair processes, and that markers far from the origin of replication have more time to be repaired.

A more likely explanation for the first result might be that during penetration or at the first replication one of the strands of the heteroduplex was chosen as a master strand, and that base correction is not as frequent as Baas and Jansz conclude. The result concerning the occurrence of mixed bursts is perplexing. From these data the authors inferred the correct origin and direction of RF DNA replication in φX. Yet the evidence suggests that for these phages the rate-limiting event in DNA synthesis is initiation, and that once initiation has occurred it takes less than half a minute to complete a round of synthesis at 37°C. Therefore, a distal marker has at most 30 extra seconds in which to be repaired, whereas the heteroduplex may wait much longer than that for replication to begin.

Similar experiments were performed with f1 by Enea and Zinder (1976). The major differences between their experiments and those of Baas and Jansz (1972a,b) were that the heteroduplexes were made by cutting with a restriction enzyme and that all of the molecules had the break in the minus strand at the same place. In addition, Enea and Zinder (1976) transfected calcium-treated *E. coli* cells rather than spheroplasts, and two markers were used to differentiate between the two strands of the heteroduplex. Here, also, the majority of the bursts yielded progeny from only one of the strands—the minus strand. However, about 6% yielded recombinant progeny, which probably reflected gene conversion. Since the recombinants occurred in either strand with equal frequency, it was concluded that there was no bias in marker correction. This frequency is analogous to that found for lambda heteroduplexes by Wildenberg and Meselson (1975). For FP, the frequency of base correction (recombination) of synthetic heteroduplexes could be increased by having the minus strand contain a gene-II mutation so that little or no replication could occur. An increase in conversion recombination of about fivefold did occur. From these experiments we conclude that gene conversion could occur should a heteroduplex structure appear as an intermediate in genetic recombination, but it is not the major source of recombinants (see below).

This experiment also helped resolve the apparent conflict between the results of the early label-transfer experiment (which suggested that both strands were involved in recombination) and the evidence for single-strand exchange by demonstrating how a recombinant viral strand might segregate from the complementary strand without gene-II function. In contrast to the case with φX, we found that the segregation of normal heteroduplexes yielded about 75% of the minus-strand type alone and only 10% of the plus-strand type; this was a significant bias in favor of the master strand, as would be predicted by the rolling-circle model for DNA replication. However, when there was a mutation in gene II in the minus strand of the heteroduplex the amount of plus-strand progeny rose to almost 50%. We envisage that the nicked minus strand was replicated off, and since it had been cut in vitro or in the recombination event at a point other than the replication terminus, it could not be circularized by the usual mechanism. Thus, we end up with a functional homoduplex, derived from the plus strand as the master strand, which can now go on to replicate.

One further experiment supports our contention that in FP, and probably in IP, genetic recombination occurs via the formation of a large heteroduplex molecule that segregates to give one parent and one recombinant while the displaced segment of DNA and the residue of the other parent are lost.

A number of large deletion mutants of f1 have been isolated (Enea and Zinder 1975; Enea et al. 1977). They contain, for the most part, about 20% of the DNA of f1 plus a duplication of about 7% of f1. The relevant point for the experiment to be described is that they have no intact phage genes and contain only a proximal marker of gene II and a distal marker of gene IV plus the origin/terminus of DNA replication. They cannot grow by themselves; they require a helper phage. A cross was performed between such deletion phage and a phage containing the proximal amber gene II. This setup mimics the gene-II-by-gene-II cross. The yield of progeny was equivalent to that obtained in a normal cross, and again about 50% of the bursts contained one parent and one recombinant. In this instance, the parent in the burst was derived from the complete genome and the *am*$^+$ gene was derived from the defective genome. If in the normal cross the recombinant had been produced by an asymmetric single-strand exchange with subsequent base correction to homozygous *am*$^+$ and if the parent that emerged with it had been produced by repair synthesis of the donated region, then in the cross described above all bursts should contain recombinants only; the donor-defective parent cannot form a plaque.

Thus, we conclude that the only model consistent with all the data is one that results in a large region of heteroduplex DNA being produced by the breakage, transfer, and joining of single strands (Zinder 1974). Recombinants are formed when the breaks are between markers. One of the parents is the recipient, the other the donor. Neither the donor nor the strand displaced in the recipient is recovered. The heteroduplex formed segregates to give a one parent-one recombinant yield. A few recombinants arise by base correction. This singularity of the small-phage systems might be a result of their size and ring nature. With linear genomes or rings sufficiently large that any given domain can be considered linear, a minor modification of the events envisaged could also give reciprocal recombination of a more classical kind. Also, these phages are utilizing their host as a system for genetic recombination. It may be no coincidence that the genetic recombination systems of *E. coli*, such as conjugation, transduction, and transformation, also involve nonreciprocal events which give rise to intermediates of heteroduplex DNA. The kinds of physical structures that are believed to be formed during recombination events are discussed by Warner and Tessman (this volume).

ACKNOWLEDGMENTS

Work in this laboratory is supported by grants from the National Science Foundation and the National Institutes of Health.

REFERENCES

Arber, W. 1966. Host specificity of DNA produced by *Escherichia coli*. *J. Mol. Biol.* **20:** 483.

Arber, W. and U. Kühnlein. 1967. Mutationeller verlust B-spezifischer restriction des Bakteriophagen fd. *Pathol. Microbiol.* **30:** 946.

Arber, W. and S. Linn. 1969. DNA modification and restriction. *Annu. Rev. Biochem.* **38:** 467.

Baas, P. D. and H. S. Jansz. 1972a. Asymmetric information transfer during φX174 DNA replication. *J. Mol. Biol.* **63:** 557.

———. 1972b. φX174 replicative form DNA replication, origin and direction. *J. Mol. Biol.* **63:** 569.

Baker, R. and I. Tessman. 1967. The circular genetic map of phage S13. *Proc. Natl. Acad. Sci.* **58:** 1438.

Baker, R., J. Doniger, and I. Tessman. 1971. Roles of parental and progeny DNA in two mechanisms of phage S13 recombination. *Nat. New Biol.* **230:** 23.

Benbow, R. M., C. A. Hutchison III, J. D. Fabricant, and R. L. Sinsheimer. 1971. Genetic map of bacteriophage φX174. *J. Virol.* **7:** 549.

Benbow, R. M., A. J. Zuccarelli, G. C. Davis, and R. L. Sinsheimer. 1974. Genetic recombination in bacteriophage φX174. *J. Virol.* **13:** 898.

Boon, T. and N. D. Zinder. 1970. Genetic recombination in bacteriophage f1: Transfer of parental DNA to the recombinant. *Virology* **41:** 444.

———. 1971. Genotypes produced by individual recombination events involving bacteriophage f1. *J. Mol. Biol.* **58:** 133.

Denhardt, D. 1975. The single-stranded DNA phages. *CRC Crit. Rev. Microbiol.* **4:** 161.

Doniger, J., R. C. Warner, and I. Tessman. 1973. Role of circular dimer DNA in the primary recombination mechanism of bacteriophage S13. *Nat. New Biol.* **242:** 9.

Enea, V. and N. D. Zinder. 1975. A deletion mutant of bacteriophage f1 containing no intact cistrons. *Virology* **68:** 105.

———. 1976. Heteroduplex DNA: A recombinational intermediate in bacteriophage f1. *J. Mol. Biol.* **101:** 25.

Enea, V., G. Vovis, and N. D. Zinder. 1975. Genetic studies with heteroduplex DNA of bacteriophage f1: Asymmetric segregation, base correction and implications for the mechanism of genetic recombination. *J. Mol. Biol.* **96:** 495.

Enea, V., K. Horiuchi, B. G. Turgeon, and N. D. Zinder. 1977. Physical map of defective interfering particles of bacteriophage f1. *J. Mol. Biol.* **111:** 395.

Fogel, S. and R. K. Mortimer. 1969. Informational transfer in meiotic gene conversion. *Proc. Natl. Acad. Sci.* **62:** 96.

Hartman, N. 1973. "The effects of B-specific restriction and modification of DNA on linkage relationships in f1 bacteriophage." Ph.D. thesis, Rockefeller University, New York, New York.

Hartman, N. and N. D. Zinder. 1974a. The effect of B specific restriction and modification of DNA on linkage relationships in f1 bacteriophage. I. Studies on the mechanism of B restriction in vivo. *J. Mol. Biol.* **85:** 345.

———. 1974b. The effect of B restriction and modification on linkage relationships in f1 bacteriophage. II. Evidence for a heteroduplex intermediate in f1 recombination. *J. Mol. Biol.* **85:** 357.

Horiuchi, K. and N. D. Zinder. 1973. Cleavage of f1 DNA by the restriction enzyme of *Escherichia coli* B. *Proc. Natl. Acad. Sci.* **69:** 3220.

Horiuchi, K., G. F. Vovis, V. Enea, and N. D. Zinder. 1975. Cleavage map of bacteriophage f1: Location of the *E. coli* B-specific modification sites. *J. Mol. Biol.* **95:** 147.

Hutchison, C. A., III and M. Edgell. 1971. Genetic assay for small fragments of bacteriophage φX174 deoxyribonucleic acid. *J. Virol.* **8:** 181.

Ihler, G. and M. Meselson. 1963. Genetic recombination in bacteriophage λ by breakage and joining of DNA molecules. *Virology* **21:** 7.

Lyons, L. and N. D. Zinder. 1972. The genetic map of the filamentous bacteriophage f1. *Virology* **49:** 45.

Meselson, M. and C. M. Radding. 1975. A general model for genetic recombination. *Proc. Natl. Acad. Sci.* **72:** 358.

Meselson, M. and J. Weigle. 1961. Chromosome breakage accompanying genetic recombination in bacteriophage λ. *Proc. Natl. Acad Sci.* **47:** 857.

Pratt, D. and W. S. Erdahl. 1968. Genetic control of bacteriophage M13 DNA synthesis. *J. Mol. Biol.* **37:** 181.

Pratt, D., H. Tzagoloff, and W. S. Erdahl. 1966. Conditional lethal mutants of the small filamentous phage M13. *Virology* **30:** 397.

Salivar, W. O., T. J. Henry, and D. Pratt. 1967. Purification and properties of diploid particles of M13. *Virology* **32:** 41.

Scott, J. R. and N. D. Zinder. 1967. Heterozygotes of phage f1. In *Molecular biology of viruses,* p. 211. Academic Press, New York.

Sugiura, M., T. Okamoto, and M. Takanami. 1969. Starting nucleotide sequences of RNA synthesized on the replicative form DNA of coliphage fd. *J. Mol. Biol.* **43:** 299.

Tessman, I. 1966. Genetic recombination of phage S13 in a recombination-deficient mutant of *Escherichia coli* K-12. *Biochem. Biophys. Res. Commun.* **22:** 1966.

Vovis, G. F. and N. D. Zinder. 1975. Methylation of f1 DNA by a restriction endonuclease from *E. coli* B. *J. Mol. Biol.* **95:** 557.

Vovis, G. F., K. Horiuchi, and N. D. Zinder. 1974. Kinetics of methylation of DNA by a restriction endonuclease from *Escherichia coli* B. *Proc. Natl. Acad Sci.* **71:** 3810.

———. 1975. Endonuclease R·*Eco* RII restriction of bacteriophage f1 DNA *in vitro*: Ordering of genes V and VII, location of an RNA promoter for gene VII. *J. Virol.* **16:** 674.

Vovis, G. F., K. Horiuchi, N. Hartman, and N. D. Zinder. 1973. Restriction endonuclease B and f1 heteroduplex DNA. *Nat. New Biol.* **246:** 13.

Wildenberg, J. and M. Meselson. 1975. Mismatch repair in heteroduplex DNA. *Proc. Natl. Acad. Sci.* **72:** 2202.

Zahler, S. A. 1958. Some biological properties of bacteriophage S13 and φX174. *J. Bacteriol.* **75:** 310.

Zinder, N. D. 1974. Recombination in bacteriophage f1. In *Mechanisms in recombination* (ed. R. Grell), p. 19. Plenum Press, New York.

Mechanism of Genetic Recombination

Robert C. Warner
Department of Molecular Biology and Biochemistry
University of California
Irvine, California 92717

Irwin Tessman
Department of Biological Sciences
Purdue University
West Lafayette, Indiana 47907

Genetic studies of the circular single-stranded DNA phages were initiated in the expectation that the smallness of these phages reflected a simplicity conducive to understanding the recombination process on the molecular level. Early extensive efforts just to find genetic recombination failed; the most serious work (Zahler 1958) showed no recombinants at the level of 10^{-3} in several crosses of plaque-type mutants of S13.

This inability to find genetic recombination was rationalized for several years by inferences from radiosensitivity studies (Tessman et al. 1957) which suggested that φX174 and S13 behaved radiologically like RNA viruses, which are notably single-stranded in their nucleic acid and do not undergo genetic recombination. In the end, however, it turned out that recombination of S13, φX, and G4 was easy to observe even without special selective techniques (Tessman and Tessman 1959). The original success was due to the stimulatory action of UV light, observed first for phage λ (Jacob and Wollman 1955); with UV light the recombination frequencies observed were of the order of 10^{-3}, just within the range needed at that time for easy observation. However, UV stimulation is not essential for observing recombinants, and the mechanism of stimulation remains a major puzzle in molecular genetics.

The early recombination experiments revealed a useful feature of this phage system: the probability of a mating event (the interaction of two genotypes leading to recombination) is around 0.01. This means that the recombinants emerging from a single cell are, with rare exceptions, the products of a single mating event. This is the only prokaryotic system known in which all the products of one mating event can be examined. Fungi provide the only eukaryotic genetic system known to have this useful property.

The techniques employing conditionally lethal mutants, introduced into phage research by Campbell (1961), were adopted by E. S. Tessman (1965) for S13 in work that marked the start of an intensive genetic analysis of S13 and φX. Her quantitative genetic mapping of genes *A*, *F*, *G*, and *H* showed that between neighboring genes the recombination frequencies are roughly additive; she also supplied the genetic landmarks for the subsequent proof that the genetic map is circular (see Fig. 4, Weisbeek and van Arkel [this volume] for an up-to-date map).

Circularity

By 1967 seven genes could be identified by mutation in S13, and they were ordered qualitatively into a circular map by a series of three-factor crosses (Baker and Tessman 1967). The positions of two genes were later seen to be reversed (R. Baker and I. Tessman, cited in Vanderbilt et al. 1971). A similar series of three-factor crosses was subsequently performed for φX (Benbow et al. 1971), and the same genetic map was observed; gene *D* was also located (no mutant of gene *D* is known for S13). In 1971 the genes of S13 and φX were renamed in alphabetical order (*A*–*H*) according to the circular order. The alphabetical order is disturbed by gene *J*, for which no mutants are known, but which codes for a small capsid protein and is located physically between the end of *D* and the start of *F* (Sanger et al. 1977; Godson et al., this volume).

A circular genetic map implies an even number of genetic exchanges in a mating. It also rules out the possibility that one of the even exchanges is always at a particular site because in that case the map would be linear, with the regions on either side of the site at opposite ends of the map. This is of special interest because of the function of the phage gene-*A* protein in nicking superhelical phage (RFI) DNA (Francke and Ray 1971; Henry and Knippers 1974). Nicking of the viral strand occurs at the origin of replication located within gene *A* (Baas and Jansz 1972b; Langeveld et al. 1978) at the site where the gene-*A* protein binds to the DNA (Eisenberg et al. 1977). This nicked DNA is an obvious choice as an intermediate in the formation of a recombinant particle, but if it really is an intermediate, the site of nicking cannot always be the locus for genetic exchange.

Uniformity of Recombination

An important question is whether genetic recombination is uniform throughout the genome. Using the nucleotide-sequence data of Sanger et al. (1977), one can now correlate recombination values with physical distance. All data presently available indicate that recombination is roughly uniform over the genome. The following quantitative estimates can be made (recombination frequencies are calculated as twice the wild-type frequency in crosses between conditionally lethal mutants). Within gene *A* of S13 the maximum recombination frequency is 2.5×10^{-3} (Tessman 1965; E. S.

Tessman and S. Germeraad, pers. comm.), which corresponds to about 3×10^{-6} per nucleotide; within gene *B* of S13 (Tessman et al. 1976) the frequency is at least 1×10^{-6} per nucleotide; in the *D–E* region of φX the frequency is 2×10^{-6} per nucleotide (T. J. Pollack, pers. comm.); gene-*F* crosses yield a frequency of 1×10^{-6} per nucleotide (Vanderbilt et al. 1971) and 2×10^{-6} per nucleotide (Vanderbilt et al. 1972).

Benbow et al. (1971) stress that recombination in φX gene *A* is unusually high, but their data do not appear to support that conclusion. Their largest recombination frequency in gene *A* corresponds to approximately 4×10^{-6} per nucleotide, which is in rough agreement with the values above. In their crosses, phage were adsorbed in cyanide, a procedure that can have a stimulating but variable effect on recombination (Chase and Doermann 1958; Tomizawa and Anraku 1964). E. S. Tessman (pers. comm.) finds about a twofold stimulation of recombination in S13 when cyanide is used during adsorption, even though it is subsequently diluted to 3×10^{-6} M. Roughly, then, recombination in S13 and φX occurs at a frequency of about 2×10^{-6} per nucleotide. This is particularly pertinent to the possible effect that the nicking activity of the gene-*A* protein might have on recombination.

Two Recombination Mechanisms

There are at least two mechanisms of genetic recombination for these phages, a primary one and a secondary one. Phage recombination in a recombination-deficient *recA* bacterial host distinguishes the two mechanisms (Tessman 1966). Recombination is markedly reduced in *recA*, which indicates that the primary phage mechanism is dependent on the host *recA* gene, but a significant amount of recombination remains, and the residual recombination is attributed to a secondary mechanism. Thus, the primary mechanism is *recA*-dependent and the secondary mechanism is *recA*-independent. It is not known whether the two mechanisms can function together because the secondary mechanism can be observed only when the primary one has been eliminated. The secondary mechanism is qualitatively distinguishable from the primary one: the primary mechanism is stimulated by UV irradiation, the secondary mechanism is not (Tessman 1968).

The relative contributions of the primary and secondary mechanisms vary according to the cross (Tessman 1966) and may reflect a variation along the genome in the operation of the two mechanisms. The *recBC* gene is responsible for the secondary mechanism (Benbow et al. 1974).

Role of Superhelical DNA

Several lines of evidence indicate that superhelical RFI DNA plays a central role in the primary recombination mechanism. The parental RFI alone can provide a full yield of recombinants; this is shown by the fact that crosses involving gene-*A* amber mutants, which cannot make progeny DNA, still produce a normal number of recombinants in a $recA^+$ host (Baker et al. 1971). In the absence of the phage gene-*A* nicking function, the DNA remains as superhelical RFI (Francke and Ray 1971).

That superhelical DNA is directly implicated in recombination was shown in transfection of spheroplasts with both φX single-stranded fragments and RFI DNA; recombinants between the fragments and the RFI were more frequent than when RFII (circular, duplex DNA in which at least one strand is not continuous) was used (Holloman and Radding 1976). Furthermore, this recombination required the host $recA^+$ function. Although there is about 30 times more progeny RF DNA than parental RF DNA in a cross in which gene *A* is expressed, the number of recombinants formed is no higher, which suggests that progeny RF DNA is not a major contributor to recombinants via the primary mechanism. However, it is quite possible that it is not progeny RF DNA per se that is ineffective but the high proportion of the relaxed RFII. Thus, the amount of RFI may be nearly the same whether or not gene *A* is expressed and that may account for the nearly identical number of recombinants. The higher efficiency of RFI relative to RFII for recombination is paralleled by its high efficiency for transcription, both in vitro (Hayashi and Hayashi 1971) and in vivo (Puga and Tessman 1973); the higher degree of single-strandedness in the underwound superhelical DNA may account for its higher efficiency for both recombination and transcription.

In contrast to the role of superhelical DNA in the primary, $recA^+$-dependent recombination mechanism, the secondary mechanism cannot function with parental RFI alone. In a cross of *amA* mutants (which make only parental RFI) in a $recA^-$ host, which permits only the secondary mechanism to function, recombination is reduced about 30-fold (Baker et al. 1971). Two possibilities are suggested: either progeny DNA (RF or single-stranded [SS]) is needed, or the nicking function of gene *A* is directly involved in the secondary mechanism.

A MODEL FOR RECOMBINATION

At about the same time that the genetic map of S13 was shown to be circular by Baker and Tessman (1967), circular dimers were found to be present in φX RF DNA (Rush et al. 1967). The two observations were consistent in that the characteristics of the map—circularity and paired exchanges leading to high negative interference—could be explained by a mechanism in which the circular dimer is an intermediate. This postulate was supported by the findings that circular dimers are infectious (Rush and Warner 1967) and that they can be heterozygous for two genetic markers (Rush and Warner 1968, 1969). An additional dimeric species in RF DNA, the figure eight, was suggested by Doniger et al. (1973) to be involved in the pathway leading to the circular dimer. It had been observed by Gordon et al. (1970) in preparations of RF DNA, but at that time the nature of its structure was not clear. Its relevance to recombination was its presumed ability to branch migrate and produce heteroduplex regions. This property led Broker and Lehman (1971), as a result of their studies on T4, to suggest that a figure-eight configuration might be an intermediate in recombination. Thompson et al. (1975, 1976) proved that figure eights are distinct from catenated dimers and that they contain a branch migrating junction. Figure eights have been observed in φX RF DNA by Benbow et al. (1975) and in ColE1 DNA by Potter and Dressler (1976).

A model for recombination based on the figure eight as the primary intermediate is shown in Figure 1. It is formulated at the outset as a specific mechanism in order to facilitate the discussion below of the evidence bearing on it. The model is a modification of those described by Thompson et al. (1975) and Doniger et al. (1973) and includes features discussed by Benbow et al. (1975).

The Holliday Structure

The figure eight is a simple prototype of the Holliday structure diagrammed in Figure 2. This configuration was suggested by Holliday (1962, 1964) to explain gene conversion in fungi by the processes of heteroduplex formation and mismatch repair. It is characterized by the feature that the initiating nicks are in strands of the same polarity, leading to the invasion of each genome by a single strand of the other. It thus contains heteroduplex regions, which, with the postulate of mismatch repair, has been the basis for its success in accounting for gene conversion in fungi (Holliday 1964, 1968; Mortimer and Fogel 1974). Modified mechanisms have been developed by several authors (Holliday 1968; Sigal and Alberts 1972; Radding 1973, 1978; Hotchkiss 1974; Meselson and Radding 1975) to explain features of recombination in various systems. The mechanism in Figure 1 thus has relevance for organisms more complex than the small DNA phages. It supports and extends the validity of mechanisms of the Holliday type because it is based on the figure eight as an isolatable intermediate rather than on a hypothetical structure.

Pathways

The evidence for pathways that permit interchanges among monomers, figure eights, and circular dimers (Fig. 1) has been reviewed recently (Thompson et al. 1978). The important points include the following: the interchange between figure eights and circular dimers is reversible, as shown by analysis of the products of coinfection by φX and G4 (Thompson et al. 1978). This is further supported by evidence from studies on plasmid sys-

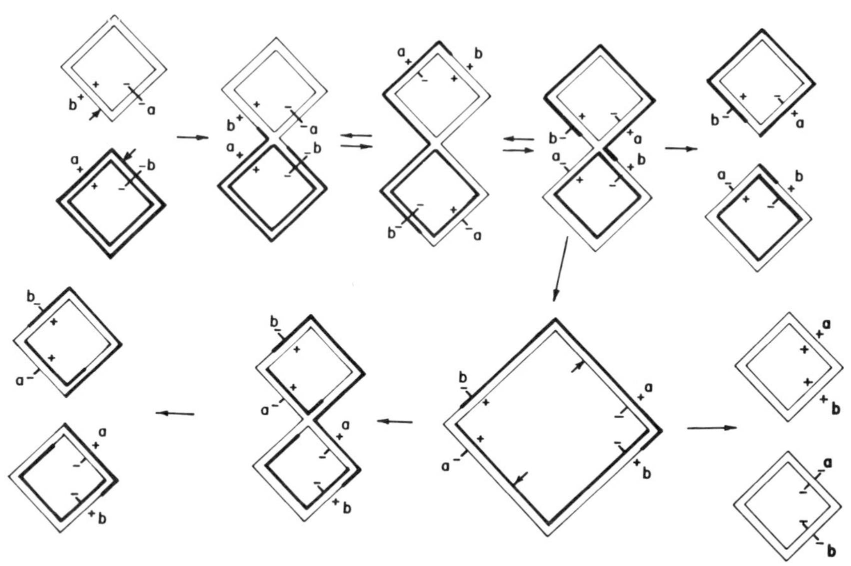

Figure 1 Model for recombination in small DNA phages. The formation of the figure eight, shown at the upper left, is initiated by nicks (arrows) in strands of the same polarity. Two possible steps of branch migration of the junction are shown, followed by two alternative paths for resolution of the figure eight. If the same strands that were nicked in the formation of the figure eight are again nicked at the junction, two monomers containing mismatches are formed (upper right). If the nicks are in the strands of the opposite polarity, a circular dimer will result. The dimer may yield monomers by a return pathway that includes the figure eight or by the replicative pathway in the lower right. The products of replication are shown as the monomer RF DNA that would be formed if replication originated between the markers on the long arcs of the dimer and if the monomers were replicas of the outside strand. Such a replicative pathway could also operate at the stage of SS DNA maturation and produce viral DNA.

The mismatches will be resolved by repair, or replication, or both (Thompson et al. 1975). These processes are not shown in the figure.

All of the reactions except those of the replicative pathway are believed to be reversible. They are not indicated to be so in order to make the model more representative of the probable pathway of in vivo recombination.

A separate replicative pathway leading to the formation of homozygous, circular dimers from monomers is assumed, but it is not shown on the diagram.

tems (Bedbrook and Ausubel 1976; Potter and Dressler 1976, 1977). The formation of circular dimers by replication was suggested by Gilbert and Dressler (1969) and Rush and Warner (1969), and this route, in contrast to that involving recombination, was clearly shown by Benbow et al. (1972) and Hofs et al. (1972, 1973) to be the major pathway. The existence of replicative formation of monomers from circular dimers is supported by the

Figure 2 The Holliday structure. The heteroduplex region shown is presumed to have been formed by hybrid overlap during initiation of the structure. Longer stretches of heteroduplex may be created by branch migration of the junction in either direction. Mechanisms for resolution of the structure to yield separate genomes still containing heteroduplex regions, with or without exchange of flanking markers, have been formulated by Sigal and Alberts (1972) and Meselson and Radding (1975).

recent electron microscope study of the replication of φX by Koths and Dressler (1978) in which replicating circular dimers with tails of up to monomer length were observed.

Figure eights have been reported to be present in the dimer fraction of φX, S13, and G4 RF DNAs to the extent of about 7% (Thompson et al. 1978). This indicates a frequency of figure eights of about 0.2% in RF DNA. A higher frequency was reported by Benbow et al. (1975). Figure eights have also been observed in filamentous phage fd RF DNA (C. Palm, F. C. Wheeler, and R. C. Warner, unpubl.). They have been shown to be recombinant by Benbow et al. (1975) by coinfection with φX wild-type and an uncharacterized deletionlike mutant, which resulted in finding figure eights to which each parent had contributed one monomer unit. This was confirmed in a different system by coinfection with φX and G4. Recombinant figure eights were identified by the difference in susceptibility of the two genomes to *Eco*RI restriction endonuclease (Thompson et al. 1978). In this study, circular dimers were also shown to be recombinant, and the dimer-length strand of the figure eight was found to be the plus or minus strand with equal probability.

Catenated dimers have not been included in the model because neither their route of formation nor their relation to recombination is understood. They have been shown to be recombinant in several studies (Benbow et al. 1972, 1975; Doniger et al. 1973; Thompson et al. 1978).

Figure eights were indicated to be on the *recA* pathway of *E. coli* by the tenfold reduction in recombinant circular dimers when infections were carried out in a *recA*$^-$ host (Doniger et al. 1973). They are absent in RF DNA prepared from *recA*$^-$ cells (Benbow et al. 1975) and in ColE1 DNA grown in *recA*$^-$ cells but not in *recBC*$^-$ cells (Potter and Dressler 1976). The requirement for the *recA* function in the formation of multimeric rings has been shown by Bedbrook and Ausubel (1976) and Potter and Dressler (1977) by recombination in plasmid systems.

Branch Migration

The process of branch migration is an essential feature of models of recombination developed from the Holliday structure. It was not part of the original mechanism, in which heteroduplex regions were considered to be formed as a result of hybrid overlap of the invading single strand. Movement of a "half-chiasma" was postulated by Fogel and Hurst (1967) as part of a variation on the Holliday model. It is now clear that movement of the junction and extension of the heteroduplex region can be accomplished by branch migration. Evidence that single-stranded branch migration is a physically possible process was first provided by Lee et al. (1970). Kim et al. (1972) extended this to double-stranded branch migration of the kind that occurs in a figure eight. The evidence in both studies was based on electron microscopic observations of structures that could have been interconverted had branch migration occurred. The relevance of branch migration to recombination was developed by Broker and Lehman (1971) and Broker (1973) in their studies on T4. They suggested a number of models, including the figure eight, in which heteroduplex regions were created by branch migration. The report of Sigal and Alberts (1972) that the double-stranded branch migrating junction could be formed without any loss of base pairing was an important step in the incorporation of branch migration into models of recombination (Sigal and Alberts 1972; Sobell 1972).

The figure eight has provided a new approach to the study of branch migration (Thompson et al. 1976). The monomer of G4 RF DNA has one site for hydrolysis by *Eco*RI restriction endonuclease, and, as a result, a figure eight made from G4 is converted to an *X*-shaped form by digestion with this enzyme. The *X* forms will branch migrate into two linear monomers, which comprise a terminal, irreversible configuration. This conversion can be used to demonstrate the occurrence of branch migration in the figure eight and to measure its rate. Thompson et al. (1976) treated branch migration as a random-walk process, and from measurements by electron microscopy of the disappearance of *X* forms they calculated a random step rate of about 6 kilobase pairs per second at 37°C. Other workers, using an assay not dependent on electron microscopy, have subsequently reported a slightly higher rate, and it has been confirmed that the kinetic course of the reaction follows the predictions of the random-walk theory (R. A. Fishel, F. C. Wheeler, and R. C. Warner, unpubl.). Because of the random movement of the junction, the probability of its progress can be predicted from a Gaussian distribution for which σ is proportional to the square root of the number of random steps. For example, there is a probability of 0.32 that a junction in a G4 figure eight will be found more than 300 nucleotides in either direction from its origin in 1 minute or more than 950 nucleotides from its origin in 10 minutes ($950 = 300\sqrt{10}$). Thus, despite the high step rate, it is a slow net movement and appears to be just fast enough to account for recombination by the mechanism of Figure 1. Measurements of the rate of single-stranded branch migration have been reported by Radding et al. (1977).

Mismatch Repair

The idea that mismatched bases, created by heteroduplex formation, may be subject to enzymatic repair was first suggested as a mechanism for gene conversion by Holliday (1962). Although there has been progress in understanding the enzymology of excision repair, no recognition system for mismatches has been identified that would lead to excision and repair. However, the occurrence of mismatch repair has been clearly shown in a number of experiments in which heteroduplexes prepared in vitro have been introduced into cells by transfection or in bacterial transformation systems (Doerfler and Hogness 1968; Spatz and Trautner 1970; Roger 1972; Baas and Jansz 1972a; Wildenberg and Meselson 1975; Enea et al. 1975). Mutants in *uvrD* and *uvrE* of *E. coli* are deficient in mismatch repair (Nevers and Spatz 1975).

The extent of mismatch repair is most conveniently expressed by the quantity a_i, as defined by Wildenberg and Meselson (1975). It is the probability of repair from mutant to wild type for a given marker, i, and will have values ranging from 0 to 0.5 for complete repair. It will be assumed that the probability of repair in the opposite direction, wild-type to mutant, is the same; this has not been tested directly, and indirect examination has indicated that it may not hold generally (Wagner and Meselson 1976). The probability of repair has been shown to be site-specific (Wildenberg and Meselson 1975; Wagner and Meselson 1976; Berger and Pardoll 1976; Roger 1977), although in the absence of specific information an average value must generally be assumed. For heteroduplex transfections, values of a_i from 0.04 to 0.2 were found in λ (Wildenberg and Meselson 1975); similar values were found in f1 (Enea et al. 1975), and up to 0.4 was found in φX (Baas and Jansz 1972a). There is no reason to suppose that the probability of repair in a normal infection will be the same as in a heteroduplex transfection. Such probability will be determined by the time that a mismatch is exposed to repair systems of a given efficiency before it is resolved by replication. It thus may vary with the experimental arrangement as well as with the physiology of the organism being studied.

Initiation of Figure-eight Formation

The introduction of homologous nicks in monomers (indicated by the arrows in Fig. 1) is purely a formal assumption, and so far nothing is known about the enzymology or even the sequence of the steps leading from monomers to the figure eight. However, the single-strand invasion of superhelical DNA suggested as the initiating step in recombination could be the first step in a figure-eight mechanism (Holloman et al. 1975; Holloman and Radding 1976; Beattie et al. 1977; Wiegand et al. 1977). These workers have shown the uptake of homologous single-stranded fragments of φX by RFI and the efficient conversion of the complex into recombinants when it is transfected into spheroplasts. This process could be a model for joining randomly nicked RFII to RFI in an homologous orientation, with the subsequent introduction of an approximately corresponding nick in the RFI and conversion to a figure eight.

GENETIC RESULTS THAT TEST THE MODEL

A model based entirely on the sequence in the upper part of Figure 1 can be shown to be in accord with the basic genetic results for small DNA phages. Mapping functions have been developed for random initiation and resolution of the figure eight, with the probability of repair and the rate of branch migration as variables (R. C. Warner, unpubl.). A relationship of frequency of recombination to distance separating two markers, similar to that characteristic of circular genomes (Steinberg and Stahl 1967), is maintained over the entire range of possible repair probabilities and for branch migration rates in the range of recent measurements (Thompson et al. 1976). Thus, the observed genetic circularity of small-phage genomes (Baker and Tessman 1967; Lyons and Zinder 1972; Benbow et al. 1974) is predicted. Additional involvement of the circular dimer with the pathways shown in the lower part of Figure 1 would modify the expected progeny but it would not change the nature of the linkage relationships.

Consistency with the requirements for mapping a circular genome, however, is only a trivial test of the model. The full advantage of small DNA phage as a genetic system can be realized in single-burst experiments, as the identified progeny are likely to be the result of a single recombinational event. A critical test of some features of the model is thus provided by the extensive analysis of single-burst experiments with the filamentous phage f1 (Boon and Zinder 1970, 1971; Hartman and Zinder 1974a; Enea et al. 1975; Enea and Zinder 1976). These results are pertinent to the model because in recent work the presence of the figure eight in phage fd RF DNA has been demonstrated (C. Palm, F. C. Wheeler, and R. C. Warner,unpubl.). A single-burst protocol for f1 was devised by Boon and Zinder (1971) and has been used in a series of studies. The basic plan is that infection is carried out under restrictive conditions with two gene-II amber mutants. Since the gene-II product is required for replication of RF DNA, there will be no DNA synthesis until there has been recombination. This assures that all phage produced are descendants of the progeny RF DNA from a single recombinational event that produced a wild-type recombinant with its functional gene-II product. The basic plan was supplemented by the observation of unselected markers, usually for two sites that determine resistance to restriction by *E. coli* B. With respect to the selected progeny, none of the bursts contained reciprocal recombinants, and the great majority consisted of either one parent and a wild-type recombinant or the wild-type recombinant only. This regular asymmetry was attributed by Hartman and Zinder (1974b) to a mechanism that involves the formation of considerable heteroduplex stretches and the operation of repair processes based on the recognition of mismatches. It has been discussed with respect to a number of specific postulates about the mechanism of recombination, most recently by Zinder (this volume). He points out that the results are most nearly consistent with a model in which heteroduplex regions are created by the invasion of a receptor molecule by a single strand of a donor, followed by loss of the donor and maturation of the receptor.

This model can be explained in a more general way by the mechanism outlined in Figure 1. The only additional assumption to those contained in the top line of Figure 1 is that upon resolution of the figure eight one of the two resulting monomers, on a random basis, is not recovered. Such an outcome could conceivably be initiated by two asymmetric nicks at the junction of the figure eight rather than by the symmetric nicks indicated in Figure 1. On the basis of this assumption, the predicted bursts would be (1) bursts containing one parent and the wild-type recombinant, with either parent being equally likely; (2) some bursts containing only wild type; (3) no bursts containing more than two types of phage; (4) no bursts containing a wild type and a double amber mutant. These are exactly the results obtained with f1, with the exception that it was not possible to prove the absence of bursts containing wild type and a double mutant (Boon and Zinder 1971).

To make a quantitative prediction about the relative frequencies of the first two classes of bursts, it is necessary to make assumptions about the frequencies of repair and corepair, i.e., the simultaneous repair of two or more mismatches. In the f1 system it is probable that the frequency of repair is high because there is no replication without recombination, and heteroduplexes may be exposed to repair systems for a longer time than in a normal infection before they are resolved by replication. The frequency with which a repair becomes a corepair when the heteroduplex is mismatched for both markers must also be high because the two gene-mutant sites are less than 300 nucleotide pairs apart. Repair stretches of 1000 to 3000 nucleotides leading to corepair have been estimated (Wildenberg and Meselson 1975; Wagner and Meselson 1976). Values for the frequency of repair, a_i, in the range 0.2 to 0.3, depending on the extent of corepair, are necessary to account for the observed fraction of the bursts containing wild type only. If the repair frequency were 0.1, no more than 10% of the bursts would be expected to be wild type only. In making calculations it has been assumed for simplicity that branch migration is very rapid, although this assumption will not have much effect on the results. The asymmetries observed in f1 crosses for the unselected markers are more complex. However, they are also predicted by the assumptions underlying the figure-eight model. They are critically dependent on the linkage distances between the markers and will be observed only when these are small fractions of the genome. Details of the calculations will be presented elsewhere (R. C. Warner, in prep.).

Another single-burst experiment, done by Benbow et al. (1974) with φX, can also be considered in relation to the model. They crossed two gene-*F* mutants in a nonpermissive host and analyzed the bursts that contained wild type. Their nine bursts all contained one parent and one wild-type recombinant. No bursts were found containing only wild type or wild type plus the double amber-opal mutant. However, so few bursts were analyzed that these last results are probably not significant. The single bursts analyzed by Doniger et al. (1973) have some similarity to those considered above. However, the resemblance is probably superficial. These bursts resulted from the transfection of *recA*$^-$ spheroplasts by purified circular dimers obtained from a four-factor cross. In the absence of the *recA* function, the pathway in Figure 1 for the reduction of the circular dimer via the figure eight would be blocked, and resolution of the circular dimers would have to proceed by the replicative pathway shown in the lower right of Figure 1.

The interconversions among monomers, figure eights, and circular dimers shown in Figure 1 are all suggested by physical and genetic evidence to be possible pathways leading to the formation of recombinants. However, the extent to which these pathways are normally utilized in vivo is not clear. The single-burst analyses discussed above provide evidence that most recombination results from the pathway in the upper line of Figure 1 and suggest that the circular dimer is less frequently involved and is certainly not an obligatory intermediate. Passage through a circular dimer would be expected to result in some bursts containing a wild type and a double mutant and probably bursts containing more than two types of phage, the latter depending on the route assumed for resolution of the dimer. A small percentage of such bursts were found by Boon and Zinder (1971) in all of their experiments.

The model (Fig. 1) for genetic recombination in the small DNA phages has been derived from a combination of genetic and physical studies. The enzymology of the process remains largely unexplored.

ACKNOWLEDGMENTS

The writing of this article was supported by grants CA 22239 and CA 12627 from the National Cancer Institute to I. Tessman and R. C. Warner, respectively.

REFERENCES

Baas, P. D. and H. S. Jansz. 1972a. Asymmetric information transfer during φX174 DNA replication. *J. Mol. Biol.* **63:** 557.

———. 1972b. φX174 replicative form DNA replication, origin and direction. *J. Mol. Biol.* **63:** 569.

Baker, R. and I. Tessman. 1967. The circular genetic map of phage S13. *Proc. Natl. Acad. Sci.* **58:** 1438.

Baker, R., J. Doniger, and I. Tessman. 1971. Roles of parental and progeny DNA in the two mechanisms of phage S13 recombination. *Nat. New Biol.* **230:** 23.

Beattie, K. L., R. C. Wiegand, and C. M. Radding. 1977. Uptake of homologous single-stranded fragments by superhelical DNA. II. Characterization of the reaction. *J. Mol. Biol.* **116:** 783.

Bedbrook, J. R. and F. M. Ausubel. 1976. Recombination between bacterial plasmids leading to the formation of plasmid multimers. *Cell* **9:** 707.

Benbow, R. M., M. Eisenberg, and R. L. Sinsheimer. 1972. Multiple length DNA molecules of bacteriophage φX174. *Nat. New Biol.* **237:** 141.

Benbow, R. M., A. J. Zuccarelli, and R. L. Sinsheimer. 1975. Recombinant DNA molecules of bacteriophage φX174. *Proc. Natl. Acad. Sci.* **72:** 235.

Benbow, R. M., C. A. Hutchison III, J. D. Fabricant, and R. L. Sinsheimer. 1971. Genetic map of bacteriophage φX174. *J. Virol.* **7:** 549.

Benbow, R. M., A. J. Zuccarelli, G. C. Davis, and R. L. Sinsheimer. 1974. Genetic recombination in bacteriophage φX174. *J. Virol.* **13:** 898.

Berger, H. and D. Pardoll. 1976. Evidence that mismatched bases in heteroduplex T4 bacteriophage are recognized in vivo. *J. Virol.* **20:** 441.

Boon, T. and N. D. Zinder. 1970. Genetic recombination in bacteriophage f1: Transfer of parental DNA to the recombinant. *Virology* **41:** 444.

———. 1971. Genotypes produced by individual recombinational events involving bacteriophage f1. *J. Mol. Biol.* **58:** 133.

Broker, T. R. 1973. An electron microscopic analysis of pathways for bacteriophage T4 DNA recombination. *J. Mol. Biol.* **81:** 1.

Broker, T. R. and I. R. Lehman. 1971. Branched DNA molecules: Intermediates in T4 recombination. *J. Mol. Biol.* **60:** 131.

Campbell, A. 1961. Sensitive mutants of bacteriophage λ. *Virology* **14:** 22.

Chase, M. and A. H. Doermann. 1958. High negative interference over short segments of the genetic structure of bacteriophage T4. *Genetics* **43:** 332.

Doerfler, W. and D. S. Hogness. 1968. Gene orientation in bacteriophage lambda as determined from the genetic activities of heteroduplex DNA formed in vitro. *J. Mol. Biol.* **33:** 661.

Doniger, J., R. C. Warner, and I. Tessman. 1973. Role of circular dimer DNA in the primary recombination mechanism of bacteriophage S13. *Nat. New Biol.* **242:** 9.

Eisenberg, S., J. Griffith, and A. Kornberg. 1977. φX174 cistron A protein is a multifunctional enzyme in DNA replication. *Proc. Natl. Acad. Sci.* **74:** 3198.

Enea, V. and N. D. Zinder. 1976. Heteroduplex DNA: A recombinational intermediate in bacteriophage f1. *J. Mol. Biol.* **101:** 25.

Enea, V., G. F. Vovis, and N. D. Zinder. 1975. Genetic studies with heteroduplex DNA of bacteriophage f1. Asymmetric segregation, base correction and implications for the mechanism of genetic recombination. *J. Mol. Biol.* **96:** 495.

Fogel, S. and D. D. Hurst. 1967. Meiotic gene conversion in yeast tetrads and the theory of gene conversion. *Genetics* **57:** 455.

Francke, B. and D. S. Ray. 1971. Formation of the parental replicative form DNA of bacteriophage φX174 and the initial events in its replication. *J. Mol. Biol.* **61:** 565.

Gilbert, W. and D. Dressler. 1969. DNA replication: The rolling circle model. *Cold Spring Harbor Symp. Quant. Biol.* **33:** 473.

Gordon, C. N., M. G. Rush, and R. C. Warner. 1970. Complex replicative form molecules of bacteriophages φX174 and S13 *su*105. *J. Mol. Biol.* **47:** 495.

Hartman, N. and N. D. Zinder. 1974a. The effect of B specific restriction and modification of DNA on linkage relationships in f1 bacteriophage. I. Studies on the mechanism of B restriction in vivo. *J. Mol. Biol.* **85:** 345.

———. 1974b. The effect of B specific restriction and modification of DNA on linkage relationships in f1 bacteriophage. II. Evidence for a heteroduplex intermediate in f1 recombination. *J. Mol. Biol.* **85:** 357.

Hayashi, Y. and M. Hayashi. 1971. Template activities of the φX174 replicative allomorphic deoxyribonucleic acids. *Biochemistry* **10:** 4212.

Henry, T. J. and R. Knippers. 1974. Isolation and function of the gene *A* initiation of bacteriophage φX174. *Proc. Natl. Acad. Sci.* **71:** 1549.

Hofs, E. B. H., J. H. van de Pol, G. A. van Arkel, and H. S. Jansz. 1972. Dimeric circular duplex DNA of bacteriophage φX174 and recombination. *Mol. Gen. Genet.* **118:** 161.

Hofs, E. B. H., G. A. van Arkel, P. D. Baas, D. J. Ellens, and H. S. Jansz. 1973. Mechanism of formation of bacteriophage φX174 circular and catenated dimer RF DNA. *Mol. Gen. Genet.* **126:** 37.

Holliday, R. 1962. Mutation and replication in *Ustilago maydis*. *Genet. Res.* **3:** 472.

———. 1964. A mechanism for gene conversion in fungi. *Genet. Res.* **5:** 282.

———. 1968. Genetic recombination in fungi. In *Replication and recombination of genetic material* (ed. W. J. Peacock and R. D. Brock), p. 157. Australian Academy of Science, Canberra.

Holloman, W. K. and C. M. Radding. 1976. Recombination promoted by superhelical DNA and the *recA* gene of *Escherichia coli*. *Proc. Natl. Acad. Sci.* **73:** 3910.

Holloman, W. K., R. Weigand, C. Hoessli, and C. M. Radding. 1975. Uptake of homologous single-stranded fragments by superhelical DNA: A possible mechanism for initiation of genetic recombination. *Proc. Natl. Acad. Sci.* **72:** 2394.

Hotchkiss, R. D. 1974. Models of genetic recombination. *Annu. Rev. Microbiol.* **28:** 445.

Jacob, F. and E.-L. Wollman. 1955. Étude génétique d'un bactériophage tempéré d'*Escherichia coli.* III. Effet du rayonnement ultraviolet sur la recombinaison génétique. *Ann. Inst. Pasteur* **88:** 724.

Kim, J.-S., P. A. Sharp, and N. Davidson. 1972. Electron microscope studies of heteroduplex DNA from a deletion mutant of bacteriophage φX174. *Proc. Natl. Acad. Sci.* **69:** 1948.

Koths, K. and D. Dressler. 1978. An analysis of the φX DNA replication cycle by electron microscopy. *Proc. Natl. Acad. Sci.* **75:** 605.

Langeveld, S. A., A. D. M. van Mansfield, P. D. Baas, H. S. Jansz, G. A. van Arkel, and P. J. Weisbeek. 1978. Nucleotide sequence of the origin of replication in bacteriophage φX174 RF DNA. *Nature* **271:** 417.

Lee, C. H., R. W. Davis, and N. Davidson. 1970. A physical study by electron microscopy of the terminally repetitious, circularly permuted DNA from the coliphage particles of *Escherichia coli* 15. *J. Mol. Biol.* **48:** 1.

Lyons, L. B. and N. D. Zinder. 1972. The genetic map of the filamentous bacteriophage f1. *Virology* **49:** 45.

Meselson, M. S. and C. M. Radding. 1975. A general model for genetic recombination. *Proc. Natl. Acad. Sci.* **72:** 358.

Mortimer, R. K. and S. Fogel. 1974. Genetical interference and gene conversion. In *Mechanisms in recombination* (ed. R. E. Grell), p. 263. Plenum Press, New York.

Nevers, P. and H.-C. Spatz. 1975. *Escherichia coli* mutants *uvrD* and *uvrE* deficient in gene conversion of λ-heteroduplexes. *Mol. Gen. Genet.* **139:** 233.

Potter, H. and D. Dressler. 1976. On the mechanism of genetic recombination: Electron microscopic observation of recombination intermediates. *Proc. Natl. Acad. Sci.* **73:** 3000.

———. 1977. On the mechanism of genetic recombination: The maturation of recombination intermediates. *Proc. Natl. Acad. Sci.* **74:** 4168.

Puga, A. and I. Tessman. 1973. Mechanism of transcription of bacteriophage S13. I. Dependence of messenger RNA synthesis on amount and configuration of DNA. *J. Mol. Biol.* **75:** 83.

Radding, C. M. 1973. Molecular mechanisms in genetic recombination. *Annu. Rev. Genet.* **7:** 87.

———. 1978. Genetic recombination: Strand transfer and mismatch repair. *Annu. Rev. Biochem.* (in press).

Radding, C. M., K. L. Beattie, W. K. Holloman, and R. C. Wiegand. 1977. Uptake of homologous single-stranded fragments by superhelical DNA. IV. Branch migration. *J. Mol. Biol.* **116:** 825.

Roger, M. 1972. Evidence for conversion of heteroduplex transforming DNAs to homoduplexes in recipient pneumococcal cells. *Proc. Natl. Acad. Sci.* **69:** 464.

———. 1977. Mismatch excision and possible polarity effects result in preferred deoxyribonucleic acid strand of integration in pneumococcal transformation. *J. Bacteriol.* **129:** 298.

Rush, M. G. and R. C. Warner. 1967. Multiple-length rings of φX174 replicative form. II. Infectivity. *Proc. Natl. Acad. Sci.* **58:** 2372.

———. 1968. Multiple-length rings of φX174 and S13 replicative forms. III. A possible intermediate in recombination. *J. Biol. Chem.* **243:** 4821.

———. 1969. Molecular recombination in a circular genome—φX174 and S13. *Cold Spring Harbor Symp. Quant. Biol.* **33:** 459.

Rush, M. G., A. T. Kleinschmidt, W. Hellmann, and R. C. Warner. 1967. Multiple-length rings in preparations of φX174 replicative form. *Proc. Natl. Acad. Sci.* **58:** 1676.

Sanger, F., G. M. Air, B. G. Barrell, N. L. Brown, A. R. Coulson, J. C. Fiddes, C. A. Hutchison III, P. M. Slocombe, and M. Smith. 1977. Nucleotide sequence of bacteriophage φX174 DNA. *Nature* **265:** 687.

Sigal, N. and B. Alberts. 1972. Genetic recombination: The nature of a crossed-strand exchange between two homologous DNA molecules. *J. Mol. Biol.* **71:** 789.

Sobell, H. M. 1972. Molecular mechanism for genetic recombination. *Proc. Natl. Acad. Sci.* **69:** 2483.

Spatz, H.-C. and T. A. Trautner. 1970. One way to do experiments in gene conversion? Transfection with heteroduplex SPP1 DNA. *Mol. Gen. Genet.* **109:** 84.

Steinberg, C. M. and F. W. Stahl. 1967. Interference in circular maps. *J. Cell. Physiol.* (Suppl. 1) **70:** 4.

Tessman, E. S. 1965. Complementation groups in phage S13. *Virology* **25:** 303.

Tessman, E. S. and I. Tessman. 1959. Genetic recombination in phage S13. *Virology* **7:** 465.

Tessman, I. 1966. Genetic recombination of phage S13 in a recombination deficient mutant of *Escherichia coli* K12. *Biochem. Biophys. Res. Commun.* **22:** 169.

———. 1968. Selective stimulation of one of the mechanisms for genetic recombination of bacteriophage S13. *Science* **161:** 481.

Tessman, I., E. S. Tessman, and G. S. Stent. 1957. The relative radiosensitivity of bacteriophages S13 and T2. *Virology* **4:** 209.

Tessman, I., E. S. Tessman, T. J. Pollock, M.-T. Borrás, A. Puga, and R. Baker. 1976. Reinitiation mutants of gene *B* of bacteriophage S13 that mimic gene *A* mutants in blocking replicative form DNA synthesis. *J. Mol. Biol.* **103:** 583.

Thompson, B. J., M. N. Camien, and R. C. Warner. 1976. Kinetics of branch migration in double-stranded DNA. *Proc. Natl. Acad. Sci.* **73:** 2299.

Thompson, B. J., C. H. Sussman, and R. C. Warner. 1978. Coinfection of *E. coli* with phages G4 and φX174: Origin of dimeric replicative form species. *Virology* (in press).

Thompson, B. J., C. Escarmis, B. Parker, W. C. Slater, J. Doniger, I. Tessman, and R. C. Warner. 1975. Figure 8 configuration of dimers of S13 and φX174 replicative form DNA. *J. Mol. Biol.* **91:** 409.

Tomizawa, J.-I. and N. Anraku. 1964. Molecular mechanisms of genetic recombination in bacteriophage. I. Effect of KCN on genetic recombination in phage T4. *J. Mol. Biol.* **8:** 508.

Vanderbilt, A. S., M.-T. Borrás, and E. S. Tessman. 1971. Direction of translation of phage S13 as determined from the sizes of polypeptide fragments of nonsense mutants. *Virology* **43:** 352.

Vanderbilt, A. S., M.-T. Borrás, S. Germeraad, I. Tessman, and E. S. Tessman. 1972. A promoter site and polarity gradients in S13. *Virology* **50:** 171.

Wagner, R., Jr. and M. Meselson. 1976. Repair tracts in mismatched DNA heteroduplexes. *Proc. Natl. Acad. Sci.* **73:** 4135.

Wiegand, R. C., K. L. Beattie, W. K. Holloman, and C. M. Radding. 1977. Uptake of homologous single-stranded fragments by superhelical DNA. III. The product and its enzymic conversion to a recombinant molecule. *J. Mol. Biol.* **116:** 805.

Wildenberg, J. and M. Meselson. 1975. Mismatch repair in heteroduplex DNA. *Proc. Natl. Acad. Sci.* **72:** 2202.

Zahler, S. A. 1958. Some biological properties of bacteriophages S13 and φX174. *J. Bacteriol.* **75:** 310.

Repair of Radiation Lesions in Single-stranded DNA Phages

Ramendra K. Poddar, Chanchal K. Dasgupta,* and Ashoke R. Thakur
Biophysics Section, Physics Department
University College of Science
Calcutta 700 009, India

φX174, S13, and related phages have evoked considerable interest among radiation biologists because of the single-stranded nature of their DNA. In fact, the radiobiological (^{32}P-suicide) experiment of Tessman et al. (1957) gave the first hint that phage S13 might contain single-stranded DNA (SS DNA); it was based on the observation that the decay of a single ^{32}P atom incorporated into S13 DNA was capable of inactivating its plaque-forming ability. It was found subsequently that the survival curves obtained by irradiating various SS DNA phages with X rays (Blok et al. 1967), γ rays (Lytle and Ginoza 1968), and UV (Yarus and Sinsheimer 1964) were all of the so-called "single-hit" type, implying that one lesion occurring somewhere in the DNA genome was sufficient to inactivate the phage. As to the nature of such lesions in SS DNA, lethality in the case of ^{32}P-suicide was attributed solely to the scissions in the SS DNA. For γ rays, however, the frequency of single-strand scission per lethal event was only 0.26 (Lytle and Ginoza 1969). Besides single-strand scissions, thymine base damages, including formation of ring-saturated compounds such as 5,6-dihydroxy-dihydrothymine, were found to be produced upon irradiation with γ rays (Swineheart and Cerruti 1975). In the UV dose ranges used in radiobiological studies, few chain scissions are likely to be produced (Poddar and Sinsheimer 1971), and David (1964) showed that thymine dimers constituted only 0.34 of the lethal lesions (the other lesions have remained uncharacterized). In contrast, there were 4.8 thymine dimers per lethal hit for phage T4v_1 (Wulf 1963).

When SS DNA phages irradiated with γ rays (Lytle and Ginoza 1969), UV (Newbold and Sinsheimer 1970), or visible light in the presence of photosensitizers (Chaudhuri and Poddar 1973) were allowed to infect normal cells, adsorption was found to be practically unaffected and some kind of replicative-form (RF)-like structures were formed in all cases. A comparative study of the loss of plaque-forming ability by infected cells harboring putative lethal lesions in the viral or complementary strand of the parental

* Present address: Department of Internal Medicine, Yale University School of Medicine, New Haven, Connecticut 06510.

RF DNA suggested a special role for the viral (+) strand in phage maturation (Denhardt and Sinsheimer 1965; Datta and Poddar 1970). Later studies, however, failed to provide any biochemical justification for this and, on the contrary, revealed that the complementary strand was more important in RF DNA replication and phage maturation (reviewed by Denhardt 1975). The inference that inside the cell an incomplete RF molecule was formed from the viral strand bearing a UV lesion received experimental support from the single-stranded regions visualized as "bushes" in the electron microscope (Benbow et al. 1974). It was found that the length of the new complementary strand decreased gradually with increasing UV dose.

Evidence for SS DNA Repair

A number of workers have investigated the possible modes of intracellular repair of the abnormal or incomplete RF molecules caused by the various irradiations. Several studies have been made with UV-irradiated φX or S13, and some illustrative experimental results are given in Figures 1–6. Yarus and Sinsheimer (1964) showed that phage φX did not undergo any host-cell reactivation of UV lesions in its DNA (Fig. 1). This can be explained by the assumption that any attempt at excision-repair of the lesions in the infecting viral strand would linearize the parental RF DNA, as the latter would have gaps in the nascent complementary strand opposite the lesions. The photoenzymatic repair process (photoreactivation) has been shown to be only marginally effective (Fig. 2) on SS DNA (Poddar and Sinsheimer 1971). The claim by Menningman (1975) that the sectors of intracellular SS DNA and RF DNA able to be photoreactivated were more or less the same seems untenable

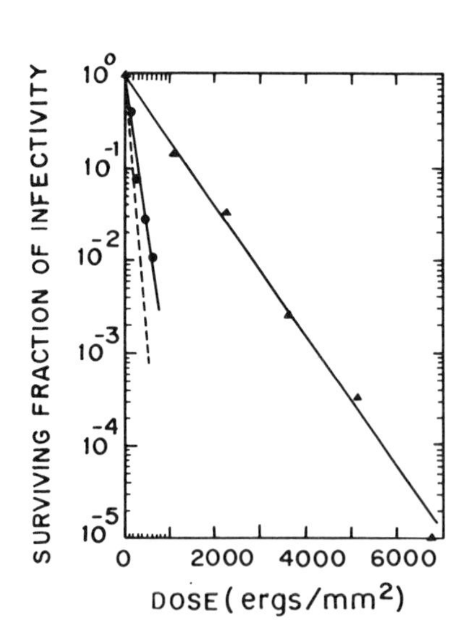

Figure 1 The UV sensitivity of RF DNA isolated from infected cells and assayed on protoplasts of *E. coli* W6 (*hcr*$^+$) (–▲–) and AB1886 (*hcr*$^-$) (–●–). SS DNA was assayed on either W6 or AB1886 (---). Cross sections of RF DNA = 1.2×10^{-16} cm² on W6 and 5.5×10^{-16} cm² on AB1886; SS DNA = ~ 9×10^{-16} cm² on W6/AB1886. (Redrawn from Yarus and Sinsheimer 1964.)

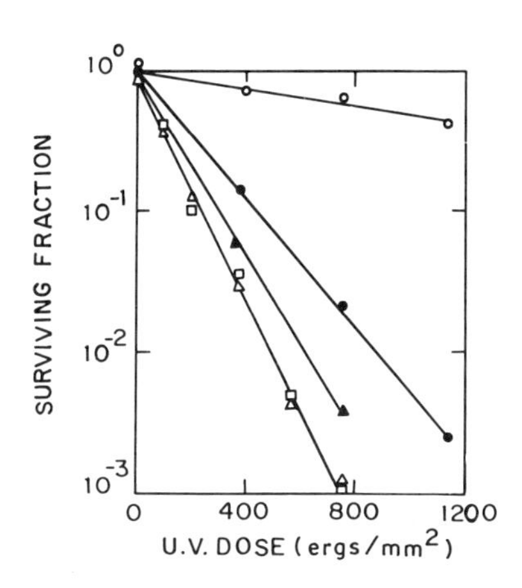

Figure 2 Inactivation curves of UV-irradiated free φX, φX SS DNA, and φX RFII DNA with and without photoreactivation. Biological activity of DNA was measured by means of spheroplast assay. Yeast photoreactivating enzyme was used for photoreactivation in vitro. RFII DNA with (-o-) and without (-●-) photoreactivation; survival curves of φX *am*3 SS DNA with (-▲-) and without (-△-) prior photoreactivation treatment; (-□-) survival of UV-irradiated φX *am*3 virus without photoreactivation. (Redrawn from Poddar and Sinsheimer 1971.)

because the inducible DNA repair processes (described below) were not taken into consideration. Neither do the phages S13 and φX exhibit any multiplicity reactivation (Tessman and Ozaki 1960; Dasgupta and Poddar 1975a), i.e., the capacity to be repaired through genetic exchange between two or more damaged phages (Fig. 3).

Thus, the prospects of repair for an infecting irradiated φX particle are marginal except for by what is known as UV reactivation (UVR). It was Weigle (1953) who first observed that irradiation of host cells with small doses of UV prior to infection with irradiated phage increased the number of surviving λ phage. Since then, reports have been published showing that

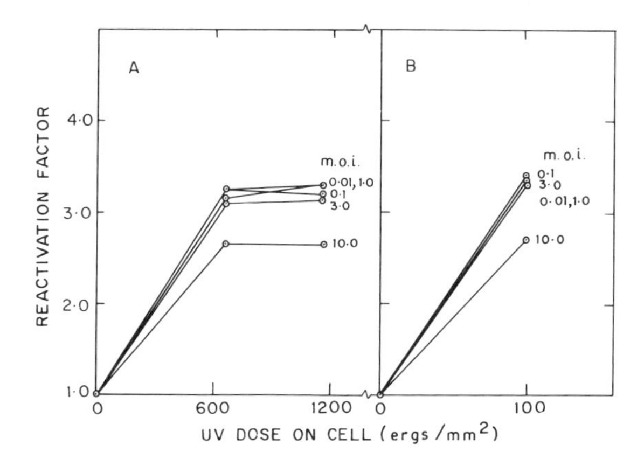

Figure 3 Absence of correlation between multiplicity of infection (m.o.i.) and UV-reactivation factor, which is defined as the relative increase in plaque-forming ability of φX irradiated with a UV dose of 1000 ergs/mm^2 on UV-irradiated versus nonirradiated (*A*) *E. coli* C and (*B*) HF4704 cells. (Redrawn from Dasgupta and Poddar 1975a.)

such repair occurs for UV-irradiated S13 (Tessman and Ozaki 1960) and φR (Ono and Shimazu 1966). It was later found that "thymineless" inactivation could also trigger this kind of repair (Thakur and Poddar 1977). The essential features of this repair process (Figs. 4 and 5) are that it occurs only in *recA*$^+$ hosts and requires de novo protein synthesis (Dasgupta and Poddar 1975a,b). This mode of repair also operates on phages inactivated photodynamically (Bandyopadhyay and Poddar 1973) or with hydroxylamine (Vizdalova 1969). Bleichrodt and Verheij (1974) demonstrated a five- to sixfold increase in the number of mutants among the survivors undergoing UVR (Fig. 6).

Repair Synthesis Past Dimers?

It was shown by Poddar and Sinsheimer (1971) that DNA polymerase I could not work past a thymine dimer on a UV-irradiated φX DNA template. G. Villani et al. (pers. comm.) also obtained similar results in a partially in vitro system using cell extracts instead of purified enzyme preparations, and they argued that the replication machinery "idles" at the dimer site on the UV-irradiated SS DNA template. Addition of an extract from UV-irradiated *recA*$^+$ cells, however, led to an increased incorporation of labeled precursors into DNA. Their interpretation was that the proofreading $3' \rightarrow 5'$ exonuclease activity of DNA polymerase I rejected repeatedly nucleotides incorporated opposite the dimer in the template, and that a protein believed to be induced in the UV-irradiated *recA*$^+$ cells incapacitated the exonuclease and allowed polymerization past the dimer by means of the insertion of any one of four nucleotides opposite the dimer.

Although the above model offers a reasonable explanation for both UVR and the concomitant mutagenesis of UV-irradiated φX (Witkin 1976), a few serious reservations should be mentioned. (1) It was not actually demon-

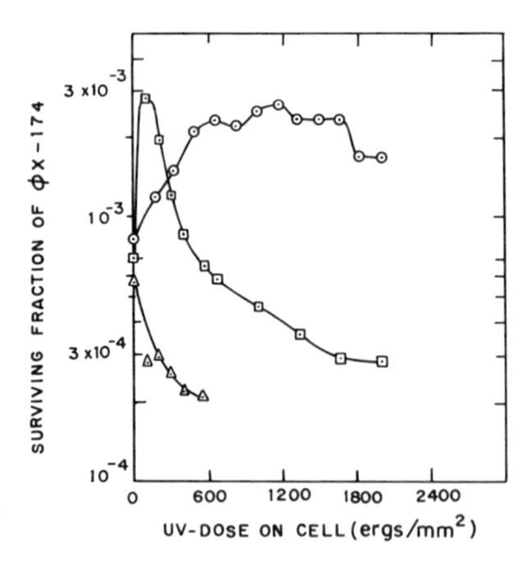

Figure 4 UV reactivation of phage φX irradiated with 1000 ergs/mm^2. The surviving fraction of φX, as measured by plaque-forming ability, is expressed as a function of the UV dose given to each bacterial strain before infection. (-o-o-) *E. coli* C (*uvrA*$^+$ *recA*$^+$); (-□-□-) HF4704 (*uvrA*$^-$ *recA*$^+$); (-△-△-) HF4712 (*uvrA*$^+$*recA*$^-$). (Redrawn from Dasgupta and Poddar 1975a.)

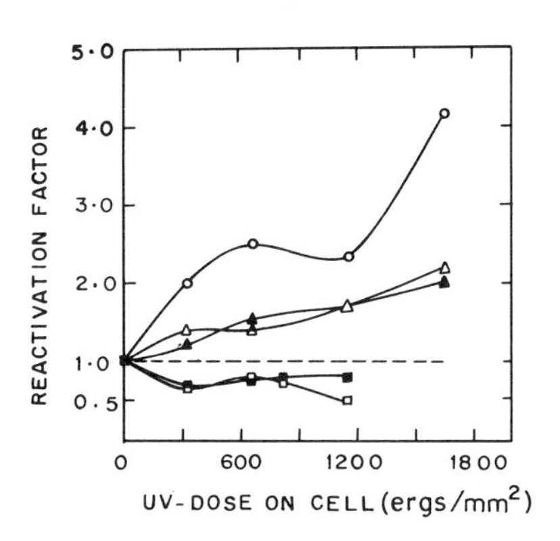

Figure 5 UV reactivation of φX as a function of the UV dose given to *E. coli* C ($uvrA^+$ $recA^+$) cells before infection when the infected cells were plated directly (–o–), grown in tryptone broth with 7.5 μg chloramphenicol (Cm)/ml for 12 min (–△–) or 30 μg Cm/ml for 30 min (–▲–), or grown in tryptone broth for 5 min followed by growth in tryptone broth with 7.5 μg Cm/ml for 5 min (–□–) or in tryptone broth with 30 μg Cm/ml for 15 min. (–■–). UV dose on φX = 1000 ergs/mm².

strated that the increased incorporation of labeled precursors on UV-irradiated φX DNA templates was due to an increase in the length of the synthesized complementary strand rather than to new initiations triggered by oligonucleotides present in the extract of UV-irradiated cells. (2) According to this model, one would expect that among the survivors of UV-irradiated φX various kinds of transition and transversion mutants should occur. Howard and Tessman (1964) found that UV mutagenesis in the related phage S13 involved mainly a C → T transition when mutants were assayed on UV-irradiated cells. This makes one wonder whether the dimers are the mutagenic lesions after all. (3) If the parental RF DNA had a completed complementary strand paired with a viral strand containing a dimer, there should not be any difficulty in obtaining a complete progeny RF molecule from such a structure according to the rolling-circle model (Fig. 7). This would very likely lead to gradual dilution of the label in the parental RF DNA,

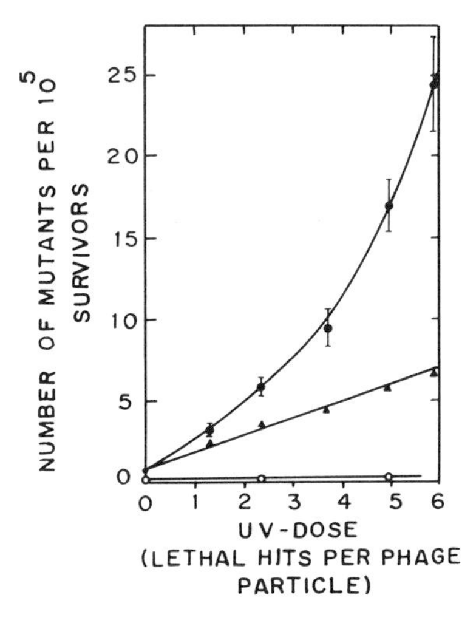

Figure 6 Induction of pseudo wild-type particles in a population of *am*18 by UV. Irradiated samples of *am*18 were assayed on nonirradiated (–o–) and UV-irradiated (700 ergs/mm²) (–●–) *E. coli* C cells. When the standard deviation based on plaque counts was larger than the dimension of the circles it has been represented by a vertical bar. The mutation induction curve for irradiated cells (–●–) was corrected for greater survival due to UVR. The corrected data are represented by (–▲–). (Redrawn from Bleichrodt and Verheij 1974.)

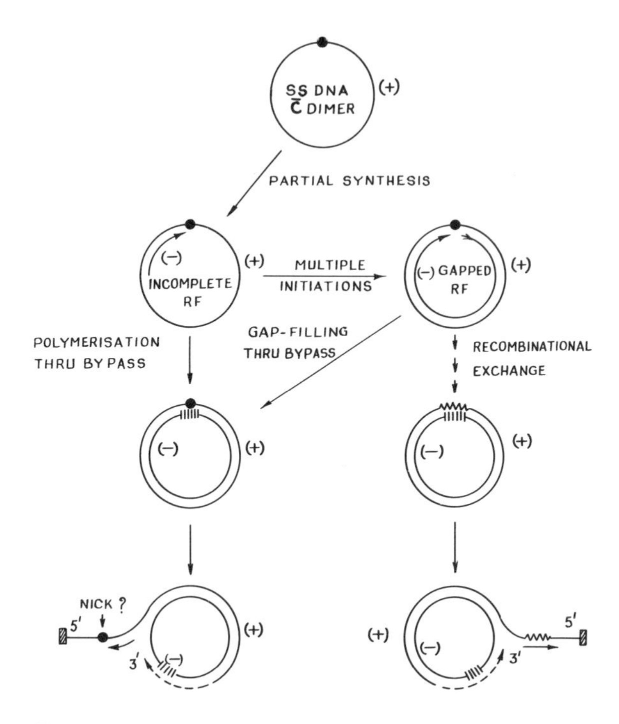

Figure 7 Probable pathways by which healthy progeny RF molecules can be obtained from parental RF DNA that has UV-induced lesions (pyrimidine dimers) in the infecting viral strand. The bypass mechanism postulated by G. Villani et al. (pers. comm.) and Witkin (1976) may lead to dilution of label in the parental RF DNA, which is not likely to happen in the recombinational pathway.

but the results of Dasgupta and Poddar (1975a,b) indicate otherwise. (4) UV-irradiated phages, when allowed to infect normal cells, have also been found to be mutagenic (Bleichrodt and Verheij 1974). If dimers are the mutagenic lesions, then one has to explain how the synthesis of the complementary strand can occur past a dimer when no UVR is operating. One explanation could be that there are other forms of lesions besides pyrimidine dimers that are mutagenic but not lethal.

Recombinational Exchange—An Alternative Model

In conclusion, it is apparent that the structure of parental RF DNA containing a UV-irradiated infecting viral strand is not yet settled. If it has a partially synthesized complementary strand, then perhaps under UVR conditions either a completed strand can be made by synthesizing past a dimer, as proposed by G. Villani et al. (pers. comm.), or a gapped RF molecule may be made, in accord with the results of the in vitro study by Poddar and Sinsheimer (1971). In the latter case, the UVR condition is imagined to supply more initiators. If this second model is true, then further replication of the incomplete RF molecules would be impossible until the gaps in the complementary strands are filled, possibly by recombinational transfer of host genetic material to the φX RF DNA mediated by the *recA*$^+$ gene product in UV-irradiated host cells.

Kornberg (1974) has suggested that φX parental RF DNA and the folded *Escherichia coli* chromosome, as well as its replication machinery, might be located at the same, or very close together, membranous sites. If this were true, then the exchange between host and phage genetic material would not be ruled out, although there is at present no experimental evidence to support such a notion. It is known that under normal conditions phage synthesis does not involve utilization of host genetic material. We have shown, however, that the transfer of ^{32}P atoms from the host DNA to the progeny phage occurs when thymine-requiring host cells are incubated in the absence of thymine (Thakur and Poddar 1977). This, of course, could be the result of rapid turnover of degraded host DNA to mature phages, rather than of genetic exchange.

DNA repair is an essential attribute of living organisms, be they phages or human beings, in their defense against environmental hazards such as radiations and mutagenic chemicals, and in the process of evolving, organisms have acquired one or more kinds of DNA-repair capacities. Among these repair processes, the enzymatic pathways of photoreactivation and excision-repair have been fairly well characterized. However, the error-prone repair process mediated by the *recA*$^+$ gene and involving induction of de novo protein synthesis is not yet fully elucidated (Witkin 1976).

ACKNOWLEDGMENTS

We wish to express our thanks and gratitude to the University Grants Commission and the Department of Atomic Energy, Government of India, for their liberal financial support of the research reported here.

REFERENCES

Bandyopadhyay, U. and R. K. Poddar. 1973. Reactivation of photoinactivated single strand DNA bacteriophage φX174 by UV-irradiated *Escherichia coli* cells. *J. Virol.* **12:** 1204.

Benbow, R. M., A. J. Zuccarelli, and R. L. Sinsheimer. 1974. A role for single strand breaks in bacteriophage φX174 genetic recombination. *J. Mol. Biol.* **88:** 629.

Bleichrodt, J. F. and W. S. D. Verheij. 1974. Mutagenesis by ultraviolet radiation in bacteriophage φX174: On the mutation stimulating processes induced by ultraviolet radiation in the host bacterium. *Mol. Gen. Genet.* **135:** 19.

Blok, J., L. H. Luthjens, and A. L. M. Roos. 1967. The radiosensitivity of bacteriophage DNA in aqueous solution. *Radiat. Res.* **30:** 468.

Chaudhuri, U. C. and R. K. Poddar. 1973. Identification of the block in the intracellular replication of SS DNA of photodynamically inactivated bacteriophage φX174. *J. Virol.* **11:** 368.

Dasgupta, C. K. and R. K. Poddar. 1975a. Ultraviolet reactivation of the single stranded DNA phage φX174. *Mol. Gen. Genet.* **139:** 77.

———. 1975b. Influence of host metabolic state on the UV reactivation (UVR) of φX174. In *Proceedings of the Symposium on Structural and Functional Aspects of Chromosomes*, p. 130. B.A.R.C., Bombay.

Datta, B. and R. K. Poddar 1970. Greater vulnerability of the infecting viral strand of replicative form DNA of bacteriophage φX174. *J. Virol.* **6:** 583.

David, C. N. 1964. UV inactivation and thymine dimerization in bacteriophage φX. *Z. Vererbungsl.* **95:** 318.

Denhardt, D. 1975. The single stranded DNA phages. *CRC Crit. Rev. Microbiol.* **4:** 161.

Denhardt, D. T. and R. L. Sinsheimer. 1965. The process of infection with bacteriophage φX174. Inactivation of the phage-bacterium complex by decay of ^{32}P incorporated in the infecting particle. *J. Mol. Biol.* **12:** 663.

Howard, B. D. and I. Tessman. 1964. Identification of the altered bases in mutated single-stranded DNA. III. Mutagenesis by ultraviolet light. *J. Mol. Biol.* **9:** 372.

Kornberg, A. 1974. *DNA synthesis*, p. 260. W. H. Freeman, San Francisco.

Lytle, C. D. and W. Ginoza. 1968. Frequency of single strand breaks per lethal γ-ray hit in φX174 DNA. *Int. J. Radiat. Biol.* **14:** 553.

———. 1969. Intracellular development of bacteriophage φX174 inactivated by γ-ray, ultraviolet light and nitrous acid. *Virology* **38:** 152.

Menningman, H. D. 1975. Photoreactivation of bacteriophage φX174. *Int. J. Radiat. Biol.* **27:** 313.

Newbold, J. E. and R. L. Sinsheimer. 1970. Process of infection with bacteriophage φX174. XXXII. Early stages in the infection process; attachment, eclipse, DNA penetration. *J. Mol. Biol.* **49:** 49.

Ono, J. and Y. Shimazu. 1966. Ultraviolet reactivation of a bacteriophage containing a single stranded deoxyribonucleic acid as a genetic element. *Virology* **29:** 295.

Poddar, R. K. and R. L. Sinsheimer. 1971. Nature of the complementary strands synthesized *in vitro* upon the single stranded circular DNA of bacteriophage φX174 after ultraviolet irradiation. *Biophys. J.* **11:** 355.

Swineheart, J. L. and P. A. Cerruti. 1975. Gamma-ray induced thymine damage in the DNA in coliphage φX174 and in *E. coli*. *Int. J. Radiat. Biol.* **27:** 83.

Tessman, E. S. and T. Ozaki. 1960. The interaction of phage S13 with ultraviolet irradiated host cells and properties of the ultraviolet irradiated phage. *Virology* **12:** 431.

Tessman, I., E. S. Tessman, and G. S. Stent. 1957. The relative radiosensitivity of bacteriophages S13 and T2. *Virology* **4:** 209.

Thakur, A. R. and R. K. Poddar. 1977. Growth and reactivation of single stranded DNA phage φX174 in *E. coli* undergoing thymineless death. *Mol. Gen. Genet.* **151:** 313.

Vizdalova, M. 1969. The inactivating effect of hydroxylamine on *E. coli* phages and the possibility of repair of the resultant damage. *Int. J. Radiat. Biol.* **16:** 147.

Weigle, J. J. 1953. Induction of mutations in a bacterial virus. *Proc. Natl. Acad. Sci.* **39:** 528.

Witkin, E. M. 1976. Ultraviolet mutagenesis and inducible DNA repair in *Escherichia coli. Bacteriol. Rev.* **40:** 869.
Wulf, D. L. 1963. The role of thymine dimer in the photo-inactivation of bacteriophage $T4v_1$. *J. Mol. Biol.* **7:** 431.
Yarus, M. and R. L. Sinsheimer. 1964. The UV resistance of double stranded φX174 DNA. *J. Mol. Biol.* **8:** 614.

In Vivo and In Vitro Recombination of Small-phage DNAs with Plasmids and Transposable Drug-resistance Elements: Introduction

Dan S. Ray
Molecular Biology Institute
and Department of Biology
University of California
Los Angeles, California 90024

The minichapters that follow have been included as an indication of some of the new directions being taken in small-phage research. The application of both recombinant DNA and transposon techniques, together with the recently developed DNA sequencing techniques, opens up many new and exciting areas of investigation related to the structure and function of small-phage genomes and allows the development of DNA cloning vectors having unique properties.

Studies of insertions into and deletions of filamentous phage genomes show clearly that the size of the virion is determined by the DNA. Miniphage containing extensive deletions of the viral genome form correspondingly smaller virions. Similarly, insertions into the viral genome lead to the production of virions of greater than unit length. Several different transposable drug-resistance elements and even the entire pSC101 plasmid have been inserted into filamentous phage DNA in vivo. Foreign DNAs have also been inserted into specific restriction sites in vitro.

Since for the filamentous phages all genes are essential for phage growth, it would be expected that most transducing phages constructed either by random in vivo insertions or by insertions into restriction sites would be defective. This appears to be the case for the recombinant phages that require a wild-type helper phage for their propagation. Analysis of helper-independent phage, on the other hand, should identify sites where the nucleotide sequence can be interrupted without interfering with the ability of the phage to propagate and to form a plaque. Such sites have already been identified within the intergenic space containing the origins of replication. Transposons conferring resistance to kanamycin and to ampicillin and the entire pSC101 plasmid have been observed to integrate into a site close to the *Hae*II restriction site in the intergenic space. Whether these insertions are at precisely the same site will have to be determined by DNA sequence analysis.

Helper-dependent transducing phage incapable of plaque formation have also been obtained. There appear to be numerous insertion sites that give rise to this particular phenotype. The observation of recombinant phage of this class suggests that the phage replicative-form (RF) DNA may be capable of replicating as a plasmid under appropriate conditions.

In the case of transposable drug-resistance elements it should also be possible to obtain deletions of the viral genome as a result of imprecise excision of the transposon. Well-defined deletions and duplications may also be obtainable as a result of recombination between viral genomes carrying identical transposable elements inserted at different sites.

The ability to insert selected genes into the filamentous phages and to package DNAs of varying sizes allows the development of novel cloning vectors. Foreign DNAs can be inserted into the duplex RF molecules of such vectors by in vitro recombinant DNA techniques. An important feature of filamentous phage vectors is that it is possible to obtain a single strand of the cloned DNA segment from phage carrying the cloned DNA. Insertion of the cloned DNA in both possible orientations also provides a source of large quantities of each of the strands of the DNA. The ready availability of the separated strands in circular form is particularly useful for DNA sequencing by the Sanger techniques. In addition, the separated strands provide hybridization probes free of contamination by the opposite strand.

Finally, the ability to clone specific small-phage genes into bacterial plasmids should provide gene-specific suppressor strains and, consequently, should allow the rapid isolation of large numbers of specific gene mutants for structure-function analyses.

The Filamentous Phages as Transducing Particles

Gerald F. Vovis and Mariko Ohsumi
The Rockefeller University
New York, New York 10021

An important property of the filamentous phages is that the length of the virion is determined by the size of the viral DNA molecule encapsulated within it. This conclusion is based on the discovery of miniphage particles, which were shown to contain the viral DNA of deletion mutants (Griffith and Kornberg 1974; Enea and Zinder 1975; Hewitt 1975). For example, the f1 deletion mutant MP-2, which consists of only about 20% of the f1 genome, is found in a phage particle whose length is about 20% that of the wild-type particle (Enea et al. 1977; see also Wheeler and Benzinger, this volume). Particles larger than unit length are also found in all filamentous stocks. Diploid particles constitute about 5% of a wild-type phage population (Scott and Zinder 1967; Salivar et al. 1967). In addition, under permissive conditions, certain amber mutants produce very large, multi-unit-length particles (Pratt et al. 1969). However, nearly all of these multi-unit-length particles, whether they be in wild-type or amber stocks, contain the corresponding number of unit-length DNA molecules. Nevertheless, the existence of greater-than-unit-length phages suggests that filamentous particles containing DNA molecules larger than that of the wild-type f1 genome should be possible. We set out to prove this supposition and, in addition, to determine whether there is a practical limit on how large a DNA molecule can be and yet still be replicated efficiently and extruded as a filamentous phage particle.

In our experiments we took advantage of the observation that recombination between apparently nonhomologous DNA molecules can occur if one of the molecules contains a transposable element (Cohen 1976). Specifically, we screened for tetracycline-transducing particles among the filamentous phages produced by an infected male strain of *Escherichia coli* that contained the nonconjugative plasmid pSC101, which confers resistance to tetracycline (Cohen and Chang 1973). Such in vivo phage recombinants between f1 and pSC101 were indeed detected. Two such particles, VO-1 and VO-2, have been characterized extensively (Vovis et al. 1977; M. Ohsumi and G. F. Vovis, unpubl.) and are the subject of the following discussion.

The two tetracycline resistance (tet^r) transducing particles copurify with wild-type f1 phage in CsCl, are sensitive to anti-f1 serum, and, except for their lengths, are indistinguishable from wild-type filamentous phage particles in the electron microscope. Both VO-1 and VO-2 are approximately 2.6 times as long as unit-length f1 phage. This is the size that would be expected if the DNA molecule within was equal in length to the sum of the f1 and pSC101 genomes. Each of the particles contains one single-stranded DNA molecule that can transform *E. coli* to *tet*^{r}. Likewise, the *tet*^{r} transduction of *E. coli* does not require the presence of f1 helper phage. However, VO-2 does not form plaques but VO-1 does. The VO-1 plaques, although very, very small on a wild-type host, are easily seen on a *recA* host.

Restriction enzyme analyses of the chimeric, double-stranded replicative-form DNA isolated from VO-1- and VO-2-transduced *E. coli* demonstrate that both chimeras contain the single *Eco*RI restriction site present in the parental pSC101 molecule (f1 DNA lacks such a restriction site). The restriction fragments obtained with *Hae*II or *Hae*III digestion of the chimeras include novel fragments in addition to almost all those that are seen with a mixture of f1 and pSC101 DNA molecules. No deletion or duplication loops are visible upon electron microscopic examination of the heteroduplexes formed between either of the two chimeras and f1 or pSC101 DNA. Thus, their sizes and the results of restriction enzyme and heteroduplex analyses suggest that these two chimeras are formed simply by the insertion of one parental molecule into the other. The transition from pSC101 DNA to f1 DNA occurs at a point approximately 20% away from the *Eco*RI restriction site as measured on the pSC101 genome in both VO-1 and VO-2, although it is not yet known whether the transition point for the two chimeras is on the same relative side of the *Eco*RI site in the pSC101 genome. The transition point from f1 DNA to pSC101 DNA on the f1 genome is different in the two chimeras. In VO-1 the transition occurs within *Hae*III F_β (Fig. 1), most probably within the IG (the intergenic space located between genes II and IV; see Horiuchi et al., this volume). In VO-2 the transition occurs within *Hae*III B (Fig. 1). DNA sequence data (J. Ravetch, unpubl.) locate it distal to the C terminus of gene VII (Fig. 1; see also Schaller et al., this volume) and most probably after the RNA-polymerase-binding site that is proximal to the N terminus of gene VIII (see Konings and Schoenmakers, this volume). Such a transition point for VO-2 is consistent with the observations that VO-2 requires an f1 helper to form phage particles and that the gene-VIII protein is not synthesized in an in vitro transcription-translation system programmed by VO-2 DNA.

The isolation of these VO phages proves that DNA molecules significantly larger than the unit-length f1 genome can be encapsulated in filamentous particles. The presence of a single *Eco*RI restriction site into which foreign DNA can be inserted without inactivating the tetracycline gene (Cohen and Chang 1973) allows one to construct even larger filamentous phage particles in vitro and then screen for them directly in vivo. Such VO derivatives should allow us to determine whether there is a limit to how large

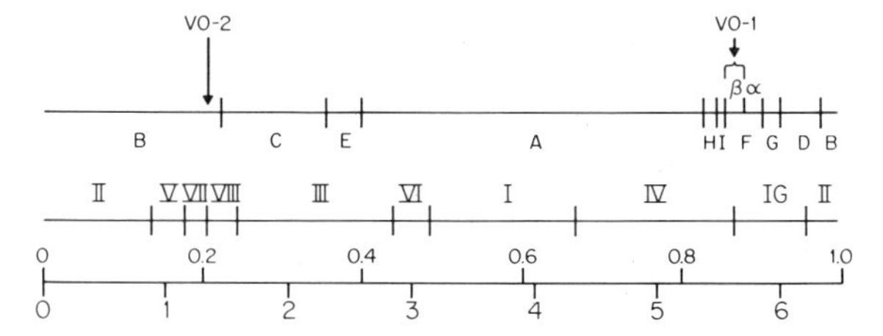

Figure 1 The VO-1 and VO-2 genomes. The upper line represents the *Hae*III restriction map of f1 (Horiuchi et al. 1975). The fragment *Hae*III F is subdivided into two fragments, a and β, by the single *Hae*II restriction site contained within *Hae*III F. The arrows indicate the region in the f1 genome where the pSC101 genome is inserted in VO-1 and VO-2. The middle line represents the genetic map of f1. IG is the intergenic space (Vovis et al. 1975; Enea et al. 1977). The bottom line is a measure of the size of the f1 genome. Map units are indicated above the line and kilobases are below it. The single *Hin*dII cleavage site in f1 serves as the reference point for the physical and genetic maps. The direction of transcription and translation is left to right.

the filamentous phage particle can be and, in addition, should provide more useful information concerning the practicality of using filamentous phages as vectors for the construction of recombinant DNA molecules.

ACKNOWLEDGMENTS

This work was supported by grants from the National Science Foundation and the National Institutes of Health to N. D. Zinder.

REFERENCES

Cohen, S. N. 1976. Transposable genetic elements and plasmid evolution. *Nature* **263:** 731.

Cohen, S. N. and A. C. Y. Chang. 1973. Recircularization and autonomous replication of a sheared R-factor DNA segment in *Escherichia coli* transformants. *Proc. Natl. Acad. Sci.* **70:** 1293.

Enea, V. and N. D. Zinder. 1975. A deletion mutant of bacteriophage f1 containing no intact cistrons. *Virology* **68:** 105.

Enea, V., K. Horiuchi, B. G. Turgeon, and N. D. Zinder. 1977. Physical map of defective interfering particles of bacteriophage f1. *J. Mol. Biol.* **111:** 395.

Griffith, J. and A. Kornberg. 1974. Mini M13 bacteriophage: Circular fragments of M13 DNA are replicated and packaged during normal infections. *Virology* **59:** 139.

Hewitt, J. A. 1975. Miniphage—A class of satellite phage to M13. *J. Gen. Virol.* **26:** 87.

Horiuchi, K., G. F. Vovis, V. Enea, and N. D. Zinder. 1975. Cleavage map of bacteriophage f1: Location of the *Escherichia coli* B-specific modification sites. *J. Mol. Biol.* **95:** 147.

Pratt, D., H. Tzagoloff, and J. Beaudoin. 1969. Conditional lethal mutants of the small filamentous coliphage M13. II. Two genes for coat proteins. *Virology* **39:** 42.

Salivar, W. O., T. J. Henry, and D. Pratt. 1967. Purification and properties of diploid particles of coliphage M13. *Virology* **32:** 41.

Scott, J. R. and N. D. Zinder. 1967. Heterozygotes of phage f1. In *The molecular biology of viruses* (ed. J. S. Colter and W. Paranchych), p. 211. Academic Press, New York.

Vovis, G. F., K. Horiuchi, and N. D. Zinder. 1975. Endonuclease R·*Eco*RII restriction of bacteriophage f1 DNA *in vitro*: Ordering of genes V and VII, location of an RNA promotor for gene VIII. *J. Virol.* **16:** 674.

Vovis, G. F., M. Ohsumi, and N. D. Zinder. 1977. Bacteriophage f1 as a vector for constructing recombinant DNA molecules. In *Molecular approaches to eucaryotic genetic systems* (ed. G. Wilcox et al.), p. 55. Academic Press, New York.

The Filamentous Phage M13 as a Carrier DNA for Operon Fusions In Vitro

Joachim Messing*
Max-Planck-Institut für Biochemie
8033 Martinsried bei München, Federal Republic of Germany

Bruno Gronenborn
Institut für Genetik der Universität zu Köln
Weyertal 121, 5000 Köln, Federal Republic of Germany

The possibility of using the filamentous single-stranded DNA phages in transduction experiments has always been an attractive speculation, albeit an unlikely one considering that integration of host genes into phage DNA, as occurs with lambda (Hershey 1971), has never been observed (Ray 1977). In considering this possibility, however, the question arose whether there was any site in the filamentous phage DNA at which integration of foreign DNA (in vivo or in vitro) could occur without loss of viability. One way to identify a nonessential region of the phage genome would be to take advantage of the generality of the translocation of transposons that is known to occur in a *rec*-independent fashion at many different sites in suitable receptor molecules (Berg et al. 1975; So et al. 1975). Once a transposon was inserted into the filamentous phage DNA, it would then provide many potential sites into which a foreign DNA fragment could be cloned. One difficulty with this approach could be that the inserted transposon is itself unstable and is lost at a significant rate (Herrmann et al. 1978).

An alternate approach would be to focus on the detection of a restriction endonuclease cleavage site within a nonessential region of the phage genome. This would require (1) the generation of linear M13 replicative-form (RF) molecules of unit length with specifically located blunt ends, preferably in a nonessential region of the phage DNA, and (2) a blunt end-to-end ligation between the linear RF molecules and a restriction endonuclease cleavage fragment containing an appropriate marker. By selecting for these specific conditions we have determined that a restriction endonuclease *Bsu*I cleavage site at 0.917 on the M13 wild-type map is located within a nonessential stretch of a nucleotide sequence in the phage genome (Messing and Gronenborn 1977; Messing et al. 1977). This has

* Present address: Department of Bacteriology, University of California, Davis, California 95616.

allowed the formation and characterization of a hybrid phage, M13 mp1, containing the *Hin*dII fragment of the *lac* regulatory region (Gilbert et al. 1975) in a functional state (Langley et al. 1975). The heteroduplex molecules of M13 mpl and M13 RF DNA that we have investigated so far show the presence of a single stranded loop of DNA of constant length at a distinct point with reference to the single *Hin*dII cleavage site on both molecules (Fig. 1).

The DNA that has been inserted into one of the *Bsu*I cleavage sites carries

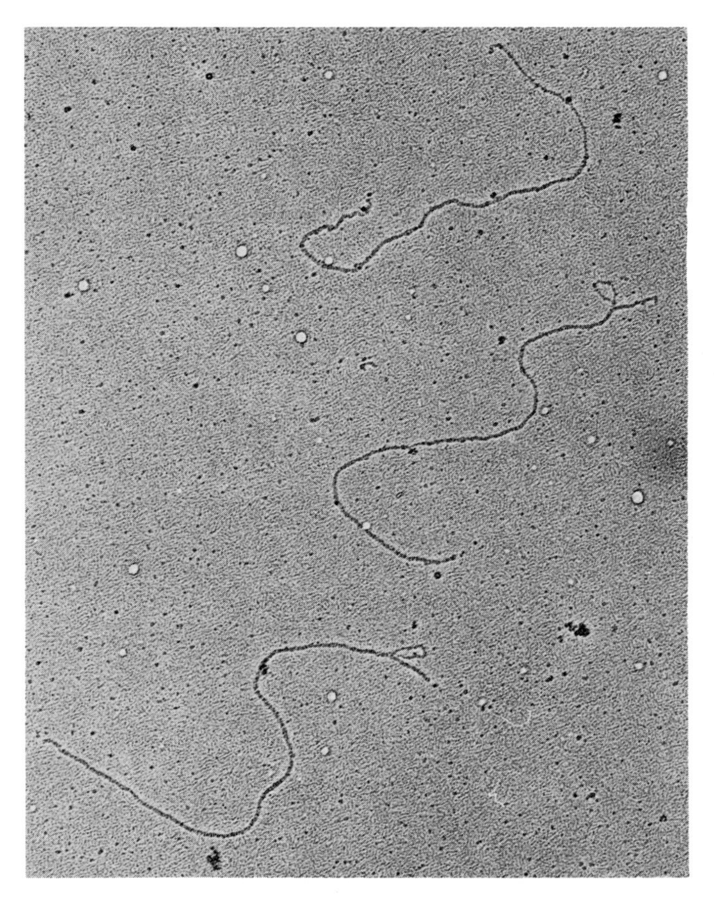

Figure 1 Heteroduplex molecules of *Hin*dII-cleaved M13 and M13 mp1 RF DNA. The replicative forms of both the wild-type and hybrid phage were digested by *Hin*dII. The DNAs were prepared for electron microscopy as described previously (Messing et al. 1977). The lengths of about 40 heteroduplex molecules were measured and the standard deviation was less than 2% of unit length. Magnification, ~ 38,000×.

the *lac* promoter-operator region and the information for the first 145 amino acid residues of the β-galactosidase gene of *Escherichia coli* (E.C. 3.2.1.23). Synthesis of this small peptide complements the function of the α region of the β-galactosidase gene in an appropriate host and, under suitable plating conditions, results in the formation of blue plaques (Malamy et al. 1972; Langley et al. 1975). Since transcription of this α peptide occurs in the same direction as the overall transcription of viral genes, as has been shown by restriction cleavage analysis (J. Messing and B. Gronenborn, in prep.), transcription from the *lac* promoter is not hampered by a counterdirected promoter of a viral gene. Therefore, this hybrid phage is a suitable vector not only for the maintenance and amplification of foreign DNA but also for the expression of an integrated gene under the control of the *lac* regulatory region.

For such operon fusions it is necessary that there be only one restriction cleavage site within the DNA sequence specifying the α peptide. The construction of such a single site was achieved by the novel method of introducing by chemical means a single base exchange in M13 mpl that leads to an *Eco*RI restriction site proximal to the *lac* promoter region (Gronenborn and Messing 1978). The change of base 13 of the β-galactosidase gene from guanine to adenine is sufficient to create such an *Eco*RI recognition site. Since M13 mpl is not sensitive to *Eco*RI, this mutation introduces a single *Eco*RI site into the phage molecule. Furthermore, a change of the fifth amino acid of β-galactosidase from aspartic acid to asparagine does not destroy its ability to mediate α complementation. This phage mutant has been denoted M13 mp2 and a combined genetic and physical map is shown in Figure 2.

Recent results have shown that the insertion of foreign DNA into the *Eco*RI site of M13 mp2 can be detected by the loss of α complementation. DNA from phages that form colorless plaques has been subjected to agarose gel electrophoresis in parallel with DNA from phages that form blue plaques. As shown in Figure 3, the DNA from the hybrid phage M13 mp2 is less mobile than that from the parental phage. The change in plaque type concomitant with the increase in size allows for the rapid detection of new hybrid phages.

Several aspects of molecular cloning with this single-stranded-phage vector currently under investigation are: the search for other mutants with different restriction cleavage sites in the *lac* region of M13 mp1; the application of the chain-termination method to sequence the cloned DNA (Sanger et al. 1977); construction of chimeric proteins still able to complement the α function; overproduction of gene products under the control of the *lac* promoter; determination of the packaging capacity of phage virions; and, finally, modification of the host range of the vector for it to meet the requirements for an authentic EK2 vector.

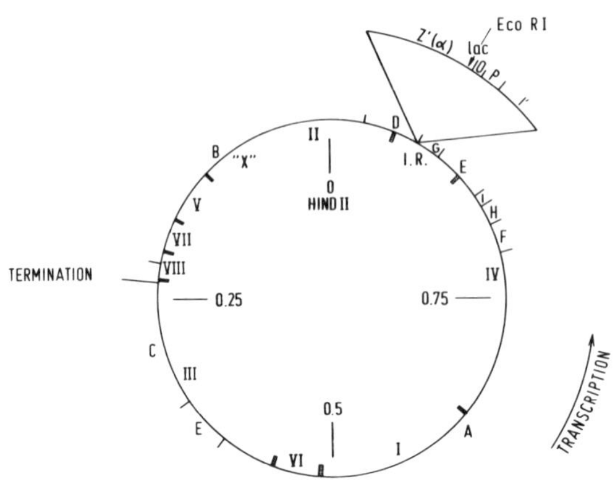

Figure 2 Combined physical and genetic map of M13 mp2. The viral genes (indicated by roman numerals) and the cleavage sites of *Bsu*I (indicated by capital letters) have been arranged on the circular map with the cleavage site of *Hin*dII used as a reference point. The inserted *lac* DNA is represented by the outer circle segment; the *Eco*RI recognition site (indicated by arrow) lies close to the operator within the *α* region. The orientation of the *lac* promoter was determined by cleavage analysis with the enzymes *Bsu*I and *Hpa*II (J. Messing and B. Gronenborn, in prep.). I′: part of the *lac* repressor gene; P: *lac* promoter; O: *lac* operator; Z′ (*α*): *α* region of the *β*-galactosidase gene.

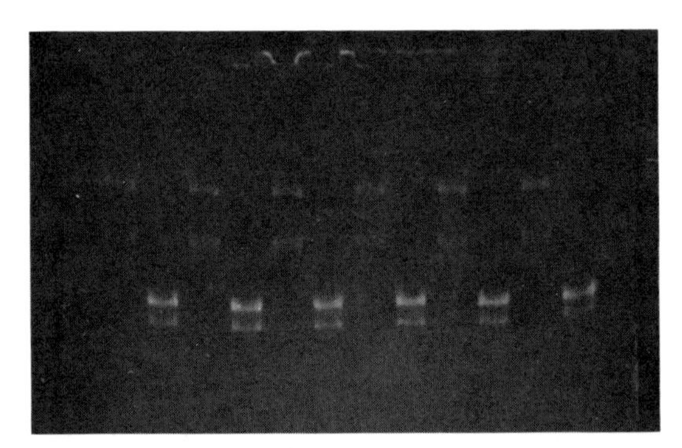

Figure 3 Analysis of the size of viral DNA by agarose gel electrophoresis. An *Eco*RI fragment of about 5600 base pairs was excised from λh80d*lac* and purified. This fragment, which carries the information for *lac* permease, was inserted into the *Eco*RI site of M13 mp2 (B. Gronenborn and J. Messing, in prep.). Insertion of the DNA was detected by the appearance of colorless plaques. Phage particles from 0.05-ml cultures made from single colorless or blue plaques were subjected directly to agarose gel electrophoresis (Messing et al. 1977). The concentration of agarose in the slab gel was 1%. The photograph of the gel shows, from left to right in an alternating manner, the viral DNAs from the white and blue plaques.

ACKNOWLEDGMENT

This work was supported by the Deutsche Forschungsgemeinschaft.

REFERENCES

Berg, D. E., J. Davies, B. Allet, and J.-D. Rochaix. 1975. Transposition of R factor genes to bacteriophage λ. *Proc. Natl. Acad. Sci.* **72:** 3628.

Gilbert, W., J. Gralla, J. Majors, and A. Maxam. 1975. Lactose operator sequences and the action of *lac* repressor. In *Protein ligand interactions* (ed. H. Sund and G. Blauer), p. 193. Walter de Gruyter, Berlin.

Gronenborn, B. and J. Messing. 1978. Methylation of single-stranded DNA in vitro introduces new restriction endonuclease cleavage sites. *Nature* **272:** 375.

Herrmann, R., K. Neugebauer, H. Zentgraf, and H. Schaller. 1978. Transposition of a DNA sequence determining kanamycin resistance into the single-stranded genome of bacteriophage fd. *Mol. Gen. Genet.* **159:** 171.

Hershey, A. D., ed. 1971. *The bacteriophage lambda.* Cold Spring Harbor Laboratory, Cold Spring Harbor, New York.

Langley, K. E., M. R. Villarejo, A. V. Fowler, P. J. Zamenhof, and I. Zabin. 1975. Molecular basis of β-galactosidase α-complementation. *Proc. Natl. Acad. Sci.* **72:** 1254.

Malamy, M. H., M. Fiandt, and W. Szybalski. 1972. Electron microscopy of polar insertions in the *lac* operon of *Escherichia coli*. *Mol. Gen. Genet.* **119:** 207.

Messing, J. and B. Gronenborn. 1977. Properties of the filamentous phage M13 as a cloning vehicle. *Hoppe-Seyler's Z. Physiol. Chem.* **358:** 278.

Messing, J., B. Gronenborn, B. Müller-Hill, and P. H. Hofschneider. 1977. Filamentous coliphage M13 as a cloning vehicle: Insertion of a *Hin*dII fragment of the *lac* regulatory region in M13 replicative form *in vitro*. *Proc. Natl. Acad. Sci.* **74:** 3642.

Ray, D. S. 1977. Replication of filamentous bacteriophages. In *Comprehensive virology* (ed. H. Fraenkel-Conrat and R. R. Wagner), vol. 7, p. 105. Plenum Press, New York.

Sanger, F., S. Nicklen, and A. R. Coulson. 1977. DNA sequencing with chain terminating inhibitors. *Proc. Natl. Acad. Sci.* **74:** 5463.

So, M., R. Gill, and S. Falkow. 1975. The generation of a ColE1-Ap^r cloning vehicle which allows detection of inserted DNA. *Mol. Gen. Genet.* **142:** 239.

van den Hondel, C. A., L. Pennings, and J. G. G. Schoenmakers. 1976. Restriction-enzyme-cleavage maps of bacteriophage M13: Existence of an intergenic region on the M13 genome. *Eur. J. Biochem.* **68:** 55.

Development of M13 as a Single-stranded Cloning Vector: Insertion of the Tn*3* Transposon into the Genome of M13

Dan S. Ray and Karin Kook
Molecular Biology Institute and Department of Biology
University of California
Los Angeles, California 90024

As a first step towards developing phage M13 as a cloning vector we have investigated the in vivo insertion of a transposable ampicillin-resistance element (Hedges and Jacob 1974; Heffron et al. 1975a,b, 1977) into its genome. Passage of M13 through *Escherichia coli* RSF2124, a strain carrying the element Tn*3* inserted into the plasmid ColE1 (So et al. 1975), yielded phage preparations capable of transducing sensitive *E. coli* strains to ampicillin resistance. By repeated colony purification of transduced cells on ampicillin plates we have isolated clones that produce homogeneous transducing phage capable of transducing ampicillin resistance with high efficiency. We describe here the properties of one of these isolates, which we have termed M13 Tn*3*-15.

To establish that the transducing activity is associated with the M13 phage produced by resistant cells, we have investigated the ability of antibodies against the M13 coat protein to inactivate both the plaque-forming ability and the transducing activity of the M13 Tn*3*-15 phage. Table 1 shows that both plaque formation and transduction to drug resistance are inactivated by purified antibodies to the viral coat protein. Gamma globulins from rabbits that had not been injected with coat protein had no effect on either plaque formation or transduction.

Electron microscopic observation of the M13 Tn*3*-15 phages shows them to have a mean length approximately 1.7 times that of the M13 unit length. Figure 1 shows M13 and M13 Tn*3*-15 in a mixture of the two phages. Replicative forms (RF) of the M13 Tn*3*-15 DNA have been isolated from infected cells and also examined by electron microscopy. Figure 2 shows electron micrographs of both M13 and M13 Tn*3*-15 RF molecules prepared by the aqueous Kleinschmidt procedure. The length of M13 Tn*3*-15 RF DNA is 1.8 times that of M13 RF DNA. The increased lengths of both the M13 Tn*3*-15 phage DNA and its duplex RF DNA agree (within experimental error) with the length predicted from the insertion of the 4.8-kb Tn*3*

Table 1 Inactivation of M13 Tn*3*-15 by antibodies against M13 coat protein

Gamma globulin (μg/ml) anti-M13	control	Plaque-forming units/ml	Ap^r colony-forming units/ml
0		2.7×10^7	2.7×10^6
0.16		1.7×10^7	2.7×10^6
1.6		2.7×10^4	1.2×10^4
16		$<10^3$	$<10^2$
160		$<10^3$	$<10^2$
	0	6.1×10^7	2.7×10^7
	2	5.1×10^7	2.4×10^7
	20	5.2×10^7	2.3×10^7
	200	6.1×10^7	1.7×10^7

M13 Tn*3*-15 plaque-forming ability and ampicillin-resistance transforming activity were assayed after incubation of M13 Tn*3*-15 with purified M13 anti-coat protein gamma globulins or gamma globulins from rabbits that had not been exposed to M13 coat protein. Incubations were for 15 min at 37°C in 0.01 M Tris (pH 8.0), 1 mM EDTA. One aliquot was diluted and assayed for infectivity by the plaque assay and a second aliquot was incubated with *E. coli* K37 for 10 min at 37°C and then spread on nutrient plates containing ampicillin at 70 μg/ml.

into the 6.4-kb M13 genome to yield an M13 Tn*3*-15 RF DNA of 11.2 kb, or 1.75 times the size of M13 RF DNA.

To determine both the site and polarity of insertion of the transposable element into the M13 genome, we have analyzed various restriction fragments of M13 Tn*3*-15 RF DNA. Restriction analyses with the enzymes *Hae*II, *Hae*III, and *Hpa*II indicate that the site of insertion must be just clockwise from the single *Hae*II site in the intergenic space (data not shown).

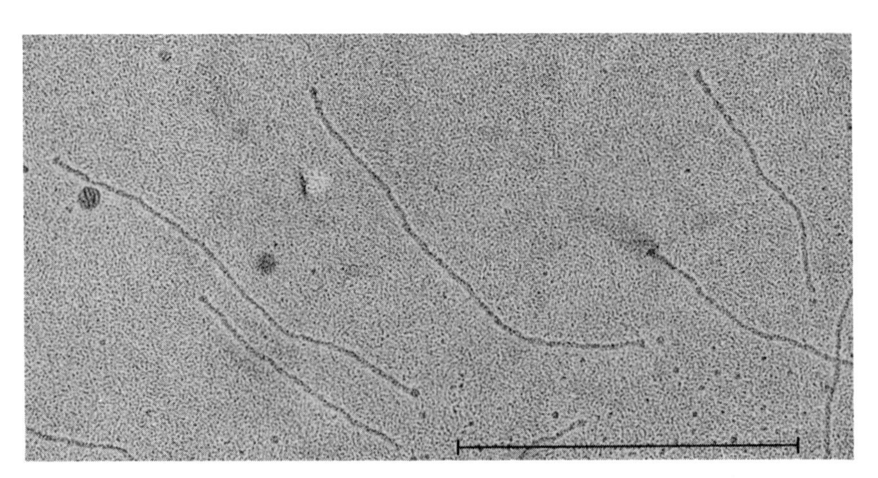

Figure 1 Electron micrograph of M13 and M13 Tn*3*-15 phage prepared by the aqueous DNA spreading technique. (Bar = 1 μm).

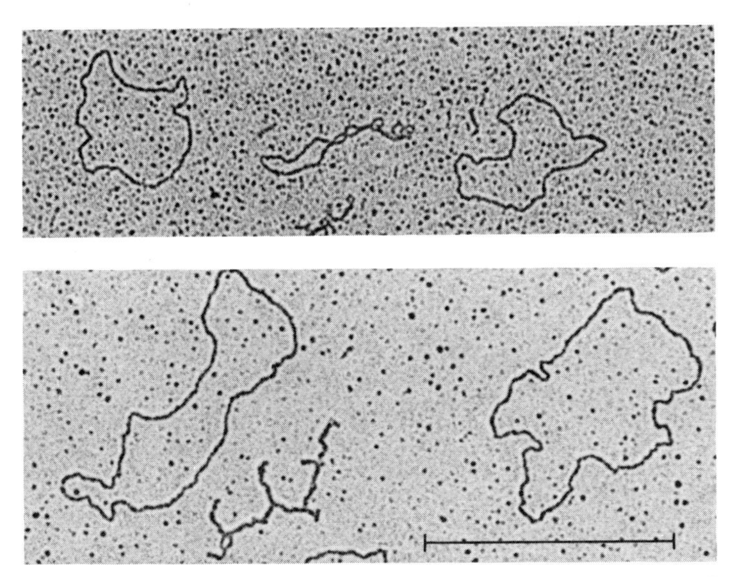

Figure 2 Electron micrographs of M13 RF DNA and M13 Tn*3*-15 RF DNA prepared by the aqueous DNA spreading technique. (Bar = 1 μm).

In each case all but one of the usual M13 fragments were observed in the digest of M13 Tn*3*-15 RF DNA. The missing fragments were *Hae*II B and one member of the *Hae*III E doublet. The *Hpa*II A fragment appeared to be slightly increased in size. Additional digestions with *Bam*I, *Hin*cII, *Hae*II, and combinations of these enzymes (Fig. 3) confirm the location indicated above and establish a unique orientation of the transposon in the viral genome (Fig. 4). *Bam*I, *Hin*cII, and *Hae*II cleave M13 Tn*3*-15 RF DNA into

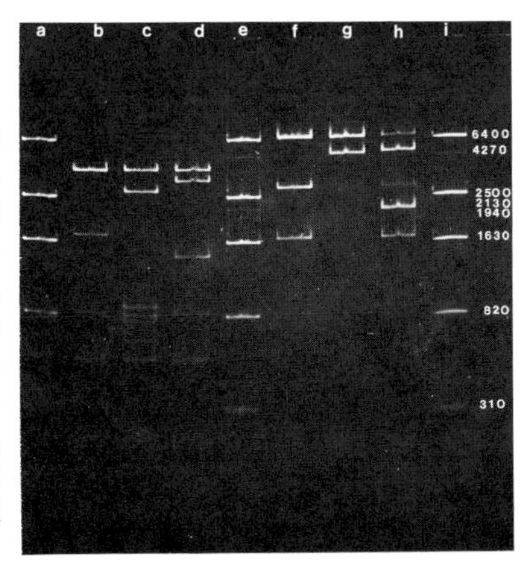

Figure 3 Gel electrophoresis of restriction fragments of M13 Tn*3*-15 RF DNA produced by treatment with (*b*) *Hae*II, (*c*) *Hae*II + *Hin*cII, (*d*) *Hae*II + *Bam*I, (*f*) *Hin*cII, (*g*) *Bam*I, and (*h*) *Hin*cII + *Bam*I. Lanes *a*, *e*, and *i* contain markers of known molecular weight: M13 RFIII (6400 bp), M13 *Bam* + *Hin* A (4270 bp), M13 *Bam* + *Hin* B (2130 bp), M13 *Hae*III A (2500 bp), M13 *Hae*III B + D (1940 bp), M13 *Hae*III B (1630 bp), M13 *Hae*III C (820 bp), and M13 *Hae*III D (310 bp).

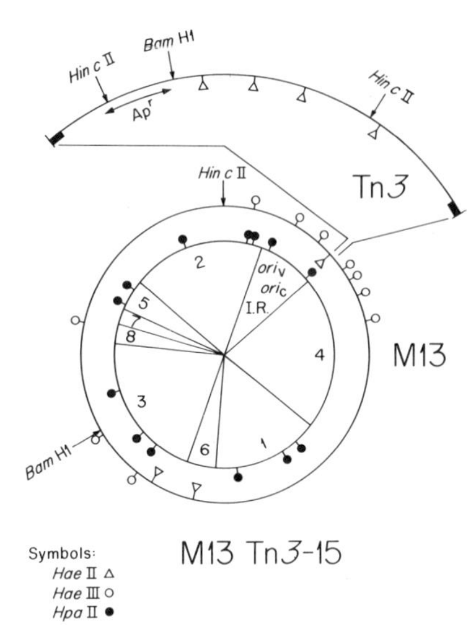

Figure 4 Restriction map of the M13 Tn*3*-15 phage genome. Numbered sectors indicate the locations of M13 genes and the intergenic region (IR) containing both the viral- and complementary-strand origins (ori_v and ori_c). The inverted repeat sequences of the Tn*3* element are indicated by black rectangles. The β-lactamase gene confers resistance to ampicillin and is indicated by Ap^r. The relative sizes of the M13 and Tn*3* portions of the M13 Tn*3*-15 genome are 6.4 kb and 4.8 kb, respectively.

two, three, and seven fragments, respectively (Table 2). In each case the number of fragments is equal to the sum of the number of restriction sites for that particular enzyme in the transposable element and in M13 RF DNA. The sum of the molecular weights of the fragments produced by each enzyme is equivalent to 11.2 kb, again in agreement with the formation of the M13 Tn*3*-15 RF DNA by insertion of Tn*3* into M13 RF DNA.

On the basis of the observations presented here and others to be described elsewhere (D. S. Ray and K. Kook, in prep.), we conclude that the ability of the M13 Tn*3*-15 phage to transduce ampicillin resistance is the result of an

Table 2 *Hin*cII, *Bam*I, and *Hae*II fragments of M13 Tn*3*-15 RF DNA

	Size (bp)		
Fragment	*Hin*cII	*Bam*I	*Hae*II
A	6700	6700	3650*
B	2860	4600	1720
C	1670		840
D			525*
E			340
	11,230	11,300	11,250

Molecular weights of M13 Tn*3*-15 restriction fragments were determined by agarose gel electrophoresis. Asterisks indicate fragments tentatively identified as doublets based on staining intensity. The *Hae*II A band has been shown to be a doublet by restriction with either *Hin*cII or *Bam*I.

insertion of the Tn*3* transposon into the intergenic region of M13 close to the terminus of gene IV. The insertion of the ampicillin-resistance gene into the viral genome provides a direct selection for infected cells. Future efforts to develop this phage as a cloning vector will include the deletion of a large fraction of the transposon so as to reduce the size of the vector and to prevent further transposition. It may be possible to use M13 as an EK2 vector system if we can obtain mutants with small deletions within viral genes corresponding to the M13 DNA fragments that have been cloned by Kaplan et al. (this volume). Such deletion mutants would be dependent on the host strain carrying the appropriate cloned gene of M13 and would therefore have a greatly reduced ability to propagate in the absence of this specially constructed host.

ACKNOWLEDGMENTS

The authors are grateful to Dr. William Wickner for providing us with purified antibodies against the M13 coat protein and to Drs. Ron Gill and Stanley Falkow for providing us with *E. coli* RSF2124. This work was supported by National Institutes of Health contract NOI AI 62505 and grant AI 10752, and by a grant from the National Science Foundation (PCM 76-02709).

REFERENCES

Hedges, R. W. and A. Jacob. 1974. Transposition of ampicillin resistance from RP4 to other replicons. *Mol. Gen. Genet.* **132:** 31.

Heffron, F., C. Rubens, and S. Falkow. 1975a. Translocation of a plasmid DNA sequence which mediates ampicillin resistance: Molecular nature and specificity of insertion. *Proc. Natl. Acad. Sci.* **72:** 3623.

Heffron, F., P. Bedinger, J. J. Champoux, and S. Falkow. 1977. Deletions affecting the transposition of an antibiotic resistance gene. *Proc. Natl. Acad. Sci.* **74:** 702.

Heffron, F., R. Sublett, R. Hedges, A. Jacob, and S. Falkow. 1975b. Origin of the TEM beta-lactamase gene found on plasmids. *J. Bacteriol.* **122:** 250.

So, M., R. Gill, and S. Falkow. 1975. The generation of a ColE1-Ap^r cloning vehicle which allows detection of inserted DNA. *Mol. Gen. Genet.* **142:** 239.

Molecular Cloning of Segments of the M13 Genome

Donald A. Kaplan, Lawrence Greenfield, and Gary Wilcox
Department of Bacteriology
and Institute of Molecular Biology
University of California
Los Angeles, California 90024

The molecular cloning of individual genes of the bacteriophage M13 (Fig. 1) provides a new approach to the study of filamentous phage gene expression and provides the potential for complementation of a defective M13 phage. This paper describes the cloning of *Bam*I-*Eco*RI* fragments of M13.[1] The cloning vehicle pBR317 was chosen for this study because it has been well characterized, can be obtained in high yield, and possesses two genes conferring antibiotic resistance on the host cell (Boyer et al. 1977). The cloning was carried out as described previously (Kaplan et al. 1977a). Two enzymes were used to digest the cloning vehicle and the M13 DNA, thus eliminating the possibility of intramolecular reactions. The plasmid pBR317 DNA was digested with *Eco*RI and *Bam*I; M13 RFI (superhelical, covalently closed, circular, duplex, replicative-form DNA) was digested with an excess of *Bam*I and a limiting amount of *Eco*RI*. Eighty Ap^r Tc^s clones were analyzed and 75 of them contained plasmid DNA ranging in molecular weight from 5.17×10^6 daltons to 5.57×10^6 daltons; the remaining five contained plasmid DNA of molecular weight greater than 5.57×10^6 daltons. Two transformants, pKG7 and pKG35, were selected for further analysis.

RESULTS

Identification and Description of the M13 Inserts: Plasmid Molecular-weight Determinations

The two plasmid DNAs, pKG7 and pKG35, were electrophoresed on a horizontal 0.7% agarose gel (Kaplan et al. 1977b) along with plasmid DNAs of known molecular weights. The molecular weight of pKG7 was determined to be 5.60×10^6 daltons and that of pKG35 to be 6.40×10^6 daltons.

Since a *Hae*III restriction map of M13 (Fig. 1) has been published (van

[1] Under conditions of low ionic strength at pH 8.5 the restriction endonuclease *Eco*RI recognizes the tetranucleotide sequence N↓AATTN (*Eco*RI* activity) rather than the hexanucleotide sequence G↓AATTC which is recognized at higher ionic strenth and pH 7.3 (Polisky et al. 1975). Since the same single-stranded ends are produced by both activities, *Eco*RI*-generated fragments can be cloned into an *Eco*RI site on a cloning vehicle.

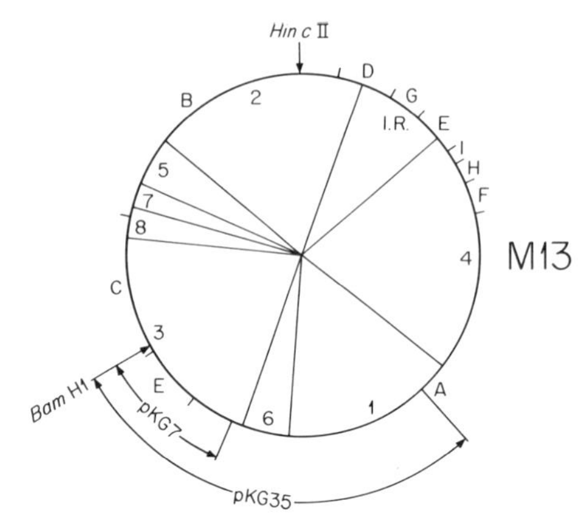

Figure 1 Combined genetic and physical map of M13 showing the *Hae*III and *Bam*I restriction sites (van den Hondel and Schoenmakers 1975; D. S. Ray, pers. comm.) and the M13 segments contained in pKG7 and pKG35.

den Hondel and Schoenmakers 1975), the *Bam*I and *Hae*III fragments generated from pBR317 and M13 RF DNAs were compared with those obtained from pKG7 and pKG35. pKG7 (10 μg), pKG35 (10 μg), M13 RF (10 μg), and pBR317 (10 μg) DNAs were digested for 18 hours at 37°C with 5 units of *Hae*III and 15 units of *Bam*I under *Hae*III restriction conditions. The cleavage products were electrophoresed either on a vertical gradient polyacrylamide gel (7.5–15%) (Fig. 2) or on a vertical step polyacrylamide gel (8% and 10%) (Jeppesen 1974). One of the two *Hae*III E fragments (0.19 $\times 10^6$ daltons) of M13 was present in both pKG7 and pKG35, along with the

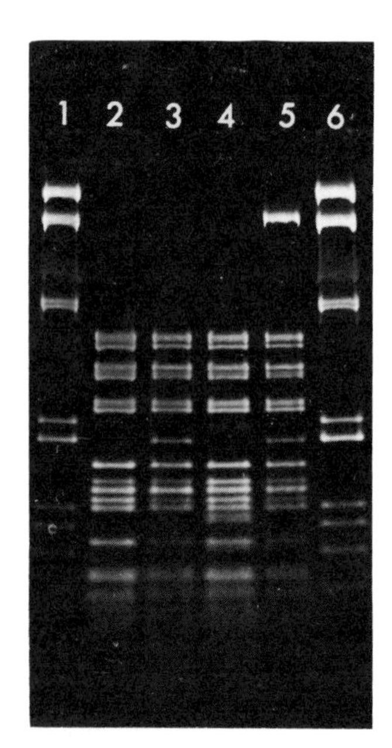

Figure 2 Gel electrophoresis of pKG7 and pKG35 DNA. DNA samples digested simultaneously with *Hae*III and *Bam*I and electrophoresed on a vertical gradient polyacrylamide gel (7.5–15%). (Lanes 1 and 6) M13 RFI; (lane 2) pBR317; (lane 3) pKG7; (lane 5) pKG35.

*Bam*I-*Hae*III fragment (0.026×10^6 daltons) of M13. It was not possible to identify definitively an additional *Hae*III fragment in pKG7, but there was an additional *Hae*III fragment (1.15×10^6 daltons) in the cleavage products of pKG35 (Fig. 2).

Molecular-weight estimates of the M13 insert in pKG35 were made by simultaneous digestion with *Bam*I and *Pst*I. pKG35 DNA was treated with *Bam*I and *Pst*I, and the cleavage products were electrophoresed, along with standard molecular-weight markers, on a horizontal 0.7% agarose gel. Three fragments were observed—two corresponded to those found in a *Bam*I-*Pst*I digest of pBR317; the other was a totally different fragment of molecular weight 1.67×10^6 daltons. To determine what fraction of this fragment was M13 DNA, we assumed that the fragment was composed of two elements—the *Pst*I-*Eco*RI fragment of pBR317 and a *Bam*I-*Eco*RI* fragment of M13. We have examined a *Pst*I-*Eco*RI digest of pBR317 by polyacrylamide gel electrophoresis. This fragment was characterized as having a molecular weight of 0.58×10^6 daltons. By subtracting this value from 1.67×10^6 daltons the size of the M13 insert was calculated to be 1.09×10^6 daltons.

Heteroduplex Formation between pKG7 and pKG35 and M13

To produce linear DNA for heteroduplex formation, M13 RFI DNA was cleaved with *Hin*dII and pKG7 or pKG35 DNA was cleaved with *Pst*I. Heteroduplexes were prepared as described by Davis et al. (1971). Eighteen hybrids between pKG7 and M13 were analyzed. The structures contained a single double-stranded region of 0.33×10^6 daltons and four single-stranded regions, two extending from one end of the double-stranded region and two from the other (Fig. 3). A similar structure was seen in an experiment where M13 was annealed to pKG35, but the size of the double-stranded region corresponded to a molecular weight of 1.10×10^6 daltons.

DISCUSSION

When these experiments were initiated, *Bam*I was the only known single restriction endonuclease site common to both phage M13 and the plasmid pBR317. Every attempt to clone the entire *Bam*I fragment of M13 failed. This was not totally unexpected, as it had been shown previously that amber mutations in every gene except gene II in M13 resulted in a phage that was lethal for *E. coli* (Hohn et al. 1971). As an alternate approach to cloning the phage DNA, M13 was digested with *Bam*I and partially with *Eco*RI*. The cleavage products were then ligated to complementary ends produced in pBR317 by treatment with *Bam*I and *Eco*RI. This type of cloning requires that at least one of the inserts be a *Bam*I-*Eco*RI* fragment. This does not exclude, however, the possibility of two inserts from different regions of

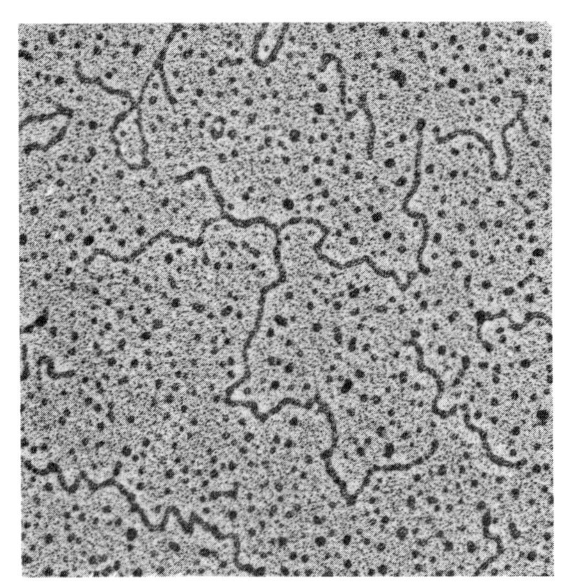

Figure 3 An electron micrograph of a heteroduplex between M13 DNA cleaved with *Hin*dII and pKG7 DNA cleaved with *Pst*I. There is a single double-stranded region (0.33×10^6 daltons) with two single-stranded regions extending from each end.

M13, one a *Bam*I-*Eco*RI* fragment and the other a *Eco*RI*-*Eco*RI* fragment. By examining the heteroduplex structures formed between M13 and pKG7 and pKG35, we have concluded that each of the two inserts analyzed represented a continuous fragment of M13. This belief is supported by the fact that a single region of double-stranded DNA with two single-stranded tails at each end is observed in heteroduplexes between M13 RFI and pKG7 and pKG35.

The molecular weight of the *Bam*I-*Eco*RI* fragment of plasmid pKG7, as determined by heteroduplex analysis, was 0.33×10^6 daltons. This may be compared with the value of 0.43×10^6 daltons obtained by subtracting the molecular weight of pBR317* (5.17×10^6 daltons) from the molecular weight of pKG7 (5.60×10^6 daltons). pBR317* represents that portion of the plasmid remaining after excision of the region between the *Bam*I and *Eco*RI sites. These data, plus the fact that pKG7 contains the *Hae*III E and *Bam*I-*Hae*III fragments of M13, define the insert: it starts in the *Bam*I site in gene III and terminates at an *Eco*RI* site in the region of gene VI (Fig. 1).

The molecular weight of the M13 insert into pKG35 was determined by three different methods: (1) The molecular weight of pKG35 DNA was determined to be 6.40×10^6. Subtracting the molecular weight of the cloning vehicle pBR317* (5.17×10^6 daltons) from that value gives a value of 1.23×10^6 daltons for the insert. (2) The *Bam*I-*Pst*I fragment of the plasmid DNA including the M13 insert was measured and found to be 1.67 $\times 10^6$ daltons. Assuming this fragment is composed of two elements, the *Pst*I-*Eco*RI (0.58 $\times 10^6$ daltons) fragment of pBR317 and the *Bam*I-*Eco*RI* M13 insert, then by subtracting 0.58 $\times 10^6$ daltons from 1.67 $\times 10^6$ daltons, the molecular weight of the insert was calculated to be 1.09 $\times 10^6$ daltons. (3) Three fragments from *Bam*I-*Hae*III cleavage of pKG35 were analyzed by gel electrophoresis. Two were identical to those obtained from a similar digestion of M13; one was the *Hae*III E fragment (0.19 $\times 10^6$ daltons) and the other the *Bam*I-*Hae*III (0.026 $\times 10^6$ daltons) fragment. The other *Hae*III fragment (1.15 $\times 10^6$ daltons) was assumed to be composed of two elements, the *Hae*III-*Eco*RI* insert of M13 and the *Eco*RI-*Hae*III (0.27 $\times 10^6$ daltons) fragment of pBR317. Therefore, by subtracting 0.27 $\times 10^6$ daltons from 1.15 $\times 10^6$ daltons, we arrive at the molecular weight of the *Hae*III-*Eco*RI* fragment of M13 (0.88 $\times 10^6$ daltons). By adding the molecular weights of the *Bam*I-*Hae*III, the *Hae*III E, and the *Hae*III-*Eco*RI* fragments, we calculate the molecular weight of the M13 insert to be 1.10 $\times 10^6$ daltons. The average value for the *Bam*I-*Eco*RI* insert of M13 is 1.14 $\times 10^6$ daltons.

From these data, from heteroduplex analyses which showed that there was a single, continuous insert into pBR317, and from the genetic map of M13 (Fig. 1; Ray 1977), we concluded that a *Bam*I-*Eco*RI* M13 fragment was cloned that starts in the *Bam*I site in gene III and ends near the end of gene I. Since the cloned fragment of pKG35 encompasses gene VI as well as an adjacent promoter, we examined the possibility that a functional gene-VI product might be produced by the clone. Preliminary data indicate that M13, with an amber mutation in gene VI, replicates on *E. coli* K38 containing the plasmid pKG35. This suggests that the product of gene VI is being synthesized by pKG35 DNA.

ACKNOWLEDGMENTS

We thank Harumi Kasamatsu for her help in preparing the heteroduplexes for electron microscopy. This research was supported by National Institutes of Health contract A16-2507 and by National Science Foundation grant PCM 76-80445. L. G. was supported by training grant CA-09056-03 from the National Cancer Institute, and G.W. by an American Cancer Society Faculty Research Award.

REFERENCES

Boyer, H., M. Betlach, F. Bolivar, R. Rodriguez, H. Heyneker, J. Shine, and H. Goodman. 1977. The construction of molecular cloning vehicles. In *Recombinant molecules: Impact on science and society* (ed. R. Beers and E. Basset), p. 9. Raven Press, New York.

Davis, R. W., M. Simon, and N. Davidson. 1971. Electron microscope heteroduplex methods for mapping regions of base sequence homology in nucleic acids. *Methods Enzymol.* **21:** 413.

Hohn, B., H. von Schütz, and D. A. Marvin. 1971. Filamentous bacterial viruses. II. Killing of bacteria by abortive infection with fd. *J. Mol. Biol.* **56:** 155.

Jeppesen, P. G. N. 1974. A method for separating DNA fragments by electrophoresis in polyacrylamide concentration gradient slab gels. *Anal. Biochem.* **58:** 195.

Kaplan, D. A., L. Greenfield, and G. Wilcox. 1977a. Molecular cloning of λh80d*ara* restriction fragments with noncomplementary ends. In *Molecular approaches to eucaryotic genetic systems* (ed. G. Wilcox et al.), p. 85. Academic Press, New York.

Kaplan, D. A., R. Russo, and G. Wilcox. 1977b. An improved horizontal slab gel electrophoresis apparatus for DNA separation. *Anal. Biochem.* **78:** 235.

Polisky, B., P. Greene, D. E. Garfin, B. J. McCarthy, H. M. Goodman, and H. W. Boyer. 1975. Specificity of substrate recognition by the *Eco*RI restriction endonuclease. *Proc. Natl. Acad. Sci.* **72:** 3310.

Ray, D. S. 1977. Replication of filamentous bacteriophages. In *Comprehensive virology* (ed. H. Fraenkel-Conrat and R. R. Wagner), vol. 7, p. 105. Plenum Press, New York.

van den Hondel, C. A. and J. G. G. Schoenmakers. 1975. Studies on bacteriophage M13 DNA. I. A cleavage map of the M13 genome. *Eur. J. Biochem.* **53:** 547.

Insertion of a Kanamycin-resistance Gene in Bacteriophage fd

Nobuo Nomura, Atsuhiro Oka, Mituru Takanami, and Hideo Yamagishi*
Institute for Chemical Research and
*Department of Biophysics, Faculty of Science
Kyoto University
Kyoto, Japan

Gene structures involved in replication of filamentous phages have been well characterized, and various derivatives, including chimeric phages, have been shown to form infectious filamentous particles (Grandis and Webster 1973; Griffith and Kornberg 1974; Enea and Zinder 1975; Herrmann et al., this volume). Thus, we can now construct appropriate vectors for cloning DNA and transducing phages using these filamentous phages. Our first interest was in isolating composite phages from wild-type fd and pML21, a plasmid containing the replication origin of ColE1 and a gene specifying kanamycin resistance (*kan*) derived from plasmid R6-5 (Hershfield et al. 1976). Phage fd was grown repeatedly on cells carrying pML21 from which *kan*-transducing phages were selected. All the *kan*-transducing phages isolated independently (fd-*kan*) contained a unique segment of the same size (3100 base pairs), but the sites of insertion apparently were different for each chimeric phage. This type of insertion is characteristic of transposable genetic elements (transposons).

All of the isolated fd-*kan* phages could transduce the *kan* marker without helper, but they could not form plaques. Upon transformation of fd-*kan* DNA into F^- cells, fd-*kan* replicative-form (RF) DNA was maintained in the cells like a plasmid, without creation of fd phage. However, F^+Kan^r cells prepared by either transduction with fd-*kan* or transformation with fd-*kan* RF DNA produced not only fd-*kan* but also wild-type fd phage, suggesting that the excision of the *kan* segment from fd-*kan* DNA is promoted in F^+ cells at the precise site where the insertion has occurred.

Isolation of *kan*-transducing Phage fd (fd-*kan*)

C600/F8 cells were infected, at a multiplicity of 2, with fd grown repeatedly on C600 (pML21)/F8, incubated for 2 hours at 37°C, and then plated at about 10^{10} cells per plate containing kanamycin (15 μg/ml). About 1000 colonies appeared per plate after incubation overnight. After treatment with chloroform vapor, the colonies were replicated onto a lawn of C600/F8 cells spread on L-agar and incubated for 3 hours at 37°C to permit some phage growth. The replica plate was then replicated onto L-agar containing kanamycin. A few large colonies appeared per plate after culture overnight. The presence of transducing phages in these colonies was confirmed by examining the transducing ability of the medium in which Kan^r cells derived from a single colony were grown (transducing-phage stock). Sixteen transducing-phage stocks were isolated independently. The characteristics of the transducing element in these stocks were analyzed and the following results were obtained: (1) Kan^r transductants were produced only when male strains were used as recipients; (2) the *kan*-transducing ability was inactivated with anti-fd serum at a rate similar to that for wild-type fd; (3) the *kan*-transducing element banded at the position where fd forms a peak in a CsCl density gradient; and (4) by electron microscopy, a filamentous structure with a length 1.5 times that of fd phage was observed in an amount comparable to the *kan*-transducing frequency (Fig. 2a). Thus, it was concluded that the transducing element in each stock was a derivative of fd, and these were designated fd-*kan*1 to fd-*kan*16. The transducing frequency in each stock was about 0.01 that of the plaque-forming units. Phages prepared from each of 1000 single plaques showed no *kan*-transducing ability. The activities of *kan* transduction and of plaque formation present in each transducing-phage stock were separable by zone centrifugation. These results indicate that each stock is a mixture of defective fd carrying *kan* (fd-*kan*) and wild-type fd. Since the transduction frequency was not affected by the addition of exogenous helper fd or by reducing the multiplicity of infection to 0.0001, fd-*kan* appears to transduce the *kan* marker without helper phages. It should be mentioned that plaque-forming phages were never eliminated from fd-*kan* preparations by successive single-colony isolation of Kan^r transductants. The result strongly suggests that plaque-forming fd (wild-type) is created from fd-*kan* DNA during intracellular multiplication.

RF DNA of fd-*kan*

DNA was extracted from Kan^r transductant cells and subjected to agarose gel electrophoresis after removal of host DNA. A typical pattern obtained from fd-*kan*1 is shown in Figure 1a: two new bands, *i* and *ii*, are seen in addition to fd RFI (superhelical, covalently closed, circular duplex DNA) and fd RFII (duplex DNA in which at least one strand is not continuous) bands. All the samples examined (fd-*kan*1 to fd-*kan*9) gave very similar patterns, although the molar ratios in the bands were different for each strain. Upon limited treatment of the sample with DNase I prior to electrophoresis, bands *I* and *i* were shifted to the positions of bands *II* and *ii*,

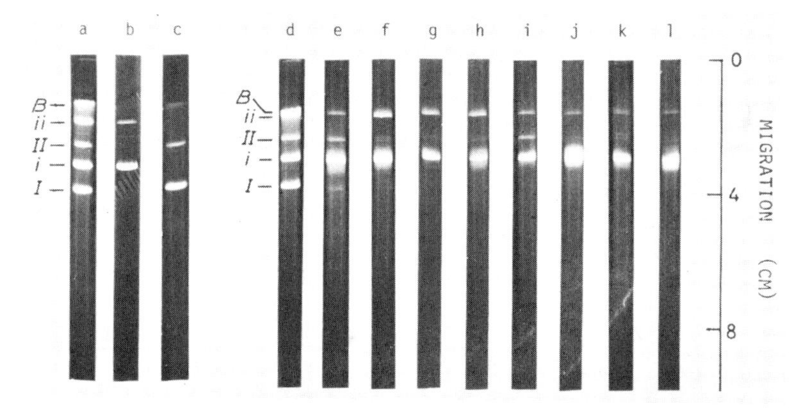

Figure 1 Agarose gel electrophoresis of RF DNA fractions from Kan^r transductant cells. RF DNA fractions were prepared from Kan^r transductant cells as has been described for the preparation of fd RF DNA (Takanami et al. 1975) and were electrophoresed on 0.7% agarose (*a–c*) or 1% agarose (*d–l*). (*a,d*) RF DNA prepared from fd-*kan* 1-transduced cells. (*c*) fd RF DNA as a reference. (*b, e–l*) RF DNA fractions were prepared from cells transduced with fd-*kan*1–9, respectively, and after resolution on 0.7% agarose gels, band-*i* regions from the respective fd-*kan* strains were reelectrophoresed. Bands *I* and *II* correspond to fd RFI and fd RFII DNA, and bands *i* and *ii* to fd-*kan* RFI and fd-*kan* RFII DNA, respectively. Band B is degraded host DNA.

respectively, suggesting that bands *i* and *ii* correspond to RFI and RFII DNA of fd-*kan*. In support of this view, DNA isolated both from band *i* and from band *ii* had the ability to transform to *kan* resistance, in contrast to DNA from bands *I* and *II* which had only plaque-forming ability. As shown in Figure 1d–l, RF DNA of all the fd-*kan* strains examined seems to have the same size: on the basis of relative mobility, the size was estimated to be about 1.5 times that of fd RF DNA.

As mentioned in the previous section, the number of phage particles of fd-*kan* and wild-type fd released into the culture medium was in a ratio of about 1:100 for all the strains. However, the molar ratio of fd-*kan* RF DNA to fd RF DNA in cells appeared to be relatively high (see Fig. 1a). This result suggests that synthesis or encapsidation of fd single-stranded (SS) DNA is more efficient than that of fd-*kan* SS DNA. This may be due to the size effect of the DNA, as small particles of f1 and M13 have been shown to be enriched during multiple passages at high multiplicities of infection (Grandis and Webster 1973; Griffith and Kornberg 1974; Enea and Zinder 1975).

Heteroduplex molecules of fd-*kan* RF DNA with fd SS DNA were visualized in the electron microscope. A typical example is shown in Figure 2b, in which a characteristic shape of a transposon can be seen. The average lengths of the inverted repeat sequence (IS) and of the *kan* loop, measured on 57 molecules, were 1070 ± 110 and 1030 ± 170 base pairs, respectively.

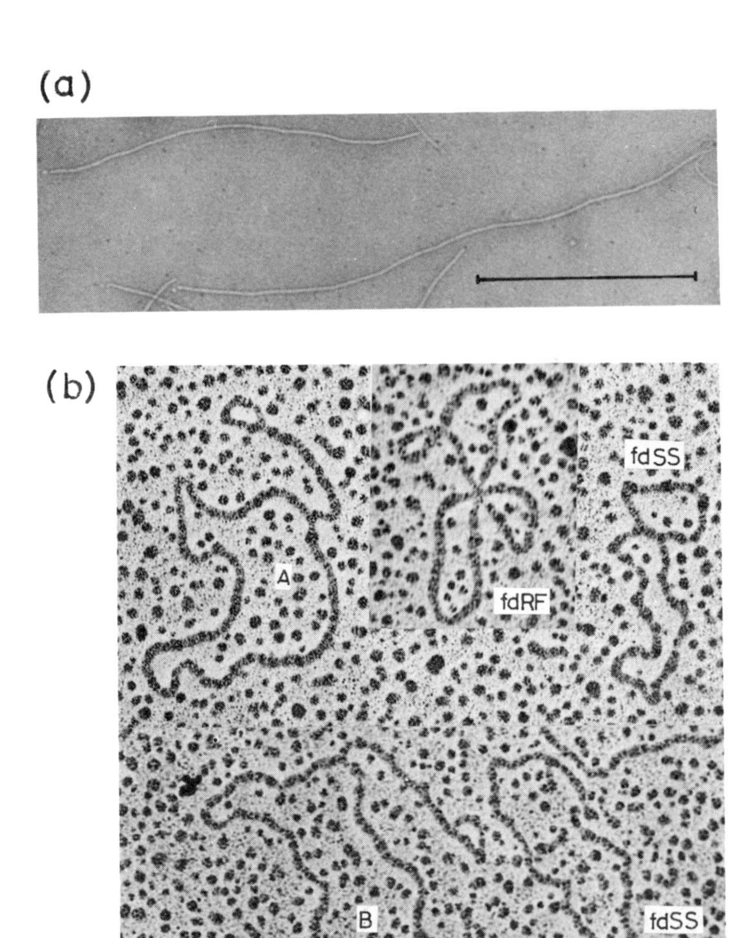

Figure 2 Electron micrographs of fd-*kan* virions and heteroduplex molecules prepared with fd-*kan* RF DNA and fd SS DNA. (*a*) Virions from fd-*kan*1 were adsorbed onto a carbon-coated and glow-discharged grid and negatively stained with 2% phosphotungstic acid. Filamentous structures with one unit length of fd (0.9 μm) and 1.5 unit length of fd-*kan* are shown. (*b*) fd-*kan*1 RF DNA was irradiated with γ-rays at a dose producing about 0.5 nick per molecule, and heteroduplexes with fd SS DNA were prepared as described previously (Yamagishi et al. 1976). Molecules A and B indicate the heteroduplex and self-annealed fd-*kan* SS DNA, respectively. Under the conditions used, fd RFII DNA has a length of 2.20 ± 0.108 μm and fd SS DNA has a length of 1.80 ± 0.104 μm. The lengths of the corresponding regions in molecules A and B, respectively, are about equal to these values. Magnification, ~ 15,000×. The bar represents 0.5 μm.

These values are consistent with those obtained for the IS region and *kan* loop on pML31 derived from plasmid R6-5 (Guyer et al. 1976). The length of the double-stranded, circular part in the heteroduplex was about equal to that of fd RFII DNA (6200 ± 390 base pairs). The results suggest that the *kan* segment recombined with the fd genome at or close to the IS region observed on pML31 without losses of the fd genome.

The stability of fd-*kan* RF DNA in cells was examined in F^-Kan^r transformant cells instead of in F^+Kan^r cells, since the latter, whether prepared by transformation or transduction, produced fd-*kan* and fd continuously by reinfecting cells during cultivation. F^-Kan^r cells were found to segregate Kan^s cells at a frequency of 3–5% per generation. When DNA was extracted from F^-Kan^r cells, DNA bands corresponding to fd-*kan* RFI and fd-*kan* RFII DNA were observed. However, fd RF DNA was not identified in these cells, and plaque-forming fd was not detected in the culture medium. The result suggests that fd-*kan* RF DNA is maintained in the F^- cells like a plasmid, without producing wild-type fd. When the fertility factor F8 was introduced into the F^-Kan^r cells, however, wild-type fd was produced as in F^+Kan^r cells. It is likely, therefore, that the excision of the *kan* segment from fd-*kan* DNA is promoted in F^+ cells at the precise site where the insertion has occurred. The result also suggests that some function coded by F8 may be involved in this excision. A similar phenomenon has been reported with another recombinant plasmid ColE1-*kan* (Bodsch 1977).

Location of Insertion Sites

RF DNA fractions prepared from fd-*kan*1–9 were digested with *Hap*II, *Hae* III, or *Hga*I, and the resulting restriction fragments were compared by polyacrylamide gel electrophoresis with those produced from fd RF DNA. Among the restriction fragments proper to fd RF DNA, the one integrated with the *kan* segment was found to be missing in the digests of fd-*kan* RF DNA. The digests of fd-*kan* RF DNA contained the common fragments derived from the internal region of the *kan* segment and two unique fragments, the lengths of which were different for each fd-*kan* RF DNA (variable fragments). The variable fragments were assigned to the junction between fd RF DNA and the *kan* segment. From the information on the missing fragments and the size of the variable fragments, it was concluded that the insertions in the nine fd-*kan* strains all occurred at different sites. This implies that the specificity of insertion is low. Although construction of the detailed insertion map is still in progress, all the sites were found to be outside of gene II, the gene required for viral DNA replication. This is consistent with the findings that all the fd-*kan* strains could transduce the *kan* marker without helper phages and that their RF DNA is maintained in the cell.

REFERENCES

Bodsch, W. 1977. Excision of a DNA sequence determining kanamycin resistance from a ColE1-Km recombinant plasmid. *Mol. Gen. Genet.* **150:** 29.

Enea, V. and N. D. Zinder. 1975. A deletion mutant of bacteriophage f1 containing no intact cistrons. *Virology* **68:** 105.

Grandis, A. S. and R. E. Webster. 1973. A new species of small covalently closed f1 DNA in *Escherichia coli* infected with an *amber* mutant of bacteriophage f1. *Virology* **55:** 14.

Griffith, J. and A. Kornberg. 1974. Mini-M13 bacteriophage: Circular fragments of M13 DNA are replicated and packaged during normal infections. *Virology* **59:** 139.

Guyer, M. S., D. Figurski, and N. Davidson. 1976. Electron microscope study of a plasmid chimera containing the replication region of the *Escherichia coli* F plasmid. *J. Bacteriol.* **127:** 988.

Hershfield, V., H. W. Boyer, L. Chow, and D. R. Helinski. 1976. Characterization of a mini-ColE1 plasmid. *J. Bacteriol.* **126:** 447.

Takanami, M., T. Okamoto, K. Sugimoto, and H. Sugisaki. 1975. Studies on bacteriophage fd DNA. I. A cleavage map of the fd genome. *J. Mol. Biol.* **95:** 21.

Yamagishi, H., H. Inokuchi, and H. Ozeki. 1976. Excision and duplication of $su3^+$-transducing fragments carried by bacteriophage φ80. I. Novel structure of $\varphi 80sus2psu3^+$ DNA molecule. *J. Virol.* **18:** 1016.

Integration of DNA Fragments Coding for Antibiotic Resistance into the Genome of Phage fd In Vivo and In Vitro

Richard Herrmann, Kristina Neugebauer, and Heinz Schaller
Mikrobiologie der Universität Heidelberg
Heidelberg, Federal Republic of Germany

Hanswalter Zentgraf
Institut für Virusforschung
Deutsches Krebsforschungszentrum
Heidelberg, Federal Republic of Germany

To test the possibility of using filamentous bacteriophages as vectors for cloning and analysis of foreign DNA, a variety of derivatives were constructed from phage fd. DNA segments of different origins and sizes were integrated into the phage genome by in vivo transposition of a transposon or by in vitro recombination between fd replicative-form (RF) DNA and DNA fragments from plasmids coding for antibiotic resistance.

The in vivo studies were carried out with transposon 5 (Tn*5*), which is known to integrate into many different sites of a DNA genome (Berg et al. 1975; Berg 1976). Phage fd grown on a host carrying Tn*5* acquired the ability to transduce kanamycin (Km) resistance, and a variety of phage clones could be isolated and analyzed (Table 1) (Herrmann et al. 1978). Integration of the intact transposon into fd DNA was accompanied by an 80% increase in the size of the DNA and phage as measured by gel electrophoresis and electron microscopy. Electron micrographs also showed the typical stem-loop structure of Tn*5* attached to the viral single-stranded (SS) DNA circle (Fig. 1). In nondefective phages the sites of integration of Tn*5* were in the intergenic region. Defective transducing phages were characterized by insertions of Tn*5* into a phage gene, or by partial deletions or duplications of phage and transposon DNA, or by both. The size of the transducing phage from different defective clones varied from 0.6 μm to 3.0 μm and was directly proportional to the DNA content. These results demonstrate that filamentous bacteriophage are able to replicate and package very different amounts of foreign DNA.

For the construction in vitro of helper-independent recombinants, fd RF DNA was linearized by partial cleavage with restriction nucleases and ligated to fragments or linear, full-length molecules of plasmid DNA bearing the same "sticky ends" as the viral DNA. After transfection of *Escherichia*

Table 1 List of fd phages carrying antibiotic resistance markers

Phage	Size of integrated DNA in kb[a]	Source of integrated DNA[b]	Antibiotic resistance marker	Single restriction cut within the resistance marker	Remarks
fdKm-16		*λbb kan*1 (Tn*5*) [1]	Km		helper-dependent; total size 3.6 kb; part of fd DNA and Tn*5* DNA deleted
fdTn*5*-13	5.3	*λbb kan*1 (Tn*5*)	Km		helper-dependent
fdTn*5*-30	5.3	*λbb kan*1 (Tn*5*)	Km		
fdKm-573	11.3	*λbb kan*1 (Tn*5*)	Km		helper-dependent
fd 177	3.6	pACYC177 [2]	Ap, Km	*Pst*I, *Hin*dIII, *Sma*I	
fd 1771	3.6	pACYC177	Km	*Hin*dIII, *Sma*I	
fd 1772	2.3	pACYC177	Ap	*Pst*I	
fd 1773	1.4	pACYC177	Km	*Xho*I, *Hin*dIII, *Sma*I	
fd 1842	1.3	pACYC184 [2]	Cm	*Eco* RI	
fd 3221	4	pBR322 [3]	Ap	*Pst*I	the restriction enzymes *Eco*RI, *Hin*dIII, and *Sal*I cut once outside of the resistance marker

[a] Data from the corresponding reference indicated in column 3 were used to calculate the size of the integrated DNA fragment. fd DNA is 6.4 kb (Schaller et al., this volume).

[b] Numbers in brackets indicate the following references: [1] Berg et al. (1975); [2] A. C. Y. Chang and S. N. Cohen (in prep); [3] Bolivar et al. (1977).

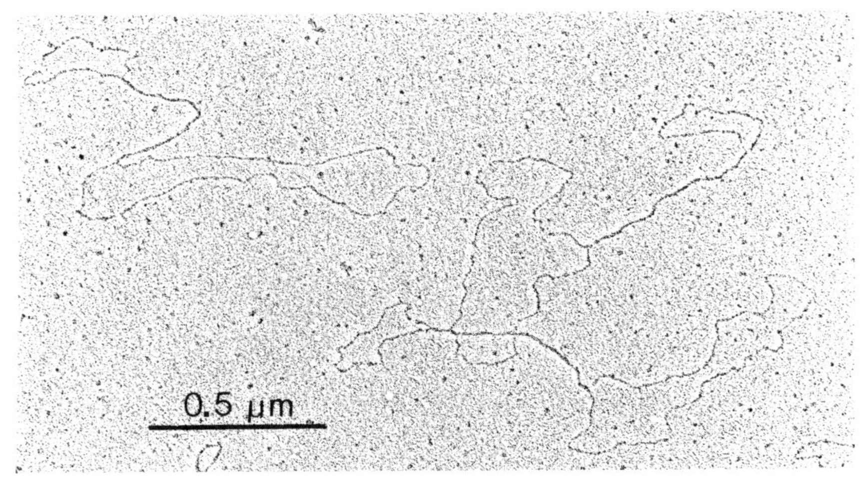

Figure 1 Electron micrograph of fd Tn*5*-30 DNA. The large, single-stranded loop represents the fd genome, the small loop with the double-stranded stem corresponds to one strand of the inserted transposon.

coli cells, phages carrying the expected antibiotic resistance (kanamycin [Km], chloramphenicol [Cm], ampicillin [Ap]) were selected by a transduction assay. A list of some of the derivatives obtained (R. Herrmann and K. Neugebauer, unpubl.) is shown in the lower part of Table 1. In these transducing phages the plasmid DNA has been integrated into the intergenic region of the fd genome either at the *Hae*II cleavage sites at map positions 5545 and 5553 (fd 177, 1771, 1772, 1773, 1842) or at the *Bam*I cleavage site at position 5626 (fd 3221) (Schaller et al., this volume).

Recently, Cm-transducing phages similar to fd 1842 have been constructed by integrating a plasmid DNA fragment into cleavage sites of restriction nucleases *Alu*I, *Hpa*II, *Hae*III, *Hha*I, *Hin*f (K. Neugebauer, unpubl.). This suggests that many positions in the intergenic region are not part of essential phage functions and thus can be modified for construction of a nondefective vector phage.

The DNA fragments used as selection markers in the construction in vitro of transducing phages all contain single cleavage sites for restriction nucleases that do not cleave the fd genome (*Eco*RI, *Hin*dIII, *Pst*I, *Sma*I, *Sal*I, *Xho*I; see Table 1). These therefore provide specific sites for selective integrations into the vector DNA molecule of additional DNA fragments carrying the identical sticky ends.

Finally, it should be noted that two of the isolates (fd 177 and fd 1771) carry the same piece of plasmid DNA in opposite orientation in the RF DNA, i.e., they contain the complementary strands of the DNA fragment in their viral DNAs. This is demonstrated by the formation of a heteroduplex between the two phage DNAs (Fig. 2). It shows that a separation of the

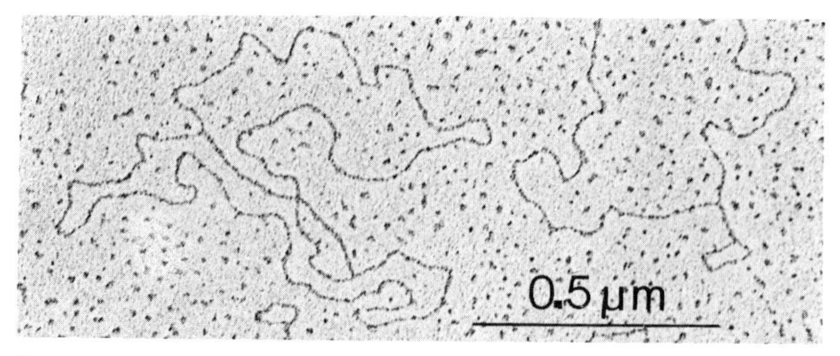

Figure 2 Heteroduplex between the SS DNAs from phages fd 177 and fd 1771. The phage DNAs contain the complementary strands of the same piece of plasmid DNA. This DNA segment exhibits a stem-loop structure (the right DNA molecule) which disappears after the formation of the duplex.

complementary strands of a cloned DNA fragment can be achieved by the use of filamentous phage vectors.

REFERENCES

Berg, D. E. 1977. Insertion and excision of the transposable kanamycin resistance determinant Tn*5* (*1*). In *DNA insertion elements, plasmids, and episomes* (ed. A. I. Bukhari et al.), p. 205. Cold Spring Harbor Laboratory, Cold Spring Harbor, New York.

Berg, D. E., J. Davies, B. Allet, and J. D. Rochaix. 1975. Transposition of R factor genes to bacteriophage λ. *Proc. Natl. Acad. Sci.* **72:** 3628.

Bolivar, F., R. L. Rodriguez, P. J. Greene, M. C. Betlach, H. L. Heyneker, H. W. Boyer, J. H. Crosa, and S. Falkow. 1977. Construction and characterization of new cloning vehicles. II. A multipurpose cloning system. *Gene* **2:** 95.

Herrmann, R., K. Neugebauer, H. Zentgraf, and H. Schaller. 1978. Transposition of a DNA sequence determining kanamycin resistance into the single-stranded genome of bacteriophage fd. *Mol. Gen. Genet.* **159:** 171.

Gene-specific Suppressor Strains of *Escherichia coli* Carrying a Functional Gene *G* from φX174 on a Plasmid

M. Zafri Humayun and Robert W. Chambers
Department of Biochemistry
New York University School of Medicine
New York, New York 10016

A majority of carcinogenic agents covalently modify DNA and produce mutations (Miller and Miller 1974). At present, no direct methods are available for determining either which of the many different kinds of modifications caused by such agents lead to mutation or the type of mutation generated by specific premutational modifications. We are exploring a direct approach to this problem in which (1) a well-characterized modification is introduced at a preselected single site in the genome of a suitable organism by a combination chemical and enzymatic synthesis; (2) the biological response to this modified DNA is followed by isolating any mutants that are produced from this modified DNA in vivo; and (3) the nature of the mutation is determined by DNA sequencing. Phage φX174 is an attractive choice for our initial studies because extensive information on its chemistry and biology is available (Sinsheimer 1968; Denhardt 1975; Sanger et al. 1977).

To test this approach, we have chosen an essential gene, gene *G*, which codes for a spike protein. However, propagation of non-conditionally lethal mutations in a virulent phage such as φX presents a problem. An attractive solution is to provide a source of the gene product in a host cell by introducing a biochemically constructed plasmid carrying a functional copy of the gene under investigation. Bacterial strains harboring such plasmids should act as gene-specific suppressors; they should be permissive for frameshift and large deletion mutations as well as base substitutions within the chosen gene. In this article we will summarize the results of our experiments to date with plasmids bearing gene *G*.

MATERIALS AND METHODS

We have used pMB9 as the cloning vehicle. This plasmid has a single *Eco*RI site and specifies tetracycline resistance (Rodriguez et al. 1976). DNA hybrids were constructed from *Eco*RI-cleaved pMB9 DNA and appropriate

φX restriction fragments by the poly(dA)-poly(dT) joining technique (Jackson et al. 1972; Wensink et al. 1974). A φX^s, Su^-, rec^+ strain of *E. coli*, H514 (Vosberg and Hoffmann-Berling 1971), was transformed with the hybrid DNA (Cohen et al. 1972; Higuchi et al. 1976). Tetracycline-resistant transformants were selected and examined by a double screening strategy involving (1) measurement of the size of the plasmid DNA by gel electrophoresis and (2) examination of the clones for their ability to suppress the nonsense mutation in *am*9, a gene-*G* mutant of φX. Selected clones were tested for their ability to suppress *am*86, a gene-*A* mutant, and *am*3, a gene-*E* mutant. In all cases the transformants were specific for the gene-*G* mutation. Plasmid DNA was isolated and characterized by electron microscopy and by restriction mapping using standard procedures which will be described in more detail elsewhere. This plasmid DNA was used to transform selected *E. coli* host cells.

RESULTS AND DISCUSSION

Using the procedure indicated above, we have constructed three types of plasmids that suppress *am*9: The pφXG100 series was prepared from *Hae*III fragment 1 (Z1) (Sanger et al. 1977). This fragment is 1353 base pairs long and contains all of gene *G* along with parts of *F* and *H* (see Fig. 1). The pφXG200 series was constructed from *Hha*I fragment 3 (H3) (Sanger et al. 1977). This fragment contains 614 base pairs. It starts in the untranslated region near the ribosome-binding site of gene *G* and extends into gene *H* (see Fig. 1). The pφXG300 series was prepared from *Taq*I fragment 1 (T1) (Sanger et al. 1977). This fragment contains 2914 base pairs and runs from gene *F* into gene *A*. It should contain genes *G* and *H* intact, but our

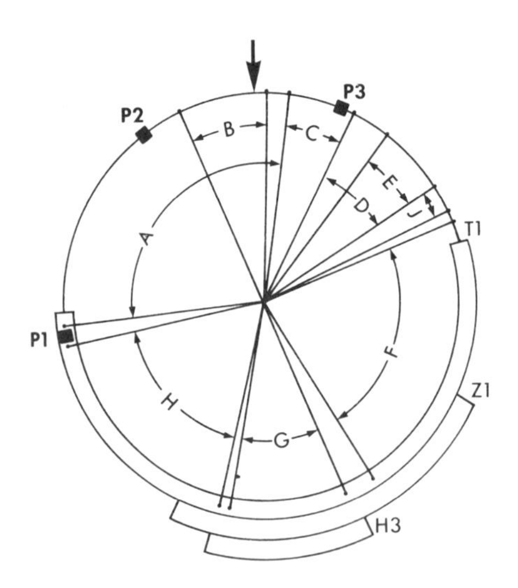

Figure 1 The genetic map of φX DNA showing the relative sizes and locations of the restriction fragments T1, Z1, and H3 (Sanger et al. 1977), which have been cloned into pMB9, giving transformants carrying functional φX gene *G*. *Pst*I cuts at a single site (↓) and provides a point of reference on the map (Sanger et al. 1977). P_1, P_2, and P_3 (■) are suggested promoter sites (Sanger et al. 1977).

characterization is not yet complete. This fragment covers 54% of the φX genome (see Fig. 1).

Table 1 shows the host cells that have been transformed with these plasmids. The 100 series, H514/pφXG109 and HF4704/pφXG109, both of which are derived from Su$^-$ *recA*$^+$ host cells, gives bursts of approximately ten virus particles per cell with *am*9 compared with approximately 200 with wild-type φX. This is about one-seventh the burst size obtained with *am*9 on the commonly used suppressor HF4714. The small burst size seems to be caused by a polar effect of *am*9 on gene *H*, another essential gene whose gene product (another spike protein) is necessary for the assembly of an infectious phage particle (Benbow et al. 1972).

Plaques produced by infection with *am*9 are tiny compared with those produced by infection with wild type. Furthermore, all of the plaques tested contained wild-type phage (up to one-third of the total) in addition to *am*9. Recombination between infecting *am*9 and the plasmid carrying gene *G* occurred in only 0.03% of the infected cells (infective-center assay). The total recombination frequency is 0.8% (single-step growth experiments) (Adams 1959). The presence of wild-type phage in plaques produced by infecting p φXG-carrying cells with *am*9 can be explained by the combined effects of the small burst size and recombination between the infecting DNA and the plasmid DNA to produce a wild-type revertant that produces a burst size 20 times that of the mutant.

Plaque purity is important, as our purpose in constructing these plasmid-bearing strains is to use them to search not only for point mutations but also for frameshift and large deletion mutations produced by different kinds of site-specific premutational lesions. Therefore, we have transformed some Su$^-$ *recA*$^-$ strains (Table 1). Strain HF4740/p φXG109 has been investigated most extensively. Two types of plaques are distinguishable after infection of this strain with *am*9: small, circular plaques which are essentially pure *am*9 ($<10^{-4}$ wild-type) and large, irregular plaques (approximately 2% of the total) which appear to be essentially pure wild-type phage. The frequency of the wild-type plaques is much higher than the wild type present in the *am*9 stocks ($<10^{-5}$), and the apparent absence of mutant phage in these wild-type plaques is not understood.

Table 1 *E. coli* strains transformed by pMB9 carrying functional genes from φX RF DNA

	pφXG109	pφXG201	pφXG303
Su$^-$ *recA*$^+$	HF514, HF4704, C600	H514	HF4704
Su$^-$ *recA*$^-$	HF4740, DD1	HF4740	—
Su$^+$ *recA*$^-$	HF4738	—	—

Host strains: H514: Su$^-$, φX^s, *end, uvrA, thyA, arg, r$^-$, m$^-$*; HF4704: F$^-$, Su$^-$, φX^s, *thy, uvrA*; HF4738: F$^-$, Su$^+$, φX^s, *recA1, gal, end, tlr*; HF4740: F$^-$, Su$^-$, φX^s, *recA1, uvrA103, tlr*; DD1 (tentative): F$^-$, Su$^-$, φX^s, *tet^r, str^r, recA, uvrA*; C600: F$^-$, Su2$^+$, φX^r, *thi, thr, leu*.

The small burst size is a serious problem with certain mutations. We are investigating two possible solutions. First, we have put both gene *G* and gene *H* (300 series) into a plasmid. Gene *G* is functional (i.e., suppresses *am*9), but we have no evidence as to whether gene *H* is functional. The plaques with *am*9 are still small, suggesting that the burst size is also small. Our second approach involves putting p φXG into a Su^+ $recA^-$ cell (HF4738, Table 1). We hope this will relieve the polarity of certain gene-*G* amber mutations, without necessarily producing a functional gene-*G* product since the normal product is furnished by the plasmid. Preliminary experiments are encouraging. Our most recent transformant (derived from a host that is Su^+ but nonpermissive for *am*9) gives a plaque purity of >99% on infection with *am*9 and a burst size similar to that obtained with the usual suppressor HF4714.

We have synthesized DNA primers with a Gln → amber mutation in the third codon of gene *G*. With these we are attempting to produce site-specific mutant replicative-form DNA by primed enzymatic synthesis (Goulian and Kornberg 1967; Goulian et al. 1967; Donelson et al. 1975). Our attempts to produce a site-specific mutant phage from this site-altered DNA (first by transfection of spheroplasts prepared from C600 Su^-/p φXG, which we have shown is permissive for *am*9 DNA, and then by propagation of the mutant phage on a rec^-/p φXG host) will provide the first test of the approach we propose for studying the molecular mechanisms of mutation by carcinogens.

Note Added in Proof

Recent experiments have shown that all six gene-*G* activity transformants that we have isolated by cloning the *Taq*I restriction fragment, T1, are non-functional for the gene-*H* mutant *am*N1. Three of the six plasmids examined have deletions or rearrangements in the gene-*H* sequence. These results suggest that the gene-*H* product may be lethal.

ACKNOWLEDGMENTS

We are indebted to the following individuals for bacterial strains and for mutants of φX: Drs. K. Backman, M. Hayashi, P. Howard-Flanders, J. Hurwitz, C. Hutchison III, S. Palchaudhuri, M. Ptashne, R. Sinsheimer, and R. Warner. The $recA^-$ mutant DD1 was constructed by Mr. David Dorfman as part of a work-study program sponsored by Princeton University at New York University School of Medicine. Excellent technical assistance was provided by Mr. Morton C. Schneider.

This work was supported by grants from the National Institutes of Health (2 ROI CA 16316–03 BIO VO 2 [M77]) and the American Cancer Society (BC-252).

REFERENCES

Adams, M. H. 1959. *Bacteriophages* (see pp. 473–490). Inter-Science, New York.

Benbow, R. M., R. F. Mayol, J. C. Picchi, and R. L. Sinsheimer. 1972. Direction of translation and size of bacteriophage φX174 cistrons. *J. Virol.* **10:** 99.

Cohen, S. N., A. C. Y. Chang, and L. Hsu. 1972. Nonchromosomal antibiotic resistance in bacteria: Genetic transformation of *Escherichia coli* by R-factor DNA. *Proc. Natl. Acad. Sci.* **69:** 2110.

Denhardt, D. T. 1975. The single-stranded DNA phages. *CRC Crit. Rev. Microbiol.* **4:** 161.

Donelson, J. E., B. G. Barrell, H. L. Weith, H. Kossel, and H. Schott. 1975. The use of primed synthesis by DNA polymerase I to study an intercistronic sequence of φX174 DNA. *Eur. J. Biochem.* **58:** 383.

Goulian, M. and A. Kornberg. 1967. Enzymatic synthesis of DNA. XXII. Synthesis of circular replicative form of phage φX174 DNA. *Proc. Natl. Acad. Sci.* **58:** 1723.

Goulian, M., A. Kornberg, and R. L. Sinsheimer. 1967. Enzymatic synthesis of DNA. XXIV. Synthesis of infectious phage φX174 DNA. *Proc. Natl. Acad. Sci.* **58:** 2321.

Higuchi, R., G. V. Paddock, R. Wall, and W. Salser. 1976. A general method for cloning eukaryotic structural gene sequences. *Proc. Natl. Acad. Sci.* **73:** 3146.

Jackson, D. A., R. H. Symons, and P. Berg. 1972. Biochemical method for inserting new genetic information into DNA of simian virus 40: Circular SV40 DNA molecules containing lambda phage genes and the galactose operon of *Escherichia coli*. *Proc. Natl. Acad. Sci.* **69:** 2904.

Miller, J. A. and E. Miller. 1974. Some current thresholds of research in chemical carcinogenesis. In *Chemical carcinogenesis* (ed. P. Ts'o and J. A. DiPaolo), part A, p. 61. Marcel Dekker, New York.

Rodriguez, R. L., F. Bolivar, H. M. Goodman, H. W. Boyer, and M. Betlach. 1976. Construction and characterization of cloning vehicles. In *Molecular mechanisms in the control of gene expression* (ed. D. P. Nierlich et al.), vol. V, p. 471. Academic Press, New York.

Sanger, F., G. M. Air, B. G. Barrell, N. L. Brown, A. R. Coulson, J. C. Fiddes, C. A. Hutchison III, P. M. Slocombe, and M. Smith. 1977. Nucleotide sequence of bacteriophage φX174 DNA. *Nature* **265:** 687.

Sinsheimer, R. L. 1968. Bacteriophage φX174 and related viruses. *Prog. Nucleic Acid Res. Mol. Biol.* **8:** 115.

Vosberg, H. P. and H. Hoffmann-Berling. 1971. DNA synthesis in nucleotide-permeable *Escherichia coli* cells. I. Preparation and properties of ether-treated cells. *J. Mol. Biol.* **58:** 739.

Wensink, P. C., D. J. Finnegan, J. E. Donelson, and D. S. Hogness. 1974. A system for mapping DNA sequences in the chromosomes of *Drosophila melanogaster*. *Cell* **3:** 315.

Transcription, Morphogenesis, and Virion Structure

Transcription of Isometric Single-stranded DNA Phage

Frank K. Fujimura and Masaki Hayashi
Department of Biology
University of California, San Diego
La Jolla, California 92093

Although the mechanisms involved are complex and varied, transcription in prokaryotic systems is characterized by the synthesis of RNA molecules of unique size and sequence that are complementary to definite regions of the particular genome. This selectivity of transcription (Chamberlin 1974) is determined by the interactions of specific sequences on the DNA template with the various components of the transcriptional apparatus. Study of the transcriptional selectivity for a particular system can proceed along several different lines: (a) characterization of the size and sequence of individual transcription units; (b) mapping of transcription units with respect to the genome; (c) determination of the regions of the genome, such as promoters and terminators, which act as transcriptional signals; (d) identification and characterization of the elements of the transcriptional machinery required for selectivity; (e) elucidation of the nature of the interactions between the transcriptional machinery and the transcriptional template.

This review summarizes progress made along these lines in characterizing transcription by the isometric single-stranded (SS) DNA phages (IP). Almost all work relevant to this area has been carried out with the two very closely related phages, φX174 and S13. Although there are differences in the nucleotide sequences of these two phages, as shown by analysis with restriction enzymes (Hayashi and Hayashi 1974; Godson and Roberts 1976), stability of heteroduplexes (Godson 1973; Compton and Sinsheimer 1977), and differences in the gel mobilities of some of the gene products (Jeng et al. 1970; Godson 1973), the basic features of transcription of φX and S13 are virtually identical, and, unless specified otherwise, results pertaining to one of these phages will be considered to be applicable to the other. Whether these statements apply to the other IP is not clear due to lack of information. A number of excellent reviews pertaining to transcription in prokaryotic systems (Chamberlin 1974; Losick and Chamberlin 1976) and others containing sections on transcription of IP (Sinsheimer 1968; Denhardt 1975; Denhardt 1977) should be consulted for additional information and different points of view.

Background

For IP, the selectivity of transcription seems to be determined solely by the interaction between the phage DNA template and the preexisting host transcriptional machinery. All phage proteins, and presumably all phage transcription units, apparently are synthesized at the same relative rates throughout the infection period, indicating the absence of controlling mechanisms involving "early" or "late" classes of genes. No evidence exists for alteration or modification of the host transcriptional apparatus from infection by IP; in fact, host transcription is not shut off and seems to proceed normally after infection. The claim by Clements and Sinsheimer (1974) that the gene-*D* protein regulates positively the rate of viral transcription in φX-infected cells could not be confirmed by Pollock (1977).

Although the virion particle of IP contains SS DNA, almost all physiological processes, such as replication and transcription, involve an intracellular duplex DNA molecule called replicative-form (RF) DNA. Upon infection of sensitive *Escherichia coli* by IP, the viral SS DNA acts as the template for synthesis of the complementary strand to form parental RF DNA, which, after semiconservative replication, yields a number of progeny RF DNA molecules. By studying φX transcription after inactivation of the viral and/or complementary strand of DNA specifically labeled with bromodeoxyuridine, Truffaut and Sinsheimer (1974) concluded that both the parental and progeny RF DNA molecules are utilized as transcriptional templates.

It has been shown by hybridization (Hayashi et al. 1963) and by mutagenesis (Vanderbilt and Tessman 1970) that in vivo all or nearly all φX and S13 mRNA is transcribed from the complementary strand of RF DNA. This asymmetry in strand selection can be reproduced in vitro in a system containing purified *E. coli* RNA polymerase holoenzyme and φX RF DNA (Hayashi et al. 1964; Warnaar et al. 1969). The presence of a limited number of nicks in the RF template does not alter the asymmetry of transcription (Warnaar et al. 1969), but extensive degradation by sonication (Hayashi et al. 1964) or denaturation by heat (Warnaar et al. 1969) does lead to transcription of both strands of RF DNA.

Because only one of the strands of IP RF DNA is transcribed, the direction of transcription with respect to the genetic map can be deduced from the direction of translation. The direction of translation can be determined by measuring the polar effects of nonsense mutations on adjacent genes and by determining the sizes of polypeptide fragments synthesized in nonpermissive hosts after infection with phage having nonsense mutations at known genetic loci. Such experiments were performed for S13 by the Tessmans and their collaborators (Tessman et al. 1967; Vanderbilt et al. 1971, 1972) and for φX by Gelfand and Hayashi (1969), Hayashi and Hayashi (1971a), and Benbow et al. (1972). Figure 1 illustrates the direction of transcription and translation and the polarities of the viral and complementary strands of RF DNA with respect to the genetic map of φX and S13.

Attempts to fractionate in vivo φX mRNA by zone sedimentation through gradients of dimethylsulfoxide or sucrose yielded a broad distribution of sizes for the phage RNA with no apparent resolution of specific species (Warnaar et al. 1970; Sedat and Sinsheimer 1971; Clements and Sinsheimer 1975). Centrifugal resolution of specific species of in vivo IP RNA has not been very successful, but separation has been accomplished by gel electrophoresis (Hayashi and Hayashi 1971a; Clements and Sinsheimer 1975). Hayashi and Hayashi (1971a) fractionated in vivo φX RNA into ten species, which ranged in size from less than one-tenth genome to greater than one complete genome. These RNA species are found after either a short (30 sec) or long (10 min) period of labeling with [^{3}H]uridine. Furthermore, no apparent precursor-product relationship that can be detected by pulse-chase experiments exists among these species. The observations of Hayashi and Hayashi (1971a) that most φX RNA species are larger than any phage gene, that the sum of the sizes of the RNA species smaller than unit genome length is much greater than the size of the RNA corresponding to transcription of the entire genome, and that RNA species larger than unit genome length can be found led these investigators to conclude that most in vivo φX RNA species are polycistronic, that some phage cistrons must be encoded in more than one species of RNA, and that RNA synthesis can occasionally continue beyond one complete round of transcription of the RF DNA template. Clements and Sinsheimer (1975) reached similar conclusions from their studies, which resolved as many as 25 size classes of in vivo φX RNA. Puga and Tessman observed that most in vivo S13 mRNA is associated with

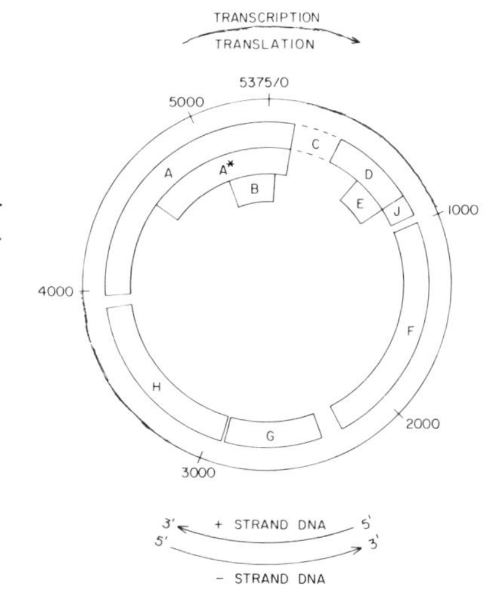

Figure 1 Direction of transcription. Arrows indicate the 5′ to 3′ direction of transcription, the N-terminal to C-terminal direction of translation, and the polarities of the viral (+) and complementary (−) strands of RF DNA. Gene locations with respect to nucleotide number are from Sanger et al. (1977). Numbering of nucleotides begins at the single *Pst*I cleavage site on φX RF DNA and proceeds in the direction of transcription around the circular genome.

polysomes (Puga and Tessman 1973a) and can be separated into at least eight species by zone sedimentation in sucrose gradients (Puga and Tessman 1973b).

Template for Isometric Phage Transcription

Two conformationally distinct forms of RF DNA can be isolated from cells infected with IP. RFI is superhelical and consists of two circular, self-continuous DNA strands. RFII is not superhelical and contains at least one discontinuity in its DNA strands. The conformational difference between RFI and RFII leads in vitro to a large difference in their template activities for transcription. Hayashi and Hayashi (1971b) found that the rate of RNA synthesis in vitro with φX RFI as template is about fivefold greater than that with RFII. This difference in template activity is observed with either "artificial" RFII, produced in vitro by random nicking of RFI with DNase I, or "natural" RFII isolated from infected cells, which probably consists mostly of relaxed molecules containing gaps in a specific region of the viral DNA strand. Although RFI and RFII differ significantly in their efficiency to serve as transcriptional templates, the transcripts synthesized in vitro from these two templates are not appreciably different when analyzed either by polyacrylamide gel electrophoresis or by translation in an in vitro coupled transcription-translation system.

The more efficient transcription of RFI over RFII by RNA polymerase in vitro begs the question as to the nature of the RF DNA template utilized for transcription of IP in vivo. It is known that intracellular RFII is the precursor for RF DNA replication and for viral SS DNA synthesis (see Dressler et al., this volume). It is possible that the involvement of RFII in replication precludes its use for transcription. That RFI is used as a template for transcription in vivo is shown by the fact that newly synthesized RNA can be isolated from φX-infected cells as a hybrid complex with RFI (Hayashi and Hayashi 1966). Furthermore, studies by Puga and Tessman (1973a,b) suggest very strongly that a majority of phage mRNA in S13-infected cells arises from transcription of RFI. These investigators (Puga and Tessman 1973b) found that low concentrations of nalidixic acid, although not affecting synthesis of *E. coli* mRNA, inhibit up to 80% of the synthesis of phage RNA. The inhibition is not due to a deficiency of RF DNA template caused by the drug or to alterations in the size or physical stability of S13 transcripts. These results led Puga and Tessman (1973b) to suggest that nalidixic acid is metabolized by sensitive bacteria into an intercalating agent that preferentially inhibits transcription of RFI. Although 80% of S13-specific transcription is inhibited by low concentrations of nalidixic acid, the remaining 20% seems relatively resistant to the drug and was postulated by Puga and Tessman (1973b) to arise through transcription of RFII. It has now been confirmed that nalidixic acid specifically inhibits formation of supercoiled DNA by inhibiting the action of a host enzyme, DNA gyrase (Gellert et al. 1977; Sugino et al. 1977).

Additional evidence that RFI is responsible for the majority of phage transcripts in vivo comes from studies of RNA synthesis in cells infected with IP gene-*A* mutants. The function of gene *A* is required for site-specific nicking of RFI, which is necessary for synthesis of progeny RF molecules (see Baas and Jansz, this volume). In the absence of gene-*A* function, phage DNA synthesis does not proceed beyond parental RF DNA, and the parental RF DNA formed under such conditions exists primarily as RFI. Puga and Tessman (1973a) found that the efficiency of phage transcription (measured as the ratio of the rate of RNA synthesis to the number of RF DNA molecules) in cells infected with S13 gene-*A* mutants (which presumably contain only RFI templates) is four- to fivefold greater than in cells infected with wild-type phage, which contain a mixture of RFI and RFII. This result agrees with that obtained in in vitro studies (Hayashi and Hayashi 1971b) and suggests that IP transcripts are copied in vivo primarily from RFI templates. It should be noted that results suggesting that transcription of parental RF DNA is more efficient than that of progeny RF DNA (Shleser et al. 1969) were apparently due to the deleterious effects of high multiplicities of infection (Puga and Tessman 1973a).

Whether intracellular RFI is actually superhelical is not known. Arguments (Denhardt 1975, 1977) suggesting that supercoiling should not be expected to occur in vivo because of its energy requirement should now be reconsidered in light of the recent discovery (Gellert et al. 1976) of an enzyme, DNA gyrase, purified from *E. coli* that in vitro is capable of introducing negative superhelical turns in closed, circular DNA. The energy for supercoiling evidently is provided by hydrolysis of ATP. The existence of DNA gyrase provides a possible mechanism by which superhelical, circular DNA molecules can be generated in vivo. Also, Ikeda et al. (1976) have shown that the purified gene-*A* protein of φX does not cleave relaxed, closed, circular RF DNA (called RFIV). Prior treatment of RFIV with DNA gyrase does allow the RF DNA to be nicked by the gene-*A* protein (Marians et al. 1977). This indicates that superhelical regions may exist in IP intracellular RFI, although the possibility that the conformational requirements for gene-*A*-protein activity in vivo are different from those in vitro cannot be ruled out.

Promoter Sites

Several approaches have been used to determine the locations of transcriptional initiation sites on φX and S13 RF DNA. The results of such experiments are consistent with the existence of three promoter sites located near the N(5′) termini of genes *A*, *B*, and *D*.

Vanderbilt et al. (1972) observed that a number of S13 nonsense mutations of genes *F* and *G* exert a polar effect on the expression of gene *H*. The polar effect of these mutations was measured both by genetic complementation and by direct analysis of protein synthesis by gel electrophoresis. These polar mutations did not affect expression of gene *A*. This observation led these investigators to conclude that a promoter site exists before gene *A* but not immediately before gene *G* or gene *H*. Recently, Pollock and Tessman (1976) found certain nonsense mutations in gene *B* that are strongly polar for synthesis of gene-*C* protein, whereas synthesis of gene-*D* protein in cells infected with these mutants remains relatively unaffected. This suggests the presence of a transcriptional initiation site between genes *C* and *D*.

Under appropriate conditions in vitro, RNA polymerase holoenzyme can bind to specific regions on native DNA and, by doing so, can either protect these regions of DNA from degradation by nucleases or allow duplex DNA fragments containing such regions to be trapped on nitrocellulose membrane filters. Many of these RNA-polymerase-binding sites have been shown to be within promoters. Chen et al. (1973) found three RNA-polymerase-binding sites on φX RF DNA. These three sites are in the regions common to the overlapping fragments of φX RF DNA *Hin*dII-4/*Hae*III-2, *Hin*dII-6/*Hae*III-3, and *Hin*dII-2/*Hae*III-1 and correspond to regions near the beginning of gene *A*, gene *D*, and gene *H*, respectively.

Hayashi et al. (1976) studied a number of in vivo φX RNA species by hybridization of isolated RNA species to restriction fragments of RF DNA. The resulting map (Fig. 2) shows two classes of RNA molecules: those with their 5′ termini in the *Hin*dII-8 fragment located near the beginning of gene *B* and those with their 5′ termini in the *Hap*II-1 *Hpa*I-2 fragment located near the beginning of gene *D*. The coincidence of the 5′ termini suggests that each of these two classes of in vivo RNA molecules is initiated at a common promoter site. With the possible exception of RNA species I, whose ends could not be determined by the methods used, there is no indication of in vivo RNA molecules with 5′ termini located at the start of gene *A*. As will be discussed below, such RNA species do seem to exist in φX-infected cells, but they are rapidly degraded and cannot be isolated as discrete species by gel electrophoresis. Consequently, they could not be mapped, as could the relatively stable φX RNA species shown in Figure 2.

The most convincing evidence concerning the promoter sites on φX RF DNA comes from the in vitro experiments of Smith and Sinsheimer 1976a,b,c) and of Axelrod (1976a,b). Although specific details are different, the basic approach taken by these two groups was to form initiation complexes of φX RFI and RNA polymerase holoenzyme, allow initiation of transcription to occur synchronously while preventing reinitiation, pulse with radioactive nucleoside triphosphate for short periods of time immediately after initiation, and chase for long periods of time to allow natural termination of transcription. These procedures result in synthesis of transcription units that are specifically labeled in their promoter-proximal (5′-

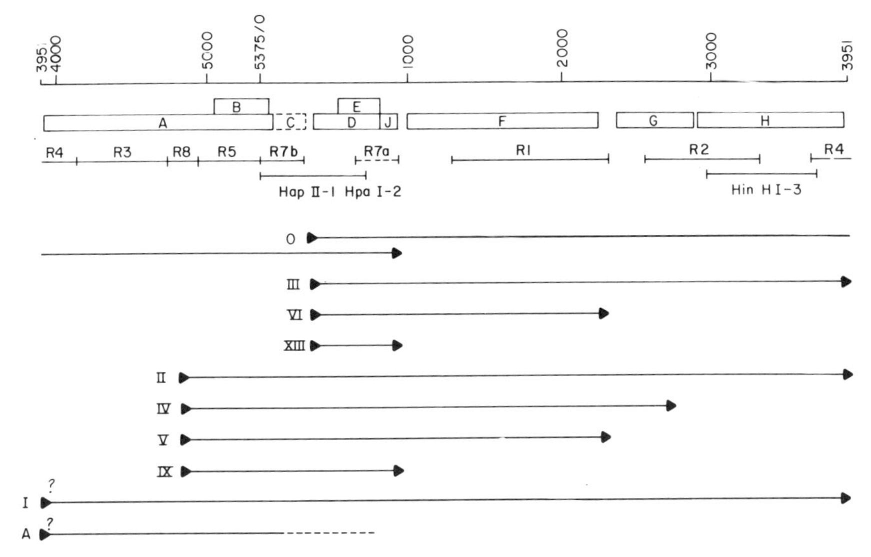

Figure 2 Map of in vivo φX RNA species. The circular genetic map of φX shown in Fig. 1 is represented linearly. Nucleotide numbering is according to Sanger et al. (1977). *Hin*dII fragments of RF DNA are represented by R. *Hap*II-1 *Hpa*I-2 indicates the second largest fragment obtained from digestion of the *Hap*II-1 fragment with the restriction enzyme *Hpa*I. The mapping of RNA species is based on their estimated molecular weights and their hybridization to the restriction fragments shown (Hayashi et al. 1976; M. N. Hayashi and M. Hayashi, in prep.). The map locations of the unit-genome-size RNA species I and the "unstable" gene-*A* mRNA have not been determined (see text). The estimated molecular weights ($\times 10^{-6}$) of these in vivo RNA species are as follows: O, 2.3; I, 1.8; II, 1.5; III, 1.3; IV, 1.1; V, 0.98; VI, 0.82; IX, 0.60; XIII, 0.23.

terminal) regions, and hybridization of these transcription units with restriction fragments of RF DNA should reveal the locations of promoter sites. The results of such experiments show the presence of three regions in φX RFI at which initiation of transcription takes place in vitro. These three regions are found near the beginning of gene *A* in the overlap of the *Hin*dII-4/*Hae*III-2/*Hpa*I+II-1 fragments, near the beginning of gene *B* in the overlap of the *Hin*dII-8/*Hae*III-3/*Hpa*I+II-1 fragments, and near the beginning of gene *D* in the overlap of the *Hin*dII-6/*Hae*III-3/*Hpa*I+II-3 fragments. Transcripts initiated at the two former sites start with pppA and those initiated at the third site begin with pppG. Axelrod called these three initiation sites $P_{A'}$, P_A, and P_G, respectively, and we will use her nomenclature.

In addition to locating the initiation sites on the φX genome, Smith and Sinsheimer (1976c) determined the nucleotide sequences of the 5′ termini of the in vitro transcripts initiated at these sites. The terminal sequences are as follows:

$P_{A'}$ pppAp(Ap)$_{2-3}$UpCpUpUpG
P_A pppApUpCpG
P_G pppGpApU and pppCpG

These four 5′-terminal sequences are found in a majority of the in vitro transcripts. Up to 80% of the RNA molecules synthesized with fresh preparations of RNA polymerase have these 5′-terminal sequences. Older preparations of RNA polymerase seem to yield relatively larger amounts of minor 5′-terminal sequences in the vitro φX transcripts.

The finding by Smith and Sinsheimer (1976c) that initiation at the P_G promoter can occur with either pppG or pppC deserves comment. Smith et al. (1974) and Grohmann et al. (1975) observed that a significant amount of 5′-pppC is found in bulk RNA synthesized in vitro with φX RFI and *E. coli* RNA polymerase. However, unpublished data from this group (cited by Grohmann et al. 1975 and Smith and Sinsheimer 1976c) indicate that only pppA and pppG are found at the 5′ termini of in vivo φX transcripts. Hayashi and Hayashi (1972) were able to show the occurrence of 5′-pppA and 5′-pppG, but they could neither detect nor rule out the presence of 5′-terminal pyrimidine triphosphates in in vivo φX RNA. It is likely that initiation at the P_G promoter with pppC is an in vitro phenomenon and initiation at this promoter in vivo probably occurs only with pppG. Axelrod (1976a) noted that the P_G promoter is less rigid in its requirements for initiation in vitro than the $P_{A'}$ or P_A promoter. By using selected dinucleotides in the presence of a low concentration of nucleoside triphosphate to allow initiation to occur at specific promoters (Downey et al. 1971), Axelrod observed that only certain oligonucleotides stimulate initiation at $P_{A'}$ and P_A, whereas a large number do so at P_G.

Sanger et al. (1977) have determined the entire genome sequence of φX, and with the information from Smith and Sinsheimer (1976b,c) and Axelrod 1976a,b) they were able to locate the exact positions of $P_{A'}$, P_A, and P_G. The locations of these promoters and some other pertinent information concerning φX promoter sites are given in Figure 3. No striking features can be observed among the nucleotide sequences of these promoters (Fig. 4). Pribnow (1975) noticed that a seven-base sequence approximately homologous to the sequence TATPuATG seems to precede the initiating nucleotide in promoters. Sequences showing varying degrees of homology with the "Pribnow box" can be found in the three φX promoter sites (Sanger et al. 1977; Fig. 4). RNA molecules initiated at these promoters have different lengths of untranslated leader sequences between their 5′ termini and the translation initiation codon of their proximal genes. The length of the leader sequence from $P_{A'}$ to its proximal gene, *A*, is 18 nucleotides; that from P_A to its proximal gene, *B*, is 176 nucleotides; that from P_G to its proximal gene, *D*, is 32 nucleotides.

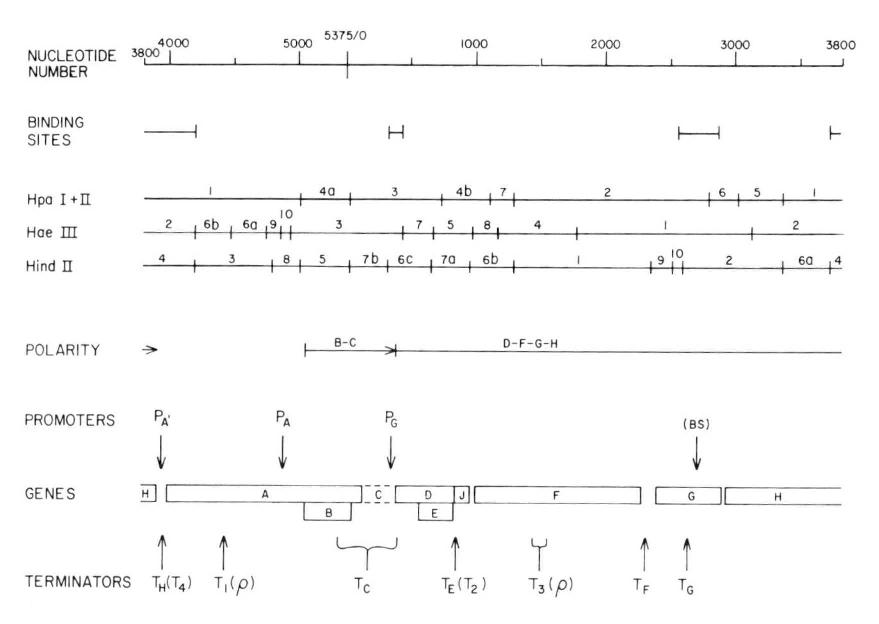

Figure 3 Genetic locations of transcriptional promoters and terminators on φX and S13 DNA. Nucleotide numbering and locations of gene boundaries are from Sanger et al. (1977) and are as shown in Fig. 1. The *Hpa*I+II, *Hae*III, and *Hin*dII restriction maps of φX RF DNA are as determined by Lee and Sinsheimer (1974) and have been aligned with respect to the φX genes as determined by Sanger et al. (1977). The three regions in φX RF DNA that bind RNA polymerase in vitro have been determined by Chen et al. (1973). Polarity from gene *B* to gene *C* was observed for S13 by Pollock and Tessman (1976). Polarity among genes *F*, *G*, and *H* of S13 was shown by Tessman et al. (1967) and by Vanderbilt et al. (1972); that among genes *D*, *F*, *G*, and *H* of φX was reported by Gelfand and Hayashi (1969), Hayashi and Hayashi (1971a), and Benbow et al. (1972). The three promoter sites, P_A , P_A, and P_G, are functional for initiation of transcripts both in vivo (Hayashi et al. 1976) and in vitro (Axelrod 1976a,b; Smith and Sinsheimer 1976a,b). Transcripts initiated in vivo at the P_A promoter are relatively unstable and are degraded rapidly. One RNA-polymerase-binding site (BS) is inactive for synthesis of RNA molecules. Transcription termination can occur at several sites. Four sites, $T_E(T_2)$, T_F, T_G, and $T_H(T_4)$, apparently are functional in vivo (Hayashi et al. 1976). Two of these sites, $T_E(T_2)$ and $T_H(T_4)$, are also functional in vitro (Axelrod 1976a,b). Three termination sites, T_1, T_C, and T_3, have been observed only in vitro (Axelrod 1976a,b; Smith and Sinsheimer 1976a,b). The termination factor *rho* is absolutely required for termination at T_1 and T_3. No evidence indicates that *rho* factor is necessary for or affects the efficiency of termination at any sites other than T_1 and T_3. Transcription termination at any termination site is not completely efficient either in vivo or in vitro. The most efficient terminator is $T_H(T_4)$; the other terminators are relatively inefficient.

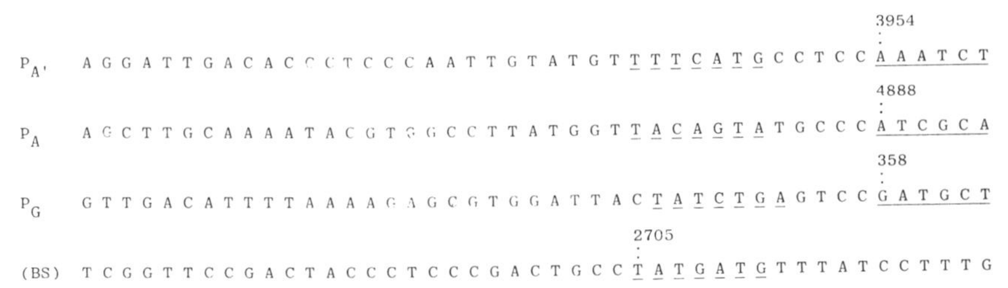

Figure 4 Nucleotide sequences of φX promoter regions. Nucleotide sequences and numbering are from Sanger et al. (1977). BS represents the possible sequence of the RNA-polymerase-binding site (Chen et al. 1973), which apparently is inactive for synthesis of RNA. Dashed underlining indicates sequences possibly corresponding to the "Pribnow box" (TATPuATG; Pribnow 1975). Solid underlining indicates sequences of the 5′ termini of transcripts initiated at $P_{A'}$, P_A, and P_G (Smith and Sinsheimer 1976c).

There is no evidence to indicate that initiation of transcription occurs either in vivo or in vitro at the RNA-polymerase-binding site in the *Hin*dII-2/*Hae*III-1 fragments observed by Chen et al. (1973). In fact, the existence of polarity among genes *F*, *G*, and *H* argues against the presence of a promoter site in this region. Sanger et al. (1977) noticed that the sequence TATGATG, which matches the "Pribnow box" sequence perfectly, is found in the overlapping region of the *Hin*dII-2/*Hae*III-1 fragments (Fig. 4). This may very well correspond to the polymerase-binding site seen by Chen et al. (1973). If so, it is interesting that synthesis of RNA chains apparently cannot be initiated at this site, even though a seemingly stable polymerase-DNA binary complex is formed.

The binding experiments of Chen et al. (1973) did not detect the P_A promoter site. Axelrod (1976b) has claimed that initiation in vitro at the P_A promoter is sensitive to variations in in vitro conditions, and she has suggested that this variability may be the reason for the failure of Chen and coworkers to detect this promoter site. On the other hand, the methods used by Chen et al. (1973) to measure binding may not have been sensitive enough to detect sites located within small restriction fragments or near restriction sites. Note that the P_A promoter is located within the *Hin*dII-8/*Hae*III-10 fragments (Sanger et al. 1977; Fig. 3). The *Hae*III-10 fragment was not used by Chen and coworkers in their studies, so it is possible that their failure to observe polymerase-binding at this site was due to technical limitations in the procedures used and not to any peculiar polymerase-binding properties of the P_A promoter.

Termination Sites

The map of in vivo RNA species (Fig. 2) suggests that there are four possible regions in the φX genome at which termination of transcription can occur.

These regions are located near the ends of genes $D(E), F, G$, and H, and they will be referred to as T_E, T_F, T_G, and T_H, respectively. The results of Axelrod's (1976a,b) in vitro studies suggest the presence of four possible termination sites on φX RF DNA. Two of these sites (T_1 and T_3) function only in the presence of the termination factor *rho*. Termination in vitro at the other two sites is independent of *rho* factor. Axelrod's two *rho*-independent in vitro termination sites, T_2 and T_4, coincide with the in vivo termination sites T_E and T_H, respectively. Smith and Sinsheimer (1976b) observed only one in vitro termination site in their experiments. This site was located near the beginning of gene C (T_C) and functioned independently of *rho* factor. Figure 3 shows the locations of these presumptive transcription termination sites.

Aside from the correspondence of the in vitro terminators T_2 and T_4 with the in vivo terminators T_E and T_H, there is ambiguity as to the locations of the φX transcription termination sites. Termination at the presumptive in vivo sites T_F and T_G has not been observed in vitro. Conversely, the *rho*-dependent in vitro termination sites T_1 and T_3 and the *rho*-independent termination site T_C do not seem to be utilized in vivo. One factor contributing to the ambiguity concerning transcription termination is that termination events are not completely efficient and the efficiency of termination at any site in vitro may vary depending on the conditions used for transcription. In the experiments of Smith and Sinsheimer (1976b), termination in vitro at T_C was quite inefficient, and most transcripts read through this termination site on their first encounter with it. Curiously, termination seemed to be relatively efficient at the second encounter of a transcript with the termination site. On the other hand, Axelrod (1976a) observed that although termination in vitro at T_2 was relatively inefficient, that at T_4 was very efficient, and few, if any, transcripts were able to read through this termination site.

From the relative numbers of molecules of the φX RNA species that were mapped, Hayashi et al. (1976) calculated the possible efficiencies of termination at the presumptive in vivo termination sites (Table 1). Although the values given in Table 1 cannot be taken too rigorously, they do suggest that termination is more efficient at T_H than at any other presumptive in vivo termination site. The efficient termination of transcription in vivo at T_H agrees with the efficiency of termination in vitro at T_4 observed by Axelrod (1976a). Sanger et al. (1977) located a sequence in the φX genome that

Table 1 Termination efficiencies at in vivo termination sites

Initiation site	Efficiency of termination			
	T_E	T_F	T_G	T_H
P_A	0.53	0.36	0.52	(1.00)
P_G	0.54	0.67	—	0.88

could correspond to the $T_H(T_4)$ termination site. This sequence consists of a GC-rich region followed by T_6A and is similar to known termination sequences recognized by *E. coli* RNA polymerase (Gilbert 1976). The surprising fact is that the T_6A sequence begins 13 nucleotides downstream from the initiating base for transcription in the $P_{A'}$ promoter. Sanger et al. (1977) speculated that this putative termination sequence may act as an "attenuator" of transcription. However, because attenuation is not consistent with the observation that efficient termination apparently occurs in this region, they suggested the alternative (as has been done for the filamentous phage fd [Sugimoto et al. 1977; see Schaller et al., this volume]) that effective termination requires the nascent RNA chain to form a hydrogen-bonded hairpin structure before the termination site (Table 2). Transcripts initiated at the $P_{A'}$ promoter cannot form such a structure upon their initial encounter with the T_6A sequence. RNA chains being synthesized upstream from and proceeding through the $P_{A'}$ promoter can form the hairpin structure and therefore are terminated at the T_6A sequence.

Recently, McMahon and Tinoco (1978) have proposed a possible method for analyzing nucleotide sequences to locate transcription termination signals. Their method is based on the hypothesis of Gilbert (1976) that termination of transcription occurs by dissociation of the transcriptional complex at sites on the DNA template at which RNA polymerase molecules pause (Maizels 1973) during the elongation process. Gilbert (1976) suggested that the probability of termination at any particular site is determined by duplex DNA stability at that site. McMahon and Tinoco (1978) analyzed known termination sequences in terms of duplex stability arising from nearest-neighbor interactions and found that these sequences exhibited a characteristic pattern of nearest-neighbor stabilities. They then looked for regions having this pattern of nearest-neighbor interactions in the Sanger et al. (1977) sequence of φX DNA and found only five such sites. Interestingly, these five theoretically possible termination sequences seem to coincide with the termination sites T_1, $T_E(T_2)$, T_F, T_G, and $T_H(T_4)$ suggested by in vivo and in vitro results (see Fig. 3). One of these sequences is located approximately half way between $P_{A'}$ and P_A and could correspond to T_1. Another is located at the end of gene *D*(*E*) within the first 20 nucleotides encoding gene *J* and could correspond to $T_E(T_2)$. The third sequence lies within the intercistronic region between genes *F* and *G* (Fiddes 1976; Sanger et al. 1977) and could correspond to T_F. The fourth sequence is located near the middle of gene *G* and may correspond to T_G. The last sequence is the same as that postulated by Sanger et al. (1977) to be the $T_H(T_4)$ termination site. The five sequences proposed by McMahon and Tinoco (1978) to be termination signals and the method they used to deduce them are summarized in Table 2.

Although *rho* factor clearly affects the size of in vitro φX RNA, its role in phage transcription in vivo is not certain. As mentioned earlier, the two *rho*-dependent in vitro terminators observed by Axelrod (1976a) appar-

Table 2 Possible nucleotide sequences of φX transcription termination sites

Terminator	Position	Sequence and stability terms
T_1	4429	T A C | C C | C | A A A A A G A A | A G G T
		5 4 1 | 1 | 1 4 | 6 6 6 6 4 4 6 | 6 4 1 4
$T_E(T_2)$	854	A A A | G G | T | A A A A A A C G | T T C T
		6 6 4 | 1 | 4 5 | 6 6 6 6 6 4 3 | 4 6 4 4
T_F	2308	A G C | G G | T | A A A A A T T T | T A A T
		4 2 3 | 1 | 4 5 | 6 6 6 6 5 6 6 | 6 5 6 5
T_G	2635	C G A | C C | C | T A A A T T T T | T T C C
		3 4 4 | 1 | 1 4 | 5 6 6 5 6 6 6 | 6 6 4 1
$T_H(T_4)$	3961	G G A | G G | C | T T T T T T A T | G G T T
		1 4 4 | 1 | 2 4 | 6 6 6 6 6 5 5 | 4 1 4 6

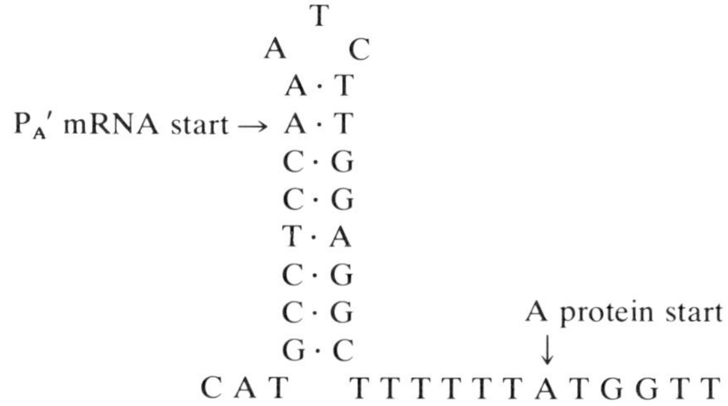

```
                       T
                     A   C
                      A · T
P_A' mRNA start →     A · T
                      C · G
                      C · G
                      T · A
                      C · G
                      C · G             A protein start
                      G · C                    ↓
                 C A T    T T T T T T A T G G T T
```

Nucleotide sequences and numbering (position) and lowermost sequence showing possible intrastrand secondary structure in the region of the $T_H(T_1)$ terminator are from Sanger et al. (1977). Stability terms (from McMahon and Tinoco 1978) give the relative stability of duplex DNA due to each pair of nearest-neighbor sequences. The relative stabilities of duplex nearest-neighbor sequences are in the order GG(CC)>GC>CG>AG(CT)~TG(CA)~AC(GT)~TC(GA)>AT~TA>AA(TT) and are numbered from 1 to 6. McMahon and Tinoco (1978) observed that termination sequences are characterized by a most-stable nearest-neighbor (1) followed by two variable terms and seven weak nearest-neighbors; of the seven weak nearest-neighbors, at least five of the seven stability terms have a value of 6. This pattern of nearest-neighbor stabilities was observed by McMahon and Tinoco (1978) to occur in the sequence of φX DNA only at the five positions shown above.

ently are not utilized for production of the in vivo RNA species. It should be noted that nine in vivo RNA species have been mapped (see Fig. 2). Although these nine species comprise about 90% of the φX RNA molecules that can be resolved as discrete bands by gel electrophoresis, the gel profiles also indicate that in addition to the RNA species that can be resolved as discrete bands, there exists a variable amount of phage-specific RNA of heterogeneous sizes. Also, Clements and Sinsheimer (1975) were able to resolve at least 18, and possibly as many as 25, discrete species of in vivo RNA by their electrophoretic techniques. Thus, the involvement of *rho* (or some other factor)-dependent termination sites and of posttranscriptional processing mechanisms in the generation of some in vivo φX RNA species cannot be ignored.

Hayashi et al. (1971) observed that transcription of φX RFI in vitro in the absence of *rho* factor led to RNA species equal to and greater than genome length. The inclusion of *rho* factor resulted in several RNA species smaller than unit genome size. Note that the results of Hayashi et al. (1971) are similar to those of Smith and Sinsheimer (1976a) in that effective termination did not seem to occur in vitro in the absence of *rho* factor. The synthesis of RNA species of discrete sizes smaller than unit genome length in the presence of *rho* factor was probably due to termination at the *rho*-dependent sites identified by Axelrod (1976a). When the products of translation of transcripts synthesized in vitro either in the presence or absence of *rho* factor were compared, Hayashi et al. (1971) observed that synthesis of the larger φX proteins (F, A, H) was severely depressed, and synthesis of low-molecular-weight material was stimulated with the *rho*-dependent RNA.

Kapitza et al. (1976) also observed synthesis of RNA molecules equal to and greater than unit genome size in an in vitro system containing φX RFI and *E. coli* RNA polymerase. However, the size distribution of RNA synthesized in the presence of both *rho* factor and RNase III, when analyzed by gel electrophoresis, was found to be similar to that of in vivo RNA. The RNA species synthesized in vitro under these conditions were not identified, but Kapitza et al. (1976) suggested that RNase III plays a role in generating in vivo φX RNA species. This suggestion, along with the known function of RNase III in processing bacteriophage T7 early mRNA (Dunn and Studier 1973), led M. N. Hayashi and M. Hayashi (in prep.) to investigate the possible involvement of RNase III in processing of in vivo φX RNA. A φX-sensitive *E. coli* strain deficient in both RNase I and RNase III was constructed, and the phage-specific RNA synthesized in this host was analyzed by gel electrophoresis. Gel electropherograms showed that these φX RNA species were identical to those synthesized in hosts having these two nucleases.

RNA Stability

Linney and Hayashi (1974) noticed that upon treatment of φX-infected cells with rifampicin, the rates of synthesis of the two gene-*A* proteins, A and A* (Linney et al. 1972), decreased more rapidly than the rates of synthesis of the other phage proteins. They interpreted this to mean that functional gene-*A* mRNA is more labile than that of other phage genes. Pollock and Tessman (1976) also observed that the functional stability of gene-*A* mRNA was relatively low (see Table 3). Hayashi et al. (1976) demonstrated by a pulse and pulse-chase experiment that a considerable amount of gene-*A*-specific RNA, as judged by hybridization to the *Hin*dII-3 and *Hin*dII-4 fragments of RF DNA, is present in bulk RNA isolated from φX-infected cells, and that this gene-*A*-specific RNA is degraded rapidly. This structural instability prevents the isolation of RNA species of discrete size encoding gene-*A* sequences. The two exceptions are RNA species O and I which do contain gene-*A*-specific sequences, but it has been shown (Hayashi and Hayashi 1971a; Hayashi et al. 1976) that cells infected with a polar mutant of gene *F* do not synthesize the larger RNA species O, I, II, and III even though they do synthesize normal amounts of A and A* proteins. This led to the suggestions that all functional gene-*A* mRNA is structurally unstable and that the gene-*A* sequences encoded in the structurally "stable" RNA

Table 3 Functional and structural stabilities of messages

mRNA for gene	Half-life (min)			
	S13[a] functional	φX174[b] functional	φX174[b] structural[c]	φX174[b] ratio[d]
A				
A	0.65	1.4	(1.1)	—
A*	2.5	2.7	—	—
B	6.0	6.5	10.1	0.64
C	6.1	—	—	—
D	4.1	5.1	8.8	0.58
F	3.7	4.0	6.8	0.59
G	3.9	4.2	7.5	0.56
H	3.7	4.3	7.5	0.57

[a] Data from Pollock and Tessman (1976). Half-lives for A and A* are for φX.

[b] Data from M. N. Hayashi and M. Hayashi (in prep.).

[c] Structural half-lives were determined as described in the text from the following information.

RNA species (relative no. of molecules)		Genes encoded	Structural half-life (min)
I	(1.0)	*B, C, D, F, G, H*	10.5
II	(2.2)	*B, C, D, F, G, H*	5.5
III	(2.3)	*D, F, G, H*	7.5
IV	(3.1)	*B, C, D, F, (G)*	7.0
V	(3.0)	*B, C, D, F*	—
VI	(5.2)	*D, F*	5.0
IX	(9.5)	*B, C, D*	13.5
XIII	(9.0)	*D*	8.5

The half-life shown in parentheses for A represents the structural half-life of bulk, infected-cell RNA hybridizing to the RF DNA fragment *Hin*dII-4.

[d] The ratio of functional to structural half-life.

species O and I may not be utilized for translation in vivo (Hayashi et al. 1976).

Synthesis of unstable gene-*A* mRNA is most probably initiated at the $P_{A'}$ promoter. Sanger et al. (1977) postulated that all transcripts initiated at $P_{A'}$ are structurally unstable and that the stable, but apparently nonfunctional, gene-*A* messages result from readthrough at the T_H terminator of transcripts initiated at either P_A or P_G. A possible exception to this hypothesis is RNA species I. Note that the locations of the termini of this RNA molecule have not been determined, but its size and hybridization behavior indicate that it is the result of one complete round of transcription of RF DNA. Given the locations of the promoters and terminators that have been identified, RNA species I should be the product of transcription initiated at $P_{A'}$ and terminated at T_H. If so, why is this particular $P_{A'}$-initiated transcript stable? The answer is unknown. Because T_H is 20 nucleotides downstream from $P_{A'}$ (Sanger et al. 1977 and see above), redundant terminal sequences may perhaps relate to the stability of such an RNA molecule. Alternatively, the ends of RNA species I many not correspond to initiation at $P_{A'}$ and termination at T_H.

Pollock and Tessman (1976) deduced that functional inactivation of gene-*A* message occurs frequently prior to the completion of transcription of that gene by RNA polymerase. Whether this functional inactivation is synonymous with physical decay of gene-*A* mRNA has not been determined. It is known that functional mRNA for synthesis of A protein is less stable than that for synthesis of A* protein (Linney and Hayashi 1974; Pollock and Tessman 1976). This could mean that utilization of transcripts initiated at the $P_{A'}$ promoter for synthesis of either A or A* protein is determined by sequential message degradation from the 5′ terminus. However, other possibilities do exist, and the reason for the functional and structural instability of gene-*A* mRNA, as well as for the relationship, if any, between message instability and the special properties of gene *A* (see Tessman and Tessman, this volume), remains unsolved.

Puga et al. (1973) showed that the structural stability of S13 RNA, as measured by hybridization of pulsed-chased bulk RNA from infected cells to RF DNA, is greater than the functional stability of messages for the B, D, and F proteins. The functional stability of message was measured by the quantity of radioactivity in the peaks of gel electropherograms of proteins isolated from infected cell cultures that had been pulse-labeled at various times after treatment with rifampicin. The half-life of approximately 10 minutes observed for structural decay of bulk phage message is considerably longer than the half-lives observed for functional decay of messages for phage proteins. Puga et al. (1973) observed that the functional message decay curves did not follow first-order kinetics, but evidently this was due to the unsuitability of the culture medium that was used (Pollock and Tessman 1976). In a later, more comprehensive set of experiments by Pollock and

Tessman (1976), first-order functional decay curves were obtained. The functional half-lives measured by these investigators for individual genes of S13 are shown in Table 3, and these confirm earlier results of Puga et al. (1973) which showed that functional mRNA for the different genes of S13 are not all the same and that the functional stabilities are lower than the structural stability of bulk, hybridizable phage RNA. In addition to the relatively low stability of functional gene-*A* message, Pollock and Tessman (1976) noticed that the functional messages for the other phage genes can be grouped into three classes. One class (genes *B* and *C*) has a relatively long functional half-life of 6.2 minutes; another class (genes *F*, *G*, and *H*) has a relatively short half-life of 3.8 minutes; the third class (gene *D*) has a half-life of 4.4 minutes. Functional half-lives for φX mRNA measured in our laboratory (M. N. Hayashi and M. Hayashi, in prep.) are also shown in Table 3. Although these values are not identical to those of Pollock and Tessman (1976), they do confirm the existence of separate classes of messages with different functional stabilities.

The differences between the functional and structural stabilities of IP messages raise the question whether the RNA species characterized by gel electrophoresis and hybridization actually represent functional RNA molecules. It seems likely that bulk, hybridizable RNA cannot consist entirely of functional messages; in fact, it is probable that a significant portion of such RNA is not functional. Furthermore, Hayashi and Hayashi (1971a) reported that the various φX RNA species seem to be structurally stable for at least 10 minutes. Recently, M. N. Hayashi and M. Hayashi (in prep.) measured the structural stabilities of the φX RNA species I, II, III, IV, V, VI, IX, and XIII. With the exception of RNA species V, the structural decay curves for these RNA species follow first-order kinetics, but the structural half-lives for these RNAs (Table 3) are not identical, ranging from 5 minutes (RNA species VI) up to 13.5 minutes (RNA species IX). These structural half-lives indicate that the φX RNA species are not as stable as believed previously (Hayashi and Hayashi 1971a) and that their structural stabilities are not adequately represented by the bulk stability of hybridizable RNA. Because the relative numbers of molecules of these RNA species are known (Hayashi et al. 1976) and because the genes encoded by these RNA species can be deduced (Fig. 2), structural decay curves for individual genes contained in these RNA species can be constructed. Curves constructed in this way have every appearance of being first-order decay curves and give the values shown in Table 3 for structural half-lives of messages for individual genes of φX. It should be noted that RNA species O, VII, VIII, X, XI, and XII were not included in these calculations, but since they comprise less than 10% of the molecules of the discrete RNA species, their omission should not affect significantly the structural half-life values given in Table 3. The half-lives for structural decay of messages for genes *B*, *C*, *D*, *F*, *G*, and *H* can be grouped into the same three classes observed by Pollock and Tessman

(1976) for the functional half-lives of messages for these phage genes. Although the functional half-lives are shorter than the corresponding structural half-lives, the ratios of functional to structural half-lives for these messages are virtually identical (Table 3). This suggests that the sequences encoding these genes found in RNA species I, II, III, IV, V, VI, IX, and XIII are functional and that they may comprise a majority of the functional sequences for these genes. This does not mean that all molecules of any particular RNA species are functional. The fact that functional decay rates are greater than structural decay rates supports the hypothesis proposed by Puga et al. (1973) that the functional inactivation event does not in itself greatly alter the structural integrity of the message. Pollock and Tessman (1976) speculated that messages for contiguous genes having identical functional decay rates (*B-C* and *F-G-H*) may have a common site (which could be comprised of sequences separated by many nucleotides) at which the functional inactivation event occurs.

ACKNOWLEDGMENTS

We gratefully acknowledge stimulating discussions with Marie N. Hayashi during the preparation of this article. Our own unpublished work presented here was supported by grants from the National Science Foundation (BM 74 10710) and the National Institute of General Medical Sciences, U.S. Public Health Service (GM 12934).

REFERENCES

Axelrod, N. 1976a. Transcription of bacteriophage φX174 *in vitro*: Selective initiation with oligonucleotides. *J. Mol. Biol.* **108:** 753.

———. 1976b. Transcription of bacteriophage φX174 *in vitro*: Analysis with restriction enzymes. *J. Mol. Biol.* **108:** 771.

Benbow, R. M., R. F. Mayol, J. C. Picchi, and R. L. Sinsheimer. 1972. Direction of translation and size of bacteriophage φX174 cistrons. *J. Virol.* **10:** 99.

Chamberlin, M. J. 1974. The selectivity of transcription. *Annu. Rev. Biochem.* **43:** 721.

Chen, C.-Y., C. A. Hutchison III, and M. H. Edgell. 1973. Isolation and genetic localization of three φX174 promoter regions. *Nat. New Biol.* **243:** 233.

Clements, J. B. and R. L. Sinsheimer. 1974. Class of φX174 mutants relatively deficient in the synthesis of viral DNA. *J. Virol.* **14:** 1630.

———. 1975. Process of infection with bacteriophage φX174. XXXVII. RNA in φX174-infected cells. *J. Virol.* **15:** 151.

Compton, J. L. and R. L. Sinsheimer. 1977. Aligning the φX174 genetic map and the φX174/S13 heteroduplex denaturation map. *J. Mol. Biol.* **109:** 217.

Denhardt, D. T. 1975. The single-stranded DNA phages. *CRC Crit. Rev. Microbiol.* **4:** 161.

———. 1977. The isometric single-stranded DNA phage. In *Comprehensive virology* (ed. H. Fraenkel-Conrat and R. Wagner), vol. 7, p. 1. Plenum Press, New York.

Downey, K. M., B. S. Jurmark, and A. G. So. 1971. Determination of nucleotide sequences at promoter regions by the use of dinucleotides. *Biochemistry* **10:** 4970.

Dunn, J. J. and F. W. Studier. 1973. T7 early RNAs are generated by site-specific cleavages. *Proc. Natl. Acad. Sci.* **70:** 1559.

Fiddes, J. C. 1976. Nucleotide sequence of the intercistronic region between genes *G* and *F* in bacteriophage φX174 DNA. *J. Mol. Biol.* **107:** 1.

Gelfand, D. H. and M. Hayashi. 1969. Electrophoretic characterization of φX174-specific proteins. *J. Mol. Biol.* **44:** 501.

Gellert, M., K. Mizuuchi, M. O'Dea, and H. A. Nash. 1976. DNA gyrase: An enzyme that introduces superhelical turns into DNA. *Proc. Natl. Acad. Sci.* **73:** 3872.

Gellert, M., K. Mizuuchi, M. H. O'Dea, T. Itoh, and J.-I. Tomizawa. 1977. Nalidixic acid resistance: A second genetic character involved in DNA gyrase activity. *Proc. Natl. Acad. Sci.* **74:** 4772.

Gilbert, W. 1976. Starting and stopping sequences for the RNA polymerase. In *RNA polymerase* (ed. R. Losick and M. Chamberlin), p. 193. Cold Spring Harbor Laboratory, Cold Spring Harbor, New York.

Godson, G. N. 1973. DNA heteroduplex analysis of the relationship between bacteriophages φX174 and S13. *J. Mol. Biol.* **77:** 467.

Godson, G. N. and R. J. Roberts. 1976. A catalogue of cleavages of φX174, S13, G4, and St-1 by 26 different restriction enzymes. *Virology* **73:** 561.

Grohmann, K., L. H. Smith, and R. L. Sinsheimer. 1975. A new method for isolation and determination of 5′-terminal regions of bacteriophage φX174 *in vitro* mRNAs. *Biochemistry* **14:** 1951.

Hayashi, M., F. K. Fujimura, and M. N. Hayashi. 1976. Mapping of bacteriophage φX174 *in vivo* messenger RNAs. *Proc. Natl. Acad. Sci.* **73:** 3519.

Hayashi, M., M. N. Hayashi, and Y. Hayashi. 1971. Size distribution of φX174 RNA synthesized *in vitro. Cold Spring Harbor Symp. Quant. Biol.* **35:** 174.

Hayashi, M., M. N. Hayashi, and S. Spiegelman. 1963. Restriction of *in vivo* transcription to one of the complementary strands of DNA. *Proc. Natl. Acad. Sci.* **50:** 664.

———. 1964. DNA circularity and the mechanism of strand selection in the generation of genetic message. *Proc. Natl. Acad. Sci.* **51:** 351.

Hayashi, M. N. and M. Hayashi. 1966. Participation of a DNA-RNA hybrid complex in *in vivo* genetic transcription. *Proc. Natl. Acad. Sci.* **55:** 635.

———. 1972. Isolation of φX174-specific messenger ribonucleic acid *in vivo* and identification of their 5′-terminal nucleotides. *J. Virol.* **9:** 207.

———. 1974. Fragment maps of φX174 replicative form DNA produced by restriction enzymes from *Haemophilus aphirophilus* and *Haemophilus influenzae* HI. *J. Virol.* **14:** 1142.

Hayashi, Y. and M. Hayashi. 1971a. Fractionation of φX174-specific mRNA. *Cold Spring Harbor Symp. Quant. Biol.* **35:** 171.

———. 1971b. Template activities of the φX174 replicative allomorphic DNAs. *Biochemistry* **10:** 4212.

Ikeda, J.-E., A. Yudelevich, and J. Hurwitz. 1976. Isolation and characterization of the protein coded by gene *A* of bacteriophage φX174 DNA. *Proc. Natl. Acad. Sci.* **73:** 2669.

Jeng, Y., D. Gelfand, M. Hayashi, R. Shleser, and E. S. Tessman. 1970. The eight genes of bacteriophages φX174 and S13 and comparison of the phage-specified proteins. *J. Mol. Biol.* **49:** 521.

Kapitza, E. L., E. A. Stukacheva, and M. F. Shemyakin. 1976. The effect of the termination rho factor and ribonuclease III on the transcription of bacteriophage φX174 DNA *in vitro*. *FEBS Lett.* **64:** 81.

Lee, A. S. and R. L. Sinsheimer. 1974. A cleavage map of bacteriophage φX174 genome. *Proc. Natl. Acad. Sci.* **71:** 2882.

Linney, E. A. and M. Hayashi. 1974. Intragenic regulation of the synthesis of φX174 gene *A* proteins. *Nature* **249:** 345.

Linney, E. A., M. N. Hayashi, and M. Hayashi. 1972. Gene *A* of φX174. I. Isolation and identification of the products. *Virology* **50:** 381.

Losick, R. and M. Chamberlin, eds. 1976. *RNA polymerase.* Cold Spring Harbor Laboratory, Cold Spring Harbor, New York.

McMahon, J. E. and I. Tinoco, Jr. 1978. Sequences and efficiencies of proposed mRNA terminators. *Nature* **271:** 275.

Maizels, N. M. 1973. The nucleotide sequence of the lactose messenger ribonucleic acid transcribed from the UV5 promoter mutant of *Escherichia coli*. *Proc. Natl. Acad. Sci.* **70:** 3585.

Marians, K. J., J. -E. Ikeda, S. Schlagman, and J. Hurwitz. 1977. Role of DNA gyrase in φX replicative-form replication *in vitro*. *Proc. Natl. Acad. Sci.* **74:** 1965.

Pollock, T. J. 1977. Gene *D* of bacteriophage φX174: Absence of transcriptional and translational regulatory properties. *J. Virol.* **21:** 468.

Pollock, T. J. and I. Tessman. 1976. Units of functional stability in bacteriophage messenger RNA. *J. Mol. Biol.* **108:** 651.

Pribnow, D. 1975. Nucleotide sequence of an RNA polymerase binding site at an early T7 promoter. *Proc. Natl. Acad. Sci.* **72:** 784.

Puga, A. and I. Tessman. 1973a. Mechanism of transcription of bacteriophage S13. I. Dependence of messenger RNA synthesis on the amount and configuration of DNA. *J. Mol. Biol.* **75:** 83.

———. 1973b. Mechanism of transcription of bacteriophage S13. II. Inhibition of phage-specific transcription by nalidixic acid. *J. Mol. Biol.* **75:** 99.

Puga, A., M. -T. Borrás, E. S. Tessman, and I. Tessman. 1973. Difference between functional and structural integrity of messenger RNA. *Proc. Natl. Acad. Sci.* **70:** 2171.

Sanger, F., G. M. Air, B. G. Barrell, N. L. Brown, A. R. Coulson, J. C. Fiddes, C. A. Hutchison III, P. M. Slocombe, and M. Smith. 1977. Nucleotide sequence of bacteriophage φX174 DNA. *Nature* **265:** 687.

Sedat, J. W. and R. L. Sinsheimer. 1971. The *in vivo* φX mRNA. *Cold Spring Harbor Symp. Quant. Biol.* **35:** 163.

Shleser, R., A. Puga, and E. S. Tessman. 1969. Synthesis of replicative form deoxyribonucleic acid and messenger ribonucleic acid by gene IV mutants of bacteriophage S13. *J. Virol.* **4:** 394.

Sinsheimer, R. L. 1968. Bacteriophage φX174 and related viruses. *Prog. Nucleic Acid Res. Mol. Biol.* **8:** 115.

Smith, L. H. and R. L. Sinsheimer. 1976a. The *in vitro* transcription units of bacteriophage φX174. I. Characterization of synthetic parameters and measurement of transcript molecular weights. *J. Mol. Biol.* **103:** 681.

———. 1976b. The *in vitro* transcription units of bacteriophage φX174. II. *In vitro* initiation sites of φX174 transcription. *J. Mol. Biol.* **103:** 699.

———. 1976c. The *in vitro* transcription units of bacteriophage φX174. III. Initiation with specific 5′-end oligonucleotides of *in vitro* φX174 DNA. *J. Mol. Biol.* **103:** 711.

Smith, L. H., K. Grohmann, and R. L. Sinsheimer. 1974. Nucleotide sequences of the 5′-termini of φX174 mRNAs synthesized *in vitro*. *Nucleic Acids Res.* **1:** 1521.

Sugimoto, K., H. Sugisaki, T. Okamoto, and M. Takanami. 1977. Studies of bacteriophage fd DNA. IV. The sequence of messenger RNA for the major coat protein gene. *J. Mol. Biol.* **110:** 487.

Sugino, A., C. L. Peebles, K. N. Kreuzer, and N. R. Cozzarelli. 1977. Mechanisms of action of nalidixic acid: Purification of *Escherichia coli nal*A gene product and its relationship to DNA gyrase and a novel nicking-closing enzyme. *Proc. Natl. Acad. Sci.* **74:** 4767.

Tessman, I., S. Kumar, and E. S. Tessman. 1967. Direction of translation in bacteriophage S13. *Science* **158:** 267.

Truffaut, N. and R. L. Sinsheimer. 1974. Use of bacteriophage φX174 replicative form progeny DNA as templates for transcription. *J. Virol.* **13:** 818.

Vanderbilt, A. S. and I. Tessman. 1970. Mutagenic method for determining which DNA strand is transcribed for individual viral genes. *Nature* **228:** 54.

Vanderbilt, A. S., M. -T. Borrás, and E. S. Tessman. 1971. Direction of translation in phage S13 as determined from the sizes of polypeptide fragments of nonsense mutants. *Virology* **43:** 352.

Vanderbilt, A. S., M. -T. Borrás, S. Germeraad, I. Tessman, and E. S. Tessman. 1972. A promoter site and polarity gradients in phage S13. *Virology* **50:** 171.

Warnaar, S. O., A. DeMol, G. Mulder, P. J. Abrahams, and J. A. Cohen. 1970. The transcription of bacteriophage φX174 DNA. *Biochim. Biophys. Acta* **99:** 340.

Warnaar, S. O., G. Mulder, I. van den Sigtenhorst-van der Sluis, L. W. van Kesteren, and J. A. Cohen. 1969. The transcription *in vitro* of various forms of φX174 DNA. *Biochim. Biophys. Acta* **174:** 239.

Transcription of the Filamentous Phage Genome

Ruud N. H. Konings and John G. G. Schoenmakers
Laboratory of Molecular Biology
University of Nijmegen
Nijmegen, The Netherlands

The F-specific filamentous coliphages, which include M13, f1, fd, ZJ/2, Ec9, AE2, HR, and δA (for reviews, see Marvin and Hohn 1969; Denhardt 1975; Ray 1977), are composed of a single-stranded DNA molecule (m.w. ca. 2×10^6) in a capsid of about 2000 molecules of a protein (m.w. 5200) encoded by gene VIII, 2–3 molecules of a protein (m.w. ca. 52,000) specified by gene III, and a few molecules of a small protein, designated C protein (m.w. ca. 3000), the genetic origin of which is still unknown (Henry and Pratt 1969; Pratt et al. 1969; G. Simons et al., unpubl.; R. Webster and W. Konigsberg, pers. comm.). The DNA of the filamentous coliphages codes for at least nine polypeptides, and their corresponding genes have been ordered into a genetic map (Lyons and Zinder 1972; Seeburg and Schaller 1975; van den Hondel et al. 1975a,b; Vovis et al. 1975).

Escherichia coli cells infected with these phages are neither lysed nor killed, but continue to grow and divide exponentially, although at a slower rate than uninfected cells, while extruding a few hundred new phage particles per cell generation. During this infection cycle, some polypeptides, such as the DNA-binding protein encoded by gene V and the major capsid protein encoded by gene VIII, are synthesized in much larger quantities than the other phage-encoded polypeptides (Henry and Pratt 1969; Lica and Ray 1976; Smits et al. 1978). These observations prompted us to initiate a study of the mechanisms which may be responsible for this diversity in gene expression. In this article we summarize our present knowledge of the regulatory mechanisms operative in transcription. The characteristics and nucleotide sequences of the regulatory elements (promoter and termination sites), which define the number and size of the RNA products made, are described by Schaller et al. (this volume). The gene order and the correlation of the genes with the polypeptides they code for are described by Horiuchi et al. (this volume).

Unless indicated otherwise, it is assumed that results obtained with one phage are also valid for the other F-specific filamentous coliphages.

EXPRESSION OF THE PHAGE GENOME IN VITRO

Template Specificity

Upon infection, the circular, single-stranded phage genome is converted into a covalently closed, double-stranded, replicative-form molecule (RFI; Ray, this volume). Hybridization studies on pulse-labeled RNA isolated from M13-infected cells (Jacob and Hofschneider 1969) or from minicells harboring M13 RF DNA (Smits et al. 1978) have indicated that during the infection cycle only the nonviral (complementary) strand of the RF molecule is transcribed. This specificity of strand selection is retained when RF DNA or fragments of this double-stranded DNA molecule (e.g., restriction fragments) are transcribed in vitro by *E. coli* RNA polymerase holoenzyme (Sugiura et al. 1969; Chan et al. 1975; Edens et al. 1975, 1976; Okamoto et al. 1975). Furthermore, transcription studies on heteroduplexes of f1 RF DNA (Pieczenik et al. 1975) have demonstrated not only that the RNA molecules synthesized in vitro are complementary to the nonviral strand of the DNA template, but also that during transcription it is this strand, not the viral strand, that determines the RNA base selection. Removal of the RNA initiation factor sigma from the RNA polymerase holoenzyme results in a loss in this specificity of template selection. In the latter case, both strands are transcribed and initiation of transcription occurs at random along both strands of the double-stranded DNA molecule (Sugiura et al. 1970; Takanami et al. 1971a).

Direction of Transcription

The direction of transcription of RF DNA has been determined by several independent methods. The earliest evidence came from observations of polarity. In genetic complementation studies it was observed that amber mutations in gene III had a polar effect on the expression of genes VI and I (Pratt et al. 1966, 1967; Lyons and Zinder 1972). Since only the complementary strand of RF DNA is transcribed by *E. coli* RNA polymerase and the order of the genes on the genetic map turned out to be III-VI-I (Lyons and Zinder 1972; Seeburg and Schaller 1975; van den Hondel et al. 1975b), these observations tended to show that transcription of RF DNA occurs in a counterclockwise direction on the conventional genetic map (Fig. 1). Proof that this was indeed the case was obtained initially from coupled protein-synthesis studies in vitro in which defined amber mutants of RF DNA and/or restriction fragments were used as templates. Model and Zinder (1974) studied in vitro protein synthesis under the direction of RF preparations bearing different amber mutations in gene II. By comparing the sizes of the polypeptides (amber protein fragments) made with the relative map positions of the corresponding amber mutations, a counterclockwise direction of transcription was deduced. An identical conclusion was reached by our group (Konings 1974; van den Hondel et al. 1975a) from coupled protein-

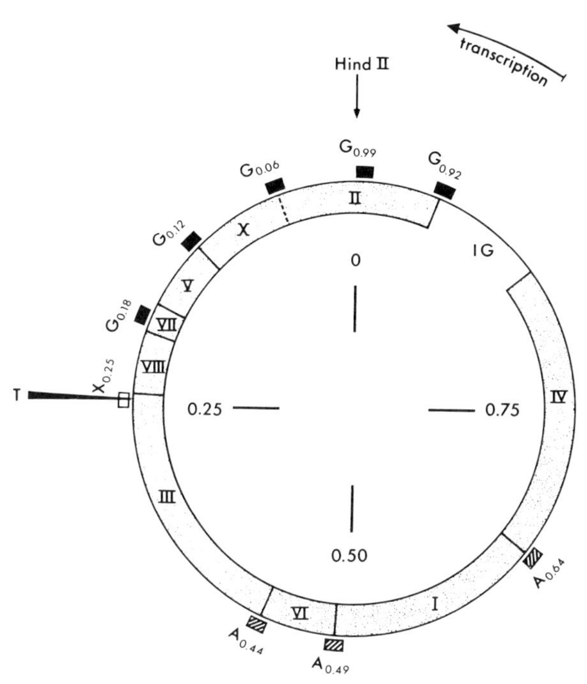

Figure 1 Circular genetic map of the filamentous bacteriophages M13, f1, and fd. The Roman numerals refer to the genes. X refers to a gene which overlaps completely the C-terminal end of gene II. The direction of transcription is as indicated. The bars refer to the promoter sites located on the M13 genome: black bars stand for G promoters and hatched bars for A promoters. The starting nucleotide of the RNA chains initiated at promoter $X_{0.25}$ is not yet known. The subscripts are the positions of the promoter sites on the physical map. T stands for the *rho*-independent termination signal for transcription, and IG refers to the intergenic region in which the replication origins for both the viral and complementary strands are located. The single site at which RF DNA is cleaved with restriction endonuclease *Hin*dII is indicated. This site serves as a reference (zero) point on the physical map (cf. Fig. 4).

synthesis studies in vitro under the direction of restriction fragments bearing either the complete or only the N-terminal end of gene IV.

Since then, these conclusions have been confirmed by in vitro transcription studies on restriction fragments (Edens et al. 1975; Okamoto et al. 1975) and more recently by nucleotide-sequence analyses of the filamentous phage genome (Takanami et al. 1976; Sugimoto et al. 1977; T. Hulsebos and P. van Wezenbeek, pers. comm.; Schaller et al., this volume).

RNA Species Formed upon Transcription of RF DNA

Transcription of RF DNA in vitro by *E. coli* RNA polymerase holoenzyme in the absence of the termination factor *rho* results in the formation of at least eight discrete RNA species ranging in size from about 8S (m.w. ca. 10^5) up to 26S (m.w. ca. 1.7×10^6; Fig. 2). In addition, there is made a heterogeneous mixture of RNA species that are equivalent in size and larger than one genome in length (Okamoto et al. 1969; Chan et al. 1975; Edens et al. 1975, 1976).

Although the activity of the DNA template is somewhat lower under low-salt (40 mM KCl) conditions than under high-salt (150 mM KCl) conditions (Takanami et al. 1971a), the patterns of the RNA species (size, number, and relative amounts) made under both ionic conditions are almost identical overall (L. Edens, pers. comm.). These observations suggest that under both conditions RNA synthesis is initiated and terminated at the same regulatory sites of the phage genome. This belief is supported by experiments which showed that, irrespective of the salt conditions used, the five smallest RNA species are initiated with pppG and the three largest are initiated with pppA. Among the RNA species with lengths equivalent to or greater than one genome length, RNA molecules have been found that are initiated with pppG as well as with pppA (Edens et al. 1975, 1976). These observations, together with the finding that the nucleotide sequences of the 5′ ends of at least three of these transcripts are different (Sugiura et al. 1969; Schaller et al., this volume), indicate that several promoter sites are located on the filamentous phage genome.

Locations of Promoter Sites

In the initial phase of transcription, RNA polymerase interacts selectively with the DNA template at the promoter sites to form binary complexes.

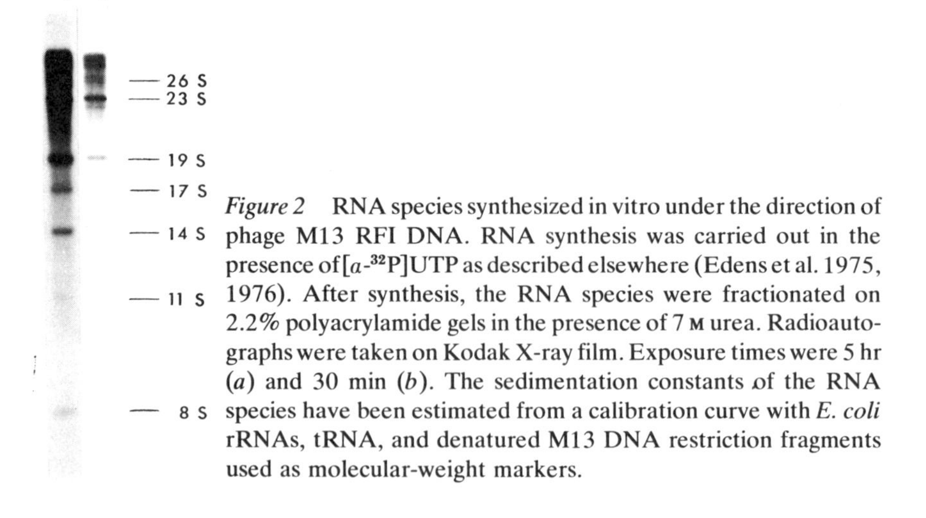

Figure 2 RNA species synthesized in vitro under the direction of phage M13 RFI DNA. RNA synthesis was carried out in the presence of [α-^{32}P]UTP as described elsewhere (Edens et al. 1975, 1976). After synthesis, the RNA species were fractionated on 2.2% polyacrylamide gels in the presence of 7 M urea. Radioautographs were taken on Kodak X-ray film. Exposure times were 5 hr (*a*) and 30 min (*b*). The sedimentation constants of the RNA species have been estimated from a calibration curve with *E. coli* rRNAs, tRNA, and denatured M13 DNA restriction fragments used as molecular-weight markers.

These complexes then bind ribonucleoside triphosphates and initiate RNA synthesis. Apart from *E. coli* RNA polymerase holoenzyme, no other regulatory elements (proteins or odd nucleotides) are required for a proper initiation of RNA synthesis on the filamentous phage genome. Proper transcription of the codogenic (nonviral) strand occurs under both low- and high-salt conditions. The ribonucleoside triphosphate concentration, also, has no remarkable influence on the size or number of RNA species produced (Takanami et al. 1971a; Edens et al. 1975, 1976). However, the ribonucleoside triphosphate concentration, as well as the ratio of enzyme to template, does influence the molar ratios in which the different RNA classes are made and the rate of chain initiation. These observations, together with the findings mentioned in the previous section, suggest that the different RNA initiation sites (promoter sites) on the filamentous phage genome have different affinities for RNA polymerase holoenzyme (cf. Seeburg et al. 1977).

A number of the strategies that have been pursued to determine the locations of these promoter sites are listed below:

1. Mapping of the DNA fragments (e.g., restriction fragments) that can form a firm complex with RNA polymerase holoenzyme under both low- and high-salt conditions and with or without ribonucleoside triphosphates.
2. Mapping of the DNA fragments that direct synthesis of discrete RNA species in vitro; the size of the RNA chains thus formed then determines the position of the promoter on the physical map.
3. Studies in a DNA-dependent, cell-free protein-synthesizing system of whether specific restriction fragments are able to direct the synthesis of phage-specified polypeptides. If this is the case, then additional information about the location of the promoter on the genetic map is gained.
4. Studies in an RNA-dependent, cell-free protein-synthesizing system of the coding information of purified in vitro transcripts. The phage-specified polypeptides made, together with the known direction of transcription, also gives information about the location of the promoter sites on the genetic map.
5. Correlation of the nucleotide sequences of in vitro transcripts with the known amino acid sequences of phage-specified polypeptides.

It should be noted that in studying promoter positions it is generally assumed that the RNA initiation sites are located within the RNA-polymerase-binding sites. Nucleotide-sequence analyses of several promoter sites and of the 5′ ends of a number of in vitro transcripts have validated this presumption (Heyden et al. 1975; Nüsslein and Schaller 1975; Pribnow 1975a; Sugimoto et al. 1975a,b; Schaller et al., this volume).

Okamoto and coworkers (Okamoto et al. 1975) and Seeburg and Schaller (1975) located promoter sites on the physical map by using the techniques described in entries 1 and 2 of the list above. Binding of RNA polymerase holoenzyme to the mixture of DNA fragments obtained after cleavage of fd RF DNA with restriction endonucleases was followed by filtration of the incubation mixture over nitrocellulose membrane filters. Retention of a particular restriction fragment was then used as a test for whether a promoter site was located on that fragment. The number and type of fragments retained were found to be dependent on the ionic conditions during binding, the molar ratio of RNA polymerase to RF DNA, and the temperature of incubation. At low ionic strength (40 mM KCl), up to six fragments formed stable complexes. For fd RF DNA, these fragments were *Hap*II (= *Hpa*II) A, B, C, D, F, and H (Fig. 3). The same set of fragments was retained by the filter when RNA polymerase was bound to fd RF DNA prior to digestion, which indicates that the ends of the fragments did not interfere with the specificity of the binding reaction and that the promoter sites are still intact after scission. Upon increasing the ionic strength of the binding buffer (to 120 mM KCl) only fragments *Hap*II A, B, D, and to some extent C were retained, but at high ionic strength (1 M KCl; Okamoto et al. 1975) only fragments *Hap*II C and D were observed. *Hap*II D was also preferentially bound by holoenzyme at low molar ratios of enzyme to RF DNA, which suggests that the strongest promoter site of the fd genome is located on this fragment (cf. Heyden et al. 1972). In the absence of ribonucleoside triphosphates and at high ionic strength (250 mM KCl) and 0°C, no fragments were bound by RNA polymerase, and preformed complexes dissociated when exposed to such treatment. Dissociation could be prevented by brief incubation of the complexes with the four ribonucleoside triphosphates, which led to the formation of a ternary complex of DNA, enzyme, and RNA. Such ternary complexes were found only with fragments that also formed promoter complexes.

From the lengths of the RNA chains formed upon transcription of fd fragments containing RNA-polymerase-binding sites, Seeburg and Schaller (1975) have calculated that the six fd promoters are located at the following approximate map positions: 0.9, 0.97, 0.07, 0.2, 0.46, and 0.67 (Fig. 3). The first four promoters initiate the synthesis of RNA chains commencing with pppG (G promoters[1]), and the RNAs initiated at the other two promoters commence with pppA (A promoters). An identical distribution of promoter sites on the genome was found for the related phages f1 and M13.

The same approach was used by Takanami and coworkers (Okamoto et al. 1975) to map four promoter sites on the fd genome. According to their data, the approximate positions of these promoter sites are 0.19, 0.06, 0.86, and 0.52. The first three sites are G promoters and the fourth is an A

[1]Nomenclature of promoter sites: Promoters that initiate the synthesis of RNA chains commencing with either pppA, pppG, or pppU are denoted by A, G, and U, respectively. The position of each promoter is given in map units and is indicated by a subscript which corresponds to the position of the promoter on the physical map.

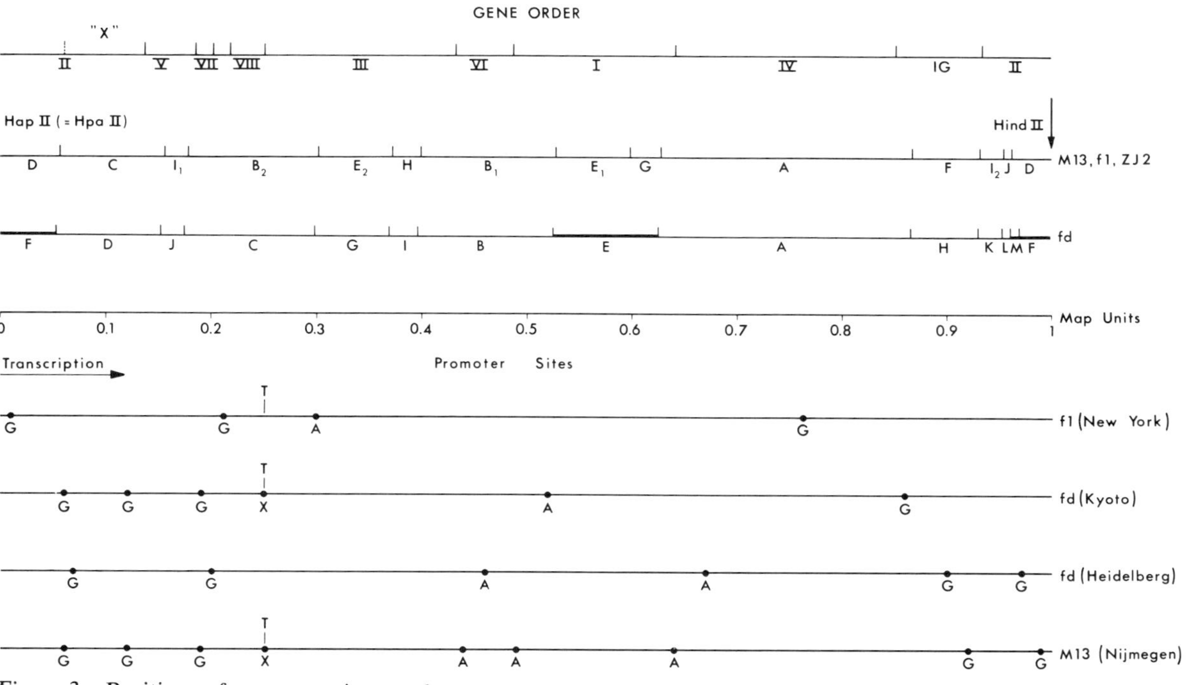

Figure 3 Positions of promoter sites on the genomes of bacteriophages f1, fd, and M13. The upper line represents the genetic map of the F-specific filamentous phages and the next two lines represent the physical maps of the *Hap*II (isoschizomer of *Hpa*II) fragments of the phages M13, f1, ZJ/2 (line 2), and fd (line 3). The black bars on the physical map of phage fd indicate fragments that differ in length from fragments located at corresponding map positions on the genomes of the phages M13, f1, and ZJ/2. The maps are divided into ten equal map units (line 4), with the *Hin*dII cleavage site as a reference (zero) point. One map unit corresponds to 6400 base pairs. On the bottom lines (lines 5–8) the promoters that have been detected by different research groups are indicated. A and G indicate promoters that initiate synthesis of RNA chains commencing with pppA and pppG, respectively. The starting nucleotide of the RNA chains initiated at promoter $X_{0\cdot 25}$ is not known. The direction of transcription is as indicated.

promoter. The G promoters were thus found at approximately the same positions as three of the G promoters detected by Seeburg and Schaller (1975) (positions 0.2, 0.07, and 0.9). Therefore, it is reasonable to assume that these G promoters are identical. However, in contrast to the results of the Heidelberg group, the A promoter detected by Okamoto et al. (1975) was at a completely different position (0.52).

Recently, Schaller and coworkers have improved their techniques for mapping RNA-polymerase-binding sites. DNA fragments protected by RNA polymerase were used as a primer for limited DNA synthesis with DNA polymerase I in the presence of [α-^{32}P]dXTP. Upon digestion of the extended DNA with the appropriate restriction enzyme, the exact distance between the binding site and the nearest restriction enzyme cleavage site could be estimated; the resulting map of the promoter sites is therefore more accurate. Furthermore, from the nucleotide sequence data available they were able to deduce tentative nucleotide sequences for these regulatory sites. The revised map positions of the promoter sites and their tentative nucleotide sequences are presented in Figure 4 of Schaller et al. (this volume). The nucleotide sequences of the promoter sites established by Takanami's group and the tentative positions of five other RNA-polymerase-binding sites detected recently by Schaller and coworkers are also included in the figure (H. Schaller, pers. comm.; Schaller et al., this volume). Interestingly, the Heidelburg group has now also found that a promoter is located near map position 0.52.

With the aid of nucleotide-sequence analyses, M. Takanami and coworkers (pers. comm.) have recently also mapped two additional promoters on the fd genome. One was found close to map position 0.12 and the other close to map position 0.25 (Fig. 3; see also Fig. 4 in Schaller et al., this volume). From the nucleotide sequences and the known amino acid sequences of phage-specified proteins, they concluded that the promoters at map positions 0.12, 0.19, and 0.25 immediately precede genes V, VIII, and III, respectively.

With respect to the genome of phage M13, we have determined, on both the physical and the genetic map, the positions of nine promoters. The techniques applied were in vitro transcription and translation studies on DNA restriction fragments obtained after cleavage of M13 RF DNA with the restriction endonucleases *Hap*II, *Hae*II, *Hae*III, *Hin*dII, *Hin*fI, *Alu*I, *Hha*I, *Taq*I, and *Bam*I (van den Hondel and Schoenmakers 1973, 1975, 1976; van den Hondel et al. 1976; T. Hulsebos and J. Schoenmakers, in prep.). The locations of these fragments on the physical and genetic maps are shown in Figure 4.

Fragments containing promoter sites were identified by means of in vitro transcription experiments. The capability of a fragment to direct, under both low- and high-salt conditions, the synthesis of a discrete RNA species was taken as evidence that the fragment contained a promoter site. The exact

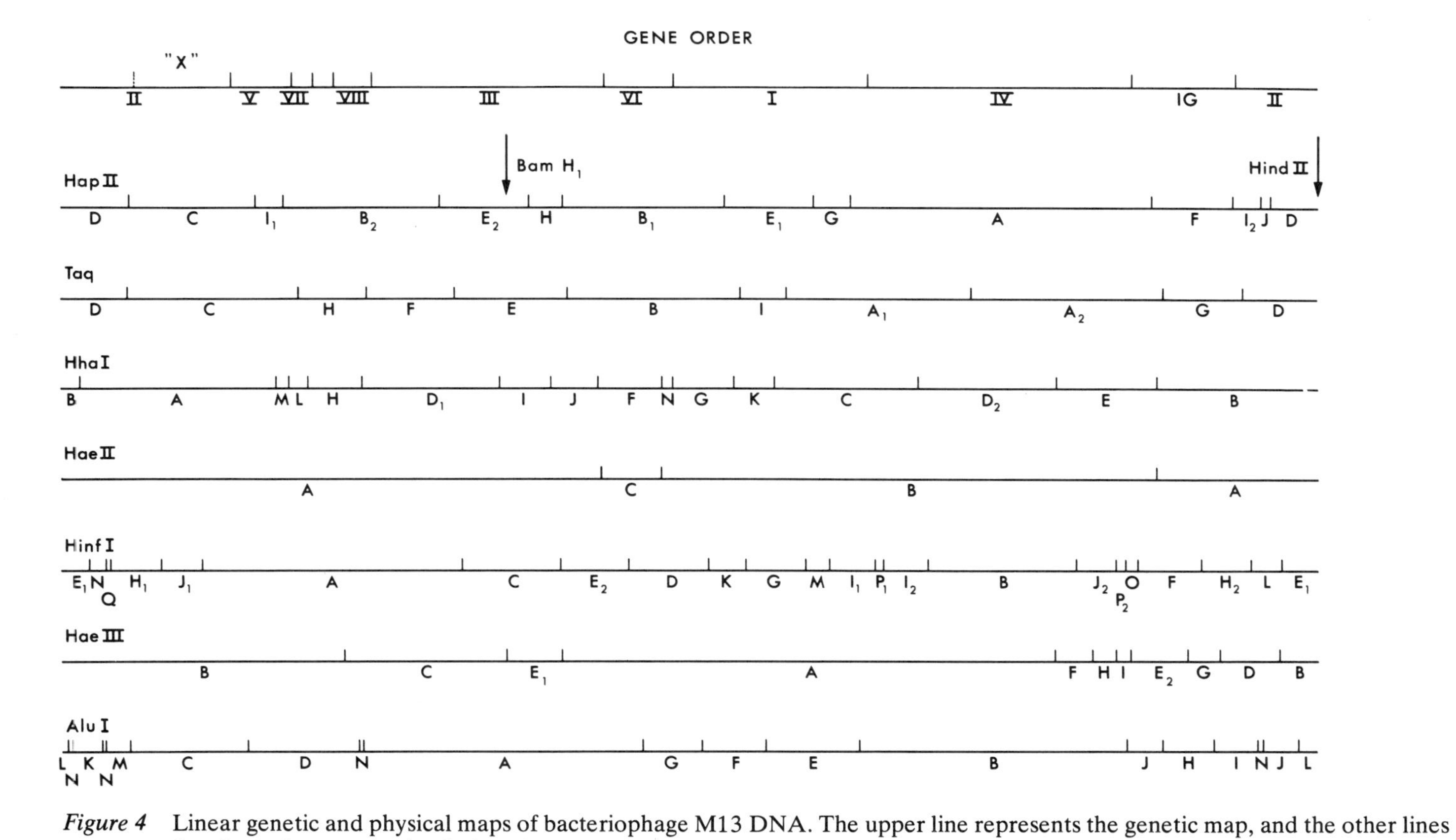

Figure 4 Linear genetic and physical maps of bacteriophage M13 DNA. The upper line represents the genetic map, and the other lines represent the physical maps obtained after cleavage of M13 RF DNA with restriction endonucleases *Hap*II, *Bam*I, *Taq*, *Hha*I, *Hae*II, *Hin*fI, *Hae*III, and *Alu*I. In all cases the site of cleavage of M13 RF DNA with restriction endonuclease *Hin*dII served as a reference point.

position of the promoter site was determined by measuring the length of the transcript made, thereby taking into account the fact that only the complementary strands of the fragments are transcribed. By means of these techniques, promoter sites were located at positions 0.18, 0.12, 0.06, 0.99, 0.92, 0.64, 0.49, 0.44, and 0.25 (Fig. 1; Edens et al. 1975, 1976, 1978).

Transcription studies in the presence of γ-^{32}P-labeled ribonucleoside triphosphates have indicated that the promoters at positions 0.18, 0.12, 0.06, 0.99, and 0.92 initiate the synthesis of RNA species that commence with pppG (G promoters) and those at positions 0.64, 0.49, and 0.44 initiate the synthesis of RNA species that commence with pppA (A promoters) (Edens et al. 1975, 1976). With which ribonucleoside triphosphate the RNA species initiated at position 0.25 (promoter $X_{0.25}$; Fig. 1) commences has not yet been established. In this connection it is of interest that M. Takanami and coworkers (pers. comm.; Schaller et al., this volume) have also recently found a promoter at position 0.25 on the genome of phage fd. Furthermore, they have established the nucleotide sequence of this promoter (see Fig. 4 in Schaller et al., this volume) and discovered that this regulatory element is within a region that also functions as a termination site for transcription. If it is assumed that RNA synthesis is initiated 6 or 7 base pairs after the so-called "Pribnow box" (TATPuATG sequence; Pribnow 1975), then the RNA transcribed from this promoter starts with either pppC or pppU. As far as we know now, these ribonucleoside triphosphates are used only rarely as starting nucleotides during the initiation of transcription. The reasons for this are unknown.

On the basis of the incorporation of γ-^{32}P-labeled ribonucleoside triphosphates into the different RNA species, it appears that the promoters have different affinities for RNA polymerase holoenzyme. The strength of G promoters decreases in the order $G_{0.06} > G_{0.18} \geqslant G_{0.92} > G_{0.99} \geqslant G_{0.12}$, and the strength of A promoters decreases in the order $A_{0.64} > A_{0.49} \geqslant A_{0.44}$. On the average, G promoters have a higher affinity for RNA polymerase holoenzyme than A promoters (Edens et al. 1975, 1976). Promoter $G_{0.06}$ is about ten times as strong as promoter $G_{0.12}$, and promoter $A_{0.64}$ is about five times as strong as promoter $A_{0.49}$. These calculations are based on the observation made by Seeburg et al. (1977) that initiation of transcription is a much faster process than binding of RNA polymerase holoenzyme to the promoter site. Although it is obvious that promoter $X_{0.25}$ has a low affinity for RNA polymerase, no quantitative data about the strength of this promoter is available so far.

As shown in Figure 1, both the G promoters and the A promoters are arranged in clusters on the physical map, with promoter $X_{0.25}$ located between these clusters. The G promoters, which are responsible for the synthesis of the five smallest in vitro RNA species (8S, 11S, 14S, 17S, and 19S; Fig. 2) are situated in a region covering only one-third of the genome. The genes whose products are needed most abundantly during phage infection, i.e., genes V and VIII, are also in this region.

With the aid of coupled and uncoupled in vitro protein-synthesis studies, most of the phage promoters have been located on the genetic map (Konings et al. 1975; van den Hondel et al. 1975a; Edens et al. 1975, 1976, 1978). The capability of a DNA fragment or of its transcript to direct the synthesis of a phage-specified polypeptide in vitro was used as an indication of whether a promoter was located in front of the gene coding for this polypeptide. Thus, it was demonstrated that the promoters $G_{0.18}$, $G_{0.12}$, $G_{0.92}$, $A_{0.64}$, $A_{0.49}$, and $X_{0.25}$ are located immediately in front of genes VIII, V, II, IV, I, and III, respectively (Fig. 1). In addition, it was found that the promoters $G_{0.06}$ and $G_{0.99}$ are intragenic RNA initiation sites within gene II. For promoter $G_{0.06}$, it was also found that this site immediately precedes a gene (gene X) whose nucleotide sequence overlaps completely the C-terminal end of gene II (Fig. 1; Schaller et al.; Horiuchi et al.; both this volume). Gene X directs the synthesis of a polypeptide, called X protein, whose formation has been observed clearly only in vitro (Konings 1973; Konings et al. 1973; Konings and Schoenmakers 1974; Model and Zinder 1974; Konings et al. 1975; van den Hondel et al. 1975a; Horiuchi et al., this volume). Although the synthesis of a phage-specified polypeptide of the size of the X protein has been observed occasionally in minicells harboring M13 RF DNA (Smits et al. 1978), at present it is not clear whether the DNA region coding for X protein is also expressed in vivo.

We have not succeeded in demonstrating, either in vitro or in vivo, the synthesis of the polypeptides encoded by genes VI and VII. Therefore, although the experimental data obtained from genetic and transcription studies favor a position for promoter $G_{0.44}$ immediately preceding gene VI, the evidence is inconclusive. It should also be emphasized that as yet no evidence has been obtained for the existence of a promoter immediately preceding gene VII. This implies that gene VII can be expressed only via the synthesis of a polycistronic mRNA molecule. This assumption is supported by the finding (Lyons and Zinder 1972; R. Benzinger, pers. comm.) that, in vivo, the adjacent genes V and VII exhibit polarity (van den Hondel et al. 1975b; Vovis et al. 1975).

As can be deduced from the genetic and physical maps (Fig. 1), promoter $G_{0.92}$ is in the intergenic region (IG) between genes IV and II, which is also the location of the origins of replication of both the viral and complementary strands (Tabak et al. 1974; Horiuchi and Zinder 1976; van den Hondel 1976; Suggs and Ray 1977). Originally it was an attractive idea to suppose that this promoter was also involved in the synthesis of the RNA primer required for the conversion of SS DNA to RF DNA (Ray, this volume), but recent studies have indicated that promoter $G_{0.92}$ and the replication origin (RNA-polymerase-binding site on SS DNA) are at different positions within this intergenic region (H. Schaller, pers. comm; Schaller et al., this volume).

The reason for the existence of more than one promoter site ($G_{0.99}$ and $G_{0.06}$) within gene II is by no means clear. Promoter $G_{0.06}$ is in fact the strongest promoter on the M13 genome. One reason may be that the genes distal to these sites need to be expressed frequently. That these promoters are situated almost proximal to the 5′ ends of two genes (genes V and VIII) whose products are needed abundantly during the phage infection cycle supports this assumption. If true, it is still surprising that these promoters are not grouped in a cluster, e.g., adjacent to promoter $G_{0.12}$, which immediately precedes gene V. It is possible, therefore, that the leading regions, which follow the promoters $G_{0.99}$ and $G_{0.06}$, code for polypeptides that have an essential role in the phage infection cycle, and that the synthesis of these polypeptides can be directed only by mRNA molecules initiated at promoter sites immediately preceding these "overlapping" genes. The observations that at least one of the intragenic regions codes in vitro for a polypeptide with a molecular weight of about 12,000 (X protein) and that relatively larger amounts of X protein are made in a DNA-dependent, cell-free protein-synthesizing system than in an RNA-dependent system (Edens et al. 1978) agree with the latter assumptions. In this connection, it is interesting that in the case of phage φX174 a promoter is found immediately preceding a gene (gene *B*) that is located within another gene (gene *A*) (Sanger et al. 1977; Smith et al. 1977). The hypothesis that translation of overlapping genes can result in functional polypeptides only if a promoter site is located immediately before these genes therefore deserves further investigation.

In general, there is excellent agreement between the results of the different research groups studying the positions of promoters on the genomes of phages M13 and fd. Strongly deviating results have been reported, however, by Chan et al. (1975), who studied the location of promoters on the genome of phage f1 (Fig. 3). The positions reported by this group are based on calculations of molecular weights of in vitro transcripts and on analyses of the protein products encoded by these RNA species. Since calculations of molecular weights of long transcripts from their sedimentation behavior or electrophoretic mobilities tend to be less accurate and RNA species obtained by density gradient centrifugation are hardly pure, we believe that the promoter positions assigned by these authors are not correct. This belief is supported by the observation of Seeburg and Schaller (1975) that at least six of the fd promoters have equivalents among the promoters on the genome of f1. Since the gene orders of all the filamentous phages studied so far (M13, f1, and fd) have been shown to be identical, and since in vitro protein-synthesis studies have indicated that the sizes of the genes located on the different phage genomes (M13, fl, fd, and ZJ/2) are also identical (Konings et al. 1975; R. N. H. Konings, unpubl.), we believe that all

F-specific filamentous coliphages contain an identical set of promoter sites.

Termination of Transcription

As outlined above, transcription of RF DNA results in the formation of at least nine RNA species that range in size from about 360 nucleotides (8S) up to 6300 nucleotides (30S). In addition, a number of RNA molecules are made that are larger than one genome length. Translation of the individual RNA species in a cell-free protein-synthesizing system has demonstrated that they all share the coding information for the polypeptide encoded by gene VIII, the gene that codes for the major capsid protein (Chan et al. 1975; Edens et al. 1978). Since the RNA species synthesized in vitro are initiated at different promoter sites and the smallest RNA product codes for only gene-VIII protein, it was concluded that, in the absence of *rho*, termination of transcription occurs at a site immediately following gene VIII. This termination site has been mapped precisely by studying RNA synthesis under the direction of a restriction fragment (fragment *Hap* II B_2; Fig. 3) that has been shown in a DNA-dependent, cell-free system to direct the synthesis of gene-VIII protein (van den Hondel et al. 1975a). Cleavage of this fragment with several restriction enzymes and subsequent measurement of the lengths of the truncated transcripts synthesized on these fragments have demonstrated that the central termination site for transcription is approximately 450 base pairs from the 5′ end of fragment *Hap* II B_2, within 60 base pairs of the last triplet of gene VIII (Edens et al. 1975). Nucleotide-sequence analyses of the 3′-terminal region of the smallest transcript (8S RNA) from fd RF DNA (Sugimoto et al. 1977) and M13 RF DNA (L. Edens and J. Schoenmakers, unpubl.) have shown that the 3′ terminus has the following nucleotide sequence: – AUACAAUUAAAGGCUC-CUUUUGGAGCCUUUUUUUU$_{OH}$. From the complete nucleotide sequence of this small 8S RNA molecule, as established by Sugimoto et al. (1977), we know now that the 3′ terminus is 42 nucleotides away from the last coding triplet of gene VIII. An interesting feature of this 3′-terminal nucleotide sequence is a stretch of eight U residues. Similar oligo(U) stretches have also been found at the 3′ end of the attenuator-terminated tryptophan messenger RNA (Bertrand et al. 1975; Lee et al. 1976), at the 3′ terminus of the 4S "oop" RNA and 6S RNA of phage λ (Lebowitz et al. 1971; Dahlberg and Blattner 1973), and at the 3′ termini of a number of other RNA species (Pieczenik et al. 1972; Ikemura and Dahlberg 1973; Sogin et al. 1976).

Another remarkable feature of this 3′-terminal region is that it has intrinsically the capacity to form a tight hairpin loop (cf. Schaller et al., this volume). Such a property is also characteristic of the 3′-terminal region of the attenuator-terminated tryptophan mRNA (Lee et al. 1976) and of the

5S ribosomal RNA precursor (Sogin et al. 1976). It is reasonable to assume, therefore, that the loop structure and the oligo(dA) stretch on the DNA molecule form the signals for the RNA polymerase molecule to stop transcription. In this connection it should be emphasized that termination of transcription at the end of a restriction fragment also occurs frequently after the formation of a small hairpin loop because of the switching of transcription from one strand to the other (Sugimoto et al. 1977).

Mechanism of Gene Expression

The results of in vitro transcription and translation studies suggest that expression of the filamentous phage genome is regulated mainly by a cascadelike mechanism of transcription, according to which RNA synthesis starts at different promoter sites but stops at a single, unique termination signal, which is located immediately following gene VIII. Most of the RNA chains initiated at a particular promoter site are thus the result of transcription through other promoters until the central termination signal is finally reached (Edens et al. 1976). Such a mechanism of gene expression leads to a stepwise gradient of transcriptional activities along the phage genome, with a maximum level immediately proximal to the 5′ end of the central termination site. Thus, the overall shape of the gradient is determined by the number, positions, and strengths of the promoters. The consequence of such a cascade mechanism is that genes with a low frequency of transcription are distal and those with a high frequency of transcription are proximal to the 5′ end of the central termination signal. The polypeptides encoded by genes V and VIII are synthesized both in vitro and in vivo in much larger quantities than the other phage-encoded polypeptides. Our finding that genes V and VIII are located immediately proximal to the central termination signal, in a region where the strongest phage promoters are also found, provides a qualitative explanation for the observed abundant synthesis of the polypeptides they encode.

There are several indications that a cascadelike mechanism of gene expression is operative not only during expression of the filamentous phage genomes but also during expression of the late genes of the phages T7, T4, and λ (E. Young; F. W. Studier; M. Pearson; W. Szybalski; all pers. comm.) and during the expression of the genome of φX174 (Hayashi et al. 1976). Whether this mechanism of gene expression is more universal awaits further investigation.

Studies of transcription of fragments containing the termination signal and one or more promoter sites, as well as of transcription of RF DNA, have indicated that termination is not stringent (Edens et al. 1976); about 10% of the RNA polymerase molecules leak through the termination signal. In this respect, the termination signal might also be considered as an attenuator which regulates the expression of the genes (genes III and VI; Fig. 1) immediately distal to its site.

Recently, solid evidence has been obtained for the existence of a promoter immediately preceding gene III (L. Edens et al., in prep.; Schaller et al., this volume). Furthermore, nucleotide-sequence analyses carried out in Takanami's laboratory (M. Takanami, pers. comm.) have demonstrated that the initiation site of the RNA transcribed from this promoter is located within the "hairpin loop" of the central termination signal. These observations suggest, but do not prove, that the RNA chains initiated at the gene-III promoter (promoter $X_{0.25}$; Fig. 1) are not prematurely terminated at the central terminator. The expression of gene III thus seems to be regulated in vitro by two different mechanisms: (1) the leakage of the RNA polymerase molecules through the central termination signal and (2) the efficiency of transcriptional initiation at the promoter located within the central terminator. It is not yet known to what extent the termination process influences the efficiency of initiation of transcription at the gene-III promoter, and vice versa. Furthermore, it is not known whether other factors in addition to RNA polymerase and the termination signal play a role in the efficiency with which termination of transcription at the termination signal occurs.

The cascade transcription model outlined above is based on results from in vitro studies performed in the absence of the termination factor *rho*. However, transcription studies in Takanami's laboratory and in ours have indicated that *rho*-dependent termination signals are also present on the filamentous phage genome (Takanami et al. 1971a,b; L. Edens et al., in prep.; M. Takanami, pers. comm.). L. Edens et al. (in prep.) observed that low concentrations of *rho* added to the transcription system had a dramatic effect on the sizes of the RNAs initiated at the A promoters, whereas only minor effects were observed on the RNAs initiated at the G promoters. Furthermore, it was observed that *rho* decreases the sizes of the "heterogeneous" RNA species that result from leakage of RNA polymerase through the (*rho*-independent) termination signal between genes VIII and III (Fig. 1). Although the existence of several weak *rho*-dependent termination signals cannot be ruled out, these observations indicate strongly that there is at least one site on the DNA genome that terminates very efficiently in the presence of *rho*. Our data suggest that this strong *rho*-dependent termination signal is most probably located at or beyond the 3′ end of gene IV. A good possibility for the location of this signal would be the intergenic region (IG) between genes IV and II (Fig. 1). No evidence was obtained that *rho* has a significant effect on the efficiency of termination at the termination signal located immediately following gene VIII. Our results suggest, therefore, that, in the presence of *rho*, two types of cascade transcription are operative in vitro, one dependent on and the other independent of *rho*. M. Takanami and coworkers (pers. comm.) have also found recently that a *rho*-dependent termination signal is located immediately following the gene coding for X protein (Fig. 1). Since, at least in our hands, *rho* has only a minor effect on the size and amount of the RNA species (14S) initiated at promoter $G_{0.06}$ (Fig. 1), which is located upstream of this termination signal, we suppose that termination of transcription at this site is very weak.

EXPRESSION OF THE PHAGE GENOME IN VIVO

Our view of the mechanisms that regulate expression of the filamentous phage genome in the infected cell is less well documented. One reason for this is that during the phage infection cycle no shutoff of the synthesis of host-specified RNA and proteins occurs. Consequently, an unambiguous characterization and identification of the phage-encoded products is hampered by the bulk of the gene products encoded by the host-cell genome.

Hofschneider and coworkers (Jaenisch et al. 1970; Jacob et al. 1973) have studied, by means of polyacrylamide gel electrophoresis and density gradient centrifugation, the size distributions of the phage-specific RNA species in cells infected with M13. The phage-specific RNA species were identified by hybridization of the different RNA fractions to M13 RF DNA. It was observed that with different fractionation techniques different size distributions were obtained. When RNA was fractionated under nondenaturing conditions on polyacrylamide gels, at least five different classes of RNA were obtained, with apparent molecular weights ranging from 1×10^5 to 1.3×10^6. A completely different size distribution was found when the RNAs were fractionated under denaturing conditions on dimethyl sulfoxide density gradients prior to hybridization. These RNA products ranged from 9×10^4 to 5×10^5 daltons. No evidence was obtained indicating that any RNA species made in vivo had a size equivalent to or larger than one genome length (m.w. $> 2\times10^6$). The majority of the RNA species had molecular weights (ca. 1×10^5) of the same order of magnitude as that of the smallest (8S) RNA species synthesized under the direction of RF DNA in vitro (Edens et al. 1975, 1976). These results were interpreted to mean that under nondenaturing conditions M13-specific RNA tends to form aggregates. Nevertheless, both results indicate strongly that in the infected cell a number of phage-specific RNA species are made that have sizes equivalent to those of the RNA chains synthesized under the direction of RF DNA in vitro. However, in view of the lack of comparative studies, we cannot say whether these RNA species are in fact identical.

Recently, we have developed a completely new approach for studying the expression of the M13 genome in vivo. To circumvent the problems of host-specific RNA and protein synthesis, we used an in vivo system prepared from minicells (Smits et al. 1978).

Minicells are small, DNA-deficient, nongrowing bodies produced by aberrant cell divisions at the polar end of rod-shaped bacteria; as such, they are anucleate, or DNA-deficient, and approximately spherical in shape (for a review, see Frazer and Curtiss 1975). When the parent strain harbors a plasmid, the plasmid is segregated into the minicells and confers on them the ability to synthesize plasmid-specific RNA and protein. When the parent strain is plasmid-free, the minicells lack DNA and are incapable of RNA synthesis. Since phage M13 neither lyses nor kills its host, its RF DNA may be considered as a nonconjugative, plasmidlike, extrachromosomal element that might be able to segregate efficiently into minicells, and that is in fact the case. Furthermore, it was shown that after infection of the proper minicell-producing strain, phage M13 RF DNA is replicated (cf. Staudenbauer and Hofschneider 1975); the minicells that are produced synthesize almost exclusively phage-specific RNA complementary to the nonviral strand (Smits et al. 1978). The ratios at which the different phage-specified polypeptides are made correspond to a large extent to the ratios at which these polypeptides are made in vitro. Also, the abundant synthesis of the polypeptides encoded by genes V and VIII is apparent (Fig. 5).

Analyses on polyacrylamide gels under denaturing conditions (90% formamide or 8 M urea) of the phage-specific RNA species revealed that at least eight RNA species, ranging in size from 8S to 20S, are made in minicells (Fig. 6). Almost all of the G-start RNAs synthesized in vitro have an equivalent among the RNA species made in minicells with respect to electrophoretic mobility. Furthermore, the RNA species (8S, 14S, and 19S) made in large amounts in vitro (Edens et al. 1976) are also the major products of the minicell system. They are not made in the same relative amounts, however. For instance, in comparison with the 14S RNA species, the 8S RNA product is produced in minicells in much larger amounts than in vitro. Furthermore, preliminary hybridization studies on restriction fragments and in vitro protein-synthesis studies have indicated that the major in vivo transcripts are products of the same region of the M13 genome that produces the major in vitro (G-start) transcripts. These observations suggest that termination of transcription occurs in vivo also at a site immediately following gene VIII (Fig. 1). A striking observation was that in minicells no RNA species are made that have sizes equivalent to those of the A-start RNAs synthesized in vitro. Also, no evidence was obtained indicating a precursor-product relationship between the various RNA species synthesized in vivo. In contrast to the in vitro system, we have never observed in minicells a heterogeneous mixture of RNA species with m.w. $> 2\times10^6$. These results can be explained in either of two ways: (1) none of the A promoters detected in vitro is operative in vivo, or (2) termination of transcription occurs in vivo not only at the termination signal located immediately following gene VIII, but also at a site or sites located on the genetic map downstream of the A promoters and upstream of the G promoters observed in vitro (Fig. 1). We favor the latter hypothesis because there are RNA species made in vivo that have sizes smaller than 20S and that have no counterpart among the RNA chains synthesized in vitro. Furthermore, as outlined above, there is evidence that *rho*-dependent termination sites are

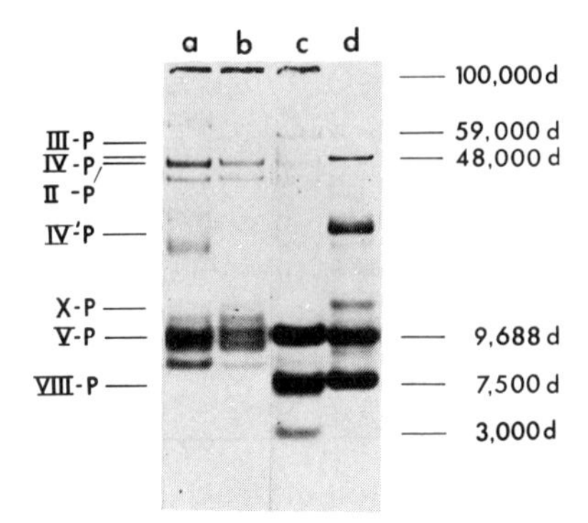

Figure 5 M13 phage-specified polypeptides synthesized either in minicells or in a DNA-dependent in vitro protein-synthesizing system. (*a*) Polypeptides synthesized in minicells produced by cells that harbor the cloacinogenic plasmid Clo DF13 (F^+ parents). (*b*) Polypeptides synthesized in minicells produced by noninfected cells (F^+ parents–M13$^-$ minicells). (*c*) Polypeptides synthesized in minicells produced by M13-infected cells (F^+ parents–M13$^+$ minicells). (*d*) Polypeptides synthesized in a DNA-dependent, cell-free system under the direction of M13 RF DNA. The identification and characterization of the M13 phage-specified polypeptides (III-P, IV-P, etc.) have been described previously (Konings et al. 1975). The polypeptides were labeled with [^{35}S]methionine and analyzed subsequently on polyacrylamide gels (Konings 1973). Note that the polypeptides encoded by genes IV and X are synthesized in much lower relative amounts in minicells than they are in vitro. Furthermore, in minicells a phage-specified polypeptide is made that has a molecular weight of about 3000. This polypeptide comigrates with the small polypeptide (C protein) present within the virion, and it is possible they are identical (G. Simons et al., unpubl.).

present on the M13 genome (cf. Takanami et al. 1971b; L. Edens, unpubl.; M. Takanami, pers. comm.).

One of the crucial questions with respect to our studies on the expression of the M13 genome is whether the mechanisms of transcription observed in vitro are also operative in the infected cell. As described in the previous paragraphs, there are several lines of evidence that suggest that this is the case. These conclusions are supported by the following observations: (1) the sum of the molecular weights of the phage-specific RNA species made in minicells is larger than the molecular weight of the codogenic (nonviral) strand, and (2) all gradient fractions obtained after fractionation on sucrose gradients (50% formamide) of the RNA species present in minicells or in infected cells contain RNA chains that can direct synthesis of the polypeptide encoded by gene VIII.

The results obtained so far suggest that the G promoters detected in vitro are also operative in vivo and that synthesis of the RNA chains initiated at these promoters is terminated at a site immediately following gene VIII. Our results suggest, furthermore, that at least one other (*rho*-dependent) termi-

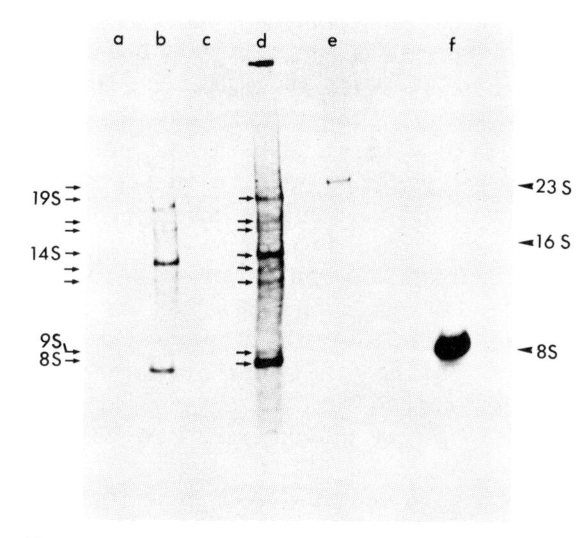

Figure 6 RNA species synthesized in minicells harboring M13 RF DNA. (*a, c*) Products synthesized in minicells produced by noninfected cells ($M13^-$ minicells). (*b, d*) Products synthesized in minicells produced by M13-infected cells ($M13^+$ minicells). (*e*) *E. coli* ribosomal RNA markers (16S and 23S). (*f*) 8S RNA marker synthesized in vitro under the direction of M13 RF DNA (Edens et al. 1976). The RNA species were labeled with [^{3}H]uridine as described elsewhere (Smits et al. 1978). In *a* and *b* the minicells were labeled for 10 min, and in *c* and *d* they were labeled for 30 min. After isolation, the RNA species were analyzed on 2.2% polyacrylamide gels in the presence of 7 M urea. Fluorographs were taken on Kodak X-ray film.

nation signal is located on the filamentous phage genome. This site is responsible for the termination of RNA chains other than the "G-start" RNAs. Thus, cascadelike mechanisms seem to be operative in the infected cell, one for the G promoters, which is *rho*-independent, and another for the other promoters, which is *rho*-dependent.

There are several indications that the mechanisms of transcription outlined above are not the only ones that regulate the synthesis of phage-specified proteins in the infected cell.

1. Genetic complementation tests have indicated that in vivo there is polarity among genes III, VI, and I and between genes V and VII (Pratt et al. 1966; Lyons and Zinder 1972; R. Benzinger, pers. comm.); expression of genes VI and I and expression of gene VII depend on the expression of gene III and gene V, respectively.
2. Translation of the individual RNA species synthesized in vitro in an uncoupled cell-free system has indicated that on a polycistronic mRNA molecule translation of the gene-VIII cistron (and of the gene-V cistron)

is favored over translation of cistrons coding for the other phage-specified polypeptides. Also, when the fractionated RNA species isolated from an infected cell were translated, they always directed the synthesis of larger amounts of gene-VIII protein (and of gene-V protein) than of the other phage-encoded polypeptides. So far, all efforts to demonstrate synthesis of gene-VII protein in vitro have been unsuccessful, despite the fact that the information coding for this polypeptide is present on all RNA chains synthesized in vitro that are larger than 8S. These observations indicate, therefore, that synthesis of phage-specified proteins is regulated at the level of translation also.

3. In vivo and in vitro protein-synthesis studies have indicated that the product of gene V can suppress specifically the synthesis of the polypeptide encoded by gene II (R. Webster and P. Model, pers. comm.; Meyer and Geider, this volume). How gene-V protein exerts this inhibitory effect is not known.
4. Studies on the degradation of phage-specified RNAs present in infected cells or in minicells have indicated that there are significant differences between the physical and functional half-lives of the mRNA molecules (Jaenisch et al. 1970; Lica and Ray 1976; LaFarina and Model 1978; M. Smits et al., unpubl.). Moreover, and what is more important, a significant difference in functional half-lives has been observed between the mRNAs coding for the gene-V and gene-VIII proteins. A transcript of gene V has a functional half-life of about 2.5 minutes. A gene-VIII transcript has a biphasic degradation pattern; part of this RNA has a half-life of about 2.5 minutes and the rest has a half-life of at least 10 minutes (Lica and Ray 1976; LaFarina and Model 1978).

FUTURE PERSPECTIVES

One of the main goals of future research will be to find out to what extent the mechanisms that have emerged from in vitro studies are valid for the transcription process in the infected cell. It has surely not escaped the notice of the readers that there are several interesting phenomena observed in vivo that have not yet been covered by the in vitro models of transcription. Undoubtedly, the control of termination by *rho* has to be studied in more detail in the near future. Other mechanisms, such as transcriptional processing (RNase III), are still poorly understood. It is also quite clear that only in vivo experiments will show definitively whether the promoters actually operate in the infected cell as postulated from in vitro studies. This is particularly interesting for those promoters that have been found to be located within other genes, such as $G_{0.06}$ and $G_{0.99}$, and probably for some others, such as those at map positions 0.75 and 0.52.

Undoubtedly, there is still a long way to go before the regulation of gene expression of the F-specific filamentous coliphages is properly understood at the molecular level. For instance, studies of the involvement of regulatory proteins and their mechanism of action have hardly begun. Other phenomena, such as the differences in the functional half-lives of phage-specific messages, the role of the host-cell membrane in this process, and the maintenance of a proper balance of proteins involved in phage assembly and extrusion of the filamentous virion through the cell envelope will be of considerable relevance to our understanding of this fundamental process of regulation of gene expression.

ACKNOWLEDGMENTS

The authors would like to thank their colleagues who provided unpublished experimental data for inclusion in this article. This research was supported in part by a grant from The Netherlands Foundation for Chemical Research (SON) with financial aid from The Netherlands Organisation for the Advancement of Pure Research (ZWO).

REFERENCES

Bertrand, K., L. Korn, F. Lee, T. Platt, C. L. Squires, C. Squires, and C. Yanofsky. 1975. New features of the regulation of the tryptophan operon. *Science* **189:** 22.

Chan, T.-S., P. Model, and N. D. Zinder. 1975. In vitro protein synthesis directed by separated transcripts of bacteriophage f1 DNA. *J. Mol. Biol.* **99:** 369.

Dahlberg, J. E. and F. R. Blattner. 1973. Sequence of a self-terminating RNA made near the origin of DNA replication of phage lambda. *Fed. Proc.* **32:** 664.

Denhardt, D. T. 1975. The single-stranded DNA phages. *CRC Crit. Rev. Microbiol.* **4:** 161.

Edens, L., R. N. H. Konings, and J. G. G. Schoenmakers. 1975. Physical mapping of the central terminator for transcription on the bacteriophage M13 genome. *Nucleic Acids Res.* **2:** 1811.

———. 1978. A cascade mechanism of transcription in bacteriophage M13 DNA. *Virology* **86:** 354.

Edens, L., P. van Wezenbeek, R. N. H. Konings, and J. G. G. Schoenmakers. 1976. Mapping of promoter sites on the genome of bacteriophage M13. *Eur. J. Biochem.* **70:** 577.

Frazer, A. C. and R. Curtiss III. 1975. Production, properties and utility of bacterial minicells. *Curr. Top. Microbiol. Immunol.* **69:** 1.

Hayashi, M., F. K. Fujimura, and M. Hayashi. 1976. Mapping of *in vivo* messenger RNAs for bacteriophage φX174. *Proc. Natl. Acad. Sci.* **73:** 717.

Henry, T. J. and D. Pratt. 1969. The proteins of bacteriophage M13. *Proc. Natl. Acad. Sci.* **62:** 800.

Heyden, B., C. Nüsslein, and H. Schaller. 1972. Single RNA polymerase binding site isolated. *Nat. New Biol.* **240:** 9.

———. 1975. Initiation of transcription within an RNA-polymerase binding site. *Eur. J. Biochem.* **55:** 147.

Horiuchi, K. and N. D. Zinder. 1976. Origin and direction of synthesis of bacteriophage f1 DNA. *Proc. Natl. Acad. Sci.* **73:** 2341.
Ikemura, T. and J. E. Dahlberg. 1973. Small ribonucleic acids of *Escherichia coli.* I. Characterization by polyacrylamide gel electrophoresis and fingerprint analysis. *J. Biol. Chem.* **248:** 5024.
Jacob, E. and P. H. Hofschneider. 1969. Replication of the single-stranded DNA bacteriophage M13: Messenger RNA synthesis directed by replicative form DNA. *J. Mol. Biol.* **46:** 359.
Jacob, E., R. Jaenisch, and P. H. Hofschneider. 1973. Replication of the single-stranded DNA bacteriophage M13. On the transcription *in vivo* of the M13 replicative-form DNA. *Eur. J. Biochem.* **32:** 432.
Jaenisch, R., E. Jacob, and P. H. Hofschneider. 1970. Replication of the small coliphage M13: Evidence for long-living M13 specific messenger RNA. *Nature* **227:** 59.
Konings, R. N. H. 1973. Synthesis of phage M13 specific proteins in a DNA-dependent cell-free system. *FEBS Lett.* **55:** 155.
———. 1974. Regulation of gene activity in bacteriophage M13 DNA. *Hoppe-Seyler's Z. Physiol. Chem.* **355:** 51.
Konings, R. N. H. and J. G. G. Schoenmakers. 1974. Bacteriophage M13 DNA-directed *in vitro* synthesis of gene 5 protein. *Mol. Biol. Rep.* **1:** 251.
Konings, R. N. H., T. Hulsebos, and C. A. van den Hondel. 1975. Identification and characterization of the *in vitro* synthesized gene products of bacteriophage M13. *J. Virol.* **15:** 570.
Konings, R. N. H., J. Jansen, T. Cuypers, and J. G. G. Schoenmakers. 1973. Synthesis of bacteriophage M13-specific proteins in a DNA-dependent cell-free system. II. In vitro synthesis of biologically active gene V protein. *J. Virol.* **12:** 1466.
LaFarina, M. and P. Model. 1978. Transcription in bacteriophage f1-infected *E. coli.* I. Translation of the RNA *in vitro*. *Virology* **86:** 368.
Lebowitz, P., S. M. Weissman, and C. M. Radding. 1971. Nucleotide sequence of a ribonucleic acid transcribed in vitro from λ phage deoxyribonucleic acid. *J. Biol. Chem.* **246:** 5120.
Lee, F., C. L. Squires, C. Squires, and C. Yanofsky. 1976. Termination of transcription *in vitro* in the *Escherichia coli* tryptophan operon leader region. *J. Mol. Biol.* **103:** 383.
Lica, L. and D. S. Ray. 1976. The functional half-lives of bacteriophage M13 gene 5 and gene 8 messages. *J. Virol.* **18:** 80.
Lyons, L. B. and N. D. Zinder. 1972. The genetic map of the filamentous bacteriophage f1. *Virology* **49:** 45.
Marvin, D. A. and B. Hohn. 1969. Filamentous bacterial viruses. *Bacteriol. Rev.* **33:** 172.
Model, P. and N. D. Zinder. 1974. In vitro synthesis of bacteriophage f1 proteins. *J. Mol. Biol.* **83:** 231.
Nüsslein, C. and H. Schaller. 1975. Stabilization of promoter complexes with a single ribonucleoside triphosphate. *Eur. J. Biochem.* **56:** 563.
Okamoto, T., M. Sugiura, and M. Takanami. 1969. Length of RNA transcribed on the replicative form DNA of coliphage fd. *J. Mol. Biol.* **45:** 101.
Okamoto, T., K. Sugimoto, H. Sugisaki, and M. Takanami. 1975. Studies on bac-

teriophage fd DNA. II. Localization of RNA initiation sites on the cleavage map of the fd genome. *J. Mol. Biol.* **95:** 33.

Pieczenik, G., B. G. Barrell, and M. L. Gefter. 1972. Bacteriophage φ80-induced low molecular weight RNA. *Arch. Biochem. Biophys.* **152:** 152.

Pieczenik, G., K. Horiuchi, P. Model, C. McGill, B. J. Mazur, G. F. Vovis, and N. D. Zinder. 1975. Is mRNA transcribed from the strand complementary to it in a DNA duplex? *Nature* **253:** 131.

Pratt, D., H. Tzagoloff, and J. Beaudoin. 1969. Conditional lethal mutants of the small filamentous coliphage M13. II. Two genes for coat proteins. *Virology* **39:** 42.

Pratt, D., H. Tzagoloff, and W. S. Erdahl. 1966. Conditional lethal mutants of the small filamentous coliphage M13. I. Isolation, complementation, cell killing, time of cistron action. *Virology* **30:** 397.

Pratt, D., H. Tzagoloff, W. S. Erdahl, and T. J. Henry. 1967. Conditional lethal mutants of coliphage M13. In *The molecular biology of viruses* (ed. J. S. Colter and W. Paranchych), p. 219. Academic Press, New York.

Pribnow, D. 1975. Nucleotide sequence of an RNA polymerase binding site at an early T7 promoter. *Proc. Natl. Acad. Sci.* **72:** 784.

Ray, D. S. 1977. Replication of filamentous bacteriophages. In *Comprehensive virology* (ed. H. Fraenkel-Conrat and R. R. Wagner), vol. 7, p. 105. Plenum Press, New York.

Sanger, F., G. M. Air, B. G. Barrell, N. L. Brown, A. R. Coulson, J. C. Fiddes, C. Hutchison III, P. M. Slocombe, and M. Smith. 1977. Nucleotide sequence of bacteriophage φX174 DNA. *Nature* **265:** 687.

Seeburg, P. and H. Schaller. 1975. Mapping and characterization of promoters in phage fd, f1, and M13. *J. Mol. Biol.* **92:** 261.

Seeburg, P. H., C. Nüsslein, and H. Schaller. 1977. Interaction of RNA polymerase with promoters from bacteriophage fd. *Eur. J. Bochem.* **74:** 107.

Smith, M., N. L. Brown, G. M. Air, B. G. Barrell, A. R. Coulson, C. A. Hutchison III, and F. Sanger. 1977. DNA sequence at the C-termini of the overlapping genes A and B in bacteriophage φX174. *Nature* **265:** 702.

Smits, M. A., G. Simons, R. N. H. Konings, and J. G. G. Schoenmakers. 1978. Expression of bacteriophage M13 DNA in vivo. I. Synthesis of phage-specific RNA and protein in minicells. *Biochim. Biophys. Acta* (in press).

Sogin, M., N. Peace, M. Rosenberg, and S. M. Weissman. 1976. Nucleotide sequence of a 5S ribosomal RNA precursor from *Bacillus subtilis. J. Biol. Chem.* **251:** 3400.

Staudenbauer, W. L. and P. H. Hofschneider. 1975. Segregation into and replication of bacteriophage M13 DNA in minicells of *Escherichia coli. Mol. Gen. Genet.* **138:** 203.

Suggs, S. V. and D. Ray. 1977. Replication of bacteriophage M13. Localization of the origin for M13 single strand synthesis. *J. Mol. Biol.* **110:** 147.

Sugimoto, K., T. Okamoto, H. Sugisaki, and M. Takanami. 1975a. The nucleotide sequence of an RNA polymerase binding site on bacteriophage fd DNA. *Nature* **253:** 410.

Sugimoto, K., H. Sugisaki, T. Okamoto, and M. Takanami. 1975b. Studies on bacteriophage fd DNA. Nucleotide sequence preceding the RNA start-site on a promoter containing fragment. *Nucleic Acids Res.* **2:** 2091.

———. 1977. Studies on bacteriophage fd DNA. IV. The sequence of messenger RNA for the major coat protein gene. *J. Mol. Biol.* **110:** 487.

Sugiura, M., T. Okamoto, and M. Takanami. 1969. Starting nucleotide sequences of RNA synthesized on the replicative form DNA of coliphage fd. *J. Mol. Biol.* **43:** 299.

———. 1970. RNA polymerase σ factor and the selection of initiation site. *Nature* **225:** 598.

Tabak, H. F., J. Griffith, K. Geider, H. Schaller, and A. Kornberg. 1974. Initiation of deoxyribonucleic acid synthesis. VII. A unique location of the gap in the M13 replicative duplex synthesized *in vitro*. *J. Biol. Chem.* **249:** 3049.

Takanami, M., T. Okamoto, and M. Sugiura. 1971a. The starting nucleotide sequences and size of RNA transcribed *in vitro* on phage DNA templates. *Cold Spring Harbor Symp. Quant. Biol.* **35:** 179.

———. 1971b. Termination of RNA transcription on the replicative form DNA of bacteriophage fd. *J. Mol. Biol.* **62:** 81.

Takanami, M., K. Sugimoto, H. Sugisaki, and T. Okamoto. 1976. Sequence of the promoter for the coat gene of bacteriophage fd. *Nature* **260:** 297.

van den Hondel, C. A. 1976. "Localization of structural genes and regulatory elements on the genome of bacteriophage M13." Ph. D. thesis, University of Nijmegen, The Netherlands.

van den Hondel, C. A. and J. G. G. Schoenmakers. 1973. Cleavage of bacteriophage M13 DNA by *Haemophilus influenzae* endonuclease R. *Mol. Biol. Rep.* **1:** 41.

———. 1975. Studies on bacteriophage M13 DNA. 1. A cleavage map of the M13 genome. *Eur. J. Biochem.* **53:** 547.

———. 1976. Cleavage maps of the filamentous bacteriophages M13, fd, f1 and ZJ/2. *J. Virol.* **18:** 1024.

van den Hondel, C. A., R. N. H. Konings, and J. G. G. Schoenmakers. 1975a. Regulation of gene activity in bacteriophage M13 DNA: Coupled transcription and translation of purified genes and gene-fragments. *Virology* **67:** 487.

van den Hondel, C. A., L. Pennings, and J. G. G. Schoenmakers. 1976. Restriction enzyme cleavage maps of bacteriophage M13: Existence of an intergenic region on the M13 genome. *Eur. J. Biochem.* **68:** 55.

van den Hondel, C. A., A. Weijers, R. N. H. Konings, and J. G. G. Schoenmakers. 1975b. Studies on bacteriophage M13 DNA. 2. The gene order of the M13 genome. *Eur. J. Biochem.* **53:** 559.

Vovis, G. F., K. Horiuchi, and N. D. Zinder. 1975. Endonuclease R·*Eco*RII restriction of bacteriophage f1 DNA in vitro: Ordering of genes V and VII, location of an RNA promoter for gene VIII. *J. Virol.* **16:** 674.

Morphogenesis of the Isometric Phages

Masaki Hayashi
Department of Biology
University of California, San Diego
La Jolla, California 92093

Virus assembly involves a wide spectrum of protein-protein and protein-nucleic acid interactions (Casjens and King 1975). Although information about morphogenetic pathways in other systems has accumulated, studies of isometric single-stranded (SS) DNA phage assembly have started only recently and references are scanty. Most information about morphogenesis in this class of phages comes from studies of φX174. We speculate that, because of their kinship, other phages in this class, such as S13 or the G phages, assemble virions in a similar way. Two basic strategies to elucidate the process of φX assembly are those that have been used in other virus systems:

1. *Identification of precursors for mature phage.*
   a. The precursors exist transiently in infected cells. They are usually pulse-labeled with radioactive material. When a pulse-labeled culture is chased with unlabeled material, the radioactivity existing in the precursors should appear in mature phage.
   b. Cells infected with conditionally lethal mutants accumulate precursors under restrictive conditions. When the infected cells are brought to permissive conditions, the accumulated precursors should be converted to mature phage.
2. *Use of in vitro systems.* Several laboratories have developed in vitro systems that synthesize viral SS DNA and infectious phage particles. Biochemical analyses of these systems provide further insight into phage morphogenesis.

Phage structures and infectious particles have been reconstituted from their separate precursors in several viral systems. Such reconstitution experiments have been useful in elucidating the sequence of events in assembly processes in other viral systems. However, such experiments have not been used extensively in the φX system.

In the case of φX, identification of precursors of mature phage in infected cells is frequently hampered by the continued synthesis of host-specific macromolecules in these cells (Lindqvist and Sinsheimer 1967a). Several methods can be used to circumvent this problem. When host cells carrying the *uvrA* mutation are irradiated with UV light prior to infection, host DNA and protein syntheses are considerably reduced. These host cells can support phage development, though at a reduced rate compared with that in nonirradiated cells (Burgess and Denhardt 1969; Gelfand and Hayashi 1969). An analogous situation occurs when host cells (*uvrA*) are treated with mitomycin. Lindqvist and Sinsheimer (1967a) found that such treatment suppresses host DNA replication without interfering with φX DNA replication and phage development.

A second problem in the identification of φX precursors is early lysis of the host cells. Lysis can be delayed either by use of the lysis-deficient (gene-*E*) mutant or by inhibition of lysis with high concentrations of mitomycin (Fukuda and Sinsheimer 1976). A combination of these procedures makes it possible to label φX-specific DNA or proteins preferentially for a prolonged period after infection.

Gene Products Required for Synthesis of Viral SS DNA

The synthesis of intact φX viral SS DNA is tightly coupled to the phage morphogenetic pathway. Several laboratories have determined the effects of conditionally lethal mutations in φX and S13 genes on the species of phage DNA synthesized in cells infected with these mutants. Table 1 summarizes the results. Two virion structural proteins, the products of genes *F* and *G*, and the nonvirion protein products of genes *A*, *B*, *C*, and *D* are required for viral DNA synthesis. Because gene-*J* mutants are not available, we do not know whether or not the gene-*J* product is required for viral DNA synthesis.

Table 1 DNA-synthesis phenotypes of φX mutants

Gene	DNA synthesis steps	Presence of gene product in 114S virion particle
A	RF → RF RF → SS	no
B	RF → SS	no
C	RF → SS	no
D	RF → SS	no
E	normal	no
F	RF → SS	yes
G	RF → SS	yes
H	SS → stable SS	yes
J	?	yes

This table is a summary of results from the following papers: Dowell and Sinsheimer (1966), Tessman (1966), Siegel and Hayashi (1969), Lindqvist and Sinsheimer (1967a,b), Iwaya and Denhardt (1971), Fujisawa and Hayashi (1976b), Tessman and Peterson (1977).

Some host functions required for φX DNA replication have been characterized (see Dumas, this volume). Viral DNA synthesis requires the products of genes *dnaE* and *dnaG* (Denhardt et al. 1973; Dumas and Miller 1973; Greenlee 1973; McFadden and Denhardt 1974). A function defined by the *rep* gene is necessary for the replication of parental replicative-form (RF) DNA (Denhardt et al. 1967). Recently, Tessman and Peterson (1976) isolated a set of *rep* mutants in which RF DNA replicates but no viral DNA is synthesized.

Structure of Replicating Intermediate DNA

Lindqvist and Sinsheimer (1968) showed that RF molecules are precursors for viral SS DNA. Subsequently, it was shown that the complementary strand of the RF molecule is conserved and that synthesis of a new viral strand in the RF molecule accompanies the displacement of the old viral strand which is encapsulated into the progeny phage (Dressler and Denhardt 1968; Knippers et al. 1968; Komano et al. 1968; Dressler 1970). The intermediate of viral SS DNA is an RF molecule with a single-stranded tail up to one genome in length. We will refer to this intermediate DNA as sigma structure (σ) (rolling-circle) DNA in this review.

Precursor Particles That Do Not Contain DNA

6S, 9S, and 12S Particles

Greenlee and Sinsheimer (1968) were the first to observe in cells infected with a lysis mutant the formation of 6S particles that were precipitable with anti-φX serum. The 6S particles can be chased into mature phage and into noninfectious 70S particles. Because the gene-*G* protein is a major antigenic determinant, they speculated that the 6S particles contain the gene-*G* protein.

Tonegawa and Hayashi (1970) reported that in host cells infected with wild-type phage and pulse-labeled with radioactive amino acids, radioactive counts are incorporated mainly into 6S, 9S, and 12S particles. These particles are chased into mature phage. Analysis of the protein compositions of these particles revealed that 6S particles contain only the gene-*G* protein, 9S particles are composed of only the gene-*F* protein, and 12S particles have both the gene-*F* and gene-*G* proteins. Tonegawa and Hayashi (1970) and Siden and Hayashi (1974) found that mutations in genes *A, D, E*, and *H* do not interfere with the formation of 6S, 9S, and 12S particles. As expected from the protein compositions of these particles, cells infected with gene-*G* mutants lack 6S and 12S particles. Mutations in gene *F* eliminate the formation of 9S and 12S particles. When infected cells lack functional gene-*B* protein, 6S and 9S particles are formed, but no 12S particles are made. Siden and Hayashi (1974) concluded that the gene-*B* protein is necessary for the association of 6S and 9S particles to form 12S particles. The gene-*B* protein seems to act as a catalyst because 12S particles do not contain gene-*B* protein. They have speculated that the gene-*B* protein may modify 6S and/or 9S particles, for example, by glycosylation, methylation, or phosphorylation, so that 6S and 9S particles can form the 12S particles.

From protein compositions and estimated molecular weights obtained by sedimentation analysis of 6S, 9S, and 12S particles, Tonegawa and Hayashi (1970) and Siden and Hayashi (1974) concluded, on the basis of icosahedral symmetry requirements, that the 6S particle is a pentamer of the gene-*G* protein, the 9S particle is a pentamer of the gene-*F* protein, and the 12S particle contains five molecules of gene-*G* protein and five molecules of gene-*F* protein. Siden and Hayashi proposed that the 12S particle could represent the structural unit of each of the twelve vertices of the icosahedral capsid of the virus (Fig. 1).

108S Particle

In sucrose gradient sedimentation analyses, Fujisawa and Hayashi (1977c) found a 108S protein complex in preparations from three different sources: (1) amber-*A*-mutant-infected cells, in which parental RF DNA is formed but neither progeny RF nor viral DNA is synthesized; (2) cells infected with an ochre mutant in gene *C*, whose product is required for synthesis of viral DNA; and (3) cells infected with a lysis-deficient mutant and starved for thymidine during the period of viral DNA synthesis. The 108S particles isolated from these cells contain the protein products of genes *F*, *G*, and *H* in the same proportions as those found in mature phage. In addition, the 108S particle contains the gene-*D* protein, which is not found in mature phage.

The 108S particle formed during thymidine starvation in cells infected with a lysis mutant is stable and maintains its structural integrity even after prolonged thymidine starvation. After release from thymidine starvation, the 108S particle is converted to 114S infectious phage and 132S infectious particles (Fujisawa and Hayashi 1977c). (The infectious 132S phage particle

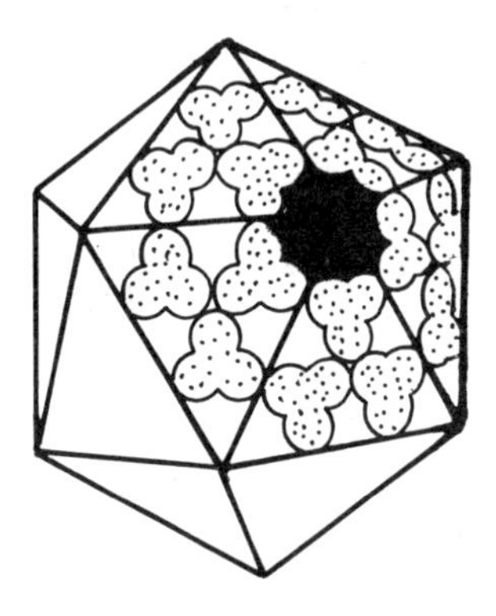

Figure 1 Schematic representation of the φX subunits in the icosahedral structure. Shaded area represents gene-*G* protein, dotted area gene-*F* protein. (Reprinted, with permission, from Edgell et al. 1969.)

is described in more detail further on in this article.) This indicates that the 108S particle is a precursor for mature phage.

Precursors That Contain DNA

50S Complex

Synthesis of viral DNA involves a replicating intermediate (sigma structure) which consists of an RF molecule with an extended tail of viral SS DNA (see above). Gilbert and Dressler (1969) proposed that viral protein aggregates with the growing viral DNA to initiate assembly of viral structure.

Fujisawa and Hayashi (1976a and unpubl.) observed that when cells infected with wild-type or lysis-deficient mutant phage are pulse-labeled with radioactive thymidine, radioactive counts are incorporated mainly into 20S and 50S particles. The radioactivity existing in these materials is chased into 132S infectious particles and 114S phage (Fig. 2). The 20S material contains RFII (circular duplex DNA in which at least one strand is not continuous). The 50S complex contains sigma-structure DNA associated with the proteins encoded by the phage genes *A* (also A*, a smaller gene-*A* product, see below), *D, F, G, H*, and *J* and several host-encoded proteins. The proportions of the products of genes *D, F, G*, and *H* in the 50S complex are the same as those found in the 108S particle. Fujisawa and Hayashi (1977c) proposed that the 50S complex is formed by the association of the 108S particle with an RFII molecule, and that synthesis of the viral DNA strand via sigma-structure DNA and the subsequent packaging of viral DNA take place in the 50S complex.

Fujisawa and Hayashi (1976b) found that when cells infected at the permissive temperature with a temperature-sensitive gene-*A* mutant are brought to the restrictive temperature, thymidine incorporation is reduced

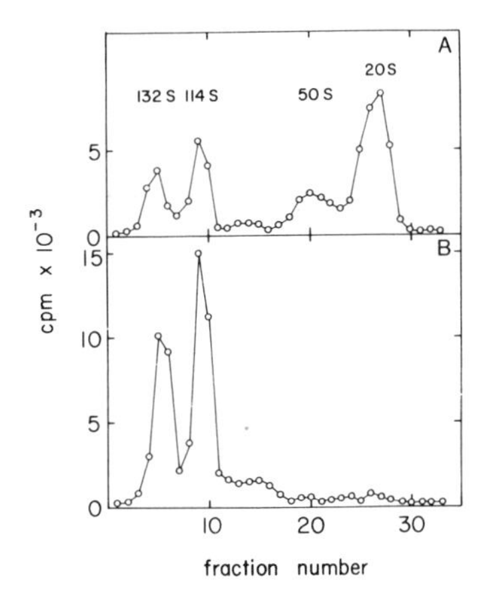

Figure 2 Sedimentation profile of precursors containing [^{3}H]thymidine-labeled DNA from extracts of cells infected with φX. Mitomycin-treated *E. coli* HF 4704 (*uvrA*) was infected with a φX lysis-deficient mutant. [^{3}H]thymidine was added to the culture 20 min after infection and chased at 20.5 min after infection by the addition of unlabeled thymidine and thymine. Samples were taken at 21 min (*a*) and 45.5 min (*b*) after infection. Cell extracts were prepared and centrifuged through a 5–30% sucrose gradient in an SW50.1 rotor at 49,000 rpm for 45 min at 4°C. Direction of sedimentation is from right to left. (Reprinted, with permission, from Fujisawa and Hayashi 1977c.)

and 50S complexes accumulate. These 50S complexes contain sigma-structure DNA with a homogeneous single-stranded tail that is one full genome in size. Upon shift-down to the permissive temperature, the 50S complexes accumulated at the restrictive temperature are converted to mature phage particles. Tessman and Peterson (1977) reported that when cells infected with a temperature-sensitive gene-*A* mutant of S13 were brought to the restrictive temperature at a later time in the infection process, production of infectious phage stopped because viral DNA synthesis ceased. These observations are consistent with the possibility that the gene-*A* product cleaves unit-genome-sized viral DNA from sigma-structure DNA in the 50S complex. In the absence of cleavage of the single-stranded tail, viral DNA replication stops after completion of one round of synthesis in the 50S complex.

Henry and Knippers (1974) purified the gene-*A* protein and showed that, in vitro, the purified preparation has an endonucleolytic activity which cleaves specifically a phosphodiester bond in the viral strand of RF DNA. They also demonstrated that φX viral SS DNA is susceptible to this endonucleolytic activity. From these experiments they proposed that gene-*A* protein functions in vivo not only to nick the viral strand of RF DNA during RF DNA replication but also to cut off unit-length pieces of viral DNA during viral DNA synthesis.

Linney et al. (1972) and Linney and Hayashi (1973) found that gene *A* codes for two proteins. The larger protein (A) has a molecular weight of about 60,000 and is the complete product of gene *A*. The smaller protein (A*) has a molecular weight of about 35,000 and seems to be synthesized in response to a start signal for protein synthesis within gene *A*. Because the 50S complex contains not only A protein but also A* protein, it is conceivable that A* protein also may be involved in the cleavage of viral DNA from sigma-structure DNA in vivo or in the circularization process of cleaved viral DNA. Extensively purified φX virions often contain a protein that comigrates with A* protein (Tonegawa and Hayashi 1970; E. Linney, pers. comm.).

*Role of Gene-*C *Protein*

Funk and Sinsheimer (1970) observed that in cells infected with an ochre mutant (*oc*6), RF molecules accumulate that sediment in preformed CsCl sedimentation gradients at the position of sigma-structure DNA. They concluded that the gene-*C* protein may function during a late stage of viral DNA synthesis, probably during the packaging of viral DNA. This conclusion was reexamined because RFI* (RFI [superhelical, covalently closed, circular, duplex DNA] complexed with φX mRNA [Hayashi 1965; Hayashi and Hayashi 1966, 1968; Fujisawa and Hayashi 1977a]) and sigma-structure DNA have similar sedimentation patterns in the gradient used by Funk and Sinsheimer in their studies (H. Fujisawa and M. Hayashi, unpubl.). Fujisawa and Hayashi (1977a) could not detect sigma-structure DNA in cells infected with an *oc*6 mutant but instead found RFI*.

Fujisawa and Hayashi (1977a) observed that in the absence of functional gene *C*, newly synthesized RFII is rapidly converted to RFI (or RFI*). This rapid conversion of RFII to RFI does not occur in cells infected with mutants of genes *B, D, F*, or *G*, suggesting that gene-*C* protein is responsible for maintaining the availability of RFII for 50S complex formation. The deficit of RFII available for 50S complex formation in cells infected with a gene-*C* mutant (*oc*6) prevents viral DNA synthesis despite the availability of the 108S particle (see above). The mechanism by which the gene-*C* protein maintains RFII as the template during viral DNA synthesis is not known. Fujisawa and Hayashi (1977a) have speculated that gene-*C* protein has DNA-binding properties and functions by attaching to RFII to prevent its conversion to RFI. Schekman et al. (1971) have reported that about 50% of the RFII molecules in infected cells have short, single-stranded tails. The gene-*C* protein might bind to these tails.

Circularization

The cleavage by the gene-*A* protein of a unit-length viral DNA strand from sigma-structure DNA in the 50S complex produces an RFII molecule and a phage precursor particle. The RFII can be recycled into further rounds of viral DNA synthesis by way of the RF pool and the 50S complex. The phage precursor released from the 50S complex might be expected to consist of linear DNA enclosed in a capsomeric structure, but such an intermediate has not been observed. This suggests that ligation of the ends of the linear SS DNA is a very rapid process, perhaps occurring instantaneously at the time of cleavage.

Denhardt (1975) has proposed a model involving strand translocation by which cleavage and ligation of a unit length of viral DNA occur simultaneously, probably via the action of the gene-*A* protein.

End Products of the φX Morphogenetic Pathway

Recently, several lines of evidence have indicated that the icosahedral 114S mature phage particle may not be the end product of intracellular maturation processes but rather is formed during isolation procedures.

Weisbeek and Sinsheimer (1974) isolated infectious particles of 140S and 114S from cells infected with φX. The 140S particle contained the products of phage genes *F, G, H*, and *J* in the same proportions as those found in the 114S phage particle. In addition, the 140S particle contained gene-*D* protein, which was not present in the 114S particle. Under appropriate conditions the 140S particle could be converted in vitro to the 114S phage particle. Fujisawa and Hayashi (1977b) found a 132S particle in phage-infected cells. This particle had virtually the same properties as those reported by Weisbeek and Sinsheimer (1974) for their 140S particle.

The 132S particle was efficiently converted to an infectious 114S particle in buffer containing Mg^{++} (or Ca^{++}, Mn^{++}). Dilution of 132S particles into 10 mM Tris buffer resulted in a loss of infectivity and the formation of defective particles of around 60S. The 114S particles were stable in this buffer (Weisbeek and Sinsheimer 1974; Fujisawa and Hayashi 1977b). In the presence of Zn^{++} the 132S particle was converted to material slightly larger than 114S. This material was not infectious. The infectivity of the 114S phage particle was not affected by Zn^{++} (Fujisawa and Hayashi 1977b).

Fujisawa and Hayashi (1977b) investigated a possible precursor-product relationship between 132S particles and 114S particles. Pulse and pulse-chase experiments demonstrated that pulse-labeled 132S particles are not converted into 114S particles after a prolonged chase (Fig. 2). These observations indicate that in vivo the 132S particle is not a precursor of the 114S particle. The 132S particle may be the mature virion made in infected cells and the 114S particle may be formed only after lysis or purification, although the remote possibility that the 132S and 114S particles are produced from a common precursor (such as the 50S complex) by two pathways cannot be eliminated.

Functions of Gene-*D* Protein

The φX gene-*D* protein is the most abundantly synthesized of all phage-encoded proteins in cells infected with wild-type phage or non–gene-*D* mutants (Burgess and Denhardt 1969; Gelfand and Hayashi 1969; Benbow et al. 1972). The gene-*D* protein is associated with several morphogenetic intermediates (108S particle, 50S complex, 132S [140S] particle). The protein compositions of these intermediates are summarized in Table 2. The relative amounts of the products of genes *F, G, H*, and *D* in these particles are very similar in each of these intermediates and in phage. Electron micrographs of 132S, 114S, and 108S particles are shown in Figure 3. The 132S particle and the 108S particle are both spherical in shape (40 nm) and are easily distinguished from the 114S phage particle, which has a smaller diameter (32 nm) and spikelike projections. Phosphotungstate stain seems to penetrate the 108S particle but not the 132S particle or the 114S phage particle, which suggests that the 108S particle has an empty-shell structure. As described above, the 108S particle is synthesized independently of viral DNA replication.

Fujisawa and Hayashi (1977c) suggested that assembly of the 108S particle is the result of aggregation of 12 of the 12S particles (capsomeric unit, see above) with the active participation of the gene-*D* product as a scaffold-

Table 2 Viral protein compositions of precursor particles and phage

Protein	50S	108S	132S	114S (phage)
F	1.00	1.00	1.00	1.00
H	0.11	0.20	0.18	0.18
G	0.38	0.36	0.37	0.36
D	0.97	1.08	1.05	0
J	0.31	0	0.20	0.20
A	0.09	0	nd[a]	0
A*	0.10	0	nd[a]	0.01 (?)

Precursor particles and phage were labeled with radioactive amino acids and analyzed by SDS-polyacrylamide gel electrophoresis as described by Gelfand and Hayashi (1969). The distribution of radioactivity is expressed in relative amounts with the radioactivity of protein F taken as 1.00. The 108S complex was isolated from cells infected with a lysis-deficient mutant under thymidine-starvation conditions. The 108S, 132S, and 114S particles were labeled with [^{14}C]lysine. The 50S complex was isolated from cells infected with a lysis-deficient mutant and labeled with [^{3}H]leucine, [^{3}H]lysine, and [^{3}H]arginine. Data from Fujisawa and Hayashi (1976a) and Fujisawa and Hayashi (1977c).

[a] nd = Not determined.

ing protein. Evidence supporting the role of gene-*D* protein in capsid assembly is the association of gene-*D* protein with several maturation intermediates and the absence of the 108S particle in cells infected with a gene-*D* mutant despite the presence of 12S particles in these cells.

The 132S infectious particles are not inactivated efficiently by antiserum against 114S phage particles (Fujisawa and Hayashi 1977a). This implies that the gene-*D* protein is on the outside of the 132S particle. The electron micrographs in Figure 3 are consistent with this interpretation concerning the location of the gene-*D* protein on the 132S particle. The similar appearance of the 108S particle and the 132S particle in electron micrographs suggests that the association of the gene-*D* protein with these two particles occurs in an identical manner.

The indications are that the 132S particle is not simply a 114S phage particle fortuitously associated with gene-*D* protein. The 114S phage particle cannot be converted to the 132S particle by incubation in infected-cell extracts containing the gene-*D* protein (Weisbeek and Sinsheimer 1974). Exposure of 132S particles to buffers of low ionic strength or to buffer containing Zn^{++} leads to loss of infectivity, whereas 114S phage particles are not affected by such treatments (see above). Evidently, the conformation of the virion particle undergoes an alteration upon removal of the gene-*D* protein from the 132S particle. "Incorrect" removal of the gene-*D* protein from the 132S particle leads to defective particles. These observations suggest an active role for gene-*D*-protein interaction with virion components during phage morphogenesis.

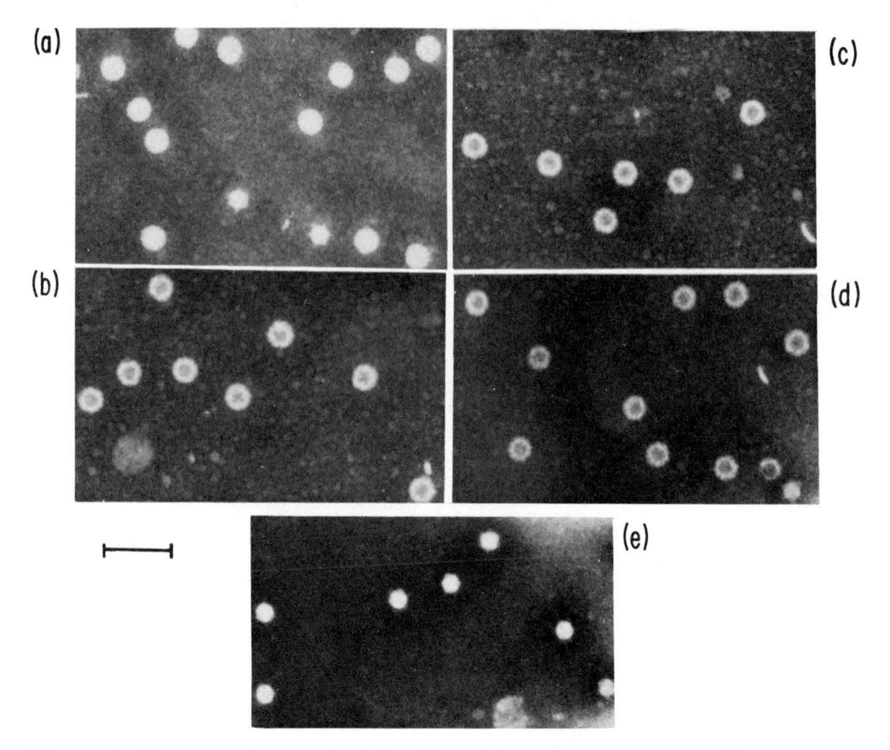

Figure 3 Electron micrographs of 114S, 132S, and 108S particles. (*a*) 132S infectious particle (114S particles are also seen in this preparation). (*b*) 108S particle from gene-*A*-mutant-infected cells. (*c*) 108S particle from gene-*C*-mutant-infected cells. (*d*) 108S particle from lysis-deficient-mutant-infected cells under thymidine starvation conditions. (*e*) 114S phage particles. All particles were negatively stained with 0.5% phosphotungstic acid (pH 7.4).

A Model for the Functions of φX Gene Products during Morphogenesis

On the basis of the information described in the preceding sections, Fujisawa and Hayashi (1977c) have proposed a model for morphogenesis in φX (Fig. 4).

Five molecules of the major capsid protein, gene-*F* protein, combine to form the 9S particle, and five molecules of the spike protein, gene-*G* protein, associate to form the 6S particle. The catalytic activity of the gene-*B* protein results in the aggregation of a 9S particle and a 6S particle to form a 12S particle. The assembly of 12 of the 12S particles into the 108S capsomeric structure requires the scaffolding function of the gene-*D* protein. The steps leading to the formation of the 108S particle occur independently of phage DNA replication. RFII molecules, which are templates for viral SS DNA synthesis, accumulate during the early stages of the infection process, either

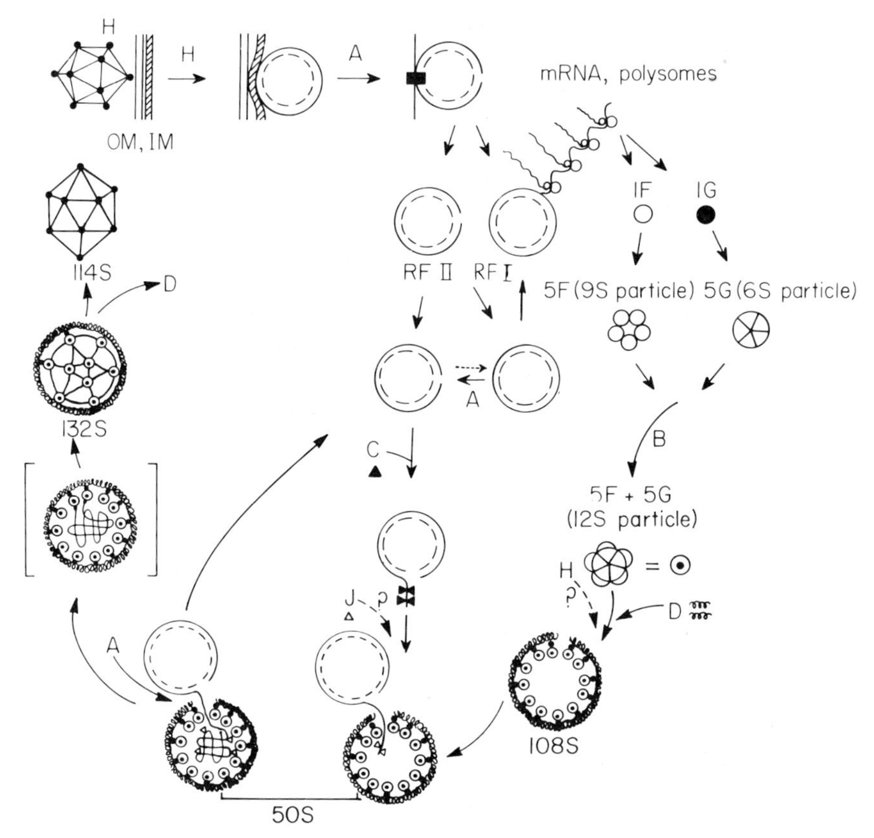

Figure 4 A model for morphogenesis in φX. This figure includes the pilot-protein model of gene-*H*-protein activity proposed by Jazwinski et al. (1975). Puga and Tessman (1973) and Hayashi and Hayashi (1966, 1968) have shown that RFI is the template for transcription in vivo, and Hayashi and Hayashi (1971) have shown that RNA polymerase has a stronger affinity for RFI than for RFII in vitro. This information is also incorporated in this figure. Additional details of the model and experimental results supporting its validity are discussed in the text. OM, outer membrane; IM, inner membrane; 1F, monomer of gene-*F* protein; 1G, monomer of gene-*G* protein; 5F, pentamer of gene-*F* protein; 5G, pentamer of gene-*G* protein. (-----) Minus strand; (——) plus (viral) strand.

directly as a result of RF DNA replication or from RFI by the endonucleolytic activity of gene-*A* protein(s). The initial stages of viral DNA replication involve RFII and the gene-*C* protein and result in the formation of RFII molecules that can associate with the 108S capsomeric structure to give rise to the 50S complex. Continuation of viral DNA synthesis by way of a sigma-structure replicating intermediate and packaging of the SS DNA into the capsomeric structure takes place in the 50S complex. As one round of viral DNA replication nears or reaches completion, the gene-*A* protein(s) cleaves the SS DNA tail of the sigma structure, resulting in the separation of RFII from a phage precursor particle. The genome of SS DNA in this precursor particle is circularized at the time of cleavage, or very soon thereafter, to form the 132S particle, which, at least in vitro, can be converted to the mature 114S virion of φX.

Functions of the Gene-*H* and Gene-*J* Products

The model of the preceding section describes possible functions for the products of φX genes *A, B, C, D, F*, and *G* during viral DNA replication and phage maturation. The roles of the products of genes *H* and *J* are not well defined. Apparently, gene-*H* protein has several functions, acting at both early and late periods of infection. Jazwinski et al. (1975) have proposed that the gene-*H* product functions during the early stages of phage infection as a pilot protein to direct the incoming viral DNA to the DNA replication site(s) on the host membrane. Siegel et al. (1968) and Siegel and Hayashi (1969) observed two kinds of defective particles in cells (sus^-) infected with amber mutants of gene *H*. One (110S particle) was not infectious, but it contained SS DNA that produced mature phage in spheroplasts prepared from suppressor-containing cells. The other (70S particle) contained linear SS DNA shorter than unit length. Neither 70S particles nor the DNA in 70S particles were infectious. K. Spindler (pers. comm.) found that both kinds of particles contained the products of genes *F*, *G*, and *J*. It is possible that phage assembly in the absence of functional gene-*H* protein leads to a defective particle whose DNA is easily attacked by nuclease, thus resulting in particles with degraded SS DNA (Iwaya and Denhardt 1971). Alternatively, a deficiency of gene-*H* protein may increase the frequency of aberrant termination of viral DNA synthesis prior to completion of one round of replication. In support of this possibility, K. Spindler (pers. comm.) has observed that the 110S particle is not a precursor of the 70S particle. In cells infected with wild-type phage the 50S complex and the 108S particle contain the gene-*H* protein. Therefore, the gene-*H* protein must enter into the proposed pathway (Fig. 4) somewhere between the 12S particle and the 108S material. However, the formation in the absence of gene-*H* protein of 110S particles containing SS DNA (Siegel and Hayashi 1969) suggests that the gene-*H* protein is not essential for the major steps in the maturation of the 110S particle. H. Fujisawa and M. Hayashi (unpubl.) have identified a particle that is identical to the 12S particle except for the presence of a small amount of gene-*H* protein. Whether or not this particle is a precursor of mature phage is not clear.

Another gene product whose function during the maturation process is unknown is the gene-*J* protein. Because of the unavailability of mutants of this gene, no function has been assigned to the gene-*J* protein. The 50S complex, but not the 108S particle, contains the gene-*J* protein. We speculate that the gene-*J* protein enters the proposed pathway (Fig. 4) somewhere between RFII and the 50S complex. The gene-*J* protein is preferentially labeled with lysine, arginine, or glycine (E. Linney, pers. comm.), which may indicate its DNA-binding (or DNA-condensing) characteristics.

φX Morphogenesis Studied in In Vitro Systems

Our understanding of the morphogenetic pathways in φX development would be greatly facilitated by the development of in vitro systems capable of producing mature virions from precursor elements. Early attempts to synthesize φX viral DNA have been reported (Denhardt and Burgess 1969; Linney and Hayashi 1970). The activity of viral DNA synthesis in these systems was low; the synthesized viral DNA could be detected only by DNA-DNA hybridization studies. Since then, our knowledge concerning the functions of the gene products has broadened as we have become more experienced in handling factors and precursor materials. We expect rapid progress in in vitro studies of viral DNA synthesis and assembly of φX.

Eisenberg et al. (1976) and Scott et al. (1977) have developed an in vitro system in which viral circular DNA (stage II[+]) is synthesized. The system consists of an RFI template and four protein components: φX-encoded gene-*A* protein, host-originated *rep* protein, and the DNA polymerase III holoenzyme, all of which act catalytically, and the *Escherichia coli* DNA-binding protein, which is needed stoichiometrically. RFI is cleaved to form RFII. *Rep*, ATP, and Mg^{++} separate the strands of the duplex DNA concurrent with the cleavage of ATP and the binding of DNA-binding protein to the viral strand of the RFII molecule. DNA polymerase III synthesizes viral DNA to fill the gap in RFII created by the peeling off of the viral strand. The intermediate in the synthesis is a sigma structure with viral DNA complexed with DNA-binding protein. The only phage-encoded protein required by this system is the gene-*A* protein. Therefore, the synthesis of sigma-structure DNA in this system does not have the same requirements found in vivo. Nevertheless, this system may provide information concerning cleavage and subsequent end-to-end joining of viral DNA.

Sumida-Yasumoto and Hurwitz (1977) reported an in vitro system that synthesized SS DNA of φX. Synthesis of SS DNA required the *E. coli* host-originated proteins *dnaB*, *dnaC*, *dnaG*, *dnaZ*, and *rep*, φX gene-*A* product, φX-encoded proteins, and RFI DNA. SS DNA synthesis in this system involved formation of a DNA-protein complex that sedimented at 50S, 60–70S, and higher. The 50S complex contained sigma-structure DNA. The DNA in the 60–70S region contained a mixture of circular and linear SS DNA. A fraction of newly synthesized SS DNA sedimented at 132S. This fraction contained infectious particles.

Mukai and Hayashi (1977a) have shown that a reaction mixture contain-

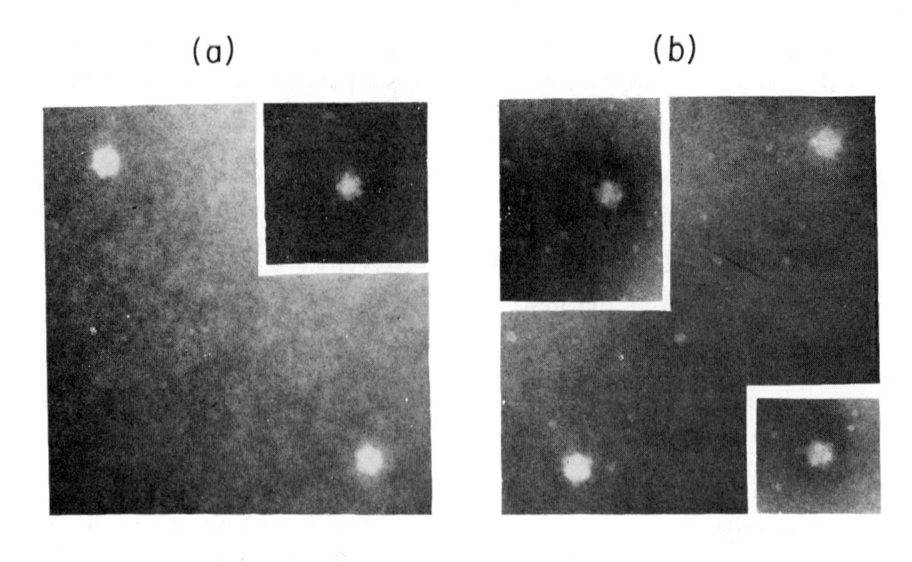

Figure 5 Electron micrographs of infectious φX phage particles formed in an in vitro system that synthesized φX circular DNA. (*a*) In vivo 114S phage particles; (*b*) in vitro infectious phage particles.

ing an extract prepared from cells infected with a φX lysis-deficient mutant is capable of synthesizing viral circular SS DNA in vitro. The DNAs synthesized in this system are associated with complexes of 20S, 50S, and 90–110S. The DNA isolated from the 20S material is RFII; the 50S material contains sigma-structure DNA; and the 90–110S material contains viral circular and linear SS DNA of full genome length. Pulse and pulse-chase experiments indicate that the 20S material is converted into 50S material, which is then converted into 90–110S material. Thus, this in vitro system mimics the in vivo mode of SS DNA synthesis. Mukai and Hayashi (1977b) showed by in vitro complementation experiments that the system required φX-encoded proteins for the synthesis of SS DNA. The reaction mixture contained extract from either gene-*B*-mutant-infected cells or gene-*C*-mutant-infected cells and synthesized material sedimenting at 20S (RFII). When the two extracts were included in the reaction mixture, the system synthesized not only 20S material but also 50S and 110S material. The 110S material contained infectious particles that clearly resembled mature φX particles as seen by electron microscopy (Fig. 5).

ACKNOWLEDGMENTS

I am grateful to Dr. Frank Fujimura and Ms. Peggy Schmitt for invaluable help with the manuscript. Our own unpublished work presented here was supported by grants from the National Science Foundation (BM74-10710) and the National Institute of General Medical Sciences, U.S. Public Health Service (GM-12934).

REFERENCES

Benbow, R. M., R. F. Mayol, J. C. Picchi, and R. L. Sinsheimer. 1972. Direction of translation and size of bacteriophage φX174 cistrons. *J. Virol.* **10:** 99.

Burgess, A. B. and D. T. Denhardt. 1969. Studies on φX174 proteins. I. Phage specific proteins synthesized after infection of *E. coli. J. Mol. Biol.* **44:** 377.

Casjens, S. and J. King. 1975. Virus assembly. *Annu. Rev. Biochem.* **44:** 555.

Denhardt, D. T. 1975. The single-stranded DNA phages. *CRC Crit. Rev. Microbiol.* **4:** 161.

Denhardt, D. T. and A. B. Burgess. 1969. DNA replication *in vitro.* I. Synthesis of single-stranded φX174 DNA in crude lysates of φX-infected *E. coli. Cold Spring Harbor Symp. Quant. Biol.* **33:** 449.

Denhardt, D. T., D. H. Dressler, and A. Hathaway. 1967. The abortive replication of φX174 DNA in a recombination-deficient mutant of *Escherichia coli. Proc. Natl. Acad. Sci.* **57:** 813.

Denhardt, D. T., M. Iwaya, G. McFadden, and G. Schochetman. 1973. The mechanism of replication of φX174 single-stranded DNA. VI. Requirements for ribonucleoside triphosphates and DNA polymerase III. *Can. J. Biochem.* **51:** 1588.

Dowell, C. E. and R. L. Sinsheimer. 1966. The process of infection with bacteriophage φX174. IX. Studies on the physiology of three φX174 temperature sensitive mutants. *J. Mol. Biol.* **16:** 374.

Dressler, D. 1970. The rolling circle for φX DNA replication. II. Synthesis of single-stranded DNA circles. *Proc. Natl. Acad. Sci.* **67:** 1934.

Dressler, D. H. and D. T. Denhardt. 1968. On the mechanisms of φX single-stranded DNA synthesis. *Nature* **219:** 346.

Dumas, L. B. and C. A. Miller. 1973. Replication of bacteriophage φX174 DNA in a temperature-sensitive *dna*E mutant of *E. coli* C. *J. Virol.* **11:** 848.

Edgell, M. H., C. A. Hutchison III, and R. L. Sinsheimer. 1969. The process of infection with bacteriophage φX174. XXVIII. Removal of the spike proteins from the phage capsid. *J. Mol. Biol.* **42:** 547.

Eisenberg, S., J. F. Scott, and A. Kornberg. 1976. Enzymatic replication of viral and complementary strands of duplex DNA of phage φX174 proceeds by separate mechanisms. *Proc. Natl. Acad. Sci.* **73:** 3151.

Fujisawa, H. and M. Hayashi. 1976a. Viral DNA synthesizing intermediate complex isolated during assembly of bacteriophage φX174. *J. Virol.* **19:** 409.

———. 1976b. Gene A product of φX174 is required for site-specific endonucleolytic cleavage during single-stranded DNA synthesis *in vivo. J. Virol.* **19:** 416.

———. 1977a. Functions of gene C and gene D products of bacteriophage φX174. *J. Virol.* **21:** 506.

———. 1977b. Two infectious forms of bacteriophage φX174. *J. Virol.* **23:** 439.

———. 1977c. Assembly of bacteriophage φX174: Identification of virion capsid precursor and proposal of a model for the functions of bacteriophage gene products during morphogenesis. *J. Virol.* **24:** 303.

Fukuda, A. and R. L. Sinsheimer. 1976. The process of infection with bacteriophage φX174. XL. Viral DNA replication of φX174 mutants blocked in progeny single-stranded DNA synthesis. *J. Virol.* **18:** 218.

Funk, F. D. and R. L. Sinsheimer. 1970. Process of infection with bacteriophage φX174. XXXV. Cistron VIII. *J. Virol.* **6:** 12.

Gelfand, D. H. and M. Hayashi. 1969. Electrophoretic characterization of φX174 specific proteins. *J. Mol. Biol.* **44:** 501.

Gilbert, W. and D. Dressler. 1969. DNA replication: The rolling circle model. *Cold Spring Harbor Symp. Quant. Biol.* **33:** 473.

Greenlee, L. L. 1973. Replication of bacteriophage φX174 in a mutant of *E. coli* defective in the *dna*E gene. *Proc. Natl. Acad. Sci.* **70:** 1757.

Greenlee, L. L. and R. L. Sinsheimer. 1968. The process of infection with bacteriophage φX174. XVIII. Intracellular antiserum-precipitable components. *J. Mol. Biol.* **32:** 321.

Hayashi, M. 1965. A DNA-RNA complex as an intermediate of *in vitro* genetic transcription. *Proc. Natl. Acad. Sci.* **54:** 1736.

Hayashi, M. N. and M. Hayashi. 1966. Participation of DNA-RNA hybrid complex in *in vivo* genetic transcription. *Proc. Natl. Acad. Sci.* **55:** 635.

———. 1968. The stability of native DNA-RNA complexes during *in vivo* φX174 transcription. *Proc. Natl. Acad. Sci.* **61:** 1107.

Hayashi, Y. and M. Hayashi. 1971. Template activities of the φX174 replicative allomorphic DNAs. *Biochemistry* **10:** 4212.

Henry, T. J. and R. Knippers. 1974. Isolation and function of the gene A initiator of bacteriophage φX174, a highly specific DNA endonuclease. *Proc. Natl. Acad. Sci.* **71:** 1549.

Iwaya, M. and D. T. Denhardt. 1971. The mechanism of replication of φX174 single-stranded DNA. II. The role of viral proteins. *J. Mol. Biol.* **57:** 159.

Jazwinski, S. M., R. Marco, and A. Kornberg. 1975. The gene H spike protein of bacteriophage φX174 and S13. II. Relation to synthesis of the parental replicative form. *Virology* **66:** 294.

Knippers, R., T. Komano, and R. L. Sinsheimer. 1968. The process of infection with bacteriophage φX174. XXI. Replication and fate of the replicative form. *Proc. Natl. Acad. Sci.* **59:** 577.

Komano, T., R. Knippers, and R. L. Sinsheimer. 1968. The process of infection with bacteriophage φX174. XXII. Synthesis of progeny single-stranded DNA. *Proc. Natl. Acad. Sci.* **59:** 911.

Lindqvist, B. H. and R. L. Sinsheimer. 1967a. Process of infection with bacteriophage φX174. XIV. Studies on macromolecular synthesis during infection with a lysis defective mutant. *J. Mol. Biol.* **28:** 87.

———. 1967b. The process of infection with bacteriophage φX174. XV. Bacteriophage DNA synthesis in abortive infections with a set of conditional lethal mutants. *J. Mol. Biol.* **30:** 69.

———. 1968. The process of infection with bacteriophage φX174. XVI. Synthesis of the replicative form and its relationship to viral single-stranded DNA synthesis. *J. Mol. Biol.* **32:** 285.

Linney, E. A. and M. Hayashi. 1970. *In vitro* synthesis of φX174 single-stranded DNA. *Biochem. Biophys. Res. Commun.* **41:** 669.

———. 1973. Two proteins of gene A of φX174. *Nat. New Biol.* **245:** 6.

Linney, E. A., M. N. Hayashi, and M. Hayashi. 1972. Gene A of φX174. I. Isolation and identification of the products. *Virology* **50:** 381.

McFadden, G. M. and D. T. Denhardt. 1974. Mechanism of replication of φX174 single-stranded DNA. IX. Requirement for the *E. coli dna*G protein. *J. Virol.* **14:** 1070.

Mukai, R. and M. Hayashi. 1977a. *In vitro* system that synthesizes circular viral DNA of bacteriophage φX174. *J. Virol.* **22:** 619.

Mukai, R. and M. Hayashi. 1977b. Synthesis of infectious φX174 bacteriophage in vitro. *Nature* **270:** 364.

Puga, A. and I. Tessman. 1973. Mechanism of transcription of bacteriophage S13. I. Dependence of mRNA on amount and configuration of DNA. *J. Mol. Biol.* **75:** 83.

Schekman, R. W., M. Iwaya, K. Bromstrup, and D. T. Denhardt. 1971. The mechanism of replication of φX174 single-stranded DNA. III. An enzymic study of the structure of replicative form II DNA. *J. Mol. Biol.* **57:** 177.

Scott, J. F., S. Eisenberg, L. L. Bertsch, and A. Kornberg. 1977. A mechanism of duplex DNA replication revealed by enzymatic studies of phage φX174: Catalytic strand separation in advance of replication. *Proc. Natl. Acad. Sci.* **74:** 193.

Siden, E. J. and M. Hayashi. 1974. Role of the gene B product in bacteriophage φX174 development. *J. Mol. Biol.* **89:** 1.

Siegel, J. E. D. and M. Hayashi. 1969. φX174 bacteriophage mutants which affect deoxyribonucleic acid synthesis. *J. Virol.* **4:** 400.

Siegel, J. E. D., M. N. Hayashi, and M. Hayashi. 1968. φX174 coat protein mutants affecting DNA synthesis. *Biochem. Biophys. Res. Commun.* **31:** 774.

Sumida-Yasumoto and J. Hurwitz. 1977. Synthesis of φX174 viral DNA in vitro depends on φX replicative form DNA. *Proc. Natl. Acad. Sci.* **74:** 4195.

Tessman, E. S. 1966. Mutants of bacteriophage S13 blocked in infectious DNA synthesis. *J. Mol. Biol.* **17:** 218.

Tessman, E. S. and P. K. Peterson. 1976. Bacterial *rep*$^-$ mutations that block development of small DNA phages late in infection. *J. Virol.* **20:** 400.

———. 1977. Gene A protein of bacteriophage S13 is required for single-stranded DNA synthesis. *J. Virol.* **21:** 806.

Tonegawa, S. and M. Hayashi. 1970. Intermediates in the assembly of φX174. *J. Mol. Biol.* **48:** 219.

Weisbeek, P. J. and R. L. Sinsheimer. 1974. A DNA-protein complex involved in bacteriophage φX174 particle formation. *Proc. Natl. Acad. Sci.* **71:** 3054.

Adsorption and Eclipse Reactions of the Isometric Phages

Nino L. Incardona
Microbiology Department
Center for the Health Sciences
University of Tennessee
Memphis, Tennessee 38163

Recent advances in the sequencing of the nucleic acid and capsid proteins of φX174 provide an opportunity to examine in more detail the early interactions between this virus and its host. The objectives of such studies are usually twofold. First, these processes involve both the formation and disruption of noncovalent interactions between proteins, nucleic acids, and, in the case of some viral receptor sites, carbohydrates. Therefore, the adsorption and nucleic-acid-injection steps of a relatively simple virus such as φX can serve as a model system for elucidating the precise structural features of these interactions and the roles they play in determining the rate of these reactions in vivo. Second, an understanding of the mechanisms by which viruses deliver their genomes to host replication systems may lead to the development of antiviral agents for prevention of and therapy for viral diseases.

Adsorption

In the earliest studies (Fujimura and Kaesberg 1962) the adsorption of infectious virus to intact cells was shown to follow pseudo first-order kinetics in the presence of excess bacteria, as is the case with most phage systems. The requirement for divalent cations and pH dependence were established. The latter seems noteworthy since the rate increased dramatically between pH 6.0 and pH 7.5, suggesting that a particular class of ionizable groups on either the viral capsid proteins or the host receptor plays an important role. The values of the rate constants in the simple aqueous solvents of this study compare reasonably well with those obtained subsequently in a complex starvation buffer (Newbold and Sinsheimer 1970b) and in nutrient broth (Bayer and Starkey 1972).

The reversibility of the adsorption step was examined by Fujimura and Kaesberg (1962) at several temperatures between 11°C and 47°C. However, their conclusions must be considered in the light of more recent work (Newbold and Sinsheimer 1970a,b; N. L. Incardona, unpubl.). Newbold and Sinsheimer (1970a,b) showed that the in vivo eclipse reaction, as defined by the molecular event that leads to the loss of infectivity of the incoming virus, results in a partial ejection of the viral DNA. Under conditions where this eclipse reaction occurs, the adsorption step will therefore appear irreversible when measured by the disappearance of infectious virus. Moreover, the large activation energy of this reaction (Newbold and Sinsheimer 1970b; Incardona 1974) is the basis for its rate approaching zero at about 15°C. As a consequence of these features, the reversibility of virus attachment must be examined at temperatures below 15°C and by methods that will not perturb the equilibrium. Preliminary results in our laboratory indicate that adsorption between 2°C and 15°C exhibits a typical approach-to-equilibrium kinetics and that the equilibrium concentration of unadsorbed virus is a function of the initial concentrations of virus and bacteria. These findings clearly indicate that the reversible nature and temperature dependence of the adsorption reaction must be reexamined.

Some insight into the precise sequence of molecular events in the initial encounter between the virus and its host has been gained from recent structural studies on the host receptor (Jazwinski et al. 1975a; Feige and Stirm 1976). The initial observation that cell-wall fragments inactivate φX (Fujimura and Kaesberg 1962) was confirmed and shown to be strain-specific (Brown et al. 1971; Beswick and Lunt 1972; Neuwald 1975). Moreover, these later studies established the fact that the inactivation at 37°C is a result of the eclipse reaction—the virus undergoes the structural transition associated with this step and a portion of its DNA becomes sensitive to DNase digestion (Newbold and Sinsheimer 1970a). A clue to the identity of the receptor component in these cell-wall fragments came from the phage sensitivity patterns of *Salmonella typhimurium* lipopolysaccharide (LPS) mutants (Lindberg 1973). That the receptor site is indeed LPS was demonstrated with purified LPS from *E. coli* C strains (Incardona and Selvidge 1973). Not only does the virus attach to the LPS and suffer the partial ejection of its DNA, but also the kinetics of these two processes are almost identical to those measured with intact cells. A comparison of φX adsorption to the core LPS mutants of *S. typhimurium* (Jazwinski et al. 1975a) and to *E. coli* C cells (Feige and Stirm 1976) reveals the requirement for a terminal (1→2)-linked hexose on the LPS core structure. If this residue is sterically blocked by the O side-chain antigen or removed from the core structure, the attachment rate is reduced tenfold. An indication of the contribution made by the capsid proteins is the difference observed with S13 phage (Jazwinski et al. 1975a). Removal of both the (1→3)-linked *N*-acetylglucosamine and the (1→2)-linked glucose from the *S. typhimurium* LPS core is required to produce the same effect for S13 as that observed with φX after the removal of only the *N*-acetylglucosamine. As far as the eclipse reaction is concerned, there is an additional structural requirement for the lipid A portion of the core LPS which is not necessary for attachment. With this structural detail as a background and the availability of well-characterized LPS mutants, it should be possible to gain more precise information about the viral capsid structure responsible for recognition of the host receptor.

A beginning has been made toward identifying the capsid proteins involved in the adsorption step. A variety of electron microscopic studies have revealed that the virus particles appear to become attached to the cell by short projections, or "spikes," located at the 12 icosahedral vertices. The particles are submerged into the cell wall to about one-half their diameter (Brown et al. 1971). Initially it was thought that the attached particles were preferentially oriented with their fivefold axis of symmetry perpendicular to the cell surface, but later stereomicrographs revealed a random distribution of all possible orientations (Bayer and Starkey 1972). More intriguing was the observation of a high frequency of adsorbed particles at the points of adhesion between the cell wall and the inner membrane. This led Bayer and Starkey (1972) to speculate that there may be either two types of receptor site or lateral movement of the virus on the cell surface prior to the DNA-injection step. The localization of the viral gene-*G* and gene-*H* proteins in these "spikes" (Edgell et al. 1969) focused attention on their role in the recognition of the host receptor site.

At the present time there is evidence implicating the major capsid proteins in host recognition. Tessman (1965) found several S13 gene-*F* mutants that were affected in their host-range and adsorption properties. In the case of gene *G*, three different host-range mutations map in or near this gene. The mapping studies were done with a DNA-fragment assay applied to fragments of varying length produced by DNase digestion (Weisbeek et al. 1973). In addition, a twofold lower adsorption rate was observed with a temperature-sensitive (ts) gene-*G* mutant (Newbold and Sinsheimer 1970b). As for the gene-*H* protein, Jazwinski et al. (1975b) demonstrated that the gene-*H* protein in crude extracts of φX-infected cells can bind to viral single-stranded DNA and increase its efficiency for transfection of spheroplasts. This enhancement is inhibited by antibodies to gene-*H* protein and by the purified LPS receptor from sensitive cells. A *ts* mutation in gene *H* produces defective 110S particles at 40°C which fail to attach to cells at 37°C in starvation buffer (Newbold and Sinsheimer 1970b). Moreover, when one considers the 5:1 molar ratio of gene-*G* protein to gene-*H* protein (Burgess 1969) in the viral "spikes," it would not be surprising to find that both proteins are involved in binding to the receptor site. To maintain the fivefold axis of symmetry at the icosahedral vertices, the simplest model would have the five gene-*G* proteins surrounding one gene-*H* protein. Since electron microscopy studies show the particles deeply submerged in the cell wall, both "spike" proteins would be in intimate contact with the cell.

Eclipse

The conditions defining the in vivo eclipse reaction have been described by Newbold and Sinsheimer (1970a) and Francke and Ray (1971). That complete injection of the viral DNA requires active metabolism was demonstrated by experiments in which infection was carried out with starved cells in buffer without an energy source. Under these conditions, the noninfectious virus particles, which were eluted from the host cells with EDTA, had a lower sedimentation coefficient as a result of extrusion of a portion of their DNA. Furthermore, little or no DNA was recovered in the parental replicative form (RF). Newbold and Sinsheimer (1970b) studied the eclipse reaction as a function of temperature. They observed that the rate of eclipse was essentially zero at 15°C and that fully infectious, unaltered virus could be eluted at this lower temperature. This difference in temperature dependence between adsorption and eclipse forms the basis for studying the kinetics of the in vivo eclipse reaction (Newbold and Sinsheimer 1970b; Incardona 1974).

As for the role of the capsid proteins in the eclipse reaction, several molecules of gene-*H* protein were found associated with the parental double-stranded RF DNA in the host when the complete injection occurred in growth medium (Jazwinski et al. 1975c). The molecules were reported to be bound to at least seven of the nine *Hin*dII restriction endonuclease fragments of φX RF DNA, thus ruling out a unique binding site for the injected gene-*H* protein on the incoming DNA. These data raise interesting questions concerning the precise mechanism of DNA injection. For example, how are the gene-*H* proteins distributed among the spikes so that more than one can be attached to the injected DNA? Does this mean that injection can occur through only one, or a few, of the 12 vertices? The combined use of mutants and in vitro systems will be required to answer these questions.

Several in vitro systems could be used for studying the eclipse reaction. These include the cell-wall-fragment systems (Brown et al. 1971; Beswick and Lunt 1972; Neuwald 1975) and the purified LPS systems (Incardona and Selvidge 1973; Jazwinski et al. 1975a) discussed above. In addition, we have developed a chemically simple eclipse reaction that consists of purified φX without any host receptor structure (Incardona et al. 1972). When the calcium ion concentration is increased 100-fold above the 10^{-3} M required for the in vivo and LPS systems, a DNA ejection reaction occurs with features that are similar, if not identical, to the features of the in vivo reaction. This in vitro inactivation reaction is accompanied by a decrease in the sedimentation coefficient of the virus without a loss of DNA; also, it has the same pH dependence, divalent ion requirement (Incardona et al. 1972), and activation energy as the in vivo reaction (Incardona 1974). However, at all temperatures the rate of the reaction is tenfold lower than that measured in vivo. In terms of the Arrhenius expression for the rate constant

$$k = A\mathrm{e}^{-\frac{E_a}{RT}},$$

the preexponential factor, A, is tenfold smaller in vitro than in vivo but the activation energy, E_a, is the same.

It is thought that in most enzyme-catalyzed reactions the catalyst increases the rate of the uncatalyzed reaction by decreasing the activation energy. If the above interpretation of the eclipse kinetics is correct, it means that this reaction represents an example where the interaction with the catalyst, in this case the host receptor site, increases the rate by increasing the preexponential factor. Since both parameters contribute to the magnitude of the rate constant, certain structural features of the reacting components would be expected to determine the numerical value of each parameter. The availability of a cold-sensitive (cs) eclipse mutant (Dowell 1967) provided a means for testing this hypothesis. A comparison of the in vitro and in vivo eclipse kinetics of *cs*70 with wild type revealed no difference in activation energy between the mutant and wild type. However, the mutant had a threefold lower preexponential factor than the wild type for the in vivo reaction but not for the in vitro reaction. Since interaction with the receptor site was not possible in the in vitro reaction, the lower value was interpreted as due to a mutational alteration in the viral protein site which involved only that portion of the molecular mechanism that determines A and not the portion that contributes to the value of E_a. Although the results with this mutant do not prove the above kinetic hypothesis for the eclipse reaction, they suggest that the interpretation can be subjected to a more critical test by the isolation of mutants with different activation energies.

The rationale behind a genetic selection technique for such eclipse mutants is based on the Arrhenius plots for the in vivo eclipse reaction of wild-type φX and for several potential A or E_a mutants (Fig. 1). Only the simplest cases are included to illustrate the principle.

A mutagenized wild-type stock is grown through one cycle of replication in order to express the mutations in the viral capsid proteins. The next step involves an enrichment for mutants with a lower eclipse rate than wild type at 37°C. The virus is adsorbed to cells at 15°C and then the virus-cell complexes are raised to 37°C for three half-times of the eclipse reaction, at which time the complexes are dissociated by dilution into EDTA at 0°C. Over 90% of the wild type and the mutants that have the same rate of eclipse as wild type at 37°C are inactivated by this treatment, whereas the slow eclipse mutants are still infectious. As can be seen from Figure 1, mutants with either a lower A or a higher E_a will be enriched by this procedure. In the last step the surviving plaques are screened for cs phenotype. The expectation of the cs phenotype is based on the results with *cs*70, where the only physiological defect observed in its replication at the nonpermissive temperature of 25°C was its lower eclipse rate (Incardona 1974; Segal and Dowell 1974). Thus, the failure to form plaques at 25°C must be because the eclipse step is rate-limiting and the cells in the surrounding lawn outgrow the virus. As in the case of the enrichment step, both lower A and higher E_a mutants will have the cs phenotype.

One result of such a selection procedure is that there is a tenfold greater frequency of cs mutants over the unenriched stock. Furthermore, over 50%

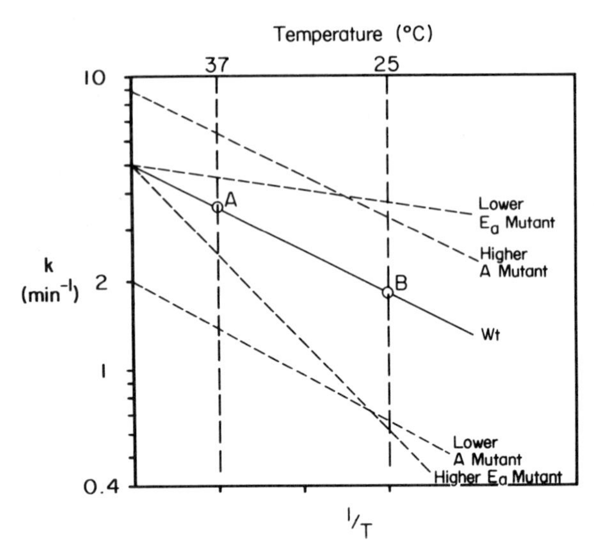

Figure 1 Arrhenius plots for the in vivo eclipse reaction of wild type (wt) and of hypothetical mutants having different values of the fundamental kinetic parameters.

of the cs mutants have a lower in vivo eclipse rate at 37°C. Preliminary kinetic measurements at 25°C suggest that some of these mutants may indeed have a higher E_a (N. L. Incardona, in prep.). Whether or not the interpretation of the kinetic parameters is correct, these new mutants are clearly defective in the virus–host-cell-receptor interactions. With the rapid DNA sequencing methods that are now available it will be possible not only to identify the viral capsid proteins involved, but also to begin to map the active site or region in the protein structure responsible for the interactions.

REFERENCES

Bayer, M. E. and T. W. Starkey. 1972. The adsorption of bacteriophage φX174 and its interaction with *Escherichia coli*, a kinetic and morphological study. *Virology* **49:** 236.

Beswick, F. M. and M. R. Lunt. 1972. Adsorption of bacteriophage φX174 to isolated bacterial cell walls. *J. Gen. Virol.* **16:** 381.

Brown, D. T., J. M. MacKenzie, and M. E. Bayer. 1971. Mode of host cell penetration by bacteriophage φX174. *J. Virol.* **7:** 836.

Burgess, A. B. 1969. Studies on the proteins of φX174. II. The protein composition of the φX coat. *Proc. Natl. Acad. Sci.* **64:** 613.

Dowell, C. E. 1967. Cold-sensitive mutants of bacteriophage φX174. I. A mutant blocked in the eclipse function at low temperature. *Proc. Natl. Acad. Sci.* **58:** 958.

Edgell, M. H., C. A. Hutchison III, and R. L. Sinsheimer. 1969. The process of infection with bacteriophage φX174. XXVIII. Removal of the spike proteins from the phage capsid. *J. Mol. Biol.* **42:** 547.

Feige, V. and S. Stirm. 1976. On the structure of the *Escherichia coli* cell wall lipopolysaccharide core and on its φX174 receptor region. *Biochem. Biophys. Res. Commun.* **71:** 566.

Francke, B. and D. S. Ray. 1971. Fate of parental φX174 DNA upon infection of starved thymine-requiring host cells. *Virology* **44:** 168.

Fujimura, R. and P. Kaesberg. 1962. Adsorption of bacteriophage φX174 to its host. *Biophys. J.* **2:** 433.

Incardona, N. L. 1974. Mechanism of adsorption and eclipse of bacteriophage φX174. III. Comparison of the activation parameters for the *in vitro* and *in vivo* eclipse reactions with mutant and wild-type virus. *J. Virol.* **14:** 469.

Incardona, N. L. and L. Selvidge. 1973. Mechanism of adsorption and eclipse of bacteriophage φX174. II. Attachment and eclipse with isolated *Escherichia coli* cell wall lipopolysaccharide. *J. Virol.* **11:** 775.

Incardona, N. L., R. Blonski, and W. Feeney. 1972. Mechanism of adsorption and eclipse of bacteriophage φX174. I. *In vitro* conformational change under conditions of eclipse. *J. Virol.* **9:** 96.

Jazwinski, S. M., A. A. Lindberg, and A. Kornberg. 1975a. The lipopolysaccharide receptor for bacteriophages φX174 and S13. *Virology* **66:** 268.

———. 1975b. The gene H spike protein of bacteriophage φX174 and S13. I. Functions in phage-receptor recognition and in transfection. *Virology* **66:** 283.

Jazwinski, S. M., R. Marco, and A. Kornberg. 1975c. The gene H spike protein of bacteriophage φX174 and S13. II. Relation to synthesis of the replicative form. *Virology* **66:** 294.

Lindberg, A. A. 1973. Bacteriophage receptors. *Annu. Rev. Microbiol.* **27:** 205.

Neuwald, P. D. 1975. *In vitro* system for the study of bacteriophage φX174 adsorption and eclipse. *J. Virol.* **15:** 497.

Newbold, J. E. and R. L. Sinsheimer. 1970a. The process of infection with bacteriophage φX174. XXXII. Early steps in the infection process: Attachment, eclipse and DNA penetration. *J. Mol. Biol.* **49:** 49.

———. 1970b. Process of infection with bacteriophage φX174. XXXIV. Kinetics of the attachment and eclipse steps of the infection. *J. Virol.* **5:** 427.

Segal, D. J. and C. E. Dowell. 1974. Cold-sensitive mutants of bacteriophage φX174. II. Comparison of two cold-sensitive mutants. *J. Virol.* **14:** 1115.

Tessman, E. S. 1965. Complementation groups in phage S13. *Virology* **25:** 303.

Weisbeek, P. J., J. H. van de Pol, and G. A. van Arkel. 1973. Mapping of host range mutants of bacteriophage φX174. *Virology* **52:** 408.

Morphogenesis of the Filamentous Single-stranded DNA Phages

Robert E. Webster and Jay S. Cashman
Department of Biochemistry
Duke University Medical Center
Durham, North Carolina 27710

One of the most fascinating aspects of the life cycle of the filamentous phages is their extraordinary morphogenesis. Whereas many other viruses, both enveloped and nonenveloped, are assembled in the cell and released by budding or lysis (Casjens and King 1975), the filamentous phages have evolved a different process for the release of their genetic material. They do not form intracellular phage particles; instead, in some way, they extrude their genomes through the bacterial membrane while assembling the capsid subunits around the genome to form the mature virion (for general reviews of the filamentous phage life cycle see Marvin and Hohn 1969; Denhardt 1975; Ray 1977). Approximately 200 phage particles are produced per cell generation without destruction of the host cell. The mature phage particle is about 6 nm in diameter by 1 μm in length and is composed of the phage single-stranded (SS) DNA encapsulated by approximately 2800 molecules of the phage-specified B, or coat (gene-VIII), protein. Approximately four molecules of the A, or adsorption, protein are at one end of the filamentous particle (Rossomando and Zinder 1968; Marco 1975; Goldsmith and Konigsberg 1977; Woolford et al. 1977). There may also be a few copies of another protein in the phage particle, possibly phage-specified and located at the other end of the particle (Pratt et al. 1969; Beaudoin 1970; Kornberg 1974). The structure of these filamentous phage particles is discussed in this volume by Marvin, by Day and Wiseman, and by Makowski and Caspar.

The membrane-associated process of phage assembly involves the gene-V protein-phage DNA complex. As shown schematically in Figure 1, the synthesis of this complex requires the products of genes II and V (Pratt et al. 1966; Pratt and Erdahl 1968). The complex can be formed in vitro and has been shown to consist basically of a phage SS DNA molecule associated with approximately 1500 molecules of the 10,000-dalton gene-V protein (Alberts et al. 1972; Oey and Knippers 1972). Correspondingly, most or all of the newly synthesized phage SS DNA isolated from infected cells is present in a complex that is extremely rich in gene-V protein and has properties (salt sensitivity, etc.) similar to those of the complex formed in vitro (Webster and Cashman 1973; Pratt et al. 1974; Lica and Ray 1977). However, it should be pointed out that some differences in fine structure, or conformation, or both exist between complexes formed in vivo and in vitro (Alberts et al. 1972; Pratt et al. 1974; Pretorius et al. 1975; Anderson et al. 1975; Lica and Ray 1977). These differences may be due to the presence in the in vivo complex of small amounts of other phage-specified proteins (see below).

The gene-V protein-DNA complex appears to associate with the bacterial inner membrane; when this membrane is isolated from infected cells by the technique of Osborn et al. (1972), the gene-V protein-DNA complex always appears to be loosely associated with it (Webster and Cashman 1973). Also, in the in vitro assembly system described by Wickner and Killick (1977), this complex sedimented with the membrane fraction. In vivo, the DNA passes from this complex through the bacterial membrane and is covered with phage capsid proteins before it emerges outside the cell in a mature phage particle (Fig. 1). During this process the gene-V protein is displaced into the cytoplasm and binds to newly synthesized phage DNA molecules (Webster and Cashman 1973; Pratt et al. 1974). The mechanism by which these events occur is not understood. The products of genes I, III, IV, VI, VII, and VIII must be required because even though bacteria infected with a phage containing mutations in any of these genes can synthesize DNA, no viable

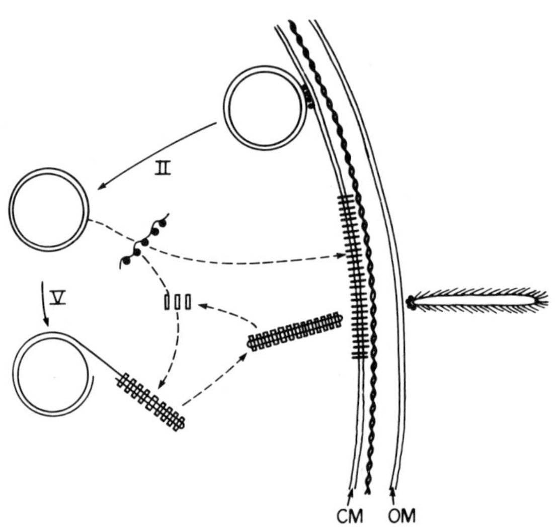

Figure 1 Life cycle of the filamentous phages. This figure shows schematically the formation of progeny replicative-form (RF) DNA, the production of gene-V protein, formation of the gene-V protein-phage DNA complex, and the production of gene-VIII protein (inserted into the membrane). The gene-V proteins are recycled as the DNA is coated by gene-VIII proteins (and approximately four copies of the gene-III protein) while passing through the membrane (see text). CM: cytoplasmic membrane; OM: outer membrane.

type of phage particle is produced (Pratt et al. 1966; Pratt and Erdahl 1968). As would be expected for such a membrane-associated assembly, the products of some of these genes (III, IV, and VIII) have been found in the cytoplasmic membrane of infected bacteria (Smilowitz et al. 1972; Webster and Cashman 1973; Lin and Pratt 1974).

The mechanism by which the phage is assembled depends a great deal on the interaction of the coat proteins with both the membrane and the phage DNA. The major coat protein is found associated only with the membrane in infected bacteria (Smilowitz et al. 1972; Webster and Cashman 1973). It is thought that it will be necessary to have detailed knowledge of its interaction with, and configuration in, the bacterial membrane to deduce its role in morphogenesis. In addition, many investigators believe that such knowledge will be helpful to the general study of how intrinsic membrane proteins are synthesized and interact with the membrane. The results of such studies form the basis for the present models of morphogenesis. Therefore, we would first like to discuss these results and to suggest a crude hypothesis of how the coat protein is involved in phage assembly. We will then discuss the possible role of the other phage-specified proteins in this process. Last, we will review recent studies, both in vitro and in vivo, that used techniques which should be useful in deducing the complicated membrane-associated events in the assembly of this phage.

Membrane-associated Major Coat Protein and Phage Assembly

Investigation of the interaction of the coat protein with the bacterial membrane was facilitated by the knowledge of its sequence. This allowed certain regions of the molecule to be identified by means of various radioactive amino acids or certain group-specific reagents. The complete sequence of the major coat protein of the filamentous phages f1, fd, and ZJ/2 has been determined (Asbeck et al. 1969; Snell and Offord 1972; Nakashima and Konigsberg 1974; Bailey et al. 1977). As can be seen in Figure 2, the coat protein of fd and f1 consists of 50 amino acids. The amino terminus is quite acidic, containing two aspartic, two glutamic, and only one lysine residue in the first 20 residues, whereas the carboxy-terminal region is quite basic, containing four lysine residues in the terminal 11 amino acids. The central core of the protein (residues 21–39) contains an uninterrupted sequence of 19 uncharged, and predominantly hydrophobic, amino acid residues.

It is logical to assume that it is this hydrophobic region that is embedded in the membrane. To test this hypothesis, coat protein radioactively labeled with specific amino acids was inserted into detergent micelles and single-walled phospholipid vesicles of varying composition and subjected to proteolysis with different proteolytic enzymes (Woolford and Webster 1975; Chamberlain et al. 1978). The products were then analyzed to determine which region of the protein was protected from enzymatic cleavage, presumably because of the region's interaction with the micelle or phospholipid vesicle. By repeating this experiment with isolated coat protein radioactively labeled with different amino acids, Chamberlain et al. (1978) determined that the 14 residues from position 26 to position 39 (underlined in Fig. 2) were probably buried in the hydrocarbon region of the membrane bilayer. Since this region is in the middle of the protein, it is theoretically possible for the protein to be inserted into the membrane in a U-shaped configuration or to span the membrane (see Fig. 3 in Tanford and Reynolds 1976). The first approach to this problem made use of antibodies directed against the amino-terminal portion of the coat protein (Wickner 1976). Wickner (1975) found that the antibody appeared to bind only to the outer surfaces of spheroplasts that contained "free coat protein,"[1] which suggests that all or almost all of the amino-terminal portion of the coat protein is exposed on the outside of the cytoplasmic membrane. This is reinforced by the accessibility to lactoperoxidase of the tyrosines of the coat protein in intact spheroplasts (R. Grant et al., in prep.). In addition, the lysine residues in the carboxy-terminal region appear to be reactive with lysine group-specific reagents only when the spheroplasts are broken, that is, when both sides of the membrane are exposed to the reagent. The data indicate that the coat protein is present in the cytoplasmic membrane with its amino-terminal 24 amino acids exposed on the outside, residues 25–39 embedded in the membrane bilayer, and the 11 carboxy-terminal residues projecting into the cytoplasm of the bacteria.

Such an asymmetric orientation of the coat protein in the membrane is consistent with three additional experimental results. First, much of the purified coat protein inserted into the bilayer during the formation of single-walled phospholipid vesicles appears to span the membrane (Wickner 1976, 1977; Chamberlain et al. 1978). This is not to say that all of the molecules are necessarily in this orientation but only that the coat protein can span the bilayer when vesicles are formed from mixtures of lipid and protein. Second, the coat protein is probably synthesized in vivo as a precursor molecule containing an additional 23 amino acids at the amino terminus (Sugimoto et al. 1977; Chang et al. 1978; Horiuchi et al., this volume). If this amino-terminal extension is necessary for the insertion of the protein into the membrane in a manner similar to that proposed by the signal hypothesis for eukaryotic secretory proteins (Blobel and Dobberstein 1975), then, after processing, the amino-terminal portion of the coat protein would be expected to be on the outside of the cytoplasmic membrane. Recent evidence supporting this idea comes from studies which indicate that some amino termini of nascent proteins in *Escherichia coli* are available to labeling by a membrane-impermeable reagent (Smith et al. 1977). Third, the coat protein

[1] "Free coat protein" is coat protein associated with the membrane of *Escherichia coli* but which is not able to produce phage because of a defect in phage SS DNA synthesis.

```
                                    10
Ala-Glu-Gly-Asp-Asp-Pro-Ala-Lys-Ala-Ala-Phe-Asp-Ser-Leu-Gln-Ala-Ser-

        20
Ala-Thr-Glu-TYR-ILE-GLY-TYR-ALA-TRP-ALA-MET-VAL-VAL-VAL-ILE-VAL-GLY-

                                   40                                50
ALA-THR-ILE-GLY-ILE-Lys-Leu-Phe-Lys-Lys-Phe-Thr-Ser-Lys-Ala-Ser
```

Figure 2 Amino acid sequence of the gene-VIII (coat) protein. The residues in capital letters are a stretch of 19 hydrophobic amino acids.

of infecting phages, which enters the cytoplasm intact and can be incorporated into newly synthesized phage (Smilowitz 1974), appears to have its amino terminus exposed on the outside of the membrane (Wickner 1975).

It is difficult to deduce the conformational state of the "free coat protein" in the membrane. The coat protein exists in the phage as a closely packed layer of helices, with essentially all of the protein in the alpha-helical conformation (Marvin et al. 1974a,b; Marvin and Wachtel 1975; Nozaki et al. 1976; Marvin; Day and Wiseman; both this volume). In contrast, circular dichroism measurements on coat protein in detergent micelles, lysophospholipid micelles, or single-walled vesicles of various phospholipid compositions show the protein to contain about 50% alpha helix and about 30% beta structure (Makino et al. 1975; Nozaki et al. 1976, 1978). This is probably the thermodynamically stable conformation in the membrane, as it can be obtained in detergent micelles prepared from either randomly coiled or fully helical protein (Williams and Dunker 1977; Nozaki et al. 1978). When the empirical procedure of Chou and Fasman (1974) is used to predict the structure from the amino acid sequence, one obtains a conformation of 46% alpha helix, 30% beta structure, and 24% unordered, in good agreement with the estimates from circular dichroism studies (Makino et al. 1975; Nozaki et al. 1976, 1978). This prediction identifies residues 5–20 and 39–45 of the hydrophilic termini of the protein as containing the alpha-helical portions and residues 26–37 of the hydrophobic portion as containing the beta structure. Although there may be a question as to the validity of using this procedure for membrane proteins (Green and Flanagan 1976), in this case the assignment of beta structure to the hydrophobic portion is probably correct. Consistent with this is the fact that when detergent-solubilized coat protein is digested with pronase, the undigested portion remaining with the detergent is at least 70% in the beta conformation, as judged by circular dichroism measurements (Chamberlain et al. 1978).

Considering all of the above and predictions from previous models (Marvin and Wachtel 1975; Wickner 1975), we suggest that the coat protein, after synthesis, insertion, and processing, may be present in the *E. coli* cytoplasmic membrane as shown schematically in Figure 3. All of the amino

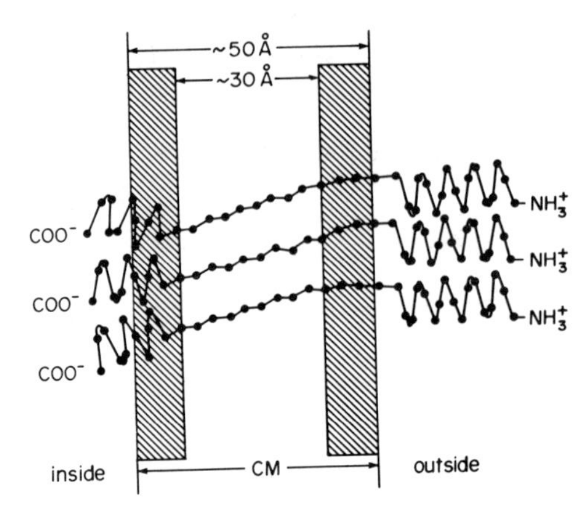

Figure 3 Coat protein conformation within the membrane. This is a schematic view of the possible spanning by the coat protein of the *E. coli* cytoplasmic membrane. The distances are approximate, as are the regions representing alpha-helical and beta structures.

termini would be exposed on the outside and all of the carboxy termini would project into the cytoplasm. A hydrophobic region consisting maximally of residues 26–39 and containing a beta structure would be embedded in the hydrocarbon portion of the bilayer. Small numbers of these chains may be associated together in a laterally hydrogen-bonded arrangement that would tie up all of the main-chain polar groups. This grouping may be analogous to the parallel β barrel structure described for the internal structural domain of certain globular proteins (Richardson 1977). Rough calculations indicate that a 14-residue-long β chain in this type of structure would have a length of 24–33 Å depending on the twist and the number of chains (J. S. Richardson, pers. comm.). This length can be easily accommodated within the hydrocarbon core of the *E. coli* cytoplasmic membrane of dimensions shown in Figure 3 (De Petris 1965; Costerton et al. 1974; Tanford and Reynolds 1976). This is especially true if one considers that the region of the protein inserted in the membrane may be slightly less than the maximum defined by the proteolysis data (residues 26–39) (Chamberlain et al. 1978).

This model for the configuration of membrane-associated coat protein suggests a hypothesis for steady-state phage assembly that is an extension of the one proposed by Marvin and Wachtel (1975). The SS DNA, during or after release of the gene-V protein, would interact with a group of basic carboxy-terminal regions of the coat proteins which protrude into the cytoplasm. This would lead to a disruption of the grouping of β chains as they assemble onto the DNA in an essentially alpha-helical configuration. This is consistent with the observations that the carboxy-terminal amino acids are present at the inner radius of the protein shell in the phage, next to the DNA (Marvin, this volume), and that the coat protein has a helical conformation in the phage particles (Marvin et al. 1974a,b; Marvin and Wachtel 1975; Nozaki et al. 1976). The next portion of the DNA would then be available to another group of carboxy-terminal ends of coat proteins; this process would be repeated until the phage particle was extruded from the cell. The protein-protein interactions of the coat protein are thermodynamically important both in the membrane and in the phage. Therefore, this model predicts that the disruption of the conformation of membrane-associated coat protein by the interaction of the protein and DNA leads to the change in the protein structure which results in the essentially 100% alpha helix found in the phage. It should be noted that the steady-state assembly model does not indicate the nature of the initial or final events of phage assembly, nor does it specify the possible roles of the other phage proteins. Instead, the model should be used only as a working hypothesis, one that can be a useful tool for designing future experiments related to the membrane-associated assembly of these phages.

Role of Phage-specified Proteins in Morphogenesis

It has been shown genetically that the maturation of phage particles requires the products of genes I, III, IV, VI, VII, and VIII (Pratt et al. 1966; Marvin and Hohn 1969). The products of genes III (A protein) and VIII (B protein) are integral parts of the phage particle, and there have been consistent reports, based on the separation of histidine- or arginine-labeled proteins from purified phage by polyacrylamide gel electrophoresis, of the presence of another minor protein in the phage (Rossomando and Zinder 1968; Henry and Pratt 1969; Model and Zinder 1974; Kornberg 1974; Makino et al. 1975). This small protein has not yet been isolated or even shown to be phage-specified. However, it has been suggested that this protein might be the product of gene VI, based on the fact that cells infected with gene-VI amber mutants produce a small number of multiple-length defective particles (Pratt et al. 1969; Beaudoin 1970). If this were found to be the case, it would partially explain the requirement for gene VI for phage assembly. It would also correlate well with the data of Mazur and Zinder (1975), which suggest that the products of genes III and VI act in the last steps of morphogenesis.

There is no real knowledge concerning the roles of the products of genes I, IV, or VII in morphogenesis. They have been identified only as 35,000-, 50,000-, and 3,600-dalton proteins, respectively (Model and Zinder 1974; Konings et al. 1975; Takanami et al. 1976; J. G. G. Schoenmakers, pers. comm.). The involvement of the gene-I protein or the gene-IV protein in processing the newly synthesized minor or major coat proteins (Mazur and Zinder 1975) appears to have been excluded; although both coat proteins are synthesized as precursors (Pieczenik et al. 1974; Konings et al. 1975; Chang et al. 1978), the gene-VIII protein can be processed correctly in vitro by inner membranes from uninfected bacteria (Chang et al. 1978). Some of these proteins may act at the level of the gene-V protein-DNA complex: Mazur and Zinder (1975) showed that the products of genes I, IV, VII, and

VIII are required for the dissociation of the gene-V protein from nascent phage SS DNA. It is possible that some of these gene products may be necessary for starting the interaction of the DNA in the gene-V protein-DNA complex with the coat protein in the membrane. In addition, it has been shown that bacteria infected with gene-VII amber-mutant phage cease division very rapidly and that the synthesis of f1 DNA is greatly reduced (Hohn et al. 1971; Grandis and Webster 1973; Mazur and Zinder 1975). This suggests that the gene-VII protein may be necessary for the formation of a functional gene-V protein-DNA complex, i.e., one that has the proper conformation to enter the next stage of morphogenesis. This is not unreasonable since there is evidence that gene-V protein-DNA complexes can have different conformations. Experiments utilizing various biochemical, biophysical, and electron microscope techniques have shown that the complexes formed in vitro (Alberts et al. 1972; Pretorius et al. 1975; Anderson et al. 1975; Y. Nakashima and W. Konigsberg, pers. comm.) must have a somewhat different conformation than those isolated from infected bacteria (Pratt et al. 1974; Pretorius et al. 1975; Lica and Ray 1977).

To summarize, little if anything is known about the nature of the products of genes I, IV, VI, and VII other than their probable molecular weights and that they are required for morphogenesis. The product of gene VII appears to be needed along with the gene-V protein early in this process (Mazur and Zinder 1975) and therefore may play a role in the formation of, and perhaps is an integral part of, a proper gene-V protein-DNA complex. The roles of the products of genes I and IV are unknown. All that is known is that they are definitely needed for assembly both in vivo and in vitro (Wickner and Killick 1977). Perhaps they are needed for the initial interaction of the gene-V protein-DNA complex with the membrane-associated coat protein at the site of phage assembly. The products of genes VI and III appear to be needed in the later part of the assembly cycle (Mazur and Zinder 1975). Last, the products of genes III and VIII are necessary if only because they are part of the mature phage particle.

Future Prospects

Our almost complete lack of knowledge about the mechanism of this fascinating process of assembly is probably the largest chasm in our understanding of the life cycle of these phages. As such, it is one of the exciting areas for future research. The most logical way to attack this problem would be to continue the strategies that others have found so useful for elucidating the assembly processes in other phage systems (Casjens and King 1975); that is, to attempt to identify precursors in infected cells by means of various conditionally lethal phage, bacterial mutants, or both. This also includes identifying all of the components and the structure of the mature phage particle. Also, attempts should be made to develop in vitro systems that can synthesize infective phage particles.

With regard to the latter strategy, a system has been developed recently in which extracts of M13-infected bacteria are able to support phage production (Wickner and Killick 1977). Extracts prepared from bacteria infected with M13 mutants carrying temperature-sensitive mutations in genes I, III, IV, and V are temperature-sensitive in this cell-free assembly reaction, which suggests that this system is representative of the in vivo process. The activity of this in vitro system resides in the membrane fraction (and requires unidentified low-molecular-weight factors from yeast extracts). What is needed now is the resolution of this system into, and its reconstitution from, its necessary components. This would certainly be aided by the isolation and characterization of more single and multiple conditionally lethal mutants of the phage. Various bacterial mutants with defects in certain processes, such as membrane biosynthesis (McIntyre et al. 1977), would also be helpful.

A more concerted effort to isolate viable precursors should also yield valuable information. Gene-V protein-phage DNA complexes isolated from bacteria infected with normal or conditionally lethal mutant phage could be examined for their protein contents and conformational differences. Certainly, recent developments in polyacrylamide gel electrophoresis and other separation techniques will allow identification of minor protein components. The use of UV-induced cross-linking (Anderson et al. 1975; Lica and Ray 1977) and group-specific reagents and electron microscopic studies should allow the detection of differences in the structures of such complexes. Similar methods could identify structures either from the cytoplasm or membrane that are intermediate between the gene-V protein-DNA complex and phage particles. As in other such steady-state studies on phage assembly, one must take pains to insure that any structure isolated is indeed a precursor. Since such studies will require the use of various conditionally lethal mutants, this emphasizes again the need for the construction of more single and multiple temperature-sensitive mutants.

Last, it might be profitable to attempt to find conditions allowing self-assembly of the phage from its component parts. The recent advances in isolating and handling membrane proteins (Helenius and Simons 1975; Tanford and Reynolds 1976) and, more specifically, the major coat protein of the phage (Nozaki et al. 1978) now make it more reasonable to pursue such an investigation. If such a self-assembly is at all possible, it would be of great help for designing further experiments for ascertaining the process of morphogenesis in vivo.

Future research on this problem will certainly be profitable. Not only will it fill an important gap in our understanding of the phage life cycle, but it will also help us to understand the complex interactions that occur between various membrane-associated components.

ACKNOWLEDGMENTS

The unpublished work from our laboratory cited in this article was supported by grants from the National Science Foundation and the National Institutes of Health.

REFERENCES

Alberts, B., L. Frey, and H. Delius. 1972. Isolation and characterization of gene V protein of filamentous bacterial viruses. *J. Mol. Biol.* **68:** 139.

Anderson, E., Y. Nakashima, and W. Konigsberg. 1975. Photo-induced cross-linkage of gene 5 protein and bacteriophage fd DNA. *Nucleic Acids Res.* **2:** 361.

Asbeck, F., K. Beyreuther, H. Köhler, G. von Wettstein, and G. Braunitzer. 1969. Virusproteine. IV. Die Konstitution des Hullproteins des Phagen fd. *Hoppe-Seyler's Z. Physiol. Chem.* **350:** 1047.

Bailey, G. S., D. Gillet, D. F. Hill, and G. B. Petersen. 1977. Automated sequencing of insoluble peptides using detergent: Bacteriophage f1 coat protein. *J. Biol. Chem.* **252:** 2218.

Beaudoin, J. 1970. "Studies on coat protein mutants and on multiple-length particles of coliphage M13." Ph. D. thesis, University of Wisconsin, Madison.

Blobel, G. and B. Dobberstein. 1975. Transfer of proteins across membranes. I. Presence of proteolytically processed and unprocessed nascent immunoglobulin light chains on membrane-bound ribosomes of murine myeloma. *J. Cell Biol.* **67:** 835.

Casjens, S. and J. King. 1975. Virus assembly. *Annu. Rev. Biochem.* **44:** 555.

Chamberlain, B. K., Y. Nozaki, C. Tanford, and R. E. Webster. 1978. Association of the major coat protein of fd bacteriophage with phospholipid vesicles. *Biochim. Biophys. Acta* **62:** 18.

Chang, C. N., G. Blobel, and P. Model. 1978. Detection of prokaryotic signal peptidase in an *Escherichia coli* membrane fraction: Endoproteolytic cleavage of nascent f1 pre-coat protein. *Proc. Natl. Acad. Sci.* **75:** 361.

Chou, P. Y. and G. D. Fasman. 1974. Conformational parameters for amino acids in helical β-sheet, and random coil regions from proteins. *Biochemistry* **13:** 211.

Costerton, J. W., J. M. Ingram, and K. J. Chen. 1974. Structure and function of the cell envelope of gram-negative bacteria. *Bacteriol. Rev.* **38:** 87.

Denhardt, D. T. 1975. The single-stranded DNA phages. *CRC Crit. Rev. Microbiol.* **4:** 161.

De Petris, S. 1965. Ultrastructure of the cell wall of *Escherichia coli. J. Ultrastruct. Res.* **12:** 247.

Goldsmith, M. E. and W. Konigsberg. 1977. Adsorption protein of the bacteriophage fd: Isolation, molecular properties and location in the virus. *Biochemistry* **16:** 2686.

Grandis, A. S. and R. E. Webster. 1973. Abortive infection of *Escherichia coli* with the bacteriophage f1: DNA synthesis associated with the membrane. *Virology* **55:** 39.

Green, N. M. and M. T. Flanagan. 1976. The prediction of the conformation of membrane proteins from the sequence of amino acids. *Biochem. J.* **153:** 729.

Helenius, A. and K. Simons. 1975. Solubilization of membranes by detergents. *Biochim. Biophys. Acta* **415:** 29.

Henry, T. J. and D. Pratt. 1969. The proteins of bacteriophage M13. *Proc. Natl. Acad. Sci.* **62:** 800.

Hohn, B., H. von Schutz, and D. A. Marvin. 1971. Filamentous bacterial viruses. II. Killing of bacteria by abortive infection with fd. *J. Mol. Biol.* **56:** 155.
Konings, R. N. H., T. Hulsebos, and C. A. van den Hondel. 1975. Identification and characterization of the *in vitro* synthesized gene products of bacteriophage M13. *J. Virol.* **15:** 570.
Kornberg, A. 1974. *DNA synthesis*. W. H. Freeman, San Francisco.
Lica, S. and D. S. Ray. 1977. Replication of bacteriophage M13. XII. *In vivo* cross-linking of a phage-specific DNA binding protein to the single-stranded DNA of bacteriophage M13 by ultraviolet irradiation. *J. Mol. Biol.* **115:** 45.
Lin, N. S.-C. and D. Pratt. 1974. Bacteriophage M13 gene 2 protein: Increasing its yield in infected cells, and identification and localization. *Virology* **61:** 334.
Makino, S., J. L. Woolford, C. Tanford, and R. E. Webster. 1975. Interaction of deoxycholate and of detergents with the coat protein of bacteriophage f1. *J. Biol. Chem.* **250:** 4327.
Marco, R. 1975. The adsorption protein in mini M13 phages. *Virology* **68:** 280.
Marvin, D. A. and B. Hohn. 1969. Filamentous bacterial viruses. *Bacteriol. Rev.* **33:** 172.
Marvin, D. A. and E. J. Wachtel. 1975. Structure and assembly of filamentous bacterial viruses. *Nature* **253:** 19.
Marvin, D. A., R. L. Wiseman, and E. J. Wachtel. 1974a. Filamentous bacterial viruses. XI. Molecular architecture of the class II (Pf1, Xf) virion. *J. Mol. Biol.* **82:** 121.
Marvin, D. A., W. J. Pigram, R. L. Wiseman, E. J. Wachtel, and F. J. Marvin. 1974b. Filamentous bacterial viruses. XII. Molecular architecture of the class I (fd, If1, Ike) virion. *J. Mol. Biol.* **88:** 581.
Mazur, B. J. and N. D. Zinder. 1975. The role of gene V protein in f1 single-stranded synthesis. *Virology* **68:** 490.
McIntyre, T. M., B. K. Chamberlain, R. E. Webster, and R. M. Bell. 1977. Mutants of *E. coli* defective in membrane phospholipid synthesis. *J. Biol. Chem.* **252:** 4487.
Model, P. and N. D. Zinder. 1974. *In vitro* synthesis of bacteriophage f1 proteins. *J. Mol. Biol.* **83:** 231.
Nakashima, Y. and W. Konigsberg. 1974. Reinvestigation of a region of the fd bacteriophage coat protein sequence. *J. Mol. Biol.* **88:** 598.
Nozaki, Y., J. A. Reynolds, and C. Tanford. 1978. Conformational states of a hydrophobic protein: The coat protein of fd bacteriophage. *Biochemistry* **17:** 1239.
Nozaki, Y., B. K. Chamberlain, R. E. Webster, and C. Tanford. 1976. Evidence for a major conformational change of coat protein in assembly of f1 bacteriophage. *Nature* **259:** 335.
Oey, J. L. and R. Knippers. 1972. Properties of the isolated gene 5 protein of bacteriophage fd. *J. Mol. Biol.* **68:** 125.
Osborn, M. J., J. E. Gander, E. Parisi, and J. Carson. 1972. Mechanism of assembly of the outer membrane of *Salmonella typhimurium*. Isolation and characterization of cytoplasmic and inner membrane. *J. Biol. Chem.* **247:** 3962.
Pieczenik, G., P. Model, and H. D. Robertson. 1974. Sequence and symmetry in ribosome binding sites of bacteriophage f1 RNA. *J. Mol. Biol.* **90:** 191.
Pratt, D. and W. S. Erdahl. 1968. Genetic control of bacteriophage M13 DNA synthesis. *J. Mol. Biol.* **37:** 181.

Pratt, D., P. Laws, and J. Griffith. 1974. Complex of bacteriophage M13 single-stranded DNA and gene 5 protein. *J. Mol. Biol.* **82:** 425.

Pratt, D., H. Tzagoloff, and J. Beaudoin. 1969. Conditional lethal mutants of the small filamentous coliphage M13. II. Two genes for coat proteins. *Virology* **39:** 42.

Pratt, D., H. Tzagoloff, and W. S. Erdahl. 1966. Conditional lethal mutants of the small filamentous coliphage M13. I. Isolation, complementation, cell killing, time of cistron action. *Virology* **30:** 397.

Pretorius, H. T., M. Klein, and L. A. Day. 1975. Gene V protein of fd bacteriophage: Dimer formation and the role of tyrosyl groups in DNA binding. *J. Biol. Chem.* **250:** 9262.

Ray, D. S. 1977. Replication of filamentous bacteriophages. In *Comprehensive virology* (ed. H. Fraenkel-Conrat and R. R. Wagner), vol. 7, p. 105. Plenum Press, New York.

Richardson, J. S. 1977. β-Sheet topology and the relatedness of proteins. *Nature* **268:** 495.

Rossomando, E. F. and N. D. Zinder. 1968. Studies on the bacteriophage f1. I. Alkali-induced disassembly of the phage into DNA and protein. *J. Mol. Biol.* **36:** 387.

Smilowitz, H. 1974. Bacteriophage f1 infection: Fate of the parental major coat protein. *J. Virol.* **13:** 94.

Smilowitz, H., J. Carson, and P. W. Robbins. 1972. Association of newly synthesized major f1 coat protein with infected host cell inner membrane. *J. Supramol. Struct.* **1:** 8.

Smith, W. P., P. C. Tai, R. C. Thompson, and B. D. Davis. 1977. Extracellular labeling of nascent polypeptides traversing the membrane of *Escherichia coli*. *Proc. Natl. Acad. Sci.* **74:** 2830.

Snell, D. T. and R. E. Offord. 1972. The amino acid sequence of the B-protein of bacteriophage ZJ-2. *Biochem. J.* **127:** 167.

Sugimoto, K., H. Sugisaki, T. Okamoto, and M. Takanami. 1977. Studies on bacteriophage fd DNA. IV. The sequence of messenger RNA for the major coat protein gene. *J. Mol. Biol.* **111:** 487.

Takanami, M., K. Sugimoto, H. Sugisaki, and T. Okamoto. 1976. Sequence of promoter for coat protein gene of bacteriophage fd. *Nature* **260:** 297.

Tanford, C. and J. A. Reynolds. 1976. Characterization of membrane proteins in detergent solutions. *Biochim. Biophys. Acta* **457:** 133.

Webster, R. E. and J. S. Cashman. 1973. Abortive infection of *Escherichia coli* with the bacteriophage f1: Cytoplasmic membrane proteins and the f1 DNA-gene 5 protein complex. *Virology* **55:** 20.

Wickner, W. 1975. Asymmetric orientation of a phage coat protein in cytoplasmic membrane of *Escherichia coli*. *Proc. Natl. Acad. Sci* **72:** 4749.

———. 1976. Asymmetric orientation of phage M13 coat protein in *Escherichia coli* cytoplasmic membranes and in synthetic lipid vesicles. *Proc. Natl. Acad. Sci.* **73:** 1159.

———. 1977. Role of hydrophobic forces in membrane protein asymmetry. *Biochemistry* **16:** 254.

Wickner, W. and T. Killick. 1977. Membrane-associated assembly of M13 phage in extracts of virus-infected *Escherichia coli*. *Proc. Natl. Acad. Sci.* **74:** 505.

Williams, R. W. and A. K. Dunker. 1977. Circular dichroism studies of fd coat protein in membrane vesicles. *J. Biol. Chem.* **252:** 6253.
Woolford, J. L., Jr. and R. E. Webster. 1975. Proteolytic digestion of the micellar complex of f1 coat protein and deoxycholate. *J. Biol. Chem.* **250:** 4333.
Woolford, J. L., Jr., H. M. Steinman, and R. E. Webster. 1977. Adsorption protein of bacteriophage f1: Solubilization in deoxycholate and localization in the f1 virion. *Biochemistry* **16:** 2694.

Abnormal Forms of Filamentous Phage Virions and Phage DNA

Frances C. Wheeler and Rolf H. Benzinger
Department of Biology
University of Virginia
Charlottesville, Virginia 22901

A careful analysis of mistakes that occur in biological systems often leads to useful insights into the normal process. Whereas the icosahedral single-stranded DNA phages are constrained to package a fixed "headful" of DNA, the filamentous phages exhibit great flexibility in the amount of DNA they accept. One of the earliest papers about these phages (Marvin and Hoffmann-Berling 1963) contained the suggestion that phage of different lengths exist in wild-type stocks. During a study of complementation between different conditionally lethal mutants, two groups (Salivar et al. 1967; Scott and Zinder 1967) discovered independently that the 1–10% turbid plaques formed on nonpermissive hosts arose from "heterozygous" phage; these plaques contained both parental phage types, which maintained themselves by a process of continuous complementation. Electron microscopy and antiserum inactivation experiments suggested that these plaques arose from double-length phage (Scott and Zinder 1967; Salivar et al. 1967; Beaudoin and Pratt 1974). A much lower frequency of trimer phage was also noted. Phage produced after infection of nonpermissive strains with amber mutants of genes III and VI alone consisted of up to 70% dimer phage as well as many trimer and even tetramer phage (Pratt et al. 1969). The products of genes III and VI are implicated in phage assembly; the gene-III protein is found at the end of the phage and is necessary for adsorption to the host cell (Pratt et al. 1969).

Early attempts to find double-length, circular phage DNA in the dimers were unsuccessful (Salivar et al. 1967). After examination of a large number of DNA molecules and partial purification of dimer DNA on CsCl velocity gradients, we were able to show that about one in ten dimeric phages do contain double-length, circular, single-stranded (SS) DNA (Wheeler et al. 1974).

Abnormal phage particles of variable, nonintegral lengths have also been observed. After Grandis and Webster (1973) found an f1 small, circular, replicative-form (RF) DNA in cells infected with amber mutants, M13 miniphage particles with lengths varying from 20% to 50% of the normal phage length were discovered in phage preparations that had been passed serially without plaque purification (Griffith and Kornberg 1974; Hewitt

1975). These particles were not infectious unless the host was infected simultaneously with a helper phage. Similar deletion mutants of phage f1 have been shown to contain the origin of replication but to lack any intact cistrons (Enea and Zinder 1975). In this article we will discuss the infectivity and propagation of filamentous phage particles of variable and multimeric lengths.

RESULTS

Multimeric Phage and Phage DNA

Our preliminary experiments showed that the frequency of abnormal fd phage depended strongly on the initial multiplicity of infection (m.o.i.). Thus, we used a well-characterized wild-type stock containing 3% double-length and 1% miniphage to infect *Escherichia coli* K38 rec^+ bacteria at an initial m.o.i. ranging from 0.01 to 100. The frequency of double-length phage particles in the yield is shown in Table 1. Except for a rise to 6.7% at the lowest m.o.i., the frequency of dimer phage among the progeny remained relatively constant at around 2–3%. When phage stocks were prepared at different m.o.i. on an *E. coli recAB* strain (with the five rec^+ stocks at the same m.o.i.), the frequency of dimer phage varied between 1% and 3%; thus, the only significant difference between rec^+ and *recAB* bacteria was observed at the m.o.i. of 0.01.

Quite different results were obtained when DNA extracted from some of these phage particles was examined. For the phage grown on the rec^+ host,

Table 1 Frequency of abnormal phage in fd phage stocks propagated on rec^+ and *recAB* cells

Type of aberrant phage	Host strain	Frequency (%) at m.o.i.				
		0.01	0.1	1.0	10	100
Dimer	rec^+	6.7	3.3	3.1	2.0	2.0
(2×)	*recAB*	2.3	2.6	1.1	1.0	3.1
Miniphage	rec^+	0.3	0.8	1.5	4.2	6.3
(0–0.6×)	*recAB*	0.3	0.3	<0.1	0.3	0.6
Midiphage	rec^+	<0.1	<0.1	<0.1	<0.1	<0.1
(1.2–1.8×)	*recAB*	1.0	3.3	1.3	2.5	12.0

Wild-type phage passed continually at an m.o.i. of 0.1 on *E. coli* K38 and containing 3% dimer and 1% miniphage were used to infect K38 at five different m.o.i. The phage were harvested 20 hr later, purified extensively (including two DNase treatments to remove extraneous DNA), and used to infect *E. coli recAB* bacteria at corresponding m.o.i. The ten phage stocks, as well as phenol-extracted DNA from some of these phage, were then prepared for electron microscopy by the Kleinschmidt or diffusion technique; electron micrographs of the phage (at least 1500 molecules for each percentage) were then evaluated quantitatively.

the frequency of double-length, circular phage DNA ranged from 0.3% in the 0.01 m.o.i. stock, to the 0.03% observed previously in a 0.1 m.o.i. stock (Wheeler et al. 1974), to less than 0.01% in the 100 m.o.i. stock. These data show that phage particles containing double-length, circular DNA are favored by a low m.o.i.

When the DNA from the phage propagated on *E. coli recAB* was examined, no double-length, circular molecules were found in either the 0.01 m.o.i. or 100 m.o.i. stock (less then 0.01%). Thus, the *recAB* function is required for the generation and/or efficient propagation of double-length, circular phage DNA.

Miniphage and Miniphage DNA

Table 1 compares the percentages of miniphage in five stocks grown with different initial m.o.i. on rec^+ and *recAB* hosts. On the rec^+ host, the frequency of miniphage increased steadily to a maximum of 6.3% at the highest multiplicity. However, the frequency of miniphage remained very low at all m.o.i. on the *recAB* host, never exceeding 0.6% (Table 1). In fact, repeated growth of fd phage at high m.o.i. on the *recAB* host decreased the frequency of miniphage to 0.1%. Therefore, the propagation of most miniphage is dependent on a process controlled by the host $recAB^+$ gene function. DNA extracted from the highest and lowest m.o.i. stocks was examined, and the same relationship was found: on the rec^+ host, mini-length DNA rose from 0.3% to 6.3%; in the corresponding *recAB* stocks, mini-length DNA remained at the low frequency of approximately 0.3%.

Midiphage

Examination of the phage propagated on the *E. coli recAB* bacteria revealed the presence of variable-length particles intermediate between monomers and dimers. Figure 1 shows an electron micrograph of such a phage. The frequency of these midi-length particles varied from 1% to 3%, with an exceptionally high frequency of 12% in a 100 m.o.i. stock (Table 1). Less than 0.1% of the phage particles propagated on the rec^+ host were of midi-length (Table 1). A histogram of the length distribution of midiphage particles gives a bell-shaped curve with a mean extra length of 0.4 (Fig. 2).

Sigma Single-stranded DNA from Midiphage

DNAs extracted from the *recAB* phage stocks containing 1% and 12% midiphage were examined by electron microscopy. Eight percent of the DNA molecules from the 100 m.o.i. stock were sigma-shaped (lariats); 0.8% of the DNA molecules from the 0.01 m.o.i. stock were lariats. An example of such a molecule is shown in Figure 3. The lengths of 220 sigma

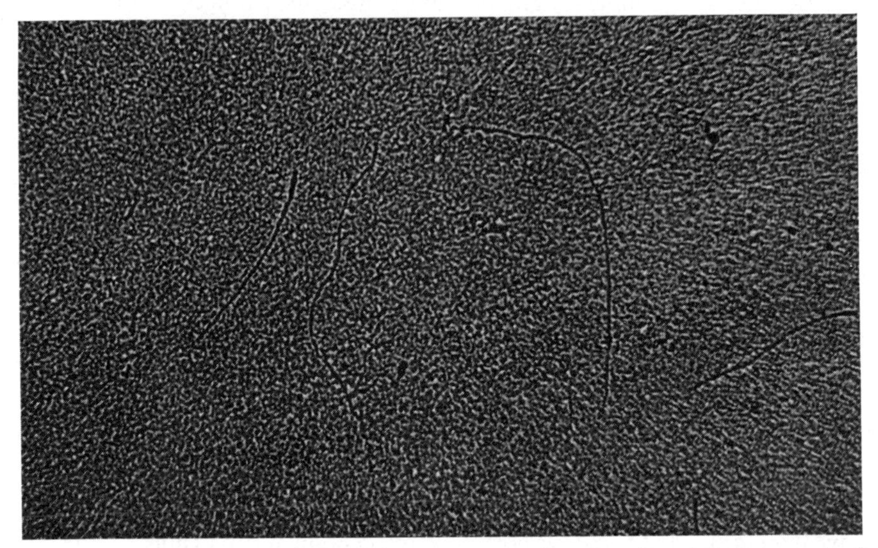

Figure 1 Electron micrograph of phage propagated on *E. coli recAB* at an m.o.i. of 100. The molecule on the left is of normal length, the middle one is of midi-length, and the one on the right is a dimer.

molecules were measured; the circular portions of the molecules were of unit-length, whereas the "tails" of sigma DNA varied between 0.04 and 0.36 times the unit length. A histogram of the sigma DNA tail lengths shows a bell-shaped curve similar to that observed for midiphage (Fig. 2); however, the extra lengths of sigma DNA are shorter than those of the phage (Fig. 2).

What is the nature of the junction holding the circular portion of the sigma DNA molecules? Treatment with 0.1 M NaOH for 10 minutes at room temperature converted all of the sigma DNA into a linear DNA of greater than unit length (Table 2). Hydrogen bonds across an inverted "terminal repetition" could stabilize the circular part of the sigma DNA (Fig. 4), whereas a protein linker seems a less likely possibility as such a protein would have to resist three phenol extractions in the presence of 1% dodecyl sulfate (Wheeler et al. 1974, and in prep.).

DISCUSSION

Multimeric Phage and Phage DNA

The frequency of dimer phage changed very little with initial m.o.i. (Table 1) in one experiment. However, continued passage of fd phage at high m.o.i. does lead to a reduction in the frequency of dimer phage to below 1%. Since

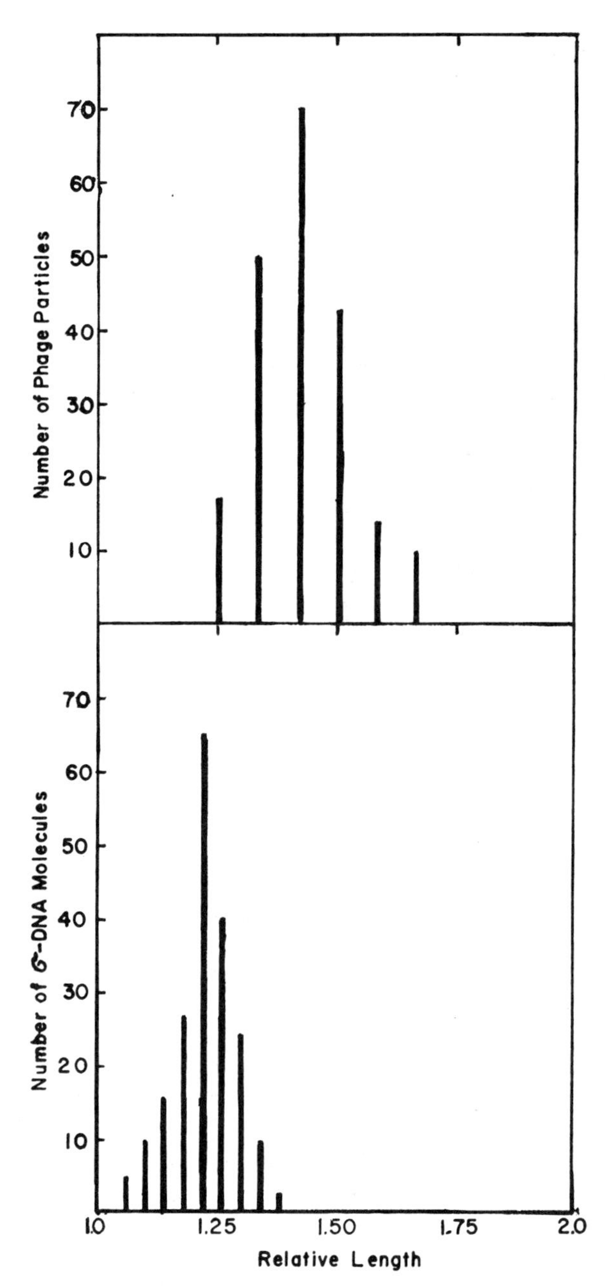

Figure 2 (*Top*) Histogram of the length distribution of midiphage particles from the fd phage stock propagated on *E. coli recAB* at an m.o.i. of 100. (*Bottom*) Histogram of the length distribution of sigma DNA extracted from the fd phage stock propagated on *E. coli recAB* at an m.o.i. of 100.

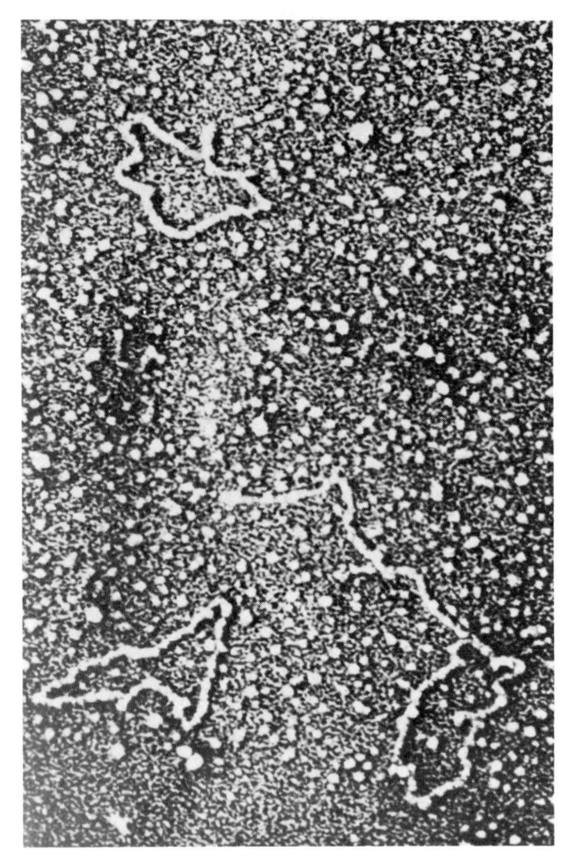

Figure 3 Electron micrograph of DNA extracted from phage propagated on *E. coli recAB* at an m.o.i. of 100. Two normal phage DNA molecules and one sigma-shaped (lariat) DNA molecule are visible.

most dimer phage contain two molecules of normal-length DNA (Wheeler et al. 1974), it is not clear why the frequency of such phage should be reduced when phage are propagated at high m.o.i. Interpretation is complicated further by our inability to distinguish propagation of preexisting dimers from newly arising dimers, as our infecting stocks contain dimer phage.

About 10% of the dimer phage contain one SS DNA circle of double length rather than two SS DNA circles (Wheeler et al. 1974). These molecules are definitely favored at low m.o.i. (see Results). Their extra length should slow their rate of replication compared with wild-type DNA, and they might therefore be at a selective disadvantage in high m.o.i. infections. Phage containing double-length DNA need to be cloned to see whether

Table 2 Effect of alkaline denaturation on the frequency of sigma-shaped DNA molecules in DNA extracted from fd phage propagated on *E. coli recAB* at an m.o.i. of 100

Type of molecule	Before NaOH (%)	After NaOH (%)
1× circle	85.2	85.6
1× linear	6.0	5.9
Sigma DNA	7.7	0
>1× linear	0.4	8.2

For each preparation, at least 1000 DNA molecules were examined.

they remain stable dimers; if they segregate normal-length particles, this would put them at a further selective disadvantage relative to wild type.

We have not observed any SS DNA circles of double length from stocks propagated on the *recAB* host (see Results); thus, either the *recBC* nuclease or the host *recA* function is required for the efficient propagation or synthesis of the circular SS DNA of double length. The nature of this requirement is unclear.

Miniphage and Miniphage DNA

Our data (Table 1) indicate that miniphage and miniphage DNA increase in frequency with increasing initial m.o.i. These results are in accord with those of Griffith and Kornberg (1974), who found miniphage in stocks propagated for many generations at high m.o.i.; they attributed this finding to the nonviability of miniphage in single infections. At high m.o.i. the required coinfection with a wild-type helper phage is much more likely. Enea and Zinder (1975) also found miniphage in stocks passed at high m.o.i.; they

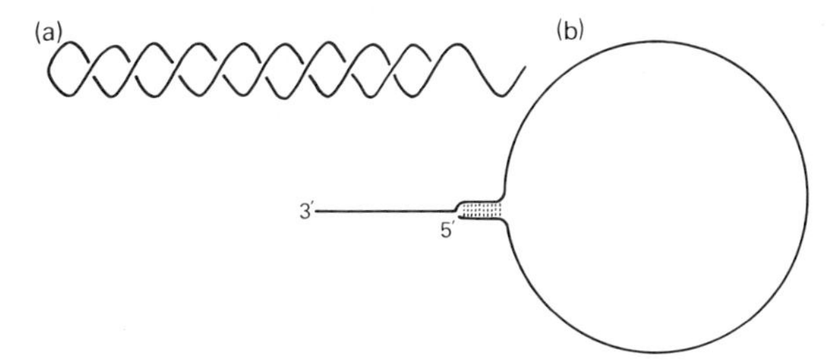

Figure 4 A model for the structure of sigma DNA. The circular part of the molecule is held by the hydrogen bonds across the hairpin loop at the 5′ end of the molecule; the 3′ end of the molecule extends beyond the hairpin and is of variable length. It is suggested that the 3′ end protrudes because it seems likely that the sigma DNA results from an abnormal replication event rather than from an abnormal recombination event. (a) In the phage; (b) extended.

observed that these miniphage actually interfered with wild-type phage replication. Quite possibly this interference may be more efficient at high m.o.i. where several miniphage infect a single cell. Hewitt (1975) found miniphage in stocks passed at an m.o.i. of 10.

Table 1 shows that the propagation of miniphage was very inefficient on the *E. coli recAB* host. It is not clear whether the host *recA* function or the *recBC* nuclease is required to propagate the miniphage, as both functions are missing in the *recAB* strain. Therefore, we have begun to compare the frequencies of miniphage in *rec*$^+$, *recA* (no recombination, active *recBC* nuclease), *recAB* (no recombination, no nuclease), and *recBC* (very weak recombination, no nuclease) strains. Measurement of several hundred phage has shown that miniphage are observed after high m.o.i. passage on the *rec*$^+$ and *recA* strains but are greatly reduced on the *recAB* and *recBC* strains. This result agrees with the findings of Enea and Zinder (1975), who also found miniphage after infection of *E. coli recA* bacteria. The data suggest that the *recBC* nuclease is required for the efficient production of miniphage; it is not obvious how the nuclease promotes miniphage DNA formation.

Midiphage

A new, variable-length phage particle was discovered in stocks prepared at high m.o.i. on the *recAB* host; the lengths of these particles varied from 1.2 to 1.8 times that of the normal phage (Fig. 2 and unpubl.) The stock considered in Figure 2 had a median length 1.4 times that of the normal phage.

The following results indicate that these midiphage contain sigma DNA: (1) The frequency of sigma DNA and midiphage agreed very well in different stocks (12% midiphage, 8% sigma DNA in the stock passed at m.o.i. 100; 1% midiphage and 0.8% sigma DNA in the stock passed at m.o.i. 0.01). (2) Both the midiphage and the sigma DNAs are of variable length and have similar bell-shaped length-distribution curves (Fig. 2). The only discrepancy between the two classes of particles is the longer average length of the midiphage (Fig. 2). This can be explained by assuming that the tails of the sigma DNA associate with more than the usual amount of phage protein; one way would be to imagine a steeper DNA helix in this tail, like the one found in Pf1 DNA, for example (Wiseman et al. 1976). According to this model, the tail of the sigma DNA would loop back to the junction of the circular portion, creating a "double-helical" structure similar to the normal phage but with a steeper pitch. A second model would extend the SS DNA tail without a "partner" strand at the normal pitch. At present, there is no way to distinguish between these models (see Day and Wiseman, this volume).

The existence of midiphage and the other abnormal forms of filamentous phage discussed above indicates a great flexibility in the packaging process for these phage. Any model of this process should take into account at least the following points:

1. Different miniphage share only a small common sequence in the intercistronic region between genes II and IV, and none of the cistrons of these phage are intact (Enea and Zinder 1975). If there is a unique start site for the packaging process, it is limited to this region, unless miniphage have other start sites different from those of wild-type phage. Some evidence for (Marco et al. 1974) and against (Tate and Peterson 1974) a unique polarity of the DNA in the phage particle leaves the question unresolved. However, polarity presumably must exist in the midiphage because the SS DNA tails can fit only at either end of the phage, not in the middle.
2. The actual length of DNA that can be packaged varies from 0.2 genome length to at least 4 genome lengths; not only are multiple-length particles formed, but also a continuum of particles ranging from 0.2 to 0.5 and 1.2 to 1.8 unit length. It is probable that the variations in phage length reflect the variations in length of the progeny DNA molecule, which seems to be the dominant force in determining the length of the particle.
3. Although the usual substrate for packaging is a circular SS DNA molecule, which is highly condensed into a double-stranded-like structure by the gene-V protein, linear SS DNA can be packaged also.
4. End-to-end joining of two phage DNA-protein complexes (to form phage dimers) can occur even when genetic recombination is blocked completely (Table 1).

Whether or not midiphage are infectious is not known. Attempts to propagate stocks containing a high frequency of midiphage on either the *rec*$^+$ or *recAB* host have been unsuccessful. Thus, midiphage probably are at a selective disadvantage compared with normal phage or they are converted back to normal phage upon reinfection of a new host.

Sigma Single-stranded DNA

This new form of SS DNA was found in phage propagated on the *E. coli recAB* strain (Fig. 3); alkaline denaturation converted the sigma DNA to linear DNA of greater than unit length (Table 2). These results are compatible with a model for sigma SS DNA that has a circular portion held together by hydrogen bonding and a SS DNA tail of variable length (Fig. 4). The constant unit length of the circular portions of the molecules suggests that the same complementary sequences stabilize all of the molecules; since two hairpin loops have been found near the origin of fd DNA replication (Schaller et al., this volume), it is tempting to equate the complementary sequences to the two halves of the stem of one of the hairpin loops.

The variable length of the tail of sigma SS DNA is puzzling. At present, we cannot even distinguish between two of the simpler models possible for these different lengths: (1) The progeny SS DNA is cut from the RF DNA with an endonuclease at a unique, abnormal site, yielding molecules about 1.5 times normal length; 3′-specific host nucleases digest varying lengths of the DNA to give the distribution observed (Fig. 2). (2) The progeny SS DNA is cut from RF DNA with an endonuclease at a great variety of sites, yielding DNA molecules 1.05–1.4 times normal length.

ACKNOWLEDGMENTS

We thank A. Klein, A. J. Clarke, and H. Drexler for their generous gifts of strains, C. Chase for some of the electron micrographs, and L. Day for helpful discussion. This work was supported by grants to R. C. Warner (U.S. Public Health Service grant CA12627) and to R. H. B. (AI-08572, a Research Career Development Award 6-K04-GM-50284 GEN, and National Science Foundation grant PCM-77-13970).

REFERENCES

Beaudoin, J. and D. Pratt. 1974. Antiserum inactivation of electrophoretically purified M13 diploid virions: Model for the F-specific filamentous bacteriophages. *J. Virol.* **13:** 466.

Enea, V. and N. Zinder. 1975. A deletion mutant of bacteriophage f1 containing no intact cistrons. *Virology* **68:** 105.

Grandis, A. S. and R. E. Webster. 1973. A new species of small covalently-closed f1 DNA in *E. coli* infected with an amber mutant of bacteriophage f1. *Virology* **55:** 14.

Griffith, J. and A. Kornberg. 1974. Mini M13 bacteriophage: Circular fragments of M13 DNA are replicated and packaged during normal infections. *Virology* **59:** 139.

Hewitt, J. A. 1975. Miniphage—A class of satellite phage to M13. *J. Gen. Virol.* **26:** 87.

Marco, R., S. M. Jazwinski, and A. Kornberg. 1974. Binding, eclipse, and penetration of the filamentous bacteriophage M13 in intact and disrupted cells. *Virology* **62:** 209.

Marvin, D. A. and H. Hoffmann-Berling. 1963. A fibrous DNA phage (fd) and a spherical RNA phage (fr) specific for male strains of *E. coli.* II. Physical characteristics. *Z. Naturforsch.* **18b:** 884.

Pratt, D., H. Tzagoloff, and J. Beaudoin. 1969. Conditional lethal mutants of the small filamentous coliphage M13. II. Two genes for coat proteins. *Virology* **39:** 42.

Salivar, W. O., T. Henry, and D. Pratt. 1967. Purification and properties of diploid particles of coliphage M13. *Virology* **32:** 41.

Scott, J. R. and N. D. Zinder. 1967. Heterozygotes of phage f1. In *The molecular biology of viruses* (ed. J. S. Colter and W. Paranchych), p. 211. Academic Press, New York.

Tate, W. P. and G. B. Peterson. 1974. Structure of the filamentous bacteriophages: Orientation of the DNA molecule within the phage particle. *Virology* **62:** 17.

Wheeler, F. C., R. H. Benzinger, and H. Bujard. 1974. Double-length, circular, single-stranded DNA from filamentous phage. *J. Virol.* **14:** 620.

Wiseman, R. L., S. A. Berkowitz, and L. A. Day. 1976. Different arrangements of protein subunits and single-stranded circular DNA in the filamentous bacterial viruses fd and Pf1. *J. Mol. Biol.* **102:** 549.

Structure of the Filamentous Phage Virion

Donald A. Marvin
European Molecular Biology Laboratory
69 Heidelberg, Federal Republic of Germany

The filamentous phages are flexible nucleoprotein rods about 60 Å in diameter and 1–2 μm long. Their circular, single-stranded DNA is encapsulated at the core of a tubular array of protein subunits. Each protein subunit contains only about 50 amino acids and is almost entirely α-helical. Concentrated gels of this phage type can be oriented into fibers that give relatively high-quality X-ray diffraction patterns. The diffraction patterns of the best-known filamentous phages (fd, f1, and M13) are difficult to analyze because there is a perturbation in the helical arrangement of subunits (Marvin et al. 1974b). The similar phage Pf1 gives a less complicated diffraction pattern (Figs. 1 and 2) and is therefore the strain that will be emphasized here.

The filamentous phages have several unique features that enable structure determination without resorting to multiple isomorphous replacement. The technique that we have used is molecular model building, a technique that has been successful for macromolecules such as the α helix and DNA. The word "model" as used here is not a synonym for "hypothesis"; the stereochemical constraints imposed by a precise molecular model are an essential part of this technique of structure determination.

METHODS

Elongated assemblies of subunits such as filamentous phage seldom form true crystals. Instead, they form oriented, semicrystalline fibers in which the assemblies lie parallel to one another and hence diffract X rays in unison. Most elongated assemblies are helical arrangements of identical subunits. There are two periodicities along the axis of a helix: the unit rise p, or axial spacing from one subunit to the next within the helix, and the pitch P, or height of one turn of the helix (Fig. 3). To make up the helical assembly, an identical copy of the subunit or monomer is associated with each of these equivalent points (Fig. 4).

The X-ray diffraction pattern of the fiber shows a series of lines of intensity, the layer lines, oriented perpendicular to the long axis of the helical assembly. The distribution of diffracted intensity on the zero layer line, or equator (which runs horizontally through the centers of the diffraction patterns in Figs. 1 and 2), gives information about the cross-sectional

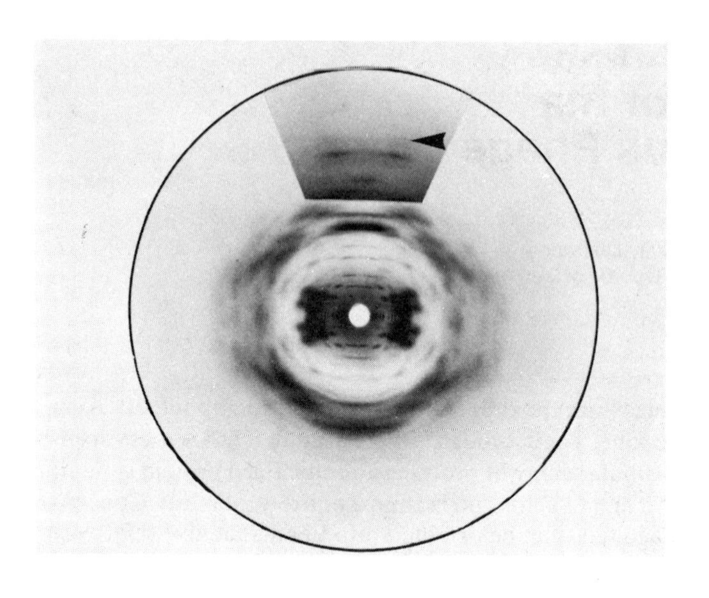

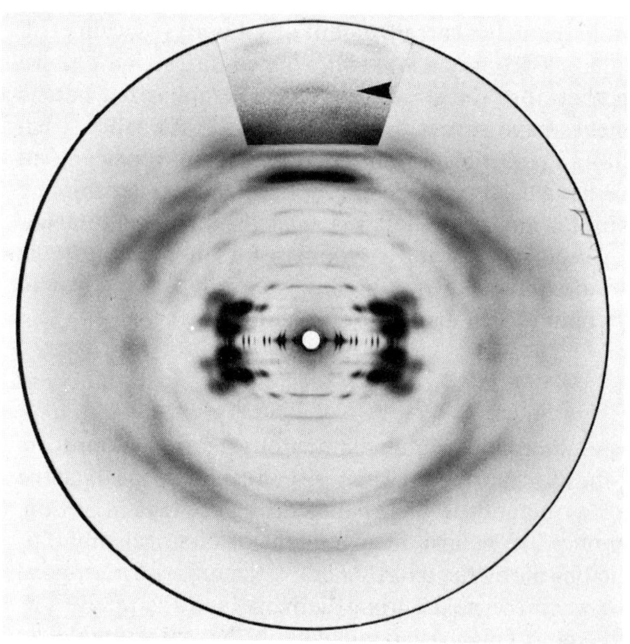

Figure 1 X-ray fiber diffraction patterns of Pf1 at 75% relative humidity. Fiber axis vertical. Fiber G128C (prepared by S. Malsey), 4.4 units per turn. (*Bottom*) Film 3231, tilted 13° away from normal to the X-ray beam to show the meridional reflection at 3.4 Å (arrow). (*Top*) Film 3247, tilted 16° away from normal to the X-ray beam to show absence of a meridional reflection at 2.8 Å (arrow). The meridional region is more heavily exposed to bring out weaker reflections.

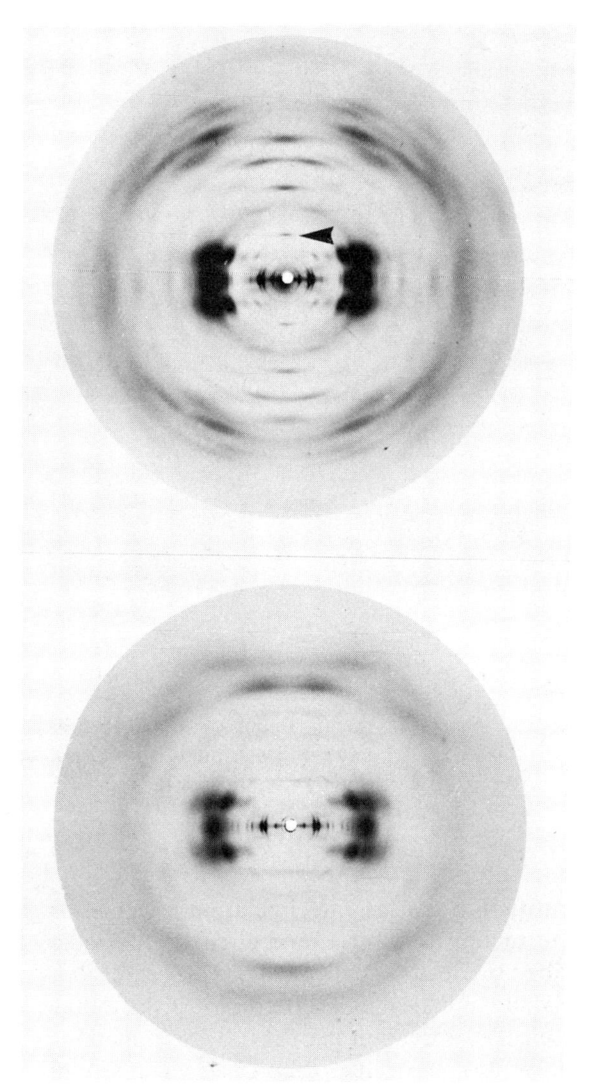

Figure 2 X-ray fiber diffraction patterns of filamentous phage, with 4.5 units per turn, at 75% relative humidity. Fiber axis vertical. (*Top*) fd: Fiber 187, film 552 (prepared by F. J. Marvin). Arrow indicates the 16-Å meridional reflection (*Bottom*) Pf1: Fiber G150A, film 3429 (prepared by S. Malsey).

structure of the helix and about the side-by-side packing of neighboring helices. The distribution of intensity on the meridian, a line running vertically through the centers of the diffraction patterns in Figures 1 and 2, gives information about periodicity along the axis of the helix (for instance, the axial spacing p between subunits).

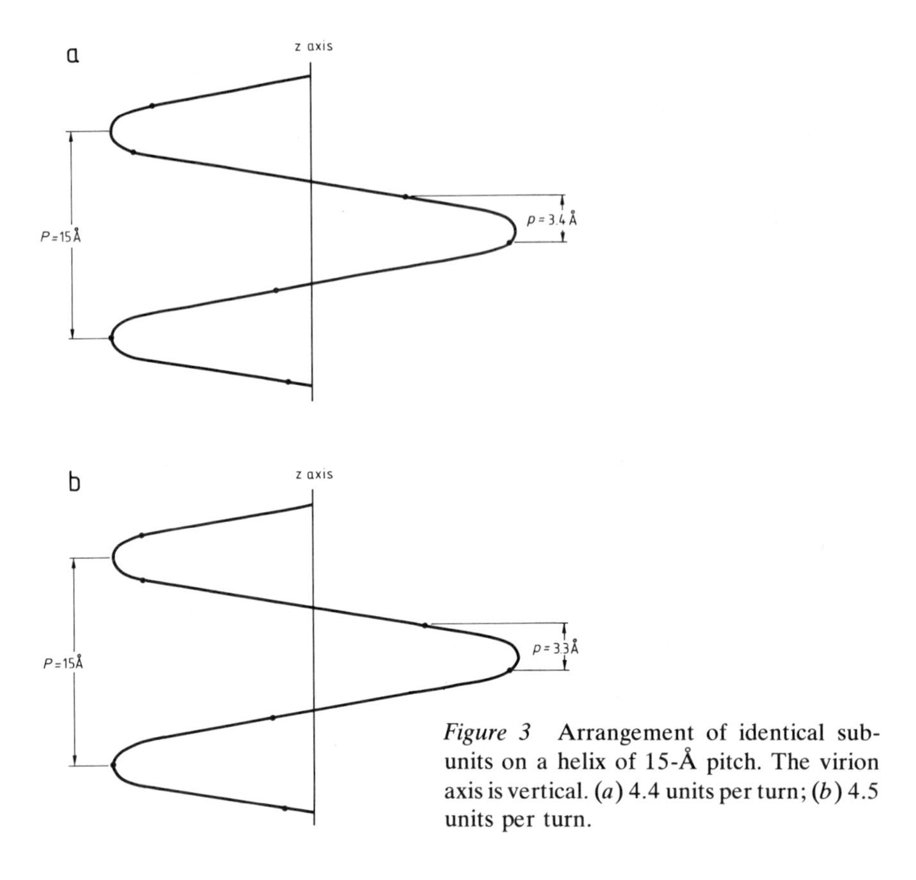

Figure 3 Arrangement of identical subunits on a helix of 15-Å pitch. The virion axis is vertical. (*a*) 4.4 units per turn; (*b*) 4.5 units per turn.

The helical character of the structure gives rise to circular periodicities as well as linear periodicities. Diffraction from an n-fold circularly periodic structure is most simply described by Bessel functions of order n, $J_n(2\pi Rr)$, where r is the radius of material in the helical molecule, and R is the radius (distance from the meridian) on the diffraction patterns. Bessel functions have the property that as n increases, the value of $2\pi Rr$ at which strong diffraction is observed also increases. Therefore, for a structure of known maximum radius r, a rough estimate of n can be made from the smallest value of R at which appreciable intensity is observed.

The positions of the layer lines are related to the periodicities within the helix by the selection rule $l = tn + um$, where l is the layer-line index, u is the number of units in t turns of the helix, n is the Bessel function index, and m is any integer. For example, for $m = 0$, a helix of 22 units in five turns gives $n = 1$ on $l = 5$. The presence of the low-order Bessel function J_1 predicts near-meridional intensity on this layer line. The same selection rule can be written as $\zeta = (n/P) + (m/p)$, where ζ is the height of the layer line above the equator, P is the pitch of the helix, and p is the unit rise. There is a reciprocal relationship between the dimensions in the helical assembly and the dimen-

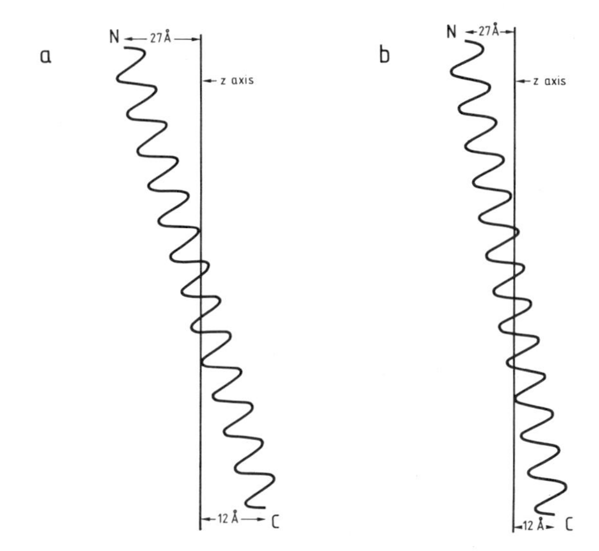

Figure 4 Orientation of the subunit in Pf1. The virion axis is vertical. Associated with each equivalent point of Fig. 3 is a single gently curved rod of α helix. The N terminus of the 46-residue length of α helix is at the 27-Å outer radius of the protein shell and the C terminus is at the 12-Å inner radius. The axis of the α helix in the virion follows a left-handed helix of decreasing radius. (a) Orientation of the subunit in the 4.4-units-per-turn virion. (b) Orientation of the subunit in the 4.5-units-per-turn virion.

sions on the diffraction pattern. For instance, if $m = 0, n = 1$, and $P = 15$ Å, then $\zeta = 0.067$ Å^{-1} ($0.067 = 1/15$).

A helical array of points can be described in several ways. The simplest description is the basic helix, which runs between points separated by p in the axial direction and by $\theta = t/u$ in the azimuthal, where θ is a fraction of 2π. A single basic helix describes the structure completely unless a rotation axis runs down the helix axis. Other helices, the n-start helices, describe the same array of points equally well. These helices have pitch of $P_n = np/|n\theta|$, the vertical lines indicating the nonintegral part of $n\theta$. For instance, if $n = 5, p = 3.4$ Å, and $t/u = 5/22$, then $P_5 = 125$ Å. To describe the structure completely, n such helices must be drawn.

An introductory discussion of the theory of X-ray fiber diffraction can be found in the book by Fraser and MacRae (1973).

RESULTS

Structure of the Virion

The first step in analyzing a helix diffraction pattern is to determine the helix parameters. The Pf1 diffraction pattern (Fig. 1) shows near-meridional intensity on a system of layer lines at orders of about 0.067 $Å^{-1}$, which gives the pitch of the basic helix as about 15 Å. There is another set of layer lines having near-meridional intensity, which are separated from each other by 0.067 $Å^{-1}$ but are displaced from the first set by 0.4 of this spacing. This shows that there are $s + 0.4$ structure units in one turn of the helix, where s is an integer limited to $2 \leqslant s \leqslant 5$ by the position of intensity on the layer lines (Marvin et al. 1974a).

Unequivocal determination of the number of units per turn (that is, of s in the above expression) is difficult, but in every X-ray experiment where we have detected a bias, $s = 4$ is favored. The most direct evidence comes from the meridional reflection (corresponding to the axial spacing p between subunits) at 3.4 Å (Fig. 1): a spacing of 3.4 Å along a helix of 15-Å pitch gives 15/3.4 = 4.4 units per turn. A second line of evidence comes from the comparison of fd with the form of Pf1 having 4.5 units per turn (Fig. 2). Analysis of the equatorial intensity distribution around $R = 0.1$ $Å^{-1}$ indicates that the order of the Bessel function contributing in this region is the same for 4.5 Pf1 as for fd (Fig. 5). Since mass-per-length calculations support

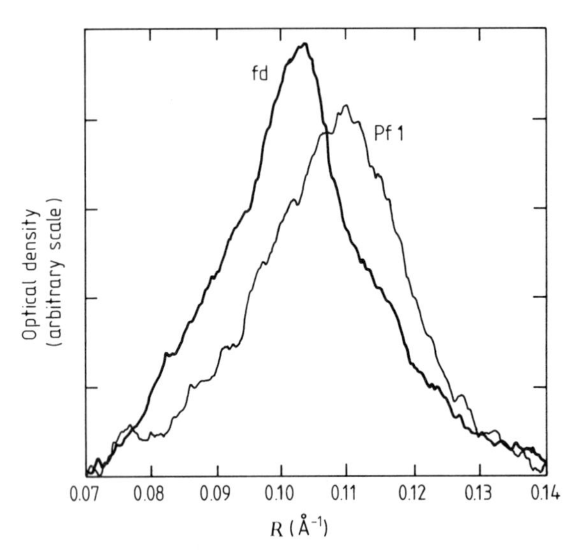

Figure 5 Densitometer tracings along the equator. (*Light line*) 4.5-units-per-turn Pf1 pattern; (*heavy line*) 4.5-units-per-turn fd pattern. The intensity maximum of fd occurs at a position R about 5% smaller than the position of the Pf1 maximum, which indicates that the Bessel function contributing in this region has the same order n for both virions since the radius r of fd is about 5% larger than that of Pf1 (making the assumption that the two virions are built according to the same general design, as discussed by Marvin et al. [1974b]). Pf1: Fiber 1439B, film 2123; fd: Fiber 177C, film 545. The experiment was performed in collaboration with E. J. Wachtel.

$s = 4$ for fd (Marvin and Hohn 1969; Berkowitz and Day 1976), this line of reasoning also supports $s = 4$ for Pf1.

On the basis of biochemical data, Wiseman et al. (1976) have suggested that models with 5.4 units per turn should be considered. However, no meridional reflection is observed at 2.8 Å, where it would be predicted for 5.4 units per turn (Fig. 1).

The next step in the analysis is to determine the general shape of the asymmetric unit. Here we are aided by the fact that most of the intensity on the Pf1 diffraction pattern is found on the first and third layer lines. The radial position of intensity and the Bessel function index n on layer line l show that intensity on the first layer line comes largely from material near a radius in the virion of 16 Å and that intensity on the third layer line comes largely from material near a radius of 12 Å. The strength of this region of intensity shows that the phase $2\pi\ (lz - n\varphi)$ of the material contributing to the diffraction on these layer lines is constant (z and φ are the fractional axial and azimuthal coordinates of atoms in the asymmetric unit of the virion). That is, the structure unit can be represented roughly by a rod of electron density centered at a radius of 16 Å and oriented with $z/\varphi = -9/1$ contiguous with another rod of electron density centered at a radius of 12 Å and oriented with $z/\varphi = -5/3$ (Marvin et al. 1974a).

Most of the protein in Pf1 is in the α-helix conformation, as shown by the typical 10-Å equatorial and 5-Å meridional intensities on the fiber diffraction pattern; by laser Raman spectroscopy, which indicates over 90% α helix (Thomas and Murphy 1975); and by theoretical analysis of the amino acid sequence (Green and Flanagan 1976). Therefore, the next step in the structure analysis was to build a model of an α-helical protein subunit having the orientation determined for the rod of electron density from consideration of the first and third layer lines and to refine this model by calculating the predicted diffraction pattern, comparing it with the observed pattern, and modifying the model to improve the fit of the calculated to the observed pattern. Early models were simply 46-residue lengths of polyalanine (Marvin et al. 1974a; Marvin and Wachtel 1975; Nakashima et al. 1975). On later models (Marvin and Wachtel 1976) side chains were added in the known sequence and a variety of constraints were used to define the orientation of the protein. General chemical considerations suggest that the collection of basic residues at the C terminus should be at the inner radius of the protein shell to neutralize the DNA phosphates and that the collection of acidic residues at the N terminus should be at the outer radius to explain the low isoelectric point of the virion. Specific chemical modification experiments (Table 1) have shown that the α-amino group at the N terminus, but not Lys-45, is accessible to succinylation and that several other predictions of the model with regard to chemical accessibility are in fact observed. Analysis of the radial electron-density distribution shows low electron density in the middle of the protein shell, which can be identified with the

Table 1 Chemical modification of the major coat protein of Pf1 virus

Modification reagent	Potentially reactive residues	Predicted by the model	Results of modification[a]
[^{125}I]chloramine T	Tyr 25	out	+
[^{125}I]lactoperoxidase	Tyr 40	in	−
[^{14}C]succinic anhydride	Gly 1	out	+
	Lys 20	in	−
	Lys 45	in	−
Phenylglyoxal	Arg 44	in	−
Carbodiimide	Asp 4	out	+
[^{14}C]glycine amide	Asp 14	out	−
	Asp 18	out	−
	Ala 46	in	−
H_2O_2 then	Met 19	out	−
[^{14}C]iodoacetic acid	Met 42	in	−

Data from Y. Nakashima and W. Konigsberg (in prep.).
[a]+: Completely reacted; −: no modification or very little.

collection of hydrophobic residues in the middle of the sequence (Marvin and Wachtel 1976). All of these considerations support a picture of the protein subunit as a gently curved rod of *a* helix roughly parallel to the long axis of the virion, but slewing round this axis and also sloping radially from the outer to the inner diameter of the virion (Fig. 6).

Can one trust a model that is not based on a multiple isomorphous replacement determination of phases? Modern protein-structure determination uses multiple isomorphous replacement only to calculate the first low-resolution electron-density map. Once a trial model has been fitted to the density, it is more accurate to replace the observed phases with the phases calculated from the model (Diamond 1976). The information derived from the intensity distribution on the first and third layer lines of Pf1 can be used instead of multiple isomorphous replacement for initial determination of the line of the *a* helix, and refinement can proceed from there by model-building. The ultimate test of the validity of any structure determination, whether multiple isomorphous replacement was used or not, is how good the fit is between the calculated and observed diffraction. The diffraction calculated from the model of Marvin and Wachtel (1976) gives a reasonably good fit to the observed data. Further refinement is in progress.

The space curve defining the axis of the *a* helix can be represented by four line segments, each with a different pitch and slope. The azimuthal orientation of the *a* helix around this axis requires a further parameter. A radius is

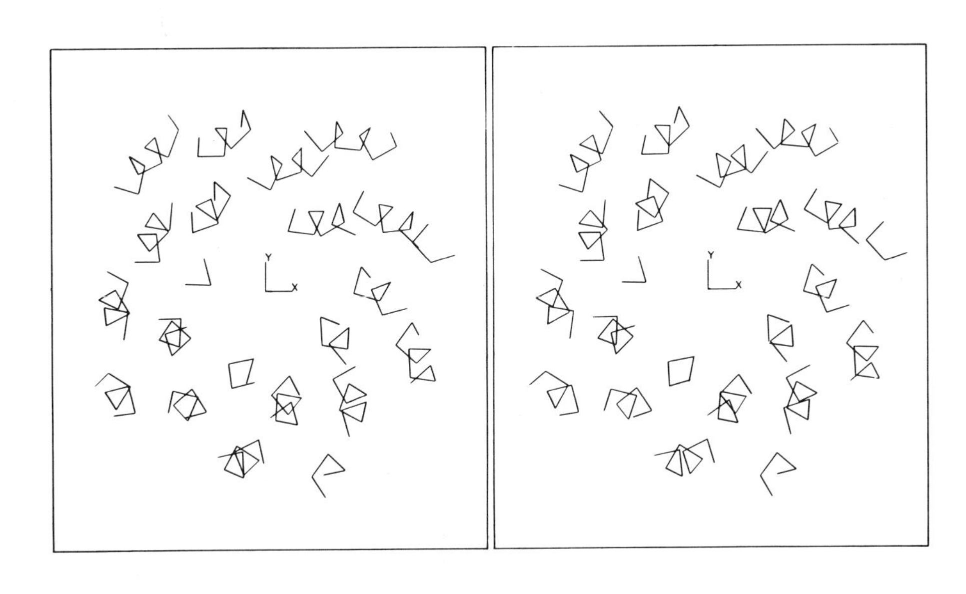

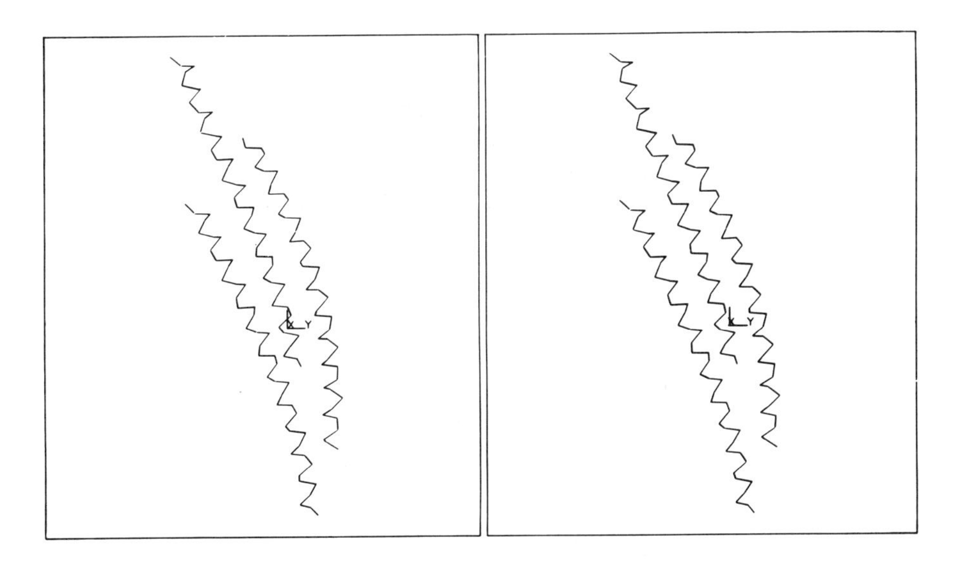

Figure 6 Stereo views of an α-carbon representation of the 4.4-units-per-turn Pf1 model described by Marvin and Wachtel (1976). (*Top*) Axial section of a 14-Å slab through one virion, viewed down the helix axis. (*Bottom*) View normal to the helix axis showing packing between a subunit (*center*) and its nearest neighbors on the 5-start (*right*) and 9-start (*left*) helicoids. The N terminus is at the upper end of the subunit. A stereo viewer may be purchased from Abrams Instrument Co., 606 E. Shiawassee St., Lansing, Michigan 48901.

necessary to define the position of the α helix in the cylindrical-polar coordinate system of the virion. These parameters are constrained by the observed diffraction intensity. The observed intensity extends to about 0.20 $Å^{-1}$ along each of 15 layer lines. The Fourier transform of a structure 60 Å in diameter can be sampled meaningfully at about 0.015-$Å^{-1}$ intervals (Bracewell 1965). Therefore, there are about 200 independent observations, which satisfactorily overdetermine the ten parameters defining the α-helix approximation to the structure. That is, if a good fit of calculated-to-observed transform can be found, then there is no chance that a different model with a linear α-helical monomer could also be found to fit the observed data. On the other hand, the number of independent diffraction observations is not sufficient to discover details such as precise orientation of all side chains or the presence of nonhelical regions. Stereochemical constraints can help to fix these aspects of the structure. Energy considerations argue against bends or breaks in the α helix within the hydrophobic interior of the protein because hydrogen bonds could not be satisfied. Even though atomic resolution is not accessible directly, structural insights can be developed from the model.

Polymorphism

Description of static structure is an important first step in understanding biological function, but real understanding can come only with a description of dynamics. One way to gain such understanding is to examine the kinetics of the process under study while the process is taking place. Another simple entry into dynamics is to examine a series of static stages along a dynamic pathway and to attempt to deduce the dynamic flow from the nature of these structures.

Pf1 was first observed in a form having 4.4 units per turn in the helix, but the virion is also found in several other forms. If fibers are prepared at 5°C instead of 20°C, there is a slight tightening of the helix to 4.5 units per turn (Wachtel et al. 1976; S. Malsey and H. Siegrist, in prep.). Other variants with 4.38 or 4.46 units per turn have also been observed. Since some diffraction patterns clearly show all virions in the fiber to be in one or the other of these forms, there must be local conformational energy minima corresponding to each of these forms. However, there are only slight differences between the conditions promoting the altered forms, so these energy minima are shallow.

Brief treatment with ether leads to a more extensive change in the structure. The virions swell to hollow cylinders of about three times normal diameter but only about 1/10 normal length (Amako and Yasunaka 1977; Fig. 7). The diffraction patterns of the swollen virions (Fig. 8) show an equatorial reflection at about 68 Å and a strong ring at about 10 Å. The equatorial reflection corresponds to a center-to-center distance between particles considerably smaller than the diameter of swollen phage observed in the electron microscope, perhaps because the first-order reflection is weak. The 10-Å ring corresponds to the spacing between close-packed α helices. The change of virions from normal to swollen may involve a change in the orientation of the α-helical subunits with respect to the virion axis from roughly parallel to roughly radial.

Many of the swollen virions appear to be rounded at one end but open at the other. The gene-III protein, which is required for adsorption of the virion to its host, is located at one end of the fd virion—and by analogy at one end of the Pf1 virion (Bradley 1973). Thus, the open end of the swollen virion most likely corresponds to the site of the gene-III protein. Consideration of how an assembly of overlapping rods might swell suggests that the C terminus of the coat protein molecule is directed toward the gene-III-protein end of the virion (Fig. 9).

Treatment with ether for a longer period of time leads to a variety of structures, including bottles open at one end, filled or empty spheres, stars, and flat sheets (Fig. 7). These structures can be envisaged to occupy a series of local energy minima along a continuous smooth change in the assembly of α-helical subunits. The presence of sheetlike structures suggests the possibility that the Pf1 virion is assembled from sheets of coat protein in the bacterial membrane by a budding process that would be the reverse of the ether-induced swelling of virions from tubes to spheres to sheets.

The effects of temperature and ether on the structure of the virion may be due to a change in the dielectric constant of the solvent which alters the internal charge distribution on the protein (Warshel and Levitt 1976); binding of ether to hydrophobic amino acids could also be involved. A small change in the shape of each protein subunit would result in a large change in the orientation of the protein needed for optimum close packing and thereby change the parameters of the helix.

The fd structure is more interesting to biologists than that of Pf1 because fd has been analyzed more extensively both genetically and biochemically than Pf1. The broad distribution of intensities on the diffraction patterns is similar for fd and the 4.5-units-per-turn form of Pf1, but there are small differences in detail (Marvin et al. 1974b; Fig. 2). Such differences are to be expected from the differences in the protein monomer: although the overall character of the protein is the same (hydrophobic central portion of the sequence, acidic N terminus, and basic C terminus), the amino acid sequences are quite different, and Pf1 has only 46 residues, compared with 50 for fd (Nakashima et al. 1975). The minimum lateral packing between virions is about 53 Å for Pf1 and about 55 Å for fd, which is consistent with the differences in the size of the monomer. Strong meridional reflections are observed at orders of 16 Å on fd, whereas the corresponding reflections are weak or absent on Pf1. This 16-Å meridional repeat is probably due to a periodic perturbation of the protein subunit away from the positions

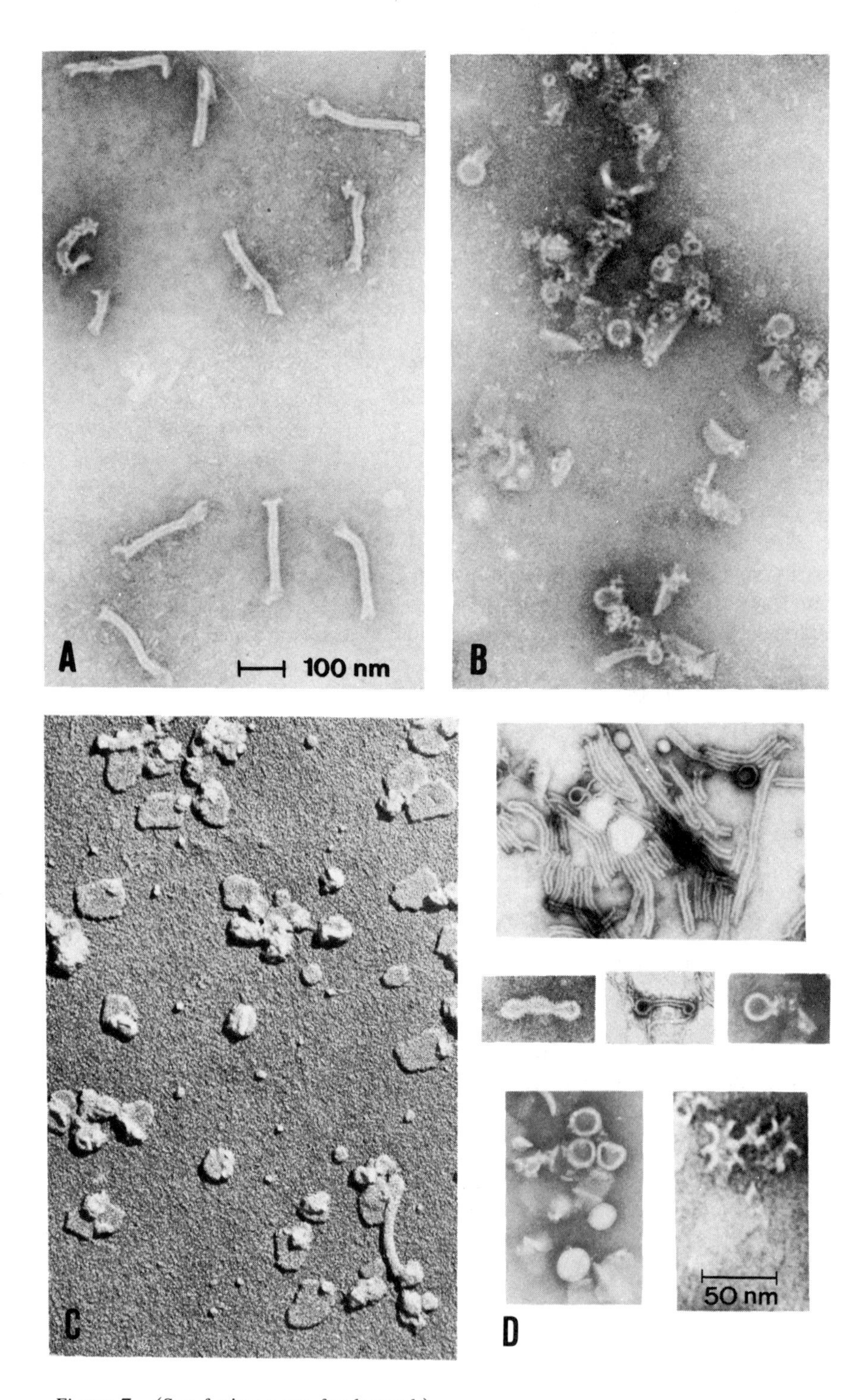

Figure 7 (See facing page for legend.)

expected for a perfect helix with a period that repeats every five subunits. The fd structure has not been analyzed in as much detail as the Pf1 structure because the perturbation complicates the diffraction pattern, but it is clear from the similarity of the diffraction patterns that the structures are closely related.

Symmetry within the Virion

Understanding the symmetry principles involved in a structure reduces the number of parameters that need to be determined experimentally and simplifies the task of understanding function. To discuss the symmetry principles involved in filamentous phage structure, it is convenient to replace each α helix in the virion with a space curve following the axis of the α helix. If one such curve is rotated about the virion axis and at the same time translated parallel to the axis, through a set of similar curves in the virion, it will sweep out a twisted ribbon, or helicoid (Fig. 10). The set of subunits comprising the virion can then be identified with the intersections of two sets of helicoids.

The closest contacts are between any subunit and its nearest neighbors along the 5-start or 9-start helicoids. In a minimum energy conformation, each subunit will be symmetrically located between its neighbors. There are two kinds of symmetry that need to be considered. The first is equivalence of neighbor-neighbor interaction in the 5-start and 9-start helicoids. This constrains the axis of the α-helical subunit to a linear relationship between z and φ and requires that r be a function of $\sinh \varphi$. The second requirement is that the distance from each subunit to its neighbor be the same along the length of the subunit. This also requires a linear relationship between z and φ but that r be a function of φ (the circle involute). The two shapes are similar ($\sinh \varphi \simeq \varphi$ for small φ), and the shape determined experimentally in fact lies between the two.

The pitch of the helicoids is coupled to the shape of the subunits. This coupling means that a small change in the shape of the α helix will yield a corresponding change in the helix parameters without destroying the overall symmetry of the assembly. A local biochemical change can thus be transduced into a macroscopic geometrical change, as in the transition from 4.4 to 4.5 units per turn or in the ether-induced swollen phage. The similarity

Figure 7 Electron micrographs of Pf1 treated with ether. Samples were gently mixed with excess ether, incubated at room temperature for the times indicated below, diluted 100-fold into 0.01% glutaraldehyde, and negatively stained according to the method of Valentine et al. (1968) (fields *A, B, D*) or shadowed at a 20° angle (field *C*). (*A*) 1 min, (*B, C*) 10 min, (*D*) selected fields. Enlargement is as given in *A* except where indicated otherwise. The experiment was performed in collaboration with T. Arad, S. Malsey, and K. Leonard.

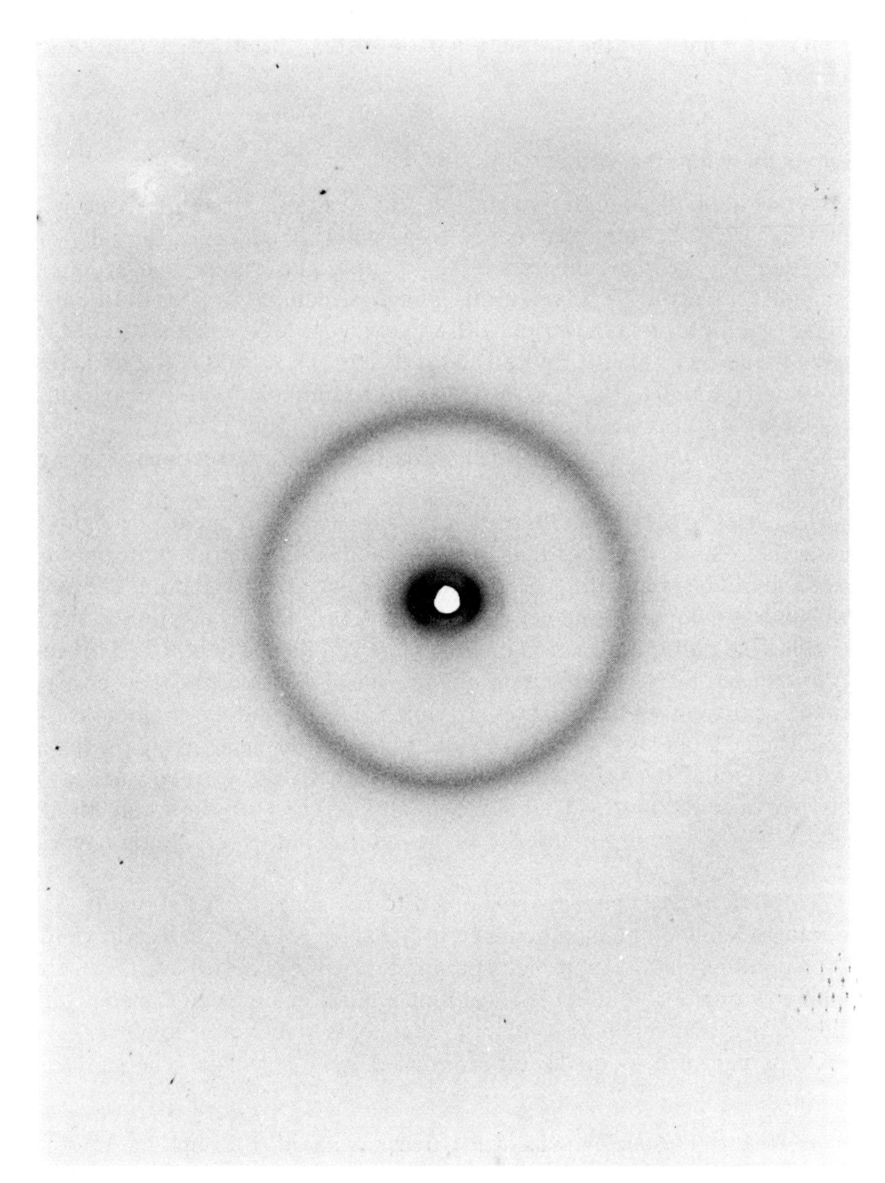

Figure 8 X-ray diffraction pattern of Pf1 treated with ether as described in the legend to Fig. 7. Fiber axis vertical. Specimen-to-film distance = 9.5 cm, enlargement 2×. The strong equatorial reflection has a spacing of 68 Å, and the strong ring has a spacing of 9.4 Å. Fiber G133A, film 3138. The experiment was performed in collaboration with S. Malsey.

between fd and a Pf1 form (Fig. 2) shows that the details of the amino acid sequence are not critical. The structure is determined by the principles of optimum packing of rodlike subunits, overlayed with second-order effects introduced by general charge or hydrophobic distribution. The simple form of the protein subunit makes this a good model system in which to study how different patterns of amino acids can generate similar protein surfaces.

A third helicoid may be drawn orthogonal to both the 5-start and the 9-start helicoids. If this helicoid is a surface of constant positive curvature, it can be mapped isometrically (with no local change in dimensions or shape) onto the surface of a sphere. The axes of the α-helical subunits that were orthogonal to this helicoid in the original structure will remain orthogonal to the surface of the sphere. This geometrical transformation describes a possible origin of the spherical structures observed after ether treatment of the virions. The symmetry of neighbor-neighbor interaction at the rounded ends of the virion has similarities to the symmetry involved in phyllotaxis, or the arrangement of leaves on a plant (Coxeter 1969). Two sets of oppositely directed spirals, the 5-start and the 9-start, can be drawn through the subunits (as shown in cross section in Fig. 11). Similar spirals can be drawn through adjacent subunits at the rounded end of the virion, as illustrated in Figure 12 for the 8-start and 13-start spirals on a pine cone. One can visualize how the packing shown in Figure 11 could be distorted to an analog of Figure 12 at the rounded end of the virion.

If the helicoid that is orthogonal to the 5-start and 9-start helicoids is a surface of constant negative curvature (in practice this means a slightly altered shape of the subunit), it can be mapped isometrically onto the surface of a pseudosphere. The pseudosphere has a shape similar to a trumpet, wide at one end and tapering at the other. The stars observed after ether treatment may be a set of pseudospheres joined at their wide ends.

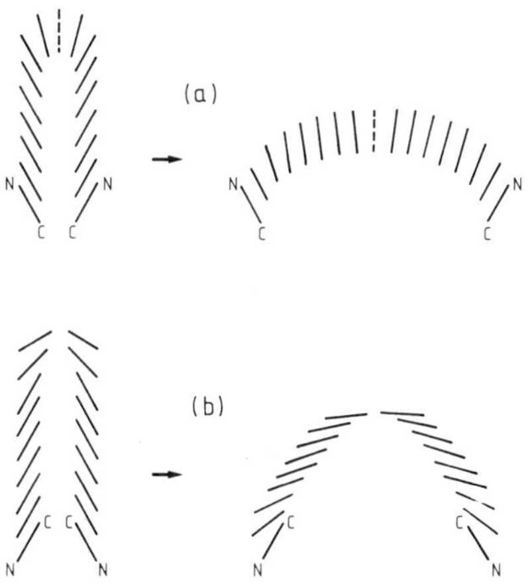

Figure 9 Models for the swelling of filamentous phage. The protein subunits are represented by short, overlapping lines, with the N terminus at the outer radius and the C terminus at the inner radius. The two alternatives for the orientation of the subunit with respect to the closed end are shown in *a* and *b*. With model *a*, the structure can swell smoothly to give the observed rounded ends; model *b* predicts a nonobserved discontinuity after swelling. The dashed subunit represents the suggested position of a minor coat protein component.

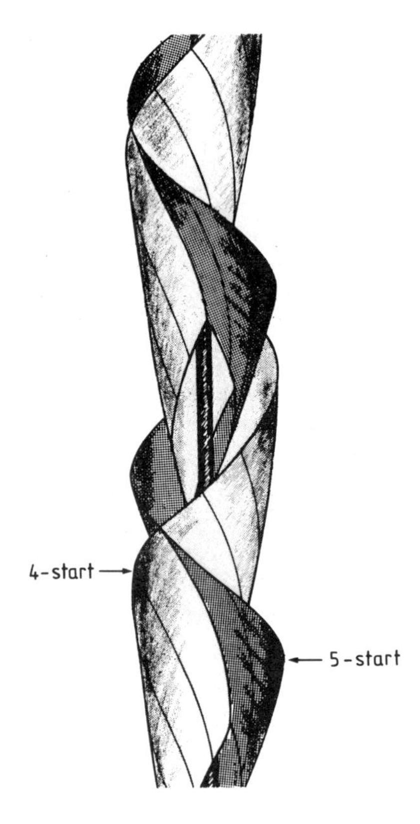

Figure 10 Representation of the symmetry of the filamentous phage virion by the intersections of two helicoids. The short curves represent the α-helix axes of protein subunits arranged along the 4-start (right-handed) and 5-start (left-handed) helicoids of the model of Marvin and Wachtel (1976). One helicoid in each of these two sets is shown; in the complete structure there are four 4-start helicoids intersecting five 5-start helicoids around the axis. The dark vertical line represents DNA in the core of the virion. The 4-start and 5-start helicoids are shown rather than the 5-start and 9-start because the intersections are easier to visualize.

The description of regular, three-dimensional structures in terms of orthogonal sets of surfaces is part of differential geometry (Struik 1961). Division of space into regular blocks, and transformation of one such space to another with a minimum of local distortion, is of interest in discussing not

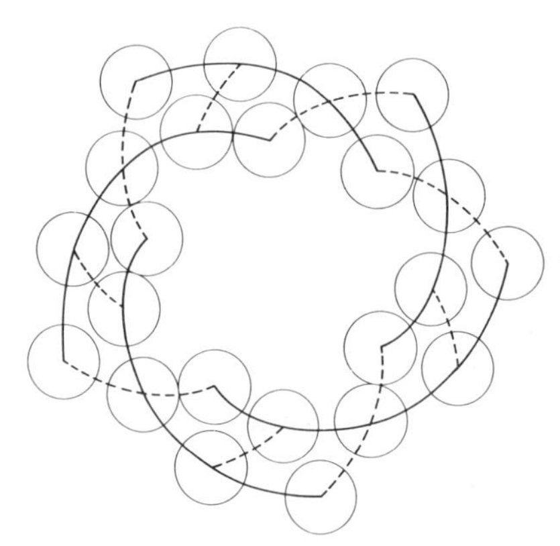

Figure 11 Symmetry of the 5-start and 9-start helicoids in Pf1. A section normal to the virion axis is shown. Each circle represents a section through one of the overlapping α-helix subunits. The solid curves represent sections through the 5-start helicoids and the dashed curves represent sections through the 9-start helicoids.

only molecular assemblies, but also assemblies of cells in developmental biology and morphogenesis.

DISCUSSION

Two general questions concerning filamentous phage structure are still under discussion. One is the question of the number of units per turn in the virion: Is it 4.4 or 5.4? We have chosen the value 4.4 for our detailed model-building, since interpretation of our X-ray data is much more difficult on the basis of 5.4 units per turn. Nevertheless, we have also built trial models with 5.4 units per turn and have found that the general picture of a protein shell made of overlapping interdigitating α helices is also valid for such models (Marvin et al. 1974a). Therefore, our general picture of the virion would be unchanged for a 5.4 model.

The other question concerns the shape of the α-helix subunit. L. Makowski and D. L. D. Caspar (in prep.) have suggested from model-independent electron-density calculations that there is a break in the middle of the α helix. We feel that an attempt to solve structures from low-resolution data in the absence of molecular model building can be misleading. In particular, the low electron density in the interior of the protein shell is best attributed to the presence of low-density hydrophobic side chains, and not to a break in the α helix. Certainly, breaks in the α helix, either in the middle or at the ends, cannot be ruled out unequivocally at this time, but one must not abandon the basic principle that the best hypothesis is always the simplest hypothesis consistent with available data.

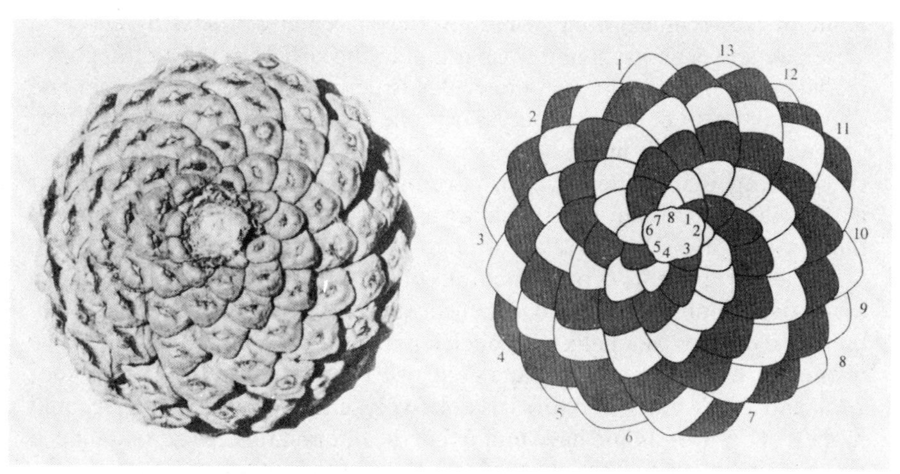

Figure 12 Symmetry of the 8-start and 13-start spirals at the rounded end of a pine cone. There is a discontinuity at the center (the stem) which might be occupied by a minor coat protein in the filamentous phage model. (Reprinted, with permission, from Hoggatt 1969.)

Knowledge of the structure of the filamentous phage virion is not an end in itself but the first step in a broad field. We chose this problem initially for the same reason that many other workers entered the field of small DNA phages: the simplest system that carries out any given function is the system most likely to reveal the basic principles of this function clearly.

One important principle illustrated by this work is that macroscopic geometry is determined by microscopic structure. Mild treatments that alter the nature of the interactions between neighboring proteins in small ways can be coupled, by the geometry of an assembly, so as to give a gross change in the structure of the assembly—a kind of allosteric amplification. Coupled structural changes of this sort may well be important in other biological polymers such as the cytoskeleton.

The absence of integral stoichiometry between DNA and protein (Marvin and Hohn 1969; Berkowitz and Day 1976) shows that specific interactions between these molecules are not possible. This discovery, although disappointing at first, illustrates the important principle that DNA and protein can interact with one another in the absence of a specific stereochemical fit. Indeed, for many purposes it is more instructive to consider them as a negatively charged rod interacting with a positively charged sheath. This is not to say that there is no specificity of interaction. On the contrary, there may be subtle variations in charge distribution along the lengths of the structures which are quite important for function but which would be overlooked if one focused on local lock-and-key stereochemistry.

NOTE ADDED IN PROOF

Some of the technical points raised in the succeeding articles by Day and Wiseman and Makowski and Caspar and in this article may be difficult for readers who are not expert in molecular structure analysis, and the general effect may be to confuse unnecessarily the current picture of filamentous phage structure. It is in the hope that we might clarify some of these points that this note is appended. We wish to emphasize in particular that the other authors do not offer any solid evidence against our proposal that all strains of filamentous phage are built according to the same general design.

The important feature of filamentous phage structure is that it consists of a set of overlapping, elongated, α-helical subunits arranged with $5s+2$ subunits in five turns of a helix with pitch length of 15 Å, where s is 4 or 5. The distinction between s equals 4 and s equals 5 (that is between 4.4 and 5.4, or 22/5 and 27/5, units per turn) is relatively unimportant. We have built models of both kinds and have found that the interaction between subunits is the same in both cases: the change in units per turn is counterbalanced by the change in rise per unit, although the 5.4 model is somewhat fatter than the 4.4 model. It is of course necessary to know the number of units per turn, and the attempts by Day and Wiseman and by Makowski and Caspar to resolve this question have contributed to narrowing the range of alternatives that need to be considered. However, the techniques that they use are not the techniques of choice for an absolute determination of mass per length. There are many instances of molecular mass having been reported from hydrodynamic measurements only to be proven in error by 5–10% when sequence data became available (cf. Edsall 1953 and Dayhoff 1972; see also Day and Berkowitz 1977).

Makowski and Caspar compare our X-ray data with various predictions calculated for the alternative units-per-turn options. We have done such comparisons ourselves, but feel they are inconclusive. To illustrate our doubts: (1) Packing ratios calculated for the 27/5 range as low as 1.01; only 1% water is predicted in the fiber, an impossibly small amount. The packing ratio calculated for fd with a lattice constant of 55.3 Å (more representative than 54.1 Å), a unit rise of 3.5 Å (calculated for 4.46 units per turn in a 15.6-Å repeat), and a molecular volume of 7428 $Å^3$ is 1.24, which is comparable to the 22/5, but not the 27/5, model for Pf1. (2) Density and water-content calculations on fibers are notoriously misleading. For instance, when applied to DNA they give a mass per length that is 10–20% too high (Langridge et al. 1960). (3) Dr. C. Nave of our laboratory has found a good molecular symmetry option for the 22/5 model.

The authors of both succeeding articles emphasize in their contributions the difference that they believe to exist between the structures of the different filamentous phage virions. Although there are certainly experimental differences between the virions of various strains of filamentous phage, the most striking features of the data are the similarities between the different strains. The coat proteins of all strains that have been sequenced have an acidic N terminus, a basic C terminus, and a hydrophobic middle region and are 46 (± 10%) residues long. All strains have about 1% of a minor protein of ≃ 50,000 m.w. The DNA of all strains is a circular, single-stranded molecule 6900 (±10%) nucleotides long (with the exception of If1, which is 1.5 times this standard length). The virions of all the strains are 55 Å (±10%) in diameter (measured by closest packing in X-ray diffraction experiments). All strains have the general distribution of intensity on X-ray fiber diffraction patterns typical of α-helical fibrous proteins. The differences between the strains are relatively small, of the degree often found between structures that are variations on a single theme. The structural similarities between biochemically different virions support the idea that geometry rather than biochemistry controls the common structure of these viruses.

ACKNOWLEDGMENTS

This review draws extensively on unpublished work of T. Arad, C. Gray, J. Ladner, R. Ladner, K. Leonard, S. Malsey, F. Marvin, C. Nave, and H.

Siegrist at the European Molecular Biology Laboratory, and of E. J. Wachtel at the Weizmann Institute, Rehovot, Israel. We are indebted to the Deutscher Krebsforschungszentrum, Heidelberg, for free use of their computing facilities.

REFERENCES

Amako, K. and K. Yasunaka. 1977. Ether-induced morphological alteration of Pf1 filamentous phage. *Nature* **267:** 862.

Berkowitz, S. A. and L. A. Day. 1976. Mass, length, composition and structure of the filamentous bacterial virus fd. *J. Mol. Biol.* **102:** 31.

Bracewell, R. 1965. *The Fourier transform and its applications.* McGraw-Hill, New York.

Bradley, D. E. 1973. The adsorption of the *Pseudomonas aeruginosa* filamentous bacteriophage Pf to its host. *Can. J. Microbiol.* **19:** 623.

Coxeter, H. S. M. 1969. *Introduction to geometry.* Wiley, New York.

Day, L. A. and S. A. Berkowitz. 1977. The number of nucleotides and the density and refractive index increments of fd virus DNA. *J. Mol. Biol.* **116:** 603.

Dayhoff, M. O. 1972. *Atlas of protein sequence and structure*, vol. 5. National Biochemical Research Foundation. Washington, D.C.

Diamond, R. 1976. Coordinate refinement with chain constraints. In *Crystallographic computing techniques* (ed. F. R. Ahmed), p. 291. Munksgaard, Copenhagen.

Edsall, J. T. 1953. The size, shape and hydration of protein molecules. In *The proteins* (ed. H. Neurath and K. Bailey), 1st Ed., vol. IB, p. 549. Academic Press, New York.

Fraser, R. D. B. and T. P. MacRae. 1973. *Conformation in fibrous proteins.* Academic Press, New York.

Green, N. M. and M. T. Flanagan. 1976. The prediction of the conformation of membrane proteins from the sequence of amino acids. *Biochem. J.* **153:** 729.

Hoggatt, V. E. 1969. *Fibonacci and Lucas numbers.* Houghton-Mifflin, Boston.

Langridge, R., H. R. Wilson, C. W. Hooper, M. H. F. Wilkins, and L. D. Hamilton. 1960. The molecular configuration of deoxyribonucleic acid. *J. Mol. Biol.* **2:** 19.

Marvin, D. A. and B. Hohn. 1969. Filamentous bacterial viruses. *Bacteriol. Rev.* **33:** 172.

Marvin, D. A. and E. J. Wachtel. 1975. Structure and assembly of filamentous bacterial viruses. *Nature* **253:** 19.

———. 1976. Structure and assembly of filamentous bacterial viruses. *Philos. Trans. R. Soc. Lond. B* **276:** 81.

Marvin, D.A., R. L. Wiseman, and E. J. Wachtel. 1974a. Filamentous bacterial viruses. XI. Molecular architecture of the class II (Pf1, Xf) virion. *J. Mol. Biol.* **82:** 121.

Marvin, D. A., W. J. Pigram, R. L. Wiseman, E. J. Wachtel, and F. J. Marvin. 1974b. Filamentous bacterial viruses. XII. Molecular architecture of the class I (fd, If1, Ike) virion. *J. Mol. Biol.* **88:** 581.

Nakashima, Y., R. L. Wiseman, W. Konigsberg, and D. A. Marvin. 1975. Primary structure and sidechain interactions of Pf1 filamentous bacterial virus coat protein. *Nature* **253:** 68.

Struik, D. J. 1961. *Differential geometry,* 2nd Ed. Addison-Wesley, Reading, Pennsylvania.

Thomas, G. J. and P. Murphy. 1975. Structure of coat proteins in Pf1 and fd virions by laser Raman spectroscopy. *Science* **188:** 1205.

Valentine, R. C., B. M. Shapiro, and E. R. Stadman. 1968. Regulation of glutamine synthetase. XII. Electron microscopy of the enzyme from *Escherichia coli. Biochemistry* **7:** 2143.

Wachtel, E. J., F. J. Marvin, and D. A. Marvin. 1976. Structural transition in a filamentous protein. *J. Mol. Biol.* **107:** 379.

Warshel, A. and M. Levitt. 1976. Theoretical studies of enzymic reactions: Dielectric, electrostatic and steric stabilization of the carbonium ion in the reaction of lysozyme. *J. Mol. Biol.* **103:** 227.

Wiseman, R. L., S. A. Berkowitz, and L. A. Day. 1976. Different arrangements of protein subunits and single-stranded circular DNA in the filamentous bacterial viruses fd and Pf1. *J. Mol. Biol.* **102:** 549.

A Comparison of DNA Packaging in the Virions of fd, Xf, and Pf1

Loren A. Day and Robert L. Wiseman
Department of Biochemistry
The Public Health Research Institute
of the City of New York, Inc.
New York, New York 10016

In this article comparisons are made among structural parameters derived from physicochemical data for fd, Pf1, and Xf viruses and, in some cases, those for the filamentous complex formed between fd DNA and the fd gene-V protein. Each of these nucleoproteins contains a circular, single-stranded (SS) DNA molecule and each is about 10^4 Å in length and 10^2 Å in diameter. In spite of these common characteristics, their structures differ significantly.

Viruses and Hosts

All known filamentous bacterial viruses have gram-negative bacteria as hosts. Those infecting strains of *Escherichia coli* carrying the F^+ transmissible sex factor include f1, fd, M13, ZJ/2, Ec9, AE2, and δA (see Marvin and Hohn 1969). Because no structural differences have yet been reported for these viruses, we can assume that the structural parameters of fd hold for all of them. According to a number of criteria, two other viruses of *E. coli*, If1 (Meynell and Lawn 1968) and Ike (Khatoon et al. 1972), have structures that are similar to that of fd, even though their host specificities are determined by different transmissible genetic factors (Lawn et al. 1967; Khatoon and Iyer 1971). Members of a second group of viruses have *Pseudomonas aeruginosa* strains as hosts. These are Pf1, Pf2 (Takeya and Amako 1966; Minamishima et al. 1968), and Pf3 (Stanisich 1974), but structural data are available only for Pf1. The third virus in our comparison is Xf, the only filamentous virus of *Xanthomonas oryzae* that has been described (Kuo et al. 1967). One filamentous virus, v6, has been found for *Vibrio parahemolyticus*, another gram-negative bacterium (Nakanishi et al. 1966), but structural studies have not been done on it.

The genetics and molecular biology of fd, M13, f1, and their close relatives have been studied extensively (see Denhardt 1975; Ray 1977; and various chapters in this volume), but little is known about the other viruses. To understand some of the comparisons made here it is important to remember that during an infection by fd (M13, f1) a large amount of gene-V protein is produced. This protein regulates the synthesis of progeny viral-strand (+) circular DNA by forming filamentous intracellular complexes with the circular SS DNA. These complexes are precursors in the assembly process of the fd virion, the final stage of which takes place at the cell membrane and involves the exchange of about one molecule of gene-V protein for two molecules of the major coat protein.

Concepts

The lengths of the filamentous structures considered here are determined by the size of the DNA molecules contained in them. In each structure the DNA molecule must pass from a point near one end to a point near the other end and back again. In accord with this is the existence of spontaneous deletion mutants, called miniphage, having DNA molecules one-third the size of the wild-type genome packaged in virions one-third the length of the wild-type virion (Grandis and Webster 1973; Griffith and Kornberg 1974; Enea and Zinder 1975). In addition, insertion mutants exist in which kanamycin-resistance genes are integrated in fd genomes that are, in some cases, three times the size of wild-type fd DNA, and these are packaged in virions three times the length of the normal virion (Herrmann et al. 1978).

A second concept is the axial separation between identical structural elements. This parameter is simply the projection upon the virus axis of the distance from one element to the next along the length of the structure. It is obtained by dividing the length by the number of elements. The average axial separations cited below for protein subunits are virion lengths divided by the total number of subunits, whereas those cited for nucleotides are virion lengths divided by half the number of nucleotides and correspond to the axial translation from one nucleotide to the next in the same DNA chain. Because they are projections upon the structural axes, the separations can be shorter than the distances between the mass centers of the elements.

The final concept introduced here is the nucleotide-protein subunit ratio, which is a way of expressing the chemical stoichiometry of a rod-shaped complex containing a large nucleic acid molecule and thousands of protein subunits. If the subunits are chemically identical, then one might expect that each subunit would interact with the same number of nucleotides, and therefore that the value of the nucleotide-subunit ratio would be a whole number. The results presented below for fd, however, show that the value of the ratio can be a noninteger quantity.

Chemical Constituents of the Structures

The Composition and Number of Bases in Each DNA

Although the base sequence of fd DNA is known almost completely (Schaller et al., this volume), no work at all has been reported on the chemical structures of the other DNAs except determinations of base compositions (Table 1). The base compositon of fd is clearly incompatible with complete Watson-Crick pairing, but for each of the others listed the adenine and thymine contents and the guanine and cytosine contents are not grossly different.

The most accurate way to count the nucleotides in viral genomes is to determine nucleotide sequences. The sequences of φX174 and fd DNAs (Sanger et al. 1977 [see also Appendix I, this volume]; Schaller et al., this volume) provide standards for relative measurements of other DNAs by physical techniques. The results for fd, Xf, and Pf1 DNAs in Table 1 are relative to φX DNA. For fd DNA, the difference of a few nucleotides between the 6389 nucleotides obtained from its sequence (Schaller et al., this volume) and the 6370 nucleotides obtained from its sedimentation rate relative to φX DNA is well within the limits of the statistical uncertainties of the methods used.

This agreement gives us confidence in the number of nucleotides found for both Xf DNA and Pf1 DNA, i.e., 7400. Although both circular SS DNA genomes have, within experimental uncertainty, the same number of nucleotides, their virion structures differ remarkably.

Amino Acid Compositions of the Proteins

Significant features of the amino acid compositions (Table 2) of the major coat proteins are the high concentrations of alanine in all of them, the high concentrations of glycine in Pf1 and Xf, the absence of histidine and cysteine in all, the absence of proline in Pf1 and If1, and the presence of five basic

Table 1 DNA base composition and number of bases

	fd	Ifl	Xf	Pf1
Base composition				
adenine	24.4	27.1	20.6	19.2
thymine	34.1	28.4	19.3	20.4
guanine	19.9	23.1	32.6	29.8
cytosine	21.7	21.7	27.5	30.6
Average nucleotide weight (Na^+ salt)	331	333	333	332
Total number of nucleotides	6370±140	—	7420±240	7390±180
Molecular weights (Na^+ form)	2.11×10^6	—	2.47×10^6	2.45×10^6

The base compositions for fd, Ifl, Xf, and Pf1 DNAs are from Hoffmann-Berling et al. (1963), Wiseman et al. (1972), Kuo et al. (1969), and Wiseman et al. (1976), respectively. The base composition for fd DNA can also be calculated from its sequence of 6389 bases (Schaller et al., this volume). Each number of bases listed in the table is from the sedimentation velocity of each DNA relative to φX DNA (Day and Berkowitz 1977; Wiseman and Day 1977).

amino acids in fd, four in Xf, and three in Pf1. The sequence of the major capsid protein (B protein) is identical for fd (Asbeck et al. 1969; Sanger et al. 1973; Nakashima and Konigsberg 1974) and f1 (Bailey et al. 1977) and differs from that for ZJ/2 by the substitution of one amino acid (Snell and Offord 1972). The sequence of the B protein of Pf1 (Nakashima et al. 1975) differs from that of Xf (recently completed in the laboratories of B. Fran-

Table 2 Amino acid compositions of the proteins

Amino acid	DNA-binding protein[a] (gene V)	Minor coat protein[b] (gene III)	Major coat proteins			
			fd,[c] (ZJ/2)[d]	If1[e]	Xf[f]	Pf1[g]
Lys	6	14	5	4	2	2
His	1	2	0	0	0	0
Arg	4	9	0	1	2	1
Asp (Asx)	3	24	3	3	2	3
Asn (Asx)	2	29	0	0	0	0
Thr	4	26	3 (2)	4	1	2
Ser	7	31	4	4	2	3
Glu (Glx)	4	26	2	1	1	0
Gln (Glx)	6	15	1	2	0	2
Pro	6	25	1	0	1	0
Gly	7	63	4	3	8	7
Ala	4	28	10 (11)	12	7	7
Cys	1	8	0	0	0	0
Val	8	20	4	6	9	5
Met	2	7	1	1	2	2
Ile	4	9	4	1	4	6
Leu	10	19	2	4	1	4
Tyr	5	21	2	1	1	2
Phe	3	20	3	3	0	0
Trp	0	4	1	1	1	0
Total residues	87	400	50	51	44	46
Molecular weight	9690	42200	5240	5278	4343	4609

[a] Nakashima et al. (1974); Cuypers et al. (1974).
[b] Read from the amino acid sequence as read from the fd DNA sequence (Schaller et al., this volume).
[c] Asbeck et al. (1969); Sanger et al. (1973); Nakashima and Konigsberg (1974); Bailey et al. (1977).
[d] Snell and Offord (1972).
[e] From the sequence determined by B. Frangione (pers. comm.). See also Wiseman et al. (1972).
[f] The amino acid composition of Xf protein is from the sequence determined by B. Frangione et al. (unpubl.).
[g] Nakashima et al. (1975).

gione and W. Konigsberg in collaboration with R.L.W.), and both sequences are different from the other three. Of particular interest is the finding that proline is near the center of the sequence of Xf coat protein. Each B protein has a high concentration of negatively charged acidic residues in the N-terminal region, a middle region with hydrophobic residues, and a positively charged C-terminal region.

The amino acid sequence of the A protein, the principal minor coat protein of fd, can be read from the DNA sequence of gene III (Schaller et al., this volume). The amino acid composition (expressed in mole fraction) is very close to that obtained for the gene-III protein purified by M. Goldsmith (Goldsmith and Konigsberg 1977). The A proteins of Xf and Pf1 migrate on gels like fd A protein (Marvin et al. 1974a) and are probably the same size, i.e., about 42,000 daltons.

The sequence of the gene-V protein (Nakashima et al. 1974; Cuypers et al. 1974) is unrelated to that of either the A or B protein. The basic amino acids and the tyrosines, which have important roles in DNA binding, are distributed throughout the chain.

Stoichiometry of Structure

The ratio of nucleotides to major coat protein subunits is a key parameter in the present approach to the DNA packing problem. Gene-V protein complexes isolated from infected cells by the method of Pratt et al. (1974) contain 5 nucleotides per subunit according to results of radiochemical tracer analysis (Pratt et al. 1974) and 4.7 ± 0.1 nucleotides per subunit according to multicomponent spectral analysis (Pretorius et al. 1975). Both values differ from the value of 4.0 obtained by most workers from titrations of isolated gene-V protein with SS DNA or polynucleotides (Alberts et al. 1972; Day 1973; Anderson et al. 1975; Pretorius et al. 1975). This difference is one of several small structural differences between gene-V complexes made in vivo and in vitro (Pratt et al. 1974; Pretorius et al. 1975).

The data for the fd virion indicate a minimum of 2.32 ± 0.07 and a maximum of 2.37 ± 0.07 for the number of nucleotides per major coat protein subunit (Table 3). The lower value is based on the assumption that the minor coat protein components, taken to constitute 2% of the virus mass, bind to the DNA but do not contribute to the virion length, whereas the higher value is based on the assumption that the minor proteins do not bind to the DNA but do contribute 2% to the length. The ranges of systematic and statistical uncertainties thus exclude an integer value for the number of nucleotides per subunit in the fd virion. The results for Pf1 and Xf, however, indicate that these structures may very well contain exactly 1 nucleotide and exactly 2 nucleotides per subunit, respectively (Table 3).

The number of A-protein molecules (gene-III product) in the fd virion appears to be five. The data of Goldsmith and Konigsberg (1977) and Woolford et al. (1977) yield an average of 0.015 for the ratio of grams of A

Table 3 Constituents of the virions

	fd	Xf	Pf1
Percent by weight DNA	12.0±0.3	12.6±0.3	6.4±0.3
Nitrogen/phosphorus (moles)	28–29	26, 29	56–60
Nucleotides/subunit along filament	2.32±0.07[a]	2.02±0.04[b]	0.97±0.05[c]
Total protein subunits[d]	2710±110	3670±190	7620±440

The data are from the literature (Marvin and Hohn 1969; Berkowitz and Day 1976; Newman et al. 1977; Wiseman and Day 1977) except for two values for the nitrogen/phosphorus mole ratio of Xf which, although inconsistent with each other, show that Xf has about twice as much DNA as Pf1. Ranges of observed nitrogen/phosphorus ratios are given for fd and Pf1. The uncertainties cited are two standard deviations from the mean values of each parameter.

[a] This value is based on the assumption that minor coat proteins (~ 2% by weight) bind to 2% of the DNA; other assumptions lead only to systematically higher values up to 2.37±0.07.

[b] Calculated on the basis of the Xf subunit molecular weight given in Table 2. The value of 1.92 (Wiseman and Day 1977) was based on an estimate of the subunit molecular weight from earlier amino acid composition data (Lin et al. 1971). Different assumptions about minor components could make the value higher by about 2%.

[c] As in notes a and b, different assumptions about minor constituents would only increase this value, in this case only 1% or 2%.

[d] The number of A-protein molecules (gene-III product) in fd is five (see text).

protein/grams of B protein in the virion. This divided by 8.05, the molecular weight ratio (Table 2), and multiplied by 2700, the number of B-protein molecules per virion (Table 3), gives 5.0 A-protein subunits. They are at one end of the fd virion (Woolford et al. 1977; Goldsmith and Konigsberg 1977) and can be removed by techniques that leave an intact filament (Ikehara and Utiyama 1975).

Some Properties of the Isolated Components

Isolated DNA

The sequences of fd DNA and φX DNA can be used to assess the earlier determinations of molecular weights obtained by physical techniques. The most extensive studies were those by Newman et al. (1974) and Berkowitz and Day (1974) of fd DNA, in which correct values were obtained for all experimental parameters except for differential density and refractive index increments. New values for these have been calculated from the earlier light scattering and sedimentation results, and methods for measuring them directly have been improved (Day and Berkowitz 1977).

Hydrodynamic properties of isolated DNA give a clue to the DNA structure in the virion. In neutral aqueous buffers, phenol-extracted DNA has the hydrodynamic behavior of a "hard sphere" (Newman et al. 1974), which means that the coiled, single-stranded chain (whether truly random or not) and solvent entrapped in its domain constitute a spherical hydrodynamic entity. However, if the DNA is isolated from the virus or from infected cells in the presence of 1 M NaCl by mild detergent techniques, it sediments more slowly and therefore must have a more asymmetric structure (Forsheit and Ray 1970). This might indicate (contrary to the results of Tate and Peterson [1974]) that there are specific orientations of M13 (fd) DNA in the intracellular gene-V complex and in the virion such that a given nucleotide sequence has the same position relative to the ends of the structures.

Isolated Coat Protein

Physical studies on major coat protein subunits have been reported for the B protein of fd (M13, f1) but not for the B protein of Pf1 or Xf. Knippers and Hoffmann-Berling (1966) showed that the major coat protein from fd tends to form aggregates that have the hydrodynamic properties of rodlike particles under certain conditions. However, their results and unpublished work of others indicate the difficulty, if not the impossibility, of getting isolated coat protein to reassemble into structures similar to the native structure in the virion. In several detergent solutions and in lipid vesicles, solubilized or suspended B-protein subunits have different secondary structures from those in the virus (Nozaki et al. 1976; Williams and Dunker 1977). The structures of isolated B proteins from Xf and Pf1, also, probably depend strongly on the environment, but in different ways from each other and from fd. The virions themselves react very differently to denaturants, especially ether (Amako and Yasunaka 1977). In general, experiments with isolated viral proteins are now best designed with knowledge of the many components required for proper virus assembly (see Webster and Cashman, this volume) including, possibly, the processing of a precursor protein species by an endopeptidase to form the mature coat protein subunit (Chang et al. 1978).

Gene-V Protein

This DNA-binding protein is soluble and functional in aqueous buffers. In the absence of DNA or oligonucleotides it shows little tendency to form multisubunit structures larger than dimers (Pretorius et al. 1975; Cavalieri et al. 1976), and a monomer-dimer equilibrium exists which is easily displaced toward monomers by NaCl and KCl at moderate concentrations (Oey and Knippers 1972; Pretorius et al. 1975). Cross-linking studies (Rasched and Pohl 1974) indicate that binding to very short oligonucleotides induces formation of aggregates as large as octamers, and recent X-ray results show 12 monomers per unit cell in crystals formed in the presence of dinucleotides and short oligonucleotides (A. McPherson, pers. comm.). Gene-V protein crystallizes as a dimer in the absence of oligonucleotides, and its X-ray structure is being determined (McPherson et al. 1976).

Axial Separations from Virus Lengths

Lengths and Diameters

When using virion lengths to derive structural parameters, one must consider problems of length heterogeneity. Normally, the passaging of wild-type fd a few times at moderate multiplicities of infection yields preparations in which the virions, when fresh, have quite uniform length distributions (Wheeler et al. 1974; Newman et al. 1977; Wheeler and Benzinger, this volume). However, in cultures of mutated f1 (M13, fd), end-to-end dimers and multimers have been observed which contain two or more unit lengths of DNA sometimes of different genotype (Notani and Zinder 1964; Bradley 1964; Salivar et al. 1967; Beaudoin et al. 1974). In work referred to above, certain culturing conditions for wild-type virus produce miniphage of only one-third the virion length. In addition, fd virions with lengths between the unit length and the dimer length have been observed; the fd DNA molecules in these phage, called midiphage, are longer than unit length and they are not covalently closed. Detailed analyses have led to the conclusion that there must be unusual DNA packaging at the ends of this type of abnormal virion (Wheeler and Benzinger, this volume). In work done together with F. C. Chen and G. Koopmans (unpubl.), we have observed, by agarose slab gel electrophoresis, end-to-end dimers and higher multimers in preparations of wild-type Xf and Pf1 viruses. A few Xf virions of less than monomer length and a few of between the monomer and dimer lengths have been observed by electron microscopy, but about 90% of the virions are of unit length. Preparations enriched to greater than 98% virions of unit length can be obtained by sucrose gradient centrifugation or preparative slab gel electrophoresis.

The overall dimensions obtained from electron microscopy and hydrodynamic measurements are given in Table 4. We have listed only two of the many lengths obtained for fd by electron microscopy. The conventional microscope used for one of these values was modified to account for variations in magnification resulting from changes in current to the objective lens on focusing (Frank and Day 1970), a problem not encountered with the scanning transmission electron microscope used for the other value (Crewe and Wall 1970; Wall 1971). Carbon replicas of diffraction gratings were used for instrument calibration in both studies. Diffraction gratings were also used for calibrations by Kuo et al. (1969; pers. comm.) in the measurement of Xf length and by Bradley (1973) in the measurement of Pf1 length. The Pf1 length measured by Wiseman et al. (1976) was calculated from the relative lengths of Pf1 and fd images on the same electron micrographs and the lengths for fd obtained by Frank and Day (1970).

The dimensions of fd and Xf in solution have been obtained from combined measurements of rotational diffusion by transient electric birefringence (TEB) and translational diffusion by intensity fluctuation spectro-

Table 4 Dimensions and axial separations

	fd	Xf	Pf1
Length			
from electron microscopy	880±30[a]	977±35[e]	1915±77[h]
	883±24[b]		1960±70[i]
from diffusion measurements	895±20[c]	990±30[f]	([j])
Diameter			
from electron microscopy	8.4±1.0[b]	—	
from diffusion measurements	9.0±1.0[c]	7.0±1.0[f]	
from X-ray analysis			
0% relative humidity	5.5[d]	—	5.5[g]
98% relative humidity	8.5[d]	6.6[g]	6.6[g]
Nucleotide axial separation			
(in one chain) from EM length	0.28[k]	0.26[k]	0.53[k]
Subunit axial separation			
from EM length	0.32[k]	0.25[k]	0.26[k]

The values are given in nanometers (1 nm = 10 Å).

The data are from [a]Frank and Day (1970); [b]Wall (1971); [c]Newman et al. (1977); [d]Dunker et al. (1974); [e]Kuo et al. (1969); [f]F. C. Chen et al. (in prep.); [g]Marvin et al. (1974a); [h]Bradley (1973); [i]Wiseman et al. (1976); [j]the length by diffusion is about 2 μm (E. Loh, pers. comm.); [k]Wiseman and Day (1977).

The uncertainties given in the table are two standard deviations from the mean.

scopy (IFS) (Newman et al. 1977; unpubl. work in collaboration with F. C. Chen, G. Koopmans, and H. L. Swinney). For TEB measurements, the viruses are partially oriented by an external electric field. When the field is turned off, there is decay to random orientation, and the rate of the decay is directly related to the rotational diffusion coefficient, D_R. In IFS measurements, the relative contributions of different molecular motions of the scatterers to the autocorrelation function vary with scattering angle. At low angles, of the order of 10° for fd and Xf, the dominant contribution to the time-dependent scattering is translational motion, so diffusion coefficients for pure translational motion, D_T, can be deduced unambiguously. The hydrodynamic lengths and diameters listed for fd and Xf in Table 4 were obtained from D_T and D_R values through simultaneous solutions of the two equations of Broersma (1960a,b). These equations were derived for the hydrodynamic behavior of stiff rods, and the exact suitability of this model for filamentous viruses is under further investigation; any systematic error in the lengths calculated by these equations is thought to be about 5% or less (Newman et al. 1977).

Axial Separation of Protein Subunits in Relation to Results from X-ray Diffraction Studies

Axial separations between protein subunits provide guides and constraints for the construction of models of the structures. The average axial subunit separation in both Xf and Pf1 is close to 2.6 Å; the average axial subunit separation in fd is close to 3.2 Å, clearly different from that for Pf1 and Xf (Table 4).

X-ray diffraction patterns of oriented fibers show that each of the virions has a helical sheath of protein subunits surrounding a cylindrical core region about 20 Å in diameter that contains the DNA (Marvin and Wachtel 1976; Wachtel et al. 1974; Marvin; Makowski and Caspar; both this volume). The patterns for Pf1 and Xf have layer-line spacings of 75 Å, which corresponds to the apparent repeat of the physical structure in the axial direction. According to physicochemical data, the helices of subunits in Xf and Pf1 would have 28 ± 2 subunits every 75 Å (Wiseman and Day 1977; Table 4). According to the major layer-line spacings in X-ray diffraction photographs, the apparent axial repeat of the fd virion is 32 Å (Marvin 1966; Marvin et al. 1974b). According to the axial separations from physicochemical data, the helix of protein subunits in fd would have exactly 10 subunits every 32 Å (Newman et al. 1977).

The Nucleotide Axial Separation

The conclusion that the DNAs are in cylindrical core regions of about 20 Å (Wachtel et al. 1974; Marvin and Wachtel 1976) is an important one. If the DNA is in a core region, then it is definite that there are two antiparallel polynucleotide chains in that core. The nucleotide axial separation for fd DNA in its virion is 2.8 Å, and for Xf DNA in its virion it is 2.6 Å. These values lie between the 2.6-Å nucleotide translation for classical A-type DNA and the 3.4-Å nucleotide translation for classical B-type DNA, both of which have diameters of 20 Å (Franklin and Gosling 1953; Watson and Crick 1953; Arnott and Hukins 1972). The correspondence of diameters and translations for the DNA structures in fd and Xf to these parameters for classical base-paired DNA structures suggests that the two noncomplementary polynucleotide chains in each virion core have sugar-phosphate backbone conformations resembling the backbone structures in classical A- or B-type DNA. That is, the phosphates would be on the outside of the core region and the bases on the inside.

The nucleotide separation in the Pf1 viral DNA is unusually long, 5.3 Å. This distance is based on particle length as determined by electron microscopy. When allowances are made for shrinkage in the axial direction on drying and for experimental uncertainty, the distance for Pf1 in solution may be as long as 5.9 Å. Pf1 viral DNA thus has the most extended natural DNA conformation known (Wiseman et al. 1976; Wiseman and Day 1977). We will discuss the possible orientation of the phosphates and bases relative to the virion axis and the protein sheath of Pf1 after we present results of spectroscopic studies.

Spectroscopic Studies

Evidence for Base-Tyrosine stacking in the Gene-V Protein-DNA Complex

Nuclear magnetic resonance (NMR) spectroscopy and chemical modification (Anderson et al. 1975), as well as solvent perturbation of the protein absorbance (Pretorius et al. 1975), show that three of five tyrosines of gene-V protein are particularly accessible to solvent. Ultraviolet difference spectra for mixed versus separated protein and DNA show that the solvent-exposed tyrosines become less exposed on DNA binding, and tyrosine fluorescence of free protein is strongly quenched by DNA and polynucleotides (Day 1973; Pretorius et al. 1975). On the basis of work by Helené and Dimicoli (1972) on model tyrosyl compounds, Pretorius et al. (1975) concluded that the bases stack with tyrosines in the complexes. In a detailed NMR study Coleman et al. (1976) observed the ring current shifts expected for the formation of tyrosine-base stacks. Garssen et al. (1977) have also concluded from NMR data that there may be base-tyrosine stacking.

Evidence for Base-Base Stacking in fd and Xf but Base-Tyrosine Stacking in Pf1

The near ultraviolet absorbance of each virion is dominated by DNA contributions. Since light scattering contributions are small, the calculated protein absorbance can be subtracted from the total virus absorbance to obtain the apparent absorbance per nucleotide (Table 5, first row). The DNAs in fd and in Xf have relatively low absorbance, similar to double-stranded DNAs in conformations with base-base stacking. However, the DNA in Pf1 has a nucleotide absorbance of 9000 at 260 nm. The DNA structure may involve a tilting of the bases to give both an orientation of transition dipoles with high absorbance and the long axial nucleotide translation. A complete interpretation, however, must account for tyrosyl spectral properties.

Two tyrosines are the only aromatic groups in the Pf1 subunit. Tyrosyls in the Pf1 virion titrate reversibly up to pH 11.7 (Fig. 1). If the portion of the curve below pH 11.7 is analyzed in terms of a single ionizable group, an apparent pK_a of 11.3 and the absorbance change expected for one tyrosine per subunit are obtained. The absorbance change for two tyrosines per subunit with normal pK_a's of about 10 is obtained if the virion is disrupted by 6 M guanidine-HCl.

Fluorescence measurements of intact and disrupted Pf1 also indicate two environments for tyrosines. The fluorescence of disrupted virus in 6 M guanidine-HCl has the features of *N*-acetyl-tyrosyl-glycine fluorescence, including its intensity at 303 nm per tyrosyl group. Guanidine-HCl has no significant effects on tyrosyl fluorescence in neutral aqueous buffers (McGuire and Feldman 1973). The fluorescence of intact Pf1 at pH 7 also

Table 5 Spectral features

	fd	Xf	Pf1	Gene-V complex
Nucleotide absorbance in native structure at 260 nm	6750	6400	9000	8400
CD of native structures				
[θ] at 208 nm	−27000	−18000	−35000	zero
[θ] at 222 nm	−41000	−15000	−32000	zero
CD of NBS-treated structures				
[θ] at 208 nm	−31000	−18000	−35000	—
[θ] at 222 nm	−35000	−18000	−32000	—
Range of estimates of α-helix content from CD (%)	80–100	40–65	80–100	0–10

Absorbances for Xf DNA and Pf1 DNA in the virions, $\epsilon(P)$, were estimated from virus extinction coefficient (see legend to Fig. 2) and handbook values for contributions from protein chromophores; the results for fd (Day 1969) and for the gene-V protein-DNA complex (Day 1973) were obtained differently. The CD amplitudes are molar ellipticities per peptide chromophore. The values for native structures are averages from measurements on at least three different virus preparations. The estimated uncertainties are ±1000 molar ellipticity units (degrees cm²/decimole). The concentrations were based on extinction coefficients of 3.84 mg^{-1} cm^2 at 269 nm for fd (Berkowitz and Day 1976), 2.07 mg^{-1} cm^2 at 260 nm for Pf1 (Wiseman et al. 1976), and 3.52 mg^{-1} cm^2 at 263 nm for Xf, all in 0.15 M KCl, 0.02 M Na phosphate, pH 7, and the data in Tables 2 and 3. The values for the NBS-modified samples were based on relative changes produced by the reagent and the values for the native structures. Three moles of NBS per mole protein were sufficient to give the change in CD. The ranges of estimated α-helix contents result principally from lack of knowledge of contributions from nonpeptide chromophores and lack of criteria for choosing reference spectra.

has a spectrum similar to *N*-acetyl-tyrosyl-glycine, but it is only half as intense at 303 nm. The data (unpubl.) seem to indicate that one tyrosyl per subunit in the intact virion fluoresces like the model compound and that the other has its fluorescence quenched completely. One way complete tyrosyl quenching might occur is through base-tyrosyl stacking (Helené and Dimicoli 1972). To account for the spectral results, we propose that one of the two tyrosyls in each Pf1 subunit is involved in base-tyrosine stacking.

Circular Dichroism of the Virions

CD spectra from 200 nm to 250 nm for the virions of fd, Xf, and Pf1 are given in Figure 2. The spectra are different. The spectrum for fd virus appears to be distorted relative to those for highly α-helical proteins. The amplitudes for Xf are much lower than those for Pf1. The shape of the fd spectrum becomes more like those of Pf1 and other α-helical structures upon oxidation of tryptophan by *N*-bromsuccinimide (NBS), as indicated by the data in Table 5. Thus, the deep minimum of the native fd virion CD at 222 nm appears to have a significant tryptophan contribution. The tryptophan reagent, at 3 moles NBS per mole subunit, has only a small effect on Xf and

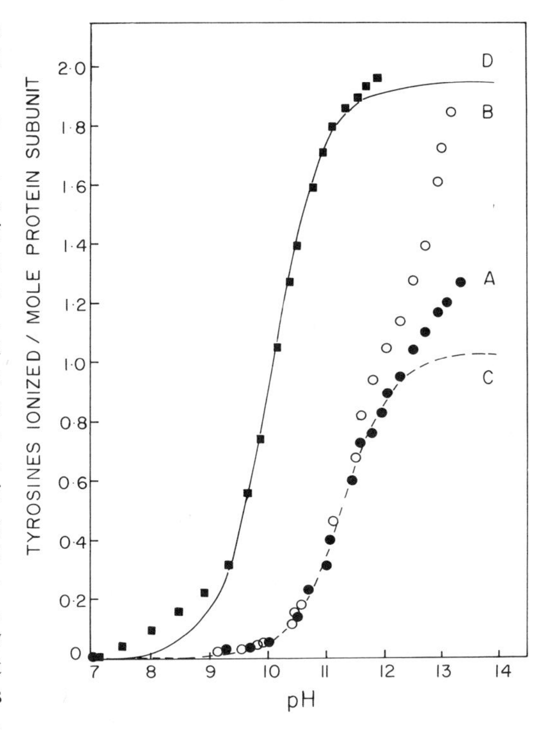

Figure 1 Spectrophotometric titrations of phenolic hydroxyl groups of Pf1 virus. Curve *A* (●) shows a titration at 245 nm for samples at 0.40 mg/ml Pf1 (1.63×10^{-4} M in total tyrosines); points are zero time values after mixing aliquots of a stock solution in 0.5 M KCl with 0.5 M KOH or with dilute buffers containing 0.5 M KCl. Curve *B* (o) is for samples in *A* 5 hr after mixing. Titration up to pH 11.7 apparently is reversible, and the data below pH 11.7 yield a derived apparent pK_a of 11.3 and ΔOD_{245} of 0.92 (dashed curve). Applying a molar extinction coefficient of 11,000 observed for tyrosinamide in 0.5 M KOH at 245 nm, we calculate 1.03 reversibly titratable tyrosines per subunit (curve *C*). Curve *D* (■) shows a titration in 6 M guanidine-HCl measured at 300 nm. With an extinction coefficient of 2320, the value obtained for tyrosinamide at its long wavelength maximum in 0.5 M KOH, the 6 M guanidine-HCl data fit reasonably well with the ratio two tyrosines (each with pK_a 10.1) per subunit (solid curve).

no effect on Pf1. CD spectra for fd have also been reported by Ikehara et al. (1973, 1975), Nozaki et al. (1976), and Williams and Dunker (1977), and the high estimates for the α-helix content in fd have agreed reasonably well with earlier estimates from optical rotatory dispersion and far ultraviolet absorbance (Day 1966, 1969; see also Asbeck et al. 1969). Possible locations of α-helical segments in fd and Pf1 subunits have been calculated by Green and Flanagan (1976). It is difficult, if not impossible, to obtain by these methods estimates of helix content for the virions that are better than $\pm 10\%$. This is because of unknown contributions from chromophores other than the peptide group and from differential light scattering and flattening effects and because of the difficulty of choosing appropriate reference spectra (Chen et al. 1974; Baker and Isenberg 1976). It appears that Xf protein has a far lower α-helix content than either fd or Pf1, although it is conceivable, but unlikely, that in Xf there are some left-handed helical segments.

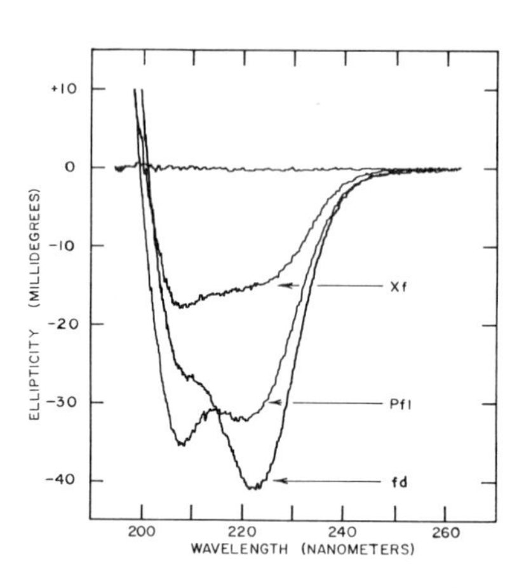

Figure 2 CD spectra for fd, Pf1, and Xf viruses in 0.1 M Na phosphate, pH 7. The optical path through each solution was 1 mm and the concentration of each solution was 10^{-3} M in peptide chromophores so that these actual spectra can be compared directly for indications of protein conformation. Spectra for isolated DNAs in aqueous buffers at the concentrations present in the virus solutions would have ellipticities of 2 millidegrees or less in this wavelength range. No corrrections for DNA were made in calculations of peptide-bond molar ellipticities (Table 5), which have numerical values that are 10^6 times the experimental ellipticities in the spectra shown here.

DNA Packing and Symmetry in the Virions

DNA Conformation in Pf1

The DNA structure in the Pf1 virion is unusual: The two antiparallel chains are confined to a core region of about 20 Å; the nucleotide translation in each chain is between 5 and 6 Å; the nucleotide-to-subunit ratio is probably unity; and there is evidence for base-tyrosine stacking. The results lead us to conclude that the bases are situated on the outside of the core in such a way that each base can interact with one tyrosine. This would mean that the phosphates are in the center of the core.

The data suggest that there is direct interaction between one nucleotide and one subunit in Pf1. This implies different structures for the subunits interacting with the up and the down chains of the DNA in the virion. Since the X-ray diffraction patterns have led to proposals of models with one protein monomer per asymmetric unit (see Marvin; Makowski and Caspar; both this volume), the two types of subunits probably have the same overall orientation relative to the virion axis. Thus, the structural differences are expected to be small and localized to the region of DNA-protein contact. One-for-one nucleotide-subunit interaction would, in addition, raise the question whether the antiparallel DNA chains, which are presumably interwound in a two-chain helix, can fit naturally into a sheath of protein subunits located on a single helix of 15-Å pitch, as proposed by Marvin et al. (1974a). Preliminary analyses indicate that a natural fit could be achieved if the protein subunits were arranged on a two-start helix of 30-Å pitch. To reconcile such an arrangement with the X-ray patterns, however, would require the introduction of a parallel dyad as a symmetry element of the helix of subunits and other modifications of the current interpretations of the X-ray data.

The Xf-Pf1 Phenomenon

The X-ray diffraction patterns of Xf and Pf1 are similar (Marvin et al. 1974a). Both circular SS DNA genomes have almost exactly 7400 nucleotides. Nevertheless, Pf1 is 2 μm long and Xf is only 1 μm long. The calculated nucleotide translations therefore differ by a factor of 2. In accord with this, the nucleotide-to-subunit ratios also differ by a factor of 2.

The factor of 2 might lead some to think that 2-μm Pf1 particles could be end-to-end dimers, that Pf1 DNA is not circular, or that there is adsorbed coat protein or empty protein sheaths in Pf1 preparations. Our results indicate that our Pf1 preparations are homogeneous in CsCl density gradients, that they contain very few 1-μm particles relative to the number of 2-μm particles, that the DNA molecules are circular, and, perhaps of most importance, that the rate of loss of Pf1 infectivity by mechanical shearing is about twice that of fd (which is 0.9-μm long) (Wiseman et al. 1976). In addition, Bradley (1973) counted only 3 1-μm particles per 87 2-μm particles in his micrographs of Pf1, and independent diffusion measurements indicate the length of Pf1 in solution is 2 μm (E. Loh, pers. comm.).

Although Xf and Pf1 X-ray patterns are quite similar, as are their DNA sizes and base compositions, their major coat protein amino acid compositions differ significantly. This genetic difference indicates that the DNA size similarity may be fortuitous or that it has a complex explanation. In any case, the twofold differences, in length and in nucleotide-to-subunit ratio, suggest that a coat protein mutation or other effect could lead to either DNA packing as in Xf or DNA packing as in Pf1.

As discussed above, we have concluded that the DNA bases in the Xf virion are in the center of the structure, whereas the DNA phosphates are in the center of the Pf1 structure. Preliminary analyses indicate that such vastly different DNA conformations could be accommodated naturally in virions having the subunits on two-start helices with 30-Å pitch, but a fully satisfactory explanation is not yet available. A phenomenon also characterized by two types of DNA packing with linear densities differing by a factor of 2 appears to occur within certain abnormal fd virions (Wheeler and Benzinger, this volume).

fd Structure and Assembly

The layer-line spacings in X-ray patterns for fd fibers are about 32 Å, and there are meridional reflections along the axis corresponding to distances of 16 Å (Marvin 1966; Marvin et al. 1974b). There are almost exactly 5 subunits per 16-Å, or 10 subunits per 32-Å, layer-line spacing (Newman et al. 1977). The noninteger value of the nucleotide to subunit ratio in the fd virion means that neither all nucleotides nor all subunits would have the same environments (Berkowitz 1975; Newman et al. 1977). The number of unique environments could be small if the nucleotide-to-subunit ratio were a ratio of small whole numbers. A ratio with a small numerator and a small denominator that is consistent with the chemical data is 12 nucleotides to 5 subunits. A structural repeat with the fewest unique subunit environments would contain 24 nucleotides and 10 subunits, or a multiple of this grouping.

The data for the gene-V protein-fd DNA complex and the fd virion indicate the replacement of 1300–1600 gene-V-protein molecules with 2700 subunits of gene-VIII-protein molecules during virus assembly. The DNA conformation with apparent base-tyrosine stacking in the gene-V protein-DNA complex is transformed in the fd virion into one with apparent base-base stacking. The repeating structural units of the gene-V protein complexed with fd DNA in vitro appear to contain 12 gene-V-protein subunits (Alberts et al. 1972), which would correspond to 48 nucleotides. In vivo, gene-V protein-DNA complexes have repeats of about 160 Å (Pratt et al. 1974), which is close to 1.5% of the length of the complex; hence they might contain 96 nucleotides (1.5% of 6400 nucleotides). Thus, the subunit exchange reaction of virus assembly may proceed in stages involving 24 nucleotides, with 5 or 6 gene-V-protein subunits being replaced by 10 gene-VIII-protein subunits.

Concluding Remarks

The data summarized in this article show that four filamentous nucleoproteins that contain circular SS DNA all have different primary structures of protein subunits, different secondary structures of proteins and DNAs, different nucleotide-to-subunit ratios, and different axial separations between structural elements. Eventually, when detailed structures are known, it may be possible to understand how the common function of packaging circular SS DNA is accomplished in so many diverse ways.

ACKNOWLEDGMENTS

We thank S. A. Berkowitz, F. C. Chen, G. Koopmans, J. Newman, H. T. Pretorius, and H. L. Swinney for their many contributions to this work, and we thank those who have communicated manuscripts and results prior to publication. The previously unpublished work from our laboratory cited here has been supported by grants AI 09049 and KO5-GM 70363 from the U.S. Public Health Service.

REFERENCES

Alberts, B., L. Frey, and H. Delius. 1972. Isolation and characterization of gene 5 protein of filamentous bacterial viruses. *J. Mol. Biol.* **68:** 139.

Amako, K. and K. Yasunaka. 1977. Ether-induced morphological alteration of Pf-1 filamentous phage. *Nature* **267:** 862.

Anderson, R. A., Y. Nakashima, and J. E. Coleman. 1975. Chemical modifications of functional residues of fd gene 5 DNA-binding protein. *Biochemistry* **14:** 907.

Arnott, S. and D. W. L. Hukins. 1972. Optimised parameters for A-DNA and B-DNA. *Biochem. Biophys. Res. Commun.* **47:** 1504.

Asbeck, F., K. Beyreuther, H. Köhler, G. von Wettstein, and G. Braunitzer. 1969. Virusproteine. IV. Die Konstitution des Hüllproteins des Phagen fd. *Hoppe-Seyler's Z. Physiol. Chem.* **350:** 1047.

Bailey, G. S., D. Gillett, D. F. Hill, and G. B. Petersen. 1977. Automated sequencing of insoluble peptides using detergent. Bacteriophage f1 coat protein. *J. Biol. Chem.* **252:** 2218.

Baker, C. C. and I. Isenberg. 1976. On the analysis of circular dichroic spectra of proteins. *Biochemistry* **15:** 629.

Beaudoin, J., T. J. Henry, and D. Pratt. 1974. Purification of single- and double-length M13 virions by polyacrylamide gel electrophoresis. *J. Virol.* **13:** 470.

Berkowitz, S. A. 1975. "Molecular weight of fd bacteriophage and fd NaDNA. Evidence for quasi-equivalent DNA-protein interactions in a filamentous virus." Ph.D. thesis, New York University, New York.

Berkowitz, S. A. and L. A. Day. 1974. Molecular weight of single-stranded fd bacteriophage DNA. High speed equilibrium sedimentation and light scattering measurements. *Biochemistry* **13:** 4825.

———. 1976. Mass, length, composition and structure of the filamentous bacterial virus fd. *J. Mol. Biol.* **102:** 531.

Bradley, D. E. 1964. The structure of some bacteriophages associated with male strains of *Escherichia coli. J. Gen. Microbiol.* **35:** 471.

———. 1973. The length of the filamentous *Pseudomonas aeruginosa* bacteriophage Pf. *J. Gen. Virol.* **20:** 249.

Broersma, S. 1960a. Rotational diffusion constant of a cylindrical particle. *J. Chem. Phys.* **32:** 1626.

———. 1960b. Viscous force constant for a closed cylinder. *J. Chem. Phys.* **32:** 1632.

Cavalieri, S. J., D. A. Goldthwait, and K. E. Neet. 1976. The isolation of a dimer of gene 8 protein of bacteriophage fd. *J. Mol. Biol.* **102:** 713.

Chang, C. N., G. Blobel, and P. Model. 1978. Detection of prokaryotic signal peptidase in an *Escherichia coli* membrane fraction: Endoproteolytic cleavage of nascent f1 pre-coat protein. *Proc. Natl. Acad. Sci.* **75:** 361.

Chen, Y.-H., J. T. Yang, and K. H. Chau. 1974. Determination of the helix and β form of proteins in aqueous solution by circular dichroism. *Biochemistry* **13:** 3350.

Coleman, J. E., R. A. Anderson, R. G. Ratcliffe, and I. M. Armitage. 1976. Structure of gene 5 protein-oligodeoxynucleotide complexes as determined by ^{1}H, ^{19}F, and ^{31}P nuclear magnetic resonance. *Biochemistry* **15:** 5419.

Crewe, A. V. and J. Wall. 1970. A scanning microscope with 5 Å resolution. *J. Mol. Biol.* **48:** 375.

Cuypers, T., F. J. Van der Ouderaa, and W. W. de Jong. 1974. The amino acid sequence of gene 5 protein of bacteriophage M13. *Biochem. Biophys. Res. Commun.* **59:** 557.

Day, L. A. 1966. Protein conformation in fd bacteriophage as investigated by optical rotary dispersion. *J. Mol. Biol.* **15:** 395.

———. 1969. Conformations of single-stranded DNA and coat protein in fd bacteriophage as revealed by ultraviolet absorption spectroscopy. *J. Mol. Biol.* **39:** 265.

———. 1973. Circular dichroism and ultraviolet absorption of a deoxyribonucleic acid binding protein of filamentous bacteriophage. *Biochemistry* **12:** 5329.

Day, L. A. and S. A. Berkowitz. 1977. The number of nucleotides and the density and refractive index increments of fd virus DNA. *J. Mol. Biol.* **116:** 603.

Denhardt, D. T. 1975. The single-stranded DNA phages. *CRC Crit. Rev. Microbiol.* **4:** 161.

Dunker, A. K., R. D. Klausner, D. A. Marvin, and R. L. Wiseman. 1974. Filamentous bacterial viruses. X. X-ray diffraction studies of the R4 A-protein mutant. *J. Mol. Biol.* **82:** 115.

Enea, V. and N. D. Zinder. 1975. A deletion mutant of bacteriophage f1 containing no intact cistrons. *Virology* **68:** 105.

Forsheit, A. B. and D. S. Ray. 1970. Conformations of the single-stranded DNA of bacteriophage M13. *Proc. Natl. Acad. Sci.* **67:** 1534.

Frank, H. and L. A. Day. 1970. Electron microscopic observations on fd bacteriophage, its alkali denaturation products and its DNA. *Virology* **42:** 144.

Franklin, R. E. and R. G. Gosling. 1953. The structure of sodium thymonucleate fibres. II. The cylindrically symmetrical Patterson function. *Acta Cryst.* **6:** 678.

Garssen, G. J., C. W. Hilbers, J. G. G. Schoenmakers, and J. H. van Boom. 1977. Studies on DNA unwinding. Proton and phosphorus nuclear-magnetic-resonance studies of gene V protein from bacteriophage M13 interacting with d(pC-G-C-G). *Eur. J. Biochem.* **81:** 453.

Goldsmith, M. E. and W. H. Konigsberg. 1977. Adsorption protein of the bacteriophage fd: Isolation, molecular properties, and location in the virus. *Biochemistry* **16:** 2686.

Grandis, A. S. and R. E. Webster. 1973. A new species of small covalently closed f1 DNA in *Escherichia coli* infected with an *amber* mutant of bacteriophage f1. *Virology* **55:** 14.

Green, N. M. and M. T. Flanagan. 1976. The prediction of the conformation of membrane proteins from the sequence of amino acids. *Biochem. J.* **153:** 729.

Griffith, J. and A. Kornberg. 1974. Mini M13 bacteriophage: Circular fragments of M13 DNA are replicated and packaged during normal infections. *Virology* **59:** 139.

Helené, C. and J. L. Dimicoli. 1972. Interaction of oligopeptides containing aromatic amino acids with nucleic acids. Fluorescence and proton magnetic resonance studies. *FEBS Lett.* **26:** 6.

Herrmann, R., K. Neugebauer, H. Zentgraf, and H. Schaller. 1978. Transposition of a DNA sequence determining kanamycin resistance into the single-stranded genome of bacteriophage fd. *Mol. Gen. Genet.* **159:** 171.

Hoffmann-Berling, H., D. A. Marvin, and H. Dürwald. 1963. Ein fädiger DNS-phage (fd) und ein sphärischer RNS-phage (fr), wirtsspezifisch für männliche Stämme von *E. coli*. I. Präparation und chemische Eigenschaften von fd und fr. *Z. Naturforsch.* **18b:** 876.

Ikehara, K. and H. Utiyama. 1975. Studies on the structure of filamentous bacteriophage fd. III. A stable intermediate of the 2-chloroethanol-induced disassembly. *Virology* **66:** 316.

Ikehara, K., H. Utiyama, and M. Kurata. 1975. Studies on the structure of filamentous bacteriophage fd. II. All-or-none disassembly in guanidine-HCl and sodium dodecyl sulfate. *Virology* **66:** 306.

Ikehara, K., Y. Obata, H. Utiyama, and M. Kurata. 1973. Studies on the structure of filamentous bacteriophage fd. I. Physicochemical properties of phage fd and its components. *Bull. Inst. Chem. Res.* **51:** 140.

Khatoon, H. and R. V. Iyer. 1971. Stable coexistence of Rfi$^-$ factors in *Escherichia coli. Can. J. Microbiol.* **17:** 669.

Khatoon, H., R. V. Iyer, and V. N. Iyer. 1972. A new filamentous bacteriophage with sex-factor specificity. *Virology* **48:** 145.

Knippers, R. and H. Hoffmann-Berling. 1966. A coat protein from bacteriophage fd. I. Hydrodynamic measurements and biological characterization. *J. Mol. Biol.* **21:** 281.

Kuo, T.-T., T.-C. Huang, and T.-Y. Chow. 1969. A filamentous bacteriophage from *Xanthomonas oryzae*. *Virology* **39:** 548.

Kuo, T.-T., T.-C. Huang, R.-Y. Wu, and C.-M. Yang. 1967. Characterization of three bacteriophages of *Xanthomonas oryzae* (uyeda et ishiyama) Dowson. *Bot. Bull. Acad. Sin.* **8:** 246.

Lawn, A. M., E. Meynell, G. G. Meynell, and N. Datta. 1967. Sex pili and the classification of sex factors in the Enterobacteriaceae. *Nature* **216:** 343.

Lin, J.-Y., C.-C. Wu, and T.-T. Kue. 1971. Amino acid analysis of the coat protein of the filamentous bacterial virus Xf from *Xanthomonas oryzae. Virology* **45:** 38.

Marvin, D. A. 1966. X-ray diffraction and electron microscopic studies on the structure of the small filamentous bacteriophage fd. *J. Mol. Biol.* **15:** 8.

Marvin, D. A. and B. Hohn. 1969. Filamentous bacterial viruses. *Bacteriol. Rev.* **33:** 172.

Marvin, D. A. and E. J. Wachtel. 1976. Structure and assembly of filamentous bacterial viruses. *Philos. Trans. R. Soc. Lond. B* **276:** 81.

Marvin, D. A., R. L. Wiseman, and E. J. Wachtel. 1974a. Filamentous bacterial viruses. XI. Molecular architecture of the class II (Pf1, Xf) virion. *J. Mol. Biol.* **82:** 121.

Marvin, D. A., W. J. Pigram, R. L. Wiseman, E. J. Wachtel, and F. J. Marvin. 1974b. Filamentous bacterial viruses. XII. Molecular architecture of the class I (fd, If1, Ike) virion. *J. Mol. Biol.* **88:** 581.

McGuire, R. and I. Feldman. 1973. The quenching of tyrosine and tryptophan fluorescence by H_2O and D_2O. *Photochem. Photobiol.* **18:** 119.

McPherson, A., I. Molineux, and A. Rich. 1976. Crystallization of a DNA unwinding protein: Preliminary X-ray analysis of fd bacteriophage gene 5 product. *J. Mol. Biol.* **106:** 1077.

Meynell, G. G. and A. M. Lawn. 1968. Filamentous phages specific for the I sex factor. *Nature* **217:** 1184.

Minamishima, Y., K. Takeya, Y. Ohnishi, and K. Amako. 1968. Physicochemical and biological properties of fibrous *Pseudomonas* bacteriophages. *J. Virol.* **2:** 208.

Nakanishi, H., Y. Iida, K. Maeshima, T. Teramoto, Y. Hosaka, and M. Ozaki. 1966. Isolation and properties of bacteriophages of *Vibrio parahaemolyticus. Biken. J.* **9:** 149.

Nakashima, Y. and W. Konigsberg. 1974. Reinvestigation of a region of the fd bacteriophage coat protein sequence. *J. Mol. Biol.* **88:** 598.

Nakashima, Y., A. K. Dunker, D. A. Marvin, and W. Konigsberg. 1974. The amino acid sequence of a DNA binding protein, the gene 5 product of fd filamentous bacteriophage. *FEBS Lett.* **43:** 125.

Nakashima, Y., R. L. Wiseman, W. Konigsberg, and D. A. Marvin. 1975. Primary structure and sidechain interactions of Pf1 filamentous bacterial virus coat protein. *Nature* **253:** 68.

Newman, J., H. L. Swinney, and L. A. Day. 1977. Hydrodynamic properties and structure of fd virus. *J. Mol. Biol.* **116:** 593.

Newman, J., H. L. Swinney, S. A. Berkowitz, and L. A. Day. 1974. Hydrodynamic properties and molecular weight of fd bacteriophage DNA. *Biochemistry* **13:** 4832.

Notani, G. W. and N. D. Zinder. 1964. Genetically heterozygous particles of the bacteriophage f1. In *Bacterial proceedings*, p. 140. American Society for Microbiology, Washington, D.C.

Nozaki, Y., B. K. Chamberlain, R. E. Webster, and C. Tanford. 1976. Evidence for a major conformational change of coat protein in assembly of f1 bacteriophage. *Nature* **259:** 335.

Oey, J. L. and R. Knippers. 1972. Properties of the isolated gene 5 protein of bacteriophage fd. *J. Mol. Biol.* **68:** 125.

Pratt, D., P. Laws, and J. Griffith. 1974. Complex of bacteriophage M13 single-stranded DNA and gene 5 protein. *J. Mol. Biol.* **82:** 425.

Pretorius, H. T., M. Klein, and L. A. Day. 1975. Gene V protein of fd bacteriophage. *J. Biol. Chem.* **250:** 9262.

Rasched, I. and F. M. Pohl. 1974. Oligonucleotides and the quaternary structure of gene-5 protein from filamentous bacteriophage. *FEBS Lett.* **46:** 115.

Ray, D. S. 1977. Replication of filamentous bacteriophages. In *Comprehensive virology* (ed. H. Fraenkel-Conrat and R. R. Wagner), vol. 7, p. 105. Plenum Press, New York.

Salivar, W. O., T. J. Henry, and D. Pratt. 1967. Purification and properties of diploid particles of coliphage M13. *Virology* **32:** 41.

Sanger, F., J. E. Donelson, A. R. Coulson, H. Kössel, and D. Fischer. 1973. Use of DNA polymerase I primed by a synthetic oligonucleotide to determine a nucleotide sequence in phage f1 DNA. *Proc. Natl. Acad. Sci.* **70:** 1209.

Sanger, F., B. G. Air, G. M. Barrell, N. L. Brown, A. R. Coulson, J. C. Fiddes, C. A. Hutchison III, P. M. Slocombe, and M. Smith. 1977. Nucleotide sequence of bacteriophage φX 174 DNA. *Nature* **265:** 687.

Snell, D. T. and R. E. Offord. 1972. The amino acid sequence of the B-protein of bacteriophage ZJ-2. *Biochem. J.* **127:** 167.

Stanisich, V. A. 1974. The properties and host range of male-specific bacteriophages of *Pseudomonas aeruginosa. J. Gen. Microbiol.* **84:** 332.

Takeya, K. and K. Amako. 1966. A rod-shaped *Pseudomonas* phage. *Virology* **28:** 163.

Tate, W. P. and G. B. Peterson. 1974. Structure of the filamentous bacteriophages: Orientation of the DNA molecule within the phage particle. *Virology* **62:** 17.

Wachtel, E. J., R. L. Wiseman, W. J. Pigram, D. A. Marvin, and L. Manuelides. 1974. Filamentous bacterial viruses. XIII. Molecular structure of the virion in projection. *J. Mol. Biol.* **88:** 601.

Wall, J. S. 1971. "A high resolution scanning electron microscope for the study of single biological molecules." Ph.D. thesis, University of Chicago, Illinois.

Watson, J. D. and F. H. C. Crick. 1953. A structure for deoxyribose nucleic acid. *Nature* **171:** 737.

Wheeler, F. C., R. H. Benzinger, and H. Bujard. 1974. Double-length, circular, single-stranded DNA from filamentous phage. *J. Virol.* **14:** 620.

Williams, R. W. and A. K. Dunker. 1977. Circular dichroism studies of fd coat protein in membrane vesicles. *J. Biol. Chem.* **252:** 6253.

Wiseman, R. L. and L. A. Day. 1977. Different packaging of DNA in the filamentous viruses Pf1 and Xf. *J. Mol. Biol.* **116:** 607.

Wiseman, R. L., S. A. Berkowitz, and L. A. Day. 1976. Different arrangements of protein subunits and single-stranded circular DNA in the filamentous bacterial viruses fd and Pf1. *J. Mol. Biol.* **102:** 549.

Wiseman, R. L., A. K. Dunker, and D. A. Marvin. 1972. Filamentous bacterial viruses. III. Physical and chemical characterization of the If1 virion. *Virology* **48:** 230.

Woolford, J. L., H. M. Steinman, and R. E. Webster. 1977. Adsorption protein of bacteriophage f1: Solubilization in deoxycholate and localization in the f1 virion. *Biochemistry* **16:** 2694.

Filamentous Bacteriophage Pf1 Has 27 Subunits in Its Axial Repeat

Lee Makowski and Donald L. D. Caspar
Rosenstiel Basic Medical Sciences Research Center
Brandeis University
Waltham, Massachusetts 02154

Molecular model building is now being applied to analyze the structure of the filamentous phage coat protein (Marvin and Wachtel 1975, 1976) by using information from low-resolution fiber diffraction and amino acid sequence without recourse to conventional methods of protein crystallography. This approach has important implications for the study of the structure of structural proteins that aggregate in their own purposeful way to build symmetric filaments, shells, or sheets rather than crystals suitable for high-resolution diffraction analysis.

The general picture of the Pf1 phage structure (Marvin et al. 1974a; Marvin and Wachtel 1975; Nakashima et al. 1975) is that the 46-residue protein subunit is largely α-helical; the subunits are interlocked in a regular, overlapping spiral arrangement to form a protein tube about 60 Å in diameter with a wall about 20 Å thick. This helical tube contains the two strands of the extended circular DNA molecule; the positively charged C-terminal end of the subunit is at the inside of the tube near the DNA and the acidic N terminus is at the outside surface.

Marvin and Wachtel (1976) have constructed a detailed molecular model of the Pf1 protein coat by interlocking α-helical subunits in a regular helical assembly. This model presumes that there is one subunit every 3.4 Å along the axis of the particle; i.e., 4.4 subunits per turn of the 15-Å pitch helix or 22 subunits in the 75-Å helix repeat period. (The repeat period varies slightly with changes in humidity, from about 72 Å for completely dry fibers to about 77 Å at high humidity.) Although the X-ray pattern appears to favor a helix with 22 units in the repeat (Marvin et al. 1974a), the evidence is inconclusive. Structures with 23, 27, or 28 subunits in the five-turn helix repeat are also compatible with the diffraction data. Determination of the correct number of units in the repeat is crucial for analysis of the molecular structure and packing interactions of the subunit. The α-helix interlocking, for example, will be substantially different in a model with a 3.4-Å axial translation per subunit compared with one having, say, a 2.8-Å unit separation.

An axial spacing per subunit of 2.57 ± 0.17 Å in the dry Pf1 particle has been calculated by Wiseman and Day (1977) from a combination of chemical, physicochemical, and electron microscopic measurements. The DNA content of 6.1% corresponds to 0.97 nucleotide of average molecular weight 309 for each protein subunit of molecular weight 4609 (Nakashima et al. 1975). Comparison of the sedimentation rate of the Pf1 DNA with those of DNA molecules of known size indicates that there are 7400 nucleotides in the chain and therefore about 7630 subunits in the particle. Dividing the 1.96-μm length of the dry particle obtained from electron microscopy by the number of subunits gives the axial separation of 2.57 Å. Since the repeat period of the dry particle obtained from X-ray diffraction is about 72 Å (Marvin et al. 1974a), the axial separations for 22, 23, 27, or 28 subunits in the repeat would be 3.27, 3.13, 2.66, or 2.57 Å. The measurements by Wiseman and Day (1977) are therefore compatible with 27 or 28 subunits in the repeat, but their estimated experimental uncertainty excludes 22- or 23-subunit models.

The distribution of intensity along the ~75-Å-period layer lines in the Pf1 diffraction pattern shows that there are $5s+2$ or $5s+3$ (where s is an integer) subunits equally spaced along five turns of a helix of 15-Å pitch (Marvin et al. 1974a). Considering the size of the subunit and the estimated volume of the particle repeat, there may be 22, 23, 27, or 28 subunits in the repeat length. Observation of a strong, apparently meridional diffraction maximum on the 22nd layer line and the absence of comparable diffraction on the 23rd, 27th, or 28th layer line suggests that there are 22 subunits in the repeat. However, this form of measurement is often unreliable, as illustrated by early attempts to determine the helical symmetry of tobacco mosaic virus (TMV). Watson (1954) showed that TMV has $3s+1$ subunits in three turns of a helix of 23-Å pitch and suggested that the number in the repeat was 31, since he observed an apparently meridional maximum on the 31st layer line. Franklin (1955) showed that this maximum was off-meridional and suggested, instead, that the repeat consisted of 37 units because of a 37th-layer-line reflection; later this also proved to be off-meridional. The correct number, 49, was first recognized in the equatorial diffraction from a mercury derivative which revealed the side-to-side separation of the heavy atoms around the circle seen in projection (Franklin and Holmes 1958). The meridional 49th-layer-line reflections expected from the 1.41-Å unit axial translation in TMV has not yet been observed and may be too weak to measure. The 1.41-Å unit axial spacing was confirmed (Klug and Caspar 1960) by comparing the TMV mass per unit length from physicochemical measurements and electron microscopy with the subunit weight.

In support of 22 subunits in the 75-Å repeat for Pf1, Marvin and Wachtel (1976) pointed out that the 3.4-Å axial separation between subunits needed for this symmetry is close to the 3.3-Å separation they had estimated from mass-per-unit-length measurements for the structurally similar strain fd (Marvin and Hohn 1969). The most recent measurements of particle length and number of coat-protein subunits for fd (Newman et al. 1977) give a subunit separation of 3.23 Å in the hydrated virus, which corresponds very closely to 5 subunits in the 16.1-Å axial period. (Like with Pf1, the length of fd expands with humidity, but by about half the percentage; the axial period of ~15.7 Å in dry fibers increases to ~16.3 Å in water.) Subunit separations for fd and Pf1 are not, however, directly comparable because both the subunit sizes and the particle diameters for these two viruses are different.

Each subunit (protein plus DNA) occupies a volume in the virus defined by the product of the subunit axial separation and the effective cross-sectional area of the particle. Cross-sectional areas can be defined objectively in the close-packed lattices of semicrystalline virus fibers. The lattice volume per subunit will be greater than the specific molecular volume of the subunit if there are crevices or cavities in the packing. Similar packing of protein and nucleic acid in different viruses will lead to similar ratios of lattice volume to molecular volume regardless of the subunit size and particle dimensions. Our comparison of the ratios of lattice volume to molecular volume in dry fibers of Pf1, fd, and TMV indicate that there are 27 subunits in the Pf1 repeat. Furthermore, measurements of protein volume from the electron-density profile and of particle hydration from the volume of moist fibers are also consistent with a 27-subunit model.

Marvin et al. (1974a) have argued that the semicrystalline sampling observed in the diffraction pattern of Pf1 at low relative humidities could not be generated by any packing of a regular helical structure with a multiple of 3 subunits per repeat. If this symmetry argument were generally valid, it would have eliminated 27 as a possibility. We show, however, that interlocking of particles with a helical symmetry of 27 subunits in five turns can explain the observed partially ordered packing.

MOLECULAR VOLUME AND PACKING DENSITY

One Pf1 protein subunit plus 0.97 nucleotide has an average molecular weight of 4909 daltons (Wiseman and Day 1977). From the subunit weight and the partial specific volume of the virus, the possible values of the physical volume of the repeating unit consisting of 22, 23, 27, or 28 subunits can be calculated. The ratio of lattice volume measured by X-ray diffraction to the correct molecular volume of the virus repeating unit should be in the same range as the ratios calculated for similar structures for which the mass, density, and lattice volume are known.

Calculation of the physical volume of a molecule from its molecular weight and partial specific volume has some uncertainty because the partial specific volume, defined thermodynamically, need not correspond exactly to the volume occupied by a unit weight of the molecule. Comparisons of molecular volumes measured for a number of protein crystals with partial specific volumes of these proteins indicate, however, that there is a close correspondence between these two quantities (Richards 1974, 1977). These measurements have led to a set of standard volumes occupied by amino acid residues inside protein molecules (Chothia 1975). The correspondence of these volumes with the volumes occupied in crystals of pure amino acids demonstrates that the packing in the interior of a protein molecule resembles a molecular crystal. Furthermore, the packing densities of residues at intersubunit contacts in protein assemblies are indistinguishable from those found inside the molecules (Chothia and Janin 1975). In contrast, the packing of protein molecules in wet crystals is quite loose—the solvent space is generally comparable to the volume occupied by the protein (Matthews 1968).

To make comparisions among different structures, we have used Chothia's (1975) volumes for amino acid residues and volumes for the nucleotides from measurements on DNA fibers (Langridge et al. 1960) to calculate the molecular volume from chemical composition. Use of mean volumes occupied by residues in the interior of proteins to estimate total protein volume gives a slight overestimate of specific volume because no correction for electrostriction by charged residues is included and there is some uncertainty in the geometrical definition of the protein-solvent interface (Richards 1974, 1977). Nevertheless, the standard residue-packing volumes will give self-consistent molecular volumes for comparison of protein structures with similar surface-to-volume ratios. The problem in estimating the number of subunits in the repeat period of the filamentous viruses from comparisons of lattice and subunit volumes is not in determining these volumes; rather, the critical problem is to decide how to interpret these measurements.

Lattice Volume of Dry Fibers

Marvin et al. (1974a) have reported three sets of X-ray measurements of lattice dimensions on two Pf1 fibers dried at zero humidity. The lattice constants for the hexagonally packed particles are 52.4, 53.7, and 54.0 Å, and the corresponding axial repeats are 72.2, 71.6, and 74.1 Å. The variation in shape of the repeating unit indicates slightly different rearrangements of the subunits on drying. These small variations are not surprising considering the polymorphism of the subunit packing observed under other conditions (Wachtel et al. 1976). The volumes of the repeat unit from these three sets of measurements are 1.72, 1.79, and 1.87×10^5 Å^3. The 9% range of variation is larger than the estimated experimental errors. Within the experimental uncertainties, the smallest volume may be close to the molecular volume of the close-packed parts.

Each Pf1 subunit (protein plus DNA) of molecular weight 4909 daltons has a molecular volume of 6290 Å^3. The molecular volumes of 22, 23, 27, or 28 subunits in the helix repeat would be 1.38, 1.45, 1.70, and 1.76×10^5 Å^3, respectively. From the X-ray measurements on dry fibers, the ratios of lattice volume to molecular volume of 22, 23, 27, or 28 subunits are in the ranges 1.24–1.35, 1.19–1.29, 1.01–1.10, and 0.97–1.06, respectively. The 28-subunit model can be ruled out because the predicted molecular volume exceeds the smallest lattice volume by more than the estimated experimental error. The range 1.01–1.10 for 27 subunits in the repeat is physically plausible as the protein and DNA could close pack so as to leave only a small fraction of the volume in dry fibers unoccupied. With 22 subunits in the repeat, the vacant spaces in the most compact dry fibers would add up to about a quarter of the particle volume. Since the lattice volume of some dry protein crystals exceeds the molecular volume by about this much (Matthews 1968), the low packing density calculated by this comparison does not necessarily rule out the 22-subunit model.

Considering the similarities in structure of Pf1 and fd, the packing densities of their fibers should be similar (Marvin et al. 1974a,b). The measurements of particle length and number of subunits for fd by Newman et al. (1977) indicate that there are 5.0±0.2 subunits in the axial repeat distance (15.7–16.3 Å for dry to wet fibers). Earlier measurements of mass per unit length indicated between 4.5 and 5.5 subunits in this repeat (Marvin and Hohn 1969). Marvin et al. (1974b) have interpreted the fd diffraction pattern in terms of a perturbed, single-start, 4.5-subunit-per-turn helix with axial displacements that repeat every 5 subunits to give the ~16.1-Å axial period. A simpler interpretation of the ~16.1-Å-spacing meridional reflections is that this period is the projected axial separation between subunits. Since there are 5 subunits in this distance, they all would have to be at the same axial level related by a fivefold rotation axis. In the 5-start helix generated by pentamers stacked ~16.1 Å apart with an approximate repeat every two layers, consecutive pentamers would be rotated by approximately 36°. The measured layer-line splitting (Marvin et al. 1974b) indicates an angle of rotation between neighboring pentamers of about 34.6°. Whether the X-ray pattern is interpreted in terms of a regularly symmetric 5-start helix or a periodically perturbed single-start helix, there are, in either case, 5 subunits in the axial period.

Each fd subunit, consisting of one protein of molecular weight 5240 (Asbeck et al. 1969; Nakashima and Konigsberg 1974) plus 2.3 nucleotides (Marvin and Hohn 1969; Newman et al. 1977), has a molecular volume of 7465 $Å^3$. The lattice volume occupied by 5 subunits in dry fibers with lattice constants of 54.1 to 55.4 Å (Marvin 1966; Milman 1968) is in the range 39,700–42,200 $Å^3$. Dividing this lattice volume by the molecular volume of 5 subunits gives a ratio of 1.06–1.13. This calculation demonstrates that the protein and DNA of fd particles pack compactly in fibers at zero humidity. Within the range of variation and uncertainty of the lattice-volume measurements, the ratio of vacant space to particle volume of 0.06–0.13 for fd fibers overlaps with the ratios 0.01, 0.05, and 0.10 calculated from the three sets of X-ray measurements on dry Pf1 fibers for which a 27-subunit repeat was assumed. Correspondence of the fd packing density with that calculated for 27 subunits in the Pf1 repeat might be fortuitous if the protein and nucleic acid of different viruses were not packed with comparable compactness.

Another comparison that can be made is with TMV. Filamentous bacteriophage particles pack in gels and fibers (Lapointe and Marvin 1973) in much the same way as particles of TMV (Bernal and Fankuchen 1941). The ratio of lattice volume to molecular volume calculated for dried TMV (for which the number and volume of subunits in the axial repeat are known [see Klug and Caspar 1960]) is 1.16. After correcting for the 40-Å-diameter hole along the axis of the particle, this ratio is reduced to 1.08. TMV, which has a structure very different from fd, nevertheless has a similarly compact packing of its protein and nucleic acid. If the packing in Pf1 is also compact, there would be 27 subunits in its axial repeat. It could be argued, however, that there are only 22 subunits, leaving about 20% more vacant space in dry fibers of Pf1 than for fd or TMV. There is only about half as much DNA in Pf1 as in fd (Wiseman et al. 1976; Wiseman and Day 1977) and the symmetry of their protein coats may be quite different. Pf1 protein could form a thin-walled coat leaving a large axial hole in which the DNA fits loosely. This possibility can be assessed by looking more closely at details of the structure which are defined unambiguously by the X-ray diffraction pattern.

Radial Density Distribution

The cylindrically averaged electron-density distribution in the particle (Fig. 1) is computed from the equatorial lattice reflections, which have been measured to a spacing of 8 Å. To this resolution, the transform of the particle is cylindrically symmetric and it is oversampled by the X-ray measurements at different swelling stages of the lattice. Thus, phase relations for this centrosymmetric projection are uniquely fixed by the minimum wavelength principle (Bragg and Perutz 1952). The profile in Figure 1 is very similar to that calculated by Marvin and Wachtel (1976) from equatorial continuous diffraction data from swollen fibers. The central peak corresponds to the viral DNA, and the peaks at the 15- and 25.5-Å radii can be identified with α-helical segments of the coat protein. From this interpretation, the protein coat is about 20 Å thick, and DNA occupies a central cavity with a radius of about 10 Å.

DNA constitutes about 6.1% of the particle mass (Wiseman and Day 1977) and therefore accounts for about 12% of the total number of effective scattering electrons above the water density. (The DNA is more electron-dense than protein and contributes proportionally more than its weight fraction to the scattering.) The integral of the electron-density profile out to 7.5 Å radius corresponds to the fraction of scattering electrons in DNA. There may be some interdigitation of nucleotides and protein side chains,

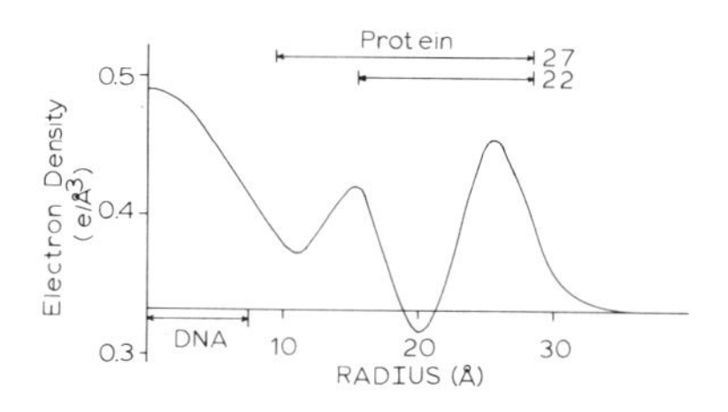

Figure 1 Radial electron-density distribution of Pf1. This cylindrically averaged electron-density profile was calculated from the equatorial lattice reflections measured out to a spacing of 8 Å for a Pf1 fiber at 92% relative humidity. The central peak can be identified with the DNA core and the peaks at the 15- and 25.5-Å radii with layers of *a* helix in the protein coat. The fraction of scattering electrons in DNA is accounted for by the integral of the electron-density profile out to 7.5 Å radius, marked by the arrow in the lower left. Calculated thicknesses for protein coats consisting of 22 and 27 subunits in the 75-Å axial repeat are marked in the upper right. Compact subunit packing in a uniform shell of external radius 28.5 Å is assumed. Protein-subunit volume was derived from amino acid composition and standard packing volumes of the residues. With 27 subunits the inner radius of the shell would be 9.5 Å, and with 22 subunits the inner radius would be 15.0 Å. The electron-density scale marked on the ordinate was evaluated for 27 subunits in the 75-Å repeat distance; for 22 subunits, the scale would be reduced relative to the water level of 0.334 electrons/Å³ by the factor 22/27.

and some portions of the DNA may extend to a somewhat larger radius. The density minimum at 11 Å radius may represent the boundary between the inner surface of the interlocked *a* helices of the coat and the DNA core. These estimates of the volume occupied by the DNA are independent of assumptions about particle symmetry.

Protein Coat Thickness

Protein extends to a maximum radius of about 31 Å in the hydrated particle (Fig. 1). A minimum packing radius of about 28.5 Å can be defined from the electron-density profile as the point where the electron density falls to half the value at the outer peak. A similar estimate for the effective packing radius of the cylindrical particle is obtained by equating the circular cross-sectional area with the hexagonal area occupied per particle in dry fibers. A smooth, cylindrical shell of compactly packed protein of outer radius 28.5 Å would have a thickness of 19 Å if there were 27 subunits in the 75-Å axial repeat; with 22 units in the repeat, the shell thickness would be only 13.5 Å. The use of different possible estimates for the mean outer radius gives a range for the calculated thickness that varies by 1–2 Å. As shown in Figure 1, the 19-Å wall thickness is consistent with the estimates of the volumes occupied by DNA and protein based on apportioning the electrons in the density profile. In contrast, if the wall thickness were only 13.5 Å, the number of effective scattering electrons within the 15-Å inner radius would total about 2.8 times the number that could be accounted for by DNA.

Strong diffraction on the first and third layer lines of the Pf1 diffraction pattern shows that the protein coat is built up of interlocked α helices with a separation of about 10 Å. If there were 22 subunits in the repeat, the 10.5-Å diffraction maximum on the third layer line would come from a major density fluctuation following a 5-start helix at a radius of about 11 Å; with 27 subunits in the repeat, this diffraction would be attributed to a 6-start helix at about 12.5 Å radius. The 15-Å inner radius calculated for a close-packed shell of protein with 22 subunits in the repeat (Fig. 1) is inconsistent with the 11-Å radius calculated for the α-helix density fluctuation from the third-layer-line diffraction; but with a 27-subunit repeat, the 12.5-Å radius calculated for this fluctuation is well within the calculated thickness of the protein shell.

The detailed 22-subunit molecular model constructed by Marvin and Wachtel (1976) for the Pf1 protein has a shell about 20 Å thick, but the outer radius is smaller than that indicated by experimental measurements. The calculated density profile of the model roughly parallels the measured profile, with an inward shift of about 2.5 Å. The 8–9% decrease in outer radius corresponds to a 15–17% smaller volume for the model than for the virus. This is just about the difference in volume for a protein shell made of 22 subunits compared with one made of 27 subunits. Packing volumes of amino acid residues in the model appear to conform generally to the standard volumes measured in proteins (Richards 1974; Chothia 1975), which gives a reasonable level for the calculated electron density. Any model with 22 compactly packed subunits will have a volume for the protein shell about 18% smaller than that indicated by the electron-density profile.

Another possibility that can be considered is that the protein is not compactly packed in the coat; e.g., 22 subunits in the repeat distance form a coat extending to the measured outer radius of the virus particle, but about 18% of the space within this radius is vacant. Interlocked α-helical segments are necessarily packed compactly. The electron-density profile (Fig. 1) suggests two layers of α-helical segments separated radially by about 10 Å. The low density at 20 Å radius can be identified with the hydrophobic residues in the middle of the chain sequence (Nakashima et al. 1975) that are distributed along the apposed surfaces of the inner and outer α-helical segments. Considering the hydrophobic nature of the central region of the protein sequence and the X-ray evidence for coiled-coil interlocking, the subunit packing inside the protein coat is likely to be as compact as in the interior of globular protein molecules. A gap-overlap arrangement of chain segments, as in fibrous protein assemblies such as collagen, might nevertheless occur, which would leave water-accessible channels or grooves between subunits. Negative staining should reveal any gaps, but the micrographs of filamentous bacteriophage show a quite smooth surface. Furthermore, there are no major changes in structure detected between wet and dry specimens by X-ray diffraction, as might be expected if there were large gaps in the protein-coat packing.

Particle Hydration

Hydration of the virus fibers equilibrated at different relative humidities can be calculated from the difference between the lattice volume V and the physical volume V_p of the repeating unit. The hydration in grams of water of unit density per gram of particle of density ρ_p is $(V-V_p)/\rho_p V_p$. This relation gives a valid estimate of water content if the partial specific volumes of protein and water are constant (as is the case in partially dried protein crystals) for water contents greater than about 10% (Low and Richards 1974). In Figure 2, the values of $(V-V_p)/\rho_p V_p$ calculated from the X-ray measurements on Pf1 fibers (Marvin et al. 1974a), for which 22 and 27 subunits in the repeat distance were assumed, are compared with hydration measurements on collagen, wool, and β-lactoglobulin (Bull 1944). Collagen has the highest hydration of the 11 proteins compared by Bull, β-lactoglobulin is typical of globular proteins, and wool takes up less water than the others at high relative humidity. The hydration of 0.11–0.21 g/g at 75–80% relative humidity calculated for 27 subunits in the Pf1 repeat

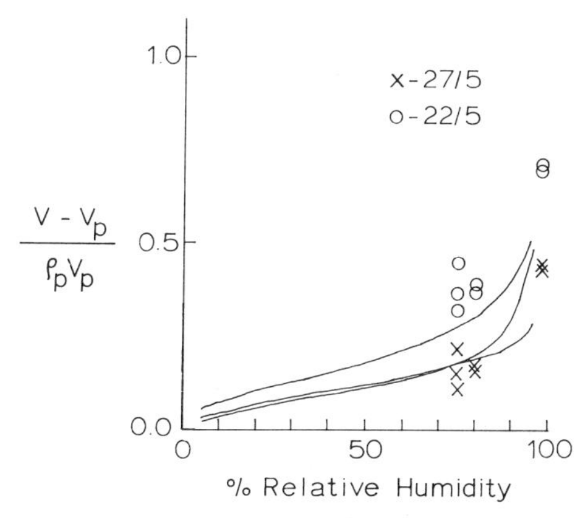

Figure 2 "Hydration" calculated from excess volume for Pf1 fibers assuming 22 (o) and 27 (x) subunits in the repeat compared with hydration measurements for three proteins. The excess volume $(V-V_p)$ is the difference between the volume V of the lattice that contains one repeat unit and the volume V_p calculated for the anhydrous particle; $\rho_p V_p$ is the anhydrous mass of the repeat unit of density ρ_p. The ratio $(V-V_p)/\rho_p V_p$ will represent hydration in grams of water per gram of particle if the excess volume is occupied by water of unit density. The lattice volume V of the repeat unit is calculated from the area of the hexagonal unit cell and the axial repeat length obtained from X-ray spacing measurements reported by Marvin et al. (1974a). V_p is equal to either 22 or 27 times the subunit volume derived from the chemical composition and standard packing volumes of the amino acid residues and nucleotides. The solid lines are water-binding isotherms measured gravimetrically by Bull (1944). Measured hydration is plotted for (top to bottom at 90% relative humidity) collagen, β-lactoglobulin, and wool.

period is typical of most proteins, whereas the 22-subunit model would require two to three times as much water in the fiber. There would have to be even more hydration than that measured for collagen, which has a large ratio of water-accessible surface to volume.

Packing-density Comparisons

All of the comparisons based on calculation of the molecular volume of the protein and nucleic acid in the five-turn axial repeat period of the Pf1 particle are consistent with a 27-subunit model. To reconcile a 22-subunit model with the experimental observations requires ad hoc assumptions to account for undetected vacant spaces that would comprise about one-fifth of the volume within the particle. A 22-subunit model is not demonstrably impossible, but the dependence on special assumptions to resolve all the apparent contradictions makes it unlikely.

PARTICLE SYMMETRY AND LATTICE SAMPLING

Lattice sampling in the X-ray patterns provides information about particle symmetry which is independent of inferences based on particle volume and packing density. X-ray diffraction patterns from fibers of Pf1 (Marvin et al. 1974a) have sharp equatorial reflections from the regular hexagonal packing of the cylindrically shaped particles. Distinct crystalline lattice sampling is evident on the third layer line at relative humidities below about 80%. This lattice sampling implies some kind of three-dimensional regularity in the particle packing. The relation between particle symmetry and lattice packing can be explored by describing the packing arrangements that can give rise to the observed interference effects in the diffraction pattern.

The crystal sampling on the third layer line indexes on a hexagonal lattice with an area three times that defined by the equatorial reflections. Systematic extinctions are observed on the third layer line for reflections $h-k=3q$, where q is any integer. The equatorial reflections indexed on this larger unit cell correspond to those missing on the third layer line. This sampling is like that for a rhombohedral lattice (*International Tables for X-ray Crystallography*, 1959), but the packing in the Pf1 fibers cannot be regularly crystalline as the continuous transform is observed on other nonequatorial layer lines. Before explaining irregularities in the packing, the kinds of regular arrangements that can account for the lattice sampling on the equator and third layer line need to be considered.

Smallest Regular Lattice

A sufficient lattice for analyzing the interference effects, as described by Marvin et al. (1974a), has a unit cell containing three hexagonally arranged particles that are in different orientations. These orientations can differ only by rotation by an angle, φ, about the particle axis measured relative to some reference orientation and/or a translation, z, parallel to the axis measured in units of the repeat distance, c, relative to an arbitrary origin. In the equatorial projection, all the particles look the same, so the repeating unit appears to be one particle. This lattice is diagrammed in Figure 3a. The numbers 0, 1, and 2 refer to the orientations of the three particles. If we assume regularity in the packing, the difference in orientation between particles 0 and 1 can be taken to be the same as that between particles 1 and 2. Only if the displacements are in multiples of $2\pi/3$ for φ or $1/3$ for z (or both types occur) will the

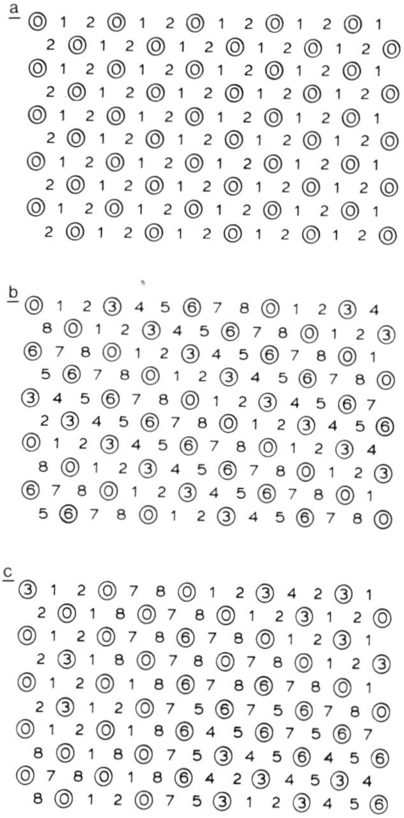

a
⓪ 1 2 ⓪ 1 2 ⓪ 1 2 ⓪ 1 2 ⓪ 1
2 ⓪ 1 2 ⓪ 1 2 ⓪ 1 2 ⓪ 1 2 ⓪
⓪ 1 2 ⓪ 1 2 ⓪ 1 2 ⓪ 1 2 ⓪ 1
2 ⓪ 1 2 ⓪ 1 2 ⓪ 1 2 ⓪ 1 2 ⓪
⓪ 1 2 ⓪ 1 2 ⓪ 1 2 ⓪ 1 2 ⓪ 1
2 ⓪ 1 2 ⓪ 1 2 ⓪ 1 2 ⓪ 1 2 ⓪
⓪ 1 2 ⓪ 1 2 ⓪ 1 2 ⓪ 1 2 ⓪ 1
2 ⓪ 1 2 ⓪ 1 2 ⓪ 1 2 ⓪ 1 2 ⓪
⓪ 1 2 ⓪ 1 2 ⓪ 1 2 ⓪ 1 2 ⓪ 1
2 ⓪ 1 2 ⓪ 1 2 ⓪ 1 2 ⓪ 1 2 ⓪

b
⓪ 1 2 ③ 4 5 ⑥ 7 8 ⓪ 1 2 ③ 4
8 ⓪ 1 2 ③ 4 5 ⑥ 7 8 ⓪ 1 2 ③
⑥ 7 8 ⓪ 1 2 ③ 4 5 ⑥ 7 8 ⓪ 1
5 ⑥ 7 8 ⓪ 1 2 ③ 4 5 ⑥ 7 8 ⓪
③ 4 5 ⑥ 7 8 ⓪ 1 2 ③ 4 5 ⑥ 7
2 ③ 4 5 ⑥ 7 8 ⓪ 1 2 ③ 4 5 ⑥
⓪ 1 2 ③ 4 5 ⑥ 7 8 ⓪ 1 2 ③ 4
8 ⓪ 1 2 ③ 4 5 ⑥ 7 8 ⓪ 1 2 ③
⑥ 7 8 ⓪ 1 2 ③ 4 5 ⑥ 7 8 ⓪ 1
5 ⑥ 7 8 ⓪ 1 2 ③ 4 5 ⑥ 7 8 ⓪

c
③ 1 2 ⓪ 7 8 ⓪ 1 2 ③ 4 2 ③ 1
2 ⓪ 1 8 ⓪ 7 8 ⓪ 1 2 ③ 1 2 ⓪
⓪ 1 2 ⓪ 7 8 ⑥ 7 8 ⓪ 1 2 ③ 1
2 ③ 1 8 ⓪ 7 8 ⓪ 7 8 ⓪ 1 2 ③
⓪ 1 2 ⓪ 1 8 ⑥ 7 8 ⑥ 7 8 ⓪ 1
2 ③ 1 2 ⓪ 7 5 ⑥ 7 5 ⑥ 7 8 ⓪
⓪ 1 2 ⓪ 1 8 ⑥ 4 5 ⑥ 7 5 ⑥ 7
8 ⓪ 1 8 ⓪ 7 5 ③ 4 5 ⑥ 4 5 ⑥
⓪ 7 8 ⓪ 1 8 ⑥ 4 2 ③ 4 5 ③ 4
8 ⓪ 1 2 ⓪ 7 5 ③ 1 2 ③ 4 5 ⑥

Figure 3 (*a*) A lattice of hexagonally arranged particles with three particles in the unit cell. The numbers 0, 1, and 2 refer to different orientations of the particles. All particles represented by a given number are arranged with the same rotation and translation relative to some reference. In the three directions of the hexagonal lattice (horizontal left to right and the two diagonals right to left) there is a regular sequence of displacements of +1 or −2 between nearest neighbors. There are twice as many nearest-neighbor displacements of +1 as there are of −2 in the lattice.

(*b*) A lattice of hexagonally arranged particles with nine particles in the unit cell. The numbers refer to different orientations of the particles. All nearest-neighbor displacements in the three directions of the lattice are either +1 or −2.

(*c*) A lattice of hexagonally arranged particles constructed from a random selection of the nearest-neighbor displacements of +1 and −2. The lattice was constructed as follows: The arrangement of particles along one horizontal line was determined by randomly choosing displacements of +1 (with probability 2/3) and −2 (with probability 1/3). A neighboring horizontal row was then added, constraining each triangle of three particles to have two displacements of +1 and one displacement of −2. A displacement of −2 occurring along one edge of a triangle determined the displacements along the other two sides. When a displacement of +1 occurred along one edge of the triangle, the edge on which a displacement of −2 occurred was chosen at random.

difference in orientation between particle 2 and 0 be the same as that for the other two. In general, though, the difference between 2 and 0 is minus twice that between 0 and 1 (or 1 and 2).

Fourier Transform of an Array of Helical Particles

The Fourier transform of a helical particle on layer-line l can be represented by the terms $G_{n,l}(R)e^{-in\psi}$, where R is the radius in reciprocal space and ψ is the angle and n the order of rotational symmetry (corresponding to an nth-order Bessel function). The intensity on the lth layer line can be written as a product of $|G_{n,l}(R)|^2$ and a packing factor (see Fuller et al. 1967) which is a function of the indices, h, k, l, and the difference in the orientations of the particles in the unit cell. If the value of the packing factor for a given reflection is zero, that reflection will not be observed.

The differences between the orientations of particles 0 and 1 can be expressed as $\Delta z_{0,1} = z_0 - z_1$ and $\Delta\varphi_{0,1} = \varphi_0 - \varphi_1$. With similar expressions for the displacements between the other pairs of particles in the unit cell, we can define $\Delta z = \Delta z_{0,1} = \Delta z_{1,2}$ such that $\Delta z_{2,0} = -2\Delta z$. Similarly, $\Delta\varphi = \Delta\varphi_{0,1} = \Delta\varphi_{1,2}$ and $\Delta\varphi_{2,0} = -2\Delta\varphi$. The necessary condition to produce the extinction of reflections $h-k = 3q$ on layer-line l for terms with n-fold rotational symmetry is $2\pi l\Delta z - n\Delta\varphi = 2\pi(p/3)$, where p is an integer not divisible by 3. Setting $l = 3$ gives the general condition for producing the extinctions observed on the third layer line of the Pf1 diffraction pattern.

Particle Arrangements Giving Observed Sampling

A solution to this relation suggested by Marvin et al. (1974a) is $\Delta z = 1/3$ and $\Delta\varphi = 2\pi/3$, which gives a lattice with trigonal symmetry in which all the displacement parameters are the same. This solution would give the observed extinctions only if n were not a multiple of 3, which would rule out the 27-subunits-in-five-turns symmetry for which $n = 6$ on the third layer line.

There are many other possible solutions for values Δz and $\Delta\varphi$ that will give the observed extinctions. Since there are a large number of subunits in the helix repeat, any translation can be approximated by a rotation and vice versa. For a helix with u subunits in the repeat, a translation by any multiple of c/u can be represented exactly by a rotation by some multiple of $2\pi/u$. We will consider possible translational packing relations for particles all in the same rotational orientation. For $\Delta\varphi = 0$, the observed extinctions will be produced for $\Delta z = p/9$, where p is any integer not divisible by 3. This condition applies for any helical symmetry of the particles regardless of the order of the rotational symmetry of the third layer line.

For $\Delta z = p/9$, a regular lattice with nine particles in the unit cell can be constructed that will still give the same extinctions on the third layer line as the three-particle lattice. This lattice (Fig. 3b) will give more closely spaced reflections on the non-third-order layer lines than the lattice in Figure 3a because the number of reflections is inversely proportional to the size of the unit cell. The diffraction on the third layer line would be unchanged because the projection of the lattice corresponding to these reflections is invariant when any particle is translated by a multiple of 1/3 in the z direction.

Since displacements by multiples of 1/3 for Δz do not affect the third-layer-line diffraction, it is possible to construct a lattice with a random combination of translational displacements of $p/9$ and $-2p/9$ that will give no interference on non-third-order layer lines but diffracts the same as the regular lattices on the third layer line. Figure 3c illustrates a random combination of displacements Δz and $-2\Delta z$ going from left to right along the horizontal lattice direction or right to left along the diagonal lattice directions. The lattice was constructed by choosing nearest-neighbor displacements of either Δz or $-2\Delta z$ randomly along the horizontal direction. Each triangle of three particles in the lattice was constrained so that along the three sides of the triangle there were two displacements of Δz and one of $-2\Delta z$. Thus, the lattice is not entirely random, since a displacement of $-2\Delta z$ along one side of a triangle determines the displacement on the other two sides. In each of the lattices in Figure 3 there are twice as many nearest-neighbor displacements of Δz as there are of $-2\Delta z$, but the relative arrangements of these displacements are different. Lattices of the type shown in Figure 3c can account for the interference effects observed in the third-layer-line diffraction and for the usual absence of these effects on other layer lines. Furthermore, the same nearest-neighbor packing relations can lead to more ordered lattices, as shown in Figure 3 a and b, which could explain the crystalline reflections seen sometimes on other layer lines.

Symmetry of Contact Relations

From considerations of symmetry, a particle with 27 subunits in the repeat would be more likely to pack in lattices such as those in Figure 3 than particles with 22, 23, or 28 subunits per repeat. A displacement of Δz between neighboring particles along each of the three directions of the hexagonal array involves the same contact relations only if the line group symmetry of the particle is of order 3. Since 27 is a multiple of 3, there will be only two kinds of nearest-neighbor contacts in these lattices, corresponding to the displacements Δz and $-2\Delta z$. With other particle symmetries there will be six types of nearest-neighbor contacts in these lattices, and the contacts in the three different directions of the lattice are different. Consequently, it is unlikely that an accurate hexagonal lattice with displacements Δz and $-2\Delta z$ would result from the packing of particles with 22, 23, or 28 subunits in the repeating unit.

For the trigonal lattice constructed from particles with 22 subunits in the repeat (or any other number not a multiple of 3) which are related by nearest-neighbor displacements of $\Delta z = 1/3$ and $\Delta\varphi = 2\pi/3$, there will be three symmetrically distinct contact relations between neighboring particles. In contrast to the lattices with $\Delta z = p/9$, there is no obvious way to randomize the trigonal lattice without altering the types of nearest-neighbor contacts. Random displacements by $\pm$ 1/3 would double the number of different kinds of contact. Thus, the absence of crystalline sampling on non-third-order layer lines cannot be accounted for reasonably by packing in a lattice with trigonal symmetry.

Packing Interactions

Analysis of the equatorial X-ray diffraction shows that the outer parts of a hydrated Pf1 virus particle extend to a diameter of about 62 Å. This is significantly larger than the packing diameter for virus particles in fibers at 0% relative humidity, where the observed lattice constants range from 52.5 to 54 Å. Protrusions on the surfaces of particles must interlock or be compressed for the particles to pack tightly. The interlocking of bumps on the surfaces of the particles is likely to lead to some regularity in the packing of virus particles in the fibers. Since any bumps on the particle surface will be arranged with the helical symmetry of the virus particle, the possible regular interlocking arrangements in a lattice are related to this helical symmetry.

A mechanical model was constructed of three cylindrical particles with bumps arranged on a helical net with 27 subunits in five turns. By rotating and translating these models relative to one another, orientations could be found in which the bumps on one particle could interlock with the bumps on another particle. To explore all possible interlocking combinations a computer model was constructed. As a measure of the interaction, a pseudo interaction energy was calculated that depends on the distance between the centers of the bumps on adjacent particles. The closer two bumps are to each other, the greater their interpenetration and the higher the interaction energy. Thus, relative orientations for which good interlocking will occur will have low calculated interaction energies. If the bumps are symmetrically shaped, as in our models, particles could interlock in both parallel and antiparallel orientations.

Figure 4 is a plot of one such calculation for helices with 27 subunits in five turns. In this plot a composite interaction energy $\mathbf{E}(\Delta z)$ is plotted as a function of the relative axial displacement of particles with the same orientation φ. The energy plotted here is a sum of two terms, $\mathbf{E}(\Delta z) = 2\,E(\Delta z) + E(-2\Delta z)$, one for an axial displacement of Δz and the other for a displacement of $-2\Delta z$. Since all the lattices in Figure 3 have twice as many pairs displaced by Δz as by $-2\Delta z$, the first term has been multiplied by two. There is a pronounced minimum in the calculated interaction energy at $\Delta z \simeq 2/9$, a somewhat less pronounced minimum at $\Delta z \simeq 4/9$, and a maximum near $\Delta z \simeq 1/9$. This calculation shows that helices with 27 subunits in five turns

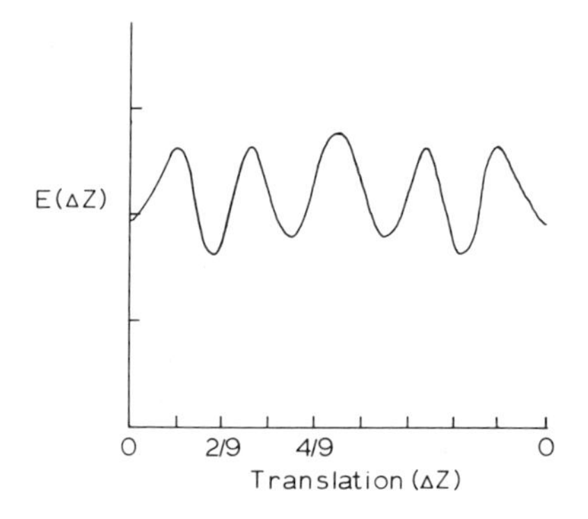

Figure 4 Interaction energy $\mathbf{E}(\Delta z)$ for a particle with 27 units in five turns plotted as a function of axial displacement. The energy plotted is the sum of two terms, one for an axial displacement of Δz and the other for a displacement of $-2\Delta z$. Since the lattices in Fig. 3 have twice as many pairs displaced by Δz as by $-2\Delta z$, the first term has been multiplied by two. Low values of the interaction energy indicate relative orientations of the particles for which interlocking of bumps on the particles will occur.

arranged in one of the lattices in Figure 3 will interlock most favorably for displacements of $\Delta z \simeq 2/9$. As we have shown above, arrangement of particles into a lattice of the type shown in Figure 3c with $\Delta z = p/9$ can account for the observed interference effects on the third layer line and the absence of these effects on non-third-order layer lines.

Similar calculations were done for helices with 22 subunits in five turns. In this case, favorable interlocking with displacements of $\Delta z \simeq 2/9$ and $\Delta z \simeq 4/9$ can be found for some lines of contact. However, interactions in directions at 60° and 120° to the favorable ones are generally unfavorable. This is due to the different contact relations in the three directions of the hexagonal lattice. The calculations indicate that it is unlikely that helices with 22 subunits in five turns would pack into lattices of the type in Figure 3 with nearest-neighbor displacements that would lead to the observed semicrystalline sampling.

CONCLUSIONS

Three independent lines of argument indicate that of the possible symmetries allowed by the X-ray patterns the repeating unit of Pf1 is composed of 27 subunits arranged on five turns of a helix. First, the length of the virus obtained from electron microscopy and the number of subunits per particle obtained from chemical data (Wiseman et al. 1976; Wiseman and Day 1977) indicate that there are 27 or 28 subunits in the axial repeat distance. Second, the observed packing volumes of the virus particle in fibers at different relative humidities are consistent with 27 subunits in the repeating unit; with 22 or 23 subunits, unaccounted-for vacant spaces are predicted within the particle, and 28 subunits would not fit in the available space. Third, the interlocking of particles with 27 subunits in five turns leads naturally to the formation of a semicrystalline lattice that accounts for the observed interference effects; particles with 22, 23, or 28 subunits in the repeating unit would be unlikely to pack into a lattice of this type.

In each of these analyses there is some uncertainty; the results of any one of these sets of data would not constitute a conclusive demonstration of the virus-particle symmetry. Since the three arguments are independent, the probability that these consistent conclusions are all wrong is of the order of the product of the probability that each individual one is invalid. The combined results from the observations on mass per unit length, packing density, and packing arrangement establish that the probability of Pf1 having 27 subunits in the axial repeat is very close to one. This conclusion provides a stringent constraint for building molecular models to represent the structure and packing interactions of the coat protein.

ACKNOWLEDGMENTS

We thank Dr. D. A. Marvin for helpful information and for providing several specimens for our X-ray studies, and Dr. L. Day for communicating unpublished results. This work was supported by National Institutes of Health grant CA15468-04 and National Science Foundation grant PCM77-16271 to D.L.D.C., and a National Institutes of Health postdoctoral fellowship to L.M.

REFERENCES

Asbeck, F., K. Beyreuther, H. Köhler, and G. Braunitzer. 1969. Die Konstitution des Hüllproteins des Phagen fd. *Hoppe-Seyler's Z. Physiol. Chem.* **350:** 1047.

Bernal, J. D. and I. Fankuchen. 1941. X-ray and crystallographic studies of plant virus preparations III. *J. Gen. Physiol.* **25:** 147.

Bragg, L. and M. F. Perutz. 1952. The structure of haemoglobin. *Proc. R. Soc. Lond.* **213A:** 425.

Bull, H. B. 1944. Adsorption of water vapor by proteins. *J. Am. Chem. Soc.* **66:** 1499.

Chothia, C. 1975. Structural invariants in protein folding. *Nature* **254:** 304.

Chothia, C. and J. Janin. 1975. Principles of protein-protein recognition. *Nature* **256:** 705.

Franklin, R. E. 1955. Structure of tobacco mosaic virus. *Nature* **175:** 379.

Franklin, R. E. and K. C. Holmes. 1958. Tobacco mosaic virus: Application of the method of isomorphous replacement to the determination of the helical parameters and radial density distribution. *Acta Cryst.* **11:** 213.

Fuller, W., F. Hutchinson, M. Spencer, and M. H. F. Wilkins. 1967. Molecular and crystal structures of double-helical RNA. *J. Mol. Biol.* **27:** 507.

Klug, A. and D. L. D. Caspar. 1960. The structure of small viruses. *Adv. Virus Res.* **7:** 225.

Langridge, R., H. R. Wilson, C. W. Hooper, M. W. Wilkins, and L. D. Hamilton. 1960. The molecular configuration of deoxyribonucleic acid. I. X-ray diffraction study of a crystalline form of the lithium salt. *J. Mol. Biol.* **2:** 19.

Lapointe, J. and D. A. Marvin. 1973. Filamentous bacterial viruses. VIII. Liquid crystals of fd. *Mol. Cryst. Liq. Cryst.* **19:** 269.

Low, B. W. and F. M. Richards. 1954. Measurement of the density, composition and related unit cell dimensions of some protein crystals. *J. Am. Chem. Soc.* **76:** 2511.

Marvin, D. A. 1966. X-ray diffraction and electron microscope studies of the small filamentous bacteriophage fd. *J. Mol. Biol.* **15:** 8.

Marvin, D. A. and B. Hohn. 1969. Filamentous bacterial viruses. *Bacteriol. Rev.* **33:** 172.

Marvin, D. A. and E. J. Wachtel. 1975. Structure and assembly of filamentous bacterial viruses. *Nature* **253:** 19.

———. 1976. Structure and assembly of filamentous bacterial viruses. *Philos. Trans. Soc. Lond. B* **276:** 81.

Marvin, D. A., R. L. Wiseman, and E. J. Wachtel. 1974a. Filamentous bacterial viruses. XI. Molecular architecture of the class II (Pf1, Xf) virion. *J. Mol. Biol.* **82:** 121.

Marvin, D. A., W. J. Pigram, R. L. Wiseman, E. J. Wachtel, and F. J. Marvin. 1974b. Filamentous bacterial viruses. XII. Molecular architecture of the class I (fd, If1, Ike) virion. *J. Mol. Biol.* **88:** 581.

Matthews, B. W. 1968. Solvent content of protein crystals. *J. Mol. Biol.* **83:** 491.

Milman, G. 1968. "X-ray diffraction and related studies of filamentous phage." Ph.D. thesis, Harvard University, Cambridge, Massachusetts.

Nakashima, Y. and W. Konigsberg. 1974. Reinvestigation of a region of the fd bacteriophage coat protein sequence. *J. Mol. Biol.* **88:** 598.

Nakashima, Y., R. L. Wiseman, W. Konigsberg, and D. A. Marvin. 1975. Primary structure and sidechain interactions of Pf1 filamentous bacterial coat protein. *Nature* **253:** 68.

Newman, J., H. L. Swinney, and L. A. Day. 1977. Hydrodynamic properties and structure of fd virus. *J. Mol. Biol.* **116:** 593.

Richards, F. M. 1974. The interpolation of protein structures: Total volume, group volume distributions and packing density. *J. Mol. Biol.* **82:** 1.

———. 1977. Area, volume, packing and protein structure. *Annu. Rev. Biophys. Bioeng.* **6:** 151.

Wachtel, E. J., F. S. Marvin, and D. A. Marvin. 1976. Structural transition in a filamentous protein. *J. Mol. Biol.* **107:** 379.

Watson, J. D. 1954. The structure of tobacco mosaic virus. I. X-ray evidence of a helical arrangement of subunits around the longitudinal axis. *Biochim. Biophys. Acta* **13:** 10.

Wiseman, R. L. and L. Day. 1977. Different packaging of DNA in the filamentous viruses Pf1 and Xf. *J. Mol. Biol.* **116:** 607.

Wiseman, R. L., S. A. Berkowitz, and L. A. Day. 1976. Different arrangements of protein subunits and single-stranded circular DNA in the filamentous bacterial virus fd and Pf1. *J. Mol. Biol.* **102:** 549.

Overview: A Comparison of the Isometric and Filamentous Phages

David T. Denhardt
Department of Biochemistry
McGill University
Montreal, Quebec, Canada H3G 1Y6

This overview is written partly as a summary and partly with the intent to make a comparison between the isometric and filamentous phages. I hope in this way to cover most of the salient points made in this book which might be of interest to the casual reader. Because these two phage types are structurally so patently dissimilar, it might seem unlikely that they are related; nevertheless, underlying similarities have led me to argue that there is a close evolutionary connection (Denhardt 1975). Inevitably, however, the argument is weakened by the fact that similarities do not necessarily mean that a close evolutionary relationship exists—they may reflect convergent evolution or they may be coincidental. It is hoped that as we learn more about these viruses the truth will become clear. I have excluded the mycoplasma viruses (Maniloff et al.)[1] from this discussion because not enough is known about them to allow a meaningful comparison with the coliphages.

The Virion

The articles by Day and Wiseman, by Marvin, and by Makowski and Caspar detail what we know about the structure of the filamentous phage virion—not only of the coliphages, but also of the *Pseudomonas* phage Pf1 and the *Xanthomonas* phage Xf. General agreement on the details of the structure of the f1-fd-M13 group of phages is lacking, and despite the similarity of their X-ray diffraction patterns, the Pf1 and Xf virions seem to be constructed on the basis of different protein-DNA interactions. In all of these viruses, circular, single-stranded (SS) DNA occupies a central core area and is surrounded by a sheath of helically arranged overlapping and interdigitating protein monomers, each in a largely alpha-helical conformation. The precise relations between the major structural protein (gene-VIII protein, or B protein) and the circular SS DNA are not yet understood. There are different ways in which the SS DNA molecules interact with the virus proteins, and these interactions are affected by the environment of the virion. The relation between the number of nucleotides and the number of protein monomers is not usually integral. In both the Pf1 virion and the

[1]Unless specified otherwise, the citations are to contributions in this book.

gene-V protein-DNA complex there is evidence for stacking between tyrosine residues and the nucleotide bases. The involvement of other proteins in the assembly of the nucleoprotein complexes in vivo is suggested by the fact that structures formed by reconstitution in vitro are often different from the naturally occurring structures. This research is currently revealing, in a dramatic manner, the manifold ways in which proteins can interact with a DNA molecule.

The structure of the isometric phages is outlined by Hayashi. From the stoichiometry of the different proteins in the phage particle and their relative affinity for the particle, the following picture emerges. Sixty molecules of the gene-*F* protein form a shell about the DNA. Interacting with each of the 12 faces of the icosahedron thus formed are five molecules of the gene-*G* protein and one molecule of the gene-*H* protein: these form 12 spikes and seem to constitute the adsorption organelles. The association of multiple copies of the gene-*J* protein, and perhaps one copy of the A* protein, with this structure is less well defined.

During morphogenesis of both types of phage particles there is a transition from an intracellular, immature form to an extracellular, mature form. Webster and Cashman describe how this occurs in the case of the filamentous phages where the gene-V protein-SS DNA complex interacts with a site at the cell surface and the DNA is extruded through the cell membrane; during this process the DNA forms a stable complex with gene-VIII-protein molecules. It is likely that one or more phage-encoded protein is required both to generate a proper gene-V protein-DNA complex and to create the appropriate membrane site. About five molecules of the A protein (gene-III product) become associated with the virus at one end.

Morphogenesis of the isometric phages is tightly coupled to synthesis of phage SS DNA. Hayashi discusses the requirement for the functions of genes *B, D, F,* and *G*. The *D, F,* and *G* proteins seem to be positive effectors of SS DNA synthesis. This is in contrast to the case with the filamentous phages where, in addition to the gene-II protein, only the gene-V protein is needed for SS DNA synthesis. The immature form of the isometric phage virion is a 132–140S structure which has 50–100 molecules of gene-*D* protein associated with it. It is possible that the function of the gene-*D* protein is to serve as a scaffold for organizing the virion proteins (particularly, F, G, and H) about the DNA. Sumida-Yasumoto et al. report the synthesis of infectious φX174 virions in an in vitro system. Lysis of the cell usually results in the loss of the gene-*D* protein and the conversion of the 140S immature particle to the 114S mature particle.

The two phage types exploit different routes of entry into the cell. The filamentous phage are brought to the surface by the retraction of the pilus, to which they are adsorbed at the tip (for further details see Denhardt 1975; Ray 1977). This involves the breakdown of pilus structure by an ATP-dependent process. Once in contact with the cell surface, a similar fate befalls the virion. Starting at the end adsorbed to the pilus (where the gene-III, or A, protein is located), the gene-VIII-protein molecules are stripped from the DNA and the DNA is intruded into the cell. One could speculate that the structures of the virus filament and the pilus filament are analogous, and therefore that the cellular mechanism used to break down the pilus would function on the virion also. One could also expect that the filamentous phages in general require "retractable" piluslike structures for entry into the cell and that each type of pilus would have a particular class of phage that would adsorb to it.

Adsorption of the isometric phages is quite different. Incardona presents some aspects of the adsorption and eclipse phenomena and a partial characterization of the lipopolysaccharide-virion interaction. Both the *F* and the *G* protein appear to be involved in recognition of the adsorption site and, together with the *H* protein, are involved in the eclipse reaction. The φX174 adsorption locus (*phx* [Bachmann et al. 1976]) is located near the xylose locus at 70 minutes on the *Escherichia coli* map.

Genome Function

Both types of viruses have a circular SS DNA molecule as their chromosome. The isometric and filamentous phage DNAs are some 5400 and 6400 nucleotides in length, respectively (Schaller et al; Godson et al.; Appendices I and II). There is a strong constraint on the size of the DNA that can be packaged by the isometric phages, and this may have favored the development of overlapping genes (Godson et al.). Godson presents a review of the various isometric phages. St-1, having some 6050 bases, is about 10% larger than φX. Although G4 and φX have markedly different contents of guanine plus thymine in their viral strands, they are about the same size and code for a very similar set of proteins. What kind of evolutionary pressure led to this difference? The S13 sequence determined by Spencer et al. suggests that S13 is very similar to φX.

Because DNA molecules of practically any length can be packaged by the morphogenetic process used by the filamentous phages, there is little constraint on the size of the genome. For this reason, filamentous phage are good cloning vectors. The insertion of DNA into various sites on filamentous phage DNA is described below. Wheeler and Benzinger summarize what we know about shorter-than-unit-length miniphage, which possess deletions of the genome, and phage which possess a linear SS DNA molecule that is longer than unit length.

Figure 1 presents a comparision of the genomes of φX and fd which are drawn to scale and oriented so that the origins of replication of the viral strands are in approximately the same position. The reason for taking the origin of viral-strand synthesis, which in φX equates with the origin of RF DNA replication, as the point of reference is that this part of the DNA sequence is likely to resist evolutionary drift, since it probably interacts with one or more bacterial proteins in addition to the phage-encoded gene-*A* or gene-II protein. Thus, for the purpose of asking whether there are similarities between these two phage types, it seems logical to take this site as a benchmark; the fact that the sequence in this region is identical in both φX and G4 supports this argument (Godson et al.). The manifold requirements of the different phages for various host-specified functions are summarized by Dumas and by Taketo and Kodaira.

Eight genes have been identified in the filamentous (I, II, III, IV, V, VI, VII, and VIII) and isometric (*A, B, C, D, E, F, G,* and *H*) phages by means of conditionally lethal mutations. Proteins A* and X have been identified as the C-terminal segments of genes *A* and II, respectively, and are translated as polypeptides in their own right in the same phase as the larger protein. φX gene *J* codes for the J protein that is found in the virion. Two additional proteins, IX (Schaller et al.) and K (Godson et al.; Tessman and Tessman), have been identified, but nothing is known about their function, if any.

Restriction enzyme mapping of the isometric phage genome is described

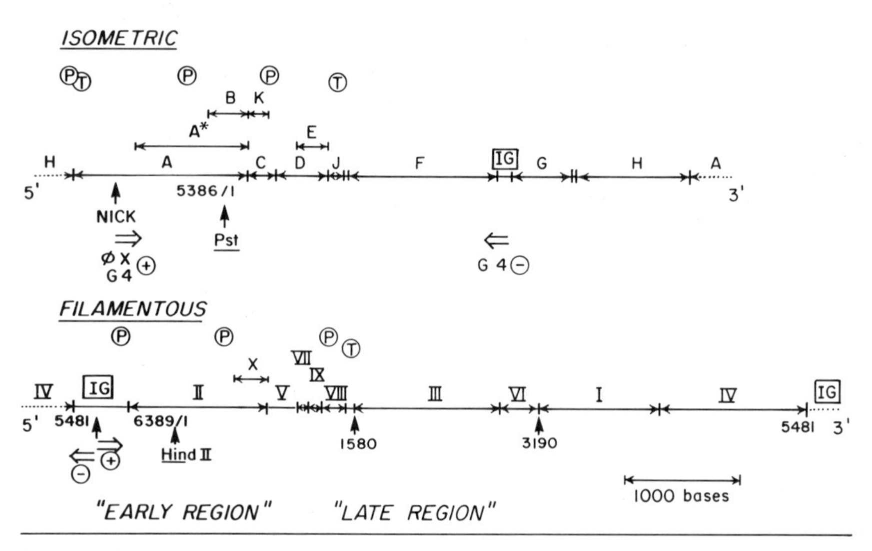

Figure 1 Comparison of the genomes of the isometric and filamentous phages. The genomes of φXand fd are depicted in a linear fashion in a 5′→3′ left-to-right direction just as the genes are transcribed. The positions of the genes and their approximate relative sizes are indicated on the same scale. P and T indicate the positions of strong promoters and terminators. IG indicates the largest intergenic space in each genome. The location of the single cleavage site of the restriction endonuclease that is used as a reference for numbering the bases is indicated below the genome. See text for further details.

by Weisbeek and van Arkel. Mutations are found more readily in the proximal and distal portions of gene *A* than in the central region, where the origin of viral-strand replication is located. The detailed mapping of various mutations onto restriction enzyme fragments has provided evidence that gene *B* overlaps with the distal portion of gene *A*.

The concept of what a gene is had to be revised as the result of the work of Barrell et al. (1976) and Weisbeek et al. (1977). We now realize that more than one gene can be contained within one stretch of nucleotides; the protein products, which define the individual genes, are coded for in different reading frames. Tessman and Tessman discuss this aspect with respect to the isometric phages and they describe the variety of protein products and their functions in depth. The filamentous phage genome appears to have one overlap of 20 base pairs between genes I and IV (Schaller et al.).

The variety of mRNA molecules produced in φX-infected cells are discussed by Fujimura and Hayashi. Promoters have been found preceding genes *A, B,* and *D*. The major promoter preceding gene *D* controls much of the transcription of the "late" genes *D, E, J, F, G,* and *H* which are needed for SS DNA synthesis, phage morphogenesis, and lysis. Terminators of transcription are found in the *A*/*H* and *J*/*F* untranslated regions (Godson et al.). The existence in vivo of other *rho*-dependent terminators is suggested by the variety of size classes of mRNA molecules produced.

Horiuchi et al. describe the structure of the filamentous phage genome and the functions of the various genes. It is remarkable how little we know about genes I, IV, VI, and VII. Gene-II protein is required for RF DNA replication and SS DNA synthesis. It has recently been purified and its endonucleolytic activity confirmed (Ray; Meyer and Geider). Its synthesis is suppressed by gene-V protein, apparently at the level of translation. The gene-III and gene-VIII proteins are part of the virion; both proteins are processed by proteolytic cleavage as they are being synthesized. The discrepancy between the molecular weight determined for the gene-III protein isolated from the virion and that predicted by the sequence may indicate the presence of carbohydrate on the virion protein. Gene-V protein is an SS DNA-binding protein necessary for the late stage of progeny SS DNA synthesis. Between genes II and IV is a particularly large intergenic space that contains the origins of viral- and complementary-strand replication. Details of the M13, f1, and fd restriction enzyme maps are given by Konings and Schoenmakers, by Horiuchi et al., and by Schaller et al., respectively. The complete sequence of the genome and the predicted amino acid sequences of the fd proteins are given by Schaller et al.

Transcription of the filamentous phage genome and the control of expression via a cascade model of transcription are described by Konings and Schoenmakers. Experiments performed in vitro suggest that almost every gene has a promoter immediately preceding it, and that some have internal promoters as well (Schaller et al.). In vivo, some of these promoters may be used more frequently than others with superhelical RF DNA as a template. This could explain the polar effect of mutations in genes III and V on genes VI (and I) and gene VII, respectively (Horiuchi et al.). Two of the proteins made in the largest amounts in vivo (V, VIII) are encoded in the greatest number of mRNA molecules. There is a strong *rho*-independent terminator, and at least one *rho*-dependent terminator, between genes VIII and III. Konings and Schoenmakers suggest that one set of overlapping mRNAs starts at G promoters and terminates at the central *rho*-independent terminator and that a second set of overlapping transcripts starts at the weaker A promoters in front of genes VI, I, and IV and is terminated by a *rho*-dependent process.

Other factors that affect the expression of the different genes are the stability of the different mRNA molecules and the affinity of the ribosome-binding sites for ribosomes. Small (approximately twofold) variations in the functional stability of different φX mRNA molecules in vivo are described by Hayashi. The event that inactivates the mRNA molecule is not understood, but, whatever it is, it does not result in the immediate destruction of the entire RNA molecule since the physical presence of the mRNA can be detected for some minutes after its functional inactivation.

DNA Replication

Three stages can be distinguished in the replication of these small SS DNA phages: (I) synthesis of a complementary strand to form parental replicative-form DNA (RF), (II) RF DNA replication, and (III) synthesis of progeny viral-strand DNA. Staudenbauer et al., Ray, and Dressler et al. have suggested that stage II is a combination of stages I and III, and that the transition to stage III involves simply the suppression of stage-I synthesis.

In the case of the filamentous phages, synthesis of the complementary strand (stage 1) on viral circular SS DNA requires an RNA primer (synthesized by RNA polymerase on DNA complexed with *Eco* DNA-binding protein) uniquely located in the intergenic region (see Fig. 1). There is also evidence that synthesis of the viral strand during progeny SS DNA synthesis (stage III) begins in the same region, at a nick inserted by the gene-II protein, and that the replicating intermediate can be isolated as a sigma structure containing a longer-than-unit-length viral strand (Ray; Staudenbauer et al.; Meyer and Geider). Ray, with some reservations, describes the following scenario for RF DNA replication (stage II): The gene-II protein nicks parental superhelical RF DNA at the origin of replication and viral-strand synthesis is initiated using the parental strand as a primer. Synthesis of the viral strand is continuous and counterclockwise ($5' \rightarrow 3'$) around the genome for one round of replication. This results in displacement of the resident viral strand from the RF DNA. When displacement uncovers the origin of complementary-strand synthesis, the synthesis of a new complementary strand on the displaced viral strand occurs.

Are the two processes (viral- and complementary-strand synthesis) that occur during RF DNA replication the same as those that occur during stages I and III? The following facts are relevant. According to Ray, complementary-strand synthesis during parental RF DNA formation is resistant to nalidixic acid, whereas it is sensitive during RF DNA replication. Replicating intermediates with a theta structure rather than a sigma structure, which is characteristic of stage III and is thus the expected structure, can be detected during RF DNA replication (Allison et al. 1977). Staudenbauer et al. describe requirements for the *dnaB* and *dnaG* functions which are not observed during stages I and III. It has been suggested (C. Snyder and S. Mitra, pers. comm.) that the data from which it was deduced that an origin of RF DNA replication exists in the intergenic region can also be interpreted in favor of an origin near the *Hpa*II H fragment in the latter third of gene III (see Fig. 1 and G4 complementary-strand origin).

The isometric phages present much the same ambiguous picture. Like the filamentous phages, G4 appears to have only one origin for complementary-strand synthesis during the first stage of parental RF DNA formation. Dressler et al. and Godson et al. have reviewed the evidence that in vivo this origin is in the *F/G* intergenic space. In vitro, also, complementary-strand synthesis is initiated there using a primer synthesized by the *dnaG* protein. Initiation does not occur in other regions of the SS DNA. During viral-strand synthesis (stage III) DNA replication begins at the gene-*A* nick and proceeds via an elongated viral strand. Sigma, or rolling-circle, structures can be isolated at this stage from both G4- and φX-infected cells. During G4 RF DNA replication, however, sigma structures are not observed. In their place, Godson (1977) has reported that D-loop structures are found, which suggests an initiation mechanism independent of the gene-*A* nick.

The replication of φX DNA is more complex than that of G4 or St-1 DNA, and φX complementary-strand synthesis requires several additional proteins. As described by McMacken et al. and by Wickner, these include *dnaB, dnaC(D)*, and other less well characterized proteins. The effect of these proteins is to allow the *dnaG* protein to initiate synthesis at many locations on the viral-strand DNA; in the absence of these proteins, the *dnaG* protein does not perform its assumed priming function. Synthesis of the complementary strand can begin at many sites on the viral-strand DNA. Whether a ribonucleotide is obligatory, especially in vivo, has not been established. Wickner's evidence suggests that the *dnaG* protein can initiate synthesis of short DNA chains in vitro with dADP.

Dressler et al. review the evidence that synthesis of the viral strand in vivo may initiate at the gene-*A* nick and then proceed continuously around the genome according to the rolling-circle model. Eisenberg et al. report that replication of RF DNA can be initiated in vitro at the gene-*A* nick; the *E. coli rep* protein unwinds the duplex in an ATP-dependent reaction, and DNA polymerase III holoenzyme then synthesizes a new viral strand by using the 3′ OH of the viral strand as a primer. In this system, the gene-*A* protein-*rep* protein complex, covalently bound to the 5′ terminus of the viral strand, moves around the circular complementary-strand template, freeing it from its association with the viral strand. The intermediate in this process is a "looped" rolling circle. When the complex reaches the origin, the viral strand is cleaved at the junction of the new and old DNA and the 5′ and 3′ ends of the original viral strand are rejoined. Razin discusses the data on the single 5-methyl cytosine in φX DNA and speculates on its possible involvement in DNA replication.

Baas and Jansz review in detail what is known about φX replication in vivo. Several virus-encoded proteins are required for stage-III synthesis of viral-strand DNA but not for stage-II synthesis of viral-strand DNA. The way in which the gene-*A* protein participates in progeny SS DNA synthesis is different from the way in which it participates in RF DNA replication because mutations in gene *A* can affect these two processes differently (Tessman and Tessman); also, the involvement of *rep* protein in these two stages is different. Takahashi et al. summarize what we know about the *rep* function. There is evidence that the synthesis of viral strands in stage III need not be continuous, as is demanded by the rolling-circle hypothesis. That segments of viral-strand DNA shorter than unit-length can be observed in M13 and G4 during RF DNA replication has been reported by Allison et al. (1977) and Godson (1977), respectively. Sumida-Yasumoto et al. report the observation of short pieces of viral-strand DNA during stage III of φX DNA replication in vitro, and Staudenbauer et al. discuss the presence of short pieces of viral strand DNA during M13 viral-strand synthesis. Hours et al. report that, in vivo, these short pieces come from all regions of the φX genome.

What these conflicting observations are apparently telling us is that *E. coli* has more than one way of replicating the DNA of these small DNA phages and that which mode is observed depends on subtleties that we do not yet appreciate. There are different mechanisms for initiating synthesis of DNA molecules, and, at least in some cases, when one pathway is blocked another may be used. Both the isometric and the filamentous phages are dependent on the *rep* function of *E. coli* for unwinding duplex RF DNA. Most of the actual DNA synthesis (elongation of a primer) is accomplished by DNA polymerase III holoenzyme (the product of the *dnaE* gene together with the necessary elongation cofactors), especially when double-stranded DNA is the template. For synthesis of complementary strands, it seems likely that DNA polymerase I or II can sometimes serve as the elongating polymerase. Dumas reviews the host-cell gene products required by the various phages for the replication of their DNAs.

The circularity of the replicative forms of these phages has contributed to our realization of the dynamic nature of the DNA molecule. Enzymes such as DNA gyrase, DNA helicases, and *rep* protein actively unwind the DNA duplex by means of the energy of hydrolysis of ATP. Sumida-Yasumoto et al. and Geider et al. review the properties of some of these proteins. The end result of gyrase action on closed, circular, duplex DNA is an underwound molecule that has acquired negative superhelical turns so that the DNA double helix can retain its approximate B conformation. The presence of the superhelical turns facilitates the action of gene-*A* protein and favors transcription and recombination. The nicking-closing enzyme ω may counteract the effect of gyrase and decrease the superhelix density. In this way, the state of the DNA and its susceptibility to various metabolic processes is tied to the energy level of the cell.

Manipulations of the Viral Genomes

Most recombination in both the filamentous and isometric phages appears to proceed via the same mechanism, which we are well on our way to understanding. Zinder describes it as a one parent–one recombinant process, as only one parent and one recombinant are recovered as products of a single recombination event. The fact that only one mating event, or round of recombination, is all that normally occurs in a cell producing recombinant small phage is one of the reasons why research on this system has been fruitful. Recombination between two parental RF molecules can occur efficiently, and the pathway includes an asymmetric single-strand exchange, usually with the loss of the donor parent.

Warner and Tessman describe some of the intermediates. An early step is the insertion of a nick in the donor RF molecule and the transfer of a portion of a strand to the recipient RF molecule, which is likely to be superhelical as this promotes uptake of a third strand. A figure-eight structure (Holliday intermediate) is formed in which branch migration can occur and can lead to the formation of a potentially quite large heteroduplex region. Eventually, the figure-eight molecule must be resolved into two monomers; one of the monomers is usually lost, perhaps because it is linear. Mismatches in the heteroduplex are subject to a repair process dependent on the *uvrD* gene.

The primary mechanism of recombination is dependent on the *recA*-gene product, whose function is unclear. The recombination frequency is approximately 10^{-3} per genome, or 2×10^{-7} per nucleotide, and fairly constant throughout the genome. It can be stimulated by UV light, presumably because *recA* function is enhanced. A "hot spot" of recombination can be observed at times in the region of gene *A*. In the absence of the primary mechanism of recombination (*recA* deficiency), a secondary mechanism can be observed. The secondary mechanism depends on gene-*A* function and cannot be stimulated by UV light.

Repair phenomena in the SS DNA phages are reviewed by Poddar et al. Intracellular RF DNA appears susceptible to the repair mechanisms operative in the host cells. Irradiation of host cells with UV light enhances the plating efficiency of UV-irradiated phages (Weigle-reactivation, induction of "SOS" functions). The single-stranded nature of φX DNA has made it especially useful for studying error-prone repair as it is not a substrate for repair processes acting on duplex DNA (Caillet-Fauquet et al. 1977). Whether the methylated bases in the filamentous (Horiuchi et al.) or isometric (Razin) phages play a role in identifying the parental strand during mismatch repair is unknown.

Having acquired an increased appreciation of the frequency with which nonhomologous recombination occurs, it became reasonable to look for the incorporation of foreign DNA into the genome of the filamentous phages. The filamentous phages were particularly attractive because cells infected with filamentous phage continue to replicate and carry the phage RF DNA intracellularly while at the same time releasing phage particles; in addition, the packaging process allows DNA molecules of various sizes to be packaged. Several groups have looked into this: Ray and Kook describe an M13 derivative, M13 Tn*3-15*, which carries the Tn*3* transposon and is resistant to ampicillin. Herrmann et al. discuss the insertion into fd of the Tn*5* transposon which codes for kanamycin resistance; they have also isolated fd derivatives that are able to transduce resistance to various antibiotics. Vovis and Ohsumi describe f1 phage capable of transducing tetracycline resistance. Nomura et al. describe fd phage capable of transducing kanamycin resistance. Messing and Gronenborn have inserted the *lac* promoter-operator region into a nonessential region of M13; creation of an *Eco*RI site in the DNA controlled by the operator-promoter produced M13mp2, which has many of the attributes desirable for a cloning vehicle. In the nondefective phages, the transposon carrying the drug-resistance marker is integrated into one of many potential sites in the intergenic space; helper-dependent phages have been isolated that have the integrated DNA at other locations in the genome.

Attempts to clone segments of viral DNA into plasmids are described by Kaplan et al. and by Humayun and Chambers. The former group present preliminary evidence that M13 gene VI has been successfully incorporated into the plasmid pBR317. The latter group has inserted a segment of φX RF DNA carrying gene *G* into the plasmid pMB9; some of the plasmids carrying gene *G* acted as gene-specific suppressors and allowed φX gene-*G* mutants to replicate under otherwise nonpermissive conditions. Chambers plans to use these gene-specific suppressor strains in a study of different kinds of premutational lesions in DNA.

REFERENCES

Allison, D. P., A. T. Ganesan, A. C. Olson, C. Synder, and S. Mitra. 1977. Electron microscope studies of bacteriophage M13 DNA replication. *J. Virol.* **24:** 673.

Bachmann, B. J., K. B. Low, and A. L. Taylor. 1976. Recalibrated linkage map of *Escherichia coli* K12. *Bacteriol. Rev.* **40:** 116.

Barrell, B. G., G. M. Air, and C. A. Hutchison III. 1976. Overlapping genes in bacteriophage φX174. *Nature* **264:** 34.

Caillet-Fauquet, P., M. Defais, and M. Radman. 1977. Molecular mechanisms of induced mutagenesis. Replication in vivo of bacteriophage φX174 single-stranded, ultraviolet-light-irradiated DNA in intact and irradiated host cells. *J. Mol. Biol.* **117:** 95.

Denhardt, D. T. 1975. The single-stranded DNA phages. *CRC Crit. Rev. Microbiol.* **4:** 161.

Godson, N. 1977. G4 DNA replication. III. Synthesis of replicative form. *J. Mol. Biol.* **117:** 353.

Ray, D. S. 1977. Replication of filamentous bacteriophages. In *Comprehensive virology* (ed. H. Fraenkel-Conrat and R. R. Wagner), vol. 7, p. 105. Plenum Press, New York.

Weisbeek, P. J., W. E. Borrias, S. A. Langeveld, P. D. Baas, and G. A. van Arkel. 1977. Bacteriophage φX174: Gene A overlaps gene B. *Proc. Natl. Acad. Sci.* **74:** 2504.

Appendices

APPENDIX I
The Nucleotide Sequence of the DNA of φX174 *cs*70 and the Amino Acid Sequence of the Proteins for Which It Codes

F. Sanger, G. M. Air, B. G. Barrell, N. L. Brown, A. R. Coulson, J. C. Fiddes, C. A. Hutchison III, P. M. Slocombe, M. Smith, J. Drouin, T. Friedmann, and A. J. H. Smith
MRC Laboratory of Molecular Biology
Cambridge CB2 2QH, England.

A provisional sequence was determined by the plus and minus method (Sanger et al. 1977).[1] It has now been completed using the termination method of Sanger et al. (1977).[2] About 30 alterations have been made and therefore the numbering is somewhat different from that used in the provisional sequence.

The DNA sequence given is that of the circular viral (plus) strand. Numbering is from the unique *Pst* site, but the sequence is written starting at the first nucleotide after the termination of the gene-*H* protein. The letters in the left-hand margin indicate the proteins whose sequence is given on the corresponding line. Restriction enzyme recognition sites are indicated by underlining. The single-letter code for the restriction enzymes is as follows: A, *Alu*I; F, *Hin*fI; H, *Hha*I; M, *Mbo*II; P, *Pst*I; Q, *Hph*I; R, *Hin*dII; T, *Taq*I; Y, *Hap*II; Z, *Hae*III.

The position of the origin of viral-strand replication is based on the work of Langeveld et al. (1977).[3] Other references are given in our earlier paper (Sanger et al. 1977)[1] and elsewhere in this volume.

The *cs*70 mutation is cold sensitive and is believed to be a point mutation in the gene-*F* protein (C. A. Hutchison III, pers. comm.).

[1] *Nature* **265:** 687.
[2] *Proc. Natl. Acad. Sci.* **74:** 5463.
[3] *Nature* **271:** 417.

```
    C C G T C A G G A T T G A C A C C C T C C C A A T T G T A T G T T T T C A T G C C T C C A A A T C T T G G A G G C T T T
                  3927                          3937                          3947                          3957                          3967                          3977
                                                                                                  mRNA start

A        MET   VAL   ARG   SER   TYR   TYR   PRO   SER   GLU   CYS   HIS   ALA   ASP   TYR   PHE   ASP   PHE   GLU   ARG
    T T T A T G G T T C G T T C T T A T T A C C C T T C T G A A T G T C A C G C T G A T T A T T T T G A C T T T G A G C G T
                  3987                          3997                          4007                          4017                          4027                          4037
  mRNA end

A   ILE   GLU   ALA   LEU   LYS   PRO   ALA   ILE   GLU   ALA   CYS   GLY   ILE   SER   THR   LEU   SER   GLN   SER   PRO
    A T C G A G G C T C T T A A A C C T G C T A T T G A G G C T T G T G G C A T T T C T A C T C T T T C T C A A T C C C C A
      T1/6        4047                          4057                          4067                          4077                          4087                          4097

A   MET   LEU   GLY   PHE   HIS   LYS   GLN   MET   ASP   ASN   ARG   ILE   LYS   LEU   LEU   GLU   GLU   ILE   LEU   SER
    A T G C T T G G C T T C C A T A A G C A G A T G G A T A A C C G C A T C A A G C T C T T G G A A G A G A T T C T G T C T
                  4107                          4117                          4127                          4137                          4147                          4157
                                                                                          A7b/7a                M5/8        F5c/3

A   PHE   ARG   MET   GLN   GLY   VAL   GLU   PHE   ASP   ASN   GLY   ASP   MET   TYR   VAL   ASP   GLY   HIS   LYS   ALA
    T T T C G T A T G C A G G G C G T T G A G T T C G A T A A T G G T G A T A T G T A T G T T G A C G G C C A T A A G G C T
                  4167                          4177                          4187                          4197                          4207                          4217
                                                        T6/2              Q2/3a                     R4/3      Z2/6b

A   ALA   SER   ASP   VAL   ARG   ASP   GLU   PHE   VAL   SER   VAL   THR   GLU   LYS   LEU   MET   ASP   GLU   LEU   ALA
    G C T T C T G A C G T T C G T G A T G A G T T T G T A T C T G T T A C T G A G A A G T T A A T G G A T G A A T T G G C A
                  4227                          4237                          4247                          4257                          4267                          4277

A   GLN   CYS   TYR   ASN   VAL   LEU   PRO   GLN   LEU   ASP   ILE   ASN   ASN   THR   ILE   ASP   HIS   ARG   PRO   GLU
    C A A T G C T A C A A T G T G C T C C C C C A A C T T G A T A T T A A T A A C A C T A T A G A C C A C C G C C C C G A A
                  4287                          4297                          4307                          4317                          4327                          4337
                                      Origin of viral strand replication

A   GLY   ASP   GLU   LYS   TRP   PHE   LEU   GLU   ASN   GLU   LYS   THR   VAL   THR   GLN   PHE   CYS   ARG   LYS   LEU
    G G G G A C G A A A A A T G G T T T T T A G A G A A C G A G A A G A C G G T T A C G C A G T T T T G C C G C A A G C T G
                  4347                          4357                          4367                          4377                          4387                          4397
                                                                  M8/6                                                    A7a/4

A   ALA   ALA   GLU   ARG   PRO   LEU   LYS   ASP   ILE   ARG   ASP   GLU   TYR   ASN   TYR   PRO   LYS   LYS   LYS   GLY
    G C T G C T G A A C G C C C T C T T A A G G A T A T T C G C G A T G A G T A T A A T T A C C C C A A A A A G A A A G G T
                  4407                          4417                          4427                          4437                          4447                          4457

A   ILE   LYS   ASP   GLU   CYS   SER   ARG   LEU   LEU   GLU   ALA   SER   THR   MET   LYS   SER   ARG   ARG   GLY   PHE
    A T T A A G G A T G A G T G T T C A A G A T T G C T G G A G G C C T C C A C T A T G A A A T C G C G T A G A G G C T T T
                  4467                          4477                          4487                          4497                          4507                          4517
                                                                  Z6b/6a
```

```
A   ALA ILE GLN ARG LEU MET ASN ALA MET ARG GLN ALA HIS ALA ASP GLY TRP PHE ILE VAL
    G C T A T T C A G C G T T T G A T G A A T G C A A T G C G A C A G G C T C A T G C T G A T G G T T G G T T T A T C G T T
              4527                4537                4547                4557                4567                4577

A   PHE ASP THR LEU THR LEU ALA ASP ASP ARG LEU GLU ALA PHE TYR ASP ASN PRO ASN ALA
    T T T G A C A C T C T C A C G T T G G C T G A C G A C C G A T T A G A G G C G T T T T A T G A T A A T C C C A A T G C T
              4587                4597                4607                4617                4627                4637

A   LEU ARG ASP TYR PHE ARG ASP ILE GLY ARG MET VAL LEU ALA ALA GLU GLY ARG LYS ALA
    T T G C G T G A C T A T T T T C G T G A T A T T G G T C G T A T G G T T C T T G C T G C C G A G G G T C G C A A G G C T
              4647                4657                4667                4677                4687                4697

A   ASN ASP SER HIS ALA ASP CYS TYR GLN TYR PHE CYS VAL PRO GLU TYR GLY THR ALA ASN
    A A T G A T T C A C A C G C C G A C T G C T A T C A G T A T T T T T G T G T G C C T G A G T A T G G T A C A G C T A A T
          F3/5a 4707                4717                4727                4737                4747      A4/11   4757

A   GLY ARG LEU HIS PHE HIS ALA VAL HIS PHE MET ARG THR LEU PRO THR GLY SER VAL ASP
    G G C C G T C T T C A T T T C C A T G C G G T G C A C T T T A T G C G G A C A C T T C C T A C A G G T A G C G T T G A C
    Z6a/9    M6/1 4767                4777                4787                4797                4807             R3/8 4817

A   PRO ASN PHE GLY ARG ARG VAL ARG ASN ARG ARG GLN LEU ASN SER LEU GLN ASN THR TRP
    C C T A A T T T T G G T C G T C G G G T A C G C A A T C G C C G C C A G T T A A A T A G C T T G C A A A A T A C G T G G
              4827                4837                4847                4857       A11/10   4867                4877
                                                                                                                 Z9/10

A   PRO TYR GLY TYR SER MET PRO ILE ALA VAL ARG TYR THR GLN ASP ALA PHE SER ARG SER
    C C T T A T G G T T A C A G T A T G C C C A T C G C A G T T C G C T A C A C G C A G G A C G C T T T T T C A C G T T C T
              4887                4897  ↑             4907                4917                4927                4937
                                 mRNA start

A   GLY TRP LEU TRP PRO VAL ASP ALA LYS GLY GLU PRO LEU LYS ALA THR SER TYR MET ALA
    G G T T G G T T G T G G C C T G T T G A T G C T A A A G G T G A G C C G C T T A A A G C T A C C A G T T A T A T G G C T
              4947    Z10/3       4957                4967                4977     A10/12b    4987                4997
                                                       Q3a/1

A   VAL GLY PHE TYR VAL ALA LYS TYR VAL ASN LYS LYS SER ASP MET ASP LEU ALA ALA LYS
    G T T G G T T T C T A T G T G G C T A A A T A C G T T A A C A A A A A G T C A G A T A T G G A C C T T G C T G C T A A A
              5007                5017                5027                5037                5047                5057
                                              R8/5
```

```
A    GLY   LEU   GLY   ALA   LYS   GLU   TRP   ASN   ASN   SER   LEU   LYS   THR   LYS   LEU   SER   LEU   LEU   PRO   LYS
B                                    MET   GLU   GLN   LEU   THR   LYS   ASN   GLN   ALA   VAL   ALA   THR   SER   GLN   GLU
    G G T C T A G G A G C T A A A G A A T G G A A C A A C T C A C T A A A A A C C A A G C T G T C G C T A C T T C C C A A G
                  5067         5077         5087         5097         5107         5117
                  A12b/15b                               A15b/17

A    LYS   LEU   PHE   ARG   ILE   ARG   MET   SER   ARG   ASN   PHE   GLY   MET   LYS   MET   LEU   THR   MET   THR   ASN
B       ALA   VAL   GLN   ASN   GLN   ASN   GLU   PRO   GLN   LEU   ARG   ASP   GLU   ASN   ALA   HIS   ASN   ASP   LYS   SER
    A A G C T G T T C A G A A T C A G A A T G A G C C G C A A C T T C G G G A T G A A A A T G C T C A C A A T G A C A A A T
                  5127         5137         5147         5157         5167         5177
     A17/12a       F5a/6

A    LEU   SER   THR   GLU   CYS   LEU   ILE   GLN   LEU   THR   LYS   LEU   GLY   TYR   ASP   ALA   THR   PRO   PHE   ASN
B       VAL   HIS   GLY   VAL   LEU   ASN   PRO   THR   TYR   GLN   ALA   GLY   LEU   ARG   ARG   ASP   ALA   VAL   GLN   PRO
    C T G T C C A C G G A G T G C T T A A T C C A A C T T A C C A A G C T G G G T T A C G A C G C G A C G C C G T T C A A C
                  5187         5197         5207         5217         5227         5237
                                                 A12a/5

A    GLN   ILE   LEU   LYS   GLN   ASN   ALA   LYS   ARG   GLU   MET   ARG   LEU   ARG   LEU   GLY   LYS   VAL   THR   VAL
B       ASP   ILE   GLU   ALA   GLU   ARG   LYS   LYS   ARG   ASP   GLU   ILE   GLU   ALA   GLY   LYS   SER   TYR   CYS   SER
    C A G A T A T T G A A G C A G A A C G C A A A A A G A G A G A T G A G A T T G A G G G T G G G A A A A G T T A C T G T A
                  5247         5257         5267         5277         5287         5297

A    ALA   ASP   VAL   LEU   ALA   ALA   GLN   PRO   VAL   THR   THR   ASN   LEU   LEU   LYS   PHE   MET   ARG   ALA   SER
B       ARG   ARG   PHE   GLY   GLY   ALA   THR   CYS   ASP   ASP   LYS   SER   ALA   GLN   ILE   TYR   ALA   ARG   PHE   ASP
    G C C G A C G T T T T G G C G G C G C A A C C T G T G A C G A C A A A T C T G C T C A A A T T T A T G C G C G C T T C G
                  5307         5317         5327         5337         5347         5357
                            H1/15                                          H15/8b    T2/7

A    ILE   LYS   MET   ILE   GLY   VAL   SER   ASN   LEU   GLN   SER   PHE   ILE   ALA   SER   MET   THR   GLN   LYS   LEU
B       LYS   ASN   ASP   TRP   ARG   ILE   GLN   PRO   ALA   GLU   PHE   TYR   ARG   PHE   HIS   ASP   ALA   GLU   VAL   ASN
    A T A A A A A T G A T T G G C G T A T C C A A C C T G C A G A G T T T T A T C G C T T C C A T G A C G C A G A A G T T A
                  5367         5377           1            11           21           31
                                            P1/1                                      R5/7b

A    THR   LEU   SER   ASP   ILE   SER   ASP   GLU   SER   LYS   ASN   TYR   LEU   ASP   LYS   ALA   GLY   ILE   THR   THR
K                                    MET   SER   ARG   LYS   ILE   ILE   LEU   ILE   LYS   GLN   GLU   LEU   LEU   LEU
B       THR   PHE   GLY   TYR   PHE   ***
    A C A C T T T C G G A T A T T T C T G A T G A G T C G A A A A T T A T C T T G A T A A A G C A G G A A T T A C T A C T
                   41           51           61           71           81           91
                                     F6/9  T7/8
```

```
A     ALA   CYS   LEU   ARG   ILE   LYS   SER   LYS   TRP   THR   ALA   GLY   GLY   LYS   ***
K       LEU   VAL   TYR   GLU   LEU   ASN   ARG   SER   GLY   LEU   LEU   ALA   GLU   ASN   GLU   LYS   ILE   ARG   PRO   ILE
C                                                                             MET   ARG   LYS   PHE   ASP   LEU   SER
    G C T T G T T T A C G A A T T A A A T C G A A G T G G A C T G C T G G C G G A A A A T G A G A A A A T T C G A C C T A T
                101                 111                 121                 131                 141                 151
                                     T8/9                                                        T9/10

K     LEU   ALA   GLN   LEU   GLU   LYS   LEU   LEU   LEU   CYS   ASP   LEU   SER   PRO   SER   THR   ASN   ASP   SER   VAL
C       LEU   ARG   SER   SER   ARG   SER   SER   TYR   PHE   ALA   THR   PHE   ARG   HIS   GLN   LEU   THR   ILE   LEU   SER
    C C T T G C G C A G C T C G A G A A G C T C T T A C T T T G C G A C C T T T C G C C A T C A A C T A A C G A T T C T G T
                161                 171                 181                 191                 201                 211
        H8b/4    A5/18  T10/4       A18/6                                                          F9/13

K     LYS   ASN   ***
C       LYS   THR   ASP   ALA   LEU   ASP   GLU   GLU   LYS   TRP   LEU   ASN   MET   LEU   GLY   THR   PHE   VAL   LYS   ASP
    C A A A A A C T G A C G C G T T G G A T G A G G A G A A G T G G C T T A A T A T G C T T G G C A C G T T C G T C A A G G
                221                 231                 241                 251                 261                 271

C       TRP   PHE   ARG   TYR   GLU   SER   HIS   PHE   VAL   HIS   GLY   ARG   ASP   SER   LEU   VAL   ASP   ILE   LEU   LYS
    A C T G G T T T A G A T A T G A G T C A C A T T T T G T T C A T G G T A G A G A T T C T C T T G T T G A C A T T T T A A
                281                 291                 301                 311                 321                 331
                               F13/17                              F17/16a             R7b/6c

D                                                                                                               MET
C       GLU   ARG   GLY   LEU   LEU   SER   GLU   SER   ASP   ALA   VAL   GLN   PRO   LEU   ILE   GLY   LYS   LYS   SER   ***
    A A G A G C G T G G A T T A C T A T C T G A G T C C G A T G C T G T T C A A C C A C T A A T A G G T A A G A A A T C A T
                341                 351                 361                 371                 381                 391
                                         F16a/16b ↑
                                         mRNA start

D     SER   GLN   VAL   THR   GLU   GLN   SER   VAL   ARG   PHE   GLN   THR   ALA   LEU   ALA   SER   ILE   LYS   LEU   ILE
    G A G T C A A G T T A C T G A A C A A T C C G T A C G T T T C C A G A C C G C T T T G G C C T C T A T T A A G C T C A T
     F16b/1     401                 411                 421                 431                 441                 451
                                                                                 Z3/7                     A6/1

D     GLN   ALA   SER   ALA   VAL   LEU   ASP   LEU   THR   GLU   ASP   ASP   PHE   ASP   PHE   LEU   THR   SER   ASN   LYS
    T C A G G C T T C T G C C G T T T T G G A T T T A A C C G A A G A T G A T T T C G A T T T T C T G A C G A G T A A C A A
                461                 471                 481                 491                 501                 511
                                                             M1/7            T4/5

D     VAL   TRP   ILE   ALA   THR   ASP   ARG   SER   ARG   ALA   ARG   ARG   CYS   VAL   GLU   ALA   CYS   VAL   TYR   GLY
E                                                                                                            MET   VAL
    A G T T T G G A T T G C T A C T G A C C G C T C T C G T G C T C G T C G C T G C G T T G A G G C T T G C G T T T A T G G
                521                 531                 541                 551                 561                 571
```

```
D      THR   LEU   ASP   PHE   VAL   GLY   TYR   PRO   ARG   PHE   PRO   ALA   PRO   VAL   GLU   PHE   ILE   ALA   ALA   VAL
E        ARG   TRP   THR   LEU   TRP   ASP   THR   LEU   ALA   PHE   LEU   LEU   LEU   LEU   SER   LEU   LEU   LEU   PRO   SER
   T A C G C T G G A C T T T G T G G G A T A C C C T C G C T T T C C T G C T C C T G T T G A G T T T A T T G C T G C C G T
                   581                           591                           601                           611                           621                           631

D      ILE   ALA   TYR   TYR   VAL   HIS   PRO   VAL   ASN   ILE   GLN   THR   ALA   CYS   LEU   ILE   MET   GLU   GLY   ALA
E        LEU   LEU   ILE   MET   PHE   ILE   PRO   SER   THR   PHE   LYS   ARG   PRO   VAL   SER   SER   TRP   LYS   ALA   LEU
   C A T T G C T T A T T A T G T T C A T C C C G T C A A C A T T C A A A C G G C C T G T C T C A T C A T G G A A G G C G C
                   641                           651   R6c/7a          661                  671 Z7/5            681                 691 H4/13

D      GLU   PHE   THR   GLU   ASN   ILE   ILE   ASN   GLY   VAL   GLU   ARG   PRO   VAL   LYS   ALA   ALA   GLU   LEU   PHE
E        ASN   LEU   ARG   LYS   THR   LEU   LEU   MET   ALA   SER   SER   VAL   ARG   LEU   LYS   PRO   LEU   ASN   CYS   SER
   T G A A T T T A C G G A A A A C A T T A T T A A T G G C G T C G A G C G T C C G G T T A A A G C C G C T G A A T T G T T
                   701                           711                           721 T5/3            731 Y1/3            741                           751

D      ALA   PHE   THR   LEU   ARG   VAL   ARG   ALA   GLY   ASN   THR   ASP   VAL   LEU   THR   ASP   ALA   GLU   GLU   ASN
E        ARG   LEU   PRO   CYS   VAL   TYR   ALA   GLN   GLU   THR   LEU   THR   PHE   LEU   LEU   THR   GLN   LYS   LYS   THR
   C G C G T T T A C C T T G C G T G T A C G C G C A G G A A A C A C T G A C G T T C T T A C T G A C G C A G A A G A A A A
                   761                           771 H13/11                    781                           791                           801    M7/3        811

J                                                                                    MET   SER   LYS   GLY   LYS   LYS   ARG   SER
D      VAL   ARG   GLN   LYS   LEU   ARG   ALA   GLU   GLY   VAL   MET   ***
E        CYS   VAL   LYS   ASN   TYR   VAL   ARG   LYS   GLU   ***
   C G T G C G T C A A A A A T T A C G T G C G G A A G G A G T G A T G T A A T G T C T A A A G G T A A A A A A C G T T C T
                   821                           831                           841                           851                           861                           871

J    GLY   ALA   ARG   PRO   GLY   ARG   PRO   GLN   PRO   LEU   ARG   GLY   THR   LYS   GLY   LYS   ARG   LYS   GLY   ALA
   G G C G C T C G C C C T G G T C G T C C G C A G C C G T T G C G A G G T A C T A A A G G C A A G C G T A A A G G C G C T
     H11/14        881                           891                           901                           911                           921          H14/12  931

J    ARG   LEU   TRP   TYR   VAL   GLY   GLY   GLN   GLN   PHE   ***
   C G T C T T T G G T A T G T A G G T G G T C A A C A A T T T T A A T T G C A G G G G C T T C G G C C C C T T A C T T G A
                   941                           951 R7a/6b                    961                           971               981 Z5/8                    991
                                                                                                                                       Minor mRNA end ↑

F                            MET   SER   ASN   ILE   GLN   THR   GLY   ALA   GLU   ARG   MET   PRO   HIS   ASP   LEU   SER   HIS
   G G A T A A A T T A T G T C T A A T A T T C A A A C T G G C G C C G A G C G T A T G C C G C A T G A C C T T T C C C A T
                   1001                          1011                          1021 H12/10                   1031                          1041                          1051
```

```
F  LEU GLY PHE LEU ALA GLY GLN ILE GLY ARG LEU ILE THR ILE SER THR THR PRO VAL ILE
   CTT GGC TTC CTT GCT GGT CAG ATT GGT CGT CTT ATT ACC ATT TCA ACT ACT CCG GTT ATC
   1061 1071 1081 1091 1101 1111
   Y3/2

F  ALA GLY ASP SER PHE GLU MET ASP ALA VAL GLY ALA LEU ARG LEU SER PRO LEU ARG ARG
   GCT GGC GAC TCC TTC GAG ATG GAC GCC GTT GGC GCT CTC CGT CTT TCT CCA TTG CGT CGT
   1121 1131 1141 1151 1161 1171
   F1/14b T3/1 H10/7

F  GLY LEU ALA ILE ASP SER THR VAL ASP ILE PHE THR PHE TYR VAL PRO HIS ARG HIS VAL
   GGC CTT GCT ATT GAC TCT ACT GTA GAC ATT TTT ACT TTT TAT GTC CCT CAT CGT CAC GTT
   1181 1191 1201 1211 1221 1231
   Z8/4 F14b/2

F  TYR GLY GLU GLN TRP ILE LYS PHE MET LYS ASP GLY VAL ASN ALA THR PRO LEU PRO THR
   TAT GGT GAA CAG TGG ATT AAG TTC ATG AAG GAT GGT GTT AAT GCC ACT CCT CTC CCG ACT
   1241 1251 1261 1271 1281 1291
   Q1/3c

F  VAL ASN THR THR GLY TYR ILE ASP HIS ALA ALA PHE LEU GLY THR ILE ASN PRO ASP THR
   GTT AAC ACT ACT GGT TAT ATT GAC CAT GCC GCT TTT CTT GGC ACG ATT AAC CCT GAT ACC
   1301 1311 1321 1331 1341 1351
   R6b/1

F  ASN LYS ILE PRO LYS HIS LEU PHE GLN GLY TYR LEU ASN ILE TYR ASN ASN TYR PHE LYS
   AAT AAA ATC CCT AAG CAT TTG TTT CAG GGT TAT TTG AAT ATC TAT AAC AAC TAT TTT AAA
   1361 1371 1381 1391 1401 1411

F  ALA PRO TRP MET PRO ASP ARG THR GLU ALA ASN PRO ASN GLU LEU ASN GLN ASP ASP ALA
   GCG CCG TGG ATG CCT GAC CGT ACC GAG GCT AAC CCT AAT GAG CTT AAT CAA GAT GAT GCT
   1421 1431 1441 1451 1461 1471
   H7/5 A1/12c

F  ARG TYR GLY PHE ARG CYS CYS HIS LEU LYS ASN ILE TRP THR ALA PRO LEU PRO PRO GLU
   CGT TAT GGT TTC CGT TGC TGC CAT CTC AAA AAC ATT TGG ACT GCT CCG CTT CCT CCT GAG
   1481 1491 1501 1511 1521 1531

F  THR GLU LEU SER ARG GLN MET THR THR SER THR THR SER ILE ASP ILE MET GLY LEU GLN
   ACT GAG CTT TCT CGC CAA ATG ACG ACT TCT ACC ACA TCT ATT GAC ATT ATG GGT CTG CAA
   1541 1551 1561 1571 1581 1591
   A12c/13

F  ALA ALA TYR ALA ASN LEU HIS THR ASP GLN GLU ARG ASP TYR PHE MET GLN ARG TYR HIS
   GCT GCT TAT GCT AAT TTG CAT ACT GAC CAA GAA CGT GAT TAC TTC ATG CAG CGT TAC CAT
   1601 1611 1621 1631 1641 1651
   A13/2
```

```
F   ASP VAL ILE SER SER PHE GLY GLY LYS THR SER TYR ASP ALA ASP ASN ARG PRO LEU LEU
    G A T G T T A T T T C T T C A T T T G G A G G T A A A A C C T C T T A T G A C G C T G A C A A C C G T C C T T T A C T T
                      1661  M3/4       1671                1681                1691                1701                1711

F   VAL MET ARG SER ASN LEU TRP ALA SER GLY TYR ASP VAL ASP GLY THR ASP GLN THR SER
    G T C A T G C G C T C T A A T C T C T G G G C A T C T G G C T A T G A T G T T G A T G G A A C T G A C C A A A C G T C G
               H5/9a 1721              1731                1741                1751                1761                1771

F   LEU GLY GLN PHE SER GLY ARG VAL GLN GLN THR TYR LYS HIS SER VAL PRO ARG PHE PHE
    T T A G G C C A G T T T T C T G G T C G T G T T C A A C A G A C C T A T A A A C A T T C T G T G C C G C G T T T C T T T
           Z4/1      1781              1791                1801                1811                1821                1831

F   VAL PRO GLU HIS GLY THR MET PHE THR LEU ALA LEU VAL ARG PHE PRO PRO THR ALA THR
    G T T C C T G A G C A T G G C A C T A T G T T T A C T C T T G C G C T T G T T C G T T T C C G C C T A C T G C G A C T
                     1841              1851                1861  H9a/8a        1871                1881                1891

F   LYS GLU ILE GLN TYR LEU ASN ALA LYS GLY ALA LEU THR TYR THR ASP ILE ALA GLY ASP
    A A A G A G A T T C A G T A C C T T A A C G C T A A A G G T G C T T T G A C T T A T A C C G A T A T T G C T G G C G A C
                     1901              1911                1921                1931                1941                1951
                  F2/11

F   PRO VAL LEU TYR GLY ASN LEU PRO PRO ARG GLU ILE SER MET LYS ASP VAL PHE ARG SER
    C C T G T T T T G T A T G G C A A C T T G C C G C C G C G T G A A A T T T C T A T G A A G G A T G T T T T C C G T T C T
                     1961              1971                1981                1991                2001                2011

F   GLY ASP SER SER LYS LYS PHE LYS ILE ALA GLU GLY GLN TRP TYR ARG TYR ALA PRO SER
    G G T G A T T C G T C T A A G A A G T T T A A G A T T G C T G A G G G T C A G T G G T A T C G T T A T G C G C C T T C G
     Q3c/6  F11/7   2021              2031                2041                2051                2061  H8a/6       2071

F   TYR VAL SER PRO ALA TYR HIS LEU LEU GLU GLY PHE PRO PHE ILE GLN GLU PRO PRO SER
    T A T G T T T C T C C T G C T T A T C A C C T T C T T G A A G G C T T C C C A T T C A T T C A G G A A C C G C C T T C T
                     2081              2091                2101                2111                2121                2131
                                        Q6/5

F   GLY ASP LEU GLN GLU ARG VAL LEU ILE ARG HIS HIS ASP TYR ASP GLN CYS PHE GLN SER
    G G T G A T T T G C A A G A A C G C G T A C T T A T T C G C C A C C A T G A T T A T G A C C A G T G T T T C C A G T C C
     Q5/3b           2141              2151                2161                2171                2181                2191

F   VAL GLN LEU LEU GLN TRP ASN SER GLN VAL LYS PHE ASN VAL THR VAL TYR ARG ASN LEU
    G T T C A G T T G T T G C A G T G G A A T A G T C A G G T T A A A T T T A A T G T G A C C G T T T A T C G C A A T C T G
                     2201              2211                2221                2231                2241                2251
```

```
F    PRO   THR   THR   ARG   ASP   SER   ILE · MET   THR   SER   ***
     C C G A C C A C T C G C G A T T C A A T C A T G A C T T C G T G A T A A A A G A T T G A G T G T G A G G T T A T A A C G
                       2261     F7/5b      2271                 2281                2291                2301                2311

     C C G A A G C G G T A A A A A T T T T A A T T T T T G C C G C T G A G G G G T T G A C C A A G C G A A G C G C G G T A G
                       2321                2331                2341                2351 R1/9           2361     H6/3       2371

G                                                MET   PHE   GLN   THR   PHE   ILE   SER   ARG   HIS   ASN   SER   ASN   PHE
     G T T T T C T G C T T A G G A G T T T A A T C A T G T T T C A G A C T T T T A T T T C T C G C C A T A A T T C A A A C T
                       2381                2391                2401                2411                2421                2431

G       PHE   SER   ASP   LYS   LEU   VAL   LEU   THR   SER   VAL   THR   PRO   ALA   SER   SER   ALA   PRO   VAL   LEU   GLN
     T T T T T T C T G A T A A G C T G G T T C T C A C T T C T G T T A C T C C A G C T T C T T C G G C A C C T G T T T T A C
                       2441   A2/16    2451                2461                2471   M4/10        2481                2491
                                                                              A16/15a

G       THR   PRO   LYS   ALA   THR   SER   SER   THR   LEU   TYR   PHE   ASP   SER   LEU   THR   VAL   ASN   ALA   GLY   ASN
     A G A C A C C T A A A G C T A C A T C G T C A A C G T T A T A T T T T G A T A G T T T G A C G G T T A A T G C T G G T A
                       2501  A15a/3    2511   R9/10        2521                2531                2541                2551

G       GLY   GLY   PHE   LEU   HIS   CYS   ILE   GLN   MET   ASP   THR   SER   VAL   ASN   ALA   ALA   ASN   GLN   VAL   VAL
     A T G G T G G T T T T C T T C A T T G C A T T C A G A T G G A T A C A T C T G T C A A C G C C G C T A A T C A G G T T G
                       2561    M10/9       2571                2581                2591 R10/2          2601                2611

G       SER   VAL   GLY   ALA   ASP   ILE   ALA   PHE   ASP   ALA   ASP   PRO   LYS   PHE   PHE   ALA   CYS   LEU   VAL   ARG
     T T T C T G T T G G T G C T G A T A T T G C T T T T G A T G C C G A C C C T A A A T T T T T T G C C T G T T T G G T T C
                       2621                2631                2641                2651                2661                2671

G       PHE   GLU   SER   SER   SER   VAL   PRO   THR   THR   LEU   PRO   THR   ALA   TYR   ASP   VAL   TYR   PRO   LEU   ASN
     G C T T T G A G T C T T C T T C G G T T C C G A C T A C C C T C C C G A C T G C C T A T G A T G T T T A T C C T T T G A
                       2681    M9/2        2691                2701                2711                2721                2731
                    F5b/8

G       GLY   ARG   HIS   ASP   GLY   GLY   TYR   TYR   THR   VAL   LYS   ASP   CYS   VAL   THR   ILE   ASP   VAL   LEU   PRO
     A T G G T C G C C A T G A T G G T G G T T A T T A T A C C G T C A A G G A C T G T G T G A C T A T T G A C G T C C T T C
                       2741                2751                2761                2771                2781                2791

G       ARG   THR   PRO   GLY   ASN   ASN   VAL   TYR   VAL   GLY   PHE   MET   VAL   TRP   SER   ASN   PHE   THR   ALA   THR
     C C C G T A C G C C G G G C A A T A A C G T T T A T G T T G G T T T C A T G G T T T G G T C T A A C T T T A C C G C T A
                       2801 Y2/5       2811                2821                2831                2841                2851
```

```
G      LYS   CYS   ARG   GLY   LEU   VAL   SER   LEU   ASN   GLN   VAL   ILE   LYS   GLU   ILE   ILE   CYS   LEU   GLN   PRO
     C T A A A T G C C G C G G A T T G G T T T C G C T G A A T C A G G T T A T T A A A G A G A T T A T T T G T C T C C A G C
                    2861                          2871                          2881                          2891                          2901                          2911
                                                                              F8/4

H                                                  MET   PHE   GLY   ALA   ILE   ALA   GLY   GLY   ILE   ALA   SER   ALA   LEU   ALA
G      LEU   LYS   ***
     C A C T T A A G T G A G G T G A T T T A T G T T T G G T G C T A T T G C T G G C G G T A T T G C T T C T G C T C T T G C
                    2921             2931                          2941                          2951                          2961                          2971
                                 Q3b/4

H      GLY   GLY   ALA   MET   SER   LYS   LEU   PHE   GLY   GLY   GLY   GLN   LYS   ALA   ALA   SER   GLY   GLY   ILE   GLN
     T G G T G G C G C C A T G T C T A A A T T G T T T G G A G G C G G T C A A A A A G C C G C C T C C G G T G G C A T T C A
                    2981                          2991                          3001                          3011                          3021                          3031
               H3/2                                                                                                                  Y5/4

H      GLY   ASP   VAL   LEU   ALA   THR   ASP   ASN   ASN   THR   VAL   GLY   MET   GLY   ASP   ALA   GLY   ILE   LYS   SER
     A G G T G A T G T G C T T G C T A C C G A T A A C A A T A C T G T A G G C A T G G G T G A T G C T G G T A T T A A A T C
                    3041                          3051                          3061                          3071                          3081                          3091
          Q4/7                                                                                      Q7/2

H      ALA   ILE   GLN   GLY   SER   ASN   VAL   PRO   ASN   PRO   ASP   GLU   ALA   ALA   PRO   SER   PHE   VAL   SER   GLY
     T G C C A T T C A A G G C T C T A A T G T T C C T A A C C C T G A T G A G G C C G C C C C T A G T T T T G T T T C T G G
                    3101                          3111                          3121                          3131                          3141                          3151
                                                                                                  Z1/2

H      ALA   MET   ALA   LYS   ALA   GLY   LYS   GLY   LEU   LEU   GLU   GLY   THR   LEU   GLN   ALA   GLY   THR   SER   ALA
     T G C T A T G G C T A A A G C T G G T A A A G G A C T T C T T G A A G G T A C G T T G C A G G C T G G C A C T T C T G C
                    3161                          3171                          3181                          3191                          3201                          3211
                             A3/9

H      VAL   SER   ASP   LYS   LEU   LEU   ASP   LEU   VAL   GLY   LEU   GLY   GLY   LYS   SER   ALA   ALA   ASP   LYS   GLY
     C G T T T C T G A T A A G T T G C T T G A T T T G G T T G G A C T T G G T G G C A A G T C T G C C G C T G A T A A A G G
                    3221                          3231                          3241                          3251                          3261                          3271

H      LYS   ASP   THR   ARG   ASP   TYR   LEU   ALA   ALA   ALA   PHE   PRO   GLU   LEU   ASN   ALA   TRP   GLU   ARG   ALA
     A A A G G A T A C T C G T G A T T A T C T T G C T G C T G C A T T T C C T G A G C T T A A T G C T T G G G A G C G T G C
                    3281                          3291                          3301                          3311                          3321                          3331
                                                                                                      A9/12d

H      GLY   ALA   ASP   ALA   SER   SER   ALA   GLY   MET   VAL   ASP   ALA   GLY   PHE   GLU   ASN   GLN   LYS   GLU   LEU
     T G G T G C T G A T G C T T C C T C T G C T G G T A T G G T T G A C G C C G G A T T T G A G A A T C A A A A A G A G C T
                    3341                          3351                          3361                          3371                          3381                          3391
                                                            R2/6a          Y4/1                  F4/14a                A12d/7c
```

```
H    THR   LYS   MET   GLN   LEU   ASP   ASN   GLN   LYS   GLU   ILE   ALA   GLU   MET   GLN   ASN   GLU   THR   GLN   LYS
    T A C T A A A A T G C A A C T G G A C A A T C A G A A A G A G A T T G C C G A G A T G C A A A A T G A G A C T C A A A A
                 3401                          3411                          3421                          3431                          3441      F14a/12      3451

H    GLU   ILE   ALA   GLY   ILE   GLN   SER   ALA   THR   SER   ARG   GLN   ASN   THR   LYS   ASP   GLN   VAL   TYR   ALA
    A G A G A T T G C T G G C A T T C A G T C G G C G A C T T C A C G C C A G A A T A C G A A A G A C C A G G T A T A T G C
                 3461                          3471                          3481                          3491                          3501                          3511

H    GLN   ASN   GLU   MET   LEU   ALA   TYR   GLN   GLN   LYS   GLU   SER   THR   ALA   ARG   VAL   ALA   SER   ILE   MET
    A C A A A A T G A G A T G C T T G C T T A T C A A C A G A A G G A G T C T A C T G C T C G C G T T G C G T C T A T T A T
                 3521                          3531                          3541      F12/10      3551                          3561                          3571

H    GLU   ASN   THR   ASN   LEU   SER   LYS   GLN   GLN   GLN   VAL   SER   GLU   ILE   MET   ARG   GLN   MET   LEU   THR
    G G A A A A C A C C A A T C T T T C C A A G C A A C A G C A G G T T T C C G A G A T T A T G C G C C A A A T G C T T A C
                 3581                          3591                          3601                          3611                3621                          3631
                                                                                                                     H2/9b

H    GLN   ALA   GLN   THR   ALA   GLY   GLN   TYR   PHE   THR   ASN   ASP   GLN   ILE   LYS   GLU   MET   THR   ARG   LYS
    T C A A G C T C A A A C G G C T G G T C A G T A T T T T A C C A A T G A C C A A A T C A A A G A A A T G A C T C G C A A
          A7c/8  3641                          3651                          3661                          3671                          3681      F10/15      3691

H    VAL   SER   ALA   GLU   VAL   ASP   LEU   VAL   HIS   GLN   GLN   THR   GLN   ASN   GLN   ARG   TYR   GLY   SER   SER
    G G T T A G T G C T G A G G T T G A C T T A G T T C A T C A G C A A A C G C A G A A T C A G C G G T A T G G C T C T T C
                 3701                R6a/4     3711                          3721                          3731      F15/5c          3741                   3751 M2/5

H    HIS   ILE   GLY   ALA   THR   ALA   LYS   ASP   ILE   SER   ASN   VAL   VAL   THR   ASP   ALA   ALA   SER   GLY   VAL
    T C A T A T T G G C G C T A C T G C A A A G G A T A T T T C T A A T G T C G T C A C T G A T G C T G C T T C T G G T G T
                 3761 H9b/1                    3771                          3781                          3791                          3801                          3811

H    VAL   ASP   ILE   PHE   HIS   GLY   ILE   ASP   LYS   ALA   VAL   ALA   ASP   THR   TRP   ASN   ASN   PHE   TRP   LYS
    G G T T G A T A T T T T T C A T G G T A T T G A T A A A G C T G T T G C C G A T A C T T G G A A C A A T T T C T G G A A
                 3821                          3831                          3841 A8/14                    3851                          3861                          3871

H    ASP   GLY   LYS   ALA   ASP   GLY   ILE   GLY   SER   ASN   LEU   SER   ARG   LYS   ***
    A G A C G G T A A A G C T G A T G G T A T T G G C T C T A A T T T G T C T A G G A A A T A A
                 3881 A14/7b                   3891                          3901                          3911
```

APPENDIX II
Comparative DNA-sequence Analysis of the G4 and φX174 Genomes

G. N. Godson
Yale University
New Haven, Connecticut 06510

The φX174 nucleotide sequence used in this gene-by-gene comparison is the updated sequence presented in Appendix I. The G4 nucleotide sequence is part of the complete G4 sequence which will be published by G. N. Godson, B. G. Barrell, and J. C. Fiddes at a later date. Alignment of the φX and G4 sequences within the genes was only made possible by comparing the total nucleotide sequences; where deletions and insertions occur in one sequence relative to the other (i.e., in genes *A, J, G,* and *H*), the missing nucleotides are represented with dashes. The exact location of the deletion, however, is not certain, and the positions chosen are only a best guess from conservation of the surrounding nucleotides and amino acids.

The mismatched nucleotides are marked with a line and the similar amino acids with hatching. The numerology of the changes are given in Godson et al. (this volume; see especially Table 2).

gene A

```
     3981
ØX   MET VAL ARG SER TYR TYR PRO SER GLU CYS HIS ALA ASP TYR PHE ASP PHE GLU ARG ILE
     ATG GTT CGT TCT TAT TAC CCT TCT GAA TGT CAC GCT GAT TAT TTT GAC TTT GAG CGT ATC
         |   ||| |||                           | |     |   |   | ||  ||| ||| |||
     ATG TTT AAA GTA --- --- --- --- --- --- CAT TCT GAC TAC TTC AGC AAA CCT AAC ATC
G4   MET PHE LYS VAL                         HIS SER ASP TYR PHE SER LYS PRO ASN ILE
     59

     GLU ALA LEU LYS PRO ALA ILE GLU ALA CYS GLY ILE SER THR LEU SER GLN SER PRO MET
     GAG GCT CTT AAA CCT GCT ATT GAG GCT TGT GGC ATT TCT ACT CTT TCT CAA TCC CCA ATG
      |    | | |         ||| | |   | |   ||    |   |     ||      |         |      ||
     GAC GCA ATC AAA CCT CTC GTC GAA ACT GCT GGT ATG TCT CTT CTT ACT CAA TCT CCA AAC
     ASP ALA ILE LYS PRO LEU VAL GLU THR ALA GLY MET SER LEU LEU THR GLN SER PRO ASN

     LEU GLY PHE HIS LYS GLN MET ASP ASN ARG ILE LYS LEU LEU GLU GLU ILE LEU SER PHE
     CTT GGC TTC CAT AAG CAG ATG GAT AAC CGC ATC AAG CTC TTG GAA GAG ATT CTG TCT TTT
       | | | |   |||   |     ||| |    ||           | | | |       |   |   | |   |||
     CTC CGT ATC TGG AAA CAG TGC AAT ACT CGC ATC AAA TTA CTG GAA GAA ATC CTT GCT CAC
     LEU ARG ILE TRP LYS GLN CYS ASN THR ARG ILE LYS LEU LEU GLU GLU ILE LEU ALA HIS

     ARG MET GLN GLY VAL GLU PHE     ASP ASN GLY ASP
     CGT ATG CAG GGC GTT GAG TTC --- GAT AAT GGT GAT --- --- --- --- --- --- --- ---
     ||   || | |   | |   ||| |||       |           |
     TAT ACA AAT GGT ATT CGC CGT GAC GAC AAT GGT GAC TTC TGG ATG AAT CCC AAC TCT CAA
     TYR THR ASN GLY ILE ARG ARG ASP ASP ASN GLY ASP PHE TRP MET ASN PRO ASN SER GLN
```

```
--- --- --- --- --- --- --- --- --- --- --- --- --- --- --- --- --- --- --- ---

CTC GCA ACT ACT ATT GCC TAT CGT GCT CAT CAC AAA GGC CAT AAC CCT AAG TTC AAA CAA
LEU ALA THR THR ILE ALA TYR ARG ALA HIS HIS LYS GLY HIS ASN PRO LYS PHE LYS GLN

                                                                        MET TYR
--- --- --- --- --- --- --- --- --- --- --- --- --- --- --- --- --- --- ATG TAT
                                                                         || |||
TAC CCA GAA ACC TTC ACT CTG GAC GAC ATT ATT ACT GGT AAA CCA ATC CCT CAA ACT GCC
TYR PRO GLU THR PHE THR LEU ASP ASP ILE ILE THR GLY LYS PRO ILE PRO GLN THR ALA

VAL ASP GLY HIS LYS ALA ALA SER ASP VAL ARG ASP GLU PHE VAL SER VAL THR GLU LYS
GTT GAC GGC CAT AAG GCT GCT TCT GAC GTT CGT GAT GAG TTT GTA TCT GTT ACT GAG AAG
||       || ||| | | ||| | | ||    | | | ||    |   |  |  ||| ||  ||| ||| ||| |||
CCT GAC GAA TTA CAA CTC TCC GAT GAA ATC GTT GAA GAC TAT CGC CTT ACA GTC CTC TCC
PRO ASP GLU LEU GLN LEU SER ASP GLU ILE VAL GLU ASP TYR ARG LEU THR VAL LEU SER

LEU MET ASP GLU LEU ALA GLN CYS TYR ASN VAL LEU PRO GLN LEU ASP ILE ASN ASN THR
TTA ATG GAT GAA TTG GCA CAA TGC TAC AAT GTG CTC CCC CAA CTT GAT ATT AAT AAC ACT
|     |   |         ||  |         | | |         |||
ATA ATC GAA GAA TTG CAA GAA TGC TAT GAC GTG CTC GGA CAA CTT GAT ATT AAT AAC ACT
ILE ILE GLU GLU LEU GLN GLU CYS TYR ASP VAL LEU GLY GLN LEU ASP ILE ASN ASN THR

ILE ASP HIS ARG PRO GLU GLY ASP GLU LYS TRP PHE LEU GLU ASN GLU LYS THR VAL THR
ATA GAC CAC CGC CCC GAA GGG GAC GAA AAA TGG TTT TTA GAG AAC GAG AAG ACG GTT ACG
            |||     |||   | | |   | | |     ||| | | ||| | |   |   | | |   | |||
ATA GAC CAC AAA CCC CTC GGT AAT GAC CAT TGG AAC CTC CTC TAT GAA AAA CCC GTA TAC
ILE ASP HIS LYS PRO LEU GLY ASN ASP HIS TRP ASN LEU LEU TYR GLU LYS PRO VAL TYR
```

```
GLN PHE CYS ARG LYS LEU ALA ALA GLU ARG PRO LEU LYS ASP ILE ARG ASP GLU TYR ASN
CAG TTT TGC CGC AAG CTG GCT GCT GAA CGC CCT CTT AAG GAT ATT CGC GAT GAG TAT AAT

AAA CAT TGG TAT CAG CTC GTT TCA AAT CGA CCC CTT AAA GAT ATC CGC GCC GAT TAC AAC
LYS HIS TRP TYR GLN LEU VAL SER ASN ARG PRO LEU LYS ASP ILE ARG ALA ASP TYR ASN

TYR PRO LYS LYS LYS GLY ILE LYS ASP GLU CYS SER ARG LEU LEU GLU ALA SER THR MET
TAC CCC AAA AAG AAA GGT ATT AAG GAT GAG TGT TCA AGA TTG CTG GAG GCC TCC ACT ATG

TAT GCA AAG GCT AAG GGT GTT AAA GAT GAG TGC TCA AAA ATC TTG GAG GAG TCA ACT ATG
TYR ALA LYS ALA LYS GLY VAL LYS ASP GLU CYS SER LYS ILE LEU GLU GLU SER THR MET

LYS SER ARG ARG GLY PHE ALA ILE GLN ARG LEU MET ASN ALA MET ARG GLN ALA HIS ALA
AAA TCG CGT AGA GGC TTT GCT ATT CAG CGT TTG ATG AAT GCA ATG CGA CAG GCT CAT GCT

AAG TCT CGA CGT GGC TTT ACC GTT CAA AGA TTA ATG AAT GCT ATG CGT ACT GCC CAT TCT
LYS SER ARG ARG GLY PHE THR VAL GLN ARG LEU MET ASN ALA MET ARG THR ALA HIS SER

ASP GLY TRP PHE ILE VAL PHE ASP THR LEU THR LEU ALA ASP ASP ARG LEU GLU ALA PHE
GAT GGT TGG TTT ATC GTT TTT GAC ACT CTC ACG TTG GCT GAC GAC CGA TTA GAG GCG TTT

AAG GGC TGG TAT GTC GTA TTC GAT ACT CTG ACT CTA GCC GAT GAC CGC TTA AAG GAT TTC
LYS GLY TRP TYR VAL VAL PHE ASP THR LEU THR LEU ALA ASP ASP ARG LEU LYS ASP PHE

TYR ASP ASN PRO ASN ALA LEU ARG ASP TYR PHE ARG ASP ILE GLY ARG MET VAL LEU ALA
TAT GAT AAT CCC AAT GCT TTG CGT GAC TAT TTT CGT GAT ATT GGT CGT ATG GTT CTT GCT

TAT GAT AAT CCC AAT GCT CTT CGT GAC TAC TTC CGT GAT ATT GGT CGT ATG GTG CTT ACT
TYR ASP ASN PRO ASN ALA LEU ARG ASP TYR PHE ARG ASP ILE GLY ARG MET VAL LEU THR
```

```
ALA GLU GLY ARG LYS ALA ASN ASP SER HIS ALA ASP CYS TYR GLN TYR PHE CYS VAL PRO
GCC GAG GGT CGC AAG GCT AAT GAT TCA CAC GCC GAC TGC TAT CAG TAT TTT TGT GTG CCT
GCC GAA GGT CGC TCG GTG CAT GAC TCA TCT TCC GAC TGC TAT CAG TAT TTT TGT GTG CCA
ALA GLU GLY ARG SER VAL HIS ASP SER SER SER ASP CYS TYR GLN TYR PHE CYS VAL PRO

GLU TYR GLY THR ALA ASN GLY ARG LEU HIS PHE HIS ALA VAL HIS PHE MET ARG THR LEU
GAG TAT GGT ACA GCT AAT GGC CGT CTT CAT TTC CAT GCG GTG CAC TTT ATG CGG ACA CTT
GAG TAT GGT ACA CAG CAC GGT CGT CTA CAT TTC CAC GCA GTG CAT CTT ATG CGC ACA CTT
GLU TYR GLY THR GLN HIS GLY ARG LEU HIS PHE HIS ALA VAL HIS LEU MET ARG THR LEU

PRO THR GLY SER VAL ASP PRO ASN PHE GLY ARG ARG VAL ARG ASN ARG ARG GLN LEU ASN
CCT ACA GGT AGC GTT GAC CCT AAT TTT GGT CGT CGG GTA CGC AAT CGC CGC CAG TTA AAT
CCT CTG GGT TCT CTG GAC CCT AAC TTC GGT AAG CTG GTA CGC ATC AAT CGG CAA ATA AAT
PRO LEU GLY SER LEU ASP PRO ASN PHE GLY LYS LEU VAL ARG ILE ASN ARG GLN ILE ASN

SER LEU GLN ASN THR TRP PRO TYR GLY TYR SER MET PRO ILE ALA VAL ARG TYR THR GLN
AGC TTG CAA AAT ACG TGG CCT TAT GGT TAC AGT ATG CCC ATC GCA GTT CGC TAC ACG CAG
AGC TTG CAA AAC ACG TGG CCT TAT GGT TAC TCT ATG CCC ATC GCA GTC CGG TAC TCC CAA
SER LEU GLN ASN THR TRP PRO TYR GLY TYR SER MET PRO ILE ALA VAL ARG TYR SER GLN

ASP ALA PHE SER ARG SER GLY TRP LEU TRP PRO VAL ASP ALA LYS GLY GLU PRO LEU LYS
GAC GCT TTT TCA CGT TCT GGT TGG TTG TGG CCT GTT GAT GCT AAA GGT GAG CCG CTT AAA
GAC GCA TTC TCT CGC GCT GGC TGG CTC TGG CCT GTG GAC TCA AAG GGT GAA CCC CTT AAA
ASP ALA PHE SER ARG ALA GLY TRP LEU TRP PRO VAL ASP SER LYS GLY GLU PRO LEU LYS
```

```
ALA THR SER TYR MET ALA VAL GLY PHE TYR VAL ALA LYS TYR VAL ASN LYS LYS SER ASP
GCT ACC AGT TAT ATG GCT GTT GGT TTC TAT GTG GCT AAA TAC GTT AAC AAA AAG TCA GAT

GCT ACC TCG TAT ATG GCT GTA GGC TTC TAC GTT GCT AAA TAC GTT AAC AAA AAA TCA GAT
ALA THR SER TYR MET ALA VAL GLY PHE TYR VAL ALA LYS TYR VAL ASN LYS LYS SER ASP

MET ASP LEU ALA ALA LYS GLY LEU GLY ALA LYS GLU TRP ASN ASN SER LEU LYS THR LYS
ATG GAC CTT GCT GCT AAA GGT CTA GGA GCT AAA GAA TGG AAC AAC TCA CTA AAA ACC AAG

ATT GAC ATG GCC GTA AAA GGC CTA GGG AAT AAA GAA TGG AAC AAT TCA CTC AAA ACC AAA
ILE ASP MET ALA VAL LYS GLY LEU GLY ASN LYS GLU TRP ASN ASN SER LEU LYS THR LYS

LEU SER LEU LEU PRO LYS LYS LEU PHE ARG ILE ARG MET SER ARG ASN PHE GLY MET LYS
CTG TCG CTA CTT CCC AAG AAG CTG TTC AGA ATC AGA ATG AGC CGC AAC TTC GGG ATG AAA

ATC AAC CTC ATA CCC AAG AAA GTG TTC AGA ATA CGA ATG TCT CGC AAT TTC GGA ATG AAA
ILE ASN LEU ILE PRO LYS LYS VAL PHE ARG ILE ARG MET SER ARG ASN PHE GLY MET LYS

MET LEU THR MET THR ASN LEU SER THR GLU CYS LEU ILE GLN LEU THR LYS LEU GLY TYR
ATG CTC ACA ATG ACA AAT CTG TCC ACG GAG TGC TTA ATC CAA CTT ACC AAG CTG GGT TAC

CTG TTA TCA ATG GCT CAC CTG TCA GCG GAA ACC CTG ATG GAA CTG ACC CAA GTG GGT TAC
LEU LEU SER MET ALA HIS LEU SER ALA GLU THR LEU MET GLU LEU THR GLN VAL GLY TYR

ASP ALA THR PRO PHE ASN GLN ILE LEU LYS GLN ASN ALA LYS ARG GLU MET ARG LEU ARG
GAC GCG ACG CCG TTC AAC CAG ATA TTG AAG CAG AAC GCA AAA AGA GAG ATG AGA TTG AGG

GAC GTG ACC CCG TTC AAC AAC ATC TTG AAG CAG AAC GCC AAG AAA GAG CTC AAA TCG AGG
ASP VAL THR PRO PHE ASN ASN ILE LEU LYS GLN ASN ALA LYS LYS GLU LEU LYS SER ARG
```

```
LEU GLY LYS VAL THR VAL ALA ASP VAL LEU ALA ALA GLN PRO VAL THR THR ASN LEU LEU
CTG GGA AAA GTT ACT GTA GCC GAC GTT TTG GCG GCG CAA CCT GTG ACG ACA AAT CTG CTC
     |    | ||| |     |                  |                        |           |
CTG GCA AAG AAA TCT GTC GCC GAC GTT TTG GAG GCG CAA CCT GTG ACG ACG AAT CTG CTA
LEU ALA LYS LYS SER VAL ALA ASP VAL LEU GLU ALA GLN PRO VAL THR THR ASN LEU LEU

LYS PHE MET ARG ALA SER ILE LYS MET ILE GLY VAL SER ASN LEU GLN SER PHE ILE ALA
AAA TTT ATG CGC GCT TCG ATA AAA ATG ATT GGC GTA TCC AAC CTG CAG AGT TTT ATC GCT
      |         ||   |   || ||   ||   |   |  ||           |
AAA TTC ATG CGC AAT TTG ACC CGA ACA ATC GGA GCG TCC AAC CTA CAG AGT TTT ATC GCT
LYS PHE MET ARG ASN LEU THR ARG THR ILE GLY ALA SER ASN LEU GLN SER PHE ILE ALA

SER MET THR GLN LYS LEU THR LEU SER ASP ILE SER ASP GLU SER LYS ASN TYR LEU ASP
TCC ATG ACG CAG AAG TTA ACA CTT TCG GAT ATT TCT GAT GAG TCG AAA AAT TAT CTT GAT
  |       | ||    |         ||  |                     | | |       |   | |    |
TCA ATG ACC ACG AAA TTA ACA AGT ACG GAT ATT TCT GAT GAA ACC AAA AAC TAC GTT GCT
SER MET THR THR LYS LEU THR SER THR ASP ILE SER ASP GLU THR LYS ASN TYR VAL ALA

LYS ALA GLY ILE THR THR ALA CYS LEU ARG ILE LYS SER LYS TRP THR ALA GLY GLY LYS
AAA GCA GGA ATT ACT ACT GCT TGT TTA CGA ATT AAA TCG AAG TGG ACT GCT GGC GGA AAA
|||             |    ||     |||                                           |
TCT GCA GGA ATT GCT GTT GCT AAC TTA CGA ATT AAA TCG AAG TGG ACT GCT GGT GGA AAA
SER ALA GLY ILE ALA VAL ALA ASN LEU ARG ILE LYS SER LYS TRP THR ALA GLY GLY LYS

  136
***
TGA

TGA
***
  1723
```

gene B

```
     5075
ØX   MET GLU GLN LEU THR LYS ASN GLN ALA VAL ALA THR SER GLN GLU ALA VAL GLN ASN GLN
     ATG GAA CAA CTC ACT AAA AAC CAA GCT GTC GCT ACT TCC CAA GAA GCT GTT CAG AAT CAG

     ATG GAA CAA TTC ACT CAA AAC CAA AAT CAA CCT CAT ACC CAA GAA AGT GTT CAG AAT ACG
G4   MET GLU GLN PHE THR GLN ASN GLN ASN GLN PRO HIS THR GLN GLU SER VAL GLN ASN THR
     1276

     ASN GLU PRO GLN LEU ARG ASP GLU ASN ALA HIS ASN ASP LYS SER VAL HIS GLY VAL LEU
     AAT GAG CCG CAA CTT CGG GAT GAA AAT GCT CAC AAT GAC AAA TCT GTC CAC GGA GTG CTT

     AAT GTC TCG CAA TTT CGG AAT GAA ACT GTT ATC AAT GGC TCA CCT GTC AGC GGA AAC CCT
     ASN VAL SER GLN PHE ARG ASN GLU THR VAL ILE ASN GLY SER PRO VAL SER GLY ASN PRO

     ASN PRO THR TYR GLN ALA GLY LEU ARG ARG ASP ALA VAL GLN PRO ASP ILE GLU ALA GLU
     AAT CCA ACT TAC CAA GCT GGG TTA CGA CGC GAC GCC GTT CAA CCA GAT ATT GAA GCA GAA

     GAT GGA ACT GAC CCA AGT GGG TTA CGA CGT GAC CCC GTT CAA CAA CAT CTT GAA GCA GAA
     ASP GLY THR ASP PRO SER GLY LEU ARG ARG ASP PRO VAL GLN GLN HIS LEU GLU ALA GLU

     ARG LYS LYS ARG ASP GLU ILE GLU ALA GLY LYS SER TYR CYS SER ARG ARG PHE GLY GLY
     CGC AAA AAG AGA GAT GAG ATT GAG GCT GGG AAA AGT TAC TGT AGC CGA CGT TTT GGC GGC

     CGC CAA GAA AGA GCT CAA ATC GAG GCT GGC AAA GAA ATC TGT CGC CGA CGT TTT GGA GGC
     ARG GLN GLU ARG ALA GLN ILE GLU ALA GLY LYS GLU ILE CYS ARG ARG ARG PHE GLY GLY
```

```
ALA THR CYS ASP ASP LYS SER ALA GLN ILE TYR ALA ARG PHE ASP LYS ASN ASP TRP ARG
GCA ACC TGT GAC GAC AAA TCT GCT CAA ATT TAT GCG CGC TTC GAT AAA AAT GAT TGG CGT

GCA ACC TGT GAC GAC GAA TCT GCT AAA ATT CAT GCG CAA TTT GAC CCG AAC AAT CGG AGC
ALA THR CYS ASP ASP GLU SER ALA LYS ILE HIS ALA GLN PHE ASP PRO ASN ASN ARG SER

ILE GLN PRO ALA GLU PHE TYR ARG PHE HIS ASP ALA GLU VAL ASN THR PHE GLY TYR PHE
ATC CAA CCT GCA GAG TTT TAT CGC TTC CAT GAC GCA GAA GTT AAC ACT TTC GGA TAT TTC

GTC CAA CCT ACA GAG TTT TAT CGC TTC AAT GAC CAC GAA ATT AAC AAG TAC GGA TAT TTC
VAL GLN PRO THR GLU PHE TYR ARG PHE ASN ASP HIS GLU ILE ASN LYS TYR GLY TYR PHE

  51
***
TGA

TGA
***
  1638
```

gene C

```
    133
ØX  MET ARG LYS PHE ASP LEU SER LEU ARG SER SER ARG SER SER TYR PHE ALA THR PHE ARG
    ATG AGA AAA TTC GAC CTA TCC TTG CGC AGC TCG AGA AGC TCT TAC TTT GCG ACC TTT CGC

    ATG AGG AAA TTC AAT CTC AAC TTA AAA AAC TCG AGG TCG TCT TAC TTT GCA ACC TTT CGC
G4  MET ARG LYS PHE ASN LEU ASN LEU LYS ASN SER ARG SER SER TYR PHE ALA THR PHE ARG
    1720
```

```
HIS GLN LEU THR ILE LEU SER LYS THR ASP ALA LEU ASP GLU GLU LYS TRP LEU ASN MET
CAT CAA CTA ACG ATT CTG TCA AAA ACT GAC GCG TTG GAT GAG GAG AAG TGG CTT AAT ATG

CAT CAT CTC AAC GTG CTG GCA AAA ACT GAC GCC CTT GAC GAG GAG AAA TAC TTA AAT ATG
HIS HIS LEU ASN VAL LEU ALA LYS THR ASP ALA LEU ASP GLU GLU LYS TYR LEU ASN MET

LEU GLY THR PHE VAL LYS ASP TRP PHE ARG TYR GLU SER HIS PHE VAL HIS GLY ARG ASP
CTT GGC ACG TTC GTC AAG GAC TGG TTT AGA TAT GAG TCA CAT TTT GTT CAT GGT AGA GAT

TTA GGT GCT CTC CTC AAG GAC TGG TTC CGG TAC GAA GAA CAT TTC GTG CAT GGT AAA CAA
LEU GLY ALA LEU LEU LYS ASP TRP PHE ARG TYR GLU GLU HIS PHE VAL HIS GLY LYS GLN

SER LEU VAL ASP ILE LEU LYS GLU ARG GLY LEU LEU SER GLU SER ASP ALA VAL GLN PRO
TCT CTT GTT GAC ATT TTA AAA GAG CGT GGA TTA CTA TCT GAG TCC GAT GCT GTT CAA CCA

TCA ATG CTT GAC ATA CTG AAA GAA CGT GGC CTA TTA TCC ACA TCG TCA ACT GAC ACA AAC
SER MET LEU ASP ILE LEU LYS GLU ARG GLY LEU LEU SER THR SER SER THR ASP THR ASN

                                393
LEU ILE GLY LYS LYS SER ***
CTA ATA GGT AAG AAA TCA TGA

CAC AAA GGA AAC TGA
HIS LYS GLY ASN ***
                       1974
```

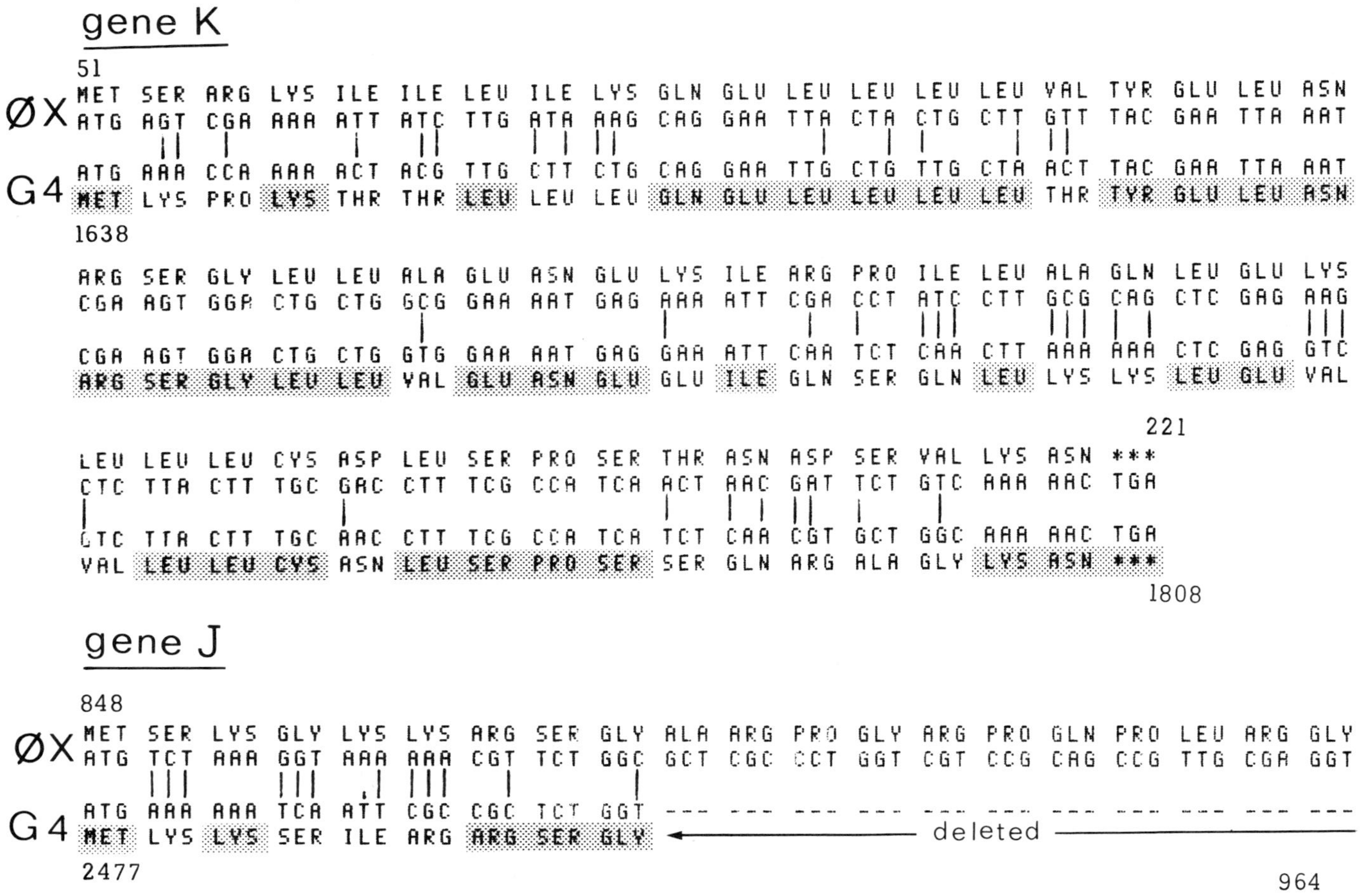

gene K

```
    51
    MET SER ARG LYS ILE ILE LEU ILE LYS GLN GLU LEU LEU LEU LEU VAL TYR GLU LEU ASN
ØX  ATG AGT CGA AAA ATT ATC TTG ATA AAG CAG GAA TTA CTA CTG CTT GTT TAC GAA TTA AAT

    ATG AAA CCA AAA ACT ACG TTG CTT CTG CAG GAA TTG CTG TTG CTA ACT TAC GAA TTA AAT
G4  MET LYS PRO LYS THR THR LEU LEU LEU GLN GLU LEU LEU LEU LEU THR TYR GLU LEU ASN
    1638

    ARG SER GLY LEU LEU ALA GLU ASN GLU LYS ILE ARG PRO ILE LEU ALA GLN LEU GLU LYS
    CGA AGT GGA CTG CTG GCG GAA AAT GAG AAA ATT CGA CCT ATC CTT GCG CAG CTC GAG AAG

    CGA AGT GGA CTG CTG GTG GAA AAT GAG GAA ATT CAA TCT CAA CTT AAA AAA CTC GAG GTC
    ARG SER GLY LEU LEU VAL GLU ASN GLU GLU ILE GLN SER GLN LEU LYS LYS LEU GLU VAL

                                                                 221
    LEU LEU LEU CYS ASP LEU SER PRO SER THR ASN ASP SER VAL LYS ASN ***
    CTC TTA CTT TGC GAC CTT TCG CCA TCA ACT AAC GAT TCT GTC AAA AAC TGA

    CTC TTA CTT TGC AAC CTT TCG CCA TCA TCT CAA CGT GCT GGC AAA AAC TGA
    VAL LEU LEU CYS ASN LEU SER PRO SER SER GLN ARG ALA GLY LYS ASN ***
                                                                 1808
```

gene J

```
    848
    MET SER LYS GLY LYS LYS ARG SER GLY ALA ARG PRO GLY ARG PRO GLN PRO LEU ARG GLY
ØX  ATG TCT AAA GGT AAA AAA CGT TCT GGC GCT CGC CCT GGT CGT CCG CAG CCG TTG CGA GGT

    ATG AAA AAA TCA ATT CGC CGC TCT GGT --- --- --- --- --- --- --- --- --- --- ---
G4  MET LYS LYS SER ILE ARG ARG SER GLY <---------- deleted ---------->
    2477                                                                     964
```

```
THR LYS GLY LYS ARG LYS GLY ALA ARG LEU TRP TYR VAL GLY GLY GLN GLN PHE ***
ACT AAA GGC AAG CGT AAA GGC GCT CGT CTT TGG TAT GTA GGT GGT CAA CAA TTT TAA
              | ||    |   |   |       |                   |   | ||      | |
--- --- GGC AAA TCT AAG GGT GCC CGT CTC TGG TAT GTA GGC GGA ACA CAA TAC TAA
  -------> GLY LYS SER LYS GLY ALA ARG LEU TRP TYR VAL GLY GLY THR GLN TYR ***
                                                                            2554
```

gene D

```
     390
ØX   MET SER GLN VAL THR GLU GLN SER VAL ARG PHE GLN THR ALA LEU ALA SER ILE LYS LEU
     ATG AGT CAA GTT ACT GAA CAA TCC GTA CGT TTC CAG ACC GCT TTG GCC TCT ATT AAG CTC
         ||  |   |||  ||     ||| | |     |||   |   |   |     | |   |       |   |   |
     ATG TCT AAA TCA AAC GAA TCT GCT GTA GCC TTT CAA ACT GCT ATC GCT TCT ATC AAA CTT
G4   MET SER LYS SER ASN GLU SER ALA VAL ALA PHE GLN THR ALA ILE ALA SER ILE LYS LEU
     1976

     ILE GLN ALA SER ALA VAL LEU ASP LEU THR GLU ASP ASP PHE ASP PHE LEU THR SER ASN
     ATT CAG GCT TCT GCC GTT TTG GAT TTA ACC GAA GAT GAT TTC GAT TTT CTG ACG AGT AAC
           |   |     | |     | |   |   |   |   |   |       |   |   | |   | | | |
     ATT CAA GCA TCT TCT GTT CTT GAC TTG ACA GAG GAC GAT TTT GAC TTC TTA ACT CGC GAC
     ILE GLN ALA SER SER VAL LEU ASP LEU THR GLU ASP ASP PHE ASP PHE LEU THR ARG ASP

     LYS VAL TRP ILE ALA THR ASP ARG SER ARG ALA ARG ARG CYS VAL GLU ALA CYS VAL TYR
     AAA GTT TGG ATT GCT ACT GAC CGC TCT CGT GCT CGT CGC TGC GTT GAG GCT TGC GTT TAT
     |||   |               |       |   |   |       |   | ||| | |               |
     CGT GTC TGG ATT GCT ACC GAC CGT TCA CGC GCT CGC CGT GCT ATC GAG GCT TGC GTA TAT
     ARG VAL TRP ILE ALA THR ASP ARG SER ARG ALA ARG ARG ALA ILE GLU ALA CYS VAL TYR
```

```
GLY THR LEU ASP PHE VAL GLY TYR PRO ARG PHE PRO ALA PRO VAL GLU PHE ILE ALA ALA
GGT ACG CTG GAC TTT GTA GGA TAC CCT CGC TTT CCT GCT CCT GTT GAG TTT ATT GCT GCC
  |   |               |   |   |                                         |
GGA ACA CTG GAC TTT GTC GGG TAT CCT CGC TTT CCT GCT CCT GTT GAG TTT ATT TCT GCC
GLY THR LEU ASP PHE VAL GLY TYR PRO ARG PHE PRO ALA PRO VAL GLU PHE ILE SER ALA

VAL ILE ALA TYR TYR VAL HIS PRO VAL ASN ILE GLN THR ALA CYS LEU ILE MET GLU GLY
GTC ATT GCT TAT TAT GTT CAT CCC GTC AAC ATT CAA ACG GCC TGT CTC ATC ATG GAA GGC
                  |               |       |       |                           |
GTC ATT GCT TAT TAC GTT CAT CCC GTT AAC ATC CAA ACC GCC TGT CTC ATC ATG GAA GGT
VAL ILE ALA TYR TYR VAL HIS PRO VAL ASN ILE GLN THR ALA CYS LEU ILE MET GLU GLY

ALA GLU PHE THR GLU ASN ILE ILE ASN GLY VAL GLU ARG PRO VAL LYS ALA ALA GLU LEU
GCT GAA TTT ACG GAA AAC ATT ATT AAT GGC GTC GAG CGT CCG GTT AAA GCC GCT GAA TTG
      |       |           | | |       |   |   |   |   |             |       |
GCT GAG TTT ACC GAA AAC ATC GTC AAT GGT GTT GAA CGC CCC GTT AAA GCC TCT GAA CTG
ALA GLU PHE THR GLU ASN ILE VAL ASN GLY VAL GLU ARG PRO VAL LYS ALA SER GLU LEU

PHE ALA PHE THR LEU ARG VAL ARG ALA GLY ASN THR ASP VAL LEU THR ASP ALA GLU GLU
TTC GCG TTT ACC TTG CGT GTA CGC GCA GGA AAC ACT GAC GTT CTT ACT GAC GCA GAA GAA
      |   |   | | |  |    |       |          ||   | |   | | ||  |     |     ||
TTC GCC TTC ACT CTT CTT GTT CGC GCC GGA AAC AAA GAT CTT ATC GGT CAC GCT GAA ACA
PHE ALA PHE THR LEU LEU VAL ARG ALA GLY ASN LYS ASP LEU ILE GLY HIS ALA GLU THR

                                                 848
ASN VAL ARG GLN LYS LEU ARG ALA GLU GLY VAL MET ***
AAC GTG CGT CAA AAA TTA CGT GCG GAA GGA GTG ATG TAA
    | |     |   | |       |   | |     |   |
AAC ATC CGT GAA CAG TTA CGC GCT CAA GGT GTC ATG TAA
ASN ILE ARG GLU GLN LEU ARG ALA GLN GLY VAL MET ***
                                                 2434
```

gene E

```
                                    wt=  trp
                                         TGG
    568
ØX  MET VAL ARG TRP THR LEU *** ASP THR LEU ALA PHE LEU LEU LEU LEU SER LEU LEU LEU
    ATG GTA CGC TGG ACT TTG TAG GAT ACC CTC GCT TTC CTG CTC CTG TTG AGT TTA TTG CTG

    ATG GAA CAC TGG ACT TTG TCG GGT ATC CTC GCT TTC CTG CTC CTG TTG AGT TTA TTT CTG
G4  MET GLU HIS TRP THR LEU SER GLY ILE LEU ALA PHE LEU LEU LEU LEU SER LEU PHE LEU
    2154

    PRO SER LEU LEU ILE MET PHE ILE PRO SER THR PHE LYS ARG PRO VAL SER SER TRP LYS
    CCG TCA TTG CTT ATT ATG TTC ATC CCG TCA ACA TTC AAA CGG CCT GTC TCA TCA TGG AAG

    CCG TCA TTG CTT ATT ACG TTC ATC CCG TTA ACA TCC AAA CCG CCT GTC TCA TCA TGG AAG
    PRO SER LEU LEU ILE THR PHE ILE PRO LEU THR SER LYS PRO PRO VAL SER SER TRP LYS

    ALA LEU ASN LEU ARG LYS THR LEU LEU MET ALA SER SER VAL ARG LEU LYS PRO LEU ASN
    GCG CTG AAT TTA CGG AAA ACA TTA TTA ATG GCG TCG AGC GTC CGG TTA AAG CCG CTG AAT

    GTG CTG AGT TTA CCG AAA ACA TCG TCA ATG GTG TTG AAC GCC CCG TTA AAG CCT CTG AAC
    VAL LEU SER LEU PRO LYS THR SER SER MET VAL LEU ASN ALA PRO LEU LYS PRO LEU ASN

    CYS SER ARG LEU PRO CYS VAL TYR ALA GLN GLU THR LEU THR PHE LEU LEU THR GLN LYS
    TGT TCG CGT TTA CCT TGC GTG TAC GCG CAG GAA ACA CTG ACG TTC TTA CTG ACG CAG AAG

    TGT TCG CCT TCA CTC TTC TTG TTC GCG CCG GAA ACA AAG ATC TTA TCG GTC ACG CTG AAA
    CYS SER PRO SER LEU PHE LEU PHE ALA PRO GLU THR LYS ILE LEU SER VAL THR LEU LYS
```

```
                                                        843
     LYS THR CYS VAL LYS ASN TYR VAL ARG LYS GLU ***
     AAA ACG TGC GTC AAA AAT TAC GTG CGG AAG GAG TGA
     |     |  |    |   |  |       |   ||      |   |
     CAA ACA TCC GTG AAC AGT TAC GCG CTC AAG GTG TCA TGT AAA GAC CTT TGA
     GLN THR SER VAL ASN SER TYR ALA LEU LYS VAL SER CYS LYS ASP LEU ***
                                                                    2444
```

gene F

```
     1001
     MET SER ASN ILE GLN THR GLY ALA GLU ARG MET PRO HIS ASP LEU SER HIS LEU GLY PHE
ØX   ATG TCT AAT ATT CAA ACT GGC GCC GAG CGT ATG CCG CAT GAC CTT TCC CAT CTT GGC TTC
               | |         |  |||   |   |     | |   |         | |   |   |      |    |
G4   ATG TCT AAC GTT CAA ACA TCT GCG GAC CGT GTA CCT CAT GAC TTA TCT CAC CTT GTC TTT
     MET SER ASN VAL GLN THR SER ALA ASP ARG VAL PRO HIS ASP LEU SER HIS LEU VAL PHE
     2600

     LEU ALA GLY GLN ILE GLY ARG LEU ILE THR ILE SER THR THR PRO VAL ILE ALA GLY ASP
     CTT GCT GGT CAG ATT GGT CGT CTT ATT ACC ATT TCA ACT ACT CCG GTT ATC GCT GGC GAC
     |||         | |       |   |   |  ||   |   |   | |||   |   |  | |||       |
     GAG GCT GGT AAA ATT GGC CGC CTC AAA ACT ATC TCT TGG ACG CCT GGT GGT GCT GGT GAC
     GLU ALA GLY LYS ILE GLY ARG LEU LYS THR ILE SER TRP THR PRO GLY GLY ALA GLY ASP

     SER PHE GLU MET ASP ALA VAL GLY ALA LEU ARG LEU SER PRO LEU ARG ARG GLY LEU ALA
     TCC TTC GAG ATG GAC GCC GTT GGC GCT CTC CGT CTT TCT CCA TTG CGT CGT GGC CTT GCT
       |         |||   | |||             | |       |   |   | | |               |
     TCT TTC GAG TGT GAT ATG GTT GGC GCT ATT CGT CTG TCC CCT CTT CGT CGT GGC CTC GCT
     SER PHE GLU CYS ASP MET VAL GLY ALA ILE ARG LEU SER PRO LEU ARG ARG GLY LEU ALA
```

```
ILE ASP SER THR VAL ASP ILE PHE THR PHE TYR VAL PRO HIS ARG HIS VAL TYR GLY GLU
ATT GAC TCT ACT GTA GAC ATT TTT ACT TTT TAT GTC CCT CAT CGT CAC GTT TAT GGT GAA
GTT GAC TCT CGC GTT GAT ATC TTT CCT TTC TAT ATC CCA CAC CGT CAT ATC TAC GGT CAG
VAL ASP SER ARG VAL ASP ILE PHE PRO PHE TYR ILE PRO HIS ARG HIS ILE TYR GLY GLN

GLN TRP ILE LYS PHE MET LYS ASP GLY VAL ASN ALA THR PRO LEU PRO THR VAL ASN THR
CAG TGG ATT AAG TTC ATG AAG GAT GGT GTT AAT GCC ACT CCT CTC CCG ACT GTT AAC ACT
CAG TGG ATT AAC TTT ATG AAG GAT GGC GTT AAT GCC TCC CCT CTT CCT CCT GTT ACA TGC
GLN TRP ILE ASN PHE MET LYS ASP GLY VAL ASN ALA SER PRO LEU PRO PRO VAL THR CYS

THR GLY TYR ILE ASP HIS ALA ALA PHE LEU GLY THR ILE ASN PRO ASP THR ASN LYS ILE
ACT GGT TAT ATT GAC CAT GCC GCT TTT CTT GGC ACG ATT AAC CCT GAT ACC AAT AAA ATC
TCC TCT GGT TGG GAC TCT GCT GCT TAT CTC GGT ACC ATT CCG TCT TCT ACC CTT AAA GTG
SER SER GLY TRP ASP SER ALA ALA TYR LEU GLY THR ILE PRO SER SER THR LEU LYS VAL

PRO LYS HIS LEU PHE GLN GLY TYR LEU ASN ILE TYR ASN ASN TYR PHE LYS ALA PRO TRP
CCT AAG CAT TTG TTT CAG GGT TAT TTG AAT ATC TAT AAC AAC TAT TTT AAA GCG CCG TGG
CCT AAA TTC TTA CAT CAG GGC TAT CTG AAT ATT TAT AAT AAC TAC TTC AAA CCG CCT TGG
PRO LYS PHE LEU HIS GLN GLY TYR LEU ASN ILE TYR ASN ASN TYR PHE LYS PRO PRO TRP

MET PRO ASP ARG THR GLU ALA ASN PRO ASN GLU LEU ASN GLN ASP ASP ALA ARG TYR GLY
ATG CCT GAC CGT ACC GAG GCT AAC CCT AAT GAG CTT AAT CAA GAT GAT GCT CGT TAT GGT
TCT GAT GAC TTA ACT TAC GCT AAC CCA TCC AAT ATG CCT TCT GAG GAT TAT AAA TGG GGC
SER ASP ASP LEU THR TYR ALA ASN PRO SER ASN MET PRO SER GLU ASP TYR LYS TRP GLY
```

```
PHE ARG CYS CYS HIS LEU LYS ASN ILE TRP THR ALA PRO LEU PRO PRO GLU THR GLU LEU
TTC CGT TGC TGC CAT CTC AAA AAC ATT TGG ACT GCT CCG CTT CCT CCT GAG ACT GAG CTT
GTA CGT GTC GCT AAC CTT AAA TCT ATC TGG ACT GCG CCA CTT CCA CCG GAT ACT CGT ACA
VAL ARG VAL ALA ASN LEU LYS SER ILE TRP THR ALA PRO LEU PRO PRO ASP THR ARG THR

SER ARG GLN MET THR THR SER THR THR SER ILE ASP ILE MET GLY LEU GLN ALA ALA TYR
TCT CGC CAA ATG ACG ACT TCT ACC ACA TCT ATT GAC ATT ATG GGT CTG CAA GCT GCT TAT
TCT GAA AAC ATG ACT ACT GGT ACA TCC ACT ATT GAC ATT ATG GGC CTT CAA GCC GCA TAC
SER GLU ASN MET THR THR GLY THR SER THR ILE ASP ILE MET GLY LEU GLN ALA ALA TYR

ALA ASN LEU HIS THR ASP GLN GLU ARG ASP TYR PHE MET GLN ARG TYR ARG ASP VAL ILE
GCT AAT TTG CAT ACT GAC CAA GAA CGT GAT TAC TTC ATG CAG CGT TAC CGT GAT GTT ATT
GCT AAA TTA CAT ACG GAA CAG GAA CGT GAT TAC TTC ATG ACC CGT TAC CGT GAC ATT ATG
ALA LYS LEU HIS THR GLU GLN GLU ARG ASP TYR PHE MET THR ARG TYR ARG ASP ILE MET

SER SER PHE GLY GLY LYS THR SER TYR ASP ALA ASP ASN ARG PRO LEU LEU VAL MET ARG
TCT TCA TTT GGA GGT AAA ACC TCT TAT GAC GCT GAC AAC CGT CCT TTA CTT GTC ATG CGC
AAA GAG TTC GGC GGT CAT ACA TCC TAT GAT GGT GAC AAT CGT CCT CTT CTG CTC ATG CGC
LYS GLU PHE GLY GLY HIS THR SER TYR ASP GLY ASP ASN ARG PRO LEU LEU LEU MET ARG

SER ASN LEU TRP ALA SER GLY TYR ASP VAL ASP GLY THR ASP GLN THR SER LEU GLY GLN
TCT AAT CTC TGG GCA TCT GGC TAT GAT GTT GAT GGA ACT GAC CAA ACG TCG TTA GGC CAG
TCT GAA TTT TGG GCA TCT GGC TAT GAC GTA GAC GGT ACT GAC CAA TCT TCT CTC GGT CAA
SER GLU PHE TRP ALA SER GLY TYR ASP VAL ASP GLY THR ASP GLN SER SER LEU GLY GLN
```

```
PHE SER GLY ARG VAL GLN GLN THR TYR LYS HIS SER VAL PRO ARG PHE PHE VAL PRO GLU
TTT TCT GGT CGT GTT CAA CAG ACC TAT AAA CAT TCT GTG CCG CGT TTC TTT GTT CCT GAG

TTC TCT GGT CGC GTT CAG CAG ACC TTC AAT CAC AAA GTC CCA CGC TTC TAT GTT CCT GAG
PHE SER GLY ARG VAL GLN GLN THR PHE ASN HIS LYS VAL PRO ARG PHE TYR VAL PRO GLU

HIS GLY THR MET PHE THR LEU ALA LEU VAL ARG PHE PRO PRO THR ALA THR LYS GLU ILE
CAT GGC ACT ATG TTT ACT CTT GCG CTT GTT CGT TTT CCG CCT ACT GCG ACT AAA GAG ATT

CAT CGC GTA ATT ATG ACT CTG GCG GTT ACT CGC TTT CCG CCT ACT CAT GAA ATG GAA ATG
HIS ARG VAL ILE MET THR LEU ALA VAL THR ARG PHE PRO PRO THR HIS GLU MET GLU MET

GLN TYR LEU ASN ALA LYS GLY ALA LEU THR TYR THR ASP ILE ALA GLY ASP PRO VAL LEU
CAG TAC CTT AAC GCT AAA GGT GCT TTG ACT TAT ACC GAT ATT GCT GGC GAC CCT GTT TTG

CAC TAT CTT GTA GGT AAA GAA AAC TTA ACT TAT ACC GAT ATT GCT TGT GAC CCT GCA CTG
HIS TYR LEU VAL GLY LYS GLU ASN LEU THR TYR THR ASP ILE ALA CYS ASP PRO ALA LEU

TYR GLY ASN LEU PRO PRO ARG GLU ILE SER MET LYS ASP VAL PHE ARG SER GLY ASP SER
TAT GGC AAC TTG CCG CCG CGT GAA ATT TCT ATG AAG GAT GTT TTC CGT TCT GGT GAT TCG

ATG GCT AAC CTG CCA CCT CGT GAA GTA TCC TTA AAA GAA TTC TTC CAC TCC TCA CCC GAT
MET ALA ASN LEU PRO PRO ARG GLU VAL SER LEU LYS GLU PHE PHE HIS SER SER PRO ASP

SER LYS LYS PHE LYS ILE ALA GLU GLY GLN TRP TYR ARG TYR ALA PRO SER TYR VAL SER
TCT AAG AAG TTT AAG ATT GCT GAG GGT CAG TGG TAT CGT TAT GCG CCT TCG TAT GTT TCT

TCT GCT AAA TTC AAA ATC GCT GAG GGC CAA TGG TAT CGT ACA CAG CCT GAC CGT GTC GCT
SER ALA LYS PHE LYS ILE ALA GLU GLY GLN TRP TYR ARG THR GLN PRO ASP ARG VAL ALA
```

```
PRO ALA TYR HIS LEU LEU GLU GLY PHE PRO PHE ILE GLN GLU PRO PRO SER GLY ASP LEU
CCT GCT TAT CAC CTT CTT GAA GGC TTC CCA TTC ATT CAG GAA CCG CCT TCT GGT GAT TTG

TTC CCA TAT AAT GCT CTG GAT GGA TTC CCG TTC TAC TCT GCT CTC CCG TCC ACG GAA CTA
PHE PRO TYR ASN ALA LEU ASP GLY PHE PRO PHE TYR SER ALA LEU PRO SER THR GLU LEU

GLN GLU ARG VAL LEU ILE ARG HIS HIS ASP TYR ASP GLN CYS PHE GLN SER VAL GLN LEU
CAA GAA CGC GTA CTT ATT CGC CAC CAT GAT TAT GAC CAG TGT TTC CAG TCC GTT CAG TTG

AAA GAC CGT GTA CTG GTT AAC ACT AAT AAC TAC GAT GAA ATC TTC CAG TCT ATG CAG CTT
LYS ASP ARG VAL LEU VAL ASN THR ASN ASN TYR ASP GLU ILE PHE GLN SER MET GLN LEU

LEU GLN TRP ASN SER GLN VAL LYS PHE ASN VAL THR VAL TYR ARG ASN LEU PRO THR THR
TTG CAG TGG AAT AGT CAG GTT AAA TTT AAT GTG ACC GTT TAT CGC AAT CTG CCG ACC ACT

GCA CAC TGG AAT ATG CAA ACT AAA TTT AAC ATT AAC GTT TAT CGT CAC ATG CCT ACG ACA
ALA HIS TRP ASN MET GLN THR LYS PHE ASN ILE ASN VAL TYR ARG HIS MET PRO THR THR
                                  2284
ARG ASP SER ILE MET THR SER ***
CGC GAT TCA ATC ATG ACT TCG TGA

CGT GAC TCA ATC ATG ACC TCG TAA
ARG ASP SER ILE MET THR SER ***
                                  3883
```

gene G

```
      2395
ØX    MET PHE GLN THR PHE ILE SER ARG HIS ASN SER ASN PHE PHE SER ASP LYS LEU VAL LEU
      ATG TTT CAG ACT TTT ATT TCT CGC CAT AAT TCA AAC TTT TTT TCT GAT AAG CTG GTT CTC

      ATG TTC CAG AAA TTC ATT TCT AAG CAC AAT GCT CCA ATT AAC TCT ACT CAG CTT GCT GCT
G4    MET PHE GLN LYS PHE ILE SER LYS HIS ASN ALA PRO ILE ASN SER THR GLN LEU ALA ALA
      4020

      THR SER VAL THR PRO ALA SER SER ALA PRO VAL LEU GLN THR PRO LYS ALA THR SER SER
      ACT TCT GTT ACT CCA GCT TCT TCG GCA CCT GTT TTA CAG ACA CCT AAA GCT ACA TCG TCA

      ACT AGA --- ACC CCA GCT GTA GCG GCA CCT GTT TTA TCT GTG CCT AAC TTA AGT CGC TCT
      THR ARG     THR PRO ALA VAL ALA ALA PRO VAL LEU SER VAL PRO ASN LEU SER ARG SER

      THR LEU TYR PHE ASP SER LEU THR VAL ASN ALA GLY ASN GLY GLY PHE LEU HIS CYS ILE
      ACG TTA TAT TTT GAT AGT TTG ACG GTT AAT GCT GGT AAT GGT GGT TTT CTT CAT TGC ATT

      ACT ATT CTA ATC AAT GCA ACA ACC ACA GCC GTT ACA ACT CAT TCA GGC TTG TGC CAT GTT
      THR ILE LEU ILE ASN ALA THR THR THR ALA VAL THR THR HIS SER GLY LEU CYS HIS VAL

      GLN MET ASP THR SER VAL ASN ALA ALA ASN GLN VAL VAL SER VAL GLY ALA ASP ILE ALA
      CAG ATG GAT ACA TCT GTC AAC GCC GCT AAT CAG GTT GTT TCT GTT GGT GCT GAT ATT GCT

      GTT CGC ATT GAT GAA ACA AAC CCA ACA AAT CAT CAC GCT CTA TCT ATT GCA GGT TCG TTA
      VAL ARG ILE ASP GLU THR ASN PRO THR ASN HIS HIS ALA LEU SER ILE ALA GLY SER LEU
```

```
PHE ASP ALA ASP PRO LYS PHE PHE ALA CYS LEU VAL ARG PHE GLU SER SER SER
TTT GAT GCC GAC CCT AAA TTT TTT GCC TGT TTG GTT CGC TTT GAG TCT TCT TCG --- ---
TCA AAC GTC CCT GCT GAT ATG ATT GCT TTT GCC ATT CGT TTT GAG GTC GCT GAT GGC GTC
SER ASN VAL PRO ALA ASP MET ILE ALA PHE ALA ILE ARG PHE GLU VAL ALA ASP GLY VAL

VAL PRO THR THR LEU PRO THR ALA TYR ASP VAL TYR PRO LEU ASN GLY ARG HIS ASP GLY
GTT CCG ACT ACC CTC CCG ACT GCC TAT GAT GTT TAT CCT TTG AAT GGT CGC CAT GAT GGT
GTT CCT ACT GCC GTC CCG GCC TTA TAC GAC GTC TAT CCA ATC GAA ACC TTC AAT AAC GGC
VAL PRO THR ALA VAL PRO ALA LEU TYR ASP VAL TYR PRO ILE GLU THR PHE ASN ASN GLY

GLY TYR TYR THR VAL LYS ASP CYS VAL THR ILE ASP VAL LEU PRO ARG THR PRO GLY ASN
GGT TAT TAT ACC GTC AAG GAC TGT GTG ACT ATT GAC GTC CTT CCC CGT ACG CCG GGC AAT
AAA GCA ATC TCA TTT AAA GAT GCT GTT ACC ATC GAC TCA CAT CCC CGC ACG GTC GGT AAT
LYS ALA ILE SER PHE LYS ASP ALA VAL THR ILE ASP SER HIS PRO ARG THR VAL GLY ASN

ASN VAL TYR VAL GLY PHE MET VAL TRP SER ASN PHE     THR ALA THR LYS CYS ARG GLY
AAC GTT TAT GTT GGT TTC ATG GTT TGG TCT AAC TTT --- ACC GCT ACT AAA TGC CGC GGA
CGC GTT TAT GCG GGA ATC ATG CTC TGG TCT AAC GCT TGG ACC GCT CTC TAC TAT CTC TGG
ARG VAL TYR ALA GLY ILE MET LEU TRP SER ASN ALA TRP THR ALA LEU TYR TYR LEU TRP
```

```
                                                                                     2922
LEU VAL SER LEU ASN GLN VAL ILE LYS GLU ILE ILE CYS LEU GLN PRO LEU LYS ***
TTG GTT TCG CTG AAT CAG GTT ATT AAA GAG ATT ATT TGT CTC CAG CCA CTT AAG TGA

CTC CTC TCT GTT AAT CAG GTA AAC CGT GAA GCA ACC GTC CTT CAA CCT CTG AAA TAA
LEU LEU SER VAL ASN GLN VAL ASN ARG GLU ALA thr val leu gln PRO LEU LYS ***
                                                                                     4553
```

gene H

```
      2931
ØX    MET PHE GLY ALA ILE ALA GLY GLY ILE ALA SER ALA LEU ALA GLY GLY ALA MET SER LYS
      ATG TTT GGT GCT ATT GCT GGC GGT ATT GCT TCT GCT CTT GCT GGT GGC GCC ATG TCT AAA

      ATG TTT GGC TCT ATC GGT GGC GGT ATC GCC TCC GCA CTT GCC GGA GGC CTT ATG GGT AAA
G4    MET PHE GLY SER ILE GLY GLY GLY ILE ALA SER ALA LEU ALA GLY GLY LEU MET GLY LYS
      4564

      LEU PHE GLY GLY GLY GLN LYS ALA ALA SER GLY GLY ILE GLN GLY ASP VAL LEU ALA THR
      TTG TTT GGA GGC GGT CAA AAA GCC GCC TCC GGT GGC ATT CAA GGT GAT GTG CTT GCT ACC

      TTA TTT GGC GGC GGT CAG TCC GCC GAC TCT ACC GGA ATC CAA GGC AAC GTC CTT GCT TCC
      LEU PHE GLY GLY GLY GLN SER ALA ASP SER THR GLY ILE GLN GLY ASN VAL LEU ALA SER

      ASP ASN ASN THR VAL GLY MET GLY ASP ALA GLY ILE LYS SER ALA ILE GLN GLY SER ASN
      GAT AAC AAT ACT GTA GGC ATG GGT GAT GCT GGT ATT AAA TCT GCC ATT CAA GGC TCT AAT

      GAC AAC AAT GTT GTA GGT GCT AAT GAT GCT GGT ATT AAA TCT GCT ATT CAG GGC TCT ACT
      ASP ASN ASN VAL VAL GLY ALA ASN ASP ALA GLY ILE LYS SER ALA ILE GLN GLY SER THR
```

```
VAL PRO ASN PRO ASP GLU ALA ALA PRO SER PHE VAL SER GLY ALA MET ALA LYS ALA GLY
GTT CCT AAC CCT GAT GAG GCC GCC CCT AGT TTT GTT TCT GGT GCT ATG GCT AAA GCT GGT

GTC CCT AAC TCT CAG GAA GCT GCT CCC TCT GCT ATC TCT GGC ATC CTT GCC GAC TCT GGC
VAL PRO ASN SER GLN GLU ALA ALA PRO SER ALA ILE SER GLY ILE LEU ALA ASP SER GLY
```

```
LYS GLY LEU LEU GLU GLY THR LEU GLN ALA GLY THR SER ALA VAL SER ASP LYS LEU LEU
AAA GGA CTT CTT GAA GGT ACG TTG CAG GCT GGC ACT TCT GCC GTT TCT GAT AAG TTG CTT

AAG --- --- --- --- CAT GCG CTC TCG TCT CTT ACC AAT GCT GGT GCA AAT AAA CTC ATG
LYS                 HIS ALA LEU SER SER LEU THR ASN ALA GLY ALA ASN LYS LEU MET
```

```
ASP LEU VAL GLY LEU GLY GLY LYS SER ALA ALA ASP LYS GLY LYS ASP THR ARG ASP TYR
GAT TTG GTT GGA CTT GGT GGC AAG TCT GCC GCT GAT AAA GGA AAG GAT ACT CGT GAT TAT

GAG GCT GTC GGC CTT --- TCT AAG TCT GCC TCT GAC AAG GGC AAG GAC ACA AAA GAC TAT
GLU ALA VAL GLY LEU     SER LYS SER ALA SER ASP LYS GLY LYS ASP THR LYS ASP TYR
```

```
LEU ALA ALA ALA PHE PRO GLU LEU ASN ALA TRP GLU ARG ALA GLY ALA ASP ALA SER SER
CTT GCT GCT GCA TTT CCT GAG CTT AAT GCT TGG GAG CGT GCT GGT GCT GAT GCT TCC TCT

CTC GCC GCC GCA TTC CCC GAG CTC AAC CCA TGG GAA CGT GCT GGT GCT GGC GCT TCT AGT
LEU ALA ALA ALA PHE PRO GLU LEU ASN PRO TRP GLU ARG ALA GLY ALA GLY ALA SER SER
```

```
ALA GLY MET VAL ASP ALA GLY PHE GLU ASN GLN LYS GLU LEU THR LYS MET GLN LEU ASP
GCT GGT ATG GTT GAC GCC GGA TTT GAG AAT CAA AAA GAG CTT ACT AAA ATG CAA CTG GAC

CCA GGC ATG CAA GAC GCT GGC TTC CAA AAT CAG AAG GAG CTA ACC AAG ATG CAG CTT GAC
PRO GLY MET GLN ASP ALA GLY PHE GLN ASN GLN LYS GLU LEU THR LYS MET GLN LEU ASP
```

```
ASN GLN LYS GLU I'.E ALA GLU MET GLN ASN GLU THR GLN LYS GLU ILE ALA GLY ILE CLN
AAT CAG AAA GAG ATT GCC GAG ATG CAA AAT GAG ACT CAA AAA GAG ATT GCT GGC ATT CAG
  |   |   |   |   |   | |             |   |       |   |   |           | |     |
AAC CAA AAG GAA ATC GCT AAG ATG CAA AAC GAA ACT CAG AAG GAA ATT GCT GGG CTT CAA
ASN GLN LYS GLU ILE ALA LYS MET GLN ASN GLU THR GLN LYS GLU ILE ALA GLY LEU GLN

SER ALA THR SER ARG GLN ASN THR LYS ASP GLN VAL TYR ALA GLN ASN GLU MET LEU ALA
TCG GCG ACT TCA CGC CAG AAT ACG AAA GAC CAG GTA TAT GCA CAA AAT GAG ATG CTT GCT
  | |||   |         | |       |   |   | |||   |       |           |       |   |
TCT ATT ACA TCA CGC GAA AAT ACT AAG GAT ACT GTC TAT GCT CAA AAT GAA ATG CTC GCG
SER ILE THR SER ARG GLU ASN THR LYS ASP THR VAL TYR ALA GLN ASN GLU MET LEU ALA

TYR GLN GLN LYS GLU SER THR ALA ARG VAL ALA SER ILE MET GLU ASN THR ASN LEU SER
TAT CAA CAG AAG GAG TCT ACT GCT CGC GTT GCG TCT ATT ATG GAA AAC ACC AAT CTT TCC
    | |   |   |          || | |          || |       | |           | ||   | |
TAT AAC CAA AAA GAG TCT ATG TCA CGC GTT GGC GCT ATT CTC GAA AAC ACA AGC CTC ACC
TYR ASN GLN LYS GLU SER MET SER ARG VAL GLY ALA ILE LEU GLU ASN THR SER LEU THR

LYS GLN GLN GLN VAL SER GLU ILE MET ARG GLN MET LEU THR GLN ALA GLN THR ALA GLY
AAG CAA CAG CAG GTT TCC GAG ATT ATG CGC CAA ATG CTT ACT CAA GCT CAA ACG GCT GGT
  |   |   |   | ||    |   |           |                   |           |   |
AAA CAG CAA CAA ACT TCT GAA ATT ATG CGG CAA ATG CTT ACT CAG GCT CAA ACC GCA GGT
LYS GLN GLN GLN THR SER GLU ILE MET ARG GLN MET LEU THR GLN ALA GLN THR ALA GLY

GLN TYR PHE THR ASN ASP GLN ILE LYS GLU MET THR ARG LYS VAL SER ALA
CAG TAT TTT ACC AAT GAC CAA ATC AAA GAA ATG ACT CGC AAG GTT AGT GCT --- --- ---
  |           |                   |   | | |   | | |   |     |
CAA TAT TTT ACA AAT GAC CAA ATC AAG GAG CTT ACG AGA AAA GTT GGT GCT GAT ATT GAT
GLN TYR PHE THR ASN ASP GLN ILE LYS GLU LEU THR ARG LYS VAL GLY ALA ASP ILE ASP
```

```
                                            GLU VAL ASP LEU VAL HIS GLN GLN THR
--- --- --- --- --- --- --- --- --- --- --- GAG GTT GAC TTA GTT CAT CAG CAA ACG
                                              | |||     ||| ||| | |     |   |||
GCT GTT CGC GCT AAC ACT GCA CGC ACC CAT GTA GAA ACC GAC CGC TCA AAA CAG GAA GTA
ALA VAL ARG ALA ASN THR ALA ARG THR HIS VAL GLU THR ASP ARG SER LYS GLN GLU VAL

GLN ASN GLN ARG TYR GLY SER SER HIS ILE GLY ALA THR ALA LYS ASP ILE SER ASN VAL
CAG AAT CAG CGG TAT GGC TCT TCT CAT ATT GGC GCT ACT GCA AAG GAT ATT TCT AAT GTC
  |   | |||   |      |            | | |   | |||   |   |       | | |   |      ||
CAA AAC TCC CGT TAT GCC TCT TCT CAG GTA GGT AAA ACC GCT AAG GAC GTG TCC AAT GCA
GLN ASN SER ARG TYR ALA SER SER GLN VAL GLY LYS THR ALA LYS ASP VAL SER ASN ALA

VAL THR ASP ALA ALA SER GLY VAL VAL ASP ILE PHE HIS GLY ILE ASP LYS ALA VAL ALA
GTC ACT GAT GCT GCT TCT GGT GTG GTT GAT ATT TTT CAT GGT ATT GAT AAA GCT GTT GCC
| |         | |     ||  ||  | |         ||        |      ||      |   |||   |   |
ATT ACT GAT ACA GCT GGT TCT ATT GTT GAT TAT TTT CGT GGT GCT GAT CAA AAA GTG GCA
ILE THR ASP THR ALA GLY SER ILE VAL ASP TYR PHE ARG GLY ALA ASP GLN LYS VAL ALA

ASP THR TRP ASN ASN PHE TRP LYS ASP GLY LYS ALA ASP GLY ILE GLY SER ASN LEU SER
GAT ACT TGG AAC AAT TTC TGG AAA GAC GGT AAA GCT GAT GGT ATT GGC TCT AAT TTG TCT
  | ||   ||       |  ||  ||       |   |     | | | |   |   |  ||   |     ||| |||
GAC GTT TAC AAC AAC TAT TTC AAA GAT GGC AAA TCG AAA GGA ATA GAA TCC AAT CAC CGC
ASP VAL TYR ASN ASN TYR PHE LYS ASP GLY LYS SER LYS GLY ILE GLU SER ASN HIS ARG

         3957
ARG LYS ***
AGG AAA TAA
|||
TCC AAA TAA
SER LYS ***
         5577
```

Reference Index

Numbers in *italics* refer to pages on which the complete references are listed; **boldface** type designates where author's article is located in this volume.

Subject Index

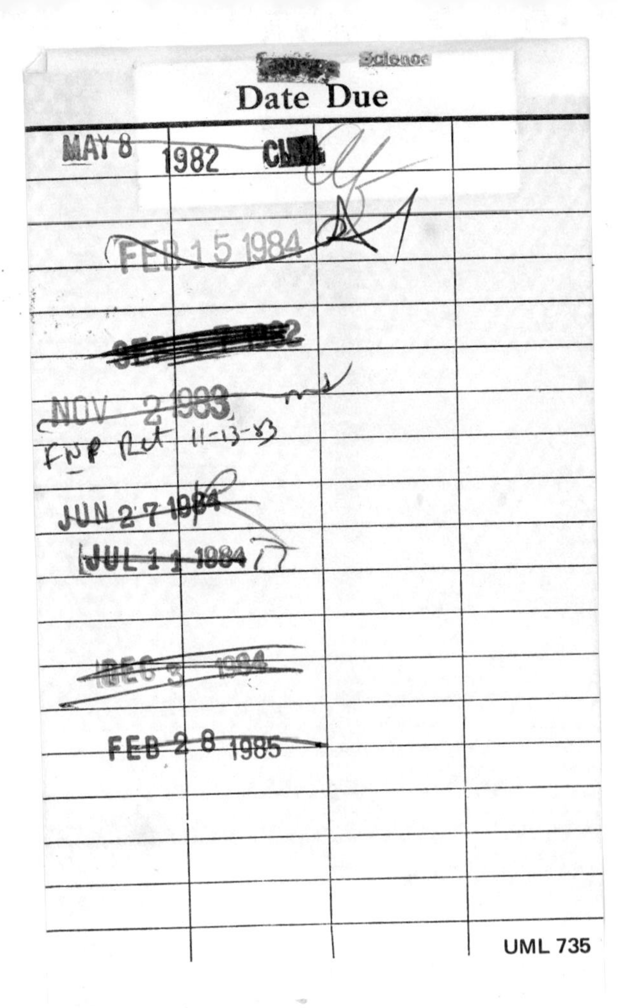
Science
Date Due
MAY 8 1982
FEB 15 1984
NOV 2 1983
FNP Ret 11-13-83
JUN 27 1984
JUL 11 1984
DEC 3 1984
FEB 28 1985
UML 735